T0178466

PRACTICAL OPTIMIZATION
Algorithms and Engineering Applications

PRACTICAL OPTIMIZATION
Algorithms and Engineering Applications

Andreas Antoniou
Wu-Sheng Lu

Department of Electrical and Computer Engineering
University of Victoria, Canada

Andreas Antoniou
Department of ECE
University of Victoria
British Columbia
Canada
aantoniou@shaw.ca

Wu-Sheng Lu
Department of ECE
University of Victoria
British Columbia
Canada
wslu@ece.uvic.ca

Practical Optimization: Algorithms and Engineering Applications
by Andreas Antoniou and Wu-Sheng Lu

ISBN 978-1-4419-4383-5

e-ISBN-10: 0-387-71107-4
e-ISBN-13: 978-0-387-71107-2

Printed on acid-free paper.

9 8 7 6 5 4 3 2 1

springer.com

*To
Lynne
and
Chi-Tang Catherine
with our love*

About the authors:

Andreas Antoniou received the Ph.D. degree in Electrical Engineering from the University of London, UK, in 1966 and is a Fellow of the IET and IEEE. He served as the founding Chair of the Department of Electrical and Computer Engineering at the University of Victoria, B.C., Canada, and is now Professor Emeritus in the same department. He is the author of Digital Filters: Analysis, Design, and Applications (McGraw-Hill, 1993) and Digital Signal Processing: Signals, Systems, and Filters (McGraw-Hill, 2005). He served as Associate Editor/Editor of IEEE Transactions on Circuits and Systems from June 1983 to May 1987, as a Distinguished Lecturer of the IEEE Signal Processing Society in 2003, as General Chair of the 2004 International Symposium on Circuits and Systems, and is currently serving as a Distinguished Lecturer of the IEEE Circuits and Systems Society. He received the Ambrose Fleming Premium for 1964 from the IEE (best paper award), the CAS Golden Jubilee Medal from the IEEE Circuits and Systems Society, the B.C. Science Council Chairman's Award for Career Achievement for 2000, the Doctor Honoris Causa degree from the Metsovio National Technical University of Athens, Greece, in 2002, and the IEEE Circuits and Systems Society 2005 Technical Achievement Award.

Wu-Sheng Lu received the B.S. degree in Mathematics from Fudan University, Shanghai, China, in 1964, the M.E. degree in Automation from the East China Normal University, Shanghai, in 1981, the M.S. degree in Electrical Engineering and the Ph.D. degree in Control Science from the University of Minnesota, Minneapolis, in 1983 and 1984, respectively. He was a post-doctoral fellow at the University of Victoria, Victoria, BC, Canada, in 1985 and Visiting Assistant Professor with the University of Minnesota in 1986. Since 1987, he has been with the University of Victoria where he is Professor. His current teaching and research interests are in the general areas of digital signal processing and application of optimization methods. He is the co-author with A. Antoniou of Two-Dimensional Digital Filters (Marcel Dekker, 1992). He served as an Associate Editor of the Canadian Journal of Electrical and Computer Engineering in 1989, and Editor of the same journal from 1990 to 1992. He served as an Associate Editor for the IEEE Transactions on Circuits and Systems, Part II, from 1993 to 1995 and for Part I of the same journal from 1999 to 2001 and from 2004 to 2005. Presently he is serving as Associate Editor for the International Journal of Multidimensional Systems and Signal Processing. He is a Fellow of the Engineering Institute of Canada and the Institute of Electrical and Electronics Engineers.

Preface

The rapid advancements in the efficiency of digital computers and the evolution of reliable software for numerical computation during the past three decades have led to an astonishing growth in the theory, methods, and algorithms of numerical optimization. This body of knowledge has, in turn, motivated widespread applications of optimization methods in many disciplines, e.g., engineering, business, and science, and led to problem solutions that were considered intractable not too long ago.

Although excellent books are available that treat the subject of optimization with great mathematical rigor and precision, there appears to be a need for a book that provides a practical treatment of the subject aimed at a broader audience ranging from college students to scientists and industry professionals. This book has been written to address this need. It treats unconstrained and constrained optimization in a unified manner and places special attention on the algorithmic aspects of optimization to enable readers to apply the various algorithms and methods to specific problems of interest. To facilitate this process, the book provides many solved examples that illustrate the principles involved, and includes, in addition, two chapters that deal exclusively with applications of unconstrained and constrained optimization methods to problems in the areas of pattern recognition, control systems, robotics, communication systems, and the design of digital filters. For each application, enough background information is provided to promote the understanding of the optimization algorithms used to obtain the desired solutions.

Chapter 1 gives a brief introduction to optimization and the general structure of optimization algorithms. Chapters 2 to 9 are concerned with unconstrained optimization methods. The basic principles of interest are introduced in Chapter 2. These include the first-order and second-order necessary conditions for a point to be a local minimizer, the second-order sufficient conditions, and the optimization of convex functions. Chapter 3 deals with general properties of algorithms such as the concepts of descent function, global convergence, and rate of convergence. Chapter 4 presents several methods for one-dimensional optimization, which are commonly referred to as line searches. The chapter also deals with inexact line-search methods that have been found to increase the efficiency in many optimization algorithms. Chapter 5 presents several basic gradient methods that include the steepest descent, Newton, and Gauss-Newton methods. Chapter 6 presents a class of methods based on the concept of conjugate directions such as the conjugate-gradient, Fletcher-Reeves, Powell, and Partan methods. An important class of unconstrained optimization methods known as quasi-Newton methods is presented in Chapter 7. Representative methods of this class such as the Davidon-Fletcher-Powell and Broydon-Fletcher-Goldfarb-Shanno methods and their properties are investigated. The chapter also includes a practical, efficient, and reliable quasi-Newton algorithm that eliminates some problems associated with the basic quasi-Newton method. Chapter 8 presents minimax methods that are used in many applications including the design of digital filters. Chapter 9 presents three case studies in which several of the unconstrained optimization methods described in Chapters 4 to 8 are applied to point pattern matching, inverse kinematics for robotic manipulators, and the design of digital filters.

Chapters 10 to 16 are concerned with constrained optimization methods. Chapter 10 introduces the fundamentals of constrained optimization. The concept of Lagrange multipliers, the first-order necessary conditions known as Karush-Kuhn-Tucker conditions, and the duality principle of convex programming are addressed in detail and are illustrated by many examples. Chapters 11 and 12 are concerned with linear programming (LP) problems. The general properties of LP and the simplex method for standard LP problems are addressed in Chapter 11. Several interior-point methods including the primal affine-scaling, primal Newton-barrier, and primal dual-path following methods are presented in Chapter 12. Chapter 13 deals with quadratic and general convex programming. The so-called active-set methods and several interior-point methods for convex quadratic programming are investigated. The chapter also includes the so-called cutting plane and ellipsoid algorithms for general convex programming problems. Chapter 14 presents two special classes of convex programming known as semidefinite and second-order cone programming, which have found interesting applications in a variety of disciplines. Chapter 15 treats general constrained optimization problems that do not belong to the class of convex programming; special emphasis is placed on several sequential quadratic programming methods that are enhanced through the use of efficient line searches and approximations of the Hessian matrix involved. Chapter 16, which concludes the book, examines several applications of constrained optimization for the design of digital filters, for the control of dynamic systems, for evaluating the force distribution in robotic systems, and in multiuser detection for wireless communication systems.

The book also includes two appendices, A and B, which provide additional support material. Appendix A deals in some detail with the relevant parts of linear algebra to consolidate the understanding of the underlying mathematical principles involved whereas Appendix B provides a concise treatment of the basics of digital filters to enhance the understanding of the design algorithms included in Chaps. 8, 9, and 16.

The book can be used as a text for a sequence of two one-semester courses on optimization. The first course comprising Chaps. 1 to 7, 9, and part of Chap. 10 may be offered to senior undergraduate or first-year graduate students. The prerequisite knowledge is an undergraduate mathematics background of calculus and linear algebra. The material in Chaps. 8 and 10 to 16 may be used as a text for an advanced graduate course on minimax and constrained optimization. The prerequisite knowledge for this course is the contents of the first optimization course.

The book is supported by online solutions of the end-of-chapter problems under password as well as by a collection of MATLAB programs for free access by the readers of the book, which can be used to solve a variety of optimization problems. These materials can be downloaded from the book's website: http://www.ece.uvic.ca/~optimization/.

We are grateful to many of our past students at the University of Victoria, in particular, Drs. M. L. R. de Campos, S. Netto, S. Nokleby, D. Peters, and Mr. J. Wong who took our optimization courses and have helped improve the manuscript in one way or another; to Chi-Tang Catherine Chang for typesetting the first draft of the manuscript and for producing most of the illustrations; to R. Nongpiur for checking a large part of the index; and to P. Ramachandran for proofreading the entire manuscript. We would also like to thank Professors M. Ahmadi, C. Charalambous, P. S. R. Diniz, Z. Dong, T. Hinamoto, and P. P. Vaidyanathan for useful discussions on optimization theory and practice; Tony Antoniou of Psicraft Studios for designing the book cover; the Natural Sciences and Engineering Research Council of Canada for supporting the research that led to some of the new results described in Chapters 8, 9, and 16; and last but not least the University of Victoria for supporting the writing of this book over a number of years.

Andreas Antoniou and Wu-Sheng Lu

ABBREVIATIONS

AWGN	additive white Gaussian noise
BER	bit-error rate
BFGS	Broyden-Fletcher-Goldfarb-Shanno
CDMA	code-division multiple access
CMBER	constrained minimum BER
CP	convex programming
DFP	Davidon-Fletcher-Powell
D-H	Denavit-Hartenberg
DNB	dual Newton barrier
DS	direct sequence
FDMA	frequency-division multiple access
FIR	finite-duration impulse response
FR	Fletcher-Reeves
GCO	general constrained optimization
GN	Gauss-Newton
IIR	infinite-duration impulse response
IP	integer programming
KKT	Karush-Kuhn-Tucker
LCP	linear complementarity problem
LMI	linear matrix inequality
LP	linear programming
LSQI	least-squares minimization with quadratic inequality
LU	lower-upper
MAI	multiple access interference
ML	maximum likelihood
MPC	model predictive control
PAS	primal affine-scaling
PCM	predictor-corrector method
PNB	primal Newton barrier
QP	quadratic programming
SD	steepest descent
SDP	semidefinite programming
SDPR-D	SDP relaxation-dual
SDPR-P	SDP relaxation-primal
SNR	signal-to-noise ratio
SOCP	second-order cone programming
SQP	sequential quadratic programming
SVD	singular-value decomposition
TDMA	time-division multiple access

Chapter 1

THE OPTIMIZATION
PROBLEM

1.1   Introduction

Throughout the ages, man has continuously been involved with the process of optimization. In its earliest form, optimization consisted of unscientific rituals and prejudices like pouring libations and sacrificing animals to the gods, consulting the oracles, observing the positions of the stars, and watching the flight of birds. When the circumstances were appropriate, the timing was thought to be auspicious (or optimum) for planting the crops or embarking on a war.

As the ages advanced and the age of reason prevailed, unscientific rituals were replaced by rules of thumb and later, with the development of mathematics, mathematical calculations began to be applied.

Interest in the process of optimization has taken a giant leap with the advent of the digital computer in the early fifties. In recent years, optimization techniques advanced rapidly and considerable progress has been achieved. At the same time, digital computers became faster, more versatile, and more efficient. As a consequence, it is now possible to solve complex optimization problems which were thought intractable only a few years ago.

The process of optimization is the process of obtaining the '*best*', if it is possible to measure and change what is 'good' or 'bad'. In practice, one wishes the 'most' or 'maximum' (e.g., salary) or the 'least' or 'minimum' (e.g., expenses). Therefore, the word '*optimum*' is taken to mean '*maximum*' or '*minimum*' depending on the circumstances; 'optimum' is a technical term which implies quantitative measurement and is a stronger word than 'best' which is more appropriate for everyday use. Likewise, the word '*optimize*', which means to achieve an optimum, is a stronger word than 'improve'. Optimization theory is the branch of mathematics encompassing the quantitative study of optima and methods for finding them. Optimization practice, on the other hand, is the collection of techniques, methods, procedures, and algorithms that can be used to find the optima.

Optimization problems occur in most disciplines like engineering, physics, mathematics, economics, administration, commerce, social sciences, and even politics. Optimization problems abound in the various fields of engineering like electrical, mechanical, civil, chemical, and building engineering. Typical areas of application are modeling, characterization, and design of devices, circuits, and systems; design of tools, instruments, and equipment; design of structures and buildings; process control; approximation theory, curve fitting, solution of systems of equations; forecasting, production scheduling, quality control; maintenance and repair; inventory control, accounting, budgeting, etc. Some recent innovations rely almost entirely on optimization theory, for example, neural networks and adaptive systems.

Most real-life problems have several solutions and occasionally an infinite number of solutions may be possible. Assuming that the problem at hand admits more than one solution, optimization can be achieved by finding the best solution of the problem in terms of some performance criterion. If the problem admits only one solution, that is, only a unique set of parameter values is acceptable, then optimization cannot be applied.

Several general approaches to optimization are available, as follows:

1. Analytical methods
2. Graphical methods
3. Experimental methods
4. Numerical methods

Analytical methods are based on the classical techniques of differential calculus. In these methods the maximum or minimum of a performance criterion is determined by finding the values of parameters x_1, x_2, ..., x_n that cause the derivatives of $f(x_1, x_2, ..., x_n)$ with respect to x_1, x_2, ..., x_n to assume zero values. The problem to be solved must obviously be described in mathematical terms before the rules of calculus can be applied. The method need not entail the use of a digital computer. However, it cannot be applied to highly nonlinear problems or to problems where the number of independent parameters exceeds two or three.

A graphical method can be used to plot the function to be maximized or minimized if the number of variables does not exceed two. If the function depends on only one variable, say, x_1, a plot of $f(x_1)$ versus x_1 will immediately reveal the maxima and/or minima of the function. Similarly, if the function depends on only two variables, say, x_1 and x_2, a set of contours can be constructed. A *contour* is a set of points in the (x_1, x_2) plane for which $f(x_1, x_2)$ is constant, and so a contour plot, like a topographical map of a specific region, will reveal readily the peaks and valleys of the function. For example, the contour plot of $f(x_1, x_2)$ depicted in Fig. 1.1 shows that the function has a minimum at point

A. Unfortunately, the graphical method is of limited usefulness since in most practical applications the function to be optimized depends on several variables, usually in excess of four.

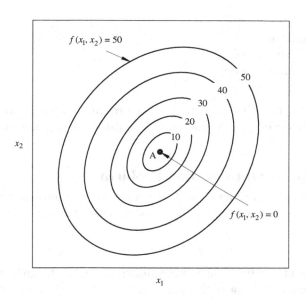

Figure 1.1. Contour plot of $f(x_1, x_2)$.

The optimum performance of a system can sometimes be achieved by direct experimentation. In this method, the system is set up and the process variables are adjusted one by one and the performance criterion is measured in each case. This method may lead to optimum or near optimum operating conditions. However, it can lead to unreliable results since in certain systems, two or more variables interact with each other, and must be adjusted simultaneously to yield the optimum performance criterion.

The most important general approach to optimization is based on numerical methods. In this approach, iterative numerical procedures are used to generate a series of progressively improved solutions to the optimization problem, starting with an initial estimate for the solution. The process is terminated when some convergence criterion is satisfied. For example, when changes in the independent variables or the performance criterion from iteration to iteration become insignificant.

Numerical methods can be used to solve highly complex optimization problems of the type that cannot be solved analytically. Furthermore, they can be readily programmed on the digital computer. Consequently, they have all but replaced most other approaches to optimization.

The discipline encompassing the theory and practice of numerical optimization methods has come to be known as *mathematical programming* [1]–[5]. During the past 40 years, several branches of mathematical programming have evolved, as follows:

1. Linear programming
2. Integer programming
3. Quadratic programming
4. Nonlinear programming
5. Dynamic programming

Each one of these branches of mathematical programming is concerned with a specific class of optimization problems. The differences among them will be examined in Sec. 1.6.

1.2 The Basic Optimization Problem

Before optimization is attempted, the problem at hand must be properly formulated. A performance criterion F must be derived in terms of n parameters $x_1, x_2, \ldots, x_n$ as

$$F = f(x_1, x_2, \ldots, x_n)$$

F is a scalar quantity which can assume numerous forms. It can be the cost of a product in a manufacturing environment or the difference between the desired performance and the actual performance in a system. Variables $x_1, x_2, \ldots, x_n$ are the parameters that influence the product cost in the first case or the actual performance in the second case. They can be independent variables, like time, or control parameters that can be adjusted.

The most basic optimization problem is to adjust variables $x_1, x_2, \ldots, x_n$ in such a way as to minimize quantity F. This problem can be stated mathematically as

$$\text{minimize } F = f(x_1, x_2, \ldots, x_n) \tag{1.1}$$

Quantity F is usually referred to as the *objective* or *cost function*.

The objective function may depend on a large number of variables, sometimes as many as 100 or more. To simplify the notation, matrix notation is usually employed. If $\mathbf{x}$ is a column vector with elements $x_1, x_2, \ldots, x_n$, the transpose of $\mathbf{x}$, namely, $\mathbf{x}^T$, can be expressed as the row vector

$$\mathbf{x}^T = [x_1 \ x_2 \ \cdots \ x_n]$$

In this notation, the basic optimization problem of Eq. (1.1) can be expressed as

$$\text{minimize } F = f(\mathbf{x}) \qquad \text{for } \mathbf{x} \in E^n$$

where E^n represents the *n-dimensional Euclidean space*.

On many occasions, the optimization problem consists of finding the maximum of the objective function. Since

$$\max[f(\mathbf{x})] = -\min[-f(\mathbf{x})]$$

the maximum of F can be readily obtained by finding the minimum of the negative of F and then changing the sign of the minimum. Consequently, in this and subsequent chapters we focus our attention on minimization without loss of generality.

In many applications, a number of distinct functions of $\mathbf{x}$ need to be optimized simultaneously. For example, if the system of nonlinear simultaneous equations

$$f_i(\mathbf{x}) = 0 \qquad \text{for} \quad i = 1, 2, \ldots, m$$

needs to be solved, a vector $\mathbf{x}$ is sought which will reduce all $f_i(\mathbf{x})$ to zero simultaneously. In such a problem, the functions to be optimized can be used to construct a vector

$$\mathbf{F}(\mathbf{x}) = [f_1(\mathbf{x}) \ f_2(\mathbf{x}) \ \cdots \ f_m(\mathbf{x})]^T$$

The problem can be solved by finding a point $\mathbf{x} = \mathbf{x}^*$ such that $\mathbf{F}(\mathbf{x}^*) = \mathbf{0}$. Very frequently, a point $\mathbf{x}^*$ that reduces all the $f_i(\mathbf{x})$ to zero simultaneously may not exist but an approximate solution, i.e., $\mathbf{F}(\mathbf{x}^*) \approx \mathbf{0}$, may be available which could be entirely satisfactory in practice.

A similar problem arises in scientific or engineering applications when the function of $\mathbf{x}$ that needs to be optimized is also a function of a continuous independent parameter (e.g., time, position, speed, frequency) that can assume an infinite set of values in a specified range. The optimization might entail adjusting variables $x_1, x_2, \ldots, x_n$ so as to optimize the function of interest over a given range of the independent parameter. In such an application, the function of interest can be sampled with respect to the independent parameter, and a vector of the form

$$\mathbf{F}(\mathbf{x}) = [f(\mathbf{x}, \ t_1) \ f(\mathbf{x}, \ t_2) \ \cdots \ f(\mathbf{x}, \ t_m)]^T$$

can be constructed, where t is the independent parameter. Now if we let

$$f_i(\mathbf{x}) \equiv f(\mathbf{x}, \ t_i)$$

we can write

$$\mathbf{F}(\mathbf{x}) = [f_1(\mathbf{x}) \ f_2(\mathbf{x}) \ \cdots \ f_m(\mathbf{x})]^T$$

A solution of such a problem can be obtained by optimizing functions $f_i(\mathbf{x})$ for $i = 1, 2, \ldots, m$ simultaneously. Such a solution would, of course, be

6

approximate because any variations in $f(\mathbf{x}, t)$ between sample points are ignored. Nevertheless, reasonable solutions can be obtained in practice by using a sufficiently large number of sample points. This approach is illustrated by the following example.

Example 1.1 The step response $y(\mathbf{x}, t)$ of an nth-order control system is required to satisfy the specification

$$y_0(\mathbf{x}, t) = \begin{cases} t & \text{for } 0 \le t < 2 \\ 2 & \text{for } 2 \le t < 3 \\ -t + 5 & \text{for } 3 \le t < 4 \\ 1 & \text{for } 4 \le t \end{cases}$$

as closely as possible. Construct a vector $\mathbf{F}(\mathbf{x})$ that can be used to obtain a function $f(\mathbf{x}, t)$ such that

$$y(\mathbf{x}, t) \approx y_0(\mathbf{x}, t) \qquad \text{for } 0 \le t \le 5$$

Solution The difference between the actual and specified step responses, which constitutes the *approximation error*, can be expressed as

$$f(\mathbf{x}, t) = y(\mathbf{x}, t) - y_0(\mathbf{x}, t)$$

and if $f(\mathbf{x}, t)$ is sampled at $t = 0, 1, 2, \ldots, 5$, we obtain

$$\mathbf{F}(\mathbf{x}) = [f_1(\mathbf{x})\ f_2(\mathbf{x})\ \cdots\ f_6(\mathbf{x})]^T$$

where

$$f_1(\mathbf{x}) = f(\mathbf{x}, 0) = y(\mathbf{x}, 0)$$
$$f_2(\mathbf{x}) = f(\mathbf{x}, 1) = y(\mathbf{x}, 1) - 1$$
$$f_3(\mathbf{x}) = f(\mathbf{x}, 2) = y(\mathbf{x}, 2) - 2$$
$$f_4(\mathbf{x}) = f(\mathbf{x}, 3) = y(\mathbf{x}, 3) - 2$$
$$f_5(\mathbf{x}) = f(\mathbf{x}, 4) = y(\mathbf{x}, 4) - 1$$
$$f_6(\mathbf{x}) = f(\mathbf{x}, 5) = y(\mathbf{x}, 5) - 1$$

The problem is illustrated in Fig. 1.2. It can be solved by finding a point $\mathbf{x} = \mathbf{x}^*$ such that $\mathbf{F}(\mathbf{x}^*) \approx \mathbf{0}$. Evidently, the quality of the approximation obtained for the step response of the system will depend on the density of the sampling points and the higher the density of points, the better the approximation.

∎

Problems of the type just described can be solved by defining a suitable objective function in terms of the element functions of $\mathbf{F}(\mathbf{x})$. The objective function

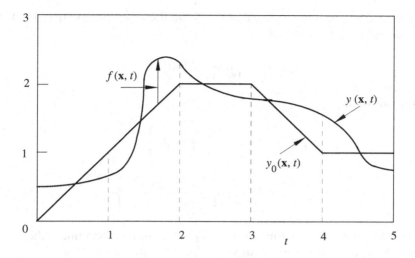

Figure 1.2. Graphical construction for Example 1.1.

must be a *scalar quantity* and its optimization must lead to the simultaneous optimization of the element functions of $\mathbf{F}(\mathbf{x})$ in some sense. Consequently, a norm of some type must be used. An objective function can be defined in terms of the L_p *norm* as

$$F \equiv L_p = \left\{ \sum_{i=1}^{m} |f_i(\mathbf{x})|^p \right\}^{1/p}$$

where p is an integer.[1]

Several special cases of the L_p norm are of particular interest. If $p = 1$

$$F \equiv L_1 = \sum_{i=1}^{m} |f_i(\mathbf{x})|$$

and, therefore, in a minimization problem like that in Example 1.1, the sum of the magnitudes of the individual element functions is minimized. This is called an L_1 *problem.*

If $p = 2$, the *Euclidean norm*

$$F \equiv L_2 = \left\{ \sum_{i=1}^{m} |f_i(\mathbf{x})|^2 \right\}^{1/2}$$

is minimized, and if the square root is omitted, the sum of the squares is minimized. Such a problem is commonly referred as a *least-squares problem.*

---

[1]See Sec. A.8 for more details on vector and matrix norms. Appendix A also deals with other aspects of linear algebra that are important to optimization.

In the case where $p = \infty$, if we assume that there is a unique maximum of $|f_i(\mathbf{x})|$ designated $\hat{F}$ such that

$$\hat{F} = \max_{1 \le i \le m} |f_i(\mathbf{x})|$$

then we can write

$$F \equiv L_\infty = \lim_{p \to \infty} \left\{ \sum_{i=1}^m |f_i(\mathbf{x})|^p \right\}^{1/p}$$

$$= \hat{F} \lim_{p \to \infty} \left\{ \sum_{i=1}^m \left[\frac{|f_i(\mathbf{x})|}{\hat{F}} \right]^p \right\}^{1/p}$$

Since all the terms in the summation except one are less than unity, they tend to zero when raised to a large positive power. Therefore, we obtain

$$F = \hat{F} = \max_{1 \le i \le m} |f_i(\mathbf{x})|$$

Evidently, if the L_∞ *norm* is used in Example 1.1, the maximum approximation error is minimized and the problem is said to be a *minimax problem*.

Often the individual element functions of $\mathbf{F}(\mathbf{x})$ are modified by using constants $w_1, w_2, \ldots, w_m$ as *weights*. For example, the least-squares objective function can be expressed as

$$F = \sum_{i=1}^m [w_i f_i(\mathbf{x})]^2$$

so as to emphasize important or critical element functions and de-emphasize unimportant or uncritical ones. If F is minimized, the residual errors in $w_i f_i(\mathbf{x})$ at the end of the minimization would tend to be of the same order of magnitude, i.e.,

$$\text{error in } |w_i f_i(\mathbf{x})| \approx \varepsilon$$

and so

$$\text{error in } |f_i(\mathbf{x})| \approx \frac{\varepsilon}{|w_i|}$$

Consequently, if a large positive weight w_i is used with $f_i(\mathbf{x})$, a small residual error is achieved in $|f_i(\mathbf{x})|$.

1.3   General Structure of Optimization Algorithms

Most of the available optimization algorithms entail a series of steps which are executed sequentially. A typical pattern is as follows:

Algorithm 1.1 General optimization algorithm

Step 1

(*a*) Set $k = 0$ and initialize $\mathbf{x}_0$.

(*b*) Compute $F_0 = f(\mathbf{x}_0)$.

Step 2

(*a*) Set $k = k + 1$.

(*b*) Compute the changes in $\mathbf{x}_k$ given by column vector $\boldsymbol{\Delta}\mathbf{x}_k$ where

$$\boldsymbol{\Delta}\mathbf{x}_k^T = [\Delta x_1 \; \Delta x_2 \; \cdots \; \Delta x_n]$$

by using an appropriate procedure.

(*c*) Set $\mathbf{x}_k = \mathbf{x}_{k-1} + \boldsymbol{\Delta}\mathbf{x}_k$

(*d*) Compute $F_k = f(\mathbf{x}_k)$ and $\Delta F_k = F_{k-1} - F_k$.

Step 3

Check if convergence has been achieved by using an appropriate criterion, e.g., by checking ΔF_k and/or $\boldsymbol{\Delta}\mathbf{x}_k$. If this is the case, continue to Step 4; otherwise, go to Step 2.

Step 4

(*a*) Output $\mathbf{x}^* = \mathbf{x}_k$ and $F^* = f(\mathbf{x}^*)$.

(*b*) Stop.

In Step 1, vector $\mathbf{x}_0$ is initialized by estimating the solution using knowledge about the problem at hand. Often the solution cannot be estimated and an arbitrary solution may be assumed, say, $\mathbf{x}_0 = \mathbf{0}$. Steps 2 and 3 are then executed repeatedly until convergence is achieved. Each execution of Steps 2 and 3 constitutes one iteration, that is, k is the number of iterations.

When convergence is achieved, Step 4 is executed. In this step, column vector

$$\mathbf{x}^* = [x_1^* \; x_2^* \; \cdots \; x_n^*]^T = \mathbf{x}_k$$

and the corresponding value of F, namely,

$$F^* = f(\mathbf{x}^*)$$

are output. The column vector $\mathbf{x}^*$ is said to be the *optimum, minimum, solution point*, or simply the *minimizer*, and F^* is said to be the optimum or minimum value of the objective function. The pair $\mathbf{x}^*$ and F^* constitute the solution of the optimization problem.

Convergence can be checked in several ways, depending on the optimization problem and the optimization technique used. For example, one might decide to stop the algorithm when the reduction in F_k between any two iterations has become insignificant, that is,

$$|\Delta F_k| = |F_{k-1} - F_k| < \varepsilon_F \tag{1.2}$$

where ε_F is an *optimization tolerance* for the objective function. Alternatively, one might decide to stop the algorithm when the changes in all variables have become insignificant, that is,

$$|\Delta x_i| < \varepsilon_x \qquad \text{for } i = 1, 2, \ldots, n \qquad (1.3)$$

where ε_x is an optimization tolerance for variables $x_1, x_2, \ldots, x_n$. A third possibility might be to check if both criteria given by Eqs. (1.2) and (1.3) are satisfied simultaneously.

There are numerous algorithms for the minimization of an objective function. However, we are primarily interested in algorithms that entail the minimum amount of effort. Therefore, we shall focus our attention on algorithms that are simple to apply, are reliable when applied to a diverse range of optimization problems, and entail a small amount of computation. A reliable algorithm is often referred to as a *'robust'* algorithm in the terminology of mathematical programming.

1.4   Constraints

In many optimization problems, the variables are interrelated by physical laws like the conservation of mass or energy, Kirchhoff's voltage and current laws, and other system equalities that must be satisfied. In effect, in these problems certain equality constraints of the form

$$a_i(\mathbf{x}) = 0 \qquad \text{for } \mathbf{x} \in E^n$$

where $i = 1, 2, \ldots, p$ must be satisfied before the problem can be considered solved. In other optimization problems a collection of inequality constraints might be imposed on the variables or parameters to ensure physical realizability, reliability, compatibility, or even to simplify the modeling of the problem. For example, the power dissipation might become excessive if a particular current in a circuit exceeds a given upper limit or the circuit might become unreliable if another current is reduced below a lower limit, the mass of an element in a specific chemical reaction must be positive, and so on. In these problems, a collection of inequality constraints of the form

$$c_j(\mathbf{x}) \geq 0 \qquad \text{for } \mathbf{x} \in E^n$$

where $j = 1, 2, \ldots, q$ must be satisfied before the optimization problem can be considered solved.

An optimization problem may entail a set of equality constraints and possibly a set of inequality constraints. If this is the case, the problem is said to be a *constrained optimization problem*. The most general constrained optimization problem can be expressed mathematically as

$$\text{minimize } f(\mathbf{x}) \qquad \text{for } \mathbf{x} \in E^n \tag{1.4a}$$
$$\text{subject to: } \quad a_i(\mathbf{x}) = 0 \qquad \text{for } i = 1, 2, \ldots, p \tag{1.4b}$$
$$c_j(\mathbf{x}) \geq 0 \qquad \text{for } j = 1, 2, \ldots, q \tag{1.4c}$$

A problem that does not entail any equality or inequality constraints is said to be an *unconstrained optimization problem.*

Constrained optimization is usually much more difficult than unconstrained optimization, as might be expected. Consequently, the general strategy that has evolved in recent years towards the solution of constrained optimization problems is to reformulate constrained problems as unconstrained optimization problems. This can be done by redefining the objective function such that the constraints are simultaneously satisfied when the objective function is minimized. Some real-life constrained optimization problems are given as Examples 1.2 to 1.4 below.

Example 1.2 Consider a control system that comprises a double inverted pendulum as depicted in Fig. 1.3. The objective of the system is to maintain the pendulum in the upright position using the minimum amount of energy. This is achieved by applying an appropriate control force to the car to damp out any displacements $\theta_1(t)$ and $\theta_2(t)$. Formulate the problem as an optimization problem.

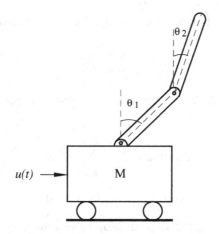

Figure 1.3.   The double inverted pendulum.

Solution The dynamic equations of the system are nonlinear and the standard practice is to apply a linearization technique to these equations to obtain a small-signal linear model of the system as [6]

$$\dot{\mathbf{x}}(t) = \mathbf{A}\mathbf{x}(t) + \mathbf{f}u(t) \tag{1.5}$$

where

$$\mathbf{x}(t) = \begin{bmatrix} \theta_1(t) \\ \dot{\theta}_1(t) \\ \theta_2(t) \\ \dot{\theta}_2(t) \end{bmatrix}, \quad \mathbf{A} = \begin{bmatrix} 0 & 1 & 0 & 0 \\ \alpha & 0 & -\beta & 0 \\ 0 & 0 & 0 & 1 \\ -\alpha & 0 & \alpha & 0 \end{bmatrix}, \quad \mathbf{f} = \begin{bmatrix} 0 \\ -1 \\ 0 \\ 0 \end{bmatrix}$$

with $\alpha > 0$, $\beta > 0$, and $\alpha \neq \beta$. In the above equations, $\dot{\mathbf{x}}(t)$, $\dot{\theta}_1(t)$, and $\dot{\theta}_2(t)$ represent the first derivatives of $\mathbf{x}(t)$, $\theta_1(t)$, and $\theta_2(t)$, respectively, with respect to time, $\ddot{\theta}_1(t)$ and $\ddot{\theta}_2(t)$ would be the second derivatives of $\theta_1(t)$ and $\theta_2(t)$, and parameters α and β depend on system parameters such as the length and weight of each pendulum, the mass of the car, etc. Suppose that at instant $t = 0$ small nonzero displacements $\theta_1(t)$ and $\theta_2(t)$ occur, which would call for immediate control action in order to steer the system back to the equilibrium state $\mathbf{x}(t) = \mathbf{0}$ at time $t = T_0$. In order to develop a digital controller, the system model in (1.5) is discretized to become

$$\mathbf{x}(k+1) = \mathbf{\Phi}\mathbf{x}(k) + \mathbf{g}u(k) \tag{1.6}$$

where $\mathbf{\Phi} = \mathbf{I} + \Delta t \mathbf{A}$, $\mathbf{g} = \Delta t \mathbf{f}$, Δt is the sampling interval, and $\mathbf{I}$ is the identity matrix. Let $\mathbf{x}(0) \neq \mathbf{0}$ be given and assume that T_0 is a multiple of Δt, i.e., $T_0 = K\Delta t$ where K is an integer. We seek to find a sequence of control actions $u(k)$ for $k = 0, 1, \ldots, K - 1$ such that the zero equilibrium state is achieved at $t = T_0$, i.e., $\mathbf{x}(T_0) = \mathbf{0}$.

Let us assume that the energy consumed by these control actions, namely,

$$J = \sum_{k=0}^{K-1} u^2(k)$$

needs to be minimized. This optimal control problem can be formulated analytically as

$$\text{minimize } J = \sum_{k=0}^{K-1} u^2(k) \tag{1.7a}$$

$$\text{subject to: } \mathbf{x}(K) = \mathbf{0} \tag{1.7b}$$

From Eq. (1.6), we know that the state of the system at $t = K\Delta t$ is determined by the initial value of the state and system model in Eq. (1.6) as

$$\mathbf{x}(K) = \mathbf{\Phi}^K \mathbf{x}(0) + \sum_{k=0}^{K-1} \mathbf{\Phi}^{K-k-1} \mathbf{g} u(k)$$

$$\equiv -\mathbf{h} + \sum_{k=0}^{K-1} \mathbf{g}_k u(k)$$

where $\mathbf{h} = -\boldsymbol{\Phi}^K \mathbf{x}(0)$ and $\mathbf{g}_k = \boldsymbol{\Phi}^{K-k-1}\mathbf{g}$. Hence constraint (1.7b) is equivalent to

$$\sum_{k=0}^{K-1} \mathbf{g}_k u(k) = \mathbf{h} \tag{1.8}$$

If we define $\mathbf{u} = [u(0) \ u(1) \ \cdots \ u(K-1)]^T$ and $\mathbf{G} = [\mathbf{g}_0 \ \mathbf{g}_1 \ \cdots \ \mathbf{g}_{K-1}]$, then the constraint in Eq. (1.8) can be expressed as $\mathbf{Gu} = \mathbf{h}$, and the optimal control problem at hand can be formulated as the problem of finding a $\mathbf{u}$ that solves the minimization problem

$$\text{minimize } \mathbf{u}^T\mathbf{u} \tag{1.9a}$$

$$\text{subject to: } \mathbf{a}(\mathbf{u}) = \mathbf{0} \tag{1.9b}$$

where $\mathbf{a}(\mathbf{u}) = \mathbf{Gu} - \mathbf{h}$. In practice, the control actions cannot be made arbitrarily large in magnitude. Consequently, additional constraints are often imposed on $|u(i)|$, for instance,

$$|u(i)| \le m \qquad \text{for } i = 0, \ 1, \ \ldots, \ K-1$$

These constraints are equivalent to

$$m + u(i) \ge 0$$
$$m - u(i) \ge 0$$

Hence if we define

$$\mathbf{c}(\mathbf{u}) = \begin{bmatrix} m + u(0) \\ m - u(0) \\ \vdots \\ m + u(K-1) \\ m - u(K-1) \end{bmatrix}$$

then the magnitude constraints can be expressed as

$$\mathbf{c}(\mathbf{u}) \ge \mathbf{0} \tag{1.9c}$$

Obviously, the problem in Eq. (1.9) fits nicely into the standard form of optimization problems given by Eq. (1.4). ∎

Example 1.3 High performance in modern optical instruments depends on the quality of components like lenses, prisms, and mirrors. These components have reflecting or partially reflecting surfaces, and their performance is limited by the reflectivities of the materials of which they are made. The surface reflectivity

14

can, however, be altered by the deposition of a thin transparent film. In fact, this technique facilitates the control of losses due to reflection in lenses and makes possible the construction of mirrors with unique properties [7][8].

As is depicted in Fig. 1.4, a typical N-layer thin-film system consists of N layers of thin films of certain transparent media deposited on a glass substrate. The thickness and refractive index of the ith layer are denoted as x_i and n_i, respectively. The refractive index of the medium above the first layer is denoted as n_0. If ϕ_0 is the angle of incident light, then the transmitted ray in the $(i-1)$th layer is refracted at an angle ϕ_i which is given by Snell's law, namely,

$$n_i \sin \phi_i = n_0 \sin \phi_0$$

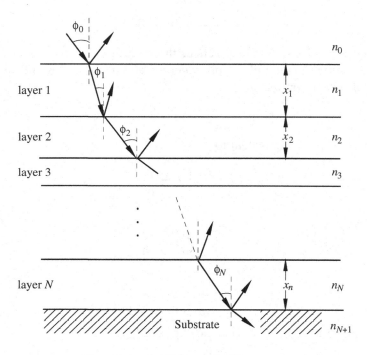

Figure 1.4. An N-layer thin-film system.

Given angle ϕ_0 and the wavelength of light, λ, the energy of the light reflected from the film surface and the energy of the light transmitted through the film surface are usually measured by the reflectance R and transmittance T which satisfy the relation

$$R + T = 1$$

For an N-layer system, R is given by (see [9] for details)

$$R(x_1, \ldots, x_N, \lambda) = \left| \frac{\eta_0 - y}{\eta_0 + y} \right|^2 \tag{1.10}$$

$$y = \frac{c}{b} \tag{1.11}$$

$$\begin{bmatrix} b \\ c \end{bmatrix} = \left\{ \prod_{k=1}^{N} \begin{bmatrix} \cos \delta_k & (j \sin \delta_k)/\eta_k \\ j\eta_k \sin \delta_k & \cos \delta_k \end{bmatrix} \right\} \begin{bmatrix} 1 \\ \eta_{N+1} \end{bmatrix} \tag{1.12}$$

where $j = \sqrt{-1}$ and

$$\delta_k = \frac{2\pi n_k x_k \cos \phi_k}{\lambda} \tag{1.13}$$

$$\eta_k = \begin{cases} n_k / \cos \phi_k & \text{for light polarized with the electric} \\ & \text{vector lying in the plane of incidence} \\[2ex] n_k \cos \phi_k & \text{for light polarized with the electric} \\ & \text{vector perpendicular to the} \\ & \text{plane of incidence} \end{cases} \tag{1.14}$$

The design of a multilayer thin-film system can now be accomplished as follows: Given a range of wavelenghs $\lambda_l \leq \lambda \leq \lambda_u$ and an angle of incidence ϕ_0, find $x_1, x_2, \ldots, x_N$ such that the reflectance $R(\mathbf{x}, \lambda)$ best approximates a desired reflectance $R_d(\lambda)$ for $\lambda \in [\lambda_l, \lambda_u]$. Formulate the design problem as an optimization problem.

Solution In practice, the desired reflectance is specified at grid points $\lambda_1, \lambda_2,$ $\ldots, \lambda_K$ in the interval $[\lambda_l, \lambda_u]$; hence the design may be carried out by selecting x_i such that the objective function

$$J = \sum_{i=1}^{K} w_i [R(\mathbf{x}, \lambda_i) - R_d(\lambda_i)]^2 \tag{1.15}$$

is minimized, where

$$\mathbf{x} = [x_1 \, x_2 \, \cdots \, x_N]^T$$

and $w_i > 0$ is a weight to reflect the importance of term $[R(\mathbf{x}, \lambda_i) - R_d(\lambda_i)]^2$ in Eq. (1.15). If we let $\boldsymbol{\eta} = [1 \, \eta_{N+1}]^T$, $\mathbf{e}_+ = [\eta_0 \, 1]^T$, $\mathbf{e}_- = [\eta_0 \, -1]^T$, and

$$\mathbf{M}(\mathbf{x}, \lambda) = \prod_{k=1}^{N} \begin{bmatrix} \cos \delta_k & (j \sin \delta_k)/\eta_k \\ j\eta_k \sin \delta_k & \cos \delta_k \end{bmatrix}$$

then $R(\mathbf{x}, \lambda)$ can be expressed as

$$R(\mathbf{x}, \lambda) = \left| \frac{b\eta_0 - c}{b\eta_0 + c} \right|^2 = \left| \frac{\mathbf{e}_-^T \mathbf{M}(\mathbf{x}, \lambda) \boldsymbol{\eta}}{\mathbf{e}_+^T \mathbf{M}(\mathbf{x}, \lambda) \boldsymbol{\eta}} \right|^2 \tag{1.16}$$

Finally, we note that the thickness of each layer cannot be made arbitrarily thin or arbitrarily large and, therefore, constraints must be imposed on the elements of $\mathbf{x}$ as

$$d_{il} \leq x_i \leq d_{iu} \qquad \text{for } i = 1, 2, \ldots, N \qquad (1.17)$$

The design problem can now be formulated as the constrained minimization problem

$$\text{minimize } J = \sum_{i=1}^{K} w_i \left[\left| \frac{\mathbf{e}_-^T \mathbf{M}(\mathbf{x}, \lambda_i) \boldsymbol{\eta}}{\mathbf{e}_+^T \mathbf{M}(\mathbf{x}, \lambda_i) \boldsymbol{\eta}} \right|^2 - R_d(\lambda_i) \right]^2 \qquad (1.18a)$$

$$\text{subject to:} \quad x_i - d_{il} \geq 0 \qquad \text{for } i = 1, 2, \ldots, N \qquad (1.18b)$$

$$d_{iu} - x_i \geq 0 \qquad \text{for } i = 1, 2, \ldots, N \qquad (1.18c)$$

∎

Example 1.4 Quantities $q_1, q_2, \ldots, q_m$ of a certain product are produced by m manufacturing divisions of a company, which are at distinct locations. The product is to be shipped to n destinations that require quantities $b_1, b_2, \ldots, b_n$. Assume that the cost of shipping a unit from manufacturing division i to destination j is c_{ij} with $i = 1, 2, \ldots, m$ and $j = 1, 2, \ldots, n$. Find the quantity x_{ij} to be shipped from division i to destination j so as to minimize the total cost of transportation, i.e.,

$$\text{minimize } C = \sum_{i=1}^{m} \sum_{j=1}^{n} c_{ij} x_{ij}$$

This is known as the *transportation problem*. Formulate the problem as an optimization problem.

Solution Note that there are several constraints on variables x_{ij}. First, each division can provide only a fixed quantity of the product, hence

$$\sum_{j=1}^{n} x_{ij} = q_i \qquad \text{for } i = 1, 2, \ldots, m$$

Second, the quantity to be shipped to a specific destination has to meet the need of that destination and so

$$\sum_{i=1}^{m} x_{ij} = b_j \qquad \text{for } j = 1, 2, \ldots, n$$

In addition, the variables x_{ij} are nonnegative and thus, we have

$$x_{ij} \geq 0 \qquad \text{for } i = 1, 2, \ldots, m \quad \text{and} \quad j = 1, 2, \ldots, n$$

If we let

$$\mathbf{c} = \begin{bmatrix} c_{11} & \cdots & c_{1n} & c_{21} & \cdots & c_{2n} & \cdots & c_{m1} & \cdots & c_{mn} \end{bmatrix}^T$$

$$\mathbf{x} = \begin{bmatrix} x_{11} & \cdots & x_{1n} & x_{21} & \cdots & x_{2n} & \cdots & x_{m1} & \cdots & x_{mn} \end{bmatrix}^T$$

$$\mathbf{A} = \begin{bmatrix} 1 & 1 & \cdots & 1 & 0 & 0 & \cdots & 0 & \cdots & \cdots & \cdots & \cdots \\ 0 & 0 & \cdots & 0 & 1 & 1 & \cdots & 1 & \cdots & \cdots & \cdots & \cdots \\ \cdots & \cdots & \cdots & \cdots & \cdots & \cdots & \cdots & \cdots & \cdots & \cdots & \cdots & \cdots \\ 0 & 0 & \cdots & 0 & 0 & 0 & \cdots & 0 & \cdots & 1 & 1 & \cdots & 1 \\ 1 & 0 & \cdots & 0 & 1 & 0 & \cdots & 0 & \cdots & 1 & 0 & \cdots & 0 \\ 0 & 1 & \cdots & 0 & 0 & 1 & \cdots & 0 & \cdots & 0 & 1 & \cdots & 0 \\ \cdots & \cdots & \cdots & \cdots & \cdots & \cdots & \cdots & \cdots & \cdots & \cdots & \cdots & \cdots \\ 0 & 0 & \cdots & 1 & 0 & 0 & \cdots & 1 & \cdots & 0 & 0 & \cdots & 1 \end{bmatrix}$$

$$\mathbf{b} = \begin{bmatrix} q_1 & \cdots & q_m & b_1 & \cdots & b_n \end{bmatrix}^T$$

then the minimization problem can be stated as

$$\text{minimize } C = \mathbf{c}^T \mathbf{x} \tag{1.19a}$$

$$\text{subject to: } \mathbf{A}\mathbf{x} = \mathbf{b} \tag{1.19b}$$

$$\mathbf{x} \geq \mathbf{0} \tag{1.19c}$$

where $\mathbf{c}^T\mathbf{x}$ is the inner product of $\mathbf{c}$ and $\mathbf{x}$. The problem in Eq. (1.19) like those in Examples 1.2 and 1.3 fits into the standard optimization problem in Eq. (1.4). Since both the objective function in Eq. (1.19a) and the constraints in Eqs. (1.19b) and (1.19c) are linear, the problem is known as a *linear programming (LP) problem* (see Sect. 1.6.1).

∎

1.5 The Feasible Region

Any point $\mathbf{x}$ that satisfies both the equality as well as the inequality constraints is said to be a *feasible point* of the optimization problem. The set of all points that satisfy the constraints constitutes the *feasible domain region* of $f(\mathbf{x})$. Evidently, the constraints define a subset of E^n. Therefore, the feasible region can be defined as a set[2]

$$\mathcal{R} = \{\mathbf{x} : a_i(\mathbf{x}) = 0 \text{ for } i = 1, 2, \ldots, p \text{ and } c_j(\mathbf{x}) \geq 0 \text{ for } j = 1, 2, \ldots, q\}$$

where $\mathcal{R} \subset E^n$.

The optimum point $\mathbf{x}^*$ must be located in the feasible region, and so the general constrained optimization problem can be stated as

$$\text{minimize } f(\mathbf{x}) \qquad \text{for } \mathbf{x} \in \mathcal{R}$$

---

[2]The above notation for a set will be used consistently throughout the book.

Any point $\mathbf{x}$ not in $\mathcal{R}$ is said to be a *nonfeasible point*.

If the constraints in an optimization problem are all inequalities, the constraints divide the points in the E^n space into three types of points, as follows:

1. Interior points
2. Boundary points
3. Exterior points

An *interior point* is a point for which $c_j(\mathbf{x}) > 0$ for all j. A *boundary point* is a point for which at least one $c_j(\mathbf{x}) = 0$, and an *exterior point* is a point for which at least one $c_j(\mathbf{x}) < 0$. Interior points are feasible points, boundary points may or may not be feasible points, whereas exterior points are nonfeasible points.

If a constraint $c_m(\mathbf{x})$ is zero during a specific iteration, the constraint is said to be *active*, and if $c_m(\mathbf{x}^*)$ is zero when convergence is achieved, the optimum point $\mathbf{x}^*$ is located on the boundary. In such a case, the optimum point is said to be constrained. If the constraints are all equalities, the feasible points must be located on the intersection of all the hypersurfaces corresponding to $a_i(\mathbf{x}) = 0$ for $i = 1, 2, \ldots, p$. The above definitions and concepts are illustrated by the following two examples.

Example 1.5 By using a graphical method, solve the following optimization problem

$$\text{minimize } f(\mathbf{x}) = x_1^2 + x_2^2 - 4x_1 + 4$$
$$\text{subject to: } \quad c_1(\mathbf{x}) = x_1 - 2x_2 + 6 \geq 0$$
$$c_2(\mathbf{x}) = -x_1^2 + x_2 - 1 \geq 0$$
$$c_3(\mathbf{x}) = x_1 \geq 0$$
$$c_4(\mathbf{x}) = x_2 \geq 0$$

Solution The objective function can be expressed as

$$(x_1 - 2)^2 + x_2^2 = f(\mathbf{x})$$

Hence the contours of $f(\mathbf{x})$ in the (x_1, x_2) plane are concentric circles with radius $\sqrt{f(\mathbf{x})}$ centered at $x_1 = 2$, $x_2 = 0$. Constraints $c_1(\mathbf{x})$ and $c_2(\mathbf{x})$ dictate that

$$x_2 \leq \tfrac{1}{2}x_1 + 3$$

and

$$x_2 \geq x_1^2 + 1$$

respectively, while constraints $c_3(\mathbf{x})$ and $c_4(\mathbf{x})$ dictate that x_1 and x_2 be positive. The contours of $f(\mathbf{x})$ and the boundaries of the constraints can be constructed as shown in Fig. 1.5.

The feasible region for this problem is the shaded region in Fig. 1.5. The solution is located at point A on the boundary of constraint $c_2(\mathbf{x})$. In effect,

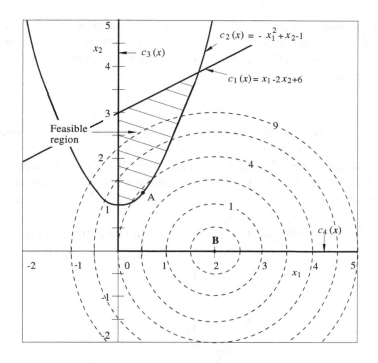

Figure 1.5. Graphical construction for Example 1.5.

the solution is a constrained optimum point. Consequently, if this problem is solved by means of mathematical programming, constraint $c_2(\mathbf{x})$ will be active when the solution is reached.

In the absence of constraints, the minimization of $f(\mathbf{x})$ would yield point B as the solution.

■

Example 1.6 By using a graphical method, solve the optimization problem

$$\text{minimize } f(\mathbf{x}) = x_1^2 + x_2^2 + 2x_2$$
$$\text{subject to: } a_1(\mathbf{x}) = x_1^2 + x_2^2 - 1 = 0$$
$$c_1(\mathbf{x}) = x_1 + x_2 - 0.5 \geq 0$$
$$c_2(\mathbf{x}) = x_1 \geq 0$$
$$c_3(\mathbf{x}) = x_2 \geq 0$$

Solution The objective function can be expressed as

$$x_1^2 + (x_2 + 1)^2 = f(\mathbf{x}) + 1$$

Hence the contours of $f(\mathbf{x})$ in the $(x_1,\ x_2)$ plane are concentric circles with radius $\sqrt{f(\mathbf{x})+1}$, centered at $x_1 = 0$, $x_2 = -1$. Constraint $a_1(\mathbf{x})$ is a circle centered at the origin with radius 1. On the other hand, constraint $c_1(\mathbf{x})$ is a straight line since it is required that

$$x_2 \geq -x_1 + 0.5$$

The last two constraints dictate that x_1 and x_2 be nonnegative. Hence the required construction can be obtained as depicted in Fig. 1.6.

In this case, the feasible region is the arc of circle $a_1(\mathbf{x}) = 0$ located in the first quadrant of the $(x_1,\ x_2)$ plane. The solution, which is again a constrained optimum point, is located at point A. There are two active constraints in this example, namely, $a_1(\mathbf{x})$ and $c_3(\mathbf{x})$.

In the absence of constraints, the solution would be point B in Fig. 1.6.

■

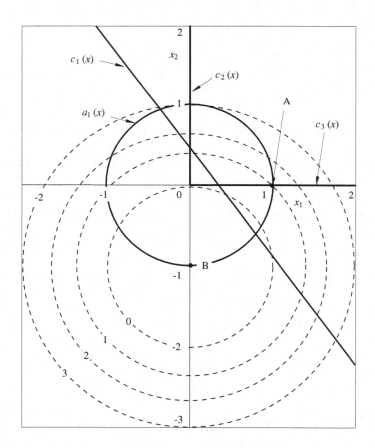

Figure 1.6. Graphical construction for Example 1.6.

In the above examples, the set of points comprising the feasible region are simply connected as depicted in Fig. 1.7a. Sometimes the feasible region may consist of two or more disjoint sub-regions, as depicted in Fig. 1.7b. If this is the case, the following difficulty may arise. A typical optimization algorithm is an iterative numerical procedure that will generate a series of progressively improved solutions, starting with an initial estimate for the solution. Therefore, if the feasible region consists of two sub-regions, say, A and B, an initial estimate for the solution in sub-region A is likely to yield a solution in sub-region A, and a better solution in sub-region B may be missed. Fortunately, however, in most real-life optimization problems, this difficulty can be avoided by formulating the problem carefully.

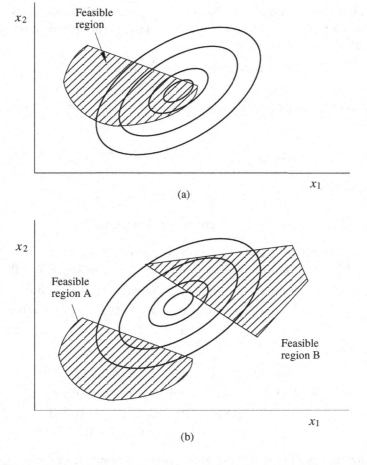

Figure 1.7. Examples of simply connected and disjoint feasible regions.

1.6 Branches of Mathematical Programming

Several branches of mathematical programming were enumerated in Sec. 1.1, namely, linear, integer, quadratic, nonlinear, and dynamic programming. Each one of these branches of mathematical programming consists of the theory and application of a collection of optimization techniques that are suited to a specific class of optimization problems. The differences among the various branches of mathematical programming are closely linked to the structure of the optimization problem and to the mathematical nature of the objective and constraint functions. A brief description of each branch of mathematical programming is as follows.

1.6.1 Linear programming

If the objective and constraint functions are linear and the variables are constrained to be positive, as in Example 1.4, the general optimization problem assumes the form

$$\text{minimize } f(\mathbf{x}) = \sum_{i=1}^{n}\alpha_i x_i$$

$$\text{subject to: } a_j(\mathbf{x}) = \sum_{i=1}^{n}\beta_{ij} x_i - \mu_j = 0 \qquad \text{for } j = 1, 2, \ldots, p$$

$$c_j(\mathbf{x}) = \sum_{i=1}^{n}\gamma_{ij} x_i - \nu_j \geq 0 \qquad \text{for } j = 1, 2, \ldots, q$$

$$x_i \geq 0 \qquad \text{for } i = 1, 2, \ldots, n$$

where α_i, β_{ij}, γ_{ij}, μ_j and ν_j are constants. For example,

$$\text{minimize } f(\mathbf{x}) = -2x_1 + 4x_2 + 7x_3 + x_4 + 5x_5$$
$$\text{subject to: } a_1(\mathbf{x}) = -x_1 + x_2 + 2x_3 + x_4 + 2x_5 - 7 = 0$$
$$a_2(\mathbf{x}) = -x_1 + 2x_2 + 3x_3 + x_4 + x_5 - 6 = 0$$
$$a_3(\mathbf{x}) = -x_1 + x_2 + x_3 + 2x_4 + x_5 - 4 = 0$$
$$x_i \geq 0 \qquad \text{for } i = 1, 2, \ldots, 5$$

or

$$\text{minimize } f(\mathbf{x}) = 3x_1 + 4x_2 + 5x_3$$
$$\text{subject to: } c_1(\mathbf{x}) = x_1 + 2x_2 + 3x_3 - 5 \geq 0$$
$$c_2(\mathbf{x}) = 2x_1 + 2x_2 + x_3 - 6 \geq 0$$
$$x_1 \geq 0, \ x_2 \geq 0, \ x_3 \geq 0$$

Optimization problems like the above occur in many disciplines. Their solution can be readily achieved by using some powerful LP algorithms as will be shown in Chaps. 11 and 12.

1.6.2   Integer programming

In certain linear programming problems, at least some of the variables are required to assume only integer values. This restriction renders the programming problem nonlinear. Nevertheless, the problem is referred to as linear since the objective and constraint functions are linear [10].

1.6.3   Quadratic programming

If the optimization problem assumes the form

$$\text{minimize } f(\mathbf{x}) = \alpha_0 + \boldsymbol{\gamma}^T \mathbf{x} + \mathbf{x}^T \mathbf{Q}\, \mathbf{x}$$

$$\text{subject to: } \boldsymbol{\alpha}^T \mathbf{x} \geq \boldsymbol{\beta}$$

where

$$\boldsymbol{\alpha} = \begin{bmatrix} \alpha_{11} & \alpha_{22} & \cdots & \alpha_{1q} \\ \alpha_{21} & \alpha_{22} & \cdots & \alpha_{2q} \\ \vdots & \vdots & & \vdots \\ \alpha_{n1} & \alpha_{n2} & \cdots & \alpha_{nq} \end{bmatrix}$$

$$\boldsymbol{\beta}^T = [\beta_1 \quad \beta_2 \quad \cdots \quad \beta_q]$$

$$\boldsymbol{\gamma}^T = [\gamma_1 \quad \gamma_2 \quad \cdots \quad \gamma_n]$$

and $\mathbf{Q}$ is a positive definite or semidefinite symmetric square matrix, then the constraints are linear and the objective function is quadratic. Such an optimization problem is said to be a quadratic programming (QP) problem (see Chap. 10 of [5]). A typical example of this type of problem is as follows:

$$\text{minimize } f(\mathbf{x}) = \tfrac{1}{2}x_1^2 + \tfrac{1}{2}x_2^2 - x_1 - 2x_2$$

$$\text{subject to: } c_1(\mathbf{x}) = 6 - 2x_1 - 3x_2 \geq 0$$
$$c_2(\mathbf{x}) = 5 - x_1 - 4x_2 \geq 0$$
$$c_3(\mathbf{x}) = x_1 \geq 0$$
$$c_4(\mathbf{x}) = x_2 \geq 0$$

1.6.4   Nonlinear programming

In nonlinear programming problems, the objective function and usually the constraint functions are nonlinear. Typical examples were given earlier as Examples 1.1 to 1.3. This is the most general branch of mathematical programming and, in effect, LP and QP can be considered as special cases of nonlinear programming. Although it is possible to solve linear or quadratic programming problems by using nonlinear programming algorithms, the specialized algorithms developed for linear or quadratic programming should be used for these problems since they are usually much more efficient.

The choice of optimization algorithm depends on the mathematical behavior and structure of the objective function. Most of the time, the objective function is a well behaved nonlinear function and all that is necessary is a general-purpose, robust, and efficient algorithm. For certain applications, however, specialized algorithms exist which are often more efficient than general-purpose ones. These are often referred to by the type of norm minimized, for example, an algorithm that minimizes an L_1, L_2, or L_∞ norm is said to by an L_1, L_2, or *minimax algorithm*.

1.6.5 Dynamic programming

In many applications, a series of decisions must be made in sequence, where subsequent decisions are influenced by earlier ones. In such applications, a number of optimizations have to be performed in sequence and a general strategy may be required to achieve an overall optimum solution. For example, a large system which cannot be optimized owing to the size and complexity of the problem can be partitioned into a set of smaller sub-systems that can be optimized individually. Often individual sub-systems interact with each other and, consequently, a general solution strategy is required if an overall optimum solution is to be achieved. Dynamic programming is a collection of techniques that can be used to develop general solution strategies for problems of the type just described. It is usually based on the use of linear, integer, quadratic or nonlinear optimization algorithms.

References

1   G. B. Dantzig, *Linear Programming and Extensions*, Princeton University Press, Princeton, N.J., 1963.

2   D. M. Himmelblau, *Applied Nonlinear Programming*, McGraw-Hill, New York, 1972.

3   P. E. Gill, W. Murray, and M. H. Wright, *Practical Optimization*, Academic Press, London, 1981.

4   D. G. Luenberger, *Linear and Nonlinear Programming*, 2nd ed., Addison-Wesley, Reading, MA, 1984.

5   R. Fletcher, *Practical Methods of Optimization*, 2nd ed., Wiley, Chichester, UK, 1987.

6   B. C. Kuo, *Automatic Control Systems*, 5th ed., Prentice Hall, Englewood Cliffs, N.J., 1987.

7   K. D. Leaver and B. N. Chapman, *Thin Films*, Wykeham, London, 1971.

8   O. S. Heavens, *Thin Film Physics*, Methuen, London, 1970.

9   Z. Knittl, *Optics of Thin Films, An Optical Multilayer Theory*, Wiley, New York, 1976.

10 G. L. Nemhauser and L. A. Wolsey, *Integer and Combinatorial Optimization*, Wiley, New York, 1988.

Problems

1.1 (*a*) Solve the following minimization problem by using a graphical method:

$$\text{minimize } f(\mathbf{x}) = x_1^2 + x_2 + 4$$

$$\text{subject to:} \quad c_1(\mathbf{x}) = -x_1^2 - (x_2 + 4)^2 + 16 \geq 0$$
$$c_2(\mathbf{x}) = x_1 - x_2 - 6 \geq 0$$

Note: An explicit numerical solution is required.

(*b*) Indicate the feasible region.

(*c*) Is the optimum point constrained?

1.2 Repeat Prob. 1(*a*) to (*c*) for the problem

$$\text{minimize } f(\mathbf{x}) = x_2 - \frac{8}{x_1}$$

$$\text{subject to:} \quad c_1(\mathbf{x}) = \tfrac{1}{5}x_1 - x_2 \geq 0$$
$$c_2(\mathbf{x}) = 16 - (x_1 - 5)^2 - x_2^2 \geq 0$$

Note: Obtain an accurate solution by using MATLAB.

1.3 Repeat Prob. 1(*a*) to (*c*) for the problem

$$\text{minimize } f(\mathbf{x}) = (x_1 - 12)x_1 + (x_2 - 6)x_2 + 45$$

$$\text{subject to:} \quad c_1(\mathbf{x}) = \tfrac{7}{5}x_1 - x_2 - \tfrac{7}{5} \geq 0$$
$$c_2(\mathbf{x}) = -x_2 - \tfrac{7}{5}x_1 + \tfrac{77}{5} \geq 0$$
$$c_3(\mathbf{x}) = x_2 \geq 0$$

1.4 Repeat Prob. 1(*a*) to (*c*) for the problem

$$\text{minimize } f(\mathbf{x}) = \tfrac{1}{4}(x_1 - 6)^2 + (x_2 - 4)^2$$

$$\text{subject to:} \quad a_1(\mathbf{x}) = x_1 - 3 = 0$$
$$c_1(\mathbf{x}) = \tfrac{80}{7} - x_2 - \tfrac{8}{7}x_1 \geq 0$$
$$c_2(\mathbf{x}) = x_2 \geq 0$$

1.5 Develop a method to determine the coordinates of point A in Example 1.5 based on the following observation: From Fig. 1.5, we see that there will be no intersection points between the contour of $f(\mathbf{x}) = r^2$ and constraint $c_2(\mathbf{x}) = 0$ if radius r is smaller than the distance A to B and there will be two distinct intersection points between them if r is larger than the distance A to B. Therefore, the solution point A can be identified by determining the value of r for which the distance between the two intersection points is sufficiently small.

1.6 Solve the constrained minimization problem

$$\text{minimize } f(\mathbf{x}) = 3x_1 + 2x_2 + x_3$$

$$\text{subject to:} \quad a_1(\mathbf{x}) = 2x_1 + 3x_2 + x_3 = 30$$
$$c_1(\mathbf{x}) = x_1 \geq 0$$
$$c_2(\mathbf{x}) = x_2 \geq 0$$
$$c_3(\mathbf{x}) = x_3 \geq 0$$

Hint: (i) Use the equality constraint to eliminate variable x_3, and (ii) use $x = \hat{x}^2$ to eliminate constraint $x \geq 0$.

1.7 Consider the constrained minimization problem

$$\text{minimize } f(\mathbf{x}) = -5\sin(x_1 + x_2) + (x_1 - x_2)^2 - x_1 - 2x_2$$

$$\text{subject to:} \quad c_1(\mathbf{x}) = 5 - x_1 \geq 0$$
$$c_2(\mathbf{x}) = 5 - x_2 \geq 0$$

(a) Plot a dense family of contours for $f(\mathbf{x})$ over the region $D = \{(x_1, x_2) : -5 < x_1 < 5, \ -5 < x_2 < 5\}$ to identify all local minimizers and local maximizers of $f(\mathbf{x})$ in D.

(b) Convert the problem in part (a) into an unconstrained minimization problem by eliminating the inequality constraints. Hint: A constraint $x \leq a$ can be eliminated by using the variable substitution $x = a - \hat{x}^2$.

Chapter 2

BASIC PRINCIPLES

2.1   Introduction

Nonlinear programming is based on a collection of definitions, theorems, and principles that must be clearly understood if the available nonlinear programming methods are to be used effectively.

This chapter begins with the definition of the gradient vector, the Hessian matrix, and the various types of extrema (maxima and minima). The conditions that must hold at the solution point are then discussed and techniques for the characterization of the extrema are described. Subsequently, the classes of convex and concave functions are introduced. These provide a natural formulation for the theory of global convergence.

Throughout the chapter, we focus our attention on the nonlinear optimization problem

$$\text{minimize } f = f(\mathbf{x})$$
$$\text{subject to: } \quad \mathbf{x} \in \mathcal{R}$$

where $f(\mathbf{x})$ is a real-valued function and $\mathcal{R} \subset E^n$ is the feasible region.

2.2   Gradient Information

In many optimization methods, gradient information pertaining to the objective function is required. This information consists of the first and second derivatives of $f(\mathbf{x})$ with respect to the n variables.

If $f(\mathbf{x}) \in C^1$, that is, if $f(\mathbf{x})$ has continuous first-order partial derivatives, the *gradient* of $f(\mathbf{x})$ is defined as

$$\mathbf{g}(\mathbf{x}) = \begin{bmatrix} \frac{\partial f}{\partial x_1} & \frac{\partial f}{\partial x_2} & \cdots & \frac{\partial f}{\partial x_n} \end{bmatrix}^T$$
$$= \nabla f(\mathbf{x}) \tag{2.1}$$

where

$$\nabla = [\ \frac{\partial}{\partial x_1} \quad \frac{\partial}{\partial x_2} \quad \cdots \quad \frac{\partial}{\partial x_n}\ ]^T \tag{2.2}$$

If $f(\mathbf{x}) \in C^2$, that is, if $f(\mathbf{x})$ has continuous second-order partial derivatives, the *Hessian*[1] of $f(\mathbf{x})$ is defined as

$$\mathbf{H}(\mathbf{x}) = \nabla \mathbf{g}^T = \nabla \{\nabla^T f(\mathbf{x})\} \tag{2.3}$$

Hence Eqs. (2.1) – (2.3) give

$$\mathbf{H}(\mathbf{x}) = \begin{bmatrix} \frac{\partial^2 f}{\partial x_1^2} & \frac{\partial^2 f}{\partial x_1 \partial x_2} & \cdots & \frac{\partial^2 f}{\partial x_1 \partial x_n} \\ \frac{\partial^2 f}{\partial x_2 \partial x_1} & \frac{\partial^2 f}{\partial x_2^2} & \cdots & \frac{\partial^2 f}{\partial x_2 \partial x_n} \\ \vdots & \vdots & & \vdots \\ \frac{\partial^2 f}{\partial x_n \partial x_1} & \frac{\partial^2 f}{\partial x_n \partial x_2} & \cdots & \frac{\partial^2 f}{\partial x_n^2} \end{bmatrix}$$

For a function $f(\mathbf{x}) \in C^2$

$$\frac{\partial^2 f}{\partial x_i \partial x_j} = \frac{\partial^2 f}{\partial x_j \partial x_i}$$

since differentiation is a linear operation and hence $\mathbf{H}(\mathbf{x})$ is an $n \times n$ square symmetric matrix.

The gradient and Hessian at a point $\mathbf{x} = \mathbf{x}_k$ are represented by $\mathbf{g}(\mathbf{x}_k)$ and $\mathbf{H}(\mathbf{x}_k)$ or by the simplified notation $\mathbf{g}_k$ and $\mathbf{H}_k$, respectively. Sometimes, when confusion is not likely to arise, $\mathbf{g}(\mathbf{x})$ and $\mathbf{H}(\mathbf{x})$ are simplified to $\mathbf{g}$ and $\mathbf{H}$.

The gradient and Hessian tend to simplify the optimization process considerably. Nevertheless, in certain applications it may be uneconomic, time-consuming, or impossible to deduce and compute the partial derivatives of $f(\mathbf{x})$. For these applications, methods are preferred that do not require gradient information.

Gradient methods, namely, methods based on gradient information may use only $\mathbf{g}(\mathbf{x})$ or both $\mathbf{g}(\mathbf{x})$ and $\mathbf{H}(\mathbf{x})$. In the latter case, the inversion of matrix $\mathbf{H}(\mathbf{x})$ may be required which tends to introduce numerical inaccuracies and is time-consuming. Such methods are often avoided.

2.3 The Taylor Series

Some of the nonlinear programming procedures and methods utilize linear or quadratic approximations for the objective function and the equality and inequality constraints, namely, $f(\mathbf{x})$, $a_i(\mathbf{x})$, and $c_j(\mathbf{x})$ in Eq. (1.4). Such

---

[1]For the sake of simplicity, the gradient vector and Hessian matrix will be referred to as the gradient and Hessian, respectively, henceforth.

approximations can be obtained by using the Taylor series. If $f(\mathbf{x})$ is a function of two variables x_1 and x_2 such that $f(\mathbf{x}) \in C^P$ where $P \to \infty$, that is, $f(\mathbf{x})$ has continuous partial derivatives of all orders, then the value of function $f(\mathbf{x})$ at point $[x_1 + \delta_1, \ x_2 + \delta_2]$ is given by the Taylor series as

$$f(x_1 + \delta_1, \ x_2 + \delta_2) = f(x_1, \ x_2) + \frac{\partial f}{\partial x_1}\delta_1 + \frac{\partial f}{\partial x_2}\delta_2$$

$$+ \frac{1}{2}\left(\frac{\partial^2 f}{\partial x_1^2}\delta_1^2 + \frac{2\partial^2 f}{\partial x_1 \partial x_2}\delta_1\delta_2 + \frac{\partial^2 f}{\partial x_2^2}\delta_2^2\right)$$

$$+ O(\|\boldsymbol{\delta}\|^3) \tag{2.4a}$$

where

$$\boldsymbol{\delta} = [\delta_1 \ \delta_2]^T$$

$O(\|\boldsymbol{\delta}\|^3)$ is the *remainder*, and $\|\boldsymbol{\delta}\|$ is the Euclidean norm of $\boldsymbol{\delta}$ given by

$$\|\boldsymbol{\delta}\| = \sqrt{\boldsymbol{\delta}^T\boldsymbol{\delta}}$$

The notation $\phi(x) = O(x)$ *denotes that* $\phi(x)$ *approaches zero at least as fast as* x *as* x *approaches zero, that is, there exists a constant* $K \geq 0$ such that

$$\left|\frac{\phi(x)}{x}\right| \leq K \quad \text{as } x \to 0$$

The remainder term in Eq. (2.4a) can also be expressed as $o(\|\boldsymbol{\delta}\|^2)$ where the notation $\phi(x) = o(x)$ *denotes that* $\phi(x)$ *approaches zero faster than* x *as* x *approaches zero*, that is,

$$\left|\frac{\phi(x)}{x}\right| \to 0 \quad \text{as } x \to 0$$

If $f(\mathbf{x})$ is a function of n variables, then the Taylor series of $f(\mathbf{x})$ at point $[x_1 + \delta_1, \ x_2 + \delta_2, \ \ldots]$ is given by

$$f(x_1 + \delta_1, \ x_2 + \delta_2, \ \ldots) = f(x_1, \ x_2, \ \ldots) + \sum_{i=1}^{n}\frac{\partial f}{\partial x_i}\delta_i$$

$$+ \frac{1}{2}\sum_{i=1}^{n}\sum_{j=1}^{n}\delta_i \frac{\partial^2 f}{\partial x_i \partial x_j}\delta_j$$

$$+ o(\|\boldsymbol{\delta}\|^2) \tag{2.4b}$$

Alternatively, on using matrix notation

$$f(\mathbf{x} + \boldsymbol{\delta}) = f(\mathbf{x}) + \mathbf{g}(\mathbf{x})^T\boldsymbol{\delta} + \tfrac{1}{2}\boldsymbol{\delta}^T\mathbf{H}(\mathbf{x})\boldsymbol{\delta} + o(\|\boldsymbol{\delta}\|^2) \tag{2.4c}$$

where $\mathbf{g}(\mathbf{x})$ is the gradient, and $\mathbf{H}(\mathbf{x})$ is the Hessian at point $\mathbf{x}$.

As $\|\boldsymbol{\delta}\| \to 0$, second- and higher-order terms can be neglected and a *linear approximation* can be obtained for $f(\mathbf{x} + \boldsymbol{\delta})$ as

$$f(\mathbf{x} + \boldsymbol{\delta}) \approx f(\mathbf{x}) + \mathbf{g}(\mathbf{x})^T \boldsymbol{\delta} \tag{2.4d}$$

Similarly, a *quadratic approximation* for $f(\mathbf{x} + \boldsymbol{\delta})$ can be obtained as

$$f(\mathbf{x} + \boldsymbol{\delta}) \approx f(\mathbf{x}) + \mathbf{g}(\mathbf{x})^T \boldsymbol{\delta} + \tfrac{1}{2}\boldsymbol{\delta}^T \mathbf{H}(\mathbf{x}) \boldsymbol{\delta} \tag{2.4e}$$

Another form of the Taylor series, which includes an expression for the remainder term, is

$$
\begin{aligned}
f(\mathbf{x} + \boldsymbol{\delta}) = \; & f(\mathbf{x}) \\
& + \sum_{1 \le k_1 + k_2 + \cdots + k_n \le P} \frac{\partial^{k_1 + k_2 + \cdots + k_n} f(\mathbf{x})}{\partial x_1^{k_1} \partial x_2^{k_2} \cdots \partial x_n^{k_n}} \prod_{i=1}^{n} \frac{\delta_i^{k_i}}{k_i!} \\
& + \sum_{k_1 + k_2 + \cdots + k_n = P+1} \frac{\partial^{P+1} f(\mathbf{x} + \alpha\boldsymbol{\delta})}{\partial x_1^{k_1} \partial x_2^{k_2} \cdots \partial x_n^{k_n}} \prod_{i=1}^{n} \frac{\delta_i^{k_i}}{k_i!}
\end{aligned} \tag{2.4f}
$$

where $0 \le \alpha \le 1$ and

$$\sum_{1 \le k_1 + k_2 + \cdots + k_n \le P} \frac{\partial^{k_1 + k_2 + \cdots + k_n} f(\mathbf{x})}{\partial x_1^{k_1} \partial x_2^{k_2} \cdots \partial x_n^{k_n}} \prod_{i=1}^{n} \frac{\delta_i^{k_i}}{k_i!}$$

is the sum of terms taken over all possible combinations of k_1, k_2, $\ldots$, k_n that add up to a number in the range 1 to P. (See Chap. 4 of Protter and Morrey [1] for proof.) This representation of the Taylor series is completely general and, therefore, it can be used to obtain cubic and higher-order approximations for $f(\mathbf{x} + \boldsymbol{\delta})$. Furthermore, it can be used to obtain linear, quadratic, cubic, and higher-order *exact closed-form* expressions for $f(\mathbf{x} + \boldsymbol{\delta})$. If $f(\mathbf{x}) \in C^1$ and $P = 0$, Eq. (2.4f) gives

$$f(\mathbf{x} + \boldsymbol{\delta}) = f(\mathbf{x}) + \mathbf{g}(\mathbf{x} + \alpha\boldsymbol{\delta})^T \boldsymbol{\delta} \tag{2.4g}$$

and if $f(\mathbf{x}) \in C^2$ and $P = 1$, then

$$f(\mathbf{x} + \boldsymbol{\delta}) = f(\mathbf{x}) + \mathbf{g}(\mathbf{x})^T \boldsymbol{\delta} + \tfrac{1}{2}\boldsymbol{\delta}^T \mathbf{H}(\mathbf{x} + \alpha\boldsymbol{\delta}) \boldsymbol{\delta} \tag{2.4h}$$

where $0 \le \alpha \le 1$. Eq. (2.4g) is usually referred to as the *mean-value theorem for differentiation*.

Yet another form of the Taylor series can be obtained by regrouping the terms in Eq. (2.4f) as

$$
\begin{aligned}
f(\mathbf{x} + \boldsymbol{\delta}) = \; & f(\mathbf{x}) + \mathbf{g}(\mathbf{x})^T \boldsymbol{\delta} + \tfrac{1}{2}\boldsymbol{\delta}^T \mathbf{H}(\mathbf{x}) \boldsymbol{\delta} + \tfrac{1}{3!} D^3 f(\mathbf{x}) \\
& + \cdots + \frac{1}{(r-1)!} D^{r-1} f(\mathbf{x}) + \cdots
\end{aligned} \tag{2.4i}
$$

where

$$D^r f(\mathbf{x}) = \sum_{i_1=1}^{n} \sum_{i_2=1}^{n} \cdots \sum_{i_r=1}^{n} \left\{ \delta_{i_1} \delta_{i_2} \cdots \delta_{i_r} \frac{\partial^r f(\mathbf{x})}{\partial x_{i_1} \partial x_{i_2} \cdots \partial x_{i_r}} \right\}$$

2.4 Types of Extrema

The *extrema* of a function are its minima and maxima. Points at which a function has minima (maxima) are said to be *minimizers* (*maximizers*). Several types of minimizers (maximizers) can be distinguished, namely, local or global and weak or strong.

Definition 2.1 A point $\mathbf{x}^* \in \mathcal{R}$, where $\mathcal{R}$ is the feasible region, is said to be a *weak local minimizer* of $f(\mathbf{x})$ if there exists a distance $\varepsilon > 0$ such that

$$f(\mathbf{x}) \geq f(\mathbf{x}^*) \tag{2.5}$$

if

$$\mathbf{x} \in \mathcal{R} \quad \text{and} \quad \|\mathbf{x} - \mathbf{x}^*\| < \varepsilon$$

∎

Definition 2.2 A point $\mathbf{x}^* \in \mathcal{R}$ is said to be a *weak global minimizer* of $f(\mathbf{x})$ if

$$f(\mathbf{x}) \geq f(\mathbf{x}^*) \tag{2.6}$$

for all $\mathbf{x} \in \mathcal{R}$.

∎

If Def. 2.2 is satisfied at $\mathbf{x}^*$, then Def. 2.1 is also satisfied at $\mathbf{x}^*$, and so a global minimizer is also a local minimizer.

Definition 2.3
If Eq. (2.5) in Def. 2.1 or Eq. (2.6) in Def. 2.2 is replaced by

$$f(\mathbf{x}) > f(\mathbf{x}^*) \tag{2.7}$$

$\mathbf{x}^*$ is said to be a *strong local* (or *global*) *minimizer*.

∎

The minimum at a weak local, weak global, etc. minimizer is called a weak local, weak global, etc. minimum.

A strong global minimum in E^2 is depicted in Fig. 2.1.

Weak or strong and local or global maximizers can similarly be defined by reversing the inequalities in Eqs. (2.5) – (2.7).

32

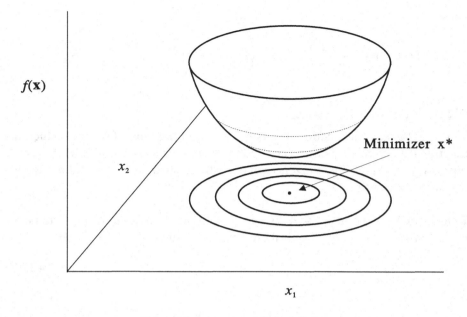

Figure 2.1. A strong global minimizer.

Example 2.1 The function of Fig. 2.2 has a feasible region defined by the set

$$\mathcal{R} = \{x : \ x_1 \leq x \leq x_2\}$$

Classify its minimizers.

Solution The function has a weak local minimum at point B, strong local minima at points A, C, and D, and a strong global minimum at point C.

■

In the general optimization problem, we are in principle seeking the global minimum (or maximum) of $f(\mathbf{x})$. In practice, an optimization problem may have two or more local minima. Since optimization algorithms in general are iterative procedures which start with an initial estimate of the solution and converge to a single solution, one or more local minima may be missed. If the global minimum is missed, a suboptimal solution will be achieved, which may or may not be acceptable. This problem can to some extent be overcome by performing the optimization several times using a different initial estimate for the solution in each case in the hope that several distinct local minima will be located. If this approach is successful, the best minimizer, namely, the one yielding the lowest value for the objective function can be selected. Although such a solution could be acceptable from a practical point of view, usually

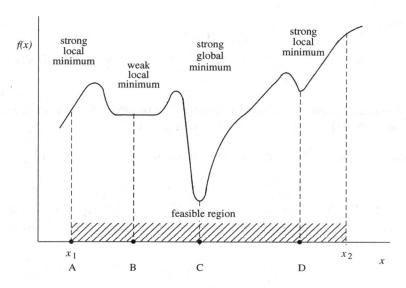

Figure 2.2. Types of minima. (Example 2.1)

there is no guarantee that the global minimum will be achieved. Therefore, for the sake of convenience, the term 'minimize $f(\mathbf{x})$' in the general optimization problem will be interpreted as 'find a local minimum of $f(\mathbf{x})$'.

In a specific class of problems where function $f(\mathbf{x})$ and set $\mathcal{R}$ satisfy certain convexity properties, any local minimum of $f(\mathbf{x})$ is also a global minimum of $f(\mathbf{x})$. In this class of problems an optimal solution can be assured. These problems will be examined in Sec. 2.7.

2.5 Necessary and Sufficient Conditions for Local Minima and Maxima

The gradient $\mathbf{g}(\mathbf{x})$ and the Hessian $\mathbf{H}(\mathbf{x})$ must satisfy certain conditions at a local minimizer $\mathbf{x}^*$, (see [2, Chap. 6]). Two sets of conditions will be discussed, as follows:

1. Conditions which are satisfied at a local minimizer $\mathbf{x}^*$. These are the necessary conditions.
2. Conditions which guarantee that $\mathbf{x}^*$ is a local minimizer. These are the sufficient conditions.

The necessary and sufficient conditions can be described in terms of a number of theorems. A concept that is used extensively in these theorems is the concept of a feasible direction.

Definition 2.4 Let $\delta = \alpha\mathbf{d}$ be a change in $\mathbf{x}$ where α is a positive constant and $\mathbf{d}$ is a direction vector. If $\mathcal{R}$ is the feasible region and a constant $\hat{\alpha} > 0$ exists

34

such that

$$\mathbf{x} + \alpha\mathbf{d} \in \mathcal{R}$$

for all α in the range $0 \leq \alpha \leq \hat{\alpha}$, then $\mathbf{d}$ is said to be a *feasible direction* at point $\mathbf{x}$.

∎

In effect, if a point $\mathbf{x}$ remains in $\mathcal{R}$ after it is moved a finite distance in a direction $\mathbf{d}$, then $\mathbf{d}$ is a feasible direction vector at $\mathbf{x}$.

Example 2.2 The feasible region in an optimization problem is given by

$$\mathcal{R} = \{\mathbf{x} : x_1 \geq 2, \ x_2 \geq 0\}$$

as depicted in Fig. 2.3. Which of the vectors $\mathbf{d}_1 = [-2 \ 2]^T$, $\mathbf{d}_2 = [0 \ 2]^T$, $\mathbf{d}_3 = [2 \ 0]^T$ are feasible directions at points $\mathbf{x}_1 = [4 \ 1]^T$, $\mathbf{x}_2 = [2 \ 3]^T$, and $\mathbf{x}_3 = [1 \ 4]^T$?

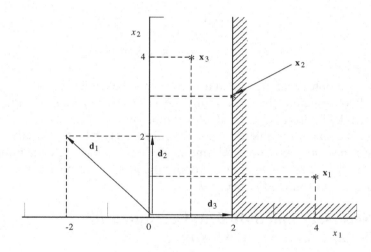

Figure 2.3. Graphical construction for Example 2.2.

Solution Since

$$\mathbf{x}_1 + \alpha\mathbf{d}_1 \in \mathcal{R}$$

for all α in the range $0 \leq \alpha \leq \hat{\alpha}$ for $\hat{\alpha} = 1$, $\mathbf{d}_1$ is a feasible direction at point $\mathbf{x}_1$; for any range $0 \leq \alpha \leq \hat{\alpha}$

$$\mathbf{x}_1 + \alpha\mathbf{d}_2 \in \mathcal{R} \quad \text{and} \quad \mathbf{x}_1 + \alpha\mathbf{d}_3 \in \mathcal{R}$$

Hence $\mathbf{d}_2$ and $\mathbf{d}_3$ are feasible directions at $\mathbf{x}_1$.
Since no constant $\hat{\alpha} > 0$ can be found such that

$$\mathbf{x}_2 + \alpha\mathbf{d}_1 \in \mathcal{R} \qquad \text{for } 0 \leq \alpha \leq \hat{\alpha}$$

$\mathbf{d}_1$ is not a feasible direction at $\mathbf{x}_2$. On the other hand, a positive constant $\hat{\alpha}$ exists such that

$$\mathbf{x}_2 + \alpha \mathbf{d}_2 \in \mathcal{R} \quad \text{and} \quad \mathbf{x}_2 + \alpha \mathbf{d}_3 \in \mathcal{R}$$

for $0 \le \alpha \le \hat{\alpha}$, and so $\mathbf{d}_2$ and $\mathbf{d}_3$ are feasible directions at $\mathbf{x}_2$.

Since $\mathbf{x}_3$ is not in $\mathcal{R}$, no $\hat{\alpha} > 0$ exists such that

$$\mathbf{x}_3 + \alpha \mathbf{d} \in \mathcal{R} \quad \text{for } 0 \le \alpha \le \hat{\alpha}$$

for any $\mathbf{d}$. Hence $\mathbf{d}_1, \mathbf{d}_2$, and $\mathbf{d}_3$ are not feasible directions at $\mathbf{x}_3$.

■

2.5.1 First-order necessary conditions

The objective function must satisfy two sets of conditions in order to have a minimum, namely, first- and second-order conditions. The first-order conditions are in terms of the first derivatives, i.e., the gradient.

Theorem 2.1 *First-order necessary conditions for a minimum*

(a) *If $f(\mathbf{x}) \in C^1$ and $\mathbf{x}^*$ is a local minimizer, then*

$$\mathbf{g}(\mathbf{x}^*)^T \mathbf{d} \ge 0$$

for every feasible direction $\mathbf{d}$ at $\mathbf{x}^*$.

(b) *If $\mathbf{x}^*$ is located in the interior of $\mathcal{R}$ then*

$$\mathbf{g}(\mathbf{x}^*) = 0$$

Proof (a) If $\mathbf{d}$ is a feasible direction at $\mathbf{x}^*$, then from Def. 2.4

$$\mathbf{x} = \mathbf{x}^* + \alpha \mathbf{d} \in \mathcal{R} \quad \text{for } 0 \le \alpha \le \hat{\alpha}$$

From the Taylor series

$$f(\mathbf{x}) = f(\mathbf{x}^*) + \alpha \mathbf{g}(\mathbf{x}^*)^T \mathbf{d} + o(\alpha \|\mathbf{d}\|)$$

If

$$\mathbf{g}(\mathbf{x}^*)^T \mathbf{d} < 0$$

then as $\alpha \to 0$

$$\alpha \mathbf{g}(\mathbf{x}^*)^T \mathbf{d} + o(\alpha \|\mathbf{d}\|) < 0$$

and so

$$f(\mathbf{x}) < f(\mathbf{x}^*)$$

This contradicts the assumption that $\mathbf{x}^*$ is a minimizer. Therefore, a necessary condition for $\mathbf{x}^*$ to be a minimizer is

$$\mathbf{g}(\mathbf{x}^*)^T \mathbf{d} \geq 0$$

(b) If $\mathbf{x}^*$ is in the interior of $\mathcal{R}$, vectors exist in all directions which are feasible. Thus from part (a), a direction $\mathbf{d} = \mathbf{d}_1$ yields

$$\mathbf{g}(\mathbf{x}^*)^T \mathbf{d}_1 \geq 0$$

Similarly, for a direction $\mathbf{d} = -\mathbf{d}_1$

$$-\mathbf{g}(\mathbf{x}^*)^T \mathbf{d}_1 \geq 0$$

Therefore, in this case, a necessary condition for $\mathbf{x}^*$ to be a local minimizer is

$$\mathbf{g}(\mathbf{x}^*) = 0$$

∎

2.5.2 Second-order necessary conditions

The second-order necessary conditions involve the first as well as the second derivatives or, equivalently, the gradient and the Hessian.

Definition 2.5

  (a) Let $\mathbf{d}$ be an arbitrary direction vector at point $\mathbf{x}$. The quadratic form $\mathbf{d}^T \mathbf{H}(\mathbf{x})\mathbf{d}$ is said to be *positive definite, positive semidefinite, negative semidefinite, negative definite* if $\mathbf{d}^T \mathbf{H}(\mathbf{x})\mathbf{d} > 0$, ≥ 0, ≤ 0, < 0, respectively, for all $\mathbf{d} \neq \mathbf{0}$ at $\mathbf{x}$. If $\mathbf{d}^T \mathbf{H}(\mathbf{x})\mathbf{d}$ can assume positive as well as negative values, it is said to be *indefinite*.
  (b) If $\mathbf{d}^T \mathbf{H}(\mathbf{x})\mathbf{d}$ is positive definite, positive semidefinite, etc., then matrix $\mathbf{H}(\mathbf{x})$ is said to be positive definite, positive semidefinite, etc.

∎

Theorem 2.2 *Second-order necessary conditions for a minimum*

  (a) *If $f(\mathbf{x}) \in C^2$ and $\mathbf{x}^*$ is a local minimizer, then for every feasible direction $\mathbf{d}$ at $\mathbf{x}^*$*
  (i) $\mathbf{g}(\mathbf{x}^*)^T \mathbf{d} \geq 0$
  (ii) *If* $\mathbf{g}(\mathbf{x}^*)^T \mathbf{d} = 0$, *then* $\mathbf{d}^T \mathbf{H}(\mathbf{x}^*)\mathbf{d} \geq 0$
  (b) *If $\mathbf{x}^*$ is a local minimizer in the interior of $\mathcal{R}$, then*
  (i) $\mathbf{g}(\mathbf{x}^*) = \mathbf{0}$
  (ii) $\mathbf{d}^T \mathbf{H}(\mathbf{x})^* \mathbf{d} \geq 0$ *for all* $\mathbf{d} \neq \mathbf{0}$

Proof Conditions (i) in parts (a) and (b) are the same as in Theorem 2.1(a) and (b).

Condition (ii) of part (a) can be proved by letting $\mathbf{x} = \mathbf{x}^* + \alpha\mathbf{d}$, where $\mathbf{d}$ is a feasible direction. The Taylor series gives

$$f(\mathbf{x}) = f(\mathbf{x}^*) + \alpha\mathbf{g}(\mathbf{x}^*)^T\mathbf{d} + \tfrac{1}{2}\alpha^2\mathbf{d}^T\mathbf{H}(\mathbf{x}^*)\mathbf{d} + o(\alpha^2\|\mathbf{d}\|^2)$$

Now if condition (i) is satisfied with the equal sign, then

$$f(\mathbf{x}) = f(\mathbf{x}^*) + \tfrac{1}{2}\alpha^2\mathbf{d}^T\mathbf{H}(\mathbf{x}^*)\mathbf{d} + o(\alpha^2\|\mathbf{d}\|^2)$$

If

$$\mathbf{d}^T\mathbf{H}(\mathbf{x}^*)\mathbf{d} < 0$$

then as $\alpha \to 0$

$$\tfrac{1}{2}\alpha^2\mathbf{d}^T\mathbf{H}(\mathbf{x}^*)\mathbf{d} + o(\alpha^2\|\mathbf{d}\|^2) < 0$$

and so

$$f(\mathbf{x}) < f(\mathbf{x}^*)$$

This contradicts the assumption that $\mathbf{x}^*$ is a minimizer. Therefore, if $\mathbf{g}(\mathbf{x}^*)^T\mathbf{d} = 0$, then

$$\mathbf{d}^T\mathbf{H}(\mathbf{x}^*)\mathbf{d} \geq 0$$

If $\mathbf{x}^*$ is a local minimizer in the interior of $\mathcal{R}$, then all vectors $\mathbf{d}$ are feasible directions and, therefore, condition (ii) of part (b) holds. This condition is equivalent to stating that $\mathbf{H}(\mathbf{x}^*)$ is positive semidefinite, according to Def. 2.5. ∎

Example 2.3 Point $\mathbf{x}^* = [\tfrac{1}{2}\ 0]^T$ is a local minimizer of the problem

$$\text{minimize } f(x_1,\ x_2) = x_1^2 - x_1 + x_2 + x_1x_2$$

$$\text{subject to}: \quad x_1 \geq 0,\ x_2 \geq 0$$

Show that the necessary conditions for $\mathbf{x}^*$ to be a local minimizer are satisfied.

Solution The partial derivatives of $f(x_1,\ x_2)$ are

$$\frac{\partial f}{\partial x_1} = 2x_1 - 1 + x_2, \quad \frac{\partial f}{\partial x_2} = 1 + x_1$$

Hence if $\mathbf{d} = [d_1\ d_2]^T$ is a feasible direction, we obtain

$$\mathbf{g}(\mathbf{x})^T\mathbf{d} = (2x_1 - 1 + x_2)d_1 + (1 + x_1)d_2$$

At $\mathbf{x} = \mathbf{x}^*$

$$\mathbf{g}(\mathbf{x}^*)^T\mathbf{d} = \tfrac{3}{2}d_2$$

and since $d_2 \geq 0$ for $\mathbf{d}$ to be a feasible direction, we have

$$\mathbf{g}(\mathbf{x}^*)^T\mathbf{d} \geq 0$$

Therefore, the first-order necessary conditions for a minimum are satisfied.
Now

$$\mathbf{g}(\mathbf{x}^*)^T\mathbf{d} = 0$$

if $d_2 = 0$. The Hessian is

$$\mathbf{H}(\mathbf{x}^*) = \begin{bmatrix} 2 & 1 \\ 1 & 0 \end{bmatrix}$$

and so

$$\mathbf{d}^T\mathbf{H}(\mathbf{x}^*)\mathbf{d} = 2d_1^2 + 2d_1d_2$$

For $d_2 = 0$, we obtain

$$\mathbf{d}^T\mathbf{H}(\mathbf{x}^*)\mathbf{d} = 2d_1^2 \geq 0$$

for every feasible value of d_1. Therefore, the second-order necessary conditions for a minimum are satisfied.

∎

Example 2.4 Points $\mathbf{p}_1 = [0\ 0]^T$ and $\mathbf{p}_2 = [6\ 9]^T$ are probable minimizers for the problem

$$\text{minimize } f(x_1,\ x_2) = x_1^3 - x_1^2x_2 + 2x_2^2$$

$$\text{subject to}: \quad x_1 \geq 0,\ x_2 \geq 0$$

Check whether the necessary conditions of Theorems 2.1 and 2.2 are satisfied.

Solution The partial derivatives of $f(x_1,\ x_2)$ are

$$\frac{\partial f}{\partial x_1} = 3x_1^2 - 2x_1x_2, \quad \frac{\partial f}{\partial x_2} = -x_1^2 + 4x_2$$

Hence if $\mathbf{d} = [d_1\ d_2]^T$, we obtain

$$\mathbf{g}(\mathbf{x})^T\mathbf{d} = (3x_1^2 - 2x_1x_2)d_1 + (-x_1^2 + 4x_2)d_2$$

At points $\mathbf{p}_1$ and $\mathbf{p}_2$

$$\mathbf{g}(\mathbf{x})^T\mathbf{d} = 0$$

i.e., the first-order necessary conditions are satisfied. The Hessian is

$$\mathbf{H}(\mathbf{x}) = \begin{bmatrix} 6x_1 - 2x_2 & -2x_1 \\ -2x_1 & 4 \end{bmatrix}$$

and if $\mathbf{x} = \mathbf{p}_1$, then

$$\mathbf{H}(\mathbf{p}_1) = \begin{bmatrix} 0 & 0 \\ 0 & 4 \end{bmatrix}$$

and so

$$\mathbf{d}^T\mathbf{H}(\mathbf{p}_1)\mathbf{d} = 4d_2^2 \geq 0$$

Hence the second-order necessary conditions are satisfied at $x = p_1$, and p_1 can be a local minimizer.

If $x = p_2$, then

$$H(p_2) = \begin{bmatrix} 18 & -12 \\ -12 & 4 \end{bmatrix}$$

and

$$d^T H(p_2)d = 18d_1^2 - 24d_1 d_2 + 4d_2^2$$

Since $d^T H(p_2)d$ is indefinite, the second-order necessary conditions are violated, that is, p_2 cannot be a local minimizer.

∎

Analogous conditions hold for the case of a local maximizer as stated in the following theorem:

Theorem 2.3 *Second-order necessary conditions for a maximum*
   (a) *If $f(x) \in C^2$, and x^* is a local maximizer, then for every feasible direction d at x^**
   (i) $g(x^*)^T d \leq 0$
   (ii) If $g(x^*)^T d = 0$, then $d^T H(x^*)d \leq 0$
   (b) *If x^* is a local maximizer in the interior of $\mathcal{R}$ then*
   (i) $g(x^*) = 0$
   (ii) $d^T H(x^*)d \leq 0$ for all $d \neq 0$

Condition (ii) of part (b) is equivalent to stating that $H(x^*)$ is negative semidefinite.

The conditions considered are necessary but not sufficient for a point to be a local extremum point, that is, a point may satisfy these conditions without being a local extremum point. We now focus our attention on a set of stronger conditions that are *sufficient* for a point to be a local extremum. We consider conditions that are applicable in the case where x^* is located in the interior of the feasible region. Sufficient conditions that are applicable to the case where x^* is located on a boundary of the feasible region are somewhat more difficult to deduce and will be considered in Chap. 10.

Theorem 2.4 *Second-order sufficient conditions for a minimum If $f(x) \in C^2$ and x^* is located in the interior of $\mathcal{R}$, then the conditions*
   (a) $g(x^*) = 0$
   (b) $H(x^*)$ *is positive definite*
are sufficient for x^* to be a strong local minimizer.

Proof For any direction d, the Taylor series yields

$$f(x^* + d) = f(x^*) + g(x^*)^T d + \tfrac{1}{2}d^T H(x^*)d + o(\|d\|^2)$$

and if condition (a) is satisfied, we have

$$f(\mathbf{x}^* + \mathbf{d}) = f(\mathbf{x}^*) + \tfrac{1}{2}\mathbf{d}^T\mathbf{H}(\mathbf{x}^*)\mathbf{d} + o(\|\mathbf{d}\|^2)$$

Now if condition (b) is satisfied, then

$$\tfrac{1}{2}\mathbf{d}^T\mathbf{H}(\mathbf{x}^*)\mathbf{d} + o(\|\mathbf{d}\|^2) > 0 \qquad \text{as } \|\mathbf{d}\| \to 0$$

Therefore,

$$f(\mathbf{x}^* + \mathbf{d}) > f(\mathbf{x}^*)$$

that is, $\mathbf{x}^*$ is a strong local minimizer.

∎

Analogous conditions hold for a maximizer as stated in Theorem 2.5 below.

Theorem 2.5 *Second-order sufficient conditions for a maximum* If $f(\mathbf{x}^*) \in C^2$ *and* $\mathbf{x}^*$ *is located in the interior of* $\mathcal{R}$*, then the conditions*
   (a) $\mathbf{g}(\mathbf{x}) = \mathbf{0}$
   (b) $\mathbf{H}(\mathbf{x}^*)$ *is negative definite*
are sufficient for $\mathbf{x}^*$ *to be a strong local maximizer.*

2.6   Classification of Stationary Points

If the extremum points of the type considered so far, namely, minimizers and maximizers, are located in the interior of the feasible region, they are called *stationary points* since $\mathbf{g}(\mathbf{x}) = \mathbf{0}$ at these points. Another type of stationary point of interest is the saddle point.

Definition 2.6 A point $\bar{\mathbf{x}} \in \mathcal{R}$, where $\mathcal{R}$ is the feasible region, is said to be a *saddle point* if
   (a) $\mathbf{g}(\bar{\mathbf{x}}) = \mathbf{0}$
   (b) point $\bar{\mathbf{x}}$ is neither a maximizer nor a minimizer.

∎

A saddle point in E^2 is illustrated in Fig. 2.4.

At a point $\mathbf{x} = \bar{\mathbf{x}} + \alpha\mathbf{d} \in \mathcal{R}$ in the neighborhood of a saddle point $\bar{\mathbf{x}}$, the Taylor series gives

$$f(\mathbf{x}) = f(\bar{\mathbf{x}}) + \tfrac{1}{2}\alpha^2\mathbf{d}^T\mathbf{H}(\bar{\mathbf{x}})\mathbf{d} + o(\alpha^2\|\mathbf{d}\|^2)$$

since $\mathbf{g}(\bar{\mathbf{x}}) = \mathbf{0}$. From the definition of a saddle point, directions $\mathbf{d}_1$ and $\mathbf{d}_2$ must exist such that

$$f(\bar{\mathbf{x}} + \alpha\mathbf{d}_1) < f(\bar{\mathbf{x}}) \quad \text{and} \quad f(\bar{\mathbf{x}} + \alpha\mathbf{d}_2) > f(\bar{\mathbf{x}})$$

Since $\bar{\mathbf{x}}$ is neither a minimizer nor a maximizer, then as $\alpha \to 0$ we have

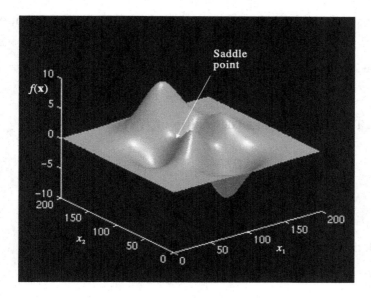

Figure 2.4. A saddle point in E^2.

$$\mathbf{d}_1^T \mathbf{H}(\bar{\mathbf{x}})\mathbf{d}_1 < 0 \quad \text{and} \quad \mathbf{d}_2^T \mathbf{H}(\bar{\mathbf{x}})\mathbf{d}_2 > 0$$

Therefore, matrix $\mathbf{H}(\bar{\mathbf{x}})$ must be indefinite.

Stationary points can be located and classified as follows:

1. Find the points $\mathbf{x}_i$ at which $\mathbf{g}(\mathbf{x}_i) = \mathbf{0}$.
2. Obtain the Hessian $\mathbf{H}(\mathbf{x}_i)$.
3. Determine the character of $\mathbf{H}(\mathbf{x}_i)$ for each point $\mathbf{x}_i$.

If $\mathbf{H}(\mathbf{x}_i)$ is positive (or negative) definite, $\mathbf{x}_i$ is a minimizer (or maximizer); if $\mathbf{H}(\mathbf{x}_i)$ is indefinite, $\mathbf{x}_i$ is a saddle point. If $\mathbf{H}(\mathbf{x}_i)$ is positive (or negative) semidefinite, $\mathbf{x}_i$ can be a minimizer (or maximizer); in the special case where $\mathbf{H}(\mathbf{x}_i) = \mathbf{0}$, $\mathbf{x}_i$ can be a minimizer or maximizer since the necessary conditions are satisfied in both cases. Evidently, if $\mathbf{H}(\mathbf{x}_i)$ is semidefinite, insufficient information is available for the complete characterization of a stationary point and further work is, therefore, necessary in such a case. A possible approach would be to deduce the third partial derivatives of $f(\mathbf{x})$ and then calculate the fourth term in the Taylor series, namely, term $D^3 f(\mathbf{x})/3!$ in Eq. (2.4i). If the fourth term is zero, then the fifth term needs to be calculated and so on. An alternative and more practical approach would be to compute $f(\mathbf{x}_i + \mathbf{e}_j)$ and $f(\mathbf{x}_i - \mathbf{e}_j)$ for $j = 1, 2, \ldots, n$ where $\mathbf{e}_j$ is a vector with elements

$$e_{jk} = \begin{cases} 0 & \text{for } k \neq j \\ \varepsilon & \text{for } k = j \end{cases}$$

42

for some small positive value of ε and then check whether the definition of a minimizer or maximizer is satisfied.

Example 2.5 Find and classify the stationary points of

$$f(\mathbf{x}) = (x_1 - 2)^3 + (x_2 - 3)^3$$

Solution The first-order partial derivatives of $f(\mathbf{x})$ are

$$\frac{\partial f}{\partial x_1} = 3(x_1 - 2)^2$$

$$\frac{\partial f}{\partial x_2} = 3(x_2 - 3)^2$$

If $\mathbf{g} = \mathbf{0}$, then

$$3(x_1 - 2)^2 = 0 \quad \text{and} \quad 3(x_2 - 3)^2 = 0$$

and so there is a stationary point at

$$\mathbf{x} = \mathbf{x}_1 = [2\ 3]^T$$

The Hessian is given by

$$\mathbf{H} = \begin{bmatrix} 6(x_1 - 2) & 0 \\ 0 & 6(x_2 - 3) \end{bmatrix}$$

and at $\mathbf{x} = \mathbf{x}_1$

$$\mathbf{H} = \mathbf{0}$$

Since $\mathbf{H}$ is semidefinite, more work is necessary in order to determine the type of stationary point.

The third derivatives are all zero except for $\partial^3 f/\partial x_1^3$ and $\partial^3 f/\partial x_2^3$ which are both equal to 6. For point $\mathbf{x}_1 + \boldsymbol{\delta}$, the fourth term in the Taylor series is given by

$$\frac{1}{3!}\left(\delta_1^3 \frac{\partial^3 f}{\partial x_1^3} + \delta_2^3 \frac{\partial^3 f}{\partial x_2^3}\right) = \delta_1^3 + \delta_2^3$$

and is positive for $\delta_1,\ \delta_2 > 0$ and negative for $\delta_1,\ \delta_2 < 0$. Hence

$$f(\mathbf{x}_1 + \boldsymbol{\delta}) > f(\mathbf{x}_1) \quad \text{for } \delta_1,\ \delta_2 > 0$$

and

$$f(\mathbf{x}_1 + \boldsymbol{\delta}) < f(\mathbf{x}_1) \quad \text{for } \delta_1,\ \delta_2 < 0$$

that is, $\mathbf{x}_1$ is neither a minimizer nor a maximizer. Therefore, $\mathbf{x}_1$ is a saddle point.

From the preceding discussion, it follows that the problem of classifying the stationary points of function $f(\mathbf{x})$ reduces to the problem of characterizing the Hessian. This problem can be solved by using the following theorems.

Theorem 2.6 *Characterization of symmetric matrices* *A real symmetric $n \times n$ matrix $\mathbf{H}$ is positive definite, positive semidefinite, etc., if for every nonsingular matrix $\mathbf{B}$ of the same order, the $n \times n$ matrix $\hat{\mathbf{H}}$ given by*

$$\hat{\mathbf{H}} = \mathbf{B}^T \mathbf{H} \mathbf{B}$$

is positive definite, positive semidefinite, etc.

Proof If $\mathbf{H}$ is positive definite, positive semidefinite etc., then for all $\mathbf{d} \neq 0$

$$
\begin{aligned}
\mathbf{d}^T \hat{\mathbf{H}} \mathbf{d} &= \mathbf{d}^T (\mathbf{B}^T \mathbf{H} \mathbf{B}) \mathbf{d} \\
&= (\mathbf{d}^T \mathbf{B}^T) \mathbf{H} (\mathbf{B} \mathbf{d}) \\
&= (\mathbf{B} \mathbf{d})^T \mathbf{H} (\mathbf{B} \mathbf{d})
\end{aligned}
$$

Since $\mathbf{B}$ is nonsingular, $\mathbf{B} \mathbf{d} = \hat{\mathbf{d}}$ is a nonzero vector and thus

$$\mathbf{d}^T \hat{\mathbf{H}} \mathbf{d} = \hat{\mathbf{d}}^T \mathbf{H} \hat{\mathbf{d}} > 0, \ \geq 0, \ \text{etc.}$$

for all $\mathbf{d} \neq 0$. Therefore,

$$\hat{\mathbf{H}} = \mathbf{B}^T \mathbf{H} \mathbf{B}$$

is positive definite, positive semidefinite, etc.

Theorem 2.7 *Characterization of symmetric matrices via diagonalization*
(a) *If the $n \times n$ matrix $\mathbf{B}$ is nonsingular and*

$$\hat{\mathbf{H}} = \mathbf{B}^T \mathbf{H} \mathbf{B}$$

is a diagonal matrix with diagonal elements $\hat{h}_1$, $\hat{h}_2$, ..., $\hat{h}_n$ then $\mathbf{H}$ is positive definite, positive semidefinite, negative semidefinite, negative definite, if $\hat{h}_i > 0$, ≥ 0, ≤ 0, < 0 for $i = 1, 2, ..., n$. Otherwise, if some $\hat{h}_i$ are positive and some are negative, $\mathbf{H}$ is indefinite.
(b) *The converse of part (a) is also true, that is, if $\mathbf{H}$ is positive definite, positive semidefinite, etc., then $\hat{h}_i > 0$, ≥ 0, etc., and if $\mathbf{H}$ is indefinite, then some $\hat{h}_i$ are positive and some are negative.*

Proof (a) For all $\mathbf{d} \neq 0$

$$\mathbf{d} \hat{\mathbf{H}} \mathbf{d} = d_1^2 \hat{h}_1 + d_2^2 \hat{h}_2 + \cdots + d_n^2 \hat{h}_n$$

Therefore, if $\hat{h}_i > 0$, ≥ 0, etc. for $i = 1, 2, \ldots, n$, then

$$\mathbf{d}^T \hat{\mathbf{H}} \mathbf{d} > 0, \geq 0, \text{ etc.}$$

that is, $\hat{\mathbf{H}}$ is positive definite, positive semidefinite etc. If some $\hat{h}_i$ are positive and some are negative, a vector $\mathbf{d}$ can be found which will yield a positive or negative $\mathbf{d}^T \hat{\mathbf{H}} \mathbf{d}$ and then $\hat{\mathbf{H}}$ is indefinite. Now since $\hat{\mathbf{H}} = \mathbf{B}^T \mathbf{H} \mathbf{B}$, it follows from Theorem 2.6 that if $\hat{h}_i > 0$, ≥ 0, etc. for $i = 1, 2, \ldots, n$, then $\mathbf{H}$ is positive definite, positive semidefinite, etc.

(b) Suppose that $\mathbf{H}$ is positive definite, positive semidefinite, etc. Since $\hat{\mathbf{H}} = \mathbf{B}^T \mathbf{H} \mathbf{B}$, it follows from Theorem 2.6 that $\hat{\mathbf{H}}$ is positive definite, positive semidefinite, etc. If $\mathbf{d}$ is a vector with element d_k given by

$$d_k = \begin{cases} 0 & \text{for } k \neq i \\ 1 & \text{for } k = i \end{cases}$$

then

$$\mathbf{d}^T \hat{\mathbf{H}} \mathbf{d} = \hat{h}_i > 0, \geq 0, \text{ etc.} \qquad \text{for } i = 1, 2, \ldots, n$$

If $\mathbf{H}$ is indefinite, then from Theorem 2.6 it follows that $\hat{\mathbf{H}}$ is indefinite, and, therefore, some $\hat{h}_i$ must be positive and some must be negative. ∎

A diagonal matrix $\hat{\mathbf{H}}$ can be obtained by performing row and column operations on $\mathbf{H}$, like adding k times a given row to another row or adding m times a given column to another column. For a symmetric matrix, these operations can be carried out by applying *elementary transformations*, that is, $\hat{\mathbf{H}}$ can be formed as

$$\hat{\mathbf{H}} = \cdots \mathbf{E}_3 \mathbf{E}_2 \mathbf{E}_1 \mathbf{H} \mathbf{E}_1^T \mathbf{E}_2^T \mathbf{E}_3^T \cdots \qquad (2.8)$$

where $\mathbf{E}_1, \mathbf{E}_2, \cdots$ are elementary matrices. Typical elementary matrices are

$$\mathbf{E}_a = \begin{bmatrix} 1 & 0 & 0 \\ 0 & 1 & 0 \\ 0 & k & 1 \end{bmatrix}$$

and

$$\mathbf{E}_b = \begin{bmatrix} 1 & m & 0 & 0 \\ 0 & 1 & 0 & 0 \\ 0 & 0 & 1 & 0 \\ 0 & 0 & 0 & 1 \end{bmatrix}$$

If E_a premultiplies a 3×3 matrix, it will cause k times the second row to be added to the third row, and if $\mathbf{E}_b$ postmultiplies a 4×4 matrix it will cause m times the first column to be added to the second column. If

$$\mathbf{B} = \mathbf{E}_1^T \mathbf{E}_2^T \mathbf{E}_3^T \cdots$$

then

$$\mathbf{B}^T = \cdots \mathbf{E}_3\mathbf{E}_2\mathbf{E}_1$$

and so Eq. (2.8) can be expressed as

$$\hat{\mathbf{H}} = \mathbf{B}^T\mathbf{H}\mathbf{B}$$

Since elementary matrices are nonsingular, B is nonsingular, and hence $\hat{\mathbf{H}}$ is positive definite, positive semidefinite, etc., if $\mathbf{H}$ is positive definite, positive semidefinite, etc.

Therefore, the characterization of $\mathbf{H}$ can be achieved by diagonalizing $\mathbf{H}$, through the use of appropriate elementary matrices, and then using Theorem 2.7.

Example 2.6 Diagonalize the matrix

$$\mathbf{H} = \begin{bmatrix} 1 & -2 & 4 \\ -2 & 2 & 0 \\ 4 & 0 & -7 \end{bmatrix}$$

and then characterize it.

Solution Add 2 times the first row to the second row as

$$\begin{bmatrix} 1 & 0 & 0 \\ 2 & 1 & 0 \\ 0 & 0 & 1 \end{bmatrix}\begin{bmatrix} 1 & -2 & 4 \\ -2 & 2 & 0 \\ 4 & 0 & -7 \end{bmatrix}\begin{bmatrix} 1 & 2 & 0 \\ 0 & 1 & 0 \\ 0 & 0 & 1 \end{bmatrix} = \begin{bmatrix} 1 & 0 & 4 \\ 0 & -2 & 8 \\ 4 & 8 & -7 \end{bmatrix}$$

Add -4 times the first row to the third row as

$$\begin{bmatrix} 1 & 0 & 0 \\ 0 & 1 & 0 \\ -4 & 0 & 1 \end{bmatrix}\begin{bmatrix} 1 & 0 & 4 \\ 0 & -2 & 8 \\ 4 & 8 & -7 \end{bmatrix}\begin{bmatrix} 1 & 0 & -4 \\ 0 & 1 & 0 \\ 0 & 0 & 1 \end{bmatrix} = \begin{bmatrix} 1 & 0 & 0 \\ 0 & -2 & 8 \\ 0 & 8 & -23 \end{bmatrix}$$

Now add 4 times the second row to the third row as

$$\begin{bmatrix} 1 & 0 & 0 \\ 0 & 1 & 0 \\ 0 & 4 & 1 \end{bmatrix}\begin{bmatrix} 1 & 0 & 0 \\ 0 & -2 & 8 \\ 0 & 8 & -23 \end{bmatrix}\begin{bmatrix} 1 & 0 & 0 \\ 0 & 1 & 4 \\ 0 & 0 & 1 \end{bmatrix} = \begin{bmatrix} 1 & 0 & 0 \\ 0 & -2 & 0 \\ 0 & 0 & 9 \end{bmatrix}$$

Since $\hat{h}_1 = 1, \hat{h}_2 = -2, \hat{h}_3 = 9$, $\mathbf{H}$ is indefinite. ∎

Example 2.7 Diagonalize the matrix

$$\mathbf{H} = \begin{bmatrix} 4 & -2 & 0 \\ -2 & 3 & 0 \\ 0 & 0 & 50 \end{bmatrix}$$

46

and determine its characterization.

Solution Add 0.5 times the first row to the second row as

$$\begin{bmatrix} 1 & 0 & 0 \\ 0.5 & 1 & 0 \\ 0 & 0 & 1 \end{bmatrix} \begin{bmatrix} 4 & -2 & 0 \\ -2 & 3 & 0 \\ 0 & 0 & 50 \end{bmatrix} \begin{bmatrix} 1 & 0.5 & 0 \\ 0 & 1 & 0 \\ 0 & 0 & 1 \end{bmatrix} = \begin{bmatrix} 4 & 0 & 0 \\ 0 & 2 & 0 \\ 0 & 0 & 50 \end{bmatrix}$$

Hence **H** is positive definite. ∎

Another theorem that can be used to characterize the Hessian is as follows:

Theorem 2.8 *Eigendecomposition of symmetric matrices*
(a) *If* **H** *is a real symmetric matrix, then there exists a real unitary (or orthogonal) matrix* **U** *such that*

$$\Lambda = \mathbf{U}^T \mathbf{H} \mathbf{U}$$

is a diagonal matrix whose diagonal elements are the eigenvalues of **H**.
(b) *The eigenvalues of* **H** *are real.*
(See Chap. 4 of Horn and Johnson [3] for proofs.)

For a real unitary matrix, we have $\mathbf{U}^T \mathbf{U} = \mathbf{I}_n$ where

$$\mathbf{I}_n = \begin{bmatrix} 1 & 0 & \cdots & 0 \\ 0 & 1 & \cdots & 0 \\ \vdots & \vdots & & \vdots \\ 0 & 0 & \cdots & 1 \end{bmatrix}$$

is the $n \times n$ identity matrix, and hence det $\mathbf{U} = \pm 1$, that is, **U** is nonsingular. From Theorem 2.6, Λ is positive definite, positive semidefinite, etc. if **H** is positive definite, positive semidefinite, etc. Therefore, **H** can be characterized by deducing its eigenvalues and then checking their signs as in Theorem 2.7.

Another approach for the characterization of a square matrix **H** is based on the evaluation of the so-called *principal minors* and *leading principal minors* of **H**, which are described in Sec. A.6. The details of this approach are summarized in terms of the following theorem.

Theorem 2.9 *Properties of matrices*
(a) *If* **H** *is positive semidefinite or positive definite, then*

$$\det \mathbf{H} \geq 0 \text{ or } > 0$$

(b) **H** *is positive definite if and only if all its leading principal minors are positive, i.e.,* det $\mathbf{H}_i > 0$ *for* $i = 1, 2, \ldots, n$.

(c) **H** *is positive semidefinite if and only if all its principal minors are nonnegative, i.e.,* det $(\mathbf{H}_i^{(l)}) \geq 0$ *for all possible selections of* $\{l_1, l_2, \ldots, l_i\}$ *for* $i = 1, 2, \ldots, n.$

(d) **H** *is negative definite if and only if all the leading principal minors of* $-\mathbf{H}$ *are positive, i.e.,* det $(-\mathbf{H}_i) > 0$ *for* $i = 1, 2, \ldots, n.$

(e) **H** *is negative semidefinite if and only if all the principal minors of* $-\mathbf{H}$ *are nonnegative, i.e.,* det $(-\mathbf{H}_i^{(l)}) \geq 0$ *for all possible selections of* $\{l_1, l_2, \ldots, l_i\}$ *for* $i = 1, 2, \ldots, n.$

(f) **H** *is indefinite if neither (c) nor (e) holds.*

Proof (a) Elementary transformations do not change the determinant of a matrix and hence

$$\det \mathbf{H} = \det \hat{\mathbf{H}} = \prod_{i=1}^{n} \hat{h}_i$$

where $\hat{\mathbf{H}}$ is a diagonalized version of **H** with diagonal elements $\hat{h}_i$. If **H** is positive semidefinite or positive definite, then $\hat{h}_i \geq 0$ or > 0 from Theorem 2.7 and, therefore,

$$\det \mathbf{H} \geq 0 \text{ or } > 0$$

(b) If

$$\mathbf{d} = [d_1 \ d_2 \ \cdots \ d_i \ 0 \ 0 \cdots \ 0]^T$$

and **H** is positive definite, then

$$\mathbf{d}^T \mathbf{H} \mathbf{d} = \mathbf{d}_0^T \mathbf{H}_i \mathbf{d}_0 > 0$$

for all $\mathbf{d}_0 \neq \mathbf{0}$ where

$$\mathbf{d}_0 = [d_1 \ d_2 \ \cdots \ d_i]^T$$

and $\mathbf{H}_i$ is the ith leading principal submatrix of **H**. The preceding inequality holds for $i = 1, 2, \ldots, n$ and, hence $\mathbf{H}_i$ is positive definite for $i = 1, 2, \ldots, n$. From part (a)

$$\det \mathbf{H}_i > 0 \qquad \text{for } i = 1, 2, \ldots, n$$

Now we prove the sufficiency of the theorem by induction. If $n = 1$, then $\mathbf{H} = a_{11}$, and det $(\mathbf{H}_1) = a_{11} > 0$ implies that **H** is positive definite. We assume that the sufficiency is valid for matrix **H** of size $(n - 1)$ by $(n - 1)$ and we shall show that the sufficiency is also valid for matrix **H** of size n by n. First, we write **H** as

$$\mathbf{H} = \begin{bmatrix} \mathbf{H}_{n-1} & \mathbf{h} \\ \mathbf{h}^T & h_{nn} \end{bmatrix}$$

where

$$\mathbf{H}_{n-1} = \begin{bmatrix} h_{11} & h_{12} & \cdots & h_{1,n-1} \\ h_{21} & h_{22} & \cdots & h_{2,n-1} \\ \vdots & \vdots & & \vdots \\ h_{n-1,1} & h_{n-1,2} & \cdots & h_{n-1,n-1} \end{bmatrix}, \quad \mathbf{h} = \begin{bmatrix} h_{1n} \\ h_{2n} \\ \vdots \\ h_{n-1,n} \end{bmatrix}$$

By assumption $\mathbf{H}_{n-1}$ is positive definite; hence there exists an $\mathbf{R}$ such that

$$\mathbf{R}^T \mathbf{H}_{n-1} \mathbf{R} = \mathbf{I}_{n-1}$$

where $\mathbf{I}_{n-1}$ is the $(n-1) \times (n-1)$ identity matrix. If we let

$$\mathbf{S} = \begin{bmatrix} \mathbf{R} & \mathbf{0} \\ \mathbf{0} & 1 \end{bmatrix}$$

we obtain

$$\mathbf{S}^T \mathbf{H} \mathbf{S} = \begin{bmatrix} \mathbf{R}^T & \mathbf{0} \\ \mathbf{0} & 1 \end{bmatrix} \begin{bmatrix} \mathbf{H}_{n-1} & \mathbf{h} \\ \mathbf{h}^T & h_{nn} \end{bmatrix} \begin{bmatrix} \mathbf{R} & \mathbf{0} \\ \mathbf{0} & 1 \end{bmatrix} = \begin{bmatrix} \mathbf{I}_{n-1} & \mathbf{R}^T \mathbf{h} \\ \mathbf{h}^T \mathbf{R} & h_{nn} \end{bmatrix}$$

If we define

$$\mathbf{T} = \begin{bmatrix} \mathbf{I}_{n-1} & -\mathbf{R}^T \mathbf{h} \\ \mathbf{0} & 1 \end{bmatrix}$$

then

$$\mathbf{T}^T \mathbf{S}^T \mathbf{H} \mathbf{S} \mathbf{T} = \begin{bmatrix} \mathbf{I}_{n-1} & \mathbf{0} \\ -\mathbf{h}^T \mathbf{R} & 1 \end{bmatrix} \begin{bmatrix} \mathbf{I}_{n-1} & \mathbf{R}^T \mathbf{h} \\ \mathbf{h}^T \mathbf{R} & h_{nn} \end{bmatrix} \begin{bmatrix} \mathbf{I}_{n-1} & -\mathbf{R}^T \mathbf{h} \\ \mathbf{0} & 1 \end{bmatrix}$$

$$= \begin{bmatrix} \mathbf{I}_{n-1} & \mathbf{0} \\ \mathbf{0} & h_{nn} - \mathbf{h}^T \mathbf{R} \mathbf{R}^T \mathbf{h} \end{bmatrix}$$

So if we let $\mathbf{U} = \mathbf{ST}$ and $\alpha = h_{nn} - \mathbf{h}^T \mathbf{R} \mathbf{R}^T \mathbf{h}$, then

$$\mathbf{U}^T \mathbf{H} \mathbf{U} = \begin{bmatrix} 1 & & & \\ & \ddots & & \\ & & 1 & \\ & & & \alpha \end{bmatrix}$$

which implies that

$$(\det \mathbf{U})^2 \det \mathbf{H} = \alpha$$

As $\det \mathbf{H} > 0$, we obtain $\alpha > 0$ and, therefore, $\mathbf{U}^T \mathbf{H} \mathbf{U}$ is positive definite which implies the positive definiteness of $\mathbf{H}$.

(c) The proof of the necessity is similar to the proof of part (b). If

$$\mathbf{d} = [0 \cdots 0 \, d_{l_1} \, 0 \cdots 0 \, d_{l_2} \, 0 \cdots 0 \, d_{l_i} \, 0 \cdots 0]^T$$

and $\mathbf{H}$ is positive semidefinite, then

$$\mathbf{d}^T \mathbf{H} \mathbf{d} = \mathbf{d}_0^T \mathbf{H}_i^{(l)} \mathbf{d}_0 \geq 0$$

for all $\mathbf{d}_0 \neq \mathbf{0}$ where

$$\mathbf{d}_0 = [d_{l_1} \ d_{l_2} \ \cdots \ d_{l_i}]^T$$

and $\mathbf{H}_i^{(l)}$ is an $i \times i$ principal submatrix. Hence $\mathbf{H}_i^{(l)}$ is positive semidefinite for all possible selections of rows (and columns) from the set $l = \{l_1, l_2, \ldots, l_i, \}$ with $1 \leq l_1 \leq l_2 < \ldots < l_i \leq n\}$ and $i = 1, 2, \ldots, n$. Now from part (a)

$$\det (\mathbf{H}_i^l) \geq 0 \qquad \text{for } 1, 2, \ldots, n.$$

The proof of sufficiency is rather lengthy and is omitted. The interested reader is referred to Chap. 7 of [3].

(d) If $\mathbf{H}_i$ is negative definite, then $-\mathbf{H}_i$ is positive definite by definition and hence the proof of part (b) applies to part (d).

(e) If $\mathbf{H}_i^{(l)}$ is negative semidefinite, then $-\mathbf{H}_i^{(l)}$ is positive semidefinite by definition and hence the proof of part (c) applies to part (e).

(f) If neither part (c) nor part (e) holds, then $\mathbf{d}^T \mathbf{H} \mathbf{d}$ can be positive or negative and hence $\mathbf{H}$ is indefinite.

■

Example 2.8 Characterize the Hessian matrices in Examples 2.6 and 2.7 by using the determinant method.

Solution Let

$$\Delta_i = \det (\mathbf{H}_i)$$

be the leading principal minors of $\mathbf{H}$. From Example 2.6, we have

$$\Delta_1 = 1, \quad \Delta_2 = -2, \quad \Delta_3 = -18$$

and if $\Delta_i' = \det (-\mathbf{H}_i)$, then

$$\Delta_1' = -1, \quad \Delta_2' = -2, \quad \Delta_3' = 18$$

since

$$\det (-\mathbf{H}_i) = (-1)^i \det (\mathbf{H}_i)$$

Hence $\mathbf{H}$ is indefinite.

From Example 2.7, we get

$$\Delta_1 = 4, \quad \Delta_2 = 8, \quad \Delta_3 = 400$$

Hence $\mathbf{H}$ is positive definite.

■

Example 2.9 Find and classify the stationary points of

$$f(\mathbf{x}) = x_1^2 + 2x_1x_2 + 2x_2^2 + 2x_1 + x_2$$

Solution The first partial derivatives of $f(\mathbf{x})$ are

$$\frac{\partial f}{\partial x_1} = 2x_1 + 2x_2 + 2$$

$$\frac{\partial f}{\partial x_2} = 2x_1 + 4x_2 + 1$$

If $\mathbf{g} = \mathbf{0}$, then

$$2x_1 + 2x_2 + 2 = 0$$
$$2x_1 + 4x_2 + 1 = 0$$

and so there is a stationary point at

$$\mathbf{x} = \mathbf{x}_1 = [-\tfrac{3}{2} \ \tfrac{1}{2}]^T$$

The Hessian is deduced as

$$\mathbf{H} = \begin{bmatrix} 2 & 2 \\ 2 & 4 \end{bmatrix}$$

and since $\Delta_1 = 2$ and $\Delta_2 = 4$, $\mathbf{H}$ is positive definite. Therefore, $\mathbf{x}_1$ is a minimizer.

∎

Example 2.10 Find and classify the stationary points of function

$$f(\mathbf{x}) = x_1^2 - x_2^2 + x_3^2 - 2x_1x_3 - x_2x_3 + 4x_1 + 12$$

Solution The first-order partial derivatives of $f(\mathbf{x})$ are

$$\frac{\partial f}{\partial x_1} = 2x_1 - 2x_3 + 4$$
$$\frac{\partial f}{\partial x_2} = -2x_2 - x_3$$
$$\frac{\partial f}{\partial x_3} = -2x_1 - x_2 + 2x_3$$

On equating the gradient to zero and then solving the simultaneous equations obtained, the stationary point $\mathbf{x}_1 = [-10 \ 4 \ -8]^T$ can be deduced. The Hessian is

$$\mathbf{H} = \begin{bmatrix} 2 & 0 & -2 \\ 0 & -2 & -1 \\ -2 & -1 & 2 \end{bmatrix}$$

and since $\Delta_1 = 2$, $\Delta_2 = -4$, $\Delta_3 = -2$, and $\Delta_1' = -2$, $\Delta_2' = -4$, $\Delta_3' = 2$, **H** is indefinite. Therefore, point $\mathbf{x}_1 = [-10 \; 4 \; - 8]^T$ is a saddle point. The solution can be readily checked by diagonalizing **H** as

$$\hat{\mathbf{H}} = \begin{bmatrix} 2 & 0 & 0 \\ 0 & -2 & 0 \\ 0 & 0 & 2\frac{1}{2} \end{bmatrix}$$

∎

2.7 Convex and Concave Functions

Usually, in practice, the function to be minimized has several extremum points and, consequently, the uncertainty arises as to whether the extremum point located by an optimization algorithm is the global one. In a specific class of functions referred to as convex and concave functions, any local extremum point is also a global extremum point. Therefore, if the objective function is convex or concave, optimality can be assured. The basic principles relating to convex and concave functions entail a collection of definitions and theorems.

Definition 2.7

A set $\mathcal{R}_c \subset E^n$ is said to be *convex* if for every pair of points $\mathbf{x}_1$, $\mathbf{x}_2 \subset \mathcal{R}_c$ and for every real number α in the range $0 < \alpha < 1$, the point

$$\mathbf{x} = \alpha\mathbf{x}_1 + (1 - \alpha)\mathbf{x}_2$$

is located in $\mathcal{R}_c$, i.e., $\mathbf{x} \in \mathcal{R}_c$.

∎

In effect, if any two points $\mathbf{x}_1$, $\mathbf{x}_2 \in \mathcal{R}_c$ are connected by a straight line, then $\mathcal{R}_c$ is convex if every point on the line segment between $\mathbf{x}_1$ and $\mathbf{x}_2$ is a member of $\mathcal{R}_c$. If some points on the line segment between $\mathbf{x}_1$ and $\mathbf{x}_2$ are not in $\mathcal{R}_c$, the set is said to be *nonconvex*. Convexity in sets is illustrated in Fig. 2.5.

The concept of convexity can also be applied to functions.

Definition 2.8

(*a*) A function $f(\mathbf{x})$ defined over a convex set $\mathcal{R}_c$ is said to be convex if for every pair of points $\mathbf{x}_1$, $\mathbf{x}_2 \in \mathcal{R}_c$ and every real number α in the range $0 < \alpha < 1$, the inequality

$$f[\alpha\mathbf{x}_1 + (1 - \alpha)\mathbf{x}_2] \leq \alpha f(\mathbf{x}_1) + (1 - \alpha)f(\mathbf{x}_2) \tag{2.9}$$

holds. If $\mathbf{x}_1 \neq \mathbf{x}_2$ and

$$f[\alpha\mathbf{x}_1 + (1 - \alpha)\mathbf{x}_2] < \alpha f(\mathbf{x}_1) + (1 - \alpha)f(\mathbf{x}_2)$$

52

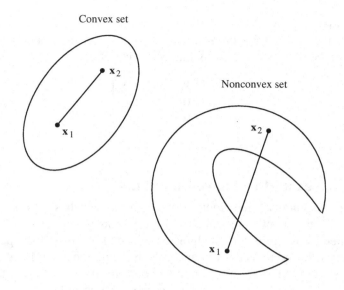

Convex set

x_2

Nonconvex set

x_2

x_1

x_1

Figure 2.5. Convexity in sets.

then $f(\mathbf{x})$ is said to be *strictly convex*.

(*b*) If $\phi(\mathbf{x})$ is defined over a convex set $\mathcal{R}_c$ and $f(\mathbf{x}) = -\phi(\mathbf{x})$ is convex, then $\phi(\mathbf{x})$ is said to be *concave*. If $f(\mathbf{x})$ is strictly convex, $\phi(\mathbf{x})$ is *strictly concave*.

∎

In the left-hand side of Eq. (2.9), function $f(\mathbf{x})$ is evaluated on the line segment joining points $\mathbf{x}_1$ and $\mathbf{x}_2$ whereas in the right-hand side of Eq. (2.9) an approximate value is obtained for $f(\mathbf{x})$ based on linear interpolation. Thus a function is convex if linear interpolation between any two points overestimates the value of the function. The functions shown in Fig. 2.6a and b are convex whereas that in Fig. 2.6c is nonconvex.

Theorem 2.10 *Convexity of linear combination of convex functions* *If*

$$f(\mathbf{x}) = af_1(\mathbf{x}) + bf_2(\mathbf{x})$$

where a, $b \geq 0$ *and* $f_1(\mathbf{x})$, $f_2(\mathbf{x})$ *are convex functions on the convex set* $\mathcal{R}_c$, *then* $f(\mathbf{x})$ *is convex on the set* $\mathcal{R}_c$.

Proof Since $f_1(\mathbf{x})$ and $f_2(\mathbf{x})$ are convex, and a, $b \geq 0$, then for $\mathbf{x} = \alpha\mathbf{x}_1 + (1 - \alpha)\mathbf{x}_2$ we have

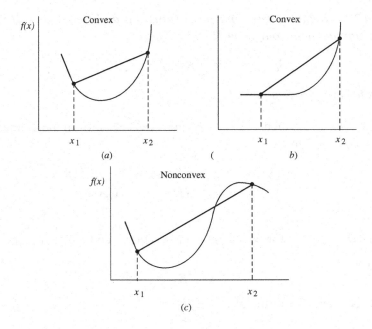

Figure 2.6. Convexity in functions.

$$af_1[\alpha \mathbf{x}_1 + (1 - \alpha)\mathbf{x}_2] \leq a[\alpha f_1(\mathbf{x}_1) + (1 - \alpha)f_1(\mathbf{x}_2)]$$

$$bf_2[\alpha \mathbf{x}_1 + (1 - \alpha)\mathbf{x}_2] \leq b[\alpha f_2(\mathbf{x}_1) + (1 - \alpha)f_2(\mathbf{x}_2)]$$

where $0 < \alpha < 1$. Hence

$$f(\mathbf{x}) = af_1(\mathbf{x}) + bf_2(\mathbf{x})$$
$$f[\alpha \mathbf{x}_1 + (1 - \alpha)\mathbf{x}_2] = af_1[\alpha \mathbf{x}_1 + (1 - \alpha)\mathbf{x}_2] + bf_2[\alpha \mathbf{x}_1 + (1 - \alpha)\mathbf{x}_2]$$
$$\leq \alpha[af_1(\mathbf{x}_1) + bf_2(\mathbf{x}_1)] + (1 - \alpha)[af_1(\mathbf{x}_2)$$
$$+ bf_2(\mathbf{x}_2)]$$

Since

$$af_1(\mathbf{x}_1) + bf_2(\mathbf{x}_1) = f(\mathbf{x}_1)$$
$$af_1(\mathbf{x}_2) + bf_2(\mathbf{x}_2) = f(\mathbf{x}_2)$$

the above inequality can be expressed as

$$f[\alpha \mathbf{x}_1 + (1 - \alpha)\mathbf{x}_2] \leq \alpha f(\mathbf{x}_1) + (1 - \alpha)f(\mathbf{x}_2)$$

that is, $f(\mathbf{x})$ is convex. ∎

54

Theorem 2.11 *Relation between convex functions and convex sets* If $f(\mathbf{x})$ is a convex function on a convex set $\mathcal{R}_c$, then the set

$$\mathcal{S}_c = \{\mathbf{x} : \mathbf{x} \in \mathcal{R}_c, \ f(\mathbf{x}) \le K\}$$

is convex for every real number K.

Proof If $\mathbf{x}_1, \ \mathbf{x}_2 \in \mathcal{S}_c$, then $f(\mathbf{x}_1) \le K$ and $f(\mathbf{x}_2) \le K$ from the definition of $\mathcal{S}_c$. Since $f(\mathbf{x})$ is convex

$$\begin{aligned} f[\alpha \mathbf{x}_1 + (1-\alpha)\mathbf{x}_2] &\le \alpha f(\mathbf{x}_1) + (1-\alpha)f(\mathbf{x}_2) \\ &\le \alpha K + (1-\alpha)K \end{aligned}$$

or

$$f(\mathbf{x}) \le K \qquad \text{for } \mathbf{x} = \alpha \mathbf{x}_1 + (1-\alpha)\mathbf{x}_2 \quad \text{and} \quad 0 < \alpha < 1$$

Therefore

$$\mathbf{x} \in \mathcal{S}_c$$

that is, $\mathcal{S}_c$ is convex by virtue of Def. 2.7.

∎

Theorem 2.11 is illustrated in Fig. 2.7, where set $\mathcal{S}_c$ is convex if $f(\mathbf{x})$ is a convex function on convex set $\mathcal{R}_c$.

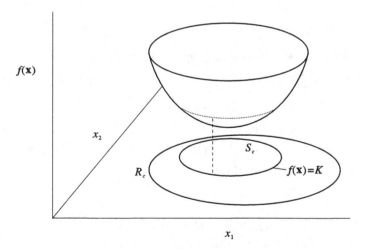

Figure 2.7. Graphical construction for Theorem 2.11.

An alternative view of convexity can be generated by examining some theorems which involve the gradient and Hessian of $f(\mathbf{x})$.

Theorem 2.12 *Property of convex functions relating to gradient* *If* $f(\mathbf{x}) \in$ C^1, *then* $f(\mathbf{x})$ *is convex over a convex set* $\mathcal{R}_c$ *if and only if*

$$f(\mathbf{x}_1) \geq f(\mathbf{x}) + \mathbf{g}(\mathbf{x})^T(\mathbf{x}_1 - \mathbf{x})$$

for all $\mathbf{x}$ *and* $\mathbf{x}_1 \in \mathcal{R}_c$, *where* $\mathbf{g}(\mathbf{x})$ *is the gradient of* $f(\mathbf{x})$.

Proof The proof of this theorem consists of two parts. First we prove that if $f(\mathbf{x})$ is convex, the inequality holds. Then we prove that if the inequality holds, $f(\mathbf{x})$ is convex. The two parts constitute the necessary and sufficient conditions of the theorem. If $f(\mathbf{x})$ is convex, then for all α in the range $0 < \alpha < 1$

$$f[\alpha\mathbf{x}_1 + (1 - \alpha)\mathbf{x}] \leq \alpha f(\mathbf{x}_1) + (1 - \alpha)f(\mathbf{x})$$

or

$$f[\mathbf{x} + \alpha(\mathbf{x}_1 - \mathbf{x})] - f(\mathbf{x}) \leq \alpha[f(\mathbf{x}_1) - f(\mathbf{x})]$$

As $\alpha \to 0$, the Taylor series of $f[\mathbf{x} + \alpha(\mathbf{x}_1 - \mathbf{x})]$ yields

$$f(\mathbf{x}) + \mathbf{g}(\mathbf{x})^T\alpha(\mathbf{x}_1 - \mathbf{x}) - f(\mathbf{x}) \leq \alpha[f(\mathbf{x}_1) - f(\mathbf{x})]$$

and so

$$f(\mathbf{x}_1) \geq f(\mathbf{x}) + \mathbf{g}(\mathbf{x})^T(\mathbf{x}_1 - \mathbf{x}) \tag{2.10}$$

Now if this inequality holds at points $\mathbf{x}$ and $\mathbf{x}_2 \in \mathcal{R}_c$, then

$$f(\mathbf{x}_2) \geq f(\mathbf{x}) + \mathbf{g}(\mathbf{x})^T(\mathbf{x}_2 - \mathbf{x}) \tag{2.11}$$

Hence Eqs. (2.10) and (2.11) yield

$$\alpha f(\mathbf{x}_1) + (1 - \alpha)f(\mathbf{x}_2) \geq \alpha f(\mathbf{x}) + \alpha\mathbf{g}(\mathbf{x})^T(\mathbf{x}_1 - \mathbf{x}) + (1 - \alpha)f(\mathbf{x})$$
$$+ (1 - \alpha)\mathbf{g}(\mathbf{x})^T(\mathbf{x}_2 - \mathbf{x})$$

or

$$\alpha f(\mathbf{x}_1) + (1 - \alpha)f(\mathbf{x}_2) \geq f(\mathbf{x}) + \mathbf{g}^T(\mathbf{x})[\alpha\mathbf{x}_1 + (1 - \alpha)\mathbf{x}_2 - \mathbf{x}]$$

With the substitution

$$\mathbf{x} = \alpha\mathbf{x}_1 + (1 - \alpha)\mathbf{x}_2$$

we obtain

$$f[\alpha\mathbf{x}_1 + (1 - \alpha)\mathbf{x}_2] \leq \alpha f(\mathbf{x}_1) + (1 - \alpha)f(\mathbf{x}_2)$$

for $0 < \alpha < 1$. Therefore, from Def. 2.8 $f(\mathbf{x})$ is convex. ∎

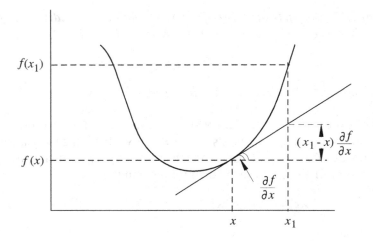

Figure 2.8.   Graphical construction for Theorem 2.12.

Theorem 2.12 states that a linear approximation of $f(\mathbf{x})$ at point $\mathbf{x}_1$ based on the derivatives of $f(\mathbf{x})$ at $\mathbf{x}$ underestimates the value of the function. This fact is illustrated in Fig. 2.8.

Theorem 2.13 *Property of convex functions relating to the Hessian A function $f(\mathbf{x}) \in C^2$ is convex over a convex set $\mathcal{R}_c$ if and only if the Hessian $H(\mathbf{x})$ of $f(\mathbf{x})$ is positive semidefinite for $\mathbf{x} \in \mathcal{R}_c$.*

Proof If $\mathbf{x}_1 = \mathbf{x} + \mathbf{d}$ where $\mathbf{x}_1$ and $\mathbf{x}$ are arbitrary points in $\mathcal{R}_c$, then the Taylor series yields

$$f(\mathbf{x}_1) = f(\mathbf{x}) + \mathbf{g}(\mathbf{x})^T(\mathbf{x}_1 - \mathbf{x}) + \tfrac{1}{2}\mathbf{d}^T\mathbf{H}(\mathbf{x} + \alpha\mathbf{d})\mathbf{d} \qquad (2.12)$$

where $0 \le \alpha \le 1$ (see Eq. (2.4h)). Now if $\mathbf{H}(\mathbf{x})$ is positive semidefinite everywhere in $\mathcal{R}_c$, then

$$\tfrac{1}{2}\mathbf{d}^T\mathbf{H}(\mathbf{x} + \alpha\mathbf{d})\mathbf{d} \ge 0$$

and so

$$f(\mathbf{x}_1) \ge f(\mathbf{x}) + \mathbf{g}(\mathbf{x})^T(\mathbf{x}_1 - \mathbf{x})$$

Therefore, from Theorem 2.12, $f(\mathbf{x})$ is convex.

If $\mathbf{H}(\mathbf{x})$ is not positive semidefinite everywhere in $\mathcal{R}_c$, then a point $\mathbf{x}$ and at least a $\mathbf{d}$ exist such that

$$\mathbf{d}^T\mathbf{H}(\mathbf{x} + \alpha\mathbf{d})\mathbf{d} < 0$$

and so Eq. (2.12) yields

$$f(\mathbf{x}_1) < f(\mathbf{x}) + \mathbf{g}(\mathbf{x})^T(\mathbf{x}_1 - \mathbf{x})$$

and $f(\mathbf{x})$ is nonconvex from Theorem 2.12. Therefore, $f(\mathbf{x})$ is convex if and only if $H(\mathbf{x})$ is positive semidefinite everywhere in $\mathcal{R}_c$.

∎

For a strictly convex function, Theorems 2.11–2.13 are modified as follows.

Theorem 2.14 *Properties of strictly convex functions*

(a) *If $f(\mathbf{x})$ is a strictly convex function on a convex set $\mathcal{R}_c$, then the set*

$$\mathcal{S}_c = \{\mathbf{x} : \mathbf{x} \in \mathcal{R}_c \ for \ f(\mathbf{x}) < K\}$$

is convex for every real number K.

(b) *If $f(\mathbf{x}) \in C^1$, then $f(\mathbf{x})$ is strictly convex over a convex set if and only if*

$$f(\mathbf{x}_1) > f(\mathbf{x}) + \mathbf{g}(\mathbf{x})^T(\mathbf{x}_1 - \mathbf{x})$$

for all $\mathbf{x}$ and $\mathbf{x}_1 \in \mathcal{R}_c$ where $\mathbf{g}(\mathbf{x})$ is the gradient of $f(\mathbf{x})$.

(c) *A function $f(\mathbf{x}) \in C^2$ is strictly convex over a convex set $\mathcal{R}_c$ if and only if the Hessian $\mathbf{H}(\mathbf{x})$ is positive definite for $\mathbf{x} \in \mathcal{R}_c$.*

If the second-order sufficient conditions for a minimum hold at $\mathbf{x}^*$ as in Theorem 2.4, in which case $\mathbf{x}^*$ is a strong local minimizer, then from Theorem 2.14(c), $f(\mathbf{x})$ must be strictly convex in the neighborhood of $\mathbf{x}^*$. Consequently, convexity assumes considerable importance even though the class of convex functions is quite restrictive.

If $\phi(\mathbf{x})$ is defined over a convex set $\mathcal{R}_c$ and $f(\mathbf{x}) = -\phi(\mathbf{x})$ is strictly convex, then $\phi(\mathbf{x})$ is strictly concave and the Hessian of $\phi(\mathbf{x})$ is negative definite. Conversely, if the Hessian of $\phi(\mathbf{x})$ is negative definite, then $\phi(\mathbf{x})$ is strictly concave.

Example 2.11 Check the following functions for convexity:

(a) $f(\mathbf{x}) = e^{x_1} + x_2^2 + 5$
(b) $f(\mathbf{x}) = 3x_1^2 - 5x_1x_2 + x_2^2$
(c) $f(\mathbf{x}) = \frac{1}{4}x_1^4 - x_1^2 + x_2^2$
(d) $f(\mathbf{x}) = 50 + 10x_1 + x_2 - 6x_1^2 - 3x_2^2$

Solution In each case the problem reduces to the derivation and characterization of the Hessian $\mathbf{H}$.

(a) The Hessian can be obtained as

$$\mathbf{H} = \begin{bmatrix} e^{x_1} & 0 \\ 0 & 2 \end{bmatrix}$$

58

For $-\infty < x_1 < \infty$, $\mathbf{H}$ is positive definite and $f(\mathbf{x})$ is strictly convex.

(b) In this case, we have

$$\mathbf{H} = \begin{bmatrix} 6 & -5 \\ -5 & 2 \end{bmatrix}$$

Since $\Delta_1 = 6$, $\Delta_2 = -13$ and $\Delta_1' = -6$, $\Delta_2' = -13$, where $\Delta_i = \det(\mathbf{H}_i)$ and $\Delta_i' = \det(-\mathbf{H}_i)$, $\mathbf{H}$ is indefinite. Thus $f(\mathbf{x})$ is neither convex nor concave.

(c) For this example, we get

$$\mathbf{H} = \begin{bmatrix} 3x_1^2 - 2 & 0 \\ 0 & 2 \end{bmatrix}$$

For $x_1 \leq -\sqrt{2/3}$ and $x_1 \geq \sqrt{2/3}$, $\mathbf{H}$ is positive semidefinite and $f(\mathbf{x})$ is convex; for $x_1 < -\sqrt{2/3}$ and $x_1 > \sqrt{2/3}$, $\mathbf{H}$ is positive definite and $f(\mathbf{x})$ is strictly convex; for $-\sqrt{2/3} < x_1 < \sqrt{2/3}$, $\mathbf{H}$ is indefinite, and $f(\mathbf{x})$ is neither convex nor concave.

(d) As before

$$\mathbf{H} = \begin{bmatrix} -12 & 0 \\ 0 & -6 \end{bmatrix}$$

In this case $\mathbf{H}$ is negative definite, and $f(\mathbf{x})$ is strictly concave.

∎

2.8 Optimization of Convex Functions

The above theorems and results can now be used to deduce the following three important theorems.

Theorem 2.15 *Relation between local and global minimizers in convex functions* *If $f(\mathbf{x})$ is a convex function defined on a convex set $\mathcal{R}_c$, then*

(a) the set of points $\mathcal{S}_c$ where $f(\mathbf{x})$ is minimum is convex;

(b) any local minimizer of $f(\mathbf{x})$ is a global minimizer.

Proof (a) If F^* is a minimum of $f(\mathbf{x})$, then $\mathcal{S}_c = \{\mathbf{x} : f(\mathbf{x}) \leq F^*, \mathbf{x} \in \mathcal{R}_c\}$ is convex by virtue of Theorem 2.11.

(b) If $\mathbf{x}^* \in \mathcal{R}_c$ is a local minimizer but there is another point $\mathbf{x}^{**} \in \mathcal{R}_c$ which is a global minimizer such that

$$f(\mathbf{x}^{**}) < f(\mathbf{x}^*)$$

then on line $\mathbf{x} = \alpha\mathbf{x}^{**} + (1 - \alpha)\mathbf{x}^*$

$$f[\alpha\mathbf{x}^{**} + (1 - \alpha)\mathbf{x}^*] \leq \alpha f(\mathbf{x}^{**}) + (1 - \alpha)f(\mathbf{x}^*)$$
$$< \alpha f(\mathbf{x}^*) + (1 - \alpha)f(\mathbf{x}^*)$$

or
$$f(\mathbf{x}) < f(\mathbf{x}^*) \qquad \text{for all } \alpha$$

This contradicts the fact that $\mathbf{x}^*$ is a local minimizer and so

$$f(\mathbf{x}) \geq f(\mathbf{x}^*)$$

for all $\mathbf{x} \in \mathcal{R}_c$. Therefore, any local minimizers are located in a convex set, and all are global minimizers.

∎

Theorem 2.16 *Existence of a global minimizer in convex functions* If $f(\mathbf{x}) \in C^1$ *is a convex function on a convex set* $\mathcal{R}_c$ *and there is a point* $\mathbf{x}^*$ *such that*

$$\mathbf{g}(x^*)^T \mathbf{d} \geq 0 \qquad \text{where } \mathbf{d} = \mathbf{x}_1 - \mathbf{x}^*$$

for all $\mathbf{x}_1 \in \mathcal{R}_c$, *then* $\mathbf{x}^*$ *is a global minimizer of* $f(\mathbf{x})$.

Proof From Theorem 2.12

$$f(\mathbf{x}_1) \geq f(\mathbf{x}^*) + \mathbf{g}(\mathbf{x}^*)^T (\mathbf{x}_1 - \mathbf{x}^*)$$

where $\mathbf{g}(\mathbf{x}^*)$ is the gradient of $f(\mathbf{x})$ at $\mathbf{x} = \mathbf{x}^*$. Since

$$\mathbf{g}(\mathbf{x}^*)^T (\mathbf{x}_1 - \mathbf{x}^*) \geq 0$$

we have

$$f(\mathbf{x}_1) \geq f(\mathbf{x}^*)$$

and so $\mathbf{x}^*$ is a local minimizer. By virtue of Theorem 2.15, $\mathbf{x}^*$ is also a global minimizer.

Similarly, if $f(\mathbf{x})$ is a strictly convex function and

$$\mathbf{g}(\mathbf{x}^*)^T \mathbf{d} > 0$$

then $\mathbf{x}^*$ is a strong global minimizer.

∎

The above theorem states, in effect, that if $f(\mathbf{x})$ is convex, then the first-order necessary conditions become sufficient for $\mathbf{x}^*$ to be a global minimizer.

Since a convex function of one variable is in the form of the letter U whereas a convex function of two variables is in the form of a bowl, there are no theorems analogous to Theorems 2.15 and 2.16 pertaining to the maximization of a convex function. However, the following theorem, which is intuitively plausible, is sometimes useful.

Theorem 2.17 *Location of maximum of a convex function* If $f(\mathbf{x})$ *is a convex function defined on a bounded, closed, convex set* $\mathcal{R}_c$, *then if* $f(\mathbf{x})$ *has a maximum over* $\mathcal{R}_c$, *it occurs at the boundary of* $\mathcal{R}_c$.

60

Proof If point $\mathbf{x}$ is in the interior of $\mathcal{R}_c$, a line can be drawn through $\mathbf{x}$ which intersects the boundary at two points, say, $\mathbf{x}_1$ and $\mathbf{x}_2$, since $\mathcal{R}_c$ is bounded and closed. Since $f(\mathbf{x})$ is convex, some α exists in the range $0 < \alpha < 1$ such that

$$\mathbf{x} = \alpha\mathbf{x}_1 + (1 - \alpha)\mathbf{x}_2$$

and

$$f(\mathbf{x}) \leq \alpha f(\mathbf{x}_1) + (1 - \alpha)f(\mathbf{x}_2)$$

If $f(\mathbf{x}_1) > f(\mathbf{x}_2)$, we have

$$f(\mathbf{x}) < \alpha f(\mathbf{x}_1) + (1 - \alpha)f(\mathbf{x}_1)$$
$$= f(\mathbf{x}_1)$$

If

$$f(\mathbf{x}_1) < f(\mathbf{x}_2)$$

we obtain

$$f(\mathbf{x}) < \alpha f(\mathbf{x}_2) + (1 - \alpha)f(\mathbf{x}_2)$$
$$= f(\mathbf{x}_2)$$

Now if

$$f(\mathbf{x}_1) = f(\mathbf{x}_2)$$

the result

$$f(\mathbf{x}) \leq f(\mathbf{x}_1) \quad \text{and} \quad f(\mathbf{x}) \leq f(\mathbf{x}_2)$$

is obtained. Evidently, in all possibilities the maximizers occur on the boundary of $\mathcal{R}_c$.

∎

This theorem is illustrated in Fig. 2.9.

References

1 M. H. Protter and C. B. Morrey, Jr., *Modern Mathematical Analysis*, Addison-Wesley, Reading, MA, 1964.
2 D. G. Luenberger, *Linear and Nonlinear Programming*, 2nd ed., Addison-Wesley, Reading, MA, 1984.
3 R. A. Horn and C. R. Johnson, *Matrix Analysis*, Cambridge, Cambridge University Press, UK, 1990.

Problems

2.1 (a) Obtain a quadratic approximation for the function

$$f(\mathbf{x}) = 2x_1^3 + x_2^2 + x_1^2 x_2^2 + 4x_1 x_2 + 3$$

at point $\mathbf{x} + \boldsymbol{\delta}$ if $\mathbf{x}^T = [1\ 1]$.

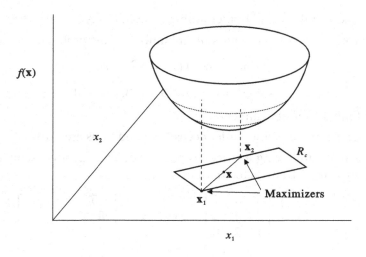

Figure 2.9. Graphical construction for Theorem 2.17.

(*b*) Now obtain a linear approximation.

2.2 An n-variable quadratic function is given by

$$f(\mathbf{x}) = a + \mathbf{b}^T \mathbf{x} + \tfrac{1}{2} \mathbf{x}^T \mathbf{Q} \mathbf{x}$$

where $\mathbf{Q}$ is an $n \times n$ symmetric matrix. Show that the gradient and Hessian of $f(\mathbf{x})$ are given by

$$\mathbf{g} = \mathbf{b} + \mathbf{Q}\mathbf{x} \quad \text{and} \quad \nabla^2 f(\mathbf{x}) = \mathbf{Q}$$

respectively.

2.3 Point $\mathbf{x}_a = [2\ 4]^T$ is a possible minimizer of the problem

$$\text{minimize } f(\mathbf{x}) = \tfrac{1}{4}[x_1^2 + 4x_2^2 - 4(3x_1 + 8x_2) + 100]$$

$$\text{subject to:} \quad x_1 = 2,\ x_2 \geq 0$$

(*a*) Find the feasible directions.

(*b*) Check if the second-order necessary conditions are satisfied.

2.4 Points $\mathbf{x}_a = [0\ 3]^T$, $\mathbf{x}_b = [4\ 0]^T$, $\mathbf{x}_c = [4\ 3]^T$ are possible maximizers of the problem

$$\text{maximize } f(\mathbf{x}) = 2(4x_1 + 3x_2) - (x_1^2 + x_2^2 + 25)$$

$$\text{subject to:} \quad x_1 \geq 0,\ x_2 \geq 0$$

(*a*) Find the feasible directions.

62

(b) Check if the second-order necessary conditions are satisfied.

2.5 Point $\mathbf{x}_a = [4 \ -1]^T$ is a possible minimizer of the problem

$$\text{minimize } f(\mathbf{x}) = \frac{16}{x_1} - x_2$$

$$\text{subject to:} \quad x_1 + x_2 = 3, \ x_1 \geq 0$$

(a) Find the feasible directions.

(b) Check if the second-order necessary conditions are satisfied.

2.6 Classify the following matrices as positive definite, positive semidefinite, etc. by using LDL^T factorization:

$$(a) \ \mathbf{H} = \begin{bmatrix} 5 & 3 & 1 \\ 3 & 4 & 2 \\ 1 & 2 & 6 \end{bmatrix}, \quad (b) \ \mathbf{H} = \begin{bmatrix} -5 & 1 & 1 \\ 1 & -2 & 2 \\ 1 & 2 & -4 \end{bmatrix}$$

$$(c) \ \mathbf{H} = \begin{bmatrix} -1 & 2 & -3 \\ 2 & 4 & 5 \\ -3 & 5 & -20 \end{bmatrix}$$

2.7 Check the results in Prob. 2.6 by using the determinant method.

2.8 Classify the following matrices by using the eigenvalue method:

$$(a) \ \mathbf{H} = \begin{bmatrix} 2 & 3 \\ 3 & 4 \end{bmatrix}, \quad (b) \ \mathbf{H} = \begin{bmatrix} 1 & 0 & 4 \\ 0 & 2 & 0 \\ 4 & 0 & 18 \end{bmatrix}$$

2.9 One of the points $\mathbf{x}_a = [1 \ -1]^T$, $\mathbf{x}_b = [0 \ 0]^T$, $\mathbf{x}_c = [1 \ 1]^T$ minimizes the function

$$f(\mathbf{x}) = 100(x_2 - x_1^2)^2 + (1 - x_1)^2$$

By using appropriate tests, identify the minimizer.

2.10 An optimization algorithm has given a solution $\mathbf{x}_a = [0.6959 \ -11.3479]^T$ for the problem

$$\text{minimize } f(\mathbf{x}) = x_1^4 + x_1 x_2 + (1 + x_2)^2$$

(a) Classify the general Hessian of $f(\mathbf{x})$ (i.e., positive definite, ..., etc.).

(b) Determine whether $\mathbf{x}_a$ is a minimizer, maximizer, or saddle point.

2.11 Find and classify the stationary points for the function

$$f(\mathbf{x}) = x_1^2 - x_2^2 + x_3^2 - 2x_1 x_3 - x_2 x_3 + 4x_1 + 12$$

2.12 Find and classify the stationary points for the following functions:

(a) $f(\mathbf{x}) = 2x_1^2 + x_2^2 - 2x_1 x_2 + 2x_1^3 + x_1^4$

(b) $f(\mathbf{x}) = x_1^2 x_2^2 - 4x_1^2 x_2 + 4x_1^2 + 2x_1 x_2^2 + x_2^2 - 8x_1 x_2 + 8x_1 - 4x_2$

2.13 Show that
$$f(\mathbf{x}) = (x_2 - x_1^2)^2 + x_1^5$$
has only one stationary point which is neither a minimizer or a maximizer.

2.14 Investigate the following functions and determine whether they are convex or concave:

(a) $f(\mathbf{x}) = x_1^2 + \cosh x_2$
(b) $f(\mathbf{x}) = x_1^2 + 2x_2^2 + 2x_3^2 + x_4^2 - x_1x_2 + x_1x_3 - 2x_2x_4 + x_1x_4$
(c) $f(\mathbf{x}) = x_1^2 - 2x_2^2 - 2x_3^2 + x_4^2 - x_1x_2 + x_1x_3 - 2x_2x_4 + x_1x_4$

2.15 A given quadratic function $f(\mathbf{x})$ is known to be convex for $\|\mathbf{x}\| < \varepsilon$. Show that it is convex for all $\mathbf{x} \in E^n$.

2.16 Two functions $f_1(\mathbf{x})$ and $f_2(\mathbf{x})$ are convex over a convex set $\mathcal{R}_c$. Show that
$$f(\mathbf{x}) = \alpha f_1(\mathbf{x}) + \beta f_2(\mathbf{x})$$
where α and β are nonnegative scalars is convex over $\mathcal{R}_c$.

2.17 Assume that functions $f_1(\mathbf{x})$ and $f_2(\mathbf{x})$ are convex and let
$$f(\mathbf{x}) = \max\{f_1(\mathbf{x}),\ f_2(\mathbf{x})\}$$
Show that $f(\mathbf{x})$ is a convex function.

2.18 Let $\gamma(t)$ be a single-variable convex function which is monotonic non-decreasing, i.e., $\gamma(t_1) \geq \gamma(t_2)$ for $t_1 > t_2$. Show that the compound function $\gamma[f(\mathbf{x})]$ is convex if $f(\mathbf{x})$ is convex [2].

Chapter 3

GENERAL PROPERTIES
OF ALGORITHMS

3.1 Introduction

In Chap. 1, an optimization algorithm has been informally introduced as a sequence of steps that can be executed repeatedly in order to obtain a series of progressively improved solutions, starting with an initial estimate of the solution. In this chapter, a more formal and mathematical description of an algorithm will be supplied and some fundamental concepts pertaining to all algorithms in general will be studied.

The chapter includes a discussion on the principle of global convergence. Specifically, a general theorem that enumerates the circumstances and conditions under which convergence can be assured in any given algorithm is proved [1]–[3].

The chapter concludes with a quantitative discussion relating to the speed of convergence of an optimization algorithm. In particular, quantitative criteria are described that can be used to compare the efficiency of different types of algorithms.

3.2 An Algorithm as a Point-to-Point Mapping

There are numerous algorithms that can be used for the solution of nonlinear programming problems ranging from some simple to some highly complex algorithms. Although different algorithms differ significantly in their structure, mathematical basis, and range of applications, they share certain common properties that can be regarded as universal. The two most fundamental common properties of nonlinear programming algorithms are

1. They are iterative algorithms.
2. They are descent algorithms.

An algorithm is *iterative* if the solution is obtained by calculating a series of points in sequence, starting with an initial estimate of the solution. On the other hand, an algorithm is a *descent* algorithm if each new point generated by the algorithm yields a reduced value of some function, possibly the objective function.

In mathematical terms, an algorithm can be regarded as a *point-to-point mapping* where a point x_k in some space, normally a subspace of the E^n vector space, is mapped onto another point x_{k+1} in the same space. The value of x_{k+1} is governed by some rule of correspondence. In effect, if point x_k is used as input to the algorithm, a point x_{k+1} is obtained as output. An algorithm can thus be represented by a block diagram as depicted in Fig. 3.1. In this representation, x_0 is an initial estimate of the solution and the feedback line denotes the iterative nature of the algorithm. The rule of correspondence between x_{k+1} and x_k, which might range from a simple expression to a large number of formulas, can be represented by the relation

$$x_{k+1} = A(x_k)$$

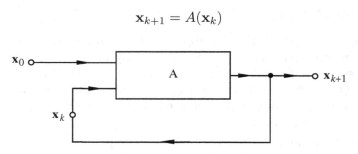

Figure 3.1. Block diagram for an iterative algorithm.

When applied iteratively to successive points, an algorithm will generate a series (or sequence) of points $\{x_0, x_1, \ldots, x_k, \ldots\}$ in space X, as depicted in Fig. 3.2. If the sequence converges to a limit $\hat{x}$, then $\hat{x}$ is the required solution.

A sequence $\{x_0, x_1, \ldots, x_k, \ldots\}$ is said to *converge to a limit* $\hat{x}$ if for any given $\varepsilon > 0$, and an integer K exists such that

$$\|x_k - \hat{x}\| < \varepsilon \qquad \text{for all } k \geq K$$

where $\| \cdot \|$ denotes the Euclidean norm. Such a sequence can be represented by the notation $\{x_k\}_{k=0}^{\infty}$ and its limit as $k \to \infty$ by $x_k \to \hat{x}$. If the sequence converges, it has a unique limit point.

Later on in this chapter, reference will be made to subsequences of a given sequence. A subsequence of $\{x_k\}_{k=0}^{\infty}$, denoted as $\{x_k\}_{k \in I}$, where I is a set of positive integers, can be obtained by deleting certain elements in $\{x_k\}_{k=0}^{\infty}$. For example, if $I = \{k : k \geq 10\}$ then $\{x_k\}_{k \in I} = \{x_{10}, x_{11}, x_{12}, \ldots\}$, if $I = \{k : k \text{ even and greater than zero}\}$ then $\{x_k\}_{k \in I} = \{x_2, x_4, x_6, \ldots\}$, and if $I = \{k : 0 \leq k \leq 100\}$, then $\{x_k\}_{k \in I} = \{x_1, x_2, \ldots, x_{100}\}$. In our notation $S = \{k : \mathcal{P}\}$, S is the set of elements such that k has property $\mathcal{P}$.

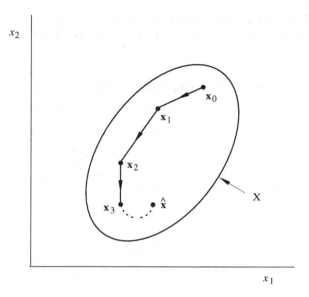

Figure 3.2. A point-to-point algorithm in E^2.

If the sequence of points generated by an algorithm A converges to a limit $\hat{x}$ as described above, then algorithm A is said to be *continuous*.

3.3 An Algorithm as a Point-to-Set Mapping

In the above discussion, an algorithm was considered as a point-to-point mapping in that for any given point x_k a corresponding unique point x_{k+1} is generated. In practice, this is the true nature of an algorithm only if a specific version of the algorithm is implemented on a specific computer. Since different implementations of an algorithm by different programmers on different computers are very likely to give slightly different results, owing to the accumulation of roundoff errors, it is advantageous to consider an algorithm as a point-to-set mapping. In this way, if any general properties of an algorithm are deduced, they will hold for all possible implementations of the algorithm. Furthermore, they may hold for similar algorithms. For these reasons, the following more general definition of an algorithm will be used throughout the rest of this chapter.

Definition 3.1 An algorithm is a *point-to-set mapping* on space X that assigns a subset of X to every point $x \in X$.

∎

According to this definition, an algorithm A will generate a sequence of points $\{x_k\}_{k=1}^{\infty}$ by assigning a set X_1 which is a subset of X, i.e., $X_1 \subset X$, to a given initial point $x_0 \in X$. Then an arbitrary point $x_1 \in X_1$ is selected and a set $X_2 \subset X$ is assigned to it, and so on, as depicted in Fig. 3.3. The rule of correspondence between $\mathbf{x}_{k+1}$ and $\mathbf{x}_k$ is, therefore, of the form

$$\mathbf{x}_{k+1} \in A(\mathbf{x}_k)$$

where $A(\mathbf{x}_k)$ is the set of all possible outputs if $\mathbf{x}_k$ is the input.

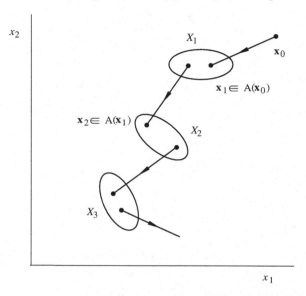

Figure 3.3. A point-to-set algorithm in E^2.

Clearly, the above definition encompasses all possible implementations of an algorithm and it would encompass a class of algorithms that are based on a similar mathematical structure. The concept of the point-to-set algorithm can be visualized by noting that in a typical algorithm

$$\mathbf{x}_{k+1} = A(\mathbf{x}_k) + \varepsilon_q$$

where ε_q is a random vector due to the quantization of numbers. Since the quantization error tends to depend heavily on the sequence in which arithmetic operations are performed and on the precision of the computer used, the exact location of $\mathbf{x}_{k+1}$ is not known. Nevertheless, it is known that $\mathbf{x}_{k+1}$ is a member of a small subset of X.

3.4   Closed Algorithms

In the above discussion, reference was made to the continuity of a point-to-point algorithm. A more general property which is applicable to point-to-point as well as to point-to-set algorithms is the property of closeness. This property reduces to continuity in a point-to-point algorithm.

Definition 3.2

    (*a*) A point-to-set algorithm A, from space X to space X_1 is said to be *closed at point* $\hat{\mathbf{x}} \in X$ if the assumptions

$$\mathbf{x}_k \to \hat{\mathbf{x}} \qquad \text{for } \mathbf{x}_k \in X$$
$$\mathbf{x}_{k+1} \to \hat{\mathbf{x}}_1 \qquad \text{for } \mathbf{x}_{k+1} \in A(\mathbf{x}_k)$$

imply that

$$\hat{\mathbf{x}}_1 \in A(\hat{\mathbf{x}})$$

The notation $\mathbf{x}_k \to \hat{\mathbf{x}}$ denotes that the sequence $\{\mathbf{x}_k\}_{k=0}^{\infty}$ converges to a limit $\hat{\mathbf{x}}$.

    (*b*) A point-to-set algorithm A is said to be *closed on* X if it is closed at each point of X.

                                                          ■

This definition is illustrated in Fig. 3.4. It states that algorithm A is closed at point $\hat{\mathbf{x}}$ if a solid line can be drawn between $\hat{\mathbf{x}}$ and $\hat{\mathbf{x}}_1$, and if a solid line can be drawn for all $\hat{\mathbf{x}} \in X$, then A is closed on X.

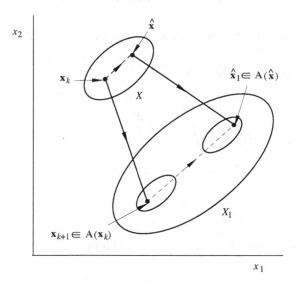

Figure 3.4. Definition of a closed algorithm in E^2.

Example 3.1 An algorithm A is defined by

$$x_{k+1} = A(x_k) = \begin{cases} \frac{1}{2}(x_k + 2) & \text{for } x_k > 1 \\[2mm] \frac{1}{4}x_k & \text{for } x_k \le 1 \end{cases}$$

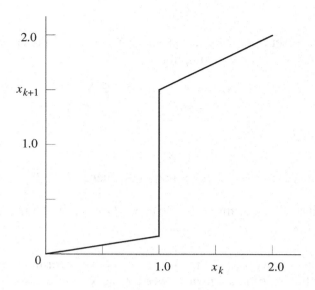

Figure 3.5. Graph for Example 3.1.

(see Fig. 3.5). Show that the algorithm is not closed at $\hat{x} = 1$.

Solution Let sequence $\{x_k\}_{k=0}^{\infty}$ be defined by

$$x_k = 1 + \frac{1}{2^{k+1}}$$

The sequence can be obtained as

$$\{x_k\}_{k=0}^{\infty} = \{1.5,\ 1.25,\ 1.125\ \ldots,\ 1\}$$

and hence

$$x_k \to \hat{x} = 1$$

The corresponding sequence $\{x_{k+1}\}_{k=0}^{\infty}$ is given by

$$x_{k+1} = A(x_k) = \tfrac{1}{2}(x_k + 2)$$

and so

$$\{x_{k+1}\}_{k=0}^{\infty} = \{1.75,\ 1.625,\ 1.5625,\ \ldots,\ 1.5\}$$

Thus

$$x_{k+1} \to \hat{x}_1 = 1.5$$

Now

$$A(\hat{x}) = \tfrac{1}{4}$$

and since $\hat{x}_1 = 1.5$, we have

$$\hat{x}_1 \neq A(\hat{x})$$

Therefore, A is not closed at $\hat{x} = 1$. The problem is due to the discontinuity of $A(x_k)$ at $x_k = 1$. ∎

Example 3.2 An algorithm A is defined by

$$x_{k+1} = A(x_k) = x_k^2 \quad \text{for } -\infty < x_k < \infty$$

Show that A is closed.

Solution Let $\{x_k\}$ be a sequence converging to $\hat{x}$, i.e., $x_k \to \hat{x}$. Then $\{x_{k+1}\} = \{A(x_k)\} = \{x_k^2\}$ is a sequence that converges to $\hat{x}^2$, i.e., $x_k^2 \to \hat{x}_1 = \hat{x}^2$. Since $\hat{x}_1 = A(\hat{x})$, we conclude that for all $\hat{x}$ in the range $-\infty < \hat{x} < \infty$, A is closed. ∎

3.5 Descent Functions

In any descent algorithm, a specific function $D(\mathbf{x})$ is utilized, which is reduced continuously throughout the optimization until convergence is achieved. $D(\mathbf{x})$ may be the objective function itself or some related function, and it is referred to as the descent function. A formal definition summarizing the required specifications for a function to be a descent function is as follows. This will be used later in Theorem 3.1.

Definition 3.3

Let $S \subset X$ be the set containing the solution points, and assume that A is an algorithm on X. A continuous real-valued function $D(\mathbf{x})$ on X is said to be a *descent function* for S and A if it satisfies the following specifications:

(a) if $\mathbf{x}_k \notin S$, then $D(\mathbf{x}_{k+1}) < D(\mathbf{x}_k)$ for all $\mathbf{x}_{k+1} \in A(\mathbf{x}_k)$
(b) if $\mathbf{x}_k \in S$, then $D(\mathbf{x}_{k+1}) \leq D(\mathbf{x}_k)$ for all $\mathbf{x}_{k+1} \in A(\mathbf{x}_k)$ ∎

Example 3.3 Obtain a descent function for the algorithm

$$x_{k+1} = A(x_k) = \tfrac{1}{4}x_k$$

Solution For an arbitrary point x_0, the sequence

$$\{x_k\}_{k=0}^\infty = \{x_0, \frac{x_0}{4}, \frac{x_0}{4^2}, \dots, 0\}$$

is generated. Therefore, $D(x_k) = |x_k|$ satisfies condition (a). The solution set is a single point at $x_\infty = 0$. Therefore, condition (b) is satisfied. Hence $D(x) = |x|$ is a descent function for the algorithm.

∎

3.6 Global Convergence

If an algorithm has the important property that an arbitrary initial point $x_0 \in X$ will lead to a converging sequence of points $\{x_k\}_{k=0}^\infty$, then the algorithm is said to be *globally convergent*. In practice, even the most efficient algorithms are likely to fail if certain conditions are violated. For example, an algorithm may generate sequences that do not converge or may converge to points that are not solutions. There are several factors that are likely to cause failure in an algorithm. However, if they are clearly understood, certain precautions can be taken which will circumvent the cause of failure. Consequently, the study of global convergence is of particular interest not only to the theorist but also the practitioner.

A large segment of the theory of global convergence deals with the circumstances and conditions that will guarantee global convergence. An important theorem in this area is as follows:

Theorem 3.1 *Convergence of an algorithm* *Let A be an algorithm on X and assume that an initial point x_0 will yield an infinite sequence $\{x_k\}_{k=0}^\infty$ where*

$$x_{k+1} \in A(x_k)$$

If a solution set S and a descent function $D(x_k)$ exist for the algorithm such that

(a) all points x_k are contained in a compact subset of X,
(b) $D(x_k)$ satisfies the specifications of Def. 3.3, and
(c) the mapping of A is closed at all points outside S,

then the limit of any convergent subsequence of $\{x_k\}_{k=0}^\infty$ is a solution point.

Proof The proof of this important theorem consists of two parts. In part (a), we suppose that $\hat{x}$ is the limit of any subsequence of $\{x_k\}_{k=0}^\infty$, say, $\{x_k\}_{k\in I}$, where I is a set of integers, and show that $D(x_k)$ converges with respect to the infinite sequence $\{x_k\}_{k=0}^\infty$. In part (b), we show that $\hat{x}$ is in the solution set S.

The second part of the proof relies heavily on the Weierstrass theorem (see [4]) which states that if W is a compact set, then the sequence $\{x_k\}_{k=0}^\infty$, where $x_k \in W$, has a limit point in W. A set W is *compact* if it is *closed*. A set W is *closed*, if all points on the boundary of W belong to W. A set W is *bounded*, if it can be circumscribed by a hypersphere of finite radius. A consequence of the Weierstrass theorem is that a subsequence $\{x_k\}_{k\in I}$ of $\{x_k\}_{k=0}^\infty$ has a limit point in set $\bar{W} = \{\mathbf{x}_k : k \in I\}$ since $\bar{W}$ is a subset of W and is, therefore, compact.

(*a*) Since $D(\mathbf{x}_k)$ is continuous on X and $\hat{\mathbf{x}}$ is assumed to be the limit of $\{\mathbf{x}_k\}_{k\in I}$, a positive number and an integer K exist such that

$$D(\mathbf{x}_k) - D(\hat{\mathbf{x}}) < \varepsilon \tag{3.1}$$

for $k \geq K$ with $k \in I$. Hence $D(\mathbf{x}_k)$ converges with respect to the subsequence $\{\mathbf{x}_k\}_{k\in I}$. We must show, however, that $D(\mathbf{x}_k)$ converges with respect to the infinite sequence $\{\mathbf{x}_k\}_{k=0}^{\infty}$.

For any $k \geq K$, we can write

$$D(\mathbf{x}_k) - D(\hat{\mathbf{x}}) = [D(\mathbf{x}_k) - D(\mathbf{x}_K)] + [D(\mathbf{x}_K) - D(\hat{\mathbf{x}})] \tag{3.2}$$

If $k = K$ in Eq. (3.1)

$$D(\mathbf{x}_K) - D(\hat{\mathbf{x}}) < \varepsilon \tag{3.3}$$

and if $k \geq K$, then $D(\mathbf{x}_k) \leq D(\mathbf{x}_K)$ from Def. 3.3 and hence

$$D(\mathbf{x}_k) - D(\mathbf{x}_K) \leq 0 \tag{3.4}$$

Now from Eqs. (3.2) – (3.4), we have

$$D(\mathbf{x}_k) - D(\hat{\mathbf{x}}) < \varepsilon$$

for *all* $k \geq K$. Therefore,

$$\lim_{k\to\infty} D(\mathbf{x}_k) = D(\hat{\mathbf{x}}) \tag{3.5}$$

that is, $D(\mathbf{x}_k)$ converges with respect to the infinite series, as $\mathbf{x}_k \to \hat{\mathbf{x}}$.

(*b*) Let us assume that $\hat{\mathbf{x}}$ is not in the solution set. Since the elements of subsequence $\{\mathbf{x}_{k+1}\}_{k\in I}$ belong to a compact set according to condition (*a*), a compact subset $\{\mathbf{x}_{k+1} : k \in \bar{I} \subset I\}$ exists such that $\mathbf{x}_{k+1}$ converges to some limit $\bar{\mathbf{x}}$ by virtue of the Weierstrass theorem. As in part (*a*), we can show that

$$\lim_{k\to\infty} D(\mathbf{x}_{k+1}) = D(\bar{\mathbf{x}}) \tag{3.6}$$

Therefore, from Eqs. (3.5) and (3.6)

$$D(\bar{\mathbf{x}}) = D(\hat{\mathbf{x}}) \tag{3.7}$$

On the other hand,

$$\mathbf{x}_k \to \hat{\mathbf{x}} \qquad \text{for } k \in \bar{I} \quad \text{(from part (}a\text{))}$$
$$\mathbf{x}_{k+1} \to \bar{\mathbf{x}} \qquad \text{for } \mathbf{x}_{k+1} \in A(\mathbf{x})$$

and since $\hat{\mathbf{x}} \notin S$ by supposition, and A is closed at points outside S according to condition (c), we have

$$\bar{\mathbf{x}} \in A(\hat{\mathbf{x}})$$

Consequently,

$$D(\bar{\mathbf{x}}) < D(\hat{\mathbf{x}}) \tag{3.8}$$

On comparing Eqs. (3.7) and (3.8), a contradiction is observed and, in effect, our assumption that point $\hat{\mathbf{x}}$ is not in the solution set S is not valid. That is, the limit of any convergent subsequence of $\{\mathbf{x}_k\}_{k=0}^{\infty}$ is a solution point.

■

In simple terms, the above theorem states that if

- (a) the points that can be generated by the algorithm are located in the *finite* E^n space,
- (b) a *descent function* can be found that satisfies the strict requirements stipulated, and
- (c) the algorithm is *closed* outside the neighborhood of the solution,

then the algorithm is globally convergent. Further, a very close approximation to the solution can be obtained in a finite number of iterations, since the limit of any convergent finite subsequence of $\{\mathbf{x}_k\}_{k=0}^{\infty}$ is a solution.

A corollary of Theorem 3.1 which is of some significance is as follows:

Corollary If under the conditions of Theorem 3.1, the solution set S consists of a single point $\hat{\mathbf{x}}$, then the sequence $\{\mathbf{x}_k\}_{k=0}^{\infty}$ converges to $\hat{\mathbf{x}}$.

Proof If we suppose that there is a subsequence $\{\mathbf{x}_k\}_{k\in I}$ that does not converge to $\hat{\mathbf{x}}$, then

$$\|\mathbf{x}_k - \hat{\mathbf{x}}\| > \varepsilon \tag{3.9}$$

for all $k \in I$ and $\varepsilon > 0$. Now set $\{\mathbf{x}_k : \in I' \subset I\}$ is compact and hence $\{\mathbf{x}_k\}_{k\in I'}$ converges to a limit point, say, $\mathbf{x}'$, by virtue of the Weierstrass theorem. From Theorem 3.1,

$$\|\mathbf{x}_k - \mathbf{x}'\| < \varepsilon \tag{3.10}$$

for all $k \geq K$. Since the solution set consists of a single point, we have $\mathbf{x}' = \hat{\mathbf{x}}$. Under these circumstances, Eqs. (3.9) and (3.10) become contradictory and, in effect, our supposition is false. That is, any subsequence of $\{\mathbf{x}_k\}_{k=0}^{\infty}$, including the sequence itself, converges to $\hat{\mathbf{x}}$.

■

If one or more of the conditions in Theorem 3.1 are violated, an algorithm may fail to converge. The possible causes of failure are illustrated in terms of the following examples.

Example 3.4 A possible algorithm for the problem

$$\text{minimize } f(x) = |x|$$

is

$$x_{k+1} = A(x_k) = \begin{cases} \frac{1}{2}(x_k + 2) & \text{for } x_k > 1 \\ \frac{1}{4}x_k & \text{for } x_k \leq 1 \end{cases}$$

Show that the algorithm is not globally convergent and explain why.

Solution If $x_0 = 4$, the algorithm will generate the sequence

$$\{x_k\}_{k=0}^\infty = \{4, 3, 2.5, 2.25, \ldots, 2\}$$

and if $x_0 = -4$, we have

$$\{x_k\}_{k=0}^\infty = \{-4, -1, -0.25, -0.0625, \ldots, 0\}$$

Since two distinct initial points lead to different limit points, the algorithm is not globally convergent. The reason is that the algorithm is not closed at point $x_k = 1$ (see Example 3.1), i.e., condition (c) of Theorem 3.1 is violated. ∎

Example 3.5 A possible algorithm for the problem

$$\text{minimize } f(x) = x^3$$

is

$$x_{k+1} = A(x_k) = -(x_k^2 + 1)$$

Show that the algorithm is not globally convergent and explain why.

Solution For an initial point x_0 the solution sequence is

$$\{x_k\}_{k=0}^\infty = \{x_0, -(x_0^2+1), -((x_0^2+1)^2+1), (((x_0^2+1)^2+1)^2+1), \ldots, -\infty\}$$

Hence the sequence does not converge, and its elements are not in a compact set. Therefore, the algorithm is not globally convergent since condition (a) of Theorem 3.1 is violated. ∎

Example 3.6 A possible algorithm for the problem

$$\text{minimize } f(x) = |x - 1|$$
$$\text{subject to: } x > 0$$

is

$$x_{k+1} = A(x_k) = \sqrt{x_k}$$

76

Show that the algorithm is globally convergent for $0 < x_0 < \infty$.

Solution For any initial point x_0 in the range $0 < x_0 < \infty$, we have

$$\{x_k\}_{k=0}^\infty = \{x_0,\ x_0^{1/2},\ x_0^{1/4},\ \ldots,\ 1\}$$
$$\{x_k\}_{k=0}^\infty = \{x_0^{1/2},\ x_0^{1/4},\ x_0^{1/8},\ \ldots,\ 1\}$$

Thus

$$x_k \to \hat{x} = 1, \quad x_{k+1} \to \hat{x}_1 = 1$$

Evidently, all points x_k belong to a compact set and so condition (a) is satisfied. The objective function $f(x)$ is a descent function since

$$|x_{k+1} - 1| < |x_k - 1| \qquad \text{for all } k < \infty$$

and so condition (b) is satisfied.
   Since

$$x_k \to \hat{x} \qquad \text{for } x_k > 0$$
$$x_{k+1} \to \hat{x}_1 \qquad \text{for } x_{k+1} = A(x_k)$$

and

$$\hat{x}_1 = A(\hat{x})$$

the algorithm is closed, and so condition (c) is satisfied. The algorithm is, therefore, globally convergent.

∎

3.7 Rates of Convergence
   The many available algorithms differ significantly in their computational efficiency. An efficient or fast algorithm is one that requires only a small number of iterations to converge to a solution and the amount of computation will be small. Economical reasons dictate that the most efficient algorithm for the application be chosen and, therefore, quantitative measures or criteria that can be used to measure the rate of convergence in a set of competing algorithms are required.
   The most basic criterion in this area is the order of convergence of a sequence. If $\{x_k\}_{k=0}^\infty$ is a sequence of real numbers, its *order of convergence* is the largest nonnegative integer p that will satisfy the relation

$$0 \le \beta < \infty$$

where

$$\beta = \lim_{k\to\infty} \frac{|x_{k+1} - \hat{x}|}{|x_k - \hat{x}|^p} \tag{3.11}$$

and $\hat{x}$ is the limit of the sequence as $k \to \infty$. Parameter β is called the *convergence ratio*.

Example 3.7 Find the order of convergence and convergence ratio of the sequence $\{x_k\}_{k=0}^{\infty}$ if

(a) $x_k = \gamma^k$ \qquad for $0 < \gamma < 1$
(b) $x_k = \gamma^{2^k}$ \qquad for $0 < \gamma < 1$

Solution (a) Since $\hat{x} = 0$, Eq. (3.11) gives

$$\beta = \lim_{k \to \infty} \gamma^{k(1-p)+1}$$

Hence for $p = 0$, 1, 2 we have $\beta = 0$, γ, ∞. Thus $p = 1$ and $\beta = \gamma$.
  (b) In this case

$$\beta = \lim_{k \to \infty} \frac{\gamma^{2^{(k+1)}}}{\gamma^{2^k p}} = \lim_{k \to \infty} \{\gamma^{2^k(2-p)}\}$$

Hence for $p = 0$, 1, 2, 3, we have $\beta = 0$, 0, 1, ∞. Thus $p = 2$ and $\beta = 1$. ∎

If the limit in Eq. (3.11) exists, then

$$\lim_{k \to \infty} |x_k - \hat{x}| = \varepsilon$$

where $\varepsilon < 1$. As a result

$$\lim_{k \to \infty} |x_{k+1} - \hat{x}| = \beta \varepsilon^p$$

Therefore, the rate of convergence is increased if p is increased and β is reduced. If $\gamma = 0.8$ in Example 3.7, the sequences in parts (a) and (b) will be

$$\{x_k\}_{k=0}^{\infty} = \{1, \ 0.8, \ 0.64, \ 0.512, \ 0.409, \ \ldots, \ 0\}$$

and

$$\{x_k\}_{k=0}^{\infty} = \{1, \ 0.64, \ 0.409, \ 0.167, \ 0.023, \ \ldots, \ 0\}$$

respectively. The rate of convergence in the second sequence is much faster since $p = 2$.

If $p = 1$ and $\beta < 1$, the sequence is said to have *linear convergence*. If $p = 1$ and $\beta = 0$ or $p \geq 2$ the sequence is said to have *superlinear convergence*.

Most of the available nonlinear programming algorithms have linear convergence and hence their comparison is based on the value of β.

78

Another measure of the rate of convergence of a sequence is the so-called *average order of convergence*. This is the lowest nonnegative integer that will satisfy the relation

$$\gamma = \lim_{k \to \infty} |x_k - \hat{x}|^{1/(p+1)^k} = 1$$

If no $p > 0$ can be found, then the order of convergence is infinity.

Example 3.8 Find the average order of convergence of the sequence $\{x_k\}_{k=0}^{\infty}$

(a) $x_k = \gamma^k$ for $0 < \gamma < 1$

(b) $x_k = \gamma^{2^k}$ for $0 < \gamma < 1$

Solution (a) Since $\hat{x} = 0$

$$\gamma = \lim_{k \to \infty} (\gamma^k)^{1/(p+1)^k} = 1$$

Hence for $p = 0,\ 1,\ 2$, we have $\gamma = 0,\ 1,\ 1$. Thus $p = 1$.

(b) In this case,

$$\gamma = \lim_{k \to \infty} (\gamma^{2^k})^{1/(p+1)^k} = 1$$

Hence for $p = 0,\ 1,\ 2,\ 3$, we have $\gamma = 0,\ \gamma,\ 1,\ 1$. Thus $p = 2$.

∎

If the average order of convergence is unity, then the sequence is said to have an *average linear convergence*. An average convergence ratio can be defined as

$$\gamma = \lim_{k \to \infty} |x_k - \hat{x}|^{1/k}$$

In the above discussion, the convergence of a sequence of numbers has been considered. Such a sequence might consist of the values of the objective function as the solution is approached. In such a case, we are measuring the rate at which the objective function is approaching its minimum. Alternatively, if we desire to know how fast the variables of the problem approach their optimum values, a sequence of numbers can be generated by considering the magnitudes or the square magnitudes of the vectors $\mathbf{x}_k - \hat{\mathbf{x}}$, namely, $\|\mathbf{x}_k - \hat{\mathbf{x}}\|$ or $\|\mathbf{x}_k - \hat{\mathbf{x}}\|^2$, as the solution is approached.

In the above measures of the rate of convergence, the emphasis is placed on the efficiency of an algorithm in the neighborhood of the solution. Usually in optimization a large percentage of the computation is used in the neighborhood of the solution and, consequently, the above measures are quite meaningful. Occasionally, however, a specific algorithm may be efficient in the neighborhood of the solution and very inefficient elsewhere. In such a case, the use of the above criteria would lead to misleading results and, therefore, other criteria should also be employed.

References

1 W. I. Zangwill, *Nonlinear Programming: A Unified Approach*, Chap. 4, Prentice-Hall, Englewood Cliffs, N.J., 1969.
2 M. S. Bazaraa and C. M. Shetty, *Nonlinear Programming*, Chap. 7, Wiley, New York, 1979.
3 D. G. Luenberger, *Linear and Nonlinear Programming*, 2nd ed., Chap. 6, Addison-Wesley, Reading, MA, 1984.
4 H. M. Edwards, *Advanced Calculus*, Houghton Mifflin, Chap. 9, Boston, MA, 1969.

Problems

3.1 Let A be a point-to-set algorithm from space E^1 to space E^1. The *graph* of A is defined as the set

$$\{(x, y) : x \in E^1, \ y \in A(x)\}$$

(*a*) Show that algorithm A defined by

$$A(x) = \{y : x/4 \leq y \leq x/2\}$$

is closed on E^1.

(*b*) Plot the graph of A.

3.2 Examine whether or not the following point-to-set mappings from E^1 to E^1 are closed:

(*a*)

$$A(x) = \begin{cases} \frac{1}{x} & \text{if } x \neq 0 \\ x & \text{if } x = 0 \end{cases}$$

(*b*)

$$A(x) = \begin{cases} \frac{1}{x} & \text{if } x \neq 0 \\ 1 & \text{if } x = 0 \end{cases}$$

(*c*)

$$A(x) = \begin{cases} x & \text{if } x \neq 0 \\ 1 & \text{if } x = 0 \end{cases}$$

3.3 Define the point-to-set mapping on E^n by

$$A(\mathbf{x}) = \{\mathbf{y} : \mathbf{y}^T\mathbf{x} \geq 1\}$$

Is A closed on E^n?

3.4 Let $\{b_k, \ k = 0, 1, \ldots\}$ and $\{c_k, \ k = 0, 1, \ldots\}$ be sequences of real numbers, where $b_k \to 0$ superlinearly in the sense that $p = 1$ and $\beta = 0$ (see Eq. (3.11)) and $c \leq c_k \leq C$ with $c > 0$. Show that $\{b_k c_k, \ k = 0, 1, \ldots\}$ converges to zero superlinearly.

Chapter 4

ONE-DIMENSIONAL OPTIMIZATION

4.1   Introduction

Three general classes of nonlinear optimization problems can be identified, as follows:

1. One-dimensional unconstrained problems
2. Multidimensional unconstrained problems
3. Multidimensional constrained problems

Problems of the first class are the easiest to solve whereas those of the third class are the most difficult. In practice, multidimensional constrained problems are usually reduced to multidimensional unconstrained problems which, in turn, are reduced to one-dimensional unconstrained problems. In effect, most of the available nonlinear programming algorithms are based on the minimization of a function of a single variable without constraints. Therefore, *efficient* one-dimensional optimization algorithms are required, if efficient multidimensional unconstrained and constrained algorithms are to be constructed.

The one-dimensional optimization problem is

$$\text{minimize } F = f(x)$$

where $f(x)$ is a function of one variable. This problem has a solution if $f(x)$ is *unimodal* in some range of x, i.e., $f(x)$ has only one minimum in some range $x_L \leq x \leq x_U$, where x_L and x_U are the lower and upper limits of the minimizer x^*.

Two general classes of one-dimensional optimization methods are available, namely, *search methods* and *approximation methods*.

In search methods, an interval $[x_L, \ x_U]$ containing x^*, known as a *bracket*, is established and is then repeatedly reduced on the basis of function evaluations until a reduced bracket $[x_{L,k}, \ x_{U,k}]$ is obtained which is sufficiently small. The minimizer can be assumed to be at the center of interval $[x_{L,k}, \; x_{U,k}]$. These methods can be applied to any function and differentiability of $f(x)$ is not essential.

In approximation methods, an approximation of the function in the form of a low-order polynomial, usually a second- or third-order polynomial, is assumed. This is then analyzed using elementary calculus and an approximate value of x^* is deduced. The interval $[x_L, \; x_U]$ is then reduced and the process is repeated several times until a sufficiently precise value of x^* is obtained. In these methods, $f(x)$ is required to be continuous and differentiable, i.e., $f(x) \in C^1$.

Several one-dimensional optimization approaches will be examined in this chapter, as follows [1]–[8]:

1. Dichotomous search
2. Fibonacci search
3. Golden-section search
4. Quadratic interpolation method
5. Cubic interpolation method
6. The Davies, Swann, and Campey method

The first three are search methods, the fourth and fifth are approximation methods, and the sixth is a practical and useful method that combines a search method with an approximation method.

The chapter will also deal with a so-called *inexact line search* due to Fletcher [9][10], which offers certain important advantages such as reduced computational effort in some optimization methods.

4.2 Dichotomous Search

Consider a unimodal function which is known to have a minimum in the interval $[x_L, \; x_U]$. This interval is said to be the *range of uncertainty*. The minimizer x^* of $f(x)$ can be located by reducing progressively the range of uncertainty until a sufficiently small range is obtained. In search methods, this can be achieved by using the values of $f(x)$ at suitable points.

If the value of $f(x)$ is known at a single point x_a in the range $x_L < x_a < x_U$, point x^* is equally likely to be in the range x_L to x_a or x_a to x_U as depicted in Fig. 4.1(a). Consequently, the information available is not sufficient to allow the reduction of the range of uncertainty. However, if the value of $f(x)$ is known at two points, say, x_a and x_b, an immediate reduction is possible. Three possibilities may arise, namely,

(a) $f(x_a) < f(x_b)$
(b) $f(x_a) > f(x_b)$
(c) $f(x_a) = f(x_b)$

In case (a), x^* may be located in range $x_L < x^* < x_a$ or $x_a < x^* < x_b$, that is, $x_L < x^* < x_b$, as illustrated in Fig. 4.1a. The possibility $x_b < x^* < x_U$

is definitely ruled out since this would imply that $f(x)$ has two minima: one to the left of x_b and one to the right of x_b. Similarly, for case (b), we must have $x_a < x^* < x_U$ as in Fig. 4.1b. For case (c), we must have $x_a < x^* < x_b$, that is, both inequalities $x_L < x^* < x_b$ and $x_a < x^* < x_U$ must be satisfied as in Fig. 4.1c.

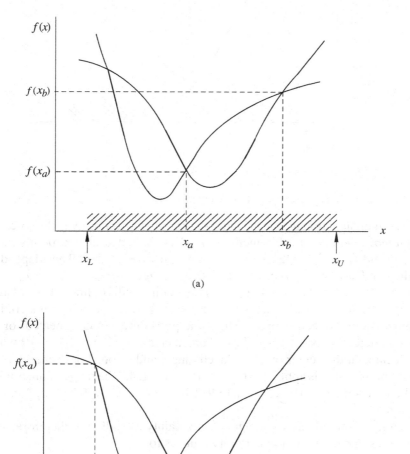

(a)

(b)

Figure 4.1. Reduction of range of uncertainty: (a) case (a), $f(x_a) < f(x_b)$, (b) case (b), $f(x_a) > f(x_b)$.

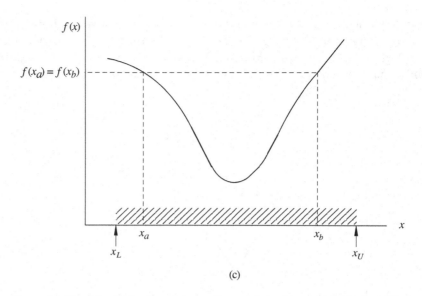

Figure 4.1 Cont'd. Reduction of range of uncertainty: (c) case $(c), f(x_a) = f(x_b)$.

A rudimentary strategy for reducing the range of uncertainty is the so-called *dichotomous search*. In this method, $f(x)$ is evaluated at two points $x_a = x_1 - \varepsilon/2$ and $x_b = x_1 + \varepsilon/2$ where ε is a small positive number. Then depending on whether $f(x_a) < f(x_b)$ or $f(x_a) > f(x_b)$, range x_L to $x_1 + \varepsilon/2$ or $x_1 - \varepsilon/2$ to x_U can be selected and if $f(x_a) = f(x_b)$ either will do fine. If we assume that $x_1 - x_L = x_U - x_1$, i.e., $x_1 = (x_L + x_U)/2$, the region of uncertainty is immediately reduced by half. The same procedure can be repeated for the reduced range, that is, $f(x)$ can be evaluated at $x_2 - \varepsilon/2$ and $x_2 + \varepsilon/2$ where x_2 is located at the center of the reduced range, and so on. For example, if the dichotomous search is applied to the function of Fig. 4.2 the range of uncertainty will be reduced from $0 < x^* < 1$ to $9/16 + \varepsilon/2 < x^* < 5/8 - \varepsilon/2$ in four iterations.

Each iteration reduces the range of uncertainty by half and, therefore, after k iterations, the interval of uncertainty reduces to

$$I_k = (\tfrac{1}{2})^k I_0$$

where $I_0 = x_U - x_L$. For example, after 7 iterations the range of uncertainty would be reduced to less than 1% of the initial interval. The corresponding computational effort would be 14 function evaluations since two evaluations are required for each iteration.

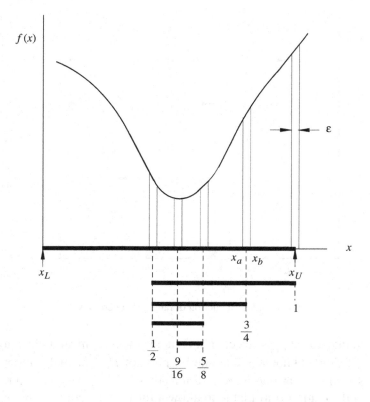

Figure 4.2. Construction for dichotomous search.

4.3 Fibonacci Search

Consider an interval of uncertainty

$$I_k = [x_{L,k}, \ x_{U,k}]$$

and assume that two points $x_{a,k}$ and $x_{b,k}$ are located in I_k, as depicted in Fig. 4.3. As in Sec. 4.2, the values of $f(x)$ at $x_{a,k}$ and $x_{b,k}$, namely, $f(x_{a,k})$ and $f(x_{b,k})$, can be used to select the left interval

$$I_{k+1}^L = [x_{L,k}, \ x_{b,k}]$$

if $f(x_{a,k}) < f(x_{b,k})$, the right interval

$$I_{k+1}^R = [x_{a,k}, \ x_{U,k}]$$

if $f(x_{a,k}) > f(x_{b,k})$, or either of I_{k+1}^R and I_{k+1}^L if

$$f(x_a, k) = f(x_b, k)$$

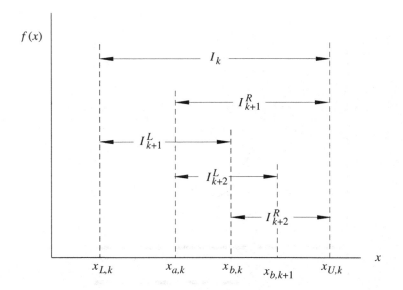

Figure 4.3. Reduction of range of uncertainty.

If the right interval I_{k+1}^R is selected, it contains the minimizer and, in addition, the value of $f(x)$ is known at one interior point of I_{k+1}^R, namely, at point $x_{b,k}$. If $f(x)$ is evaluated at one more interior point of I_{k+1}^R, say, at point $x_{b,k+1}$, sufficient information is available to allow a further reduction in the region of uncertainty, and the above cycle of events can be repeated. One of the two new sub-intervals I_{k+2}^L and I_{k+2}^R, shown in Fig. 4.3, can be selected as before, and so on. In this way, only one function evaluation is required per iteration, and the amount of computation will be reduced relative to that required in the dichotomous search.

From Fig. 4.3

$$I_k = I_{k+1}^L + I_{k+2}^R \tag{4.1}$$

and if, for the sake of convenience, we assume equal intervals, then

$$I_{k+1}^L = I_{k+1}^R = I_{k+1}$$
$$I_{k+2}^L = I_{k+2}^R = I_{k+2}$$

Eq. (4.1) gives the recursive relation

$$I_k = I_{k+1} + I_{k+2} \tag{4.2}$$

If the above procedure is repeated a number of times, a sequence of intervals $\{I_1, I_2, \ldots, I_n\}$ will be generated as follows:

$$I_1 = I_2 + I_3$$

$$I_2 = I_3 + I_4$$

$$\vdots$$

$$I_n = I_{n+1} + I_{n+2}$$

In the above set of n equations, there are $n + 2$ variables and if I_1 is the given initial interval, $n + 1$ variables remain. Therefore, an infinite set of sequences can be generated by specifying some additional rule. Two specific sequences of particular interest are the Fibonacci sequence and the golden-section sequence. The Fibonacci sequence is considered below and the golden-section sequence is considered in Sec. 4.4.

The Fibonacci sequence is generated by assuming that the interval for iteration $n + 2$ vanishes, that is, $I_{n+2} = 0$. If we let $k = n$ in Eq. (4.2), we can write

$$I_{n+1} = I_n - I_{n+2} = I_n \equiv F_0 I_n$$
$$I_n = I_{n+1} + I_{n+2} = I_n \equiv F_1 I_n$$
$$I_{n-1} = I_n + I_{n+1} = 2I_n \equiv F_2 I_n$$
$$I_{n-2} = I_{n-1} + I_n = 3I_n \equiv F_3 I_n$$
$$I_{n-3} = I_{n-2} + I_{n-1} = 5I_n \equiv F_4 I_n$$
$$I_{n-4} = I_{n-3} + I_{n-2} = 8I_n \equiv F_5 I_n$$

$$\vdots \qquad\qquad \vdots$$

$$I_k = I_{k+1} + I_{k+2} = F_{n-k+1} I_n \tag{4.3a}$$

$$\vdots \qquad\qquad \vdots$$

$$I_1 = I_2 + I_3 = F_n I_n \tag{4.3b}$$

The sequence generated, namely,

$$\{1,\ 1,\ 2,\ 3,\ 5,\ 8,\ 13,\ \ldots\} = \{F_0,\ F_1,\ F_2,\ F_3,\ F_4,\ F_5,\ F_6\ \ldots\}$$

is the well-known *Fibonacci sequence* which occurs in various branches of mathematics. It can be generated by using the recursive relation

$$F_k = F_{k-1} + F_{k-2} \qquad \text{for } k \geq 2 \tag{4.4}$$

where $F_0 = F_1 = 1$. Its application in one-dimensional optimization gives rise to the *Fibonacci search method*. The method is illustrated in Fig. 4.4 for $n = 6$ and $I_1 = 100$ for the case where the left interval is consistently selected, i.e., the minimum occurs in the neighborhood of $x = 0$.

If the number of iterations is assumed to be n, then from Eqn. (4.3b) the Fibonacci search reduces the interval of uncertainty to

$$I_n = \frac{I_1}{F_n} \tag{4.5}$$

88

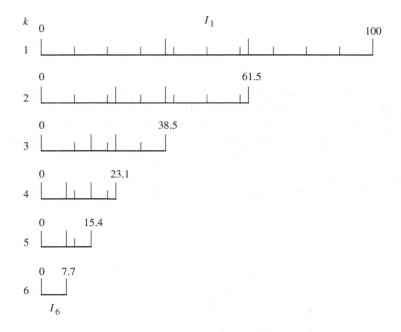

Figure 4.4. Fibonacci search for $n = 6$.

For example, if $n = 11$ then $F_n = 144$ and so I_n is reduced to a value less than 1% the value of I_1. This would entail 11 iterations and since one function evaluation is required per iteration, a total of 11 function evaluations would be required as opposed to the 14 required by the dichotomous search to achieve the same precision. In effect, the Fibonacci search is more efficient than the dichotomous search. Indeed, it can be shown, that it achieves the largest interval reduction relative to the other search methods and it is, therefore, the most efficient in terms of computational effort required.

The Fibonacci sequence of intervals can be generated only if n is known. If the objective of the optimization is to find x^* to within a prescribed tolerance, the required n can be readily deduced by using Eq. (4.5). However, if the objective is to determine the minimum of $f(x)$ to within a prescribed tolerance, difficulty will be experienced in determining the required n without solving the problem. The only available information is that n will be low if the minimum of $f(x)$ is shallow and high if $f(x)$ varies rapidly in the neighborhood of the solution.

The above principles can be used to implement the Fibonacci search. Let us assume that the initial bounds of the minimizer, namely, $x_{L,1}$ and $x_{U,1}$, and the value of n are given, and a mathematical description of $f(x)$ is available.

The implementation consists of computing the successive intervals, evaluating $f(x)$, and selecting the appropriate intervals.

At the kth iteration, the quantities $x_{L,k}$, $x_{a,k}$, $x_{b,k}$, $x_{U,k}$, I_{k+1} and

$$f_{a,k} = f(x_{a,k}), \quad f_{b,k} = f(x_{b,k})$$

are known, and the quantities $x_{L,k+1}$, $x_{a,k+1}$, $x_{b,k+1}$, $x_{U,k+1}$, I_{k+2}, $f_{a,k+1}$, and $f_{b,k+1}$ are required. Interval I_{k+2} can be obtained from Eq. (4.3a) as

$$I_{k+2} = \frac{F_{n-k-1}}{F_{n-k}} I_{k+1} \qquad (4.6)$$

The remaining quantities can be computed as follows.

If $f_{a,k} > f_{b,k}$, then x^* is in interval $[x_{a,k}, x_{U,k}]$ and so the new bounds of x^* can be updated as

$$x_{L,k+1} = x_{a,k} \qquad (4.7)$$
$$x_{U,k+1} = x_{U,k} \qquad (4.8)$$

Similarly, the two interior points of the new interval, namely, $x_{a,k+1}$ and $x_{b,k+1}$ will be $x_{b,k}$ and $x_{L,k+1} + I_{k+2}$, respectively. We can thus assign

$$x_{a,k+1} = x_{b,k} \qquad (4.9)$$
$$x_{b,k+1} = x_{L,k+1} + I_{k+2} \qquad (4.10)$$

as illustrated in Fig. 4.5. The value $f_{b,k}$ is retained as the value of $f(x)$ at $x_{a,k+1}$, and the value of $f(x)$ at $x_{b,k+1}$ is calculated, i.e.,

$$f_{a,k+1} = f_{b,k} \qquad (4.11)$$
$$f_{b,k+1} = f(x_{b,k+1}) \qquad (4.12)$$

On the other hand, if $f_{a,k} < f_{b,k}$, then x^* is in interval $[x_{L,k}, x_{b,k}]$. In this case, we assign

$$x_{L,k+1} = x_{L,k} \qquad (4.13)$$
$$x_{U,k+1} = x_{b,k} \qquad (4.14)$$
$$x_{a,k+1} = x_{U,k+1} - I_{k+2} \qquad (4.15)$$
$$x_{b,k+1} = x_{a,k} \qquad (4.16)$$
$$f_{b,k+1} = f_{a,k} \qquad (4.17)$$

and calculate

$$f_{a,k+1} = f(x_{a,k+1}) \qquad (4.18)$$

as depicted in Fig. 4.6. In the unlikely event that $f_{a,k} = f_{b,k}$, either of the above sets of assignments can be used since x^* is contained by both intervals $[x_{L,k}, x_{b,k}]$ and $[x_{a,k}, x_{U,k}]$.

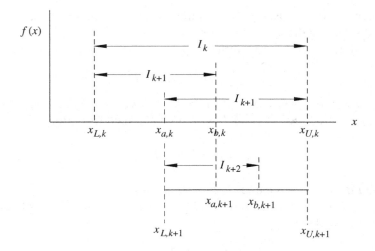

Figure 4.5. Assignments in kth iteration of the Fibonacci search if $f_{a,k} > f_{b,k}$.

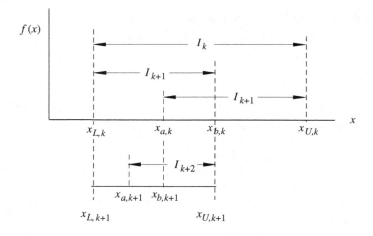

Figure 4.6. Assignments in kth iteration of the fibonacci search if $f_{a,k} < f_{b,k}$.

The above procedure is repeated until $k = n - 2$ in which case

$$I_{k+2} = I_n$$

and

$$x^* = x_{a,k+1} = x_{b,k+1}$$

as depicted in Fig. 4.7. Evidently, the minimizer is determined to within a tolerance $\pm 1/F_n$.

The error in x^* can be divided by two by applying one stage of the dichotomous search. This is accomplished by evaluating $f(x)$ at point $x = x_{a,k+1} + \varepsilon$

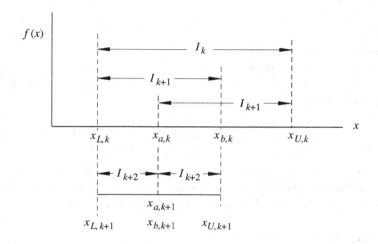

Figure 4.7. Assignments in iteration $n-2$ of the Fibonacci search if $f_{a,k} < f_{b,k}$.

where $|\varepsilon| < 1/F_n$ and then assigning

$$
x^* = \begin{cases} x_{a,k+1} + \frac{1}{2F_n} & \text{if } f(x_{a,k+1} + \varepsilon) < f(x_{a,k+1}) \\ x_{a,k+1} + \frac{\varepsilon}{2} & \text{if } f(x_{a,k+1} + \varepsilon) = f(x_{a,k+1}) \\ x_{a,k+1} - \frac{1}{2F_n} & \text{if } f(x_{a,k+1} + \varepsilon) > f(x_{a,k+1}) \end{cases}
$$

If n is very large, the difference between $x_{a,k}$ and $x_{b,k}$ can become very small, and it is possible for $x_{a,k}$ to exceed $x_{b,k}$, owing to roundoff errors. If this happens, unreliable results will be obtained. In such applications, checks should be incorporated in the algorithm for the purpose of eliminating the problem, if it occurs. One possibility would be to terminate the algorithm since, presumably, sufficient precision has been achieved if $x_{a,k} \approx x_{b,k}$.

The above principles can be used to construct the following algorithm.

Algorithm 4.1 Fibonacci search
Step 1
Input $x_{L,1}$, $x_{U,1}$, and n.
Step 2
Compute F_1, F_2, ..., F_n using Eq. (4.4).
Step 3
Assign $I_1 = x_{U,1} - x_{L,1}$ and compute

$$
I_2 = \frac{F_{n-1}}{F_n} I_1 \quad \text{(see Eq. (4.6))}
$$
$$
x_{a,1} = x_{U,1} - I_2, \quad x_{b,1} = x_{L,1} + I_2
$$
$$
f_{a,1} = f(x_{a,1}), \quad f_{b,1} = f(x_{b,1})
$$

Set $k = 1$.

Step 4

Compute I_{k+2} using Eq. (4.6).

If $f_{a,k} \geq f_{b,k}$, then update $x_{L,k+1}$, $x_{U,k+1}$, $x_{a,k+1}$, $x_{b,k+1}$, $f_{a,k+1}$, and $f_{b,k+1}$ using Eqs. (4.7) to (4.12). Otherwise, if $f_{a,k} < f_{b,k}$, update information using Eqs. (4.13) to (4.18).

Step 5

If $k = n - 2$ or $x_{a,k+1} > x_{b,k+1}$, output $x^* = x_{a,k+1}$ and $f^* = f(x^*)$, and stop. Otherwise, set $k = k + 1$ and repeat from Step 4.

The condition $x_{a,k+1} > x_{b,k+1}$ implies that $x_{a,k+1} \approx x_{b,k+1}$ within the precision of the computer used, as was stated earlier, or that there is an error in the algorithm. It is thus used as an alternative stopping criterion.

4.4   Golden-Section Search

The main disadvantage of the Fibonacci search is that the number of iterations must be supplied as input. A search method in which iterations can be performed until the desired accuracy in either the minimizer or the minimum value of the objective function is achieved is the so-called *golden-section search*. In this approach, as in the Fibonacci search, a sequence of intervals $\{I_1, I_2, I_3, \ldots\}$ is generated as illustrated in Fig. 4.8 by using the recursive relation of Eq. (4.2). The rule by which the lengths of successive intervals are generated is that the ratio of any two adjacent intervals is constant, that is

$$\frac{I_k}{I_{k+1}} = \frac{I_{k+1}}{I_{k+2}} = \frac{I_{k+2}}{I_{k+3}} = \cdots = K \tag{4.19}$$

so that

$$\frac{I_k}{I_{k+2}} = K^2 \tag{4.20}$$

$$\frac{I_k}{I_{k+3}} = K^3$$

and so on.

Upon dividing Eq. (4.2) by I_{k+2}, we obtain

$$\frac{I_k}{I_{k+2}} = \frac{I_{k+1}}{I_{k+2}} + 1 \tag{4.21}$$

and from Eqs. (4.19) to (4.21)

$$K^2 = K + 1 \tag{4.22}$$

Now solving for K, we get

$$K = \frac{1 \pm \sqrt{5}}{2} \tag{4.23}$$

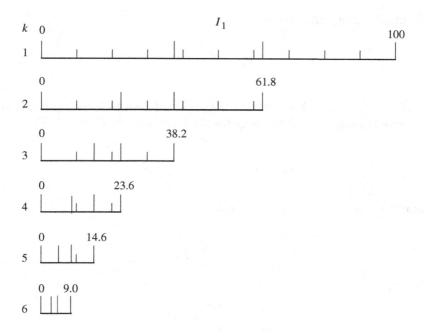

Figure 4.8. Golden section search.

The negative value of K is irrelevant and so $K = 1.618034$. This constant is known as the *golden ratio*. The terminology has arisen from the fact that in classical Greece, a rectangle with sides bearing a ratio $1 : K$ was considered the most pleasing rectangle and hence it came to be known as the golden rectangle. In turn, the sequence $\{I_1, I_1/K, I_1/K^2, \ldots, I_1/K^{n-1}\}$ came to be known as the *golden-section sequence*.

The golden-section search is illustrated in Fig. 4.8 for the case where the left interval is consistently selected. As can be seen, this search resembles the Fibonacci search in most respects. The two exceptions are:

1. Successive intervals are independent of n. Consequently, iterations can be performed until the range of uncertainty or the change in the value of the objective function is reduced below some tolerance ε.
2. The ratio between successive intervals, namely, F_{n-k-1}/F_{n-k}, is replaced by the ratio $1/K$ where

$$\frac{1}{K} = K - 1 = 0.618034$$

according to Eqs. (4.22) – (4.23).

The efficiency of the golden-section search can be easily compared with that of the Fibonacci search. A known relation between F_n and K which is

94

applicable for large values of n is

$$F_n \approx \frac{K^{n+1}}{\sqrt{5}} \qquad (4.24)$$

(e.g., if $n = 11$, $F_n = 1.44$ and $K^{n+1}/\sqrt{5} \approx 144.001$). Thus Eqs. (4.5) and (4.24) give the region of uncertainty for the Fibonacci search as

$$\Lambda_F = I_n = \frac{I_1}{F_n} \approx \frac{\sqrt{5}}{K^{n+1}} I_1$$

Similarly, for the golden-section search

$$\Lambda_{GS} = I_n = \frac{I_1}{K^{n-1}}$$

and hence

$$\frac{\Lambda_{GS}}{\Lambda_F} = \frac{K^2}{\sqrt{5}} \approx 1.17$$

Therefore, if the number of iterations is the same in the two methods, the region of uncertainty in the golden-section search is larger by about 17% relative to that in the Fibonacci search. Alternatively, the golden-section search will require more iterations to achieve the same precision as the Fibonacci search. However, this disadvantage is offset by the fact that the total number of iterations need not be supplied at the start of the optimization.

An implementation of the golden-section search is as follows:

Algorithm 4.2 Golden-section search
Step 1
Input $x_{L,1}$, $x_{U,1}$, and ε.
Step 2
Assign $I_1 = x_{U,1} - x_{L,1}$, $K = 1.618034$ and compute

$$I_2 = I_1/K$$
$$x_{a,1} = x_{U,1} - I_2, \quad x_{b,1} = x_{L,1} + I_2$$
$$f_{a,1} = f(x_{a,1}), \quad f_{b,1} = f(x_{b,1})$$

Set $k = 1$.
Step 3
Compute

$$I_{k+2} = I_{k+1}/K$$

If $f_{a,k} \geq f_{b,k}$, then update $x_{L,k+1}$, $x_{U,k+1}$, $x_{a,k+1}$, $x_{b,k+1}$, $f_{a,k+1}$, and $f_{b,k+1}$ using Eqs. (4.7) to (4.12). Otherwise, if $f_{a,k} < f_{b,k}$, update information using Eqs. (4.13) to (4.18).

Step 4
If $I_k < \varepsilon$ or $x_{a,k+1} > x_{b,k+1}$, then do:
  If $f_{a,k+1} > f_{b,k+1}$, compute

$$x^* = \tfrac{1}{2}(x_{b,k+1} + x_{U,k+1})$$

If $f_{a,k+1} = f_{b,k+1}$, compute

$$x^* = \tfrac{1}{2}(x_{a,k+1} + x_{b,k+1})$$

If $f_{a,k+1} < f_{b,k+1}$, compute

$$x^* = \tfrac{1}{2}(x_{L,k+1} + x_{a,k+1})$$

Compute $f^* = f(x^*)$.
Output x^* and f^*, and stop.
Step 5
Set $k = k + 1$ and repeat from Step 3.

4.5 Quadratic Interpolation Method

In the approximation approach to one-dimensional optimization, an approximate expression for the objective function is assumed, usually in the form of a low-order polynomial. If a second-order polynomial of the form

$$p(x) = a_0 + a_1 x + a_2 x^2 \qquad (4.25)$$

is assumed, where a_0, a_1, and a_2 are constants, a quadratic interpolation method is obtained.

Let

$$p(x_i) = a_0 + a_1 x_i + a_2 x_i^2 = f(x_i) = f_i \qquad (4.26)$$

for $i = 1$, 2, and 3 where $[x_1, x_3]$ is a bracket on the minimizer x^* of $f(x)$. Assuming that the values f_i are known, the three constants a_0, a_1, and a_2 can be deduced by solving the three simultaneous equations in Eq. (4.26). Thus a polynomial $p(x)$ can be deduced which is an approximation for $f(x)$. Under these circumstances, the plots of $p(x)$ and $f(x)$ will assume the form depicted in Fig. 4.9. As can be seen, the minimizer $\bar{x}$ of $p(x)$ is close to x^*, and if $f(x)$ can be accurately represented by a second-order polynomial, then $\bar{x} \approx x^*$. If $f(x)$ is a quadratic function, then $p(x)$ becomes an exact representation of $f(x)$ and $\bar{x} = x^*$.

The first derivative of $p(x)$ with respect to x is obtained from Eq. (4.25) as

$$p'(x) = a_1 + 2a_2 x$$

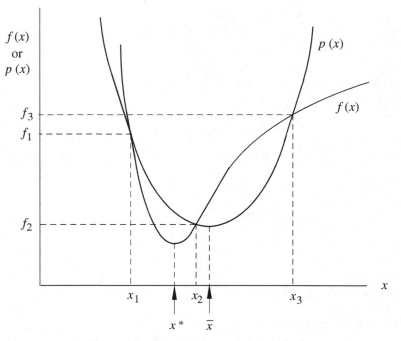

Figure 4.9. Quadratic interpolation method.

and if

$$p'(x) = 0$$

and $a_2 \neq 0$, then the minimizer of $p(x)$ can be deduced as

$$\bar{x} = -\frac{a_1}{2a_2} \qquad (4.27)$$

By solving the simultaneous equations in Eq. (4.26), we find that

$$a_1 = -\frac{(x_2^2 - x_3^2)f_1 + (x_3^2 - x_1^2)f_2 + (x_1^2 - x_2^2)f_3}{(x_1 - x_2)(x_1 - x_3)(x_2 - x_3)} \qquad (4.28)$$

$$a_2 = \frac{(x_2 - x_3)f_1 + (x_3 - x_1)f_2 + (x_1 - x_2)f_3}{(x_1 - x_2)(x_1 - x_3)(x_2 - x_3)} \qquad (4.29)$$

and from Eqs. (4.27) – (4.29), we have

$$\bar{x} = \frac{(x_2^2 - x_3^2)f_1 + (x_3^2 - x_1^2)f_2 + (x_1^2 - x_2^2)f_3}{2[(x_2 - x_3)f_1 + (x_3 - x_1)f_2 + (x_1 - x_2)f_3]} \qquad (4.30)$$

The above approach constitutes one iteration of the quadratic interpolation method. If $f(x)$ cannot be represented accurately by a second-order polynomial, a number of such iterations can be performed. The appropriate strategy is to attempt to reduce the interval of uncertainty in each iteration as was done in the search methods of Secs. 4.2–4.4. This can be achieved by rejecting either x_1 or x_3 and then using the two remaining points along with point $\bar{x}$ for a new interpolation.

After a number of iterations, the three points will be in the neighborhood of x^*. Consequently, the second-order polynomial $p(x)$ will be an accurate representation of $f(x)$ by virtue of the Taylor series, and x^* can be determined to within any desired accuracy.

An algorithm based on the above principles is as follows:

Algorithm 4.3 Quadratic interpolation search
Step 1
Input x_1, x_3, and ε.
Set $\bar{x}_0 = 10^{99}$.
Step 2
Compute
$x_2 = \frac{1}{2}(x_1 + x_3)$ and $f_i = f(x_i)$ and $i = 1, 2, 3$.
Step 3
Compute $\bar{x}$ from Eq. (4.30) and $\bar{f} = f(\bar{x})$.
If $|\bar{x} - \bar{x}_0| < \varepsilon$, then output $x^* = \bar{x}$ and $f(x^*) = \bar{f}$, and stop.
Step 4
If $x_1 < \bar{x} < x_2$, then do:
  If $\bar{f} \le f_2$, assign $x_3 = x_2$, $f_3 = f_2$, $x_2 = \bar{x}$, $f_2 = \bar{f}$;
  otherwise, if $\bar{f} > f_2$, assign $x_1 = \bar{x}$, $f_1 = \bar{f}$.
If $x_2 < \bar{x} < x_3$, then do:
  If $\bar{f} \le f_2$, assign $x_1 = x_2$, $f_1 = f_2$, $x_2 = \bar{x}$, $f_2 = \bar{f}$;
  otherwise, if $\bar{f} > f_2$, assign $x_3 = \bar{x}$, $f_3 = \bar{f}$.
Set $\bar{x}_0 = \bar{x}$, and repeat from Step 3.

In Step 4, the bracket on x^* is reduced judiciously to $[x_1, x_2]$ or $[\bar{x}, x_3]$ if $x_1 < \bar{x} < x_2$; or to $[x_2, x_3]$ or $[x_1, \bar{x}]$ if $x_2 < \bar{x} < x_3$ by using the principles developed in Sec. 4.2. The algorithm entails one function evaluation per iteration (see Step 3) except for the first iteration in which three additional function evaluations are required in Step 2.

An implicit assumption in the above algorithm is that interval $[x_1, x_3]$ is a bracket on x^*. If it is not, one can be readily established by varying x in the direction of decreasing $f(x)$ until $f(x)$ begins to increase.

A simplified version of the interpolation formula in Eq. (4.30) can be obtained by assuming that points x_1, x_2, and x_3 are equally spaced. If we let

$$x_1 = x_2 - \delta \quad \text{and} \quad x_3 = x_2 + \delta$$

then Eq. (4.30) becomes

$$\bar{x} = x_2 + \frac{(f_1 - f_3)\delta}{2(f_1 - 2f_2 + f_3)} \tag{4.31}$$

Evidently, this formula involves less computation than that in Eq. (4.30) and, if equal spacing is allowed, it should be utilized. The minimum of the function can be deduced as

$$f_{\min} = f_2 - \frac{(f_1 - f_3)^2}{8(f_1 - 2f_2 + f_3)}$$

(see Prob. 4.10).

4.5.1 Two-point interpolation

The interpolation formulas in Eqs. (4.30) and (4.31) are said to be *three-point formulas* since they entail the values of $f(x)$ at three distinct points. *Two-point interpolation formulas* can be obtained by assuming that the values of $f(x)$ and its first derivatives are available at two distinct points. If the values of $f(x)$ at $x = x_1$ and $x = x_2$ and the first derivative of $f(x)$ at $x = x_1$ are available, we can write

$$p(x_1) = a_0 + a_1 x_1 + a_2 x_1^2 = f(x_1) \equiv f_1$$
$$p(x_2) = a_0 + a_1 x_2 + a_2 x_2^2 = f(x_2) \equiv f_2$$
$$p'(x_1) = a_1 + 2a_2 x_1 = f'(x_1) \equiv f_1'$$

The solution of these equations gives a_1 and a_2, and thus from Eq. (4.27), the two-point interpolation formula

$$\bar{x} = x_1 + \frac{f_1'(x_2 - x_1)^2}{2[f_1 - f_2 + f_1'(x_2 - x_1)]}$$

can be obtained.

An alternative two-point interpolation formula of the same class can be generated by assuming that the first derivative of $f(x)$ is known at two points x_1 and x_2. If we let

$$p'(x_1) = a_1 + 2a_2 x_1 = f'(x_1) \equiv f_1'$$
$$p'(x_2) = a_1 + 2a_2 x_2 = f'(x_2) \equiv f_2'$$

we deduce

$$\bar{x} = \frac{x_2 f_1' - x_1 f_2'}{f_1' - f_2'} = \frac{x_2 f_1' - x_2 f_2' + x_2 f_2' - x_1 f_2'}{f_1' - f_2'}$$
$$= x_2 + \frac{(x_2 - x_1)f_2'}{f_1' - f_2'}$$

4.6   Cubic Interpolation

Another one-dimensional optimization method, which is sometimes quite useful, is the *cubic interpolation method*. This is based on the third-order polynomial

$$p(x) = a_0 + a_1 x + a_2 x^2 + a_3 x^3 \qquad (4.32)$$

As in the quadratic interpolation method, the coefficients a_i can be determined such that $p(x)$ and/or its derivatives at certain points are equal to $f(x)$ and/or its derivatives. Since there are four coefficients in Eq. (4.32), four equations are needed for the complete characterization of $p(x)$. These equations can be chosen in a number of ways and several cubic interpolation formulas can be generated.

The plot of $p(x)$ can assume either of the forms depicted in Fig. 4.10 and, in effect, $p(x)$ can have a maximum as well as a minimum. By equating the first derivative of $p(x)$ to zero, that is,

$$p'(x) = a_1 + 2a_2 x + 3a_3 x^2 = 0 \qquad (4.33)$$

and then solving for x, the extremum points of $p(x)$ can be determined as

$$x = \frac{1}{3a_3} \left(-a_2 \pm \sqrt{a_2^2 - 3a_1 a_3} \right) \qquad (4.34)$$

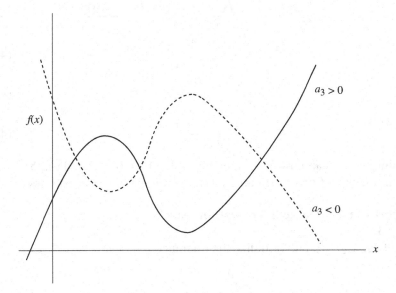

Figure 4.10.   Possible forms of third-order polynomial.

At the minimizer $\bar{x}$, the second derivative of $p(x)$ is positive, and thus Eq. (4.33) gives

$$p''(\bar{x}) = 2a_2 + 6a_3\bar{x} > 0$$

or

$$\bar{x} > -\frac{a_2}{3a_3} \tag{4.35}$$

Thus, the solution in Eq. (4.34) that corresponds to the minimizer of $p(x)$ can be readily selected.

Polynomial $p(x)$ will be an approximation for $f(x)$ if four independent equations are chosen which interrelate $p(x)$ with $f(x)$. One of many possibilities is to let

$$p(x_i) = a_0 + a_1 x_i + a_2 x_i^2 + a_3 x_i^3 = f(x_i)$$

for $i = 1,\ 2,$ and 3 and

$$p'(x_1) = a_1 + 2a_2 x_1 + 3a_3 x_1^2 = f'(x_1)$$

By solving this set of equations, coefficients a_1 and a_3 can be determined as

$$a_3 = \frac{\beta - \gamma}{\theta - \psi} \tag{4.36}$$

$$a_2 = \beta - \theta a_3 \tag{4.37}$$

$$a_1 = f'(x_1) - 2a_2 x_1 - 3a_3 x_1^2 \tag{4.38}$$

where

$$\beta = \frac{f(x_2) - f(x_1) + f'(x_1)(x_1 - x_2)}{(x_1 - x_2)^2} \tag{4.39}$$

$$\gamma = \frac{f(x_3) - f(x_1) + f'(x_1)(x_1 - x_3)}{(x_1 - x_3)^2} \tag{4.40}$$

$$\theta = \frac{2x_1^2 - x_2(x_1 + x_2)}{(x_1 - x_2)} \tag{4.41}$$

$$\psi = \frac{2x_1^2 - x_3(x_1 + x_3)}{(x_1 - x_3)} \tag{4.42}$$

The minimizer $\bar{x}$ can now be obtained by using Eqs. (4.34) and (4.35).

An implementation of the cubic interpolation method is as follows:

Algorithm 4.4 Cubic interpolation search
Step 1
Input $x_1,\ x_2,\ x_3$, and initialize the tolerance ε.
Step 2
Set $\bar{x}_0 = 10^{99}$.
Compute $f_1' = f'(x_1)$ and $f_i = f(x_i)$ for $i = 1, 2, 3$.

Step 3
Compute constants β, γ, θ, and ψ using Eqs. (4.39) – (4.42).
Compute constants a_3, a_2 and a_1 using Eqs. (4.36) – (4.38).
Compute the extremum points of $p(x)$ using Eq. (4.34), and select the minimizer $\bar{x}$ using Eq. (4.35).
Compute $\bar{f} = f(\bar{x})$.
Step 4
If $|\bar{x} - \bar{x}_0| < \varepsilon$, then output $x^* = \bar{x}$ and $f(x^*) = \bar{f}$, and stop.
Step 5
Find m such that $f_m = \max(f_1, f_2, f_3)$.
Set $\bar{x}_0 = \bar{x}$, $x_m = \bar{x}$, $f_m = \bar{f}$.
If $m = 1$, compute $f_1' = f'(\bar{x})$.
Repeat from Step 3.

In this algorithm, a bracket is maintained on x^* by replacing the point that yields the largest value in $f(x)$ by the new estimate of the minimizer $\bar{x}$ in Step 5. If the point that is replaced is x_1, the first derivative $f'(x_1)$ is computed since it is required for the calculation of a_1, β, and γ.

As can be seen in Eqs. (4.36) – (4.42), one iteration of cubic interpolation entails a lot more computation than one iteration of quadratic interpolation. Nevertheless, the former can be more efficient. The reason is that a third-order polynomial is a more accurate approximation for $f(x)$ than a second-order one and, as a result, convergence will be achieved in a smaller number of iterations. For the same reason, the method is more tolerant to an inadvertent loss of the bracket.

4.7 The Algorithm of Davies, Swann, and Campey

The methods described so far are either search methods or approximation methods. A method due to Davies, Swann, and Campey [8] will now be described, which combines a search method with an approximation method. The search method is used to establish and maintain a bracket on x^*, whereas the approximation method is used to generate estimates of x^*.

In this method, $f(x)$ is evaluated for increasing or decreasing values of x until x^* is bracketed. Then the quadratic interpolation formula for equally-spaced points is used to predict x^*. This procedure is repeated several times until sufficient accuracy in the solution is achieved, as in previous methods.

The input to the algorithm consists of an initial point $x_{0,1}$, an initial increment δ_1, a scaling constant K, and the optimization tolerance ε.

At the kth iteration, an initial point $x_{0,k}$ and an initial increment δ_k are available, and a new initial point $x_{0,k+1}$ as well as a new increment δ_{k+1} are required for the next iteration.

Initially, $f(x)$ is evaluated at points $x_{0,k} - \delta_k$, $x_{0,k}$, and $x_{0,k} + \delta_k$. Three possibilities can arise, namely,

(a) $f(x_{0,k} - \delta_k) > f(x_{0,k}) > f(x_{0,k} + \delta_k)$
(b) $f(x_{0,k} - \delta_k) < f(x_{0,k}) < f(x_{0,k} + \delta_k)$
(c) $f(x_{0,k} - \delta_k) \geq f(x_{0,k}) \leq f(x_{0,k} + \delta_k)$

In case (a), the minimum of $f(x)$ is located in the positive direction and so $f(x)$ is evaluated for increasing values of x until a value of $f(x)$ is obtained, which is larger than the previous one. If this occurs on the nth function evaluation, the interval $[x_{0,k}, x_{n,k}]$ is a bracket on x^*. The interval between successive points is increased geometrically, and so this procedure will yield the sequence of points

$$x_{0,k}$$
$$x_{1,k} = x_{0,k} + \delta_k$$
$$x_{2,k} = x_{1,k} + 2\delta_k$$
$$x_{3,k} = x_{2,k} + 4\delta_k$$
$$\vdots$$
$$x_{n,k} = x_{n-1,k} + 2^{n-1}\delta_k \tag{4.43}$$

as illustrated in Fig. 4.11. Evidently, the most recent interval is twice as long as the previous one and if it is divided into two equal sub-intervals at point

$$x_{m,k} = x_{n,-1,k} + 2^{n-2}\delta_k \tag{4.44}$$

then four equally-spaced points are available, which bracket the minimizer.

If $f(x)$ is evaluated at point $x_{m,k}$, the function values

$$f_{n-2,k} \equiv f(x_{n-2,k}) \tag{4.45}$$
$$f_{n-1,k} \equiv f(x_{n-1,k}) \tag{4.46}$$
$$f_{m,k} \equiv f(x_{m,k}) \tag{4.47}$$
$$f_{n,k} \equiv f(x_{n,k}) \tag{4.48}$$

will be available. If $f_{m,k} \geq f_{n-1,k}$, x^* is located in the interval $[x_{n-2,k}, x_{m,k}]$ (see Fig. 4.12) and so the use of Eqs. (4.31) and (4.45) – (4.48) yields an estimate for x^* as

$$x_{0,k+1} = x_{n-1,k} + \frac{2^{n-2}\delta_k(f_{n-2,k} - f_{m,k})}{2(f_{n-2,k} - 2f_{n-1,k} + f_{m,k})} \tag{4.49}$$

Similarly, if $f_{m,k} < f_{n-1,k}$, x^* is located in the interval $[x_{n-1,k}, x_{n,k}]$ (see Fig. 4.13) then an estimate for x^* is

$$x_{0,k+1} = x_{m,k} + \frac{2^{n-2}\delta_k(f_{n-1,k} - f_{n,k})}{2(f_{n-1,k} - 2f_{m,k} + f_{n,k})} \tag{4.50}$$

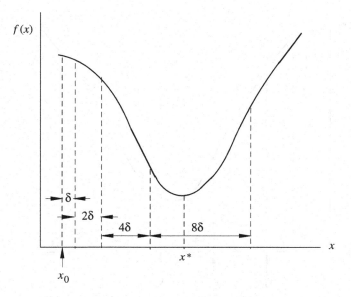

Figure 4.11.   Search method used in the Davies, Swann, and Campey algorithm.

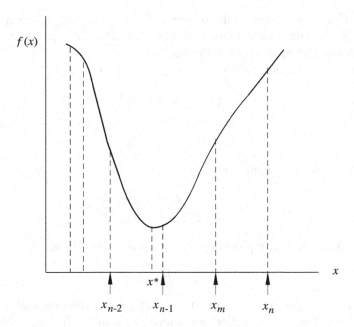

Figure 4.12.   Reduction of range of uncertainty in Davies, Swann, and Campey algorithm if $f_m \geq f_{n-1}$.

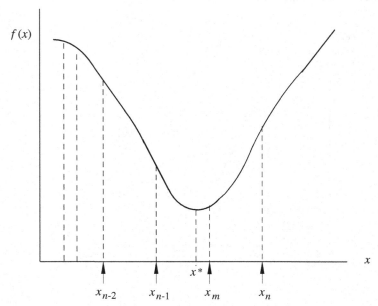

Figure 4.13. Reduction of range of uncertainty in Davies, Swann, and Campey algorithm if $f_m < f_{n-1}$.

In case (b), x^* is located in the negative direction, and so x is decreased in steps δ_k, $2\delta_k$, ... until the minimum of $f(x)$ is located. The procedure is as in case (a) except that δ_k is negative in Eqs. (4.49) and (4.50).

In case (c), x^* is bracketed by $x_{0,k} - \delta_k$ and $x_{0,k} + \delta_k$ and if

$$f_{-1,k} = f(x_{0,k} - \delta_k)$$
$$f_{0,k} = f(x_{0,k})$$
$$f_{1,k} = f(x_{0,k} + \delta_k)$$

Eq. (4.31) yields an estimate for x^* as

$$x_{0,k+1} = x_{0,k} + \frac{\delta_k(f_{-1,k} - f_{1,k})}{2(f_{-1,k} - 2f_{0,k} + f_{1,k})} \tag{4.51}$$

The kth iteration is completed by defining a new increment

$$\delta_{k+1} = K\delta_k$$

where K is a constant in the range 0 to 1. The motivation for this scaling is that as the solution is approached, a reduced range of x will be searched and, therefore, the resolution of the algorithm needs to be increased. A suitable value for K might be 0.1.

The above principles can be used to construct the following algorithm:

Algorithm 4.5 Davies, Swann, and Campey search
Step 1
Input $x_{0,1}$, δ_1, K, and initialize the tolerance ε.
Set $k = 0$.
Step 2
Set $k = k + 1$, $x_{-1,k} = x_{0,k} - \delta_k$, $x_{1,k} = x_{0,k} + \delta_k$.
Compute $f_{0,k} = f(x_{0,k})$ and $f_{1,k} = f(x_{1,k})$.
Step 3
If $f_{0,k} > f_{1,k}$, set $p = 1$ and go to Step 4; otherwise, compute $f_{-1,k} = f(x_{-1,k})$.
If $f_{-1,k} < f_{0,k}$, set $p = -1$ and go to Step 4.
Otherwise, if $f_{-1,k} \geq f_{0,k} \leq f_{1,k}$ go to Step 7.
Step 4
For $n = 1, 2, \ldots$ compute $f_{n,k} = f(x_{n-1,k} + 2^{n-1}p\delta_k)$ until $f_{n,k} > f_{n-1,k}$.
Step 5
Compute $f_{m,k} = f(x_{n-1,k} + 2^{n-2}p\delta_k)$.
Step 6
If $f_{m,k} \geq f_{n-1,k}$, compute

$$x_{0,k+1} = x_{n-1,k} + \frac{2^{n-2}p\delta_k(f_{n-2,k} - f_{m,k})}{2(f_{n-2,k} - 2f_{n-1,k} + f_{m,k})}$$

Otherwise, if $f_{m,k} < f_{n-1,k}$, compute

$$x_{0,k+1} = x_{m,k} + \frac{2^{n-2}p\delta_k(f_{n-1,k} - f_{n,k})}{2(f_{n-1,k} - 2f_{m,k} + f_{n,k})}$$

(see Eqs. (4.49) and (4.50)).
If $2^{n-2}\delta_k \leq \varepsilon$ go to Step 8; otherwise, set $\delta_{k+1} = K\delta_k$ and repeat from Step 2.
Step 7 Compute

$$x_{0,k+1} = x_{0,k} + \frac{\delta_k(f_{-1,k} - f_{1,k})}{2(f_{-1,k} - 2f_{0,k} + f_{1,k})}$$

(see Eq. (4.51)).
If $\delta_k \leq \varepsilon$ go to Step 8; otherwise, set $\delta_{k+1} = K\delta_k$ and repeat from Step 2.
Step 8
Output $x^* = x_{0,k+1}$ and $f(x^*) = f_{0,k+1}$, and stop.

Parameter δ_1 is a small positive constant that would depend on the problem, say, $0.1x_{0,1}$. Constant p in Steps 3 to 6, which can be 1 or -1, is used to render the formulas in Eqs. (4.49) and (4.50) applicable for increasing as well as decreasing values of x. Constant ε in Step 1 determines the precision of the solution. If ε is very small, say, less than 10^{-6}, then as the solution is approached, we have

$$f_{n-2,k} \approx f_{n-1,k} \approx f_{m,k} \approx f_{n,k}$$

Consequently, the distinct possibility of dividing by zero may arise in the evaluation of $x_{0,k+1}$. However, this problem can be easily prevented by using appropriate checks in Steps 6 and 7.

An alternative form of the above algorithm can be obtained by replacing the quadratic interpolation formula for equally-spaced points by the general formula of Eq. (4.30). If this is done, the mid-interval function evaluation of Step 5 is unnecessary. Consequently, if the additional computation required by Eq. (4.31) is less than one complete evaluation of $f(x)$, then the modified algorithm is likely to be more efficient.

Another possible modification is to use the cubic interpolation of Sec. 4.6 instead of quadratic interpolation. Such an algorithm is likely to reduce the number of function evaluations. However, the amount of computation could increase owing to the more complex formulation in the cubic interpolation.

4.8 Inexact Line Searches

In the multidimensional algorithms to be studied, most of the computational effort is spent in performing function and gradient evaluations in the execution of line searches. Consequently, the amount of computation required tends to depend on the efficiency and precision of the line searches used. If high precision line searches are necessary, the amount of computation will be large and if inexact line searches do not affect the convergence of an algorithm, a small amount of computation might be sufficient.

Many optimization methods have been found to be quite tolerant to line-search imprecision and, for this reason, inexact line searches are usually used in these methods.

Let us assume that

$$\mathbf{x}_{k+1} = \mathbf{x}_k + \alpha \mathbf{d}_k$$

where $\mathbf{d}_k$ is a given direction vector and α is an independent search parameter, and that function $f(\mathbf{x}_{k+1})$ has a unique minimum for some positive value of α. The linear approximation of the Taylor series in Eq. (2.4d) gives

$$f(\mathbf{x}_{k+1}) = f(\mathbf{x}_k) + \mathbf{g}_k^T \mathbf{d}_k \alpha \qquad (4.52)$$

where

$$\mathbf{g}_k^T \mathbf{d}_k = \left. \frac{df(\mathbf{x}_k + \alpha \mathbf{d}_k)}{d\alpha} \right|_{\alpha=0}$$

Eq. (4.52) represents line A shown in Fig. 4.14a. The equation

$$f(\mathbf{x}_{k+1}) = f(\mathbf{x}_k) + \rho \mathbf{g}_k^T \mathbf{d}_k \alpha \qquad (4.53)$$

where $0 \leq \rho < \frac{1}{2}$ represents line B in Fig. 4.14a whose slope ranges from 0 to $\frac{1}{2} \mathbf{g}_k^T \mathbf{d}_k$ depending on the value of ρ, as depicted by shaded area B in Fig. 4.14a. On the other hand, the equation

$$f(\mathbf{x}_{k+1}) = f(\mathbf{x}_k) + (1 - \rho) \mathbf{g}_k^T \mathbf{d}_k \alpha \qquad (4.54)$$

represents line C in Fig. 4.14a whose slope ranges from $\mathbf{g}_k^T \mathbf{d}_k$ to $\frac{1}{2} \mathbf{g}_k^T \mathbf{d}_k$ as depicted by shaded area C in Fig. 4.14a. The angle between lines C and B, designated as θ, is given by

$$\theta = \tan^{-1} \left[\frac{-(1 - 2\rho) \mathbf{g}_k^T \mathbf{d}_k}{1 + \rho(1 - \rho)(\mathbf{g}_k^T \mathbf{d}_k)^2} \right]$$

as illustrated in Fig. 4.14b. Evidently by adjusting ρ in the range 0 to $\frac{1}{2}$, the slope of θ can be varied in the range $-\mathbf{g}_k^T \mathbf{d}_k$ to 0. By fixing ρ at some value in the permissible range, two values of α are defined by the intercepts of the lines in Eqs. (4.53) and (4.54) and the curve for $f(\mathbf{x}_{k+1})$, say, α_1 and α_2, as depicted in Fig. 4.14b.

Let α_0 be an estimate of the value of α that minimizes $f(\mathbf{x}_k + \alpha \mathbf{d}_k)$. If $f(\mathbf{x}_{k+1})$ for $\alpha = \alpha_0$ is equal to or less than the corresponding value of $f(\mathbf{x}_{k+1})$ given by Eq. (4.53), and is equal to or greater than the corresponding value of $f(\mathbf{x}_{k+1})$ given by Eq. (4.54), that is, if

$$f(\mathbf{x}_{k+1}) \leq f(\mathbf{x}_k) + \rho \mathbf{g}_k^T \mathbf{d}_k \alpha_0 \qquad (4.55)$$

and

$$f(\mathbf{x}_{k+1}) \geq f(\mathbf{x}_k) + (1 - \rho) \mathbf{g}_k^T \mathbf{d}_k \alpha_0 \qquad (4.56)$$

then α_0 may be deemed to be an acceptable estimate of α^* in that it will yield a sufficient reduction in $f(\mathbf{x})$. Under these circumstances, we have $\alpha_1 \leq \alpha_0 \leq \alpha_2$, as depicted in Fig. 4.14b, i.e., α_1 and α_2 constitute a bracket of the estimated minimizer α_0. Eqs. (4.55) and (4.56), which are often referred to as the *Goldstein conditions*, form the basis of a class of *inexact line searches*. In these methods, an estimate α_0 is generated by some means, based on available information, and the conditions in Eqs. (4.55) and (4.56) are checked. If both conditions are satisfied, then the reduction in $f(\mathbf{x}_{k+1})$ is deemed to be acceptable, and the procedure is terminated. On the other hand, if either Eq. (4.55) or Eq. (4.56) is violated, the reduction in $f(\mathbf{x}_{k+1})$ is deemed to be insufficient and an improved estimate of α^*, say, $\breve{\alpha}_0$, can be obtained. If Eq. (4.55) is violated, then $\alpha_0 > \alpha_2$ as depicted in Fig. 4.15a and since $\alpha_L < \alpha^* < \alpha_0$, the new

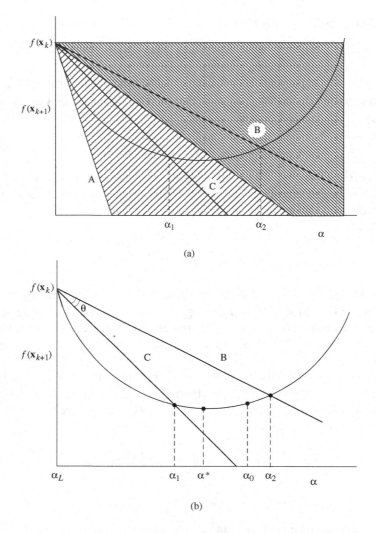

Figure 4.14. (a) The Goldstein tests. (b) Goldstein tests satisfied.

estimate $\breve{\alpha}_0$ can be determined by using interpolation. On the other hand, if Eq. (4.56) is violated, $\alpha_0 < \alpha_1$ as depicted in Fig. 4.15b, and since α_0 is likely to be in the range $\alpha_L < \alpha_0 < \alpha^*$, $\breve{\alpha}_0$ can be determined by using extrapolation.

If the value of $f(\mathbf{x}_k + \alpha\mathbf{d}_k)$ and its derivative with respect to α are known for $\alpha = \alpha_L$ and $\alpha = \alpha_0$, then for $\alpha_0 > \alpha_2$ a good estimate for $\breve{\alpha}_0$ can be deduced by using the interpolation formula

$$\breve{\alpha}_0 = \alpha_L + \frac{(\alpha_0 - \alpha_L)^2 f_L'}{2[f_L - f_0 + (\alpha_0 - \alpha_L)f_L']} \tag{4.57}$$

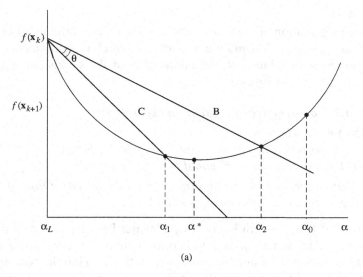

(a)

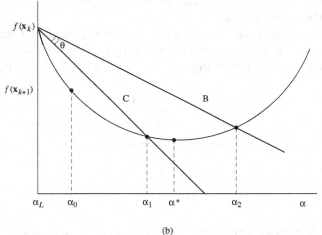

(b)

Figure 4.15. Goldstein tests violated: (a) with $\alpha_0 > \alpha_2$, (b) with $\alpha_0 < \alpha_1$.

and for $\alpha_0 < \alpha_1$ the extrapolation formula

$$\breve{\alpha}_0 = \alpha_0 + \frac{(\alpha_0 - \alpha_L)f_0'}{(f_L' - f_0')} \qquad (4.58)$$

can be used, where

$$f_L = f(\mathbf{x}_k + \alpha_L\mathbf{d}_k), \quad f_L' = f'(\mathbf{x}_k + \alpha_L\mathbf{d}_k) = \mathbf{g}(\mathbf{x}_k + \alpha_L\mathbf{d}_k)^T\mathbf{d}_k$$
$$f_0 = f(\mathbf{x} + \alpha_0\mathbf{d}_k), \quad f_0' = f'(\mathbf{x}_k + \alpha_0\mathbf{d}_k) = \mathbf{g}(\mathbf{x}_k + \alpha_0\mathbf{d}_k)^T\mathbf{d}_k$$

(see Sec. 4.5).

Repeated application of the above procedure will eventually yield a value of $\breve{\alpha}_0$ such that $\alpha_1 < \breve{\alpha}_0 < \alpha_2$ and the inexact line search is terminated.

A useful theorem relating to the application of the Goldstein tests in an inexact line search is as follows:

Theorem 4.1 *Convergence of inexact line search* *If*

   (a) $f(\mathbf{x}_k)$ *has a lower bound,*
   (b) $\mathbf{g}_k$ *is uniformly continuous on set* $\{\mathbf{x} : f(\mathbf{x}) < f(\mathbf{x}_0)\}$,
   (c) *directions* $\mathbf{d}_k$ *are not orthogonal to* $-\mathbf{g}_k$ *for all* k,

then a descent algorithm using an inexact line search based on Eqs. (4.55) and (4.56) will converge to a stationary point as $k \to \infty$.

The proof of this theorem is given by Fletcher [9]. The theorem does not guarantee that a descent algorithm will converge to a minimizer since a saddle point is also a stationary point. Nevertheless, the theorem is of importance since it demonstrates that inaccuracies due to the inexactness of the line search are not detrimental to convergence.

Conditions (a) and (b) of Theorem 4.1 are normally satisfied but condition (c) may be violated. Nevertheless, the problem can be avoided in practice by changing direction $\mathbf{d}_k$. For example, if θ_k is the angle between $\mathbf{d}_k$ and $-\mathbf{g}_k$ and

$$\theta_k = \cos^{-1} \frac{-\mathbf{g}_k^T \mathbf{d}_k}{\|\mathbf{g}_k\| \, \|\mathbf{d}_k\|} = \frac{\pi}{2}$$

then $\mathbf{d}_k$ can be modified slightly to ensure that

$$\theta_k = \frac{\pi}{2} - \mu$$

where $\mu > 0$.

The Goldstein conditions sometimes lead to the situation illustrated in Fig. 4.16, where α^* is not in the range $[\alpha_1, \alpha_2]$. Evidently, in such a case a value α_0 in the interval $[\alpha^*, \alpha_1]$ will not terminate the line search even though the reduction in $f(\mathbf{x}_k)$ would be larger than that for any α_0 in the interval $[\alpha_1, \alpha_2]$. Although the problem is not serious, since convergence is assured by Theorem 4.1, the amount of computation may be increased. The problem can be eliminated by replacing the second Goldstein condition, namely, Eq. (4.56), by the condition

$$\mathbf{g}_{k+1}^T \mathbf{d}_k \geq \sigma \mathbf{g}_k^T \mathbf{d}_k \tag{4.59}$$

where $0 < \sigma < 1$ and $\sigma \geq \rho$. This modification to the second Goldstein condition was proposed by Fletcher [10]. It is illustrated in Fig. 4.17. The scalar $\mathbf{g}_k^T \mathbf{d}_k$ is the derivative of $f(\mathbf{x}_k + \alpha \mathbf{d}_k)$ at $\alpha = 0$, and since $0 < \sigma < 1$,

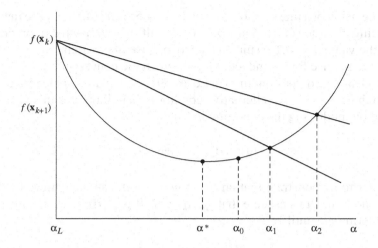

Figure 4.16. Goldstein tests violated with $\alpha^* < \alpha_1$.

$\sigma \mathbf{g}_k^T \mathbf{d}_k$ is the derivative of $f(\mathbf{x}_k + \alpha \mathbf{d}_k)$ at some value of α, say, α_1, such that $\alpha_1 < \alpha^*$. Now if the condition in Eq. (4.59) is satisfied at some point

$$\mathbf{x}_{k+1} = \mathbf{x}_k + \alpha_0 \mathbf{d}_k$$

then the slope of $f(\mathbf{x}_k + \alpha \mathbf{d}_k)$ at $\alpha = \alpha_0$ is less negative (more positive) than the slope at $\alpha = \alpha_1$ and, consequently, we conclude that $\alpha_1 \leq \alpha_0$. Now if Eq. (4.55) is also satisfied, then we must have $\alpha_1 < (\alpha^* \text{ or } \alpha_0) < \alpha_2$, as depicted in Fig. 4.17.

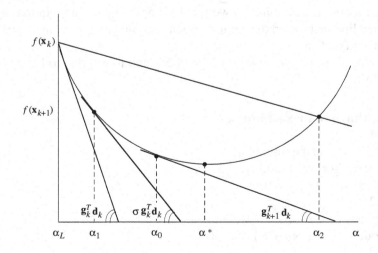

Figure 4.17. Fletcher's modification of the Goldstein tests.

The precision of a line search based on Eqs. (4.55) and (4.59) can be increased by reducing the value of σ. While $\sigma = 0.9$ results in a somewhat imprecise line search, the value $\sigma = 0.1$ results in a fairly precise line search. Note, however, that a more precise line search could slow down the convergence.

A disadvantage of the condition in Eq. (4.59) is that it does not lead to an exact line search as $\sigma \to 0$. An alternative condition that eliminates this problem is obtained by modifying the condition in Eq. (4.59) as

$$|\mathbf{g}_{k+1}^T \mathbf{d}_k| \leq -\sigma \mathbf{g}_k^T \mathbf{d}_k$$

In order to demonstrate that an exact line search can be achieved with the above condition, let us assume that $\mathbf{g}_k^T \mathbf{d}_k < 0$. If $\mathbf{g}_{k+1}^T \mathbf{d}_k < 0$, the line search will not terminate until

$$-|\mathbf{g}_{k+1}^T \mathbf{d}_k| \geq \sigma \mathbf{g}_k^T \mathbf{d}_k$$

and if $\mathbf{g}_{k+1}^T \mathbf{d}_k > 0$, the line search will not terminate until

$$|\mathbf{g}_{k+1}^T \mathbf{d}_k| \leq -\sigma \mathbf{g}_k^T \mathbf{d}_k \tag{4.60}$$

Now if $\sigma \mathbf{g}_k^T \mathbf{d}_k$, $\mathbf{g}_{k+1}^T \mathbf{d}_k$, and $-\sigma \mathbf{g}_k^T \mathbf{d}_k$ are the derivatives of $f(\mathbf{x}_k + \alpha \mathbf{d}_k)$ at points $\alpha = \alpha_1$, $\alpha = \alpha_0$, and $\alpha = \alpha_2$, respectively, we have $\alpha_1 \leq \alpha_0 \leq \alpha_2$ as depicted in Fig. 4.18. In effect, Eq. (4.60) overrides both of the Goldstein conditions in Eqs. (4.55) and (4.56). Since interval $[\alpha_1, \alpha_2]$ can be reduced as much as desired by reducing σ, it follows that α^* can be determined as accurately as desired, and as $\sigma \to 0$, the line search becomes exact. In such a case, the amount of computation would be comparable to that required by any other exact line search and the computational advantage of using an inexact line search would be lost.

An inexact line search based on Eqs. (4.55) and (4.59) due to Fletcher [10] is as follows:

Algorithm 4.6 Inexact line search
Step 1
Input $\mathbf{x}_k$, $\mathbf{d}_k$, and compute $\mathbf{g}_k$.
Initialize algorithm parameters ρ, σ, τ, and χ.
Set $\alpha_L = 0$ and $\alpha_U = 10^{99}$.
Step 2
Compute $f_L = f(\mathbf{x}_k + \alpha_L \mathbf{d}_k)$.
Compute $f_L' = \mathbf{g}(\mathbf{x}_k + \alpha_L \mathbf{d}_k)^T \mathbf{d}_k$.
Step 3
Estimate α_0.

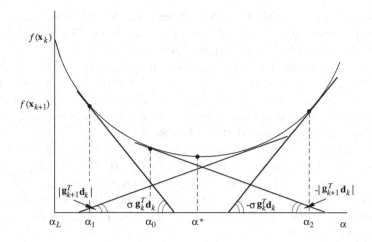

Figure 4.18. Conversion of inexact line search into an exact line search.

Step 4
Compute $f_0 = f(\mathbf{x}_k + \alpha_0 \mathbf{d}_k)$.
Step 5 (Interpolation)
If $f_0 > f_L + \rho(\alpha_0 - \alpha_L)f_L'$, then do:
  a. If $\alpha_0 < \alpha_U$, then set $\alpha_U = \alpha_0$.
  b. Compute $\breve{\alpha}_0$ using Eq. (4.57).
  c. If $\breve{\alpha}_0 < \alpha_L + \tau(\alpha_U - \alpha_L)$ then set $\breve{\alpha}_0 = \alpha_L + \tau(\alpha_U - \alpha_L)$.
  d. If $\breve{\alpha}_0 > \alpha_U - \tau(\alpha_U - \alpha_L)$ then set $\breve{\alpha}_0 = \alpha_U - \tau(\alpha_U - \alpha_L)$.
  e. Set $\alpha_0 = \breve{\alpha}_0$ and go to Step 4.
Step 6
Compute $f_0' = \mathbf{g}(\mathbf{x}_k + \alpha_0 \mathbf{d}_k)^T \mathbf{d}_k$.
Step 7 (Extrapolation)
If $f_0' < \sigma f_L'$, then do:
  a. Compute $\Delta\alpha_0 = (\alpha_0 - \alpha_L)f_0'/(f_L' - f_0')$ (see Eq. (4.58)).
  b. If $\Delta\alpha_0 < \tau(\alpha_0 - \alpha_L)$, then set $\Delta\alpha_0 = \tau(\alpha_0 - \alpha_L)$.
  c. If $\Delta\alpha_0 > \chi(\alpha_0 - \alpha_L)$, then set $\Delta\alpha_0 = \chi(\alpha_0 - \alpha_L)$.
  d. Compute $\breve{\alpha}_0 = \alpha_0 + \Delta\alpha_0$.
  e. Set $\alpha_L = \alpha_0$, $\alpha_0 = \breve{\alpha}_0$, $f_L = f_0$, $f_L' = f_0'$, and go to Step 4.
Step 8
Output α_0 and $f_0 = f(\mathbf{x}_k + \alpha_0 \mathbf{d}_k)$, and stop.

The precision to which the minimizer is determined depends on the values of ρ and σ. Small values like $\rho = \sigma = 0.1$ will yield a relatively precise line search whereas values like $\rho = 0.3$ and $\sigma = 0.9$ will yield a somewhat imprecise line search. The values $\rho = 0.1$ and $\sigma = 0.7$ give good results.

An estimate of α_0 in Step 3 can be determined by assuming that $f(\mathbf{x})$ is a convex quadratic function and using $\alpha_0 = \|\mathbf{g}_0\|^2/(\mathbf{g}_0^T \mathbf{H}_0 \mathbf{g}_0)$ which is the minimum point for a convex quadratic function.

In Step 5, $\breve{\alpha}_0$ is checked and if necessary it is adjusted through a series of interpolations to ensure that $\alpha_L < \breve{\alpha}_0 < \alpha_U$. A suitable value for τ is 0.1. This assures that $\breve{\alpha}_0$ is no closer to α_L or α_U than 10 percent of the permissible range. A similar check is applied in the case of extrapolation, as can be seen in Step 7. The value for χ suggested by Fletcher is 9.

The algorithm maintains a running bracket (or range of uncertainty) $[\alpha_L, \alpha_U]$ that contains the minimizer which is initially set to $[0, 10^{99}]$ in Step 1. This is gradually reduced by reducing α_U in Step 5a and increasing α_L in Step 7e.

In Step 7e, known data that can be used in the next iteration are saved, i.e., α_0, f_0, and f_0' become α_L, f_L, and f_L', respectively. This keeps the amount of computation to a minimum.

Note that the Goldstein condition in Eq. (4.55) is modified as in Step 5 to take into account the fact that α_L assumes a value greater than zero when extrapolation is applied at least once.

References

1 D. M. Himmelblau, *Applied Nonlinear Programming*, McGraw-Hill, New York, 1972.
2 B. S. Gottfried and J. Weisman, *Introduction to Optimization Theory*, Prentice-Hall, Englewood Cliffs, N.J., 1973.
3 P. R. Adby and M. A. H. Dempster, *Introduction to Optimization Methods*, Chapman and Hall, London, 1974.
4 C. S. Beightler, D. T. Phillips, and D. J. Wilde, *Foundations of Optimization*, Prentice-Hall, Englewood Cliffs, N.J., 1979.
5 M. S. Bazaraa and C. M. Shetty, *Nonlinear Programming, Theory and Algorithms*, Wiley, New York, 1979.
6 P. E. Gill, W. Murray, and M. H. Wright, *Practical Optimization*, Academic Press, London, 1981.
7 G. P. McCormick, *Nonlinear Programming*, Wiley, New York, 1983.
8 M. J. Box, D. Davies, and W. H. Swann, *Nonlinear Optimization Techniques*, Oliver and Boyd, London, 1969.
9 R. Fletcher, *Practical Methods of Optimization*, 2nd ed., Wiley, New York, 1987.
10 R. Fletcher, *Practical Methods of Optimization*, vol. 1, Wiley, New York, 1980.

Problems

4.1 (a) Assuming that the ratio of two consecutive Fibonacci numbers, F_{k-1}/F_k, converges to a finite limit α, use Eq. (4.4) to show that

$$\lim_{k \to \infty} \frac{F_{k-1}}{F_k} = \alpha = \frac{2}{\sqrt{5}+1} \approx 0.6180$$

(b) Use MATLAB to verify the value of α in part (a).

4.2 The 5th-order polynomial

$$f(x) = -5x^5 + 4x^4 - 12x^3 + 11x^2 - 2x + 1$$

is known to be a unimodal function on interval $[-0.5,\ 0.5]$.

(a) Use the dichotomous search to find the minimizer of $f(x)$ on $[-0.5,\ 0.5]$ with the range of uncertainty less than 10^{-5}.

(b) Solve the line search problem in part (a) using the Fibonacci search.

(c) Solve the line search problem in part (a) using the golden-section search.

(d) Solve the line search problem in part (a) using the quadratic interpolation method of Sec. 4.5.

(e) Solve the line search problem in part (a) using the cubic interpolation method of Sec. 4.6.

(f) Solve the line search problem in part (a) using the algorithm of Davies, Swann, and Campey.

(g) Compare the computational efficiency of the methods in (a) – (f) in terms of number of function evaluations.

4.3 The function[1]

$$f(x) = \ln^2(x - 2) + \ln^2(10 - x) - x^{0.2}$$

is known to be a unimodal function on $[6, 9.9]$. Repeat Prob. 4.2 for the above function.

4.4 The function

$$f(x) = -3x \sin 0.75x + e^{-2x}$$

is known to be a unimodal function on $[0,\ 2\pi]$. Repeat Prob. 4.2 for the above function.

4.5 The function

$$f(x) = e^{3x} + 5e^{-2x}$$

is known to be a unimodal function on $[0, 1]$. Repeat Prob. 4.2 for the above function.

4.6 The function

$$f(x) = 0.2x \ln x + (x - 2.3)^2$$

is known to be a unimodal function on $[0.5, 2.5]$. Repeat Prob. 4.2 for the above function.

---

[1]Here and and the rest of the book, the logarithms of x to the base e and 10 will be denoted as $\ln(x)$ and $\log_{10}(x)$, respectively.

4.7 Let $f_1(x)$ and $f_2(x)$ be two convex functions such that $f_1(-0.4) = 0.36$, $f_1(0.6) = 2.56$, $f_2(-0.4) = 3.66$, and $f_2(1) = 2$ and define the function

$$f(x) = \max\{f_1(x),\ f_x(x)\}$$

Identify the smallest interval in which the minimizer of $f(x)$ is guaranteed to exist.

4.8 The values of a function $f(x)$ at points $x = x_1$ and $x = x_2$ are f_1 and f_2, respectively, and the derivative of $f(x)$ at point x_1 is f_1'. Show that

$$\bar{x} = x_1 + \frac{f_1'(x_2 - x_1)^2}{2[f_1 - f_2 + f_1'(x_2 - x_1)]}$$

is an estimate of the minimizer of $f(x)$.

4.9 By letting $x_1 = x_2 - \delta$ and $x_3 = x_2 + \delta$ in Eq. (4.30), show that the minimizer $\bar{x}$ can be computed using Eq. (4.31).

4.10 A convex quadratic function $f(x)$ assumes the values f_1, f_2, and f_3 at $x = x_1$, x_2, and x_3, respectively, where $x_1 = x_2 - \delta$ and $x_3 = x_2 + \delta$. Show that the minimum of the function is given by

$$f_{\min} = f_2 - \frac{(f_1 - f_3)^2}{8(f_1 - 2f_2 + f_3)}$$

4.11 (a) Use MATLAB to plot

$$f(\mathbf{x}) = 0.7x_1^4 - 8x_1^2 + 6x_2^2 + \cos(x_1 x_2) - 8x_1$$

over the region $-\pi \le x_1,\ x_2 \le \pi$. A MATLAB command for plotting the surface of a two-variable function is mesh.

(b) Use MATLAB to generate a contour plot of $f(\mathbf{x})$ over the same region as in (a) and 'hold' it.

(c) Compute the gradient of $f(\mathbf{x})$, and prepare MATLAB function files to evaluate $f(\mathbf{x})$ and its gradient.

(d) Use Fletcher's inexact line search algorithm to update point $\mathbf{x}_0$ along search direction $\mathbf{d}_0$ by solving the problem

$$\underset{\alpha \ge 0}{\text{minimize}}\ f(\mathbf{x}_0 + \alpha \mathbf{d}_0)$$

where

$$\mathbf{x}_0 = \begin{bmatrix} -\pi \\ \pi \end{bmatrix}, \quad \mathbf{d}_0 = \begin{bmatrix} 1.0 \\ -1.3 \end{bmatrix}$$

This can be done in several steps:

- Record the numerical values of α^* obtained.
- Record the updated point $\mathbf{x}_1 = \mathbf{x}_0 + \alpha^* \mathbf{d}_0$.
- Evaluate $f(\mathbf{x}_1)$ and compare it with $f(\mathbf{x}_0)$.
- Plot the line search result on the contour plot generated in (*b*).
- Plot $f(\mathbf{x}_0 + \alpha \mathbf{d}_0)$ as a function of α over the interval $[0, 4.8332]$. Based on the plot, comment on the precision of Fletcher's inexact line search.

(*e*) Repeat Part (*d*) for

$$\mathbf{x}_0 = \begin{bmatrix} -\pi \\ \pi \end{bmatrix}, \quad \mathbf{d}_0 = \begin{bmatrix} 1.0 \\ -1.1 \end{bmatrix}$$

The interval of α for plotting $f(\mathbf{x}_0 + \alpha \mathbf{d}_0)$ in this case is $[0, 5.7120]$.

Chapter 5

BASIC MULTIDIMENSIONAL GRADIENT METHODS

5.1 Introduction

In Chap. 4, several methods were considered that can be used for the solution of one-dimensional unconstrained problems. In this chapter, we consider the solution of multidimensional unconstrained problems.

As for one-dimensional optimization, there are two general classes of multidimensional methods, namely, search methods and gradient methods. In search methods, the solution is obtained by using only function evaluations. The general approach is to explore the parameter space in an organized manner in order to find a trajectory that leads progressively to reduced values of the objective function. A rudimentary method of this class might be to adjust all the parameters at a specific starting point, one at a time, and then select a new point by comparing the calculated values of the objective function. The same procedure can then be repeated at the new point, and so on. Multidimensional search methods are thus analogous to their one-dimensional counterparts, and like the latter, they are not very efficient. As a result, their application is restricted to problems where gradient information is unavailable or difficult to obtain, for example, in applications where the objective function is not continuous.

Gradient methods are based on gradient information. They can be grouped into two classes, first-order and second-order methods. First-order methods are based on the linear approximation of the Taylor series, and hence they entail the gradient $\mathbf{g}$. Second-order methods, on the other hand, are based on the quadratic approximation of the Taylor series. They entail the gradient $\mathbf{g}$ as well as the Hessian $\mathbf{H}$.

Gradient methods range from some simple to some highly sophisticated methods. In this chapter, we focus our attention on the most basic ones which are as follows:

1. Steepest-descent method
2. Newton method
3. Gauss-Newton method

Some more advanced gradient methods will be considered later in Chaps. 6 and 7.

5.2 Steepest-Descent Method

Consider the optimization problem

$$\text{minimize } F = f(\mathbf{x}) \qquad \text{for } \mathbf{x} \in E^n$$

From the Taylor series

$$F + \Delta F = f(\mathbf{x} + \boldsymbol{\delta}) \approx f(\mathbf{x}) + \mathbf{g}^T \boldsymbol{\delta} + \tfrac{1}{2} \boldsymbol{\delta}^T \mathbf{H} \boldsymbol{\delta}$$

and as $\|\boldsymbol{\delta}\| \to 0$, the change in F due to change $\boldsymbol{\delta}$ is obtained as

$$\Delta F \approx \mathbf{g}^T \boldsymbol{\delta}$$

The product at the right-hand side is the *scalar* or *dot product* of vectors $\mathbf{g}$ and $\boldsymbol{\delta}$. If

$$\mathbf{g} = [g_1 \ g_2 \ \cdots \ g_n]^T$$

and

$$\boldsymbol{\delta} = [\delta_1 \ \delta_2 \ \cdots \ \delta_n]^T$$

then

$$\Delta F \approx \sum_{i=1}^{n} g_i \delta_i = \|\mathbf{g}\| \, \|\boldsymbol{\delta}\| \cos \theta$$

where θ is the angle between vectors $\mathbf{g}$ and $\boldsymbol{\delta}$, and

$$\|\mathbf{g}\| = (\mathbf{g}^T \mathbf{g})^{1/2} = \left(\sum_{i=1}^{n} g_i^2 \right)^{1/2}$$

5.2.1 Ascent and descent directions

Consider the contour plot of Fig. 5.1. If $\mathbf{x}$ and $\mathbf{x} + \boldsymbol{\delta}$ are adjacent points on contour A, then as $\|\boldsymbol{\delta}\| \to 0$

$$\Delta F \approx \|\mathbf{g}\| \, \|\boldsymbol{\delta}\| \, \cos \theta = 0$$

since F is constant on a contour. We thus conclude that the angle θ between vectors $\mathbf{g}$ and $\boldsymbol{\delta}$ is equal to 90°. In effect, the gradient at point $\mathbf{x}$ is orthogonal to contour A, as depicted in Fig. 5.1. Now for any vector $\boldsymbol{\delta}$, ΔF assumes a maximum positive value if $\theta = 0$, that is, $\boldsymbol{\delta}$ must be in the direction $\mathbf{g}$. On the

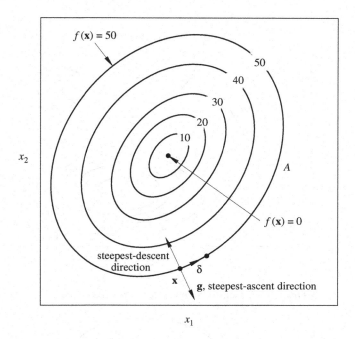

Figure 5.1.   Steepest-descent and steepest-ascent directions.

other hand, ΔF assumes a maximum negative value if $\theta = \pi$, that is, δ must in the direction $-\mathbf{g}$. The gradient $\mathbf{g}$ and its negative $-\mathbf{g}$ are thus said to be the *steepest-ascent* and *steepest-descent directions*, respectively. These concepts are illustrated in Figs. 5.1 and 5.2.

5.2.2   Basic method

Assume that a function $f(x)$ is continuous in the neighborhood of point $\mathbf{x}$. If $\mathbf{d}$ is the steepest-descent direction at point $\mathbf{x}$, i.e.,

$$\mathbf{d} = -\mathbf{g}$$

then a change δ in $\mathbf{x}$ given by

$$\delta = \alpha\mathbf{d}$$

where α is a small positive constant, will decrease the value of $f(\mathbf{x})$. Maximum reduction in $f(\mathbf{x})$ can be achieved by solving the one-dimensional optimization problem

$$\underset{\alpha}{\text{minimize}} \; F = f(\mathbf{x} + \alpha\mathbf{d}) \tag{5.1}$$

as depicted in Fig. 5.3.

122

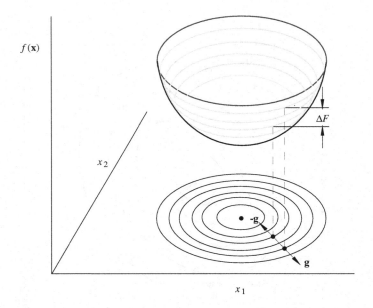

Figure 5.2. Construction for steepest-descent method.

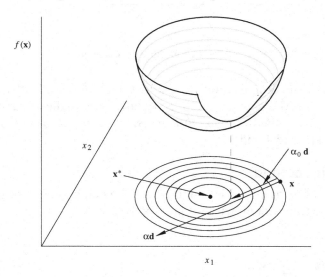

Figure 5.3. Line search in steepest-descent direction.

If the steepest-descent direction at point $\mathbf{x}$ happens to point towards the minimizer $\mathbf{x}^*$ of $f(\mathbf{x})$, then a value of α exists that minimizes $f(\mathbf{x} + \alpha\mathbf{d})$ with respect to α and $f(\mathbf{x})$ with respect to $\mathbf{x}$. Consequently, in such a case the multidimensional problem can be solved by solving the one-dimensional problem in Eq. (5.1) once. Unfortunately, in most real-life problems, $\mathbf{d}$ does not point in the direction of $\mathbf{x}^*$ and, therefore, an iterative procedure must be used for the solution. Starting with an initial point $\mathbf{x}_0$, a direction $\mathbf{d} = \mathbf{d}_0 = -\mathbf{g}$ can be calculated, and the value of α that minimizes $f(\mathbf{x}_0 + \alpha \mathbf{d}_0)$, say, α_0, can be determined. Thus a point $\mathbf{x}_1 = \mathbf{x}_0 + \alpha_0 \mathbf{d}_0$ is obtained. The minimization can be performed by using one of the methods of Chap. 4 as a line search. The same procedure can then be repeated at points

$$\mathbf{x}_{k+1} = \mathbf{x}_k + \alpha_k \mathbf{d}_k \tag{5.2}$$

for $k = 1, 2, \ldots$ until convergence is achieved. The procedure can be terminated when $\|\alpha_k \mathbf{d}_k\|$ becomes insignificant or if $\alpha_k \leq K\alpha_0$ where K is a sufficiently small positive constant. A typical solution trajectory for the steepest-descent method is illustrated in Fig. 5.4. A corresponding algorithm is as follows.

Algorithm 5.1 Steepest-descent algorithm
Step 1
Input $\mathbf{x}_0$ and initialize the tolerance ε.
Set $k = 0$.
Step 2
Calculate gradient $\mathbf{g}_k$ and set $\mathbf{d}_k = -\mathbf{g}_k$.
Step 3
Find α_k, the value of α that minimizes $f(\mathbf{x}_k + \alpha \mathbf{d}_k)$, using a line search.
Step 4
Set $\mathbf{x}_{k+1} = \mathbf{x}_k + \alpha_k \mathbf{d}_k$ and calculate $f_{k+1} = f(\mathbf{x}_{k+1})$.
Step 5
If $\|\alpha_k \mathbf{d}_k\| < \varepsilon$, then do:
    Output $\mathbf{x}^* = \mathbf{x}_{k+1}$ and $f(\mathbf{x}^*) = f_{k+1}$, and stop.
Otherwise, set $k = k + 1$ and repeat from Step 2.

5.2.3   Orthogonality of directions

In the steepest-descent method, the trajectory to the solution follows a zig-zag pattern, as can be seen in Fig. 5.4. If α is chosen such that $f(\mathbf{x}_k + \alpha \mathbf{d}_k)$ is minimized in each iteration, then successive directions are orthogonal. To demonstrate this fact, we note that

$$\frac{df(\mathbf{x}_k + \alpha \mathbf{d}_k)}{d\alpha} = \sum_{i=1}^{n} \frac{\partial f(\mathbf{x}_k + \alpha \mathbf{d}_k)}{\partial x_{ki}} \frac{d(x_{ki} + \alpha d_{ki})}{d\alpha}$$

$$= \sum_{i=1}^{n} g_i(\mathbf{x}_k + \alpha \mathbf{d}_k) d_{ki}$$

$$= \mathbf{g}(\mathbf{x}_k + \alpha \mathbf{d}_k)^T \mathbf{d}_k$$

124

where $\mathbf{g}(\mathbf{x}_k + \alpha \mathbf{d}_k)$ is the gradient at point $\mathbf{x}_k + \alpha \mathbf{d}_k$. If α^* is the value of α that minimizes $f(\mathbf{x}_k + \alpha \mathbf{d}_k)$, then

$$\mathbf{g}(\mathbf{x}_k + \alpha^* \mathbf{d}_k)^T \mathbf{d}_k = 0$$

or

$$\mathbf{d}_{k+1}^T \mathbf{d}_k = 0$$

where

$$\mathbf{d}_{k+1} = -\mathbf{g}(\mathbf{x}_k + \alpha^* \mathbf{d}_k)$$

is the steepest-descent direction at point $\mathbf{x}_k + \alpha^* \mathbf{d}_k$. In effect, successive directions $\mathbf{d}_k$ and $\mathbf{d}_{k+1}$ are orthogonal as depicted in Fig. 5.4.

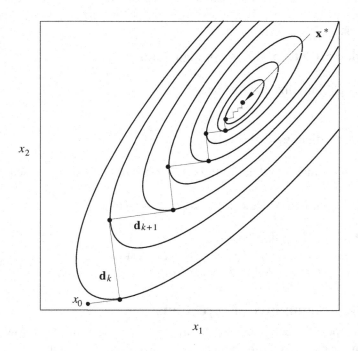

Figure 5.4. Typical solution trajectory in steepest-descent algorithm.

5.2.4 Elimination of line search

If the Hessian of $f(\mathbf{x})$ is available, the value of α that minimizes $f(\mathbf{x}_k + \alpha \mathbf{d})$, namely, α_k, can be determined by using an analytical method. If $\boldsymbol{\delta}_k = \alpha \mathbf{d}_k$, the Taylor series yields

$$f(\mathbf{x}_k + \boldsymbol{\delta}_k) \approx f(\mathbf{x}_k) + \boldsymbol{\delta}_k^T \mathbf{g}_k + \tfrac{1}{2} \boldsymbol{\delta}_k^T \mathbf{H}_k \boldsymbol{\delta}_k \tag{5.3}$$

and if $\mathbf{d}_k$ is the steepest-descent direction, i.e.,

$$\delta_k = -\alpha \mathbf{g}_k$$

we obtain

$$f(\mathbf{x}_k - \alpha \mathbf{g}_k) \approx f(\mathbf{x}_k) - \alpha \mathbf{g}_k^T \mathbf{g}_k + \tfrac{1}{2}\alpha^2 \mathbf{g}_k^T \mathbf{H}_k \mathbf{g}_k \qquad (5.4)$$

By differentiating and setting the result to zero, we get

$$\frac{df(\mathbf{x}_k - \alpha \mathbf{g}_k)}{d\alpha} \approx -\mathbf{g}_k^T \mathbf{g}_k + \alpha \mathbf{g}_k^T \mathbf{H}_k \mathbf{g}_k = 0$$

or

$$\alpha = \alpha_k \approx \frac{\mathbf{g}_k^T \mathbf{g}_k}{\mathbf{g}_k^T \mathbf{H}_k \mathbf{g}_k} \qquad (5.5)$$

Now if we assume that $\alpha = \alpha_k$ minimizes $f(\mathbf{x}_k + \alpha \mathbf{d}_k)$, Eq. (5.2) can be expressed as

$$\mathbf{x}_{k+1} = \mathbf{x}_k - \frac{\mathbf{g}_k^T \mathbf{g}_k}{\mathbf{g}_k^T \mathbf{H}_k \mathbf{g}_k} \mathbf{g}_k$$

The accuracy of α_k will depend heavily on the magnitude of δ_k since the quadratic approximation of the Taylor series is valid only in the neighborhood of point $\mathbf{x}_k$. At the start of the optimization, $\|\delta_k\|$ will be relatively large and so α_k will be inaccurate. Nevertheless, reduction will be achieved in $f(\mathbf{x})$ since $f(\mathbf{x}_k + \alpha \mathbf{d}_k)$ is minimized in the steepest-descent direction. As the solution is approached, $\|\delta_k\|$ is decreased and, consequently, the accuracy of α_k will progressively be improved, and the maximum reduction in $f(\mathbf{x})$ will eventually be achieved in each iteration. Convergence will thus be achieved. For quadratic functions, Eq. (5.3) is satisfied with the equal sign and hence $\alpha = \alpha_k$ yields maximum reduction in $f(\mathbf{x})$ in every iteration.

If the Hessian is not available, the value of α_k can be determined by calculating $f(\mathbf{x})$ at points $\mathbf{x}_k$ and $\mathbf{x}_k - \hat{\alpha} \mathbf{g}_k$ where $\hat{\alpha}$ is an estimate of α_k. If

$$f_k = f(\mathbf{x}_k) \quad \text{and} \quad \hat{f} = f(\mathbf{x}_k - \hat{\alpha} \mathbf{g}_k)$$

Eq. (5.4) gives

$$\hat{f} \approx f_k - \hat{\alpha} \mathbf{g}_k^T \mathbf{g}_k + \tfrac{1}{2}\hat{\alpha}^2 \mathbf{g}_k^T \mathbf{H}_k \mathbf{g}_k$$

or

$$\mathbf{g}_k^T \mathbf{H}_k \mathbf{g}_k \approx \frac{2(\hat{f} - f_k + \hat{\alpha} \mathbf{g}_k^T \mathbf{g}_k)}{\hat{\alpha}^2} \qquad (5.6)$$

Now from Eqs. (5.5) and (5.6)

$$\alpha_k \approx \frac{\mathbf{g}_k^T \mathbf{g}_k \hat{\alpha}^2}{2(\hat{f} - f_k + \hat{\alpha} \mathbf{g}_k^T \mathbf{g}_k)} \qquad (5.7)$$

A suitable value for $\hat{\alpha}$ is α_{k-1}, namely, the optimum α in the previous iteration. For the first iteration, the value $\hat{\alpha} = 1$ can be used.

An algorithm that eliminates the need for line searches is as follows:

Algorithm 5.2 Steepest-descent algorithm without line search
Step 1
Input $\mathbf{x}_1$ and initialize the tolerance ε.
Set $k = 1$ and $\alpha_0 = 1$.
Compute $f_1 = f(\mathbf{x}_1)$.
Step 2
Compute $\mathbf{g}_k$.
Step 3
Set $\mathbf{d}_k = -\mathbf{g}_k$ and $\hat{\alpha} = \alpha_{k-1}$.
Compute $\hat{f} = f(\mathbf{x}_k - \hat{\alpha}\mathbf{g}_k)$.
Compute α_k from Eq. (5.7).
Step 4
Set $\mathbf{x}_{k+1} = \mathbf{x}_k + \alpha_k\mathbf{d}_k$ and calculate $f_{k+1} = f(\mathbf{x}_{k+1})$.
Step 5
If $\|\alpha_k\mathbf{d}_k\| < \varepsilon$, then do:
    Output $\mathbf{x}^* = \mathbf{x}_{k+1}$ and $f(\mathbf{x}^*) = f_{k+1}$, and stop.
Otherwise, set $k = k + 1$ and repeat from Step 2.

The value of α_k in Step 3 is an accurate estimate of the value of α that minimizes $f(\mathbf{x}_k + \alpha\mathbf{d}_k)$ to the extent that the quadratic approximation of the Taylor series is an accurate representation of $f(\mathbf{x})$. Thus, as was argued earlier, the reduction in $f(\mathbf{x})$ per iteration tends to approach the maximum possible as $\mathbf{x}^*$ is approached, and if $f(\mathbf{x})$ is quadratic the maximum possible reduction is achieved in every iteration.

5.2.5   Convergence

If a function $f(\mathbf{x}) \in C^2$ has a local minimizer $\mathbf{x}^*$ and its Hessian is positive definite at $\mathbf{x} = \mathbf{x}^*$, then it can be shown that if $\mathbf{x}_k$ is sufficiently close to $\mathbf{x}^*$, we have

$$f(\mathbf{x}_{k+1}) - f(\mathbf{x}^*) \leq \left(\frac{1-r}{1+r}\right)^2 [f(\mathbf{x}_k) - f(\mathbf{x}^*)] \qquad (5.8)$$

where

$$r = \frac{\text{smallest eigenvalue of } \mathbf{H}_k}{\text{largest eigenvalue of } \mathbf{H}_k}$$

Furthermore, if $f(\mathbf{x})$ is a quadratic function then the inequality in Eq. (5.8) holds for all k (see [1] for proof). In effect, subject to the conditions stated, the steepest-descent method converges linearly (see Sec. 3.7) with a convergence ratio

$$\beta = \left(\frac{1-r}{1+r}\right)^2$$

Evidently, the rate of convergence is high if the eigenvalues of $\mathbf{H}_k$ are all nearly equal, or low if at least one eigenvalue is small relative to the largest eigenvalue.

The eigenvalues of $\mathbf{H}$, namely, λ_i for 1, 2, ..., n, determine the geometry of the surface

$$\mathbf{x}^T\mathbf{H}\mathbf{x} = \text{constant}$$

This equation gives the contours of $\mathbf{x}^T\mathbf{H}\mathbf{x}$ and if $\mathbf{H}$ is positive definite, the contours are ellipsoids with axes proportional to $1/\sqrt{\lambda_i}$. If the number of variables is two, the contours are ellipses with axes proportional to $1/\sqrt{\lambda_1}$ and $1/\sqrt{\lambda_2}$. Consequently, if the steepest-descent method is applied to a two-dimensional problem, convergence will be fast if the contours are nearly circular, as is to be expected, and if they are circular, i.e., $r = 1$, convergence will be achieved in one iteration. On the other hand, if the contours are elongated ellipses or if the function exhibits long narrow valleys, progress will be very slow, in particular as the solution is approached. The influence of r on convergence can be appreciated by comparing Figs. 5.4 and 5.5.

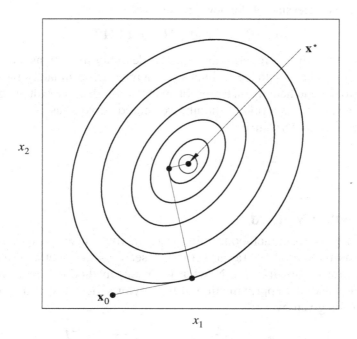

Figure 5.5. Solution trajectory in steepest-descent algorithm if $r \approx 1$.

128

The steepest-descent method attempts, in effect, to reduce the gradient to zero. Since at a saddle point, the gradient is zero, it might be questioned whether such a point is a likely solution. It turns out that such a solution is highly unlikely, in practice, for two reasons. First, the probability of locating a saddle point exactly as the next iteration point is infinitesimal. Second, there is always a descent direction in the neighborhood of a saddle point.

5.2.6   Scaling

The eigenvalues of **H** in a specific optimization problem and, in turn, the performance of the steepest-descent method tend to depend to a large extent on the choice of variables. For example, in one and the same two-dimensional problem, the contours may be nearly circular or elliptical depending on the choice of units. Consequently, the rate of convergence can often be improved by scaling the variables through variable transformations.

A possible approach to scaling might be to let

$$\mathbf{x} = \mathbf{T}\mathbf{y}$$

where **T** is an $n \times n$ diagonal matrix, and then solve the problem

$$\underset{\mathbf{y}}{\text{minimize}} \ h(\mathbf{y}) = f(\mathbf{x})|_{\mathbf{x}=\mathbf{T}\mathbf{y}}$$

The gradient and Hessian of the new problem are

$$\mathbf{g}_h = \mathbf{T}\mathbf{g}_x \quad \text{and} \quad \mathbf{H}_h = \mathbf{T}^T\mathbf{H}\mathbf{T}$$

respectively, and, therefore, both the steepest-descent direction as well as the eigenvalues associated with the problem are changed. Unfortunately, the choice of **T** tends to depend heavily on the problem at hand and, as a result, no general rules can be stated. As a rule of thumb, we should strive to as far as possible equalize the second derivatives

$$\frac{\partial^2 f}{\partial x_i^2} \quad \text{for } i = 1, 2, \ldots, n.$$

5.3   Newton Method

The steepest-descent method is a first-order method since it is based on the linear approximation of the Taylor series. A second-order method known as the Newton (also known as the Newton-Raphson) method can be developed by using the quadratic approximation of the Taylor series. If $\boldsymbol{\delta}$ is a change in **x**, $f(\mathbf{x} + \boldsymbol{\delta})$ is given by

$$f(\mathbf{x} + \boldsymbol{\delta}) \approx f(\mathbf{x}) + \sum_{j=1}^{n} \frac{\partial f}{\partial x_i}\delta_i + \frac{1}{2}\sum_{i=1}^{n}\sum_{j=1}^{n} \frac{\partial^2 f}{\partial x_i \partial x_j}\delta_i\delta_j \qquad (5.9)$$

Assuming that this is an accurate representation of the function at point $\mathbf{x} + \boldsymbol{\delta}$, then differentiating $f(\mathbf{x} + \boldsymbol{\delta})$ with respect to δ_k for $k = 1, 2, \ldots, n$ and setting the result to zero will give the values of δ_k that minimize $f(\mathbf{x} + \boldsymbol{\delta})$. This approach yields

$$\frac{\partial f}{\partial x_k} + \sum_{i=1}^{n} \frac{\partial^2 f}{\partial x_i \partial x_k} \delta_i = 0 \qquad \text{for } k = 1, 2, \ldots, n$$

or in matrix form

$$\mathbf{g} = -\mathbf{H}\boldsymbol{\delta}$$

Therefore, the optimum change in $\mathbf{x}$ is

$$\boldsymbol{\delta} = -\mathbf{H}^{-1}\mathbf{g} \tag{5.10}$$

This solution exists if and only if the following conditions hold:

   (a) The Hessian is nonsingular.
   (b) The approximation in Eq. (5.9) is valid.

Assuming that the second-order sufficiency conditions for a minimum hold at point $\mathbf{x}^*$, then $\mathbf{H}$ is positive definite at $\mathbf{x}^*$ and also in the neighborhood of the solution i.e., for $\|\mathbf{x} - \mathbf{x}^*\| < \varepsilon$. This means that $\mathbf{H}$ is nonsingular and has an inverse for $\|\mathbf{x} - \mathbf{x}^*\| < \varepsilon$. Since any function $f(x) \in C^2$ can be accurately represented in the neighborhood of $\mathbf{x}^*$ by the quadratic approximation of the Taylor series, the solution in Eq. (5.10) exists. Furthermore, for any point $\mathbf{x}$ such that $\|\mathbf{x} - \mathbf{x}^*\| < \varepsilon$ one iteration will yield $\mathbf{x} \approx \mathbf{x}^*$.

Any quadratic function has a Hessian which is constant for any $\mathbf{x} \in E^n$, as can be readily demonstrated. If the function has a minimum, and the second-order sufficiency conditions for a minimum hold, then $\mathbf{H}$ is positive definite and, therefore, nonsingular at any point $\mathbf{x} \in E^n$. Since any quadratic function is represented exactly by the quadratic approximation of the Taylor series, the solution in Eq. (5.10) exists. Furthermore, for any point $\mathbf{x} \in E^n$ one iteration will yield the solution.

If a general nonquadratic function is to be minimized and an arbitrary point $\mathbf{x}$ is assumed, condition (a) and/or condition (b) may be violated. If condition (a) is violated, Eq. (5.10) may have an infinite number of solutions or no solution at all. If, on the other hand, condition (b) is violated, then $\boldsymbol{\delta}$ may not yield the solution in one iteration and, if $\mathbf{H}$ is not positive definite, $\boldsymbol{\delta}$ may not even yield a reduction in the objective function.

The first problem can be overcome by forcing $\mathbf{H}$ to become positive definite by means of some manipulation. The second problem, on the other hand, can be overcome by using an iterative procedure which incorporates a line search for the calculation of the change in $\mathbf{x}$. The iterative procedure will counteract the fact that one iteration will not yield the solution, and the line search can be used to achieve maximum reduction in $f(\mathbf{x})$ along the direction predicted by Eq. (5.10). This approach can be implemented by selecting the next point $\mathbf{x}_{k+1}$ as

$$\mathbf{x}_{k+1} = \mathbf{x}_k + \boldsymbol{\delta}_k = \mathbf{x}_k + \alpha_k \mathbf{d}_k \qquad (5.11)$$

where

$$\mathbf{d}_k = -\mathbf{H}_k^{-1} \mathbf{g}_k \qquad (5.12)$$

and α_k is the value of α that minimizes $f(\mathbf{x}_k + \alpha \mathbf{d}_k)$. The vector $\mathbf{d}_k$ is referred to as the *Newton direction* at point $\mathbf{x}_k$. In the case where conditions (a) and (b) are satisfied, the first iteration will yield the solution with $\alpha_k = 1$.

At the start of the minimization, progress might be slow for certain types of functions. Nevertheless, continuous reduction in $f(\mathbf{x})$ will be achieved through the choice of α. As the solution is approached, however, both conditions (a) and (b) will be satisfied and, therefore, convergence will be achieved. The order of convergence can be shown to be two (see [1, Chap. 7]). In effect, the Newton method has convergence properties that are complementary to those of the steepest-descent method, namely, it can be slow away from the solution and fast close to the solution.

The above principles lead readily to the basic Newton algorithm summarized below.

Algorithm 5.3 Basic Newton algorithm
Step 1
Input $\mathbf{x}_0$ and initialize the tolerance ε.
Set $k = 0$.
Step 2
Compute $\mathbf{g}_k$ and $\mathbf{H}_k$.
If $\mathbf{H}_k$ is not positive definite, force it to become positive definite.
Step 3
Compute $\mathbf{H}_k^{-1}$ and $\mathbf{d}_k = -\mathbf{H}_k^{-1} \mathbf{g}_k$.
Step 4
Find α_k, the value of α that minimizes $f(\mathbf{x}_k + \alpha \mathbf{d}_k)$, using a line search.
Step 5
Set $\mathbf{x}_{k+1} = \mathbf{x}_k + \alpha_k \mathbf{d}_k$.
Compute $f_{k+1} = f(\mathbf{x}_{k+1})$.
Step 6
If $\|\alpha_k \mathbf{d}_k\| < \varepsilon$, then do:
    Output $\mathbf{x}^* = \mathbf{x}_{k+1}$ and $f(\mathbf{x}^*) = f_{k+1}$, and stop.

Otherwise, set $k = k + 1$ and repeat from Step 2.

5.3.1 Modification of the Hessian

If the Hessian is not positive definite in any iteration of Algorithm 5.3, it is forced to become positive definite in Step 2 of the algorithm. This modification of $\mathbf{H}_k$ can be accomplished in one of several ways.

One approach proposed by Goldfeld, Quandt, and Trotter [2] is to replace $\mathbf{H}_k$ by the $n \times n$ identity matrix $\mathbf{I}_n$ wherever it becomes nonpositive definite. Since $\mathbf{I}_n$ is positive definite, the problem of a nonsingular $\mathbf{H}_k$ is eliminated. This approach can be implemented by letting

$$\hat{\mathbf{H}}_k = \frac{\mathbf{H}_k + \beta \mathbf{I}_n}{1 + \beta} \tag{5.13}$$

where β is set to a large value if $\mathbf{H}_k$ is nonpositive definite, or to a small value if $\mathbf{H}_k$ is positive definite.

If β is large, then

$$\hat{\mathbf{H}}_k \approx \mathbf{I}_n$$

and from Eq. (5.12)

$$\mathbf{d}_k \approx -\mathbf{g}_k$$

In effect, the modification in Eq. (5.13) converts the Newton method into the steepest-descent method.

A nonpositive definite $\mathbf{H}_k$ is likely to arise at points far from the solution where the steepest-descent method is most effective in reducing the value of $f(\mathbf{x})$. Therefore, the modification in Eq. (5.13) leads to an algorithm that combines the complementary convergence properties of the Newton and steepest-descent methods.

A second possibility due to Zwart [3] is to form a modified matrix

$$\hat{\mathbf{H}}_k = \mathbf{U}^T \mathbf{H}_k \mathbf{U} + \boldsymbol{\varepsilon}$$

where $\mathbf{U}$ is a real unitary matrix (i.e., $\mathbf{U}^T \mathbf{U} = \mathbf{I}_n$) and $\boldsymbol{\varepsilon}$ is a diagonal $n \times n$ matrix with diagonal elements ε_i. It can be shown that a matrix $\mathbf{U}$ exists such that $\mathbf{U}^T \mathbf{H}_k \mathbf{U}$ is diagonal with diagonal elements λ_i for $i = 1, 2, \ldots, n$, where λ_i are the eigenvalues of $\mathbf{H}_k$ (see Theorem 2.8). In effect, $\hat{\mathbf{H}}_k$ is diagonal with elements $\lambda_i + \varepsilon_i$. Therefore, if

$$\varepsilon_i = \begin{cases} 0 & \text{if} \quad \lambda_i > 0 \\ \delta - \lambda_i & \text{if} \quad \lambda_i \leq 0 \end{cases}$$

where δ is a positive constant, then $\hat{\mathbf{H}}_k$ will be positive definite. With this modification, changes in the components of $\mathbf{x}_k$ in Eq. (5.12) due to negative eigenvalues are ignored. Matrix $\mathbf{U}^T \mathbf{H}_k \mathbf{U}$ can be formed by solving the equation

$$\det(\mathbf{H}_k - \lambda \mathbf{I}_n) = 0 \tag{5.14}$$

This method entails minimal disturbance of $\mathbf{H}_k$, and hence the convergence properties of the Newton method are largely preserved. Unfortunately, however, the solution of Eq. (5.14) involves the determination of the n roots of the characteristic polynomial of $\mathbf{H}_k$ and is, therefore, time-consuming.

A third method for the manipulation of $\mathbf{H}_k$, which is attributed to Matthews and Davies [4], is based on the Gaussian elimination. This method leads simultaneously to the modification of $\mathbf{H}_k$ and the computation of the Newton direction $\mathbf{d}_k$ and is, therefore, one of the most practical to use. As was shown in Sec. 2.6, given a matrix $\mathbf{H}_k$, a diagonal matrix $\mathbf{D}$ can be deduced as

$$\mathbf{D} = \mathbf{L}\mathbf{H}_k\mathbf{L}^T \qquad (5.15)$$

where

$$\mathbf{L} = \mathbf{E}_{n-1} \cdots \mathbf{E}_2\mathbf{E}_1$$

is a unit lower triangular matrix, and $\mathbf{E}_1$, $\mathbf{E}_2$, ... are elementary matrices. If $\mathbf{H}_k$ is positive definite, then $\mathbf{D}$ is positive definite and vice-versa (see Theorem 2.7). If $\mathbf{D}$ is not positive definite, then a positive definite diagonal matrix $\hat{\mathbf{D}}$ can be formed by replacing each zero or negative element in $\mathbf{D}$ by a positive element. In this way a positive definite matrix $\hat{\mathbf{H}}_k$ can be formed as

$$\hat{\mathbf{H}}_k = \mathbf{L}^{-1}\hat{\mathbf{D}}(\mathbf{L}^T)^{-1} \qquad (5.16)$$

Now from Eq. (5.12)

$$\hat{\mathbf{H}}_k\mathbf{d}_k = -\mathbf{g}_k \qquad (5.17)$$

and hence Eqs. (5.16) and (5.17) yield

$$\mathbf{L}^{-1}\hat{\mathbf{D}}(\mathbf{L}^T)^{-1}\mathbf{d}_k = -\mathbf{g}_k \qquad (5.18)$$

If we let

$$\hat{\mathbf{D}}(\mathbf{L}^T)^{-1}\mathbf{d}_k = \mathbf{y}_k \qquad (5.19)$$

then Eq. (5.18) can be expressed as

$$\mathbf{L}^{-1}\mathbf{y}_k = -\mathbf{g}_k$$

Therefore,

$$\mathbf{y}_k = -\mathbf{L}\mathbf{g}_k \qquad (5.20)$$

and from Eq. (5.19)

$$\mathbf{d}_k = \mathbf{L}^T\hat{\mathbf{D}}^{-1}\mathbf{y}_k \qquad (5.21)$$

The computation of $\mathbf{d}_k$ can thus be carried out by generating the unit lower triangular matrix $\mathbf{L}$ and the corresponding positive definite diagonal matrix $\hat{\mathbf{D}}$. If

$$\mathbf{H}_k = \begin{bmatrix} h_{11} & h_{12} & \cdots & h_{1n} \\ h_{21} & h_{22} & \cdots & h_{2n} \\ \vdots & \vdots & & \vdots \\ h_{n1} & h_{n2} & \cdots & h_{nn} \end{bmatrix}$$

then

$$
\mathbf{L} = \begin{bmatrix} l_{11} & 0 & \cdots & 0 \\ l_{21} & l_{22} & \cdots & 0 \\ \vdots & \vdots & & \vdots \\ l_{n1} & l_{n2} & \cdots & l_{nn} \end{bmatrix}
$$

and

$$
\hat{\mathbf{D}} = \begin{bmatrix} \hat{d}_{11} & 0 & \cdots & 0 \\ 0 & \hat{d}_{22} & \cdots & 0 \\ \vdots & \vdots & & \vdots \\ 0 & 0 & \cdots & \hat{d}_{nn} \end{bmatrix}
$$

can be computed by using the following algorithm.

Algorithm 5.4 Matthews and Davies algorithm
Step 1
Input $\mathbf{H}_k$ and n.
Set $\mathbf{L} = \mathbf{0}, \hat{\mathbf{D}} = \mathbf{0}$.
If $h_{11} > 0$, then set $h_{00} = h_{11}$, else set $h_{00} = 1$.
Step 2
For $k = 2, 3, \ldots, n$ do:
  Set $m = k - 1$, $l_{mm} = 1$.
  If $h_{mm} \leq 0$, set $h_{mm} = h_{00}$.
  Step 2.1
  For $i = k, k + 1, \ldots, n$ do:
    Set $l_{im} = -h_{im}/h_{mm}$, $h_{im} = 0$.
    Step 2.1.1
    For $j = k, k + 1, \ldots, n$ do:
      Set $h_{ij} = h_{ij} + l_{im}h_{mj}$
  If $0 < h_{kk} < h_{00}$, set $h_{00} = h_{kk}$.
Step 3
Set $l_{nn} = 1$. If $h_{nn} \leq 0$, set $h_{nn} = h_{00}$.
For $i = 1, 2, \ldots, n$ set $\hat{d}_{ii} = h_{ii}$.
Stop.

This algorithm will convert $\mathbf{H}$ into an upper triangular matrix with positive diagonal elements, and will then assign the diagonal elements obtained to $\hat{\mathbf{D}}$. Any zero or negative elements of $\mathbf{D}$ are replaced by the most recent lowest positive element of $\hat{\mathbf{D}}$, except if the first element is zero or negative, which is replaced by unity.

If $\mathbf{H}_k$ is a 4×4 matrix, k and m are initially set to 2 and 1, respectively, and l_{11} is set to unity; h_{11} is checked and if it is zero or negative it is changed to unity. The execution of Step 2.1 yields

$$\mathbf{L} = \begin{bmatrix} 1 & 0 & 0 & 0 \\ -\dfrac{h_{21}}{h_{11}} & 0 & 0 & 0 \\ -\dfrac{h_{31}}{h_{11}} & 0 & 0 & 0 \\ -\dfrac{h_{41}}{h_{11}} & 0 & 0 & 0 \end{bmatrix}$$

In addition, the elements in column 1 of $\mathbf{H}_k$ other than h_{11} are set to zero and the elements of rows 2 to 4 and columns 2 to 4 are updated to give

$$\mathbf{H}_k = \begin{bmatrix} h_{11} & h_{12} & h_{13} & h_{14} \\ 0 & h'_{22} & h'_{23} & h'_{24} \\ 0 & h'_{32} & h'_{33} & h'_{34} \\ 0 & h'_{42} & h'_{43} & h'_{44} \end{bmatrix}$$

If $0 < h'_{22} < h_{00}$, then h'_{22} is used to update h_{00}. Indices k and m are then set to 3 and 2, respectively , and l_{22} is set to unity. If $h'_{22} \leq 0$, it is replaced by the most recent value of h_{00}. The execution of Step 2.1 yields

$$\mathbf{L} = \begin{bmatrix} 1 & 0 & 0 & 0 \\ -\dfrac{h_{21}}{h_{11}} & 1 & 0 & 0 \\ -\dfrac{h_{31}}{h_{11}} & -\dfrac{h'_{32}}{h'_{22}} & 0 & 0 \\ -\dfrac{h_{41}}{h_{11}} & -\dfrac{h'_{42}}{h'_{22}} & 0 & 0 \end{bmatrix}$$

and

$$\mathbf{H}_k = \begin{bmatrix} h_{11} & h_{12} & h_{13} & h_{14} \\ 0 & h'_{22} & h'_{23} & h'_{24} \\ 0 & 0 & h''_{33} & h''_{34} \\ 0 & 0 & h''_{43} & h''_{44} \end{bmatrix}$$

If $0 < h''_{33} < h_{00}$, h''_{33} is assigned to h_{00}, and so on. In Step 3, h'''_{44} is checked and is changed to h_{00} if found to be zero or negative, and l_{44} is set to unity. Then the diagonal elements of $\mathbf{H}_k$ are assigned to $\hat{\mathbf{D}}$.

With $\hat{\mathbf{D}}$ known, $\hat{\mathbf{D}}^{-1}$ can be readily obtained by replacing the diagonal elements of $\hat{\mathbf{D}}$ by their reciprocals. The computation of $\mathbf{y}_k$ and $\mathbf{d}_k$ can be completed by using Eqs. (5.20) and (5.21). Algorithm 5.4 is illustrated by the following example.

Example 5.1 Compute $\mathbf{L}$ and $\hat{\mathbf{D}}$ for the 4×4 matrix

$$\mathbf{H}_k = \begin{bmatrix} h_{11} & h_{12} & h_{13} & h_{14} \\ h_{21} & h_{22} & h_{23} & h_{24} \\ h_{31} & h_{32} & h_{33} & h_{34} \\ h_{41} & h_{42} & h_{43} & h_{44} \end{bmatrix}$$

using Algorithm 5.4.

Solution The elements of $\mathbf{L}$ and $\hat{\mathbf{D}}$, namely, l_{ij} and $\hat{d}_{ii}$, can be computed as follows:

Step 1

Input $\mathbf{H}_k$ and set $n = 4$. Initialize $\mathbf{L}$ and $\hat{\mathbf{D}}$ as

$$\mathbf{L} = \begin{bmatrix} 0 & 0 & 0 & 0 \\ 0 & 0 & 0 & 0 \\ 0 & 0 & 0 & 0 \\ 0 & 0 & 0 & 0 \end{bmatrix} \quad \hat{\mathbf{D}} = \begin{bmatrix} 0 & 0 & 0 & 0 \\ 0 & 0 & 0 & 0 \\ 0 & 0 & 0 & 0 \\ 0 & 0 & 0 & 0 \end{bmatrix}$$

Step 2

If $h_{11} > 0$, then set $h_{00} = h_{11}$, else set $h_{00} = 1$.
$k = 2$;
  $m = 1, l_{11} = 1$;
  if $h_{11} \leq 0$, set $h_{11} = h_{00}$;
Step 2.1
$i = 2$;
  $l_{21} = -h_{21}/h_{11}, \ h_{21} = 0$;
  Step 2.1.1
  $j = 2$;
    $h_{22} = h_{22} + l_{21}h_{12} = h_{22} - h_{21}h_{12}/h_{11} \ (= h'_{22})$;
  $j = 3$;
    $h_{23} = h_{23} + l_{21}h_{13} = h_{23} - h_{21}h_{13}/h_{11} \ (= h'_{23})$;
  $j = 4$;
    $h_{24} = h_{24} + l_{21}h_{14} = h_{24} - h_{21}h_{14}/h_{11} \ (= h'_{24})$;
$i = 3$;
  $l_{31} = -h_{31}/h_{11}, \ h_{31} = 0$;
  $j = 2$;
    $h_{32} = h_{32} + l_{31}h_{12} = h_{32} - h_{31}h_{12}/h_{11} \ (= h'_{32})$;
  $j = 3$;
    $h_{33} = h_{33} + l_{31}h_{13} = h_{33} - h_{31}h_{13}/h_{11} \ (= h'_{33})$;
  $j = 4$;
    $h_{34} = h_{34} + l_{31}h_{14} = h_{34} - h_{31}h_{14}/h_{11} \ (= h'_{34})$;
$i = 4$;
  $l_{41} = -h_{41}/h_{11}, \ h_{41} = 0$;

136

$j = 2;$

$\quad h_{42} = h_{42} + l_{41}h_{12} = h_{42} - h_{41}h_{12}/h_{11} \ (= h'_{42});$

$j = 3;$

$\quad h_{43} = h_{43} + l_{41}h_{13} = h_{43} - h_{41}h_{13}/h_{11} \ (= h'_{43});$

$j = 4;$

$\quad h_{44} = h_{44} + l_{41}h_{14} = h_{44} - h_{41}h_{14}/h_{11} \ (= h'_{44});$

if $0 < h_{22} < h_{00}$, set $h_{00} = h_{22};$

$k = 3;$

$\quad m = 2, l_{22} = 1;$

$\quad$if $h_{22} < 0$, set $h_{22} = h_{00};$

$\quad i = 3;$

$\quad\quad l_{32} = -h_{32}/h_{22}, \ h_{32} = 0;$

$\quad\quad j = 3;$

$\quad\quad\quad h_{33} = h_{33} + l_{32}h_{23} = h_{33} - h_{32}h_{23}/h_{22} \ (= h''_{33});$

$\quad\quad j = 4;$

$\quad\quad\quad h_{34} = h_{34} + l_{32}h_{24} = h_{34} - h_{32}h_{24}/h_{22} \ (= h''_{34});$

$\quad i = 4;$

$\quad\quad l_{42} = -h_{42}/h_{22}, \ h_{42} = 0;$

$\quad\quad j = 3;$

$\quad\quad\quad h_{43} = h_{43} + l_{42}h_{23} = h_{43} - h_{42}h_{23}/h_{22} \ (= h''_{43});$

$\quad\quad j = 4;$

$\quad\quad\quad h_{44} = h_{44} + l_{42}h_{24} = h_{43} - h_{42}h_{24}/h_{22} \ (= h''_{44});$

if $0 < h_{33} < h_{00}$, set $h_{00} = h_{33}.$

$k = 4;$

$\quad m = 3, l_{33} = 1;$

$\quad$if $h_{33} \leq 0$, set $h_{33} = h_{00};$

$\quad i = 4;$

$\quad\quad l_{34} = -h_{43}/h_{33}, \ h_{43} = 0;$

$\quad\quad j = 4;$

$\quad\quad\quad h_{44} = h_{44} + l_{44}h_{34} = h_{44} - h_{43}h_{34}/h_{33} = h''''_{44}.$

Step 3

$l_{44} = 1;$

if $h_{44} \leq 0$, set $h_{44} = h_{00};$

set $\hat{d}_{ii} = h_{ii}$ for $i = 1, 2, \ldots, n.$

Example 5.2 The gradient and Hessian are given by

$$\mathbf{g}_k^T = [-\tfrac{1}{5} \ 2], \quad \mathbf{H}_k = \begin{bmatrix} 3 & -6 \\ -6 & \frac{59}{5} \end{bmatrix}$$

Deduce a Newton direction $\mathbf{d}_k$.

Solution If

$$\mathbf{L} = \begin{bmatrix} 1 & 0 \\ 2 & 1 \end{bmatrix}$$

Eq. (5.15) gives

$$\mathbf{D} = \begin{bmatrix} 1 & 0 \\ 2 & 1 \end{bmatrix} \begin{bmatrix} 3 & -6 \\ -6 & \frac{59}{5} \end{bmatrix} \begin{bmatrix} 1 & 2 \\ 0 & 1 \end{bmatrix} = \begin{bmatrix} 3 & 0 \\ 0 & -\frac{1}{5} \end{bmatrix}$$

A positive definite diagonal matrix is

$$\hat{\mathbf{D}} = \begin{bmatrix} 3 & 0 \\ 0 & \frac{1}{5} \end{bmatrix}$$

Hence

$$\hat{\mathbf{D}}^{-1} = \begin{bmatrix} \frac{1}{3} & 0 \\ 0 & 5 \end{bmatrix}$$

From Eq. (5.20), we get

$$\mathbf{y}_k = - \begin{bmatrix} 1 & 0 \\ 2 & 1 \end{bmatrix} \begin{bmatrix} -\frac{1}{5} \\ 2 \end{bmatrix} = \begin{bmatrix} \frac{1}{5} \\ -\frac{8}{5} \end{bmatrix}$$

Therefore, from Eq. (5.21)

$$\mathbf{d}_k = \begin{bmatrix} 1 & 2 \\ 0 & 1 \end{bmatrix} \begin{bmatrix} \frac{1}{3} & 0 \\ 0 & 5 \end{bmatrix} \begin{bmatrix} \frac{1}{5} \\ -\frac{8}{5} \end{bmatrix} = \begin{bmatrix} -\frac{239}{15} \\ -8 \end{bmatrix}$$

■

5.3.2   Computation of the Hessian

The main disadvantage of the Newton method is that the second derivatives of the function are required so that the Hessian may be computed. If exact formulas are unavailable or are difficult to obtain, the second derivatives can be computed by using the numerical formulas

$$\frac{\partial f}{\partial x_1} = \lim_{\delta \to 0} \frac{f(\mathbf{x} + \boldsymbol{\delta}_1) - f(\mathbf{x})}{\delta} = f'(\mathbf{x}) \quad \text{with } \boldsymbol{\delta}_1 = [\delta \; 0 \; 0 \; \cdots \; 0]^T$$

$$\frac{\partial^2 f}{\partial x_1 \partial x_2} = \lim_{\delta \to 0} \frac{f'(\mathbf{x} + \boldsymbol{\delta}_2) - f'(\mathbf{x})}{\delta} \quad \text{with } \boldsymbol{\delta}_2 = [0 \; \delta \; 0 \; \cdots \; 0]^T$$

5.4 Gauss-Newton Method

In many optimization problems, the objective function is in the form of a vector of functions given by

$$\mathbf{f} = [f_1(\mathbf{x})\ f_2(\mathbf{x})\ \cdots\ f_m(\mathbf{x})]^T$$

where $f_p(\mathbf{x})$ for $p = 1, 2, \ldots, m$ are independent functions of $\mathbf{x}$ (see Sec. 1.2). The solution sought is a point $\mathbf{x}$ such that all $f_p(\mathbf{x})$ are reduced to zero simultaneously.

In problems of this type, a real-valued function can be formed as

$$F = \sum_{p=1}^{m} f_p(\mathbf{x})^2 = \mathbf{f}^T \mathbf{f} \tag{5.22}$$

If F is minimized by using a multidimensional unconstrained algorithm, then the individual functions $f_p(\mathbf{x})$ are minimized in the least-squares sense (see Sec. 1.2).

A method for the solution of the above class of problems, known as the *Gauss-Newton method*, can be readily developed by applying the Newton method of Sec. 5.3.

Since there are a number of functions $f_p(\mathbf{x})$ and each one depends on x_i for $i = 1, 2, \ldots, n$ a gradient matrix, referred to as the *Jacobian*, can be formed as

$$\mathbf{J} = \begin{bmatrix} \frac{\partial f_1}{\partial x_1} & \frac{\partial f_1}{\partial x_2} & \cdots & \frac{\partial f_1}{\partial x_n} \\ \frac{\partial f_2}{\partial x_1} & \frac{\partial f_2}{\partial x_2} & \cdots & \frac{\partial f_2}{\partial x_n} \\ \vdots & \vdots & & \vdots \\ \frac{\partial f_m}{\partial x_1} & \frac{\partial f_m}{\partial x_2} & \cdots & \frac{\partial f_m}{\partial x_n} \end{bmatrix}$$

The number of functions m may exceed the number of variables n, that is, the Jacobian need not be a square matrix.

By differentiating F in Eq. (5.22) with respect to x_i, we obtain

$$\frac{\partial F}{\partial x_i} = \sum_{p=1}^{m} 2 f_p(\mathbf{x}) \frac{\partial f_p}{\partial x_i} \tag{5.23}$$

for $i = 1, 2, \ldots, n$. Alternatively, in matrix form

$$\begin{bmatrix} \frac{\partial F}{\partial x_1} \\ \frac{\partial F}{\partial x_2} \\ \vdots \\ \frac{\partial F}{\partial x_n} \end{bmatrix} = 2 \begin{bmatrix} \frac{\partial f_1}{\partial x_1} & \frac{\partial f_2}{\partial x_1} & \cdots & \frac{\partial f_m}{\partial x_1} \\ \frac{\partial f_1}{\partial x_2} & \frac{\partial f_2}{\partial x_2} & \cdots & \frac{\partial f_m}{\partial x_2} \\ \vdots & \vdots & & \vdots \\ \frac{\partial f_1}{\partial x_n} & \frac{\partial f_2}{\partial x_n} & \cdots & \frac{\partial f_m}{\partial x_n} \end{bmatrix} \begin{bmatrix} f_1(\mathbf{x}) \\ f_2(\mathbf{x}) \\ \vdots \\ f_m(\mathbf{x}) \end{bmatrix}$$

Hence the gradient of F, designated by $\mathbf{g}_F$, can be expressed as

$$\mathbf{g}_F = 2\mathbf{J}^T\mathbf{f} \tag{5.24}$$

Assuming that $f_p(\mathbf{x}) \in C^2$, Eq. (5.23) yields

$$\frac{\partial^2 F}{\partial x_i \partial x_j} = 2\sum_{p=1}^{m} \frac{\partial f_p}{\partial x_i}\frac{\partial f_p}{\partial x_j} + 2\sum_{p=1}^{m} f_p(\mathbf{x})\frac{\partial^2 f_p}{\partial x_i \partial x_j}$$

for $i,\ j = 1, 2, \ldots, n$. If the second derivatives of $f_p(\mathbf{x})$ are neglected, we have

$$\frac{\partial^2 F}{\partial x_i \partial x_j} \approx 2\sum_{p=1}^{m} \frac{\partial f_p}{\partial x_i}\frac{\partial f_p}{\partial x_j}$$

Thus the Hessian of F, designated by $\mathbf{H}_F$, can be deduced as

$$\mathbf{H}_F \approx 2\mathbf{J}^T\mathbf{J} \tag{5.25}$$

Since the gradient and Hessian of F are now known, the Newton method can be applied for the solution of the problem. The necessary recursive relation is given by Eqs. (5.11) – (5.12) and (5.24) – (5.25) as

$$\begin{aligned}\mathbf{x}_{k+1} &= \mathbf{x}_k - \alpha_k(2\mathbf{J}^T\mathbf{J})^{-1}(2\mathbf{J}^T\mathbf{f}) \\ &= \mathbf{x}_k - \alpha_k(\mathbf{J}^T\mathbf{J})^{-1}(\mathbf{J}^T\mathbf{f})\end{aligned}$$

where α_k is the value of α that minimizes $F(\mathbf{x}_k + \alpha\mathbf{d}_k)$. As k is increased, successive line searches bring about reductions in F_k and $\mathbf{x}_k$ approaches $\mathbf{x}^*$. When $\mathbf{x}_k$ is in the neighborhood of $\mathbf{x}^*$, functions $f_p(\mathbf{x}_k)$ can be accurately represented by the linear approximation of the Taylor series, the matrix in Eq. (5.25) becomes an accurate representation of the Hessian of F_k, and the method converges very rapidly. If functions $f_p(\mathbf{x})$ are linear, F is quadratic, the matrix in Eq. (5.25) is the Hessian, and the problem is solved in one iteration.

The method breaks down if $\mathbf{H}_F$ becomes singular, as in the case of Newton method. However, the remedies described in Sec. 5.3 can also be applied to the Gauss-Newton method.

An algorithm based on the above principles is as follows:

Algorithm 5.5 Gauss-Newton algorithm
Step 1
Input $\mathbf{x}_0$ and initialize the tolerance ε.
Set $k = 0$.
Step 2
Compute $f_{pk} = f_p(\mathbf{x}_k)$ for $p = 1, 2, \ldots, m$ and F_k.
Step 3
Compute $\mathbf{J}_k$, $\mathbf{g}_k = 2\mathbf{J}_k^T \mathbf{f}_k$, and $\mathbf{H}_k = 2\mathbf{J}_k^T \mathbf{J}_k$.
Step 4
Compute $\mathbf{L}_k$ and $\hat{\mathbf{D}}_k$ using Algorithm 5.4.
Compute $\mathbf{y}_k = -\mathbf{L}_k \mathbf{g}_k$ and $\mathbf{d}_k = \mathbf{L}_k^T \hat{\mathbf{D}}_k^{-1} \mathbf{y}_k$.
Step 5
Find α_k, the value of α that minimizes $F(\mathbf{x}_k + \alpha \mathbf{d}_k)$.
Step 6
Set $\mathbf{x}_{k+1} = \mathbf{x}_k + \alpha_k \mathbf{d}_k$.
Compute $f_{p(k+1)}$ for $p = 1, 2, \ldots, m$ and F_{k+1}.
Step 7
If $|F_{k+1} - F_k| < \varepsilon$, then do:
    Output $\mathbf{x}^* = \mathbf{x}_{k+1}$, $f_{p(k+1)}(\mathbf{x}^*)$ for $p = 1, 2, \ldots, m$, and F_{k+1}.
    Stop.
Otherwise, set $k = k + 1$ and repeat from Step 3.

The factors 2 in Step 3 can be discarded since they cancel out in the calculation of $\mathbf{d}_k$ (see Eq. (5.12)). In Step 4, $\mathbf{H}_k$ is forced to become positive definite, if it is not positive definite, and, further, the Newton direction $\mathbf{d}_k$ is calculated without the direct inversion of $\mathbf{H}_k$.

References

1 D. G. Luenberger, *Linear and Nonlinear Programming*, Chap. 7, Addison-Wesley, MA, 1984.

2 S. M. Goldfeld, R. E. Quandt, and H. F. Trotter, "Maximization by quadratic hill-climbing," *Econometrica*, vol. 34, pp. 541–551, 1966.

3 P. B. Zwart, *Nonlinear Programming: A Quadratic Analysis of Ridge Paralysis*, Washington University, Report COO-1493-21, St. Louis, Mo., Jan. 1969.

4 A. Matthews and D. Davies, "A comparison of modified Newton methods for unconstrained optimization," *Computer Journal*, vol. 14, pp. 293–294, 1971.

Problems

5.1 The steepest-descent method is applied to solve the problem

$$\text{minimize } f(\mathbf{x}) = 2x_1^2 - 2x_1 x_2 + x_2^2 + 2x_1 - 2x_2$$

and a sequence $\{\mathbf{x}_k\}$ is generated.

(a) Assuming that

$$\mathbf{x}_{2k+1} = \left[0 \ \left(1 - \frac{1}{5^k}\right)\right]^T$$

show that

$$\mathbf{x}_{2k+3} = \left[0 \ \left(1 - \frac{1}{5^{k+1}}\right)\right]^T$$

(b) Find the minimizer of $f(\mathbf{x})$ using the result in part (a).

5.2 The problem

$$\text{minimize } f(\mathbf{x}) = x_1^2 + 2x_2^2 + 4x_1 + 4x_2$$

is to be solved by using the steepest-descent method with an initial point $\mathbf{x}_0 = [0 \ 0]^T$.

(a) By means of induction, show that

$$\mathbf{x}_{k+1} = \left[\frac{2}{3^k} - 2 \ \left(-\frac{1}{3}\right)^k - 1\right]^T$$

(b) Deduce the minimizer of $f(\mathbf{x})$.

5.3 Consider the minimization problem

$$\text{minimize } x_1^2 + x_2^2 - 0.2x_1x_2 - 2.2x_1 + 2.2x_2 + 2.2$$

(a) Find a point satisfying the first-order necessary conditions for a minimizer.

(b) Show that this point is the global minimizer.

(c) What is the rate of convergence of the steepest-descent method for this problem?

(d) Starting at $\mathbf{x} = [0 \ 0]^T$, how many steepest-descent iterations would it take (at most) to reduce the function value to 10^{-10}?

5.4 (a) Solve the problem

$$\text{minimize } f(\mathbf{x}) = 5x_1^2 - 9x_1x_2 + 4.075x_2^2 + x_1$$

by applying the steepest-descent method with $\mathbf{x}_0 = [1 \ 1]^T$ and $\varepsilon = 3 \times 10^{-6}$.

(b) Give a convergence analysis on the above problem to explain why the steepest-decent method requires a large number of iterations to reach the solution.

5.5 Solve the problem

$$\text{minimize } f(\mathbf{x}) = (x_1 + 5)^2 + (x_2 + 8)^2 + (x_3 + 7)^2$$
$$+ 2x_1^2x_2^2 + 4x_1^2x_3^2$$

by applying Algorithm 5.1.

(a) Start with $\mathbf{x}_0 = [1\ 1\ 1]^T$ and $\varepsilon = 10^{-6}$. Verify the solution point using the second-order sufficient conditions.

(b) Repeat (a) using $\mathbf{x}_0 = [-2.3\ 0\ 0]^T$.

(c) Repeat (a) using $\mathbf{x}_0 = [0\ 2\ -12]^T$.

5.6 Solve the problem in Prob. 5.5 by applying Algorithm 5.2. Try the same initial points as in Prob. 5.5 (a)–(c). Compare the solutions obtained and the amount of computation required with that of Algorithm 5.1.

5.7 Solve the problem

$$\text{minimize } f(\mathbf{x}) = (x_1^2 + x_2^2 - 1)^2 + (x_1 + x_2 - 1)^2$$

by applying Algorithm 5.1. Use $\varepsilon = 10^{-6}$ and try the following initial points: $[4\ 4]^T$, $[4\ -4]^T$, $[-4\ 4]^T$, $[-4\ -4]^T$. Examine the solution points obtained.

5.8 Solve the problem in Prob. 5.7 by applying Algorithm 5.2. Compare the computational efficiency of Algorithm 5.2 with that of Algorithm 5.1.

5.9 Solve the problem

$$\text{minimize } f(\mathbf{x}) = -x_2^2 e^{1-x_1^2-20(x_1-x_2)^2}$$

by applying Algorithm 5.1.

(a) Start with $\mathbf{x}_0 = [0.1\ 0.1]^T$ and $\varepsilon = 10^{-6}$. Examine the solution obtained.

(b) Start with $\mathbf{x}_0 = [0.8\ 0.1]^T$ and $\varepsilon = 10^{-6}$. Examine the solution obtained.

(c) Start with $\mathbf{x}_0 = [1.1\ 0.1]^T$ and $\varepsilon = 10^{-6}$. Examine the solution obtained.

5.10 Solve the problem in Prob. 5.9 by applying Algorithm 5.2. Try the 3 initial points specified in Prob. 5.9 (a)–(c) and examine the solutions obtained.

5.11 Solve the problem

$$\text{minimize } f(\mathbf{x}) = x_1^3 e^{x_2-x_1^2-10(x_1-x_2)^2}$$

by applying Algorithm 5.1. Use $\varepsilon = 10^{-6}$ and try the following initial points: $[-3\ -3]^T$, $[3\ 3]^T$, $[3\ -3]^T$, and $[-3\ 3]^T$. Examine the solution points obtained.

5.12 Solve Prob. 5.11 by applying Algorithm 5.2. Examine the solution points obtained.

5.13 Solve the minimization problem in Prob. 5.1 with $\mathbf{x}_0 = [0\ 0]^T$ by using Newton method.

5.14 Solve the minimization problem in Prob. 5.2 with $\mathbf{x}_0 = [0\ 0]^T$ by using Newton method.

5.15 Modify the Newton algorithm described in Algorithm 5.3 by incorporating Eq. (5.13) into the algorithm. Give a step-by-step description of the modified algorithm.

5.16 Solve Prob. 5.5 by applying the algorithm in Prob. 5.15. Examine the solution points obtained and compare the algorithm's computational complexity with that of Algorithm 5.1.

5.17 Solve Prob. 5.7 by applying the algorithm in Prob. 5.15. Examine the solution points obtained and compare the amount of computation required with that of Algorithm 5.1.

5.18 Solve Prob. 5.9 by applying the algorithm in Prob. 5.15. Examine the solutions obtained and compare the algorithm's computational complexity with that of Algorithm 5.1.

5.19 Solve Prob. 5.11 by applying the algorithm in Prob. 5.15. Examine the solutions obtained and compare the amount of computation required with that of Algorithm 5.1.

5.20 (a) Find the global minimizer of the objective function

$$f(\mathbf{x}) = (x_1 + 10x_2)^2 + 5(x_3 - x_4)^2 + (x_2 - 2x_3)^4 \\ + 100(x_1 - x_4)^4$$

by using the fact that each term in the objective function is nonnegative.

(b) Solve the problem in part (a) using the steepest-descent method with $\varepsilon = 10^{-6}$ and try the initial points $[-2\ -1\ 1\ 2]^T$ and $[200\ -200\ 100\ -100]^T$.

(c) Solve the problem in part (a) using the modified Newton method in Prob. 5.15 with the same termination tolerance and initial points as in (b).

(d) Solve the problem in part (a) using the Gauss-Newton method with the same termination tolerance and initial points as in (b).

(e) Based on the results of (b)–(d), compare the computational efficiency and solution accuracy of the three methods.

5.21 Solve Prob. 5.5 by applying the Gauss-Newton method. Examine the solutions obtained and compare the results with those obtained first by using Algorithm 5.1 and then by using the algorithm in Prob. 5.15.

5.22 Solve Prob. 5.7 by applying the Gauss-Newton method. Examine the solutions obtained and compare the results with those obtained first by using Algorithm 5.1 and then by using the algorithm in Prob. 5.15.

Chapter 6

CONJUGATE-DIRECTION METHODS

6.1   Introduction

In the multidimensional optimization methods described so far, the direction of search in each iteration depends on the local properties of the objective function. Although a relation may exist between successive search directions, such a relation is incidental. In this chapter, methods are described in which the optimization is performed by using sequential search directions that bear a strict mathematical relationship to one another. An important class of methods of this type is a class based on a set of search directions known as *conjugate directions*.

Like the Newton method, conjugate-direction methods are developed for the quadratic optimization problem and are then extended to the general optimization problem. For a quadratic problem, convergence is achieved in a finite number of iterations.

Conjugate-direction methods have been found to be very effective in many types of problems and have been used extensively in the past. The four most important methods of this class are as follows:

1. Conjugate-gradient method
2. Fletcher-Reeves method
3. Powell's method
4. Partan method

The principles involved and specific algorithms based on these methods form the subject matter of this chapter.

6.2 Conjugate Directions

If $f(\mathbf{x}) \in C^1$ where $\mathbf{x} = [x_1 \; x_2 \; \cdots \; x_n]^T$, the problem

$$\min_{\mathbf{x}} \; F = f(\mathbf{x})$$

can be solved by using the following algorithm:

Algorithm 6.1 Coordinate-descent algorithm
Step 1
Input $\mathbf{x}_1$ and initialize the tolerance ε.
Set $k = 1$.
Step 2
Set $\mathbf{d}_k = [0 \; 0 \; \cdots \; 0 \; d_k \; 0 \; \cdots \; 0]^T$.
Step 3
Find α_k, the value of α that minimizes $f(\mathbf{x}_k + \alpha \mathbf{d}_k)$, using a line search.
Set $\mathbf{x}_{k+1} = \mathbf{x}_k + \alpha_k \mathbf{d}_k$
Calculate $f_{k+1} = f(\mathbf{x}_{k+1})$.
Step 4
If $\|\alpha_k \mathbf{d}_k\| < \varepsilon$ then output $\mathbf{x}^* = \mathbf{x}_{k+1}$ and $f(\mathbf{x}^*) = f_{k+1}$, and stop.
Step 5
If $k = n$, set $\mathbf{x}_1 = \mathbf{x}_{k+1}$, $k = 1$ and repeat from Step 2;
Otherwise, set $k = k + 1$ and repeat from Step 2.

In this algorithm, an initial point $\mathbf{x}_1$ is assumed, and $f(\mathbf{x})$ is minimized in direction $\mathbf{d}_1$ to obtain a new point $\mathbf{x}_2$. The procedure is repeated for points $\mathbf{x}_2, \mathbf{x}_3, \ldots$ and when $k = n$, the algorithm is reinitialized and the procedure is repeated until convergence is achieved. Evidently, this algorithm differs from those in Chap. 5 in that $f(\mathbf{x})$ is minimized repeatedly using a set of directions which bear a strict relationship to one another. The relationship among the various directions is that they form a set of *coordinate directions* since only one element of $\mathbf{x}_k$ is allowed to vary in each line search.

Algorithm 6.1, which is often referred to as a coordinate-descent algorithm, is not very effective or reliable in practice, since an oscillatory behavior can sometimes occur. However, by using another class of interrelated directions known as *conjugate directions*, some quite effective algorithms can be developed.

Definition 6.1

(a) Two distinct nonzero vectors $\mathbf{d}_1$ and $\mathbf{d}_2$ are said to be conjugate with respect to a real symmetric matrix $\mathbf{H}$, if

$$\mathbf{d}_1^T \mathbf{H} \mathbf{d}_2 = 0$$

(b) A finite set of distinct nonzero vectors $\{\mathbf{d}_0, \mathbf{d}_1, \ldots, \mathbf{d}_k\}$ is said to be *conjugate* with respect to a real symmetric matrix $\mathbf{H}$, if

$$\mathbf{d}_i^T \mathbf{H} \mathbf{d}_j = 0 \qquad \text{for all } i \neq j \qquad (6.1)$$

∎

If $\mathbf{H} = \mathbf{I}_n$, where $\mathbf{I}_n$ is the $n \times n$ identity matrix, then Eq. (6.1) can be expressed as

$$\mathbf{d}_i^T \mathbf{H} \mathbf{d}_j = \mathbf{d}_i^T \mathbf{I}_n \mathbf{d}_j = \mathbf{d}_i^T \mathbf{d}_j = 0 \qquad \text{for } i \neq j$$

This is the well known condition for *orthogonality* between vectors $\mathbf{d}_i$ and $\mathbf{d}_j$ and, in effect, conjugacy is a generalization of orthogonality.

If $\mathbf{d}_j$ for $j = 0, 1, \ldots, k$ are *eigenvectors* of $\mathbf{H}$ then

$$\mathbf{H} \mathbf{d}_j = \lambda_j \mathbf{d}_j$$

where the λ_j are the eigenvalues of $\mathbf{H}$. Hence, we have

$$\mathbf{d}_i^T \mathbf{H} \mathbf{d}_j = \lambda_j \mathbf{d}_i^T \mathbf{d}_j = 0 \qquad \text{for } i \neq j$$

since $\mathbf{d}_i$ and $\mathbf{d}_j$ for $i \neq j$ are orthogonal [5]. In effect, the set of eigenvectors $\mathbf{d}_j$ constitutes a set of conjugate directions with respect to $\mathbf{H}$.

Theorem 6.1 *Linear independence of conjugate vectors If nonzero vectors* $\mathbf{d}_0, \mathbf{d}_1, \ldots, \mathbf{d}_k$ *form a conjugate set with respect to a positive definite matrix* $\mathbf{H}$, *then they are linearly independent.*

Proof Consider the system

$$\sum_{j=0}^{k} \alpha_j \mathbf{d}_j = \mathbf{0}$$

On premultiplying by $\mathbf{d}_i^T \mathbf{H}$, where $0 \leq i \leq k$, and then using Def. 6.1, we obtain

$$\sum_{j=0}^{k} \alpha_j \mathbf{d}_i^T \mathbf{H} \mathbf{d}_j = \alpha_i \mathbf{d}_i^T \mathbf{H} \mathbf{d}_i = 0$$

Since $\mathbf{H}$ is positive definite, we have $\mathbf{d}_i^T \mathbf{H} \mathbf{d}_i > 0$. Therefore, the above system has a solution if and only if $\alpha_j = 0$ for $j = 0, 1, \ldots, k$, that is, vectors $\mathbf{d}_i$ are linearly independent.

∎

The use of conjugate directions in the process of optimization can be demonstrated by considering the quadratic problem

$$\underset{x}{\text{minimize}} \ f(\mathbf{x}) = a + \mathbf{x}^T \mathbf{b} + \tfrac{1}{2} \mathbf{x}^T \mathbf{H} \mathbf{x} \qquad (6.2)$$

where $a = f(0)$, $\mathbf{b}$ is the gradient of $f(\mathbf{x})$ at $\mathbf{x} = \mathbf{0}$, and $\mathbf{H}$ is the Hessian. The gradient of $f(\mathbf{x})$ at any point can be deduced as

$$\mathbf{g} = \mathbf{b} + \mathbf{H}\mathbf{x}$$

At the minimizer $\mathbf{x}^*$ of $f(\mathbf{x})$, $\mathbf{g} = \mathbf{0}$ and thus

$$\mathbf{H}\mathbf{x}^* = -\mathbf{b} \tag{6.3}$$

If $\mathbf{d}_0$, $\mathbf{d}_1$, $\ldots$, $\mathbf{d}_{n-1}$ are distinct conjugate directions in E^n, then they form a basis of E^n since they are linearly independent and span the E^n space. This means that all possible vectors in E^n can be expressed as linear combinations of directions $\mathbf{d}_0$, $\mathbf{d}_1$, $\ldots$, $\mathbf{d}_{n-1}$. Hence $\mathbf{x}^*$ can be expressed as

$$\mathbf{x}^* = \sum_{i=0}^{n-1} \alpha_i \mathbf{d}_i \tag{6.4}$$

where α_i for $i = 0, 1, \ldots, n-1$ are constants. If $\mathbf{H}$ is positive definite, then from Def. 6.1 we can write

$$\mathbf{d}_k^T \mathbf{H} \mathbf{x}^* = \sum_{i=0}^{n-1} \alpha_i \mathbf{d}_k^T \mathbf{H} \mathbf{d}_i = \alpha_k \mathbf{d}_k^T \mathbf{H} \mathbf{d}_k$$

and thus

$$\alpha_k = \frac{\mathbf{d}_k^T \mathbf{H} \mathbf{x}^*}{\mathbf{d}_k^T \mathbf{H} \mathbf{d}_k} \tag{6.5}$$

Now from Eq. (6.3)

$$\alpha_k = -\frac{\mathbf{d}_k^T \mathbf{b}}{\mathbf{d}_k^T \mathbf{H} \mathbf{d}_k} = -\frac{\mathbf{b}^T \mathbf{d}_k}{\mathbf{d}_k^T \mathbf{H} \mathbf{d}_k}$$

Therefore, Eq. (6.4) gives the minimizer as

$$\mathbf{x}^* = -\sum_{k=0}^{n-1} \frac{\mathbf{d}_k^T \mathbf{b}}{\mathbf{d}_k^T \mathbf{H} \mathbf{d}_k} \mathbf{d}_k \tag{6.6}$$

In effect, if n conjugate directions are known, an explicit expression for $\mathbf{x}^*$ can be obtained.

The significance of conjugate directions can be demonstrated by attempting to obtain $\mathbf{x}^*$ using a set of n nonzero orthogonal directions $\mathbf{p}_0$, $\mathbf{p}_1$, $\ldots$, $\mathbf{p}_{n-1}$. Proceeding as above, we can show that

$$\mathbf{x}^* = \sum_{k=0}^{n-1} \frac{\mathbf{p}_k^T \mathbf{x}^*}{\|\mathbf{p}_k\|^2} \mathbf{p}_k$$

Evidently, in this case, $\mathbf{x}^*$ depends on itself and, in effect, there is a distinct advantage in using conjugate directions.

6.3   Basic Conjugate-Directions Method

The computation of $\mathbf{x}^*$ through the use of Eq. (6.6) can be regarded as an iterative computation whereby n successive adjustments $\alpha_k \mathbf{d}_k$ are made to an initial point $\mathbf{x}_0 = \mathbf{0}$. Alternatively, the sequence generated by the recursive relation

$$\mathbf{x}_{k+1} = \mathbf{x}_k + \alpha \mathbf{d}_k$$

where

$$\alpha_k = -\frac{\mathbf{b}^T \mathbf{d}_k}{\mathbf{d}_k^T \mathbf{H} \mathbf{d}_k}$$

and $\mathbf{x}_0 = \mathbf{0}$ converges when $k = n - 1$ to

$$\mathbf{x}_n = \mathbf{x}^*$$

A similar result can be obtained for an arbitrary initial point $\mathbf{x}_0$ as is demonstrated by the following theorem.

Theorem 6.2 *Convergence of conjugate-directions method If* $\{\mathbf{d}_0, \mathbf{d}_1, \ldots, \mathbf{d}_{n-1}\}$ *is a set of nonzero conjugate directions,* $\mathbf{H}$ *is an* $n \times n$ *positive definite matrix, and the problem*

$$\underset{\mathbf{x}}{minimize} \ f(\mathbf{x}) = a + \mathbf{x}^T \mathbf{b} + \tfrac{1}{2} \mathbf{x}^T \mathbf{H} \mathbf{x} \tag{6.7}$$

is quadratic, then for any initial point $\mathbf{x}_0$ *the sequence generated by the relation*

$$\mathbf{x}_{k+1} = \mathbf{x}_k + \alpha_k \mathbf{d}_k \qquad for \ \ k \ge 0 \tag{6.8}$$

where

$$\alpha_k = -\frac{\mathbf{g}_k^T \mathbf{d}_k}{\mathbf{d}_k^T \mathbf{H} \mathbf{d}_k}$$

and

$$\mathbf{g}_k = \mathbf{b} + \mathbf{H} \mathbf{x}_k \tag{6.9}$$

converges to the unique solution $\mathbf{x}^*$ *of the quadratic problem in* n *iterations, i.e.,* $\mathbf{x}_n = \mathbf{x}^*$.

Proof Vector $\mathbf{x}^* - \mathbf{x}_0$ can be expressed as a linear combination of conjugate directions as

$$\mathbf{x}^* - \mathbf{x}_0 = \sum_{i=0}^{n-1} \alpha_i \mathbf{d}_i \tag{6.10}$$

Hence as in Eq. (6.5)

$$\alpha_k = \frac{\mathbf{d}_k^T \mathbf{H}(\mathbf{x}^* - \mathbf{x}_0)}{\mathbf{d}_k^T \mathbf{H} \mathbf{d}_k} \tag{6.11}$$

150

The iterative procedure in Eq. (6.8) will yield

$$\mathbf{x}_k - \mathbf{x}_0 = \sum_{i=0}^{k-1} \alpha_i \mathbf{d}_i \tag{6.12}$$

and so

$$\mathbf{d}_k^T \mathbf{H}(\mathbf{x}_k - \mathbf{x}_0) = \sum_{i=0}^{k-1} \alpha_i \mathbf{d}_k^T \mathbf{H} \mathbf{d}_i = 0$$

since $i \neq k$. Evidently,

$$\mathbf{d}_k^T \mathbf{H} \mathbf{x}_k = \mathbf{d}_k^T \mathbf{H} \mathbf{x}_0 \tag{6.13}$$

and thus Eqs. (6.11) and (6.13) give

$$\alpha_k = \frac{\mathbf{d}_k^T (\mathbf{H} \mathbf{x}^* - \mathbf{H} \mathbf{x}_k)}{\mathbf{d}_k^T \mathbf{H} \mathbf{d}_k} \tag{6.14}$$

From Eq. (6.9)

$$\mathbf{H} \mathbf{x}_k = \mathbf{g}_k - \mathbf{b} \tag{6.15}$$

and since $\mathbf{g}_k = \mathbf{0}$ at minimizer $\mathbf{x}_k$, we have

$$\mathbf{H} \mathbf{x}^* = -\mathbf{b} \tag{6.16}$$

Therefore, Eqs. (6.14) – (6.16) yield

$$\alpha_k = -\frac{\mathbf{d}_k^T \mathbf{g}_k}{\mathbf{d}_k^T \mathbf{H} \mathbf{d}_k} = -\frac{\mathbf{g}_k^T \mathbf{d}_k}{\mathbf{d}_k^T \mathbf{H} \mathbf{d}_k} \tag{6.17}$$

Now for $k = n$ Eqs. (6.12) and (6.10) yield

$$\mathbf{x}_n = \mathbf{x}_0 + \sum_{i=0}^{n-1} \alpha_i \mathbf{d}_i = \mathbf{x}^*$$

and, therefore, the iterative relation in Eq. (6.8) converges to $\mathbf{x}^*$ in n iterations. ∎

By using Theorem 6.2 in conjunction with various techniques for the generation of conjugate directions, a number of distinct conjugate-direction methods can be developed.

Methods based on Theorem 6.2 have certain common properties. Two of these properties are given in the following theorem.

Theorem 6.3 *Orthogonality of gradient to a set of conjugate directions*
(a) The gradient $\mathbf{g}_k$ is orthogonal to directions $\mathbf{d}_i$ for $0 \leq i < k$, that is,

$$\mathbf{g}_k^T \mathbf{d}_i = \mathbf{d}_i^T \mathbf{g}_k = 0 \qquad for \;\; 0 \leq i < k$$

(b) The assignment $\alpha = \alpha_k$ in Theorem 6.2 minimizes $f(\mathbf{x})$ on each line

$$\mathbf{x} = \mathbf{x}_{k-1} + \alpha \mathbf{d}_i \qquad for\ 0 \leq i < k$$

Proof

(a) We assume that

$$\mathbf{g}_k^T \mathbf{d}_i = 0 \qquad for\ \ 0 \leq i < k \tag{6.18}$$

and show that

$$\mathbf{g}_{k+1}^T \mathbf{d}_i = 0 \qquad for\ \ 0 \leq i < k+1$$

From Eq. (6.9)

$$\mathbf{g}_{k+1} - \mathbf{g}_k = \mathbf{H}(\mathbf{x}_{k+1} - \mathbf{x}_k)$$

and from Eq. (6.8)

$$\mathbf{g}_{k+1} = \mathbf{g}_k + \alpha_k \mathbf{H} \mathbf{d}_k \tag{6.19}$$

Hence

$$\mathbf{g}_{k+1}^T \mathbf{d}_i = \mathbf{g}_k^T \mathbf{d}_i + \alpha_k \mathbf{d}_k^T \mathbf{H} \mathbf{d}_i \tag{6.20}$$

For $i = k$, Eqs. (6.20) and (6.17) give

$$\mathbf{g}_{k+1}^T \mathbf{d}_k = \mathbf{g}_k^T \mathbf{d}_k + \alpha_k \mathbf{d}_k^T \mathbf{H} \mathbf{d}_k = 0 \tag{6.21}$$

For $0 \leq i < k$, Eq. (6.18) gives

$$\mathbf{g}_k^T \mathbf{d}_i = 0$$

and since $\mathbf{d}_i$ and $\mathbf{d}_k$ are conjugate

$$\mathbf{d}_k^T \mathbf{H} \mathbf{d}_i = 0$$

Hence Eq. (6.20) gives

$$\mathbf{g}_{k+1}^T \mathbf{d}_i = 0 \qquad for\ \ 0 \leq i < k \tag{6.22}$$

By combining Eqs. (6.21) and (6.22), we have

$$\mathbf{g}_{k+1}^T \mathbf{d}_i = 0 \qquad for\ \ 0 \leq i < k+1 \tag{6.23}$$

Now if $k = 0$, Eq. (6.23) gives $\mathbf{g}_1^T \mathbf{d}_i = 0$ for $0 \leq i < 1$ and from Eqs. (6.18) and (6.23), we obtain

$$\mathbf{g}_2^T \mathbf{d}_i = 0 \qquad for\ \ 0 \leq i < 2$$
$$\mathbf{g}_3^T \mathbf{d}_i = 0 \qquad for\ \ 0 \leq i < 3$$
$$\vdots \qquad\qquad\qquad \vdots$$
$$\mathbf{g}_k^T \mathbf{d}_i = 0 \qquad for\ \ 0 \leq i < k$$

152

(b) Since

$$\mathbf{g}_k^T\mathbf{d}_i \equiv \mathbf{g}^T(\mathbf{x}_k)\mathbf{d}_i = \mathbf{g}(\mathbf{x}_{k-1}+\alpha\mathbf{d}_i)^T\mathbf{d}_i$$
$$= \frac{df(\mathbf{x}_{k-1}+\alpha\mathbf{d}_i)}{d\alpha} = 0$$

$f(\mathbf{x})$ is minimized on each line

$$\mathbf{x} = \mathbf{x}_{k-1} + \alpha\mathbf{d}_i \qquad \text{for } 0 \le i < k$$

∎

The implication of the second part of the above theorem is that $\mathbf{x}_k$ minimizes $f(\mathbf{x})$ with respect to the subspace spanned by the set of vectors $\{\mathbf{d}_0, \mathbf{d}_1, \dots, \mathbf{d}_{k-1}\}$. Therefore, $\mathbf{x}_n$ minimizes $f(\mathbf{x})$ with respect to the space spanned by the set of vectors $\{\mathbf{d}_0, \mathbf{d}_1, \dots, \mathbf{d}_{n-1}\}$, namely, E^n. This is another way of stating that $\mathbf{x}_n = \mathbf{x}^*$.

6.4 Conjugate-Gradient Method

An effective method for the generation of conjugate directions proposed by Hestenes and Stiefel [1] is the so-called conjugate-gradient method. In this method, directions are generated sequentially, one per iteration. For iteration $k+1$, a new point $\mathbf{x}_{k+1}$ is generated by using the previous direction $\mathbf{d}_k$. Then a new direction $\mathbf{d}_{k+1}$ is generated by adding a vector $\beta_k\mathbf{d}_k$ to $-\mathbf{g}_{k+1}$, the negative of the gradient at the new point.

The conjugate-gradient method is based on the following theorem. This is essentially the same as Theorem 6.2 except that the method of generating conjugate directions is now defined.

Theorem 6.4 *Convergence of conjugate-gradient method*

(a) *If $\mathbf{H}$ is a positive definite matrix, then for any initial point $\mathbf{x}_0$ and an initial direction*

$$\mathbf{d}_0 = -\mathbf{g}_0 = -(\mathbf{b}+\mathbf{H}\mathbf{x}_0)$$

the sequence generated by the recursive relation

$$\mathbf{x}_{k+1} = \mathbf{x}_k + \alpha_k\mathbf{d}_k \tag{6.24}$$

where

$$\alpha_k = -\frac{\mathbf{g}_k^T\mathbf{d}_k}{\mathbf{d}_k^T\mathbf{H}\mathbf{d}_k} \tag{6.25}$$

$$\mathbf{g}_k = \mathbf{b}+\mathbf{H}\mathbf{x}_k \tag{6.26}$$

$$\mathbf{d}_{k+1} = -\mathbf{g}_{k+1} + \beta_k\mathbf{d}_k \tag{6.27}$$

$$\beta_k = \frac{\mathbf{g}_{k+1}^T\mathbf{H}\mathbf{d}_k}{\mathbf{d}_k^T\mathbf{H}\mathbf{d}_k} \tag{6.28}$$

converges to the unique solution $\mathbf{x}^*$ of the problem given in Eq. (6.2).

(b) *The gradient $\mathbf{g}_k$ is orthogonal to $\{\mathbf{g}_0, \mathbf{g}_1, \ldots, \mathbf{g}_{k-1}\}$, i.e.,*

$$\mathbf{g}_k^T \mathbf{g}_i = 0 \qquad \text{for } 0 \le i < k$$

Proof

(a) The proof of convergence is the same as in Theorem 6.2. What remains to prove is that directions $\mathbf{d}_0, \mathbf{d}_1, \ldots, \mathbf{d}_{n-1}$ form a conjugate set, that is,

$$\mathbf{d}_k^T \mathbf{H} \mathbf{d}_i = 0 \qquad \text{for } 0 \le i < k \text{ and } 1 \le k \le n$$

The proof is by induction. We assume that

$$\mathbf{d}_k^T \mathbf{H} \mathbf{d}_i = 0 \qquad \text{for } 0 \le i < k \tag{6.29}$$

and show that

$$\mathbf{d}_{k+1}^T \mathbf{H} \mathbf{d}_i = 0 \qquad \text{for } 0 \le i < k + 1$$

Let $S(\mathbf{v}_0, \mathbf{v}_1, \ldots, \mathbf{v}_k)$ be the subspace spanned by vectors $\mathbf{v}_0, \mathbf{v}_1, \ldots, \mathbf{v}_k$. From Eq. (6.19)

$$\mathbf{g}_{k+1} = \mathbf{g}_k + \alpha_k \mathbf{H} \mathbf{d}_k \tag{6.30}$$

and hence for $k = 0$, we have

$$\mathbf{g}_1 = \mathbf{g}_0 + \alpha_0 \mathbf{H} \mathbf{d}_0 = \mathbf{g}_0 - \alpha_0 \mathbf{H} \mathbf{g}_0$$

since $\mathbf{d}_0 = -\mathbf{g}_0$. In addition, Eq. (6.27) yields

$$\mathbf{d}_1 = -\mathbf{g}_1 + \beta_0 \mathbf{d}_0 = -(1 + \beta_0)\mathbf{g}_0 + \alpha_0 \mathbf{H} \mathbf{g}_0$$

that is, $\mathbf{g}_1$ and $\mathbf{d}_1$ are linear combinations of $\mathbf{g}_0$ and $\mathbf{H}\mathbf{g}_0$, and so

$$S(\mathbf{g}_0, \mathbf{g}_1) = S(\mathbf{d}_0, \mathbf{d}_1) = S(\mathbf{g}_0, \mathbf{H}\mathbf{g}_0)$$

Similarly, for $k = 2$, we get

$$\mathbf{g}_2 = \mathbf{g}_0 - [\alpha_0 + \alpha_1(1 + \beta_0)]\mathbf{H}\mathbf{g}_0 + \alpha_0\alpha_1\mathbf{H}^2\mathbf{g}_0$$
$$\mathbf{d}_2 = -[1 + (1 + \beta_0)\beta_1]\mathbf{g}_0 + [\alpha_0 + \alpha_1(1 + \beta_0) + \alpha_0\beta_1]\mathbf{H}\mathbf{g}_0$$
$$-\alpha_0\alpha_1\mathbf{H}^2\mathbf{g}_0$$

and hence

$$S(\mathbf{g}_0, \mathbf{g}_1, \mathbf{g}_2) = S(\mathbf{g}_0, \mathbf{H}\mathbf{g}_0, \mathbf{H}^2\mathbf{g}_0)$$
$$S(\mathbf{d}_0, \mathbf{d}_1, \mathbf{d}_2) = S(\mathbf{g}_0, \mathbf{H}\mathbf{g}_0, \mathbf{H}^2\mathbf{g}_0)$$

By continuing the induction, we can show that

$$S(\mathbf{g}_0, \mathbf{g}_1, \ldots, \mathbf{g}_k) = S(\mathbf{g}_0, \mathbf{H}\mathbf{g}_0, \ldots, \mathbf{H}^k\mathbf{g}_0) \qquad (6.31)$$
$$S(\mathbf{d}_0, \mathbf{d}_1, \ldots, \mathbf{d}_k) = S(\mathbf{g}_0, \mathbf{H}\mathbf{g}_0, \ldots, \mathbf{H}^k\mathbf{g}_0) \qquad (6.32)$$

Now from Eq. (6.27)

$$\mathbf{d}_{k+1}^T \mathbf{H}\mathbf{d}_i = -\mathbf{g}_{k+1}^T \mathbf{H}\mathbf{d}_i + \beta_k \mathbf{d}_k^T \mathbf{H}\mathbf{d}_i \qquad (6.33)$$

For $i = k$, the use of Eq. (6.28) gives

$$\mathbf{d}_{k+1}^T \mathbf{H}\mathbf{d}_k = -\mathbf{g}_{k+1}^T \mathbf{H}\mathbf{d}_k + \beta_k \mathbf{d}_k^T \mathbf{H}\mathbf{d}_k = 0 \qquad (6.34)$$

For $i < k$, Eq. (6.32) shows that

$$\mathbf{H}\mathbf{d}_i \in S(\mathbf{d}_0, \mathbf{d}_1, \ldots, \mathbf{d}_k)$$

and thus $\mathbf{H}\mathbf{d}_i$ can be represented by the linear combination

$$\mathbf{H}\mathbf{d}_i = \sum_{i=0}^{k} a_i \mathbf{d}_i \qquad (6.35)$$

where a_i for $i = 0, 1, \ldots, k$ are constants. Now from Eqs. (6.33) and (6.35)

$$\mathbf{d}_{k+1}^T \mathbf{H}\mathbf{d}_i = -\sum_{i=0}^{k} a_i \mathbf{g}_{k+1}^T \mathbf{d}_i + \beta_k \mathbf{d}_k^T \mathbf{H}\mathbf{d}_i$$
$$= 0 \qquad \text{for} \;\; i < k \qquad (6.36)$$

The first term is zero from the orthogonality property of Theorem 6.3(a) whereas the second term is zero from the assumption in Eq. (6.29). By combining Eqs. (6.34) and (6.36), we have

$$\mathbf{d}_{k+1}^T \mathbf{H}\mathbf{d}_i = 0 \qquad \text{for} \;\; 0 \le i < k + 1 \qquad (6.37)$$

For $k = 0$, Eq. (6.37) gives

$$\mathbf{d}_1^T \mathbf{H}\mathbf{d}_i = 0 \qquad \text{for} \;\; 0 \le i < 1$$

and, therefore, from Eqs. (6.29) and (6.37), we have

$$\mathbf{d}_2^T \mathbf{H}\mathbf{d}_i = 0 \qquad \text{for} \;\; 0 \le i < 2$$
$$\mathbf{d}_3^T \mathbf{H}\mathbf{d}_i = 0 \qquad \text{for} \;\; 0 \le i < 3$$
$$\vdots \qquad\qquad\qquad \vdots$$
$$\mathbf{d}_k^T \mathbf{H}\mathbf{d}_i = 0 \qquad \text{for} \;\; 0 \le i < k$$

(*b*) From Eqs. (6.31) – (6.32), $\mathbf{g}_0$, $\mathbf{g}_1$, ..., $\mathbf{g}_k$ span the same subspace as $\mathbf{d}_0$, $\mathbf{d}_1$, ..., $\mathbf{d}_k$ and, consequently, they are linearly independent. We can write

$$\mathbf{g}_i = \sum_{j=0}^{i} a_j \mathbf{d}_j$$

where a_j for $j = 0, 1, \ldots, i$ are constants. Therefore, from Theorem 6.3

$$\mathbf{g}_k^T \mathbf{g}_i = \sum_{j=0}^{i} a_j \mathbf{g}_k^T \mathbf{d}_j = 0 \qquad \text{for} \quad 0 \le i < k$$

∎

The expressions for α_k and β_k in the above theorem can be simplified somewhat. From Eq. (6.27)

$$-\mathbf{g}_k^T \mathbf{d}_k = \mathbf{g}_k^T \mathbf{g}_k - \beta_{k-1} \mathbf{g}_k^T \mathbf{d}_{k-1}$$

where

$$\mathbf{g}_k^T \mathbf{d}_{k-1} = 0$$

according to Theorem 6.3(*a*). Hence

$$-\mathbf{g}_k^T \mathbf{d}_k = \mathbf{g}_k^T \mathbf{g}_k$$

and, therefore, the expression for α_k in Eq. (6.25) is modified as

$$\alpha_k = \frac{\mathbf{g}_k^T \mathbf{g}_k}{\mathbf{d}_k^T \mathbf{H} \mathbf{d}_k} \tag{6.38}$$

On the other hand, from Eq. (6.19)

$$\mathbf{H}\mathbf{d}_k = \frac{1}{\alpha_k}(\mathbf{g}_{k+1} - \mathbf{g}_k)$$

and so

$$\mathbf{g}_{k+1}^T \mathbf{H} \mathbf{d}_k = \frac{1}{\alpha_k}(\mathbf{g}_{k+1}^T \mathbf{g}_{k+1} - \mathbf{g}_{k+1}^T \mathbf{g}_k) \tag{6.39}$$

Now from Eqs. (6.31) and (6.32)

$$\mathbf{g}_k \in S(\mathbf{d}_0, \mathbf{d}_1, \ldots, \mathbf{d}_k)$$

or

$$\mathbf{g}_k = \sum_{i=0}^{k} a_i \mathbf{d}_i$$

and as a result

$$\mathbf{g}_{k+1}^T \mathbf{g}_k = \sum_{i=0}^{k} a_i \mathbf{g}_{k+1}^T \mathbf{d}_i = 0 \tag{6.40}$$

by virtue of Theorem 6.3(a). Therefore, Eqs. (6.28) and (6.38) – (6.40) yield

$$\beta_k = \frac{\mathbf{g}_{k+1}^T \mathbf{g}_{k+1}}{\mathbf{g}_k^T \mathbf{g}_k}$$

The above principles and theorems lead to the following algorithm:

Algorithm 6.2 Conjugate-gradient algorithm
Step 1
Input $\mathbf{x}_0$ and initialize the tolerance ε.
Step 2
Compute $\mathbf{g}_0$ and set $\mathbf{d}_0 = -\mathbf{g}_0$, $k = 0$.
Step 3
Input $\mathbf{H}_k$, i.e., the Hessian at $\mathbf{x}_k$.
Compute

$$\alpha_k = \frac{\mathbf{g}_k^T \mathbf{g}_k}{\mathbf{d}_k^T \mathbf{H}_k \mathbf{d}_k}$$

Set $\mathbf{x}_{k+1} = \mathbf{x}_k + \alpha_k \mathbf{d}_k$ and calculate $f_{k+1} = f(\mathbf{x}_{k+1})$.
Step 4
If $\|\alpha_k \mathbf{d}_k\| < \varepsilon$, output $\mathbf{x}^* = \mathbf{x}_{k+1}$ and $f(\mathbf{x}^*) = f_{k+1}$, and stop.
Step 5
Compute $\mathbf{g}_{k+1}$.
Compute

$$\beta_k = \frac{\mathbf{g}_{k+1}^T \mathbf{g}_{k+1}}{\mathbf{g}_k^T \mathbf{g}_k}$$

Generate new direction

$$\mathbf{d}_{k+1} = -\mathbf{g}_{k+1} + \beta_k \mathbf{d}_k$$

Set $k = k + 1$, and repeat from Step 3.

A typical solution trajectory for the above algorithm for a 2-dimensional convex quadratic problem is illustrated in Fig. 6.1. Note that $\mathbf{x}_1 = \mathbf{x}_0 - \alpha_0 \mathbf{g}_0$, where α_0 is the value of α that minimizes $f(\mathbf{x}_0 - \alpha \mathbf{g}_0)$, as in the steepest-descent algorithm.

The main advantages of the conjugate-gradient algorithm are as follows:

1. The gradient is always finite and linearly independent of all previous direction vectors, except when the solution is reached.
2. The computations are relatively simple and only slightly more complicated by comparison to the computations in the steepest-descent method.
3. No line searches are required.
4. For convex quadratic problems, the algorithm converges in n iterations.

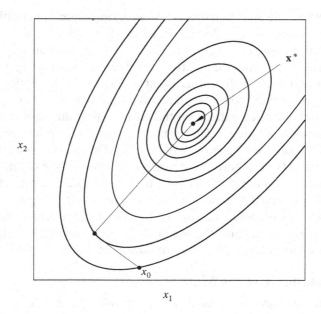

Figure 6.1. Typical solution trajectory in conjugate-gradient algorithm for a quadratic problem.

5. The first direction is a steepest-descent direction and it thus leads to a good reduction in $f(\mathbf{x})$ during the first iteration.
6. The algorithm has good convergence properties when applied for the solution of nonquadratic problems since the directions are based on gradient information.
7. Problems associated with the inversion of the Hessian are absent.

The disadvantages of the algorithm are:

1. The Hessian must be supplied, stored, and manipulated.
2. For nonquadratic problems convergence may not be achieved in rare occasions.

6.5 Minimization of Nonquadratic Functions

Like the Newton method, conjugate-direction methods are developed for the convex quadratic problem but are then applied for the solution of quadratic as well as nonquadratic problems. The fundamental assumption is made that if a steady reduction is achieved in the objective function in successive iterations, the neighborhood of the solution will eventually be reached. If $\mathbf{H}$ is positive definite near the solution, then convergence will, in principle, follow in at most n iterations. For this reason, conjugate-direction methods, like the

Newton method, are said to have *quadratic termination*. In addition, the rate of convergence is quadratic, that is, the order of convergence is two.

The use of conjugate-direction methods for the solution of nonquadratic problems may sometimes be relatively inefficient in reducing the objective function, in particular if the initial point is far from the solution. In such a case, unreliable previous data are likely to accumulate in the current direction vector, since they are calculated on the basis of past directions. Under these circumstances, the solution trajectory may wander through suboptimal areas of the parameter space, and progress will be slow. This problem can be overcome by re-initializing these algorithms periodically, say, every n iterations, in order to obliterate previous unreliable information, and in order to provide new vigor to the algorithm through the use of a steepest-descent step. Most of the time, the information accumulated in the current direction is quite reliable and throwing it away is likely to increase the amount of computation. Nevertheless, this seems to be a fair price to pay if the robustness of the algorithm is increased.

6.6 Fletcher-Reeves Method

The Fletcher-Reeves method [2] is a variation of the conjugate-gradient method. Its main feature is that parameters α_k for $k = 0, 1, 2, \ldots$ are determined by minimizing $f(\mathbf{x} + \alpha \mathbf{d}_k)$ with respect to α using a line search as in the case of the steepest-descent or the Newton method. The difference between this method and the steepest-descent or the Newton method is that $\mathbf{d}_k$ is a conjugate direction with respect to $\mathbf{d}_{k-1}, \mathbf{d}_{k-2}, \ldots, \mathbf{d}_0$ rather than the steepest-descent or Newton direction.

If the problem to be solved is convex and quadratic and the directions are selected as in Eq. (6.27) with β_k given by Eq. (6.28), then

$$\frac{df(\mathbf{x}_k + \alpha \mathbf{d}_k)}{d\alpha} = \mathbf{g}_{k+1}^T \mathbf{d}_k = 0$$

for $k = 0, 1, 2, \ldots$. Further, the conjugacy of the set of directions assures that

$$\frac{df(\mathbf{x}_k + \alpha \mathbf{d}_i)}{d\alpha} = \mathbf{g}_{k+1}^T \mathbf{d}_i = 0 \qquad \text{for } 0 \leq i \leq k$$

or

$$\mathbf{g}_k^T \mathbf{d}_i = 0 \qquad \text{for } 0 \leq i < k$$

as in Theorem 6.3. Consequently, the determination of α_k through a line search is equivalent to using Eq. (6.25). Since a line search entails more computation than Eq. (6.25), the Fletcher-Reeves modification would appear to be a retrograde step. Nevertheless, two significant advantages are gained as follows:

1. The modification renders the method more amenable to the minimization of nonquadratic problems since a larger reduction can be achieved in $f(\mathbf{x})$

along $\mathbf{d}_k$ at points outside the neighborhood of the solution. This is due to the fact that Eq. (6.25) will not yield the minimum along $\mathbf{d}_k$ in the case of a nonquadratic problem.

2. The modification obviates the derivation and calculation of the Hessian.

The Fletcher-Reeves algorithm can be shown to converge subject to the condition that the algorithm is re-initialized every n iterations. An implementation of the algorithm is as follows:

Algorithm 6.3 Fletcher-Reeves algorithm
Step 1
Input $\mathbf{x}_0$ and initialize the tolerance ε.
Step 2
Set $k = 0$.
Computer $\mathbf{g}_0$ and set $\mathbf{d}_0 = -\mathbf{g}_0$.
Step 3
Find α_k, the value of α that minimizes $f(\mathbf{x}_k + \alpha \mathbf{d}_k)$.
Set $\mathbf{x}_{k+1} = \mathbf{x}_k + \alpha_k \mathbf{d}_k$.
Step 4
If $\|\alpha_k \mathbf{d}_k\| < \varepsilon$, output $\mathbf{x}^* = \mathbf{x}_{k+1}$ and $f(\mathbf{x}^*) = f_{k+1}$, and stop.
Step 5
If $k = n - 1$, set $\mathbf{x}_0 = \mathbf{x}_{k+1}$ and go to Step 2.
Step 6
Compute $\mathbf{g}_{k+1}$.
Compute

$$\beta_k = \frac{\mathbf{g}_{k+1}^T \mathbf{g}_{k+1}}{\mathbf{g}_k^T \mathbf{g}_k}$$

Set $\mathbf{d}_{k+1} = -\mathbf{g}_{k+1} + \beta_k \mathbf{d}_k$.
Set $k = k + 1$ and repeat from Step 3.

6.7 Powell's Method

A conjugate-direction method which has been used extensively in the past is one due to Powell [3]. This method, like the conjugate-gradient method, is developed for the convex quadratic problem but it can be applied successfully to nonquadratic problems.

The distinctive feature of Powell's method is that conjugate directions are generated through a series of line searches. The technique used is based on the following theorem:

Theorem 6.5 *Generation of conjugate directions in Powell's method* *Let* $\mathbf{x}_a^*$ *and* $\mathbf{x}_b^*$ *be the minimizers obtained if the convex quadratic function*

$$f(\mathbf{x}) = a + \mathbf{x}^T \mathbf{b} + \tfrac{1}{2} \mathbf{x}^T \mathbf{H} \mathbf{x}$$

160

is minimized with respect to α on lines

$$\mathbf{x} = \mathbf{x}_a + \alpha \mathbf{d}_a$$

and

$$\mathbf{x} = \mathbf{x}_b + \alpha \mathbf{d}_b$$

respectively, as illustrated in Fig. 6.2.

If $\mathbf{d}_b = \mathbf{d}_a$, then vector $\mathbf{x}_b^* - \mathbf{x}_a^*$ is conjugate with respect to $\mathbf{d}_a$ (or $\mathbf{d}_b$).

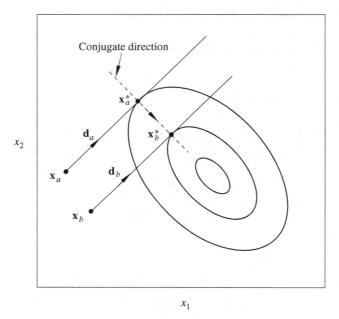

Figure 6.2. Generation of a conjugate direction.

Proof If $f(\mathbf{x}_a + \alpha \mathbf{d}_a)$ and $f(\mathbf{x}_b + \alpha \mathbf{d}_b)$ are minimized with respect to α, then

$$\frac{df(\mathbf{x}_a + \alpha \mathbf{d}_a)}{d\alpha} = \mathbf{d}_a^T \mathbf{g}(\mathbf{x}_a^*) = 0 \qquad (6.41a)$$

$$\frac{df(\mathbf{x}_b + \alpha \mathbf{d}_b)}{d\alpha} = \mathbf{d}_b^T \mathbf{g}(\mathbf{x}_b^*) = 0 \qquad (6.41b)$$

as in the case of a steepest-descent step (see Sec. 5.2). Since

$$\mathbf{g}(\mathbf{x}_a^*) = \mathbf{b} + \mathbf{H}\mathbf{x}_a^* \qquad (6.42a)$$

$$\mathbf{g}(\mathbf{x}_b^*) = \mathbf{b} + \mathbf{H}\mathbf{x}_b^* \qquad (6.42b)$$

then for $\mathbf{d}_b = \mathbf{d}_a$, Eqs. (6.41) – (6.42) yield

$$\mathbf{d}_a^T \mathbf{H}(\mathbf{x}_b^* - \mathbf{x}_a^*) = 0$$

and, therefore, vector $\mathbf{x}_b^* - \mathbf{x}_a^*$ is conjugate with respect to direction $\mathbf{d}_a$ (or $\mathbf{d}_b$).
∎

In Powell's algorithm, an initial point $\mathbf{x}_{00}$ as well as n linearly independent directions $\mathbf{d}_{01}$, $\mathbf{d}_{02}$, ..., $\mathbf{d}_{0n}$ are assumed and a series of line searches are performed in each iteration. Although any set of initial linearly independent directions can be used, it is convenient to use a set of coordinate directions.

In the first iteration, $f(\mathbf{x})$ is minimized sequentially in directions $\mathbf{d}_{01}$, $\mathbf{d}_{02}$, ..., $\mathbf{d}_{0n}$ starting from point $\mathbf{x}_{00}$ to yield points $\mathbf{x}_{01}$, $\mathbf{x}_{02}$, ..., $\mathbf{x}_{0n}$, respectively, as depicted in Fig. 6.3a. Then a new direction $\mathbf{d}_{0(n+1)}$ is generated as

$$\mathbf{d}_{0(n+1)} = \mathbf{x}_{0n} - \mathbf{x}_0$$

and $f(\mathbf{x})$ is minimized in this direction to yield a new point $\mathbf{x}_{0(n+1)}$. The set of directions is then updated by letting

$$\mathbf{d}_{11} = \mathbf{d}_{02}$$
$$\mathbf{d}_{12} = \mathbf{d}_{03}$$
$$\vdots$$
$$\mathbf{d}_{1(n-1)} = \mathbf{d}_{0n}$$
$$\mathbf{d}_{1n} = \mathbf{d}_{0(n+1)} \tag{6.43}$$

The effect of the first iteration is to reduce $f(\mathbf{x})$ by an amount $\Delta f = f(\mathbf{x}_{00}) - f(\mathbf{x}_{0(n+1)})$ and simultaneously to delete $\mathbf{d}_{01}$ from and add $\mathbf{d}_{0(n+1)}$ to the set of directions.

The same procedure is repeated in the second iteration. Starting with point

$$\mathbf{x}_{10} = \mathbf{x}_{0(n+1)}$$

$f(\mathbf{x})$ is minimized sequentially in directions $\mathbf{d}_{11}$, $\mathbf{d}_{12}$, ..., $\mathbf{d}_{1n}$ to yield points $\mathbf{x}_{11}$, $\mathbf{x}_{12}$, ..., $\mathbf{x}_{1n}$, as depicted in Fig. 6.3b. Then a new direction $\mathbf{d}_{1(n+1)}$ is generated as

$$\mathbf{d}_{1(n+1)} = \mathbf{x}_{1n} - \mathbf{x}_{10}$$

and $f(\mathbf{x})$ is minimized in direction $\mathbf{d}_{1(n+1)}$ to yield point $\mathbf{x}_{1(n+1)}$. Since

$$\mathbf{d}_{1n} = \mathbf{d}_{0(n+1)}$$

by assignment (see Eq. (6.43)), $\mathbf{d}_{1(n+1)}$ is conjugate to $\mathbf{d}_{1n}$, according to Theorem 6.5. Therefore, if we let

$$\mathbf{d}_{21} = \mathbf{d}_{12}$$
$$\mathbf{d}_{22} = \mathbf{d}_{13}$$
$$\vdots$$
$$\mathbf{d}_{2(n-1)} = \mathbf{d}_{1n}$$
$$\mathbf{d}_{2n} = \mathbf{d}_{1(n+1)}$$

162

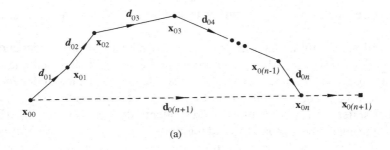

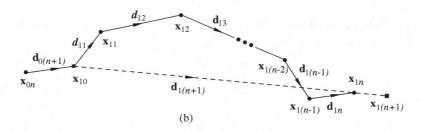

Figure 6.3. First and second iterations in Powell's algorithm.

the new set of directions will include a pair of conjugate directions, namely, $\mathbf{d}_{2(n-1)}$ and $\mathbf{d}_{2n}$.

Proceeding in the same way, each new iteration will increase the number of conjugate directions by one, and since the first two iterations yield two conjugate directions, n iterations will yield n conjugate directions. Powell's method will thus require $n(n+1)$ line searches since each iteration entails $(n+1)$ line searches. An implementation of Powell's algorithm is as follows:

Algorithm 6.4 Powell's algorithm
Step 1
Input $\mathbf{x}_{00} = [x_{01}\ x_{02}\ \cdots\ x_{0n}]^T$ and initialize the tolerance ε.
Set

$$\mathbf{d}_{01} = [\,x_{01}\ \ 0\ \ \cdots\ \ 0\,]^T$$
$$\mathbf{d}_{02} = [\,0\ \ x_{02}\ \ \cdots\ \ 0\,]^T$$
$$\vdots$$
$$\mathbf{d}_{0n} = [\,0\ \ 0\ \ \cdots\ \ x_{0n}\,]^T$$

Set $k = 0$.

Step 2
For $i = 1$ to n do:
Find α_{ki}, the value of α that minimizes $f(\mathbf{x}_{k(i-1)} + \alpha \mathbf{d}_{ki})$.
Set $\mathbf{x}_{ki} = \mathbf{x}_{k(i-1)} + \alpha_{ki}\mathbf{d}_{ki}$.
Step 3
Generate a new direction

$$\mathbf{d}_{k(n+1)} = \mathbf{x}_{kn} - \mathbf{x}_{k0}$$

Find $\alpha_{k(n+1)}$, the value of α that minimizes $f(\mathbf{x}_{k0} + \alpha \mathbf{d}_{k(n+1)})$.
Set

$$\mathbf{x}_{k(n+1)} = \mathbf{x}_{k0} + \alpha_{k(n+1)}\mathbf{d}_{k(n+1)}$$

Calculate $f_{k(n+1)} = f(\mathbf{x}_{k(n+1)})$.
Step 4
If $\|\alpha_{k(n+1)}\mathbf{d}_{k(n+1)}\| < \varepsilon$, output $\mathbf{x}^* = \mathbf{x}_{k(n+1)}$ and $f(\mathbf{x}^*) = f_{k(n+1)}$,
and stop.
Step 5
Update directions by setting

$$\mathbf{d}_{(k+1)1} = \mathbf{d}_{k2}$$
$$\mathbf{d}_{(k+1)2} = \mathbf{d}_{k3}$$
$$\vdots$$
$$\mathbf{d}_{(k+1)n} = \mathbf{d}_{k(n+1)}$$

Set $\mathbf{x}_{(k+1)0} = \mathbf{x}_{k(n+1)}$, $k = k + 1$, and repeat from Step 2.

In Step 1, $\mathbf{d}_{01}$, $\mathbf{d}_{02}$, $\ldots$, $\mathbf{d}_{0n}$ are assumed to be a set of coordinate directions. In Step 2, $f(\mathbf{x})$ is minimized along the path $\mathbf{x}_{k0}$, $\mathbf{x}_{k1}$, $\ldots$, $\mathbf{x}_{kn}$. In Step 3, $f(\mathbf{x})$ is minimized in the new conjugate direction. The resulting search pattern for the case of a quadratic 2-dimensional problem is illustrated in Fig. 6.4.

The major advantage of Powell's algorithm is that the Hessian need not be supplied, stored or manipulated. Furthermore, by using a 1-D algorithm that is based on function evaluations for line searches, the need for the gradient can also be eliminated.

A difficulty associated with Powell's method is that linear dependence can sometimes arise, and the method may fail to generate a complete set of linearly independent directions that span E^n, even in the case of a convex quadratic problem. This may happen if the minimization of $f(\mathbf{x}_{k(j-1)} + \alpha \mathbf{d}_{kj})$ with respect to α in Step 2 of the algorithm yields $\alpha_{kj} = 0$ for some j. In such a case, Step 3 will yield

$$\mathbf{d}_{k(n+1)} = \sum_{\substack{i=1 \\ i \neq j}}^{n} \alpha_{ki}\mathbf{d}_{ki}$$

164

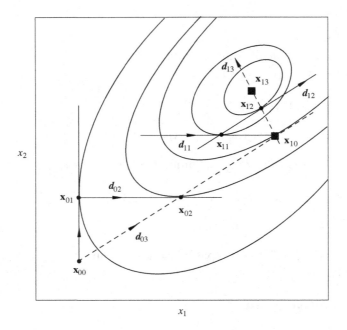

Figure 6.4. Solution trajectory in Powell's algorithm for a quadratic problem.

that is, the new direction generated will not have a component in direction $\mathbf{d}_{kj}$, and since $\mathbf{d}_{kj}$ will eventually be dropped, a set of n directions will result that does not span E^n. This means that at least two directions will be linearly dependent and the algorithm will not converge to the solution.

The above problem can be avoided by discarding $\mathbf{d}_{kn}$ if linear dependence is detected in the hope that the use of the same set of directions in the next iteration will be successful in generating a new conjugate direction. This is likely to happen since the next iteration will start with a new point $\mathbf{x}_k$.

In principle, linear dependence would occur if at least one α_{ki} becomes zero, as was demonstrated above. Unfortunately, however, owing to the finite precision of computers, zero is an improbable value for α_{ki} and, therefore, checking the value of α_{ki} is an unreliable test for linear dependence. An alternative approach due to Powell is as follows.

If the direction vectors $\mathbf{d}_{ki}$ for $i = 1, 2, \ldots, n$ are normalized such that

$$\mathbf{d}_{ki}^T \mathbf{H} \mathbf{d}_{ki} = 1 \qquad \text{for} \quad i = 1, 2, \ldots, n$$

then the determinant of matrix

$$\mathbf{D} = [\mathbf{d}_{k1} \ \mathbf{d}_{k2} \ \cdots \ \mathbf{d}_{kn}]$$

assumes a maximum value if and only if the directions $\mathbf{d}_{ki}$ belong to a conjugate set. Thus if a nonconjugate direction $\mathbf{d}_{1k}$ is dropped and conjugate direction

$\mathbf{d}_{k(n+1)}$ is added to $\mathbf{D}$, the determinant of $\mathbf{D}$ will increase. On the other hand, if the addition of $\mathbf{d}_{k(n+1)}$ results in linear dependence in $\mathbf{D}$, the determinant of $\mathbf{D}$ will decrease. On the basis of these principles, Powell developed a modified algorithm in which a test is used to determine whether the new direction generated should or should not be used in the next iteration. The test also identifies which one of the n old directions should be replaced by the new direction so as to achieve the maximum increase in the determinant, and thus reduce the risk of linear dependence.

An alternative but very similar technique for the elimination of linear dependence in the set of directions was proposed by Zangwill [4]. This technique is more effective and more economical in terms of computation than Powell's modification and, therefore, it deserves to be considered in detail.

Zangwill's technique can be implemented by applying the following modifications to Powell's algorithm.

1. The initial directions in Step 1 are chosen to be the coordinate set of vectors of unit length such that

$$\begin{aligned} \mathbf{D}_0 &= \begin{bmatrix} \mathbf{d}_{01} & \mathbf{d}_{02} & \cdots & \mathbf{d}_{0n} \end{bmatrix} \\ &= \begin{bmatrix} 1 & 0 & \cdots & 0 \\ 0 & 1 & \cdots & 0 \\ \vdots & \vdots & & \vdots \\ 0 & 0 & \cdots & 1 \end{bmatrix} \end{aligned}$$

and the determinant of $\mathbf{D}_0$, designated as Δ_0, is set to unity.

2. In Step 2, constants α_{ki} for $i = 1, 2, \ldots, n$ are determined as before, and the largest α_{ki} is then selected, i.e.,

$$\alpha_{km} = \max\{\alpha_{k1}, \alpha_{k2}, \ldots, \alpha_{kn}\}$$

3. In Step 3, a new direction is generated as before, and is then normalized to unit length so that

$$\mathbf{d}_{k(n+1)} = \frac{1}{\lambda_k}(\mathbf{x}_{kn} - \mathbf{x}_{k0})$$

where

$$\lambda_k = \|\mathbf{x}_{kn} - \mathbf{x}_{k0}\|$$

4. Step 4 is carried out as before. In Step 5, the new direction in item (3) is used to replace direction $\mathbf{d}_{km}$ provided that this substitution will maintain the determinant of

$$\mathbf{D}_k = \begin{bmatrix} \mathbf{d}_{k1} & \mathbf{d}_{k2} & \cdots & \mathbf{d}_{kn} \end{bmatrix}$$

finite and larger than a constant ε_1 in the range $0 < \varepsilon_1 \leq 1$, namely,

$$0 < \varepsilon_1 < \Delta_k = \det \mathbf{D}_k \leq 1$$

Otherwise, the most recent set of directions is used in the next iteration. Since

$$\Delta_k = \det[\mathbf{d}_{k1} \cdots \mathbf{d}_{k(m-1)} \ \mathbf{d}_{km} \ \mathbf{d}_{k(m+1)} \cdots \mathbf{d}_{kn}]$$

and

$$\mathbf{d}_{k(n+1)} = \frac{1}{\lambda_k} \sum_{i-1}^{n} \alpha_{ki} \mathbf{d}_{ki}$$

replacing $\mathbf{d}_{km}$ by $\mathbf{d}_{k(n+1)}$ yields

$$\Delta'_k = \frac{\alpha_{km}}{\lambda_k} \Delta_k$$

This result follows readily by noting that
- (a) if a constant multiple of a column is added to another column, the determinant remains unchanged, and
- (b) if a column is multiplied by a constant, the determinant is multiplied by the same constant.

From (a), the summation in Δ'_k can be eliminated and from (b) constant α_{km}/λ_k can be factored out. In this way, the effect of the substitution of $\mathbf{d}_{km}$ on the determinant of $\mathbf{D}_k$ is known. If

$$\frac{\alpha_{km}}{\lambda_k} \Delta_k > \varepsilon_1$$

we let

$$\mathbf{d}_{(k+1)m} = \mathbf{d}_{k(n+1)}$$

and

$$\mathbf{d}_{(k+1)i} = \mathbf{d}_{ki}$$

for $i = 1, 2, \ldots, m-1, m+1, \ldots, n$. Otherwise, we let

$$\mathbf{d}_{(k+1)i} = \mathbf{d}_{ki}$$

for $i = 1, 2, \ldots, n$. Simultaneously, the determinant Δ_k can be updated as

$$\delta_{k+1} = \begin{cases} \dfrac{\alpha_{km}}{\lambda_k} \Delta_k & \text{if } \dfrac{\alpha_{km}}{\lambda_k} \Delta_k > \varepsilon_1 \\ \Delta_k & \text{otherwise} \end{cases}$$

The net result of the above modifications is that the determinant of the direction matrix will always be finite and positive, which implies that the directions will always be linearly independent. The strategy in item (2) above of replacing the direction $\mathbf{d}_{ki}$ that yields the maximum α_{ki} ensures that the value of the determinant Δ_k is kept as large as possible so as to prevent linear dependence from arising in subsequent iterations.

The modified algorithm, which is often referred to as Zangwill's algorithm, can be shown to converge in the case of a convex quadratic problem. Its implementation is as follows:

Algorithm 6.5 Zangwill's algorithm
Step 1
Input $\mathbf{x}_{00}$ and initialize the tolerances ε and ε_1.
Set

$$\mathbf{d}_{01} = [1 \quad 0 \quad \cdots \quad 0]^T$$
$$\mathbf{d}_{02} = [0 \quad 1 \quad \cdots \quad 0]^T$$
$$\vdots$$
$$\mathbf{d}_{0n} = [0 \quad 0 \quad \cdots \quad 1]^T$$

Set $k = 0$, $\Delta_0 = 1$.
Step 2
For $i = 1$ to n do:
   Find α_{ki}, the value of α that minimizes $f(\mathbf{x}_{k(i-1)} + \alpha\mathbf{d}_{ki})$.
   Set $\mathbf{x}_{ki} = \mathbf{x}_{k(i-1)} + \alpha_{ki}\mathbf{d}_{ki}$.
Determine

$$\alpha_{km} = \max\{\alpha_{k1}, \ \alpha_{k2}, \ \ldots, \ \alpha_{kn}\}$$

Step 3
Generate a new direction

$$\mathbf{d}_{k(n+1)} = \mathbf{x}_{kn} - \mathbf{x}_{k0}$$

Find $\alpha_{k(n+1)}$, the value of α that minimizes $f(\mathbf{x}_{k0} + \alpha\mathbf{d}_{k(n+1)})$.
Set

$$\mathbf{x}_{k(n+1)} = \mathbf{x}_{k0} + \alpha_{k(n+1)}\mathbf{d}_{k(n+1)}$$

Calculate $f_{k(n+1)} = f(\mathbf{x}_{k(n+1)})$.
Calculate $\lambda_k = \|\mathbf{x}_{kn} - \mathbf{x}_{k0}\|$.

168

Step 4

If $\|\alpha_{k(n+1)}\mathbf{d}_{k(n+1)}\| < \varepsilon$, output $\mathbf{x}^* = \mathbf{x}_{k(n+1)}$ and $f(\mathbf{x}^*) = f_{k(n+1)}$, and stop.

Step 5

If $\alpha_{km}\Delta_k/\lambda_k > \varepsilon_1$, then do:

Set $\mathbf{d}_{(k+1)m} = \mathbf{d}_{k(n+1)}$ and $\mathbf{d}_{(k+1)i} = \mathbf{d}_{ki}$ for $i = 1, 2, \ldots, m-1$, $m+1, \ldots, n$.

Set $\Delta_{k+1} = \dfrac{\alpha_{km}}{\lambda_k}\Delta_k$.

Otherwise, set

$\mathbf{d}_{(k+1)i} = \mathbf{d}_{ki}$ for $i = 1, 2, \ldots, n$, and $\Delta_{k+1} = \Delta_k$.

Set $\mathbf{x}_{(k+1)0} = \mathbf{x}_{k(n+1)}$, $k = k + 1$, and repeat from Step 2.

6.8 Partan Method

In the early days of optimization, experimentation with two-variable functions revealed the characteristic zig-zag pattern in the solution trajectory in the steepest-descent method. It was noted that in well-behaved functions, successive solution points tend to coincide on two lines which intersect in the neighborhood of the minimizer, as depicted in Fig. 6.5. Therefore, an obvious strategy to attempt was to perform two steps of steepest descent followed by a search along the line connecting the initial point to the second solution point, as shown in Fig. 6.5. An iterative version of this approach was tried and found to converge to the solution. Indeed, for convex quadratic functions, convergence could be achieved in n iterations. The method has come to be known as the *parallel tangent method*, or *partan* for short, because of a special geometric property of the tangents to the contours in the case of quadratic functions.

The partan algorithm is illustrated in Fig. 6.6. An initial point $\mathbf{x}_0$ is assumed and two successive steepest-descent steps are taken to yield points $\mathbf{x}_1$ and $\mathbf{y}_1$. Then a line search is performed in the direction $\mathbf{y}_1 - \mathbf{x}_0$ to yield a point $\mathbf{x}_2$. This completes the first iteration. In the second iteration, a steepest-descent step is taken from point $\mathbf{x}_2$ to yield point $\mathbf{y}_2$, and a line search is performed along direction $\mathbf{y}_2 - \mathbf{x}_1$ to yield point $\mathbf{x}_3$, and so on. In effect, points $\mathbf{y}_1, \mathbf{y}_2, \ldots$, in Fig. 6.6 are obtained by steepest-descent steps and points $\mathbf{x}_2, \mathbf{x}_3, \ldots$ are obtained by line searches along the directions $\mathbf{y}_2 - \mathbf{x}_1, \mathbf{y}_3 - \mathbf{x}_2, \ldots$.

In the case of a convex quadratic problem, the lines connecting $\mathbf{x}_1, \mathbf{x}_2, \ldots, \mathbf{x}_k$, which are not part of the algorithm, form a set of conjugate-gradient directions. This property can be demonstrated by assuming that $\mathbf{d}_0, \mathbf{d}_1, \ldots, \mathbf{d}_{k-1}$ form a set of conjugate-gradient directions and then showing that $\mathbf{d}_k$ is a conjugate-gradient direction with respect to $\mathbf{d}_0, \mathbf{d}_1, \ldots, \mathbf{d}_{k-1}$.

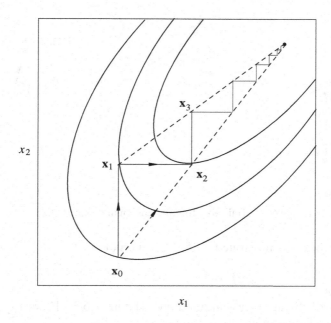

Figure 6.5. Zig-zag pattern of steepest-descent algorithm.

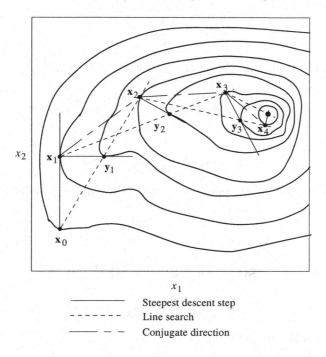

      ——————   Steepest descent step

      – – – – – –   Line search

      —— – – ——   Conjugate direction

Figure 6.6. Solution trajectory for partan method for a nonquadratic problem.

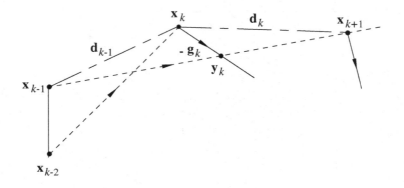

Figure 6.7.   Trajectory for kth iteration in partan method.

Consider the steps illustrated in Fig. 6.7 and note that

$$\mathbf{g}_k^T \mathbf{d}_i = 0 \qquad \text{for } 0 \leq i < k \qquad (6.44)$$

on the basis of the above assumption and Theorem 6.3. From Eqs. (6.31) – (6.32), the gradient at point $\mathbf{x}_{k-1}$ can be expressed as

$$\mathbf{g}_{k-1} = \sum_{i=0}^{k-1} a_i \mathbf{d}_i$$

where a_i for $i = 0, 1, \ldots, k-1$ are constants, and hence

$$\mathbf{g}_k^T \mathbf{g}_{k-1} = \mathbf{g}_k^T (\mathbf{b} + \mathbf{H}\mathbf{x}_{k-1}) = \sum_{i=0}^{k-1} a_i \mathbf{g}_k^T \mathbf{d}_i = 0 \qquad (6.45)$$

or

$$\mathbf{g}_k^T \mathbf{b} = -\mathbf{g}_k^T \mathbf{H}\mathbf{x}_{k-1} \qquad (6.46)$$

Since $\mathbf{y}_k$ is obtained by a steepest-descent step at point $\mathbf{x}_k$, we have

$$\mathbf{y}_k - \mathbf{x}_k = -\mathbf{g}_k$$

and

$$-\mathbf{g}(\mathbf{y}_k)^T \mathbf{g}_k = \mathbf{g}_k^T (\mathbf{b} + \mathbf{H}\mathbf{y}_k) = 0$$

or

$$\mathbf{g}_k^T \mathbf{b} = -\mathbf{g}_k^T \mathbf{H}\mathbf{y}_k \qquad (6.47)$$

Hence Eqs. (6.46) – (6.47) yield

$$\mathbf{g}_k^T \mathbf{H}(\mathbf{y}_k - \mathbf{x}_{k-1}) = 0 \qquad (6.48)$$

Since

$$\mathbf{y}_k - \mathbf{x}_{k-1} = \beta(\mathbf{x}_{k+1} - \mathbf{x}_{k-1})$$

where β is a constant, Eq. (6.48) can be expressed as

$$\mathbf{g}_k^T \mathbf{H}(\mathbf{x}_{k+1} - \mathbf{x}_{k-1}) = 0$$

or

$$\mathbf{g}_k^T \mathbf{H} \mathbf{x}_{k+1} = \mathbf{g}_k^T \mathbf{H} \mathbf{x}_{k-1} \tag{6.49}$$

We can now write

$$\mathbf{g}_k^T \mathbf{g}_{k+1} = \mathbf{g}_k^T (\mathbf{b} + \mathbf{H} \mathbf{x}_{k+1}) \tag{6.50}$$

and from Eqs. (6.45) and (6.49) – (6.50), we have

$$\mathbf{g}_k^T \mathbf{g}_{k+1} = \mathbf{g}_k^T (\mathbf{b} + \mathbf{H} \mathbf{x}_{k-1})$$
$$= \mathbf{g}_k^T \mathbf{g}_{k-1} = 0 \tag{6.51}$$

Point $\mathbf{x}_{k+1}$ is obtained by performing a line search in direction $\mathbf{x}_{k+1} - \mathbf{y}_k$, and hence

$$\mathbf{g}_{k+1}^T (\mathbf{x}_{k+1} - \mathbf{y}_k) = 0 \tag{6.52}$$

From Fig. 6.7

$$\mathbf{x}_{k+1} = \mathbf{x}_k + \mathbf{d}_k \tag{6.53}$$

and

$$\mathbf{y}_k = \mathbf{x}_k - \alpha_k \mathbf{g}_k \tag{6.54}$$

where α_k is the value of α that minimizes $f(\mathbf{x}_k - \alpha \mathbf{g}_k)$. Thus Eqs. (6.52) – (6.54) yield

$$\mathbf{g}_{k+1}^T (\mathbf{d}_k + \alpha_k \mathbf{g}_k) = 0$$

or

$$\mathbf{g}_{k+1}^T \mathbf{d}_k + \alpha_k \mathbf{g}_k^T \mathbf{g}_{k+1} = 0 \tag{6.55}$$

Now from Eqs. (6.51) and (6.55)

$$\mathbf{g}_{k+1}^T \mathbf{d}_k = 0$$

and on combining Eqs. (6.44) and (6.56), we obtain

$$\mathbf{g}_{k+1}^T \mathbf{d}_i = 0 \qquad \text{for } 0 \le i < k+1$$

that is, $\mathbf{x}_k$ satisfies Theorem 6.3.

172

References

1 M. R. Hestenes and E. L. Stiefel, "Methods of conjugate gradients for solving linear systems," *J. Res. Natl. Bureau Standards*, vol. 49, pp. 409–436, 1952.

2 R. Fletcher and C. M. Reeves, "Function minimization by conjugate gradients," *Computer J.*, vol. 7, pp. 149–154, 1964.

3 M. J. D. Powell, "An efficient method for finding the minimum of a function of several variables without calculating derivatives," *Computer J.*, vol. 7, pp. 155–162, 1964.

4 W. I. Zangwill, "Minimizing a function without calculating derivatives," *Computer J.*, vol. 10, pp. 293–296, 1968.

5 R. A. Horn and C. R. Johnson, *Matrix Analysis*, Cambridge University Press, New York, 1985.

Problems

6.1 Use the conjugate-gradient method to solve the optimization problem

$$\text{minimize } f(x) = \tfrac{1}{2}\mathbf{x}^T\mathbf{Q}\mathbf{x} + \mathbf{b}^T\mathbf{x}$$

where $\mathbf{Q}$ is given by

$$\mathbf{Q} = \begin{bmatrix} \mathbf{Q}_1 & \mathbf{Q}_2 & \mathbf{Q}_3 & \mathbf{Q}_4 \\ \mathbf{Q}_2 & \mathbf{Q}_1 & \mathbf{Q}_2 & \mathbf{Q}_3 \\ \mathbf{Q}_3 & \mathbf{Q}_2 & \mathbf{Q}_1 & \mathbf{Q}_2 \\ \mathbf{Q}_4 & \mathbf{Q}_3 & \mathbf{Q}_2 & \mathbf{Q}_1 \end{bmatrix} \quad \text{with } \mathbf{Q}_1 = \begin{bmatrix} 12 & 8 & 7 & 6 \\ 8 & 12 & 8 & 7 \\ 7 & 8 & 12 & 8 \\ 6 & 7 & 8 & 12 \end{bmatrix}$$

$$\mathbf{Q}_2 = \begin{bmatrix} 3 & 2 & 1 & 0 \\ 2 & 3 & 2 & 1 \\ 1 & 2 & 3 & 2 \\ 0 & 1 & 2 & 3 \end{bmatrix}, \quad \mathbf{Q}_3 = \begin{bmatrix} 2 & 1 & 0 & 0 \\ 1 & 2 & 1 & 0 \\ 0 & 1 & 2 & 1 \\ 0 & 0 & 1 & 2 \end{bmatrix}, \quad \mathbf{Q}_4 = \mathbf{I}_4$$

and $\mathbf{b} = -[1\,1\,1\,1\,0\,0\,0\,0\,1\,1\,1\,1\,0\,0\,0\,0]^T$.

6.2 Use the Fletcher-Reeves algorithm to find the minimizer of the Rosenbrock function

$$f(\mathbf{x}) = 100(x_2 - x_1^2)^2 + (1 - x_1)^2$$

Use $\varepsilon = 10^{-6}$ and try three initial points $\mathbf{x}_0 = [-2\ 2]^T$, $\mathbf{x}_0 = [2\ -2]^T$, and $\mathbf{x}_0 = [-2\ -2]^T$ and observe the results.

6.3 Solve Prob. 5.4 by applying the conjugate-gradient algorithm (Algorithm 6.2).

 (a) With $\varepsilon = 3 \times 10^{-7}$ and $\mathbf{x}_0 = [1\ 1]^T$, perform two iterations by following the steps described in Algorithm 6.2.

 (b) Compare the results of the first iteration obtained by using the conjugate-gradient algorithm with those obtained by using the steepest-descent method.

 (c) Compare the results of the second iteration obtained by using the conjugate-gradient algorithm with those obtained by using the steepest-descent method.

6.4 Solve Prob. 5.5 by applying the Fletcher-Reeves algorithm.

(*a*) Examine the solution obtained and the amount of computation required.

(*b*) Compare the results obtained in part (*a*) with those of Probs. 5.5, 5.16, and 5.21.

6.5 Solve Prob. 5.7 by applying the Fletcher-Reeves algorithm.

(*a*) Examine the solution obtained and the amount of computation required.

(*b*) Compare the results obtained in part (*a*) with those of Probs. 5.7, 5.17, and 5.22.

6.6 Solve Prob. 5.9 by applying the Fletcher-Reeves algorithm.

(*a*) Examine the solution obtained and the amount of computation required.

(*b*) Compare the results obtained in part (*a*) with those of Probs. 5.9 and 5.18.

6.7 Solve Prob. 5.4 by applying Powell's algorithm (Algorithm 6.4) and compare the results with those obtained in Probs. 5.4 and 6.3.

6.8 Solve Prob. 5.5 by applying Powell's algorithm and compare the results with those obtained in Probs. 5.5, 5.16, 5.21 and 6.4.

6.9 Solve Prob. 5.7 by applying Powell's algorithm and compare the results with those obtained in Probs. 5.7, 5.17, 5.22, and 6.5.

6.10 Solve Prob. 5.4 by applying Zangwill's algorithm and compare the results with those obtained in Probs. 5.4, 6.3, and 6.7.

Chapter 7

QUASI-NEWTON METHODS

7.1 Introduction

In Chap. 6, multidimensional optimization methods were considered in which the search for the minimizer is carried out by using a set of conjugate directions. An important feature of some of these methods (e.g., the Fletcher-Reeves and Powell's methods) is that explicit expressions for the second derivatives of $f(\mathbf{x})$ are not required. Another class of methods that do not require explicit expressions for the second derivatives is the class of *quasi-Newton methods*. These are sometimes referred to as *variable metric methods*.

As the name implies, the foundation of these methods is the classical Newton method described in Sec. 5.3. The basic principle in quasi-Newton methods is that the direction of search is based on an $n \times n$ direction matrix $\mathbf{S}$ which serves the same purpose as the inverse Hessian in the Newton method. This matrix is generated from available data and is contrived to be an approximation of $\mathbf{H}^{-1}$. Furthermore, as the number of iterations is increased, $\mathbf{S}$ becomes progressively a more accurate representation of $\mathbf{H}^{-1}$, and for convex quadratic objective functions it becomes identical to $\mathbf{H}^{-1}$ in $n + 1$ iterations.

Quasi-Newton methods, like most other methods, are developed for the convex quadratic problem and are then extended to the general problem. They rank among the most efficient methods available and are, therefore, used very extensively in numerous applications.

Several distinct quasi-Newton methods have evolved in recent years. In this chapter, we discuss in detail the four most important methods of this class which are:

1. Rank-one method
2. Davidon-Fletcher-Powell method
3. Broyden-Fletcher-Goldfarb-Shanno method

4. Fletcher method

We then discuss briefly a number of alternative approaches and describe two interesting generalizations, one due to Broyden and the other due to Huang.

7.2 The Basic Quasi-Newton Approach

In the methods of Chap. 5, the point generated in the kth iteration is given by

$$\mathbf{x}_{k+1} = \mathbf{x}_k - \alpha_k \mathbf{S}_k \mathbf{g}_k \tag{7.1}$$

where

$$\mathbf{S}_k = \begin{cases} \mathbf{I}_n & \text{for the steepest-descent method} \\ \mathbf{H}_k^{-1} & \text{for the Newton method} \end{cases}$$

Let us examine the possibility of using some arbitrary $n \times n$ positive definite matrix $\mathbf{S}_k$ for the solution of the quadratic problem

$$\text{minimize } f(\mathbf{x}) = a + \mathbf{b}^T\mathbf{x} + \tfrac{1}{2}\mathbf{x}^T\mathbf{H}\mathbf{x}$$

By differentiating $f(\mathbf{x}_k - \alpha\mathbf{S}_k\mathbf{g}_k)$ with respect to α and then setting the result to zero, the value of α that minimizes $f(\mathbf{x}_k - \alpha\mathbf{S}_k\mathbf{g}_k)$ can be deduced as

$$\alpha_k = \frac{\mathbf{g}_k^T\mathbf{S}_k\mathbf{g}_k}{\mathbf{g}_k^T\mathbf{S}_k\mathbf{H}\mathbf{S}_k\mathbf{g}_k} \tag{7.2}$$

where

$$\mathbf{g}_k = \mathbf{b} + \mathbf{H}\mathbf{x}_k$$

is the gradient of $f(\mathbf{x})$ at $\mathbf{x} = \mathbf{x}_k$.

It can be shown that

$$f(\mathbf{x}_{k+1}) - f(\mathbf{x}^*) \le \left(\frac{1-r}{1+r}\right)^2 [f(\mathbf{x}_k) - f(\mathbf{x}^*)]$$

where r is the ratio of the smallest to the largest eigenvalue of $\mathbf{S}_k\mathbf{H}$ (see [1] for proof). In effect, an algorithm based on Eqs. (7.1) and (7.2) would converge linearly with a convergence ratio

$$\beta = \left(\frac{1-r}{1+r}\right)^2$$

for any positive definite $\mathbf{S}_k$ (see Sec. 3.7). Convergence is fastest if $r = 1$, that is, if the eigenvalues of $\mathbf{S}_k\mathbf{H}$ are all equal. This means that the best results can be achieved by choosing

$$\mathbf{S}_k\mathbf{H} = \mathbf{I}_n$$

or
$$\mathbf{S}_k = \mathbf{H}^{-1}$$
Similarly, for the general optimization problem, we should choose some positive definite $\mathbf{S}_k$ which is equal to or, at least, approximately equal to $\mathbf{H}_k^{-1}$.

Quasi-Newton methods are methods that are motivated by the preceding observation. The direction of search is based on a positive definite matrix $\mathbf{S}_k$ which is generated from available data, and which is contrived to be an approximation for $\mathbf{H}_k^{-1}$. Several approximations are possible for $\mathbf{H}_k^{-1}$ and, consequently, a number of different quasi-Newton methods can be developed.

7.3   Generation of Matrix $\mathbf{S}_k$

Let $f(\mathbf{x}) \in C^2$ be a function in E^n and assume that the gradients of $f(\mathbf{x})$ at points $\mathbf{x}_k$ and $\mathbf{x}_{k+1}$ are designated as $\mathbf{g}_k$ and $\mathbf{g}_{k+1}$, respectively. If

$$\mathbf{x}_{k+1} = \mathbf{x}_k + \boldsymbol{\delta}_k \tag{7.3}$$

then the Taylor series gives the elements of $\mathbf{g}_{k+1}$ as

$$g_{(k+1)m} = g_{km} + \sum_{i=1}^{n} \frac{\partial g_{km}}{\partial x_{ki}} \delta_{ki} + \frac{1}{2} \sum_{i=1}^{n} \sum_{j=1}^{n} \frac{\partial^2 g_{km}}{\partial x_{ki} \partial x_{kj}} \delta_{ki} \delta_{kj} + \cdots$$

for $m = 1, 2, \ldots, n$. Now if $f(\mathbf{x})$ is quadratic, the second derivatives of $f(\mathbf{x})$ are constant and, in turn, the second derivatives of g_{km} are zero. Thus

$$g_{(k+1)m} = g_{km} + \sum_{i=1}^{n} \frac{\partial g_{km}}{\partial x_{ki}} \delta_{ki}$$

and since

$$g_{km} = \frac{\partial f_k}{\partial x_{km}}$$

we have

$$g_{(k+1)m} = g_{km} + \sum_{i=1}^{n} \frac{\partial^2 f_k}{\partial x_{ki} \partial x_{km}} \delta_{ki}$$

for $m = 1, 2, \ldots, n$. Therefore, $\mathbf{g}_{k+1}$ is given by

$$\mathbf{g}_{k+1} = \mathbf{g}_k + \mathbf{H}\boldsymbol{\delta}_k$$

where $\mathbf{H}$ is the Hessian of $f(\mathbf{x})$. Alternatively, we can write

$$\boldsymbol{\gamma}_k = \mathbf{H}\boldsymbol{\delta}_k \tag{7.4}$$

where

$$\boldsymbol{\delta}_k = \mathbf{x}_{k+1} - \mathbf{x}_k \tag{7.5}$$

$$\boldsymbol{\gamma}_k = \mathbf{g}_{k+1} - \mathbf{g}_k \tag{7.6}$$

The above analysis has shown that if the gradient of $f(\mathbf{x})$ is known at two points $\mathbf{x}_k$ and $\mathbf{x}_{k+1}$, a relation can be deduced that provides a certain amount of information about $\mathbf{H}$. Since there are n^2 unknowns in $\mathbf{H}$ (or $n(n+1)/2$ unknowns if $\mathbf{H}$ is assumed to be a real symmetric matrix) and Eq. (7.4) provides only n equations, $\mathbf{H}$ cannot be determined uniquely. However, if the gradient is evaluated sequentially at $n+1$ points, say, $\mathbf{x}_0, \mathbf{x}_1, \ldots, \mathbf{x}_n$ such that the changes in $\mathbf{x}$, namely,

$$\boldsymbol{\delta}_0 = \mathbf{x}_1 - \mathbf{x}_0$$
$$\boldsymbol{\delta}_1 = \mathbf{x}_2 - \mathbf{x}_1$$
$$\vdots$$
$$\boldsymbol{\delta}_{n-1} = \mathbf{x}_n - \mathbf{x}_{n-1}$$

form a set of linearly independent vectors, then sufficient information is obtained to determine $\mathbf{H}$ uniquely. To demonstrate this fact, n equations of the type given by Eq. (7.4) can be re-arranged as

$$[\boldsymbol{\gamma}_0 \, \boldsymbol{\gamma}_1 \, \cdots \, \boldsymbol{\gamma}_{n-1}] = \mathbf{H}[\boldsymbol{\delta}_0 \, \boldsymbol{\delta}_1 \, \cdots \, \boldsymbol{\delta}_{n-1}] \qquad (7.7)$$

and, therefore,

$$\mathbf{H} = [\boldsymbol{\gamma}_0 \, \boldsymbol{\gamma}_1 \, \cdots \, \boldsymbol{\gamma}_{n-1}][\boldsymbol{\delta}_0 \, \boldsymbol{\delta}_1 \, \cdots \, \boldsymbol{\delta}_{n-1}]^{-1}$$

The solution exists if $\boldsymbol{\delta}_0, \boldsymbol{\delta}_1, \ldots, \boldsymbol{\delta}_{n-1}$ form a set of linearly independent vectors.

The above principles can be used to construct the following algorithm:

Algorithm 7.1 Alternative Newton algorithm
Step 1
Input $\mathbf{x}_{00}$ and initialize the tolerance ε.
Set $k = 0$.
Input a set of linearly independent vectors $\boldsymbol{\delta}_0, \boldsymbol{\delta}_1, \ldots, \boldsymbol{\delta}_{n-1}$.
Step 2
Compute $\mathbf{g}_{00}$.
Step 3
For $i = 0$ to $n - 1$ do:
    Set $\mathbf{x}_{k(i+1)} = \mathbf{x}_{ki} + \boldsymbol{\delta}_i$.
    Compute $\mathbf{g}_{k(i+1)}$.
    Set $\boldsymbol{\gamma}_{ki} = \mathbf{g}_{k(i+1)} - \mathbf{g}_{ki}$.
Step 4
Compute $\mathbf{H}_k = [\boldsymbol{\gamma}_{k0} \, \boldsymbol{\gamma}_{k1} \, \cdots \, \boldsymbol{\gamma}_{k(n-1)}][\boldsymbol{\delta}_0 \, \boldsymbol{\delta}_1 \, \cdots \, \boldsymbol{\delta}_{n-1}]^{-1}$.
Compute $\mathbf{S}_k = \mathbf{H}_k^{-1}$.

Step 5
Set $\mathbf{d}_k = -\mathbf{S}_k \mathbf{g}_{k0}$.
Find α_k, the value of α that minimizes $f(\mathbf{x}_{k0} + \alpha \mathbf{d}_k)$.
Set $\mathbf{x}_{(k+1)0} = \mathbf{x}_{k0} + \alpha_k \mathbf{d}_k$.
Step 6
If $\|\alpha_k \mathbf{d}_k\| < \varepsilon$, output $\mathbf{x}_k^* = \mathbf{x}_{(k+1)0}$ and $f(\mathbf{x}^*) = f(\mathbf{x}_{(k+1)0})$, and stop.
Step 7
Set $k = k + 1$ and repeat from Step 3.

The above algorithm is essentially an implementation of the Newton method except that a mechanism is incorporated for the generation of $\mathbf{H}^{-1}$ using computed data. For a convex quadratic problem, the algorithm will yield the solution in one iteration and it will thus be quite effective. For a nonquadratic problem, however, the algorithm has the same disadvantages as any other algorithm based on the Newton method (e.g., Algorithm 5.3). First, matrix inversion is required, which is undesirable; second, matrix $\mathbf{H}_k$ must be checked for positive definiteness and rendered positive definite, if necessary, in every iteration.

A strategy that leads to the elimination of matrix inversion is as follows. We assume that a positive definite real symmetric matrix $\mathbf{S}_k$ is available, which is an approximation of $\mathbf{H}^{-1}$, and compute a quasi-Newton direction

$$\mathbf{d}_k = -\mathbf{S}_k \mathbf{g}_k \tag{7.8}$$

We then find α_k, the value of α that minimizes $f(\mathbf{x}_k + \alpha \mathbf{d}_k)$, as in the Newton method. For a convex quadratic problem, Eq. (7.2) gives

$$\alpha_k = \frac{\mathbf{g}_k^T \mathbf{S}_k \mathbf{g}_k}{(\mathbf{S}_k \mathbf{g}_k)^T \mathbf{H}(\mathbf{S}_k \mathbf{g}_k)} \tag{7.9}$$

where $\mathbf{S}_k$ and $\mathbf{H}$ are positive definite. Evidently, α_k is greater than zero provided that $\mathbf{x}_k$ is not the solution point $\mathbf{x}^*$. We then determine a change in $\mathbf{x}$ as

$$\boldsymbol{\delta}_k = \alpha_k \mathbf{d}_k \tag{7.10}$$

and deduce a new point $\mathbf{x}_{k+1}$ using Eq. (7.3). By computing the gradient at points $\mathbf{x}_k$ and $\mathbf{x}_{k+1}$, the change in the gradient, $\boldsymbol{\gamma}_k$, can be determined using Eq. (7.6). We then apply a correction to $\mathbf{S}_k$ and generate

$$\mathbf{S}_{k+1} = \mathbf{S}_k + \mathbf{C}_k \tag{7.11}$$

where $\mathbf{C}_k$ is an $n \times n$ *correction matrix* which can be computed from available data. On applying the above procedure iteratively starting with an initial point $\mathbf{x}_0$ and an initial positive definite matrix $\mathbf{S}_0$, say, $\mathbf{S}_0 = \mathbf{I}_n$, the sequences $\boldsymbol{\delta}_0, \boldsymbol{\delta}_1, \ldots, \boldsymbol{\delta}_k, \boldsymbol{\gamma}_0, \boldsymbol{\gamma}_1, \ldots, \boldsymbol{\gamma}_k$ and $\mathbf{S}_1, \mathbf{S}_2, \ldots, \mathbf{S}_{k+1}$ can be generated. If

$$\mathbf{S}_{k+1} \boldsymbol{\gamma}_i = \boldsymbol{\delta}_i \qquad \text{for } 0 \leq i \leq k \tag{7.12}$$

then for $k = n - 1$, we can write

$$\mathbf{S}_n[\gamma_0 \; \gamma_1 \; \cdots \; \gamma_{n-1}] = [\delta_0 \; \delta_1 \; \cdots \; \delta_{n-1}]$$

or

$$\mathbf{S}_n = [\delta_0 \; \delta_1 \; \cdots \; \delta_{n-1}][\gamma_0 \; \gamma_1 \; \cdots \; \gamma_{n-1}]^{-1} \qquad (7.13)$$

and from Eqs. (7.7) and (7.13), we have

$$\mathbf{S}_n = \mathbf{H}^{-1}$$

Now if $k = n$, Eqs. (7.8) – (7.10) yield

$$\mathbf{d}_n = -\mathbf{H}^{-1}\mathbf{g}_n$$
$$\alpha_n = 1$$
$$\delta_n = -\mathbf{H}^{-1}\mathbf{g}_n$$

respectively, and, therefore, from Eq. (7.3)

$$\mathbf{x}_{n+1} = \mathbf{x}_n - \mathbf{H}^{-1}\mathbf{g}_n = \mathbf{x}^*$$

as in the Newton method.

The above procedure leads to a family of quasi-Newton algorithms which have the fundamental property that they terminate in $n + 1$ iterations ($k = 0, 1, \ldots, n$) in the case of a convex quadratic problem. The various algorithms of this class differ from one another in the formula used for the derivation of the correction matrix $\mathbf{C}_n$.

In any derivation of $\mathbf{C}_n$, $\mathbf{S}_{k+1}$ must satisfy Eq. (7.12) and the following properties are highly desirable:

1. Vectors $\delta_0, \delta_1, \ldots, \delta_{n-1}$ should form a set of conjugate directions (see Chap. 6).
2. A positive definite $\mathbf{S}_k$ should give rise to a positive definite $\mathbf{S}_{k+1}$.

The first property will ensure that the excellent properties of conjugate-direction methods apply to the quasi-Newton method as well. The second property will ensure that $\mathbf{d}_k$ is a descent direction in every iteration, i.e., for $k = 0, 1, \ldots$. To demonstrate this fact, consider the point $\mathbf{x}_k + \delta_k$, and let

$$\delta_k = \alpha \mathbf{d}_k$$

where

$$\mathbf{d}_k = -\mathbf{S}_k \mathbf{g}_k$$

For $\alpha > 0$, the Taylor series in Eq. (2.4h) gives

$$f(\mathbf{x}_k + \delta_k) = f(\mathbf{x}_k) + \mathbf{g}_k^T \delta_k + \tfrac{1}{2}\delta_k^T \mathbf{H}(\mathbf{x}_k + c\delta_k)\delta_k$$

where c is a constant in the range $0 \leq c < 1$. On eliminating $\boldsymbol{\delta}_k$, we obtain

$$
\begin{aligned}
f(\mathbf{x}_k + \boldsymbol{\delta}_k) &= f(\mathbf{x}_k) - \alpha \mathbf{g}_k^T \mathbf{S}_k \mathbf{g}_k + o(\alpha \|\mathbf{d}_k\|) \\
&= f(\mathbf{x}_k) - [\alpha \mathbf{g}_k^T \mathbf{S}_k \mathbf{g}_k - o(\alpha \|\mathbf{d}_k\|)]
\end{aligned}
$$

where $o(\alpha \|\mathbf{d}_k\|)$ is the remainder which approaches zero faster than $\alpha \|\mathbf{d}_k\|$. Now if $\mathbf{S}_k$ is positive definite, then for a sufficiently small $\alpha > 0$, we have

$$
\alpha \mathbf{g}_k^T \mathbf{S}_k \mathbf{g}_k - o(\alpha \|\mathbf{d}_k\|) > 0
$$

since $\alpha > 0$, $\mathbf{g}_k^T \mathbf{S}_k \mathbf{g}_k > 0$, and $o(\alpha \|\mathbf{d}_k\|) \to 0$. Therefore,

$$
f(\mathbf{x}_k + \boldsymbol{\delta}_k) < f(\mathbf{x}_k) \tag{7.14}
$$

that is, *if $\mathbf{S}_k$ is positive definite, then $\mathbf{d}_k$ is a descent direction.*

The importance of property (2) should, at this point, be evident. A positive definite $\mathbf{S}_0$ will give a positive definite $\mathbf{S}_1$ which will give a positive definite $\mathbf{S}_2$, and so on. Consequently, directions $\mathbf{d}_0$, $\mathbf{d}_1$, $\mathbf{d}_2$, ... will all be descent directions, and this will assure the convergence of the algorithm.

7.4   Rank-One Method

The rank-one method owes its name to the fact that correction matrix $\mathbf{C}_k$ in Eq. (7.11) has a rank of unity. This correction was proposed independently by Broyden [2], Davidon [3], Fiacco and McCormick [4], Murtagh and Sargent [5], and Wolfe [6]. The derivation of the rank-one formula is as follows.

Assume that

$$
\mathbf{S}_{k+1} \boldsymbol{\gamma}_k = \boldsymbol{\delta}_k \tag{7.15}
$$

and let

$$
\mathbf{S}_{k+1} = \mathbf{S}_k + \beta_k \boldsymbol{\xi}_k \boldsymbol{\xi}_k^T \tag{7.16}
$$

where $\boldsymbol{\xi}_k$ is a column vector and β_k is a constant. The correction matrix $\beta_k \boldsymbol{\xi}_k \boldsymbol{\xi}_k^T$ is symmetric and has a rank of unity as can be demonstrated (see Prob. 7.1). From Eqs. (7.15) – (7.16)

$$
\boldsymbol{\delta}_k = \mathbf{S}_k \boldsymbol{\gamma}_k + \beta_k \boldsymbol{\xi}_k \boldsymbol{\xi}_k^T \boldsymbol{\gamma}_k \tag{7.17}
$$

and hence

$$
\begin{aligned}
\boldsymbol{\gamma}_k^T (\boldsymbol{\delta}_k - \mathbf{S}_k \boldsymbol{\gamma}_k) &= \beta_k \boldsymbol{\gamma}_k^T \boldsymbol{\xi}_k \boldsymbol{\xi}_k^T \boldsymbol{\gamma}_k \\
&= \beta_k (\boldsymbol{\xi}_k^T \boldsymbol{\gamma}_k)^2
\end{aligned} \tag{7.18}
$$

Alternatively, from Eq. (7.17)

$$
\begin{aligned}
(\boldsymbol{\delta}_k - \mathbf{S}_k \boldsymbol{\gamma}_k) &= \beta_k \boldsymbol{\xi}_k \boldsymbol{\xi}_k^T \boldsymbol{\gamma}_k = \beta_k (\boldsymbol{\xi}_k^T \boldsymbol{\gamma}_k) \boldsymbol{\xi}_k \\
(\boldsymbol{\delta}_k - \mathbf{S}_k \boldsymbol{\gamma}_k)^T &= \beta_k \boldsymbol{\gamma}_k^T \boldsymbol{\xi}_k \boldsymbol{\xi}_k^T = \beta_k (\boldsymbol{\xi}_k^T \boldsymbol{\gamma}_k) \boldsymbol{\xi}_k^T
\end{aligned}
$$

since $\xi_k^T \gamma_k$ is a scalar. Hence

$$(\delta_k - \mathbf{S}_k \gamma_k)(\delta_k - \mathbf{S}_k \gamma_k)^T = \beta_k (\xi_k^T \gamma_k)^2 \beta_k \xi_k \xi_k^T \qquad (7.19)$$

and from Eqs. (7.18) – (7.19), we have

$$\beta_k \xi_k \xi_k^T = \frac{(\delta_k - \mathbf{S}_k \gamma_k)(\delta_k - \mathbf{S}_k \gamma_k)^T}{\beta_k (\xi_k^T \gamma_k)^2}$$

$$= \frac{(\delta_k - \mathbf{S}_k \gamma_k)(\delta_k - \mathbf{S}_k \gamma_k)^T}{\gamma_k^T (\delta_k - \mathbf{S}_k \gamma_k)}$$

With the correction matrix known, $\mathbf{S}_{k+1}$ can be deduced from Eq. (7.16) as

$$\mathbf{S}_{k+1} = \mathbf{S}_k + \frac{(\delta_k - \mathbf{S}_k \gamma_k)(\delta_k - \mathbf{S}_k \gamma_k)^T}{\gamma_k^T (\delta_k - \mathbf{S}_k \gamma_k)} \qquad (7.20)$$

For a convex quadratic problem, this formula will generate $\mathbf{H}^{-1}$ on iteration $n-1$ provided that Eq. (7.12) holds. This indeed is the case as will be demonstrated by the following theorem.

Theorem 7.1 *Generation of inverse Hessian If* $\mathbf{H}$ *is the Hessian of a convex quadratic problem and*

$$\gamma_i = \mathbf{H} \delta_i \qquad for \ \ 0 \le i \le k \qquad (7.21)$$

where $\delta_1, \delta_2, \ldots, \delta_k$ *are given linearly independent vectors, then for any initial symmetric matrix* $\mathbf{S}_0$

$$\delta_i = \mathbf{S}_{k+1} \gamma_i \qquad for \ \ 0 \le i \le k \qquad (7.22)$$

where

$$\mathbf{S}_{i+1} = \mathbf{S}_i + \frac{(\delta_i - \mathbf{S}_i \gamma_i)(\delta_i - \mathbf{S}_i \gamma_i)^T}{\gamma_i^T (\delta_i - \mathbf{S}_i \gamma_i)} \qquad (7.23)$$

Proof We assume that

$$\delta_i = \mathbf{S}_k \gamma_i \qquad for \ 0 \le i \le k - 1 \qquad (7.24)$$

and show that

$$\delta_i = \mathbf{S}_{k+1} \gamma_i \qquad for \ 0 \le i \le k$$

If $0 \le i \le k - 1$, Eq. (7.20) yields

$$\mathbf{S}_{k+1} \gamma_i = \mathbf{S}_k \gamma_i + \zeta_k (\delta_k - \mathbf{S}_k \gamma_k)^T \gamma_i$$

where

$$\zeta_k = \frac{\delta_k - \mathbf{S}_k \gamma_k}{\gamma_k^T (\delta_k - \mathbf{S}_k \gamma_k)}$$

Since $\mathbf{S}_k$ is symmetric, we can write

$$\mathbf{S}_{k+1} \gamma_i = \mathbf{S}_k \gamma_i + \zeta_k (\delta_k^T \gamma_i - \gamma_k^T \mathbf{S}_k \gamma_i)$$

and if Eq. (7.24) holds, then

$$\mathbf{S}_{k+1} \gamma_i = \delta_i + \zeta_k (\delta_k^T \gamma_i - \gamma_k^T \delta_i) \qquad (7.25)$$

For $0 \le i \le k$

$$\gamma_i = \mathbf{H} \delta_i$$

and

$$\gamma_k^T = \delta_k^T \mathbf{H}$$

Hence for $0 \le i \le k - 1$, we have

$$\delta_k^T \gamma_i - \gamma_k^T \delta_i = \delta_k^T \mathbf{H} \delta_i - \delta_k^T \mathbf{H} \delta_i = 0$$

and from Eq. (7.25)

$$\delta_i = \mathbf{S}_{k+1} \gamma_i \qquad \text{for } 0 \le i \le k - 1 \qquad (7.26)$$

By assignment (see Eq. (7.15))

$$\delta_k = \mathbf{S}_{k+1} \gamma_k \qquad (7.27)$$

and on combining Eqs. (7.26) and (7.27), we obtain

$$\delta_i = \mathbf{S}_{k+1} \gamma_i \qquad \text{for } 0 \le i \le k \qquad (7.28)$$

To complete the induction, we note that

$$\delta_i = \mathbf{S}_1 \gamma_i \qquad \text{for } 0 \le i \le 0$$

by assignment, and since Eq. (7.28) holds if Eq. (7.24) holds, we can write

$$\delta_i = \mathbf{S}_2 \gamma_i \qquad \text{for } 0 \le i \le 1$$
$$\delta_i = \mathbf{S}_3 \gamma_i \qquad \text{for } 0 \le i \le 2$$
$$\vdots \qquad\qquad \vdots$$
$$\delta_i = \mathbf{S}_{k+1} \gamma_i \qquad \text{for } 0 \le i \le k$$

∎

These principles lead to the following algorithm:

Algorithm 7.2 Basic quasi-Newton algorithm
Step 1
Input $\mathbf{x}_0$ and initialize the tolerance ε.
Set $k = 0$ and $\mathbf{S}_0 = \mathbf{I}_n$.
Compute $\mathbf{g}_0$.
Step 2
Set $\mathbf{d}_k = -\mathbf{S}_k \mathbf{g}_k$.
Find α_k, the value of α that minimizes $f(\mathbf{x}_k + \alpha \mathbf{d}_k)$, using a line search.
Set $\delta_k = \alpha_k \mathbf{d}_k$ and $\mathbf{x}_{k+1} = \mathbf{x}_k + \delta_k$.
Step 3
If $\|\delta_k\| < \varepsilon$, output $\mathbf{x}^* = \mathbf{x}_{k+1}$ and $f(\mathbf{x}^*) = f(\mathbf{x}_{k+1})$, and stop.
Step 4
Compute $\mathbf{g}_{k+1}$ and set
$$\gamma_k = \mathbf{g}_{k+1} - \mathbf{g}_k$$
Compute $\mathbf{S}_{k+1}$ using Eq. (7.20).
Set $k = k + 1$ and repeat from Step 2.

In Step 2, the value of α_n is obtained by using a line search in order to render the algorithm more amenable to nonquadratic problems. However, for convex quadratic problems, α_n should be calculated by using Eq. (7.2) which should involve a lot less computation than a line search.

There are two serious problems associated with the rank-one method. First, a positive definite $\mathbf{S}_k$ may not yield a positive definite $\mathbf{S}_{k+1}$, even for a convex quadratic problem, and in such a case the next direction will not be a descent direction. Second, the denominator in the correction formula may approach zero and may even become zero. If it approaches zero, numerical ill-conditioning will occur, and if it becomes zero the method will break down since $\mathbf{S}_{k+1}$ will become undefined.

From Eq. (7.20), we can write

$$
\begin{aligned}
\gamma_i^T \mathbf{S}_{k+1} \gamma_i &= \gamma_i^T \mathbf{S}_k \gamma_i + \frac{\gamma_i^T (\delta_k - \mathbf{S}_k \gamma_k)(\delta_k^T - \gamma_k^T \mathbf{S}_k)\gamma_i}{\gamma_k^T (\delta_k - \mathbf{S}_k \gamma_k)} \\
&= \gamma_i^T \mathbf{S}_k \gamma_i + \frac{(\gamma_i^T \delta_k - \gamma_i^T \mathbf{S}_k \gamma_k)(\delta_k^T \gamma_i - \gamma_k^T \mathbf{S}_k \gamma_i)}{\gamma_k^T (\delta_k - \mathbf{S}_k \gamma_k)} \\
&= \gamma_i^T \mathbf{S}_k \gamma_i + \frac{(\gamma_i^T \delta_k - \gamma_i^T \mathbf{S}_k \gamma_k)^2}{\gamma_k^T (\delta_k - \mathbf{S}_k \gamma_k)}
\end{aligned}
$$

Therefore, if $\mathbf{S}_k$ is positive definite, a sufficient condition for $\mathbf{S}_{k+1}$ to be positive definite is

$$\gamma_k^T (\delta_k - \mathbf{S}_k \gamma_k) > 0$$

The problems associated with the rank-one method can be overcome by checking the denominator of the correction formula in Step 4 of the algorithm. If it becomes zero or negative, $\mathbf{S}_{k+1}$ can be discarded and $\mathbf{S}_k$ can be used for the subsequent iteration. However, if this problem occurs frequently the possibility exists that $\mathbf{S}_{k+1}$ may not converge to $\mathbf{H}^{-1}$. Then the expected rapid convergence may not materialize.

7.5   Davidon-Fletcher-Powell Method

An alternative quasi-Newton method is one proposed by Davidon [3] and later developed by Fletcher and Powell [7]. Although similar to the rank-one method, the Davidon-Fletcher-Powell (DFP) method has an important advantage. If the initial matrix $\mathbf{S}_0$ is positive definite, the updating formula for $\mathbf{S}_{k+1}$ will yield a sequence of positive definite matrices $\mathbf{S}_1$, $\mathbf{S}_2$, ..., $\mathbf{S}_n$. Consequently, the difficulty associated with the second term of the rank-one formula given by Eq. (7.20) will not arise. As a result every new direction will be a descent direction.

The updating formula for the DFP method is

$$\mathbf{S}_{k+1} = \mathbf{S}_k + \frac{\delta_k \delta_k^T}{\delta_k^T \gamma_k} - \frac{\mathbf{S}_k \gamma_k \gamma_k^T \mathbf{S}_k}{\gamma_k^T \mathbf{S}_k \gamma_k} \tag{7.29}$$

where the correction is an $n \times n$ symmetric matrix of *rank two*. The validity of this formula can be demonstrated by post-multiplying both sides by γ_k, that is,

$$\mathbf{S}_{k+1} \gamma_k = \mathbf{S}_k \gamma_k + \frac{\delta_k \delta_k^T \gamma_k}{\delta_k^T \gamma_k} - \frac{\mathbf{S}_k \gamma_k \gamma_k^T \mathbf{S}_k \gamma_k}{\gamma_k^T \mathbf{S}_k \gamma_k}$$

Since $\delta_k^T \gamma_k$ and $\gamma_k^T \mathbf{S}_k \gamma_k$ are scalars, they can be cancelled out and so we have

$$\mathbf{S}_{k+1} \gamma_k = \delta_k \tag{7.30}$$

as required.

The implementation of the DFP method is the same as in Algorithm 7.2 except that the rank-two formula of Eq. (7.29) is used in Step 4.

The properties of the DFP method are summarized by the following theorems.

Theorem 7.2 *Positive definiteness of* **S** *matrix. If* $\mathbf{S}_k$ *is positive definite, then the matrix* $\mathbf{S}_{k+1}$ *generated by the DFP method is also positive definite.*

Proof For any nonzero vector $\mathbf{x} \in E^n$, Eq. (7.29) yields

$$\mathbf{x}^T \mathbf{S}_{k+1} \mathbf{x} = \mathbf{x}^T \mathbf{S}_k \mathbf{x} + \frac{\mathbf{x}^T \delta_k \delta_k^T \mathbf{x}}{\delta_k^T \gamma_k} - \frac{\mathbf{x}^T \mathbf{S}_k \gamma_k \gamma_k^T \mathbf{S}_k \mathbf{x}}{\gamma_k^T \mathbf{S}_k \gamma_k} \tag{7.31}$$

For a real symmetric matrix $\mathbf{S}_k$, we can write

$$\mathbf{U}^T \mathbf{S}_k \mathbf{U} = \mathbf{\Lambda}$$

where $\mathbf{U}$ is a unitary matrix such that

$$\mathbf{U}^T\mathbf{U} = \mathbf{U}\mathbf{U}^T = \mathbf{I}_n$$

and $\boldsymbol{\Lambda}$ is a diagonal matrix whose diagonal elements are the eigenvalues of $\mathbf{S}_k$ (see Theorem 2.8). We can thus write

$$\begin{aligned} \mathbf{S}_k &= \mathbf{U}\boldsymbol{\Lambda}\mathbf{U}^T = \mathbf{U}\boldsymbol{\Lambda}^{1/2}\boldsymbol{\Lambda}^{1/2}\mathbf{U}^T \\ &= (\mathbf{U}\boldsymbol{\Lambda}^{1/2}\mathbf{U}^T)(\mathbf{U}\boldsymbol{\Lambda}^{1/2}\mathbf{U}^T) \\ &= \mathbf{S}_k^{1/2}\mathbf{S}_k^{1/2} \end{aligned}$$

If we let

$$\mathbf{u} = \mathbf{S}_k^{1/2}\mathbf{x} \quad\text{and}\quad \mathbf{v} = \mathbf{S}_k^{1/2}\boldsymbol{\gamma}_k$$

then Eq. (7.31) can be expressed as

$$\mathbf{x}^T\mathbf{S}_{k+1}\mathbf{x} = \frac{(\mathbf{u}^T\mathbf{u})(\mathbf{v}^T\mathbf{v}) - (\mathbf{u}^T\mathbf{v})^2}{\mathbf{v}^T\mathbf{v}} + \frac{(\mathbf{x}^T\boldsymbol{\delta}_k)^2}{\boldsymbol{\delta}_k^T\boldsymbol{\gamma}_k} \tag{7.32}$$

From Step 2 of Algorithm 7.2, we have

$$\boldsymbol{\delta}_k = \alpha_k\mathbf{d}_k = -\alpha_k\mathbf{S}_k\mathbf{g}_k \tag{7.33}$$

where α_k is the value of α that minimizes $f(\mathbf{x}_k + \alpha\mathbf{d}_k)$ at point $\mathbf{x} = \mathbf{x}_{k+1}$. Since $\mathbf{d}_k = -\mathbf{S}_k\mathbf{g}_k$ is a descent direction (see Eq. (7.14)), we have $\alpha_k > 0$. Furthermore,

$$\left.\frac{f(\mathbf{x}_k + \alpha\mathbf{d}_k)}{d\alpha}\right|_{\alpha=\alpha_k} = \mathbf{g}(\mathbf{x}_k + \alpha_k\mathbf{d}_k)^T\mathbf{d}_k = \mathbf{g}_{k+1}^T\mathbf{d}_k = 0$$

(see Sec. 5.2.3) and thus

$$\alpha_k\mathbf{g}_{k+1}^T\mathbf{d}_k = \mathbf{g}_{k+1}^T\alpha_k\mathbf{d}_k = \mathbf{g}_{k+1}^T\boldsymbol{\delta}_k = \boldsymbol{\delta}_k^T\mathbf{g}_{k+1} = 0$$

Hence from Eq. (7.6), we can write

$$\boldsymbol{\delta}_k^T\boldsymbol{\gamma}_k = \boldsymbol{\delta}_k^T\mathbf{g}_{k+1} - \boldsymbol{\delta}_k^T\mathbf{g}_k = -\boldsymbol{\delta}_k^T\mathbf{g}_k$$

Now from Eq. (7.33), we get

$$\boldsymbol{\delta}_k^T\boldsymbol{\gamma}_k = -\boldsymbol{\delta}_k^T\mathbf{g}_k = -[-\alpha_k\mathbf{S}_k\mathbf{g}_k]^T\mathbf{g}_k = \alpha_k\mathbf{g}_k^T\mathbf{S}_k\mathbf{g}_k \tag{7.34}$$

and hence Eq. (7.32) can be expressed as

$$\mathbf{x}^T\mathbf{S}_{k+1}\mathbf{x} = \frac{(\mathbf{u}^T\mathbf{u})(\mathbf{v}^T\mathbf{v}) - (\mathbf{u}^T\mathbf{v})^2}{\mathbf{v}^T\mathbf{v}} + \frac{(\mathbf{x}^T\boldsymbol{\delta}_k)^2}{\alpha_k\mathbf{g}_k^T\mathbf{S}_k\mathbf{g}_k} \tag{7.35}$$

Since

$$\mathbf{u}^T\mathbf{u} = \|\mathbf{u}\|^2, \quad \mathbf{v}^T\mathbf{v} = \|\mathbf{v}^2\|, \quad \mathbf{u}^T\mathbf{v} = \|\mathbf{u}\| \, \|\mathbf{v}\| \cos\theta$$

where θ is the angle between vectors $\mathbf{u}$ and $\mathbf{v}$, Eq. (7.35) gives

$$\mathbf{x}^T\mathbf{S}_{k+1}\mathbf{x} = \frac{\|\mathbf{u}\|^2\|\mathbf{v}\|^2 - (\|\mathbf{u}\| \, \|\mathbf{v}\| \cos\theta)^2}{\|\mathbf{v}\|^2} + \frac{(\mathbf{x}^T\delta_k)^2}{\alpha_k \mathbf{g}_k^T\mathbf{S}_k\mathbf{g}_k}$$

The minimum value of the right-hand side of the above equation occurs when $\theta = 0$. In such a case, we have

$$\mathbf{x}^T\mathbf{S}_{k+1}\mathbf{x} = \frac{(\mathbf{x}^T\delta_k)^2}{\alpha_k \mathbf{g}_k^T\mathbf{S}_k\mathbf{g}_k} \tag{7.36}$$

Since vectors $\mathbf{u}$ and $\mathbf{v}$ point in the same direction, we can write

$$\mathbf{u} = \mathbf{S}_k^{1/2}\mathbf{x} = \beta\mathbf{v} = \beta\mathbf{S}_k^{1/2}\gamma_k = \mathbf{S}_k^{1/2}\beta\gamma_k$$

and thus

$$\mathbf{x} = \beta\gamma_k$$

where β is a positive constant. On eliminating $\mathbf{x}$ in Eq. (7.36) and then eliminating $\gamma_k^T\delta_k = \delta_k^T\gamma_k$ using Eq. (7.34), we get

$$\mathbf{x}^T\mathbf{S}_{k+1}\mathbf{x} = \frac{(\beta\gamma_k^T\delta_k)^2}{\alpha_k \mathbf{g}_k^T\mathbf{S}_k\mathbf{g}_k} = \alpha_k\beta^2\mathbf{g}_k^T\mathbf{S}_k\mathbf{g}_k$$

Now for any $\theta \geq 0$, we have

$$\mathbf{x}^T\mathbf{S}_{k+1}\mathbf{x} \geq \alpha_k\beta^2\mathbf{g}_k^T\mathbf{S}_k\mathbf{g}_k \tag{7.37}$$

Therefore, if $\mathbf{x} = \mathbf{x}_k$ is not the minimizer $\mathbf{x}^*$ (i.e., $\mathbf{g}_k \neq \mathbf{0}$), we have

$$\mathbf{x}^T\mathbf{S}_{k+1}\mathbf{x} > 0 \qquad \text{for } \mathbf{x} \neq \mathbf{0}$$

since $\alpha_k > 0$ and $\mathbf{S}_k$ is positive definite. In effect, *a positive definite $\mathbf{S}_k$ will yield a positive definite $\mathbf{S}_{k+1}$.*

∎

It is important to note that the above result holds for any $\alpha_k > 0$ for which

$$\delta_k^T\gamma_k = \delta_k^T\mathbf{g}_{k+1} - \delta_k^T\mathbf{g}_k > 0 \tag{7.38}$$

even if $f(\mathbf{x})$ is not minimized at point $\mathbf{x}_{k+1}$, as can be verified by eliminating $\mathbf{x}$ in Eq. (7.32) and then using the inequality in Eq. (7.38) (see Prob. 7.2). Consequently, if $\delta_k^T\mathbf{g}_{k+1} > \delta_k^T\mathbf{g}_k$, the positive definiteness of $\mathbf{S}_{k+1}$ can be assured even in the case where the minimization of $f(\mathbf{x}_k + \alpha\mathbf{d}_k)$ is inexact.

188

The inequality in Eq. (7.38) will be put to good use later in the construction of a practical quasi-Newton algorithm (see Algorithm 7.3).

Theorem 7.3 *Conjugate directions in DFP method*
(a) If the line searches in Step 2 of the DFP algorithm are exact and $f(\mathbf{x})$ is a convex quadratic function, then the directions generated $\boldsymbol{\delta}_0, \boldsymbol{\delta}_1, \ldots, \boldsymbol{\delta}_k$ form a conjugate set, i.e.,

$$\boldsymbol{\delta}_i^T \mathbf{H} \boldsymbol{\delta}_j = 0 \qquad for\ 0 \le i < j \le k \tag{7.39}$$

(b) If

$$\boldsymbol{\gamma}_i = \mathbf{H}\boldsymbol{\delta}_i \qquad for\ 0 \le i \le k \tag{7.40}$$

then

$$\boldsymbol{\delta}_i = \mathbf{S}_{k+1}\boldsymbol{\gamma}_i \qquad for\ 0 \le i \le k \tag{7.41}$$

Proof As for Theorem 7.1, the proof is by induction. We assume that

$$\boldsymbol{\delta}_i^T \mathbf{H} \boldsymbol{\delta}_j = 0 \qquad for\ 0 \le i < j \le k - 1 \tag{7.42}$$
$$\boldsymbol{\delta}_i = \mathbf{S}_k \boldsymbol{\gamma}_i \qquad for\ 0 \le i \le k - 1 \tag{7.43}$$

and show that Eqs. (7.39) and (7.41) hold.
  (*a*) From Eqs. (7.4) and (7.6), we can write

$$\begin{aligned}
\mathbf{g}_k &= \mathbf{g}_{k-1} + \mathbf{H}\boldsymbol{\delta}_{k-1} \\
&= \mathbf{g}_{k-2} + \mathbf{H}\boldsymbol{\delta}_{k-2} + \mathbf{H}\boldsymbol{\delta}_{k-1} \\
&= \mathbf{g}_{k-3} + \mathbf{H}\boldsymbol{\delta}_{k-3} + \mathbf{H}\boldsymbol{\delta}_{k-2} + \mathbf{H}\boldsymbol{\delta}_{k-1} \\
&\vdots \\
&= \mathbf{g}_{i+1} + \mathbf{H}(\boldsymbol{\delta}_{i+1} + \boldsymbol{\delta}_{i+2} + \cdots + \boldsymbol{\delta}_{k-1})
\end{aligned}$$

Thus for $0 \le i \le k - 1$, we have

$$\boldsymbol{\delta}_i^T \mathbf{g}_k = \boldsymbol{\delta}_i^T \mathbf{g}_{i+1} + \boldsymbol{\delta}_i^T \mathbf{H}(\boldsymbol{\delta}_{i+1} + \boldsymbol{\delta}_{i+2} + \cdots + \boldsymbol{\delta}_{k-1}) \tag{7.44}$$

If an exact line search is used in Step 2 of Algorithm 7.2, then $f(\mathbf{x})$ is minimized exactly at point $\mathbf{x}_{i+1}$, and hence

$$\boldsymbol{\delta}_i^T \mathbf{g}_{i+1} = 0 \tag{7.45}$$

(see proof of Theorem 7.2). Now for $0 \le i \le k - 1$, Eq. (7.42) gives

$$\boldsymbol{\delta}_i^T \mathbf{H}(\boldsymbol{\delta}_{i+1} + \boldsymbol{\delta}_{i+2} + \cdots + \boldsymbol{\delta}_{k-1}) = 0 \tag{7.46}$$

and from Eqs. (7.44) – (7.46), we get

$$\delta_i^T \mathbf{g}_k = 0$$

Alternatively, from Eqs. (7.43) and (7.40) we can write

$$\delta_i^T \mathbf{g}_k = (\mathbf{S}_k \boldsymbol{\gamma}_i)^T \mathbf{g}_k = (\mathbf{S}_k \mathbf{H} \delta_i)^T \mathbf{g}_k$$
$$= \delta_i^T \mathbf{H} \mathbf{S}_k \mathbf{g}_k = 0$$

Further, on eliminating $\mathbf{S}_k \mathbf{g}_k$ using Eq. (7.33)

$$\delta_i^T \mathbf{g}_k = -\frac{1}{\alpha_k} \delta_i^T \mathbf{H} \delta_k = 0$$

and since $\alpha_k > 0$, we have

$$\delta_i^T \mathbf{H} \delta_k = 0 \qquad \text{for } 0 \le i \le k-1 \tag{7.47}$$

Now on combining Eqs. (7.42) and (7.47)

$$\delta_i^T \mathbf{H} \delta_j = 0 \qquad \text{for } 0 \le i < j \le k \tag{7.48}$$

To complete the induction, we can write

$$\delta_0^T \mathbf{g}_1 = (\mathbf{S}_1 \boldsymbol{\gamma}_0)^T \mathbf{g}_1 = (\mathbf{S}_1 \mathbf{H} \delta_0)^T \mathbf{g}_1$$
$$= \delta_0^T \mathbf{H} \mathbf{S}_1 \mathbf{g}_1$$
$$= -\frac{1}{\alpha_1} \delta_0^T \mathbf{H} \delta_1$$

and since $f(\mathbf{x})$ is minimized exactly at point $\mathbf{x}_1$, we have $\delta_0^T \mathbf{g}_1 = 0$ and

$$\delta_i^T \mathbf{H} \delta_j = 0 \qquad \text{for } 0 \le i < j \le 1$$

Since Eq. (7.48) holds if Eq. (7.42) holds, we can write

$$\delta_i^T \mathbf{H} \delta_j = 0 \qquad \text{for } 0 \le i < j \le 2$$
$$\delta_i^T \mathbf{H} \delta_j = 0 \qquad \text{for } 0 \le i < j \le 3$$
$$\vdots \qquad\qquad \vdots$$
$$\delta_i^T \mathbf{H} \delta_j = 0 \qquad \text{for } 0 \le i < j \le k$$

that is, *the directions $\delta_1, \delta_2, \ldots, \delta_k$ form a conjugate set.*
   (b) From Eq. (7.43)

$$\boldsymbol{\gamma}_k^T \delta_i = \boldsymbol{\gamma}_k^T \mathbf{S}_k \boldsymbol{\gamma}_i \qquad \text{for } 0 \le i \le k-1 \tag{7.49}$$

On the other hand, Eq. (7.40) yields

$$\gamma_k^T \delta_i = \delta_k^T \mathbf{H} \delta_i \qquad \text{for } 0 \le i \le k - 1 \tag{7.50}$$

and since $\delta_0, \; \delta_1, \; \ldots, \; \delta_k$ form a set of conjugate vectors from part (a), Eqs. (7.49) – (7.50) yield

$$\gamma_k^T \delta_i = \gamma_k^T \mathbf{S}_k \gamma_i = \delta_k^T \mathbf{H} \delta_i = 0 \qquad \text{for } 0 \le i \le k - 1 \tag{7.51}$$

By noting that

$$\delta_k^T = \gamma_k^T \mathbf{S}_{k+1} \quad \text{and} \quad \mathbf{H} \delta_i = \gamma_i$$

Eq. (7.51) can be expressed as

$$\gamma_k^T \delta_i = \gamma_k^T \mathbf{S}_k \gamma_i = \gamma_k^T \mathbf{S}_{k+1} \gamma_i = 0 \qquad \text{for } 0 \le i \le k - 1$$

and, therefore,

$$\delta_i = \mathbf{S}_k \gamma_i = \mathbf{S}_{k+1} \gamma_i \qquad \text{for } 0 \le i \le k - 1 \tag{7.52}$$

Now from Eq. (7.30)

$$\delta_k = \mathbf{S}_{k+1} \gamma_k \tag{7.53}$$

and on combining Eqs. (7.52) and (7.53), we obtain

$$\delta_i = \mathbf{S}_{k+1} \gamma_i \qquad \text{for } 0 \le i \le k$$

The induction can be completed as in Theorem 7.1.

∎

For $k = n - 1$, Eqs. (7.40) and (7.41) can be expressed as

$$[\mathbf{S}_n \mathbf{H} - \lambda \mathbf{I}] \delta_i = 0 \qquad \text{for } 0 \le i \le n - 1$$

with $\lambda = 1$. In effect, vectors δ_i are eigenvectors that correspond to the unity eigenvalue for matrix $\mathbf{S}_n \mathbf{H}$. Since they are linearly independent, we have

$$\mathbf{S}_n = \mathbf{H}^{-1}$$

that is, *in a quadratic problem $\mathbf{S}_{k+1}$ becomes the Hessian on iteration $n - 1$.*

7.5.1 Alternative form of DFP formula

An alternative form of the DFP formula can be generated by using the Sherman-Morrison formula (see [8][9] and Sec. A.4) which states that an $n \times n$ matrix

$$\hat{\mathbf{U}} = \mathbf{U} + \mathbf{V} \mathbf{W} \mathbf{X}^T$$

where $\mathbf{U}$ and $\mathbf{X}$ are $n \times m$ matrices, $\mathbf{W}$ is an $m \times m$ matrix, and $m \le n$, has an inverse

$$\hat{\mathbf{U}}^{-1} = \mathbf{U}^{-1} - \mathbf{U}^{-1} \mathbf{V} \mathbf{Y}^{-1} \mathbf{X}^T \mathbf{U}^{-1} \tag{7.54}$$

where

$$\mathbf{Y} = \mathbf{W}^{-1} + \mathbf{X}^T \mathbf{U}^{-1} \mathbf{V}$$

The DFP formula can be written as

$$\mathbf{S}_{k+1} = \mathbf{S}_k + \mathbf{XWX}^T$$

where

$$\mathbf{X} = \left[\frac{\delta_k}{\sqrt{(\delta_k^T \gamma_k)}} \quad \frac{\mathbf{S}_k \gamma_k}{\sqrt{(\gamma_k^T \mathbf{S}_k \gamma_k)}} \right] \quad \text{and} \quad \mathbf{W} = \begin{bmatrix} 1 & 0 \\ 0 & -1 \end{bmatrix}$$

and hence Eq. (7.54) yields

$$\mathbf{S}_{k+1}^{-1} = \mathbf{S}_k^{-1} - \mathbf{S}_k^{-1} \mathbf{X} \mathbf{Y}^{-1} \mathbf{X}^T \mathbf{S}_k^{-1} \tag{7.55}$$

where

$$\mathbf{Y} = \mathbf{W}^{-1} + \mathbf{X}^T \mathbf{S}_k^{-1} \mathbf{X}$$

By letting

$$\mathbf{S}_{k+1}^{-1} = \mathbf{P}_{k+1}, \quad \mathbf{S}_k^{-1} = \mathbf{P}_k$$

and then deducing $\mathbf{Y}^{-1}$, Eq. (7.55) yields

$$\mathbf{P}_{k+1} = \mathbf{P}_k + \left(1 + \frac{\delta_k^T \mathbf{P}_k \delta_k}{\delta_k^T \gamma_k}\right) \frac{\gamma_k \gamma_k^T}{\delta_k^T \gamma_k} - \frac{(\gamma_k \delta_k^T \mathbf{P_k} + \mathbf{P}_k \delta_k \gamma_k^T)}{\delta_k^T \gamma_k} \tag{7.56}$$

This formula can be used to generate a sequence of approximations for the Hessian $\mathbf{H}$.

7.6 Broyden-Fletcher-Goldfarb-Shanno Method

Another recursive formula for generating a sequence of approximations for $\mathbf{H}^{-1}$ is one proposed by Broyden [2], Fletcher [10], Goldfarb [11] and Shanno [12] at about the same time. This is referred to as the *BFGS updating formula* [13][14] and is given by

$$\mathbf{S}_{k+1} = \mathbf{S}_k + \left(1 + \frac{\gamma_k^T \mathbf{S}_k \gamma_k}{\gamma_k^T \delta_k}\right) \frac{\delta_k \delta_k^T}{\gamma_k^T \delta_k} - \frac{(\delta_k \gamma_k^T \mathbf{S}_k + \mathbf{S}_k \gamma_k \delta_k^T)}{\gamma_k^T \delta_k} \tag{7.57}$$

This formula is said to be *the dual of the DFP formula* given in Eq. (7.29) and it can be obtained by letting

$$\mathbf{P}_{k+1} = \mathbf{S}_{k+1}, \quad \mathbf{P}_k = \mathbf{S}_k$$
$$\gamma_k = \delta_k, \quad \delta_k = \gamma_k$$

in Eq. (7.56). As may be expected, for convex quadratic functions, the BFGS formula has the following properties:

1. $\mathbf{S}_{k+1}$ becomes identical to $\mathbf{H}^{-1}$ for $k = n - 1$.
2. Directions $\boldsymbol{\delta}_0$, $\boldsymbol{\delta}_1$, ..., $\boldsymbol{\delta}_{n-1}$ form a conjugate set.
3. $\mathbf{S}_{k+1}$ is positive definite if $\mathbf{S}_k$ is positive definite.
4. The inequality in Eq. (7.38) applies.

An alternative form of the BFGS formula can be obtained as

$$\mathbf{P}_{k+1} = \mathbf{P}_k + \frac{\gamma_k \gamma_k^T}{\gamma_k^T \boldsymbol{\delta}_k} - \frac{\mathbf{P}_k \gamma_k \gamma_k^T \mathbf{P}_k}{\boldsymbol{\delta}_k^T \mathbf{P}_k \boldsymbol{\delta}_k}$$

by letting

$$\mathbf{S}_{k+1} = \mathbf{P}_{k+1}, \quad \mathbf{S}_k = \mathbf{P}_k$$
$$\boldsymbol{\delta}_k = \gamma_k, \quad \gamma_k = \boldsymbol{\delta}_k$$

in Eq. (7.29) or by applying the Sherman-Morrison formula to Eq. (7.57). This is the dual of Eq. (7.56).

7.7 Hoshino Method

The application of the principle of duality (i.e., the application of the Sherman-Morrison formula followed by the replacement of $\mathbf{P}_k$, $\mathbf{P}_{k+1}$, γ_k, and $\boldsymbol{\delta}_k$ by $\mathbf{S}_k$, $\mathbf{S}_{k+1}$, $\boldsymbol{\delta}_k$, and γ_k) to the rank-one formula results in one and the same formula. For this reason, the rank-one formula is said to be *self-dual*. Another self-dual formula, which was found to give good results, is one due to Hoshino [15]. Like the DFP and BFGS formulas, the Hoshino formula is of *rank two*. It is given by

$$\mathbf{S}_{k+1} = \mathbf{S}_k + \theta_k \boldsymbol{\delta}_k \boldsymbol{\delta}_k^T - \psi_k (\boldsymbol{\delta}_k \gamma_k^T \mathbf{S}_k + \mathbf{S}_k \gamma_k \boldsymbol{\delta}_k^T + \mathbf{S}_k \gamma_k \gamma_k^T \mathbf{S}_k)$$

where

$$\theta_k = \frac{\gamma_k^T \boldsymbol{\delta}_k + 2\gamma_k^T \mathbf{S}_k \gamma_k}{\gamma_k^T \boldsymbol{\delta}_k (\gamma_k^T \boldsymbol{\delta}_k + \gamma_k^T \mathbf{S}_k \gamma_k)} \quad \text{and} \quad \psi_k = \frac{1}{(\gamma_k^T \boldsymbol{\delta}_k + \gamma_k^T \mathbf{S}_k \gamma_k)}$$

The inverse of $\mathbf{S}_{k+1}$, designated as $\mathbf{P}_{k+1}$, can be obtained by applying the Sherman-Morrison formula.

7.8 The Broyden Family

An updating formula which is of significant theoretical as well as practical interest is one due to Broyden. This formula entails an independent parameter ϕ_k and is given by

$$\mathbf{S}_{k+1} = (1 - \phi_k)\mathbf{S}_{k+1}^{DFP} + \phi_k \mathbf{S}_{k+1}^{BFGS} \tag{7.58}$$

Evidently, if $\phi_k = 1$ or 0 the Broyden formula reduces to the BFGS or DFP formula, and if

$$\phi_k = \frac{\delta_k^T \gamma_k}{\delta_k^T \gamma_k \pm \gamma_k^T \mathbf{S}_k \gamma_k}$$

the rank-one or Hoshino formula is obtained.

If the formula of Eq. (7.58) is used in Step 4 of Algorithm 7.2, a Broyden method is obtained which has the properties summarized in Theorems 7.4 – 7.6 below. These are generic properties that apply to all the methods described so far.

Theorem 7.4A *Properties of Broyden method If a Broyden method is applied to a convex quadratic function and exact line searches are used, it will terminate after $m \leq n$ iterations. The following properties apply for all $k = 0, 1, \ldots, m$:*

  (a) $\delta_i = \mathbf{S}_{k+1} \gamma_i$ for $0 \leq i \leq k$
  (b) $\delta_i^T \mathbf{H} \delta_j = 0$ for $0 \leq i < j \leq k$
  (c) If $m = n - 1$, then $\mathbf{S}_m = \mathbf{H}^{-1}$

Theorem 7.4B *If $\mathbf{S}_0 = \mathbf{I}_n$, then a Broyden method with exact line searches is equivalent to the Fletcher-Reeves conjugate gradient method (see Sec. 6.6) provided that $f(\mathbf{x})$ is a convex quadratic function. Integer m in Theorem 7.4A is the least number of independent vectors in the sequence*

$$\mathbf{g}_0, \ \mathbf{H}\mathbf{g}_0, \ \mathbf{H}^2\mathbf{g}_0, \ \ldots$$

Theorem 7.4C *If $f(\mathbf{x}) \in C^1$, a Broyden method with exact line searches has the property that $\mathbf{x}_{k+1}$ and the BFGS component of the Broyden formula are independent of $\phi_0, \ \phi_1, \ \ldots, \ \phi_{k-1}$ for all $k \geq 1$.*

The proofs of these theorems are given by Fletcher [14].

7.8.1 Fletcher switch method

A particularly successful method of the Broyden family is one proposed by Fletcher [13]. In this method, parameter ϕ_k in Eq. (7.58) is switched between zero and unity throughout the optimization. The choice of ϕ_k in any iteration is based on the rule

$$\phi_k = \begin{cases} 0 & \text{if } \delta_k^T \mathbf{H} \delta_k > \delta_k^T \mathbf{P}_{k+1} \delta_k \\ \\ 1 & \text{otherwise} \end{cases}$$

where $\mathbf{H}$ is the Hessian of $f(\mathbf{x})$, and $\mathbf{P}_{k+1}$ is the approximation of $\mathbf{H}$ generated by the updating formula. In effect, Fletcher's method compares $\mathbf{P}_{k+1}$ with $\mathbf{H}$ in direction δ_k, and if the above condition is satisfied then the DFP formula is used. Alternatively, the BFGS formula is used. The Hessian is not available in

194

quasi-Newton methods but on assuming a convex quadratic problem, it can be eliminated. From Eq. (7.4).

$$\mathbf{H}\delta_k = \gamma_k$$

and, therefore, the above test becomes

$$\delta_k^T \gamma_k > \delta_k^T \mathbf{P}_{k+1} \delta_k \qquad (7.59)$$

This test is convenient to use when an approximation for $\mathbf{H}$ is to be used in the implementation of the algorithm. An alternative, but equivalent, test which is applicable to the case where an approximation for $\mathbf{H}^{-1}$ is to be used can be readily obtained from Eq. (7.59). We can write

$$\delta_k^T \gamma_k > \delta_k^T \mathbf{S}_{k+1}^{-1} \delta_k$$

and since

$$\delta_k = \mathbf{S}_{k+1} \gamma_k$$

according to Eq. (7.12), we have

$$\delta_k^T \gamma_k > \gamma_k^T \mathbf{S}_{k+1} \mathbf{S}_{k+1}^{-1} \mathbf{S}_{k+1} \gamma_k$$
$$> \gamma_k^T \mathbf{S}_{k+1} \gamma_k$$

since $\mathbf{S}_{k+1}$ is symmetric.

7.9 The Huang Family

Another family of updating formulas is one due to Huang [16]. This is a more general family which encompasses the rank-one, DFP, BFGS as well as some other formulas. It is of the form

$$\mathbf{S}_{k+1} = \mathbf{S}_k + \frac{\delta_k(\theta\delta_k + \phi\mathbf{S}_k^T\gamma_k)^T}{(\theta\delta_k + \phi\mathbf{S}_k^T\gamma_k)^T\gamma_k} - \frac{\mathbf{S}_k\gamma_k(\psi\delta_k + \omega\mathbf{S}_k^T\gamma_k)^T}{(\psi\delta_k + \omega\mathbf{S}_k^T\gamma_k)^T\gamma_k}$$

where θ, ϕ, ψ, and ω are independent parameters. The formulas that can be generated from the Huang formula are given in Table 7.1. The McCormick formula [17] is

$$\mathbf{S}_{k+1} = \mathbf{S}_k + \frac{(\delta_k - \mathbf{S}_k\gamma_k)\delta_k^T}{\delta_k^T\gamma_k}$$

whereas that of Pearson [18] is given by

$$\mathbf{S}_{k+1} = \mathbf{S}_k + \frac{(\delta_k - \mathbf{S}_k\gamma_k)\gamma_k^T\mathbf{S}_k}{\gamma_k^T\mathbf{S}_k\gamma_k}$$

7.10 Practical Quasi-Newton Algorithm

A practical quasi-Newton algorithm that eliminates the problems associated with Algorithms 7.1 and 7.2 is detailed below. This is based on Algorithm 7.2 and uses a slightly modified version of Fletcher's inexact line search (Algorithm 4.6). The algorithm is flexible, efficient, and very reliable, and has been found to be very effective for the design of digital filters and equalizers (see [19, Chap. 16]).

Table 7.1 The Huang Family

Formula	Parameters
Rank-one	$\theta = 1,\ \phi = -1,\ \psi = 1,\ \omega = -1$
DFP	$\theta = 1,\ \phi = 0,\ \psi = 0,\ \omega = 1$
BFGS	$\begin{cases} \dfrac{\phi}{\theta} = \dfrac{-\boldsymbol{\delta}_k^T \boldsymbol{\gamma}_k}{\boldsymbol{\delta}_k^T \boldsymbol{\gamma}_k + \boldsymbol{\gamma}_k^T \mathbf{S}_k \boldsymbol{\gamma}_k}, \\ \psi = 1,\ \omega = 0 \end{cases}$
McCormick	$\theta = 1,\ \phi = 0,\ \psi = 1,\ \omega = 0$
Pearson	$\theta = 0,\ \phi = 1,\ \psi = 0,\ \omega = 1$

Algorithm 7.3 Practical quasi-Newton algorithm
Step 1 (Initialize algorithm)
a. Input $\mathbf{x}_0$ and ε_1.
b. Set $k = m = 0$.
c. Set $\rho = 0.1,\ \sigma = 0.7,\ \tau = 0.1,\ \chi = 0.75,\ \hat{M} = 600$, and $\varepsilon_2 = 10^{-10}$.
d. Set $\mathbf{S}_0 = \mathbf{I}_n$.
e. Compute f_0 and $\mathbf{g}_0$, and set $m = m + 2$. Set $f_{00} = f_0$ and $\Delta f_0 = f_0$.
Step 2 (Initialize line search)
a. Set $\mathbf{d}_k = -\mathbf{S}_k \mathbf{g}_k$.
b. Set $\alpha_L = 0$ and $\alpha_U = 10^{99}$.
c. Set $f_L = f_0$ and compute $f_L' = \mathbf{g}(\mathbf{x}_k + \alpha_L \mathbf{d}_k)^T \mathbf{d}_k$.
d. (Estimate α_0)
If $|f_L'| > \varepsilon_2$, then compute $\alpha_0 = -2\Delta f_0 / f_L'$; otherwise, set $\alpha_0 = 1$.
If $\alpha_0 \leq 0$ or $\alpha_0 > 1$, then set $\alpha_0 = 1$.
Step 3
a. Set $\boldsymbol{\delta}_k = \alpha_0 \mathbf{d}_k$ and compute $f_0 = f(\mathbf{x}_k + \boldsymbol{\delta}_k)$.
b. Set $m = m + 1$.
Step 4 (Interpolation)
If $f_0 > f_L + \rho\,(\alpha_0 - \alpha_L) f_L'$ and $|f_L - f_0| > \varepsilon_2$ and $m < \hat{M}$, then do:
a. If $\alpha_0 < \alpha_U$, then set $\alpha_U = \alpha_0$.

b. Compute $\breve{\alpha}_0$ using Eq. (4.57).

c. Compute $\breve{\alpha}_{0L} = \alpha_L + \tau(\alpha_U - \alpha_L)$; if $\breve{\alpha}_0 < \breve{\alpha}_{0L}$, then set $\breve{\alpha}_0 = \breve{\alpha}_{0L}$.

d. Compute $\breve{\alpha}_{0U} = \alpha_U - \tau(\alpha_U - \alpha_L)$; if $\breve{\alpha}_0 > \breve{\alpha}_{0U}$, then set $\breve{\alpha}_0 = \breve{\alpha}_{0U}$.

e. Set $\alpha_0 = \breve{\alpha}_0$ and go to Step 3.

Step 5

Compute $f_0' = \mathbf{g}(\mathbf{x}_k + \alpha_0 \mathbf{d}_k)^T \mathbf{d}_k$ and set $m = m + 1$.

Step 6 (Extrapolation)

If $f_0' < \sigma f_L'$ and $|f_L - f_0| > \varepsilon_2$ and $m < \hat{M}$, then do:

a. Compute $\Delta \alpha_0 = (\alpha_0 - \alpha_L) f_0' / (f_L' - f_0')$ (see Eq. (4.58)).

b. If $\Delta \alpha_0 \leq 0$, then set $\breve{\alpha}_0 = 2\alpha_0$; otherwise, set $\breve{\alpha}_0 = \alpha_0 + \Delta \alpha_0$.

c. Compute $\breve{\alpha}_{0U} = \alpha_0 + \chi(\alpha_U - \alpha_0)$; if $\breve{\alpha}_0 > \breve{\alpha}_{0U}$, then set $\breve{\alpha}_0 = \breve{\alpha}_{0U}$.

d. Set $\alpha_L = \alpha_0, \alpha_0 = \breve{\alpha}_0, f_L = f_0, f_L' = f_0'$ and go to Step 3.

Step 7 (Check termination criteria and output results)

The

a. Set $\mathbf{x}_{k+1} = \mathbf{x}_k + \boldsymbol{\delta}_k$.

b. Set $\Delta f_0 = f_{00} - f_0$.

c. If $(\|\boldsymbol{\delta}_k\|_2 < \varepsilon_1$ and $|\Delta f_0| < \varepsilon_1)$ or $m \geq \hat{M}$, then output $\breve{\mathbf{x}} = \mathbf{x}_{k+1}$, $f(\breve{\mathbf{x}}) = f_{k+1}$, and stop.

d. Set $f_{00} = f_0$.

Step 8 (Prepare for the next iteration)

a. Compute $\mathbf{g}_{k+1}$ and set $\boldsymbol{\gamma}_k = \mathbf{g}_{k+1} - \mathbf{g}_k$.

b. Compute $D = \boldsymbol{\delta}_k^T \boldsymbol{\gamma}_k$; if $D \leq 0$, then set $\mathbf{S}_{k+1} = \mathbf{I}_n$; otherwise, compute $\mathbf{S}_{k+1}$ using Eq. (7.29) for the DFP method or Eq. (7.57) for the BFGS method.

c. Set $k = k + 1$ and go to Step 2.

computational complexity of an algorithm can be determined by estimating the amount of computation required, which is not always an easy task. In optimization algorithms of the type described in Chaps. 5–7, most of the computational effort is associated with function and gradient evaluations and by counting the function and gradient evaluations, a measure of the computational complexity of the algorithm can be obtained. In Algorithm 7.3, this is done through index m which is increased by one for each evaluation of f_0, $\mathbf{g}_0$, or f_0' in Steps 1, 3, and 5. Evidently, we assume here that a function evaluation requires the same computational effort as a gradient evaluation which may not be valid, since each gradient evaluation involves the evaluation of n first derivatives. A more precise measure of computational complexity could be obtained by finding the number of additions, multiplications, and divisions associated with each function and each gradient evaluation and then modifying Steps 1, 3, and 5 accordingly.

Counting the number of function evaluations can serve another useful purpose. An additional termination mechanism can be incorporated in the algorithm that can be used to abort the search for a minimum if the number of function evaluations becomes unreasonably large and exceeds some upper limit, say, $\hat{M}$. In Algorithm 7.3, interpolation is performed in Step 4 and extrapolation is performed in Step 5 only if $m < \hat{M}$, and if $m \geq \hat{M}$ the algorithm is terminated in Step 7c. This additional termination mechanism is useful when the problem being solved does not have a well defined local minimum.

Although a positive definite matrix $\mathbf{S}_k$ will ensure that $\mathbf{d}_k$ is a descent direction for function $f(\mathbf{x})$ at point $\mathbf{x}_k$, sometimes the function $f(\mathbf{x}_k + \alpha \mathbf{d}_k)$ may have a very shallow minimum with respect to α and finding such a minimum can waste a large amount of computation. The same problem can sometimes arise if $f(\mathbf{x}_k + \alpha \mathbf{d}_k)$ does not have a well-defined minimizer or in cases where $|f_L - f_0|$ is very small and of the same order of magnitude as the roundoff errors. To avoid these problems, interpolation or extrapolation is carried out only if the expected reduction in the function $f(\mathbf{x}_k + \alpha \mathbf{d}_k)$ is larger than ε_2. In such a case, the algorithm continues with the next iteration unless the termination criteria in Step 7c are satisfied.

The estimate of α_0 in Step 2d can be obtained by assuming that the function $f(\mathbf{x}_k + \alpha \mathbf{d}_k)$ can be represented by a quadratic polynomial of α and that the reduction achieved in $f(\mathbf{x}_k + \alpha \mathbf{d}_k)$ by changing α from 0 to α_0 is equal to Δf_0, the total reduction achieved in the previous iteration. Under these assumptions, we can write

$$f_L - f_0 = \Delta f_0 \tag{7.60}$$

and from Eq. (4.57)

$$\alpha_0 = \breve{\alpha} \approx \alpha_L + \frac{(\alpha_0 - \alpha_L)^2 f_L'}{2[f_L - f_0 + (\alpha_0 - \alpha_L)f_L']} \tag{7.61}$$

Since $\alpha_L = 0$, Eqns. (7.60) and (7.61) give

$$\alpha_0 \approx \frac{\alpha_0^2 f_L'}{2[\Delta f_0 + \alpha_0 f_L']}$$

Now solving for α_0, we get

$$\alpha_0 \approx -\frac{2\Delta f_0}{f_L'}$$

This estimate of α is reasonable for points far away from the solution but can become quite inaccurate as the minimizer is approached and could even become negative due to numerical ill-conditioning. For these reasons, if the estimate is equal to or less than zero or greater than unity, it is replaced by unity in Step 2d, which is the value of α that would minimize $f(\mathbf{x}_k + \alpha \mathbf{d}_k)$ in the case of a convex quadratic problem. Recall that practical problems tend to become convex and quadratic in the neighborhood of a local minimizer.

The most important difference between the inexact line search in Algorithm 4.6 and that used in Algorithm 7.3 is related to a very real problem that can arise in practice. The first derivatives f_0' and f_L' may on occasion satisfy the inequalities

$$\alpha_0 f_L' < \alpha_L f_0' \quad \text{and} \quad f_L' > f_0'$$

and the quadratic extrapolation in Step 6 would yield

$$\breve{\alpha} = \alpha_0 + \frac{(\alpha_0 - \alpha_L)f_0'}{f_L' - f_0'} = \frac{\alpha_0 f_L' - \alpha_L f_0'}{f_L' - f_0'} < 0$$

that is, it will predict a negative α. This would correspond to a maximum of $f(\mathbf{x}_k + \alpha \mathbf{d}_k)$ since $\alpha \mathbf{d}_k$ is a descent direction only if α is positive. In such a case, $\Delta \alpha_0 = \breve{\alpha} - \alpha_0$ would assume a negative value in Step 6b and to ensure that α is changed in the descent direction, the value $2\alpha_0$ is assigned to $\breve{\alpha}_0$. This new value could turn out to be unreasonably large and could exceed the most recent upper bound α_U. Although this is not catastrophic, unnecessary computations would need to be performed to return to the neighborhood of the solution, and to avoid the problem a new and more reasonable value of $\breve{\alpha}_0$ in the current bracket is used in Step 6c. A value of $\chi = 0.75$ will ensure that $\breve{\alpha}_0$ is no closer to α_U than 25 percent of the permissible range. Note that under the above circumstances, the inexact search of Algorithm 4.6 may fail to exit Step 7.

If the DFP or BFGS updating formula is used in Step 8b and the condition in Eq. (7.38) is satisfied, then a positive definite matrix $\mathbf{S}_k$ will result in a positive definite $\mathbf{S}_{k+1}$, as was discussed just after the proof of Theorem 7.2. We will now demonstrate that if the Fletcher inexact line search is used and the search is not terminated until the inequality in Eq. (4.59) is satisfied, then Eq. (7.38) is, indeed, satisfied. When the search is terminated in the kth iteration, we have $\alpha_0 \equiv \alpha_k$ and from Step 3 of the algorithm $\boldsymbol{\delta}_k = \alpha_k \mathbf{d}_k$. Now from Eqs. (7.38) and (4.59), we obtain

$$\boldsymbol{\delta}_k^T \boldsymbol{\gamma}_k = \boldsymbol{\delta}_k^T \mathbf{g}_{k+1} - \boldsymbol{\delta}_k^T \mathbf{g}_k$$
$$= \alpha_k(\mathbf{g}_{k+1}^T \mathbf{d}_k - \mathbf{g}_k^T \mathbf{d}_k)$$
$$\geq \alpha_k(\sigma - 1)\mathbf{g}_k^T \mathbf{d}_k$$

If $\mathbf{d}_k$ is a descent direction, then $\mathbf{g}_k^T \mathbf{d}_k < 0$ and $\alpha_k > 0$. Since $\sigma < 1$ in Fletcher's inexact line search, we conclude that

$$\boldsymbol{\delta}_k^T \boldsymbol{\gamma}_k > 0$$

and, in effect, the positive definiteness of $\mathbf{S}_k$ is assured. In exceptional circumstances, the inexact line search may not force the condition in Eq. (4.59), for example, when interpolation or extrapolation is aborted, if $|f_L - f_0| < \varepsilon_2$, and a nonpositive definite $\mathbf{S}_{k+1}$ matrix may occur. To safeguard against this possibility and ensure that a descent direction is achieved in every iteration, the quantity $\delta_k^T \gamma_k$ is checked in Step $8b$ and if it is found to be negative or zero, the identity matrix $\mathbf{I}_n$ is assigned to $\mathbf{S}_{k+1}$. This is not catastrophic and it may actually be beneficial since the next change in $\mathbf{x}$ will be in the steepest-descent direction.

The algorithm will be terminated in Step $7c$ if the distance between two successive points and the reduction in the objective function $f(\mathbf{x})$ are less than ε_1. One could, of course, use different tolerances for $\mathbf{x}$ and $f(\mathbf{x})$ and, depending on the problem, one of the two conditions may not even be required.

As may be recalled, the DFP and BFGS updating formulas are closely interrelated through the principle of *duality* and one can be obtained from the other and vice versa through the use of the Sherman-Morrison formula (see Sec. 7.6). Consequently, there are no clear theoretical advantages that apply to the one and not the other formula. Nevertheless, extensive experimental results reported by Fletcher [13] show that the use of the BFGS formula tends to yield algorithms that are somewhat more efficient in a number of different problems. This is consistent with the experience of the authors.

References

1 D. G. Luenberger, *Linear and Nonlinear Programming*, 2nd ed., Addison-Wesley, Reading, MA, 1984.
2 C. G. Broyden, "Quasi-Newton methods and their application to function minimization," *Maths. Comput.*, vol. 21, pp. 368–381, 1965.
3 W. C. Davidon, "Variable metric method for minimization," AEC Res. and Dev. Report ANL-5990, 1959.
4 A. V. Fiacco and G. P. McCormick, *Nonlinear Programming*, Wiley, New York, 1968.
5 B. A. Murtagh and R. W. H. Sargent, "A constrained minimization method with quadratic convergence," *Optimization*, ed. R. Fletcher, pp. 215-246, Academic Press, London, 1969.
6 P. Wolfe, "Methods of nonlinear programming," *Nonlinear Programming*, ed. J. Abadie, pp. 97–131, Interscience, Wiley, New York, 1967.
7 R. Fletcher and M. J. D. Powell, "A rapidly convergent descent method for minimization," *Computer J.*, vol. 6, pp. 163–168, 1963.
8 T. Kailath, *Linear Systems*, Prentice Hall, Englewood Cliffs, N.J., 1980.
9 P. E. Gill, W. Murray, and W. H. Wright, *Numerical Linear Algebra and Optimization*, vol. 1, Addison Wesley, Reading, MA, 1991.
10 R. Fletcher, "A new approach to variable metric algorithms," *Computer J.*, vol. 13, pp. 317–322, 1970.
11 D. Goldfarb, "A family of variable metric methods derived by variational means," *Maths. Comput.*, vol. 24, pp. 23–26, 1970.
12 D. F. Shanno, "Conditioning of quasi-Newton methods for function minimization," *Maths. Comput.*, vol. 24, pp. 647–656, 1970.
13 R. Fletcher, *Practical Methods of Optimization*, vol. 1, Wiley, New York, 1980.
14 R. Fletcher, *Practical Methods of Optimization*, 2nd ed., Wiley, New York, 1987.
15 S. Hoshino, "A formulation of variable metric methods," *J. Inst. Maths. Applns.* vol. 10, pp. 394–403, 1972.

16 H. Y. Huang, "Unified approach to quadratically convergent algorithms for function minimization," *J. Opt. Theo. Applns.*, vol. 5, pp. 405–423, 1970.

17 G. P. McCormick and J. D. Pearson, "Variable metric methods and unconstrained optimization," in *Optimization*, ed. R. Fletcher, Academic Press, London, 1969.

18 J. D. Pearson, "Variable metric methods of minimization," *Computer J.*, vol. 12, pp. 171–178,
1969.

19 A. Antoniou, *Digital Signal Processing: Signals, Systems, and Filters*, McGraw-Hill, New
York, 2005.

Problems

7.1 Let ξ be a nonzero column vector. Show that matrix $\mathbf{M} = \xi\xi^T$ has a rank
of one and is symmetric and positive semidefinite.

7.2 In a quasi-Newton algorithm, $\mathbf{S}_{k+1}$ is obtained from a positive definite
matrix $\mathbf{S}_k$ by using the DFP updating formula. Show that the condition

$$\delta_k^T \gamma_k > 0$$

will ensure that $\mathbf{S}_{k+1}$ is positive definite.

7.3 Minimize the objective function in Prob. 5.4 by applying the DFP algo-
rithm (e.g., Algorithm 7.3 with the DFP updating formula) using $\mathbf{x}_0 = [0\ 0]^T$ and $\varepsilon = 3 \times 10^{-7}$. Compare the results with those obtained in
Prob. 5.4.

7.4 Minimize the objective function in Prob. 5.5 by applying the DFP algo-
rithm using $\mathbf{x}_0 = [1\ 1\ 1]^T$ and $\varepsilon = 10^{-6}$. Compare the results with those
obtained in Probs. 5.5 and 6.4.

7.5 Minimize the objective function in Prob. 5.7 by applying the DFP algo-
rithm using $\varepsilon = 10^{-6}$, $\mathbf{x}_0 = [4\ -4]^T$, and $\mathbf{x}_0 = [-4\ 4]^T$. Compare the
results with those obtained in Probs. 5.7 and 6.5.

7.6 Minimize the objective function in Prob. 5.9 by applying the DFP algorithm using $\mathbf{x}_0 = [0.1\ 0.1]^T$ and $\varepsilon = 10^{-6}$. Compare the results with
those obtained in Probs. 5.9 and 6.6.

7.7 Implement a quasi-Newton algorithm based on the DFP formula in a computer language of your choice and use it to minimize

$$f(\mathbf{x}) = 100(x_2 - x_1^2)^2 + (1 - x_1)^2$$

(a) Try three different initial points and observe the results.

(b) Compare the results with those obtained in Prob. 6.2.

7.8 Minimize the objective function in Prob. 5.4 by applying the BFGS algorithm (e.g., Algorithm 7.3 with the BFGS updating formula) using
$\mathbf{x}_0 = [0\ 0]^T$ and $\varepsilon = 3 \times 10^{-7}$. Compare the results with those obtained in Probs. 5.4 and 7.3.

7.9 Minimize the objective function in Prob. 5.5 by applying the BFGS algorithm using $\mathbf{x}_0 = [1\ 1\ 1]^T$ and $\varepsilon = 10^{-6}$. Compare the results with those obtained in Probs. 5.5, 6.4, and 7.4.

7.10 Minimize the objective function in Prob. 5.7 by applying the BFGS algorithm using $\varepsilon = 10^{-6}$, $\mathbf{x}_0 = [4\ -4]^T$, and $\mathbf{x}_0 = [-4\ 4]^T$. Compare the results with those obtained in Probs. 5.7, 6.5, and 7.5.

7.11 Minimize the objective function in Prob. 5.9 by applying the BFGS algorithm using $\mathbf{x}_0 = [0.1\ 0.1]^T$ and $\varepsilon = 10^{-6}$. Compare the results with those obtained in Probs. 5.9, 6.6, and 7.6.

7.12 Implement a quasi-Newton algorithm based on the BFGS formula in a computer language of your choice and use it to minimize function $f(\mathbf{x})$ given in Prob. 7.7.

(a) Try three different initial points and observe the results.

(b) Compare the results with those obtained in Probs. 7.7 and 6.2.

7.13 Using the program constructed in Prob. 7.7, minimize the function

$$f(\mathbf{x}) = 100[(x_3 - 10\theta)^2 + (r - 1)^2] + x_3^2$$

where

$$\theta = \begin{cases} \dfrac{1}{2\pi} \tan^{-1}\left(\dfrac{x_2}{x_1}\right) & \text{for } x_1 > 0 \\ 0.25 & \text{for } x_1 = 0 \\ 0.5 + \dfrac{1}{2\pi} \tan^{-1}\left(\dfrac{x_2}{x_1}\right) & \text{for } x_1 < 0 \end{cases}$$

and

$$r = \sqrt{(x_1^2 + x_2^2)}$$

Repeat with the program constructed in Prob. 7.12 and compare the results obtained.

7.14 Using the program constructed in Prob. 7.6, minimize the function

$$f(\mathbf{x}) = (x_1 + 10x_2)^2 + 5(x_3 - x_4)^2 + (x_2 - 2x_3)^4 + 100(x_1 - x_4)^4$$

Repeat with the program constructed in Prob. 7.12 and compare the results obtained.

7.15 Using the program constructed in Prob. 7.7, minimize the function

$$f(\mathbf{x}) = \sum_{i=2}^{5}[100(x_i - x_{i-1}^2)^2 + (1 - x_i)^2]$$

Repeat with the program constructed in Prob. 7.12 and compare the results obtained.

202

7.16 An interesting variant of the BFGS method is to modify the formula in
Eq. (7.57) by replacing $\mathbf{S}_k$ by the identify matrix, which gives

$$\mathbf{S}_{k+1} = \mathbf{I} + \left(1 + \frac{\boldsymbol{\gamma}_k^T \boldsymbol{\gamma}_k}{\boldsymbol{\gamma}_k^T \boldsymbol{\delta}_k}\right) \frac{\boldsymbol{\delta}_k \boldsymbol{\delta}_k^T}{\boldsymbol{\gamma}_k^T \boldsymbol{\delta}_k} - \frac{\boldsymbol{\delta}_k \boldsymbol{\gamma}_k^T + \boldsymbol{\gamma}_k \boldsymbol{\delta}_k^T}{\boldsymbol{\gamma}_k^T \boldsymbol{\delta}_k}$$

Since $\mathbf{S}_{k+1}$ is now determined without reference to $\mathbf{S}_k$, the above updating
formula is known as a *memoryless* BFGS formula [1].

Verify that the memoryless BFGS method can be implemented without
explicitly updating matrix $\mathbf{S}_k$. Instead, point $\mathbf{x}_k$ is updated as

$$\mathbf{x}_{k+1} = \mathbf{x}_k + \alpha_k \mathbf{d}_k$$

where α_k is determined by using a line search, and $\mathbf{d}_k$ is updated using
the formula

$$\mathbf{d}_{k+1} = -\mathbf{g}_{k+1} + \eta_1 \boldsymbol{\gamma}_k + (\eta_2 - \eta_3)\boldsymbol{\delta}_k$$

where

$$\eta_1 = \boldsymbol{\delta}_k^T \mathbf{g}_{k+1}/\eta_4, \quad \eta_2 = \boldsymbol{\gamma}_k^T \mathbf{g}_{k+1}/\eta_4$$

$$\eta_3 = \left(1 + \frac{\boldsymbol{\gamma}_k^T \boldsymbol{\gamma}_k}{\eta_4}\right)\eta_1, \quad \eta_4 = \boldsymbol{\gamma}_k^T \boldsymbol{\delta}_k$$

7.17 Minimize the objective function in Prob. 5.7 by applying the memoryless
BFGS method using $\varepsilon = 10^{-6}$, $\mathbf{x}_0 = [4 \ -4]^T$, and $\mathbf{x}_0 = [-4 \ 4]^T$.
Compare the results with those obtained in Probs. 5.7, 6.5, 7.5, and 7.10.

Chapter 8

MINIMAX METHODS

8.1 Introduction

In many scientific and engineering applications it is often necessary to minimize the maximum of some quantity with respect to one or more independent variables. Algorithms that can be used to solve problems of this type are said to be *minimax algorithms*. In the case where the quantity of interest depends on a real-valued parameter w that belongs to a set $\mathcal{S}$, the objective function can be represented by $f(\mathbf{x}, w)$ and the solution of the minimax problem pertaining to $f(\mathbf{x}, w)$ amounts to finding a vector variable $\mathbf{x}$ that minimizes the maximum of $f(\mathbf{x}, w)$ over $w \in \mathcal{S}$. There is also a discrete version of this problem in which the continuous parameter w is sampled to obtain discrete values $\mathcal{S}_d = \{w_i : i = 1, \ldots, L\} \subset \mathcal{S}$ and the corresponding minimax optimization problem is to find a vector $\mathbf{x}$ that minimizes the maximum of $f(\mathbf{x}, w_i)$ over $w_i \in \mathcal{S}_d$.

This chapter is concerned with efficient minimax algorithms. In Sec. 8.2, we illustrate minimax optimization using an example from digital signal processing. Two minimax algorithms due to Charalambous [1][2] are studied in Sec. 8.3 and improved versions of these algorithms using a technique of nonuniform variable sampling [3] are presented in Sec. 8.4.

8.2 Problem Formulation

A minimax problem pertaining to objective function $f(\mathbf{x}, w)$ can be formally stated as

$$\underset{\mathbf{x}}{\text{minimize}} \, \underset{w \in \mathcal{S}}{\max} \, f(\mathbf{x}, w) \tag{8.1a}$$

where $\mathcal{S}$ is a compact set on the w axis, and if $f(\mathbf{x}, w)$ is sampled with respect to w we have

$$\underset{\mathbf{x}}{\text{minimize}} \, \underset{w_i \in \mathcal{S}_d}{\max} \, f(\mathbf{x}, w_i) \tag{8.1b}$$

where $S_d = \{w_i : i = 1, 2, \ldots, L\}$ is a discrete version of set S. Obviously, the problems in Eqs. (8.1a) and (8.1b) are closely interrelated, and subject to the condition that the sampling of S is sufficiently dense, an approximate solution of the problem in Eq. (8.1a) can be obtained by solving the discrete problem in Eq. (8.1b).

As an illustrative example, let us consider a problem encountered in the field of digital signal processing whereby a digital filter needs to be designed [4, Chap. 16].[1] In this design problem, we require a transfer function of the form

$$H(z) = \frac{\displaystyle\sum_{i=0}^{N} a_i z^{-i}}{1 + \displaystyle\sum_{i=1}^{N} b_i z^{-i}} \tag{8.2}$$

where z is a complex variable and a_i, b_i are real coefficients (see Sec. B.5.1) such that the amplitude response of the filter

$$M(\mathbf{x}, \omega) = |H(e^{j\omega T})| \tag{8.3}$$

approximates a specified amplitude response $M_0(\omega)$. Vector $\mathbf{x}$ in Eq. (8.3) is defined as

$$\mathbf{x} = [a_0 \; a_1 \; \cdots \; a_N \; b_1 \; \cdots \; b_N]^T$$

and ω denotes the frequency that can assume values in the range of interest Ω. In the case of a lowpass digital filter, the desired amplitude response, $M_0(\omega)$, is assumed to be a piecewise constant function, as illustrated in Fig. 8.1 (see Sec. B.9.1). The difference between $M(\mathbf{x}, \omega)$ and $M_0(\omega)$, which is, in effect, the approximation error, can be expressed as

$$e(\mathbf{x}, \omega) = M(\mathbf{x}, \omega) - M_0(\omega) \tag{8.4}$$

(see Sec. B.9.3).

The design of a digital filter can be accomplished by minimizing one of the norms described in Sec. A.8.1. If the L_1 or L_2 norm is minimized, then the sum of the magnitudes or the sum of the squares of the elemental errors is minimized. The minimum error thus achieved usually turns out to be unevenly distributed with respect to frequency and may exhibit large peaks which are often objectionable. If prescribed amplitude response specifications are to be met, the magnitude of the largest elemental error should be minimized and, therefore, the L_∞ norm of the error function should be used. Since the L_∞ norm of the error function $e(\mathbf{x}, \omega)$ in Eq. (8.4) is numerically equal to $\max\limits_{\omega \in \Omega} |e(\mathbf{x}, \omega)|$, the minimization of the L_∞ norm can be expressed as

$$\underset{\mathbf{x}}{\text{minimize}} \; \max_{\omega \in \Omega} |e(\mathbf{x}, \omega)| \tag{8.5}$$

---

[1] See Appendix B for a brief summary of the basics of digital filters.

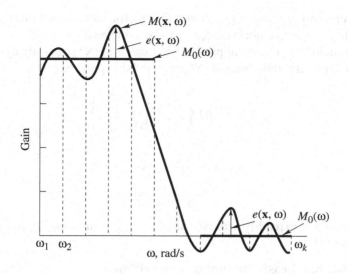

Figure 8.1. Formulation of objective function.

This is a minimax problem of the type stated in Eq. (8.1a) where the objective function is the magnitude of the approximation error, i.e., $f(\mathbf{x}, \omega) = |e(\mathbf{x}, \omega)|$.

The application of minimax algorithms for the design of digital filters usually yields designs in which the error is uniformly distributed with respect to frequency.

8.3 Minimax Algorithms

The most fundamental algorithm for the minimax optimization problem in Eq. (8.5) is the so-called *least-pth* algorithm, which involves minimizing an objective function in the form of a sum of elemental error functions, each raised to the pth power, for increasing values of p, say, $p = 2, 4, 8, \ldots$, etc.

Let $\omega_1, \omega_2, \ldots, \omega_K$ be K frequencies in Ω and define vector

$$\mathbf{e}(\mathbf{x}) = [e_1(\mathbf{x}) \, e_2(\mathbf{x}) \, \cdots \, e_n(\mathbf{x})]^T$$

where $e_i(\mathbf{x}) \equiv e(\mathbf{x}, \omega_i)$ is evaluated using Eq. (8.4). If we denote the L_p norm of vector $\mathbf{e}(\mathbf{x})$ at $\mathbf{x} = \mathbf{x}_k$ as $\Psi_k(\mathbf{x})$, i.e.,

$$\Psi_k(\mathbf{x}) = \|\mathbf{e}(\mathbf{x})\|_p = \left[\sum_{i=1}^{K} |e_i(\mathbf{x})|^p \right]^{1/p}$$

then we have

$$\lim_{p \to \infty} \Psi_k(\mathbf{x}) = \lim_{p \to \infty} \|\mathbf{e}(\mathbf{x})\|_p = \|\mathbf{e}(\mathbf{x})\|_\infty = \max_{1 \le i \le K} |e_i(\mathbf{x})| \equiv \widehat{E}(\mathbf{x})$$

In other words, by minimizing function $\Psi_k(\mathbf{x})$ for increasing power of p, the minimization of the L_∞ norm of $\mathbf{e}(\mathbf{x})$ can be achieved.

In a practical design, the approximation error $\|\mathbf{e}(\mathbf{x})\|_\infty$ is always strictly greater than zero and thus function $\Psi_k(\mathbf{x})$ can be expressed as

$$\Psi_k(\mathbf{x}) = \widehat{E}(\mathbf{x}) \left\{ \sum_{i=1}^{K} \left[\frac{|e_i(\mathbf{x})|}{\widehat{E}(\mathbf{x})} \right]^p \right\}^{1/p} \tag{8.6a}$$

where

$$e_i(\mathbf{x}) \equiv e(\mathbf{x}, \omega_i) \tag{8.6b}$$

$$\widehat{E}(\mathbf{x}) = \max_{1 \le i \le K} |e_i(\mathbf{x})| \tag{8.6c}$$

These principles lead readily to the so-called *least-pth minimax algorithm* which is as follows [1]:

Algorithm 8.1 Least-pth minimax algorithm
Step 1
Input $\breve{\mathbf{x}}_0$ and ε_1. Set $k = 1$, $p = 2$, $\mu = 2$, and $\widehat{E}_0 = 10^{99}$.
Step 2
Initialize frequencies ω_1, ω_2, ..., ω_K.
Step 3
Using $\breve{\mathbf{x}}_{k-1}$ as initial point, minimize $\Psi_k(\mathbf{x})$ in Eq. (8.6a) with respect to $\mathbf{x}$, to obtain $\breve{\mathbf{x}}_k$. Set $\widehat{E}_k = \widehat{E}(\breve{\mathbf{x}})$.
Step 4
If $|\widehat{E}_{k-1} - \widehat{E}_k| < \varepsilon_1$, then output $\breve{\mathbf{x}}_k$ and $\widehat{E}_k$, and stop. Otherwise, set $p = \mu p$ and $k = k + 1$, and go to step 3.

∎

The underlying principle for the above algorithm is that the minimax problem is solved by solving a sequence of closely related problems whereby the solution of one problem renders the solution of the next one more tractable. Parameter μ in step 1, which must obviously be an integer, should not be too large in order to avoid numerical ill-conditioning. A value of 2 gives good results.

The minimization in step 3 can be carried out by using any unconstrained optimization algorithm, for example, Algorithm 7.3 described in Sec. 7.10. The gradient of $\Psi_k(\mathbf{x})$ is given by [1]

$$\nabla \Psi_k(\mathbf{x}) = \left\{ \sum_{i=1}^{K} \left[\frac{|e_i(\mathbf{x})|}{\widehat{E}(\mathbf{x})} \right]^p \right\}^{(1/p)-1} \sum_{i=1}^{K} \left[\frac{|e_i(\mathbf{x})|}{\widehat{E}(\mathbf{x})} \right]^{p-1} \nabla |e_i(\mathbf{x})| \tag{8.7}$$

The preceding algorithm works very well, except that it requires a considerable amount of computation. An alternative and much more efficient minimax algorithm is one described in [5], [6]. This algorithm is based on principles developed by Charalambous [2] and involves the minimization of the objective function

$$\Psi(\mathbf{x}, \boldsymbol{\lambda}, \xi) = \sum_{i \in I_1} \tfrac{1}{2} \lambda_i [\phi_i(\mathbf{x}, \xi)]^2 + \sum_{i \in I_2} \tfrac{1}{2} [\phi_i(\mathbf{x}, \xi)]^2 \qquad (8.8)$$

where ξ and λ_i for $i = 1, 2, \ldots, K$ are constants and

$$\phi_i(\mathbf{x}, \xi) = |e_i(\mathbf{x})| - \xi$$
$$I_1 = \{i : \ \phi_i(\mathbf{x}, \xi) > 0 \text{ and } \lambda_i > 0\} \qquad (8.9a)$$
$$I_2 = \{i : \ \phi_i(\mathbf{x}, \xi) > 0 \text{ and } \lambda_i = 0\} \qquad (8.9b)$$

The halves in Eq. (8.8) are included for the purpose of simplifying the expression for the gradient (see Eq. (8.11)).

If

(a) the second-order sufficiency conditions for a minimum of $\widehat{E}(\mathbf{x})$ hold at $\breve{\mathbf{x}}$,

(b) $\lambda_i = \breve{\lambda}_i$ for $i = 1, 2, \ldots, K$ where $\breve{\lambda}_i$ are the minimax multipliers corresponding to the minimum point $\breve{\mathbf{x}}$ of $\widehat{E}(\mathbf{x})$, and

(c) $\widehat{E}(\breve{\mathbf{x}} - \xi)$ is sufficiently small

then it can be proved that $\breve{\mathbf{x}}$ is a *strong* local minimum point of function $\Psi(\mathbf{x}, \boldsymbol{\lambda}, \xi)$ given by Eq. (8.8) (see [2] for details). In practice, the conditions in (a) are satisfied for most practical problems. Consequently, if multipliers λ_i are forced to approach the minimax multipliers $\breve{\lambda}_i$ and ξ is forced to approach $\widehat{E}(\breve{\mathbf{x}})$, then the minimization of $\widehat{E}(\mathbf{x})$ can be accomplished by minimizing $\Psi(\mathbf{x}, \boldsymbol{\lambda}, \xi)$ with respect to $\mathbf{x}$. A minimax algorithm based on these principles is as follows:

Algorithm 8.2 Charalambous minimax algorithm
Step 1
Input $\breve{\mathbf{x}}_0$ and ε_1. Set $k = 1$, $\xi_1 = 0$, $\lambda_{11} = \lambda_{12} = \cdots = \lambda_{1K} = 1$, and $\widehat{E}_0 = 10^{99}$.
Step 2
Initialize frequencies $\omega_1, \omega_2, \ldots, \omega_K$.
Step 3
Using $\breve{\mathbf{x}}_{k-1}$ as initial point, minimize $\Psi(\mathbf{x}, \boldsymbol{\lambda}_k, \xi_k)$ with respect to $\mathbf{x}$ to obtain $\breve{\mathbf{x}}_k$. Set

$$\widehat{E}_k = \widehat{E}(\breve{\mathbf{x}}_k) = \max_{1 \leq i \leq K} |e_i(\breve{\mathbf{x}}_k)| \qquad (8.10)$$

Step 4
Compute

$$\Phi_k = \sum_{i \in I_1} \lambda_{ki} \phi_i(\breve{\mathbf{x}}_k, \, \xi_k) + \sum_{i \in I_2} \phi_i(\breve{\mathbf{x}}_k, \, \xi_k)$$

and update

$$\lambda_{(k+1)i} = \begin{cases} \lambda_{ki} \phi_i(\breve{\mathbf{x}}_k, \, \xi_k)/\Phi_k & \text{for } i \in I_1 \\ \phi_i(\breve{\mathbf{x}}_k, \, \xi_k)/\Phi_k & \text{for } i \in I_2 \\ 0 & \text{for } i \in I_3 \end{cases}$$

for $i = 1, 2, \ldots, K$ where

$$I_1 = \{i : \phi_i(\breve{\mathbf{x}}_k, \, \xi_k) > 0 \text{ and } \lambda_{ki} > 0\}$$
$$I_2 = \{i : \phi_i(\breve{\mathbf{x}}_k, \, \xi_k) > 0 \text{ and } \lambda_{ki} = 0\}$$

and

$$I_3 = \{i : \phi_i(\breve{\mathbf{x}}_k, \, \xi_k) \leq 0\}$$

Step 5
Compute

$$\xi_{k+1} = \sum_{i=1}^{K} \lambda_{(k+1)i} |e_i(\breve{\mathbf{x}})|$$

Step 6
If $|\widehat{E}_{k-1} - \widehat{E}_k| < \varepsilon_1$, then output $\breve{\mathbf{x}}_k$ and $\widehat{E}_k$, and stop. Otherwise, set $k = k + 1$ and go to step 3.

The gradient of $\Psi(\mathbf{x}, \lambda_k, \xi_k)$, which is required in step 3 of the algorithm, is given by

$$\nabla \Psi(\mathbf{x}, \lambda_k, \xi_k) = \sum_{i \in I_1} \lambda_{ki} \phi_i(\mathbf{x}, \xi_k) \nabla |e_i(\mathbf{x})| + \sum_{i \in I_2} \phi_i(\mathbf{x}, \xi_k) \nabla |e_i(\mathbf{x})| \quad (8.11)$$

Constant ξ is a lower bound of the minimum of $\widehat{E}(\mathbf{x})$ and as the algorithm progresses, it approaches $\widehat{E}(\breve{\mathbf{x}})$ from below. Consequently, the number of functions $\phi_i(\mathbf{x}, \xi)$ that do not satisfy either Eq. (8.9a) or Eq. (8.9b) increases rapidly with the number of iterations. Since the derivatives of these functions are unnecessary in the minimization of $\Psi(\mathbf{x}, \lambda, \xi)$, they need not be evaluated. This increases the efficiency of the algorithm quite significantly.

As in Algorithm 8.1, the minimization in step 3 of Algorithm 8.2 can be carried out by using Algorithm 7.3.

Example 8.1 Consider the overdetermined system of linear equations

$$
\begin{aligned}
3x_1 - 4x_2 + 2x_3 - x_4 &= -17.4 \\
-2x_1 + 3x_2 + 6x_3 - 2x_4 &= -1.2 \\
x_1 + 2x_2 + 5x_3 + x_4 &= 7.35 \\
-3x_1 + x_2 - 2x_3 + 2x_4 &= 9.41 \\
7x_1 - 2x_2 + 4x_3 + 3x_4 &= 4.1 \\
10x_1 - x_2 + 8x_3 + 5x_4 &= 12.3
\end{aligned}
$$

which can be expressed as

$$\mathbf{Ax} = \mathbf{b} \tag{8.12a}$$

where

$$
\mathbf{A} =
\begin{bmatrix}
3 & -4 & 2 & -1 \\
-2 & 3 & 6 & -2 \\
1 & 2 & 5 & 1 \\
-3 & 1 & -2 & 2 \\
7 & -2 & 4 & 3 \\
10 & -1 & 8 & 5
\end{bmatrix},
\quad
\mathbf{b} =
\begin{bmatrix}
-17.4 \\
-1.2 \\
7.35 \\
9.41 \\
4.1 \\
12.3
\end{bmatrix}
\tag{8.12b}
$$

(*a*) Find the least-squares solution of Eq. (8.12a), $\mathbf{x}_{ls}$, by solving the minimization problem

$$\text{minimize } \|\mathbf{Ax} - \mathbf{b}\| \tag{8.13}$$

(*b*) Find the minimax solution of Eq. (8.12a), $\mathbf{x}_{minimax}$, by applying Algorithm 8.2 to solve the minimization problem

$$\text{minimize } \|\mathbf{Ax} - \mathbf{b}\|_\infty \tag{8.14}$$

(*c*) Compare the magnitudes of the equation errors for the solutions $\mathbf{x}_{ls}$ and $\mathbf{x}_{minimax}$.

Solution

(*a*) The square of the L_2 norm $\|\mathbf{Ax} - \mathbf{b}\|$ is found to be

$$\|\mathbf{Ax} - \mathbf{b}\|^2 = \mathbf{x}^T \mathbf{A}^T \mathbf{Ax} - 2\mathbf{x}^T \mathbf{A}^T \mathbf{b} + \mathbf{b}^T \mathbf{b}$$

It is easy to verify that matrix $\mathbf{A}^T\mathbf{A}$ is positive definite; hence $\|\mathbf{Ax} - \mathbf{b}\|^2$ is a strictly globally convex function whose unique minimizer is given by

$$\mathbf{x}_{ls} = (\mathbf{A}^T\mathbf{A})^{-1}\mathbf{A}^T\mathbf{b} = \begin{bmatrix} 0.6902 \\ 3.6824 \\ -0.7793 \\ 3.1150 \end{bmatrix} \qquad (8.15)$$

(b) By denoting

$$\mathbf{A} = \begin{bmatrix} \mathbf{a}_1^T \\ \mathbf{a}_2^T \\ \vdots \\ \mathbf{a}_6^T \end{bmatrix}, \quad \mathbf{b} = \begin{bmatrix} b_1 \\ b_2 \\ \vdots \\ b_6 \end{bmatrix}$$

we can write

$$\mathbf{Ax} - \mathbf{b} = \begin{bmatrix} \mathbf{a}_1^T\mathbf{x} - b_1 \\ \mathbf{a}_2^T\mathbf{x} - b_2 \\ \vdots \\ \mathbf{a}_6^T\mathbf{x} - b_6 \end{bmatrix}$$

and the L_∞ norm $\|\mathbf{Ax} - \mathbf{b}\|_\infty$ can be expressed as

$$\|\mathbf{Ax} - \mathbf{b}\|_\infty = \max_{1 \le i \le 6} |\mathbf{a}_i^T\mathbf{x} - b_i|$$

Hence the problem in Eq. (8.14) becomes

$$\underset{\mathbf{x}}{\text{minimize}} \max_{1 \le i \le 6} |e_i(\mathbf{x})| \qquad (8.16)$$

where

$$e_i(\mathbf{x}) = \mathbf{a}_i^T\mathbf{x} - b_i$$

which is obviously a minimax problem. The gradient of $e_i(\mathbf{x})$ is simply given by

$$\nabla e_i(\mathbf{x}) = \mathbf{a}_i$$

By using the least-squares solution $\mathbf{x}_{ls}$ obtained in part (a) as the initial point and $\varepsilon_1 = 4 \times 10^{-6}$, it took Algorithm 8.2 four iterations to converge to the solution

$$\mathbf{x}_{minimax} = \begin{bmatrix} 0.7592 \\ 3.6780 \\ -0.8187 \\ 3.0439 \end{bmatrix} \qquad (8.17)$$

In this example as well as Examples 8.2 and 8.3, the unconstrained optimization required is Step 3 was carried out using a quasi-Newton BFGS algorithm which was essentially Algorithm 7.3 with a slightly modified version of Step 8b as follows:

Step 8b′
Compute $D = \delta_k^T \gamma_k$. If $D \leq 0$, then set $\mathbf{S}_{k+1} = \mathbf{I}_n$, otherwise, compute $\mathbf{S}_{k+1}$ using Eq. (7.57).

(c) Using Eqs. (8.15) and (8.17), the magnitudes of the equation errors for solutions $\mathbf{x}_{ls}$ and $\mathbf{x}_{minimax}$ were found to be

$$|\mathbf{A}\mathbf{x}_{ls} - \mathbf{b}| = \begin{bmatrix} 0.0677 \\ 0.0390 \\ 0.0765 \\ 0.4054 \\ 0.2604 \end{bmatrix} \quad \text{and} \quad |\mathbf{A}\mathbf{x}_{minimax} - \mathbf{b}| = \begin{bmatrix} 0.2844 \\ 0.2844 \\ 0.2843 \\ 0.2844 \\ 0.2844 \\ 0.2844 \end{bmatrix}$$

As can be seen, the minimax algorithm tends to equalize the equation errors. ∎

8.4   Improved Minimax Algorithms

To achieve good results with the above minimax algorithms, the sampling of the objective function $f(\mathbf{x}, w)$ with respect to w must be dense; otherwise, the error in the objective function may develop spikes in the intervals between sampling points during the minimization. This problem is usually overcome by using a fairly large value of K of the order of 20 to 30 times the number of variables, depending on the type of optimization problem. For example, if a 10th-order digital filter is to be designed, i.e., $N = 10$ in Eq. (8.2), the objective function depends on 21 variables and a value of K as high as 630 may be required. In such a case, each function evaluation in the minimization of the objective function would involve computing the gain of the filter as many as 630 times. A single optimization may sometimes necessitate 300 to 600 function evaluations, and a minimax algorithm like Algorithm 8.1 or 8.2 may require 5 to 10 unconstrained optimizations to converge. Consequently, up to 3.8 million function evaluations may be required to complete a design.

A technique will now be described that can be used to suppress spikes in the error function without using a large value of K [3]. The technique entails the application of *nonuniform variable sampling* and it is described in terms of the filter-design problem considered earlier. The steps involved are as follows:

1. Evaluate the error function in Eq. (8.4) with respect to a dense set of uniformly-spaced frequencies that span the frequency band of interest, say, $\bar{\omega}_1, \bar{\omega}_2, \ldots, \bar{\omega}_L$, where L is fairly large of the order of $10 \times K$.
2. Segment the frequency band of interest into K intervals.

3. For each of the K intervals, find the frequency that yields maximum error. Let these frequencies be $\widehat{\omega}_i$ for $i = 1, 2, \ldots, K$.
4. Use frequencies $\widehat{\omega}_i$ as sample frequencies in the evaluation of the objective function, i.e., set $\omega_i = \widehat{\omega}_i$ for $i = 1, 2, \ldots, K$.

By applying the above nonuniform sampling technique before the start of the second and subsequent optimizations, *frequency points at which spikes are beginning to form are located and are used as sample points in the next optimization.* In this way, the error at these frequencies is reduced and the formation of spikes is prevented.

Assume that a digital filter is required to have a specified amplitude response with respect to a frequency band B which extends from $\bar{\omega}_1$ to $\bar{\omega}_L$, and let $\bar{\omega}_1$, $\bar{\omega}_2$, $\ldots$, $\bar{\omega}_L$ be uniformly-spaced frequencies such that

$$\bar{\omega}_i = \bar{\omega}_{i-1} + \Delta\omega$$

for $i = 2, 3, \ldots, L$ where

$$\Delta\omega = \frac{\bar{\omega}_L - \bar{\omega}_1}{L - 1} \tag{8.18}$$

These frequency points may be referred to as *virtual sample points*. Band B can be segmented into K intervals, say, Ω_1 to Ω_K such that Ω_1 and Ω_K are of width $\Delta\omega/2$, Ω_2 and Ω_{K-1} are of width $l\Delta\omega$, and Ω_i for $i = 3, 4, \ldots, K-2$ are of width $2l\Delta\omega$ where l is an integer. These requirements can be satisfied by letting

$$\Omega_1 = \left\{ \omega : \bar{\omega}_1 \le \omega < \bar{\omega}_1 + \tfrac{1}{2}\Delta\omega \right\}$$

$$\Omega_2 = \left\{ \omega : \bar{\omega}_1 + \tfrac{1}{2}\Delta\omega \le \omega < \bar{\omega}_1 + (l + \tfrac{1}{2})\Delta\omega \right\}$$

$$\Omega_i = \left\{ \omega : \bar{\omega}_1 + \left[(2i - 5)l + \tfrac{1}{2} \right] \Delta\omega \le \omega < \bar{\omega}_1 + \left[(2i - 3)l + \tfrac{1}{2} \right] \Delta\omega \right\}$$

for $i = 3, 4, \ldots, K - 2$

$$\Omega_{K-1} =$$
$$\left\{ \omega : \bar{\omega}_1 + \left[(2K - 7)l + \tfrac{1}{2} \right] \Delta\omega \le \omega < \bar{\omega}_1 + \left[(2K - 6)l + \tfrac{1}{2} \right] \Delta\omega \right\}$$

and

$$\Omega_K = \left\{ \omega : \bar{\omega}_1 + \left[(2K - 6)l + \tfrac{1}{2} \right] \Delta\omega \le \omega \le \bar{\omega}_L \right\}$$

where

$$\bar{\omega}_L = \bar{\omega}_1 + [(2K - 6)l + 1]\Delta\omega. \tag{8.19}$$

The scheme is feasible if

$$L = (2K - 6)l + 2 \tag{8.20}$$

according to Eqs. (8.18) and (8.19), and is illustrated in Fig. 8.2 for the case where $K = 8$ and $l = 5$.

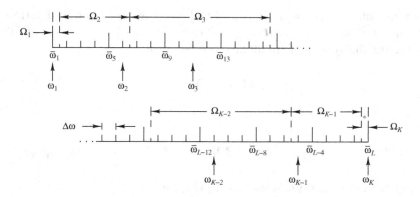

Figure 8.2. Segmentation of frequency axis.

In the above segmentation scheme, there is only one sample in each of intervals Ω_1 and Ω_K, l samples in each of intervals Ω_2 and Ω_{K-1}, and $2l$ samples in each of intervals $\Omega_3, \Omega_4, \ldots, \Omega_{K-2}$, as can be seen in Fig. 8.2. Thus step 3 of the technique will yield $\widehat{\omega}_1 = \bar{\omega}_1$ and $\widehat{\omega}_K = \bar{\omega}_L$, i.e., the lower and upper band edges are forced to remain sample frequencies throughout the optimization. This strategy leads to two advantages: (*a*) the error at the band edges is always minimized, and (*b*) a somewhat higher sampling density is maintained near the band edges where spikes are more likely to occur.

In the above technique, the required amplitude response, $M_0(\omega)$, needs to be specified with respect to a *dense* set of frequency points. If $M_0(\omega)$ is piecewise constant as in Fig. 8.1, then the required values of $M_0(\omega)$ can be easily obtained. If, on the other hand, $M_0(\omega)$ is specified by an array of numbers, the problem can be overcome through the use of interpolation. Let us assume that the amplitude response is specified at frequencies $\tilde{\omega}_1$ to $\tilde{\omega}_S$, where $\tilde{\omega}_1 = \bar{\omega}_1$ and $\tilde{\omega}_S = \bar{\omega}_L$. The required amplitude response for any frequency interval spanned by four successive specification points, say, $\tilde{\omega}_j \leq \omega \leq \tilde{\omega}_{j+3}$, can be represented by a third-order polynomial of ω of the form

$$M_0(\omega) = a_{0j} + a_{1j}\omega + a_{2j}\omega^2 + a_{3j}\omega^3 \qquad (8.21)$$

and by varying j from 1 to $S - 3$, a set of $S - 3$ third-order polynomials can be obtained which can be used to interpolate the amplitude response to any desired degree of resolution. To achieve maximum interpolation accuracy, each of these polynomials should as far as possible be used at the center of the frequency range of its validity. Hence the first and last polynomials should be used for the frequency ranges $\tilde{\omega}_1 \leq \omega < \tilde{\omega}_3$ and $\tilde{\omega}_{S-2} \leq \omega \leq \tilde{\omega}_S$, respectively, and the jth polynomial for $2 \leq j \leq S - 4$ should be used for the frequency range $\tilde{\omega}_{j+1} \leq \omega < \tilde{\omega}_{j+2}$.

Coefficients a_{ij} for $i = 0, 1, \ldots, 3$ and $j = 1$ to $S - 3$ can be determined by computing $\tilde{\omega}_m$, $(\tilde{\omega}_m)^2$, and $(\tilde{\omega}_m)^3$ for $m = j, j+1, \ldots, j+3$, and then constructing the system of simultaneous equations

$$\tilde{\mathbf{\Omega}}_j \mathbf{a}_j = \mathbf{M}_{0j} \tag{8.22}$$

where

$$\mathbf{a}_j = [\, a_{0j} \quad \cdots \quad a_{3j} \,] \quad \text{and} \quad \mathbf{M}_{0j} = [\, M_0(\tilde{\omega}_j) \quad \cdots \quad M_0(\tilde{\omega}_{j+3}) \,]^T$$

are column vectors and $\tilde{\mathbf{\Omega}}_j$ is the 4×4 matrix given by

$$\tilde{\mathbf{\Omega}}_j = \begin{bmatrix} 1 & \tilde{\omega}_j & (\tilde{\omega}_j)^2 & (\tilde{\omega}_j)^3 \\ 1 & \tilde{\omega}_{j+1} & (\tilde{\omega}_{j+1})^2 & (\tilde{\omega}_{j+1})^3 \\ 1 & \tilde{\omega}_{j+2} & (\tilde{\omega}_{j+2})^2 & (\tilde{\omega}_{j+2})^3 \\ 1 & \tilde{\omega}_{j+3} & (\tilde{\omega}_{j+3})^2 & (\tilde{\omega}_{j+3})^3 \end{bmatrix}$$

Therefore, from Eq. (8.22) we have

$$\mathbf{a}_j = \tilde{\mathbf{\Omega}}_j^{-1} \mathbf{M}_{0j}. \tag{8.23}$$

The above nonuniform sampling technique can be incorporated in Algorithm 8.1 by replacing steps 1, 2, and 4 as shown below. The filter to be designed is assumed to be a single-band filter, for the sake of simplicity, although the technique is applicable to filters with an arbitrary number of bands.

Algorithm 8.3 Modified version of Algorithm 8.1
Step 1

　　a. Input $\breve{\mathbf{x}}_0$ and ε_1. Set $k = 1, p = 2, \mu = 2$, and $\widehat{E}_0 = 10^{99}$. Initialize K.

　　b. Input the required amplitude response $M_0(\tilde{\omega}_m)$ for $m = 1, 2, \ldots, S$.

　　c. Compute L and $\Delta\omega$ using Eqs. (8.20) and (8.18), respectively.

　　d. Compute coefficients a_{ij} for $i = 0, 1, \ldots, 3$ and $j = 1$ to $S - 3$ using Eq. (8.23).

　　e. Compute the required ideal amplitude response for $\bar{\omega}_1, \bar{\omega}_2, \ldots, \bar{\omega}_L$ using Eq. (8.21).

Step 2
Set $\omega_1 = \bar{\omega}_1, \omega_2 = \bar{\omega}_{1+l}, \omega_i = \bar{\omega}_{2(i-2)l+1}$ for $i = 3, 4, \ldots, K - 2$, $\omega_{K-1} = \bar{\omega}_{L-l}$, and $\omega_K = \bar{\omega}_L$.

Step 3

Using $\breve{\mathbf{x}}_{k-1}$ as initial value, minimize $\Psi_k(\mathbf{x})$ in Eq. (8.6a) with respect to $\mathbf{x}$, to obtain $\breve{\mathbf{x}}_k$. Set $\widehat{E}_k = \widehat{E}(\breve{\mathbf{x}})$.

Step 4

a. Compute $|e_i(\breve{\mathbf{x}}_k)|$ for $i = 1, 2, \ldots, L$ using Eqs. (8.4) and (8.6b).

b. Determine frequencies $\widehat{\omega}_i$ for $i = 1, 2, \ldots, K$ and

$$\widehat{P}_k = \widehat{P}(\breve{\mathbf{x}}_k) = \max_{1 \le i \le L} |e_i(\breve{\mathbf{x}}_k)| \qquad (8.24)$$

c. Set $\widehat{\omega}_i$ for $i = 1, 2, \ldots, K$.

d. If $|\widehat{E}_{k-1} - \widehat{E}_k| < \varepsilon_1$ and $|\widehat{P}_k - \widehat{E}_k| < \varepsilon_1$, then output $\breve{\mathbf{x}}_k$ and $\widehat{E}_k$, and stop. Otherwise, set $p = \mu p$, $k = k+1$ and go to step 3.

The above nonuniform variable sampling technique can be applied to Algorithm 8.2 by replacing steps 1, 2, and 6 as follows:

Algorithm 8.4 Modified version of Algorithm 8.2

Step 1

a. Input $\breve{\mathbf{x}}_0$ and ε_1. Set $k = 1$, $\xi_1 = 0$, $\lambda_{11} = \lambda_{12} = \cdots = \lambda_{1K} = 1$, and $\widehat{E}_0 = 10^{99}$. Initialize K.

b. Input the required amplitude response $M_0(\bar{\omega}_m)$ for $m = 1, 2, \ldots, S$.

c. Compute L and $\Delta\omega$ using Eqs. (8.20) and (8.18), respectively.

d. Compute coefficients a_{ij} for $i = 0, 1, \ldots, 3$ and $j = 1$ to $S - 3$ using Eq. (8.23).

e. Compute the required ideal amplitude response for $\bar{\omega}_1, \bar{\omega}_2, \ldots, \bar{\omega}_L$ using Eq. (8.21).

Step 2

Set $\omega_1 = \bar{\omega}_1$, $\omega_2 = \bar{\omega}_{1+l}$, $\omega_i = \bar{\omega}_{2(i-2)l+1}$ for $i = 3, 4, \ldots, K - 2$, $\omega_{K-1} = \bar{\omega}_{L-l}$, and $\omega_K = \bar{\omega}_L$.

Step 3

Using $\breve{\mathbf{x}}_{k-1}$ as initial value, minimize $\Psi(\mathbf{x}, \boldsymbol{\lambda}_k, \xi_k)$ with respect to $\mathbf{x}$ to obtain $\breve{\mathbf{x}}_k$. Set

$$\widehat{E}_k = \widehat{E}(\breve{\mathbf{x}}_k) = \max_{1 \le i \le K} |e_i(\breve{\mathbf{x}}_k)|$$

Step 4

Compute

$$\Phi_k = \sum_{i \in I_1} \lambda_{ki}\phi_i(\breve{\mathbf{x}}_k, \xi_k) + \sum_{i \in I_2} \phi_i(\breve{\mathbf{x}}_k, \xi_k)$$

216

and update

$$
\lambda_{(k+1)i} = \begin{cases} \lambda_{ki}\phi_i(\breve{\mathbf{x}}_k,\, \xi_k)/\Phi_k & \text{for } i \in I_1 \\ \phi_i(\breve{\mathbf{x}}_k,\, \xi_k)/\Phi_k & \text{for } i \in I_2 \\ 0 & \text{for } i \in I_3 \end{cases}
$$

for $i = 1,\, 2,\, \ldots,\, K$ where

$$I_1 = \{i : \phi_i(\breve{\mathbf{x}}_k,\, \xi_k) > 0 \text{ and } \lambda_{ki} > 0\}$$
$$I_2 = \{i : \phi_i(\breve{\mathbf{x}}_k,\, \xi_k) > 0 \text{ and } \lambda_{ki} = 0\}$$

and

$$I_3 = \{i : \phi_i(\breve{\mathbf{x}}_k,\, \xi_k) \leq 0\}$$

Step 5
Compute

$$\xi_{k+1} = \sum_{i=1}^{K} \lambda_{(k+1)i}|e_i(\breve{\mathbf{x}})|$$

Step 6

a. Compute $|e_i(\breve{\mathbf{x}}_k)|$ for $i = 1,\, 2,\, \ldots,\, L$ using Eqs. (8.4) and (8.6b).
b. Determine frequencies $\widehat{\omega}_i$ for $i = 1,\, 2,\, \ldots,\, K$ and

$$\widehat{P}_k = \widehat{P}(\breve{\mathbf{x}}_k) = \max_{1 \leq i \leq L} |e_i(\breve{\mathbf{x}}_k)|$$

c. Set $\omega_i = \widehat{\omega}_i$ for $i = 1,\, 2,\, \ldots,\, K$.
d. If $|\widehat{E}_{k-1} - \widehat{E}_k| < \varepsilon_1$ and $|\widehat{P}_k - \widehat{E}_k| < \varepsilon_1$, then output $\breve{\mathbf{x}}_k$ and $\widehat{E}_k$, and stop. Otherwise, set $k = k + 1$ and go to step 3.

In step 2, the initial sample frequencies ω_1 and ω_K are assumed to be at the left-hand and right-hand band edges, respectively; ω_2 and ω_{K-1} are taken to be the last and first frequencies in intervals Ω_2 and Ω_{K-1}, respectively; and each of frequencies $\omega_3,\, \omega_4,\, \ldots,\, \omega_{K-2}$ is set near the center of each of intervals $\Omega_3,\, \Omega_4,\, \ldots,\, \Omega_{K-2}$. This assignment is illustrated in Fig. 8.2 for the case where $K = 8$ and $l = 5$.

Without the nonuniform sampling technique, the number of samples K should be chosen to be of the order of 20 to 30 times the number of variables, depending on the selectivity of the filter, as was mentioned in the first paragraph of Sec. 8.4. If the above technique is used, the number of virtual sample points is approximately equal to $2l \times K$, according to Eq. (8.20). As l is increased above unity, the frequencies of maximum error, $\widehat{\omega}_i$, become progressively more precise, owing to the increased resolution; however, the amount of computation required in step 4 of Algorithm 8.3 or step 6 of Algorithm 8.4 is proportionally increased. Eventually, a situation of diminishing returns is reached whereby further increases in l bring about only slight improvements in the precision of the $\widehat{\omega}_i$'s. With $l = 5$, a value of K in the range of 2 to 6 times the number of variables was found to give good results for a diverse range of designs. In effect, the use of the nonuniform sampling technique in the minimax algorithms described would lead to a reduction in the amount of computation of the order of 75 percent.

Example 8.2

(a) Applying Algorithm 8.1, design a 10th-order lowpass digital filter assuming a transfer function of the form given in Eq. (8.2). The desired amplitude response is

$$M_0(\omega) = \begin{cases} 1 & \text{for } 0 \le \omega \le \omega_p \text{ rad/s} \\ 0 & \text{for } \omega_a \le \omega \le \pi \text{ rad/s} \end{cases} \qquad (8.25)$$

where $\omega_p = 0.4\pi$, $\omega_a = 0.5\pi$, and the sampling frequency is 2π.

(b) Applying Algorithm 8.3, design the digital filter specified in part (a).

Solution (a) Using Eqs. (8.2) and (8.3), the amplitude response of the filter is obtained as

$$M(\mathbf{x}, \omega) = \left| \frac{a_0 + a_1 e^{-j\omega} + \cdots + a_N e^{-jN\omega}}{1 + b_1 e^{-j\omega} + \cdots + b_N e^{-jN\omega}} \right| \qquad (8.26)$$

If we denote

$$\mathbf{a} = \begin{bmatrix} a_0 \\ a_1 \\ \vdots \\ a_N \end{bmatrix}, \quad \mathbf{b} = \begin{bmatrix} b_1 \\ b_2 \\ \vdots \\ b_N \end{bmatrix}, \quad \mathbf{c}(\omega) = \begin{bmatrix} 1 \\ \cos \omega \\ \vdots \\ \cos N\omega \end{bmatrix}, \quad \text{and } \mathbf{s}(\omega) = \begin{bmatrix} 0 \\ \sin \omega \\ \vdots \\ \sin N\omega \end{bmatrix}$$

then $\mathbf{x} = [\mathbf{a}^T \ \mathbf{b}^T]^T$. Thus the error function in Eq. (8.4) can be expressed as

$$\begin{aligned} e_i(\mathbf{x}) &= M(\mathbf{x}, \omega_i) - M_0(\omega_i) \\ &= \frac{\{[\mathbf{a}^T \mathbf{c}(\omega_i)]^2 + [\mathbf{a}^T \mathbf{s}(\omega_i)]^2\}^{1/2}}{\{[1 + \mathbf{b}^T \hat{\mathbf{c}}(\omega_i)]^2 + [\mathbf{b}^T \hat{\mathbf{s}}(\omega_i)^2\}^{1/2}} - M_0(\omega_i) \end{aligned} \qquad (8.27)$$

where

$$\hat{\mathbf{c}}(\omega) = \begin{bmatrix} \cos \omega \\ \vdots \\ \cos N\omega \end{bmatrix} \quad \text{and} \quad \hat{\mathbf{s}}(\omega) = \begin{bmatrix} \sin \omega \\ \vdots \\ \sin N\omega \end{bmatrix}$$

The gradient of the objective function $\Psi_k(\mathbf{x})$ can be obtained as

$$\nabla|e_i(\mathbf{x})| = \text{sgn}\,[e_i(\mathbf{x})]\nabla e_i(\mathbf{x}) \tag{8.28a}$$

by using Eqs. (8.7) and (8.27), where

$$\nabla e_i(\mathbf{x}) = \begin{bmatrix} \frac{\partial e_i(\mathbf{x})}{\partial \mathbf{a}} \\ \frac{\partial e_i(\mathbf{x})}{\partial \mathbf{b}} \end{bmatrix} \tag{8.28b}$$

$$\frac{\partial e_i(\mathbf{x})}{\partial \mathbf{a}} = \frac{M(\mathbf{x},\,\omega_i)\{[\mathbf{a}^T\mathbf{c}(\omega_i)]\mathbf{c}(\omega_i) + [\mathbf{a}^T\mathbf{s}(\omega_i)]\mathbf{s}(\omega_i)\}}{[\mathbf{a}^T\mathbf{c}(\omega_i)]^2 + [\mathbf{a}^T\mathbf{s}(\omega_i)]^2} \tag{8.28c}$$

$$\frac{\partial e_i(\mathbf{x})}{\partial \mathbf{b}} = \frac{M(\mathbf{x},\,\omega_i)\{[1 + \mathbf{b}^T\hat{\mathbf{c}}(\omega_i)]\hat{\mathbf{c}}(\omega_i) + [\mathbf{b}^T\hat{\mathbf{s}}(\omega_i)]\hat{\mathbf{s}}(\omega_i)\}}{[1 + \mathbf{b}^T\hat{\mathbf{c}}(\omega_i)]^2 + [\mathbf{b}^T\hat{\mathbf{s}}(\omega_i)]^2}$$

$$\tag{8.28d}$$

The above minimax problem was solved by using a MATLAB program that implements Algorithm 8.1. The program accepts the parameters ω_p, ω_a, K, and ε_1, as inputs and produces the filter coefficient vectors $\mathbf{a}$ and $\mathbf{b}$ as output. Step 3 of the algorithm was implemented using the quasi-Newton BFGS algorithm alluded to in Example 8.1 with a termination tolerance ε_2. The program also generates plots for the approximation error $|e(\mathbf{x},\,\omega)|$ and the amplitude response of the filter designed. The initial point was taken to be $\mathbf{x}_0 = [\mathbf{a}_0^T\ \mathbf{b}_0^T]^T$ where $\mathbf{a}_0 = [1\ 1\ \cdots\ 1]^T$ and $\mathbf{b}_0 = [0\ 0\ \cdots\ 0]^T$. The number of actual sample points, K, was set to 600, i.e., 267 and 333 in the frequency ranges $0 \le \omega \le 0.4\pi$ and $0.5\pi \le \omega \le \pi$, respectively, and ε_1 and ε_2 were set to 10^{-6} and 10^{-9}, respectively. The algorithm required seven iterations and 198.20 s of CPU time on a 3.1 GHz Pentium 4 PC to converge to the solution point $\mathbf{x} = [\mathbf{a}^T\ \mathbf{b}^T]^T$ where

$$\mathbf{a} = \begin{bmatrix} 0.00735344 \\ 0.02709762 \\ 0.06800724 \\ 0.12072224 \\ 0.16823049 \\ 0.18671705 \\ 0.16748698 \\ 0.11966157 \\ 0.06704789 \\ 0.02659087 \\ 0.00713664 \end{bmatrix}, \quad \mathbf{b} = \begin{bmatrix} -3.35819120 \\ 8.39305902 \\ -13.19675182 \\ 16.35127992 \\ -14.94617828 \\ 10.68550651 \\ -5.65665532 \\ 2.15596724 \\ -0.52454530 \\ 0.06260344 \end{bmatrix}$$

Note that design problems of this type have multiple possible solutions and the designer would often need to experiment with different initial points as well as different values of K, ε_1, and ε_2, in order to achieve a good design.

The transfer function of a digital filter must have poles inside the unit circle of the z plane to assure the stability of the filter (see Sec. B.7). Since the minimax algorithms of this chapter are unconstrained, no control can be exercised on the pole positions and, therefore, a transfer function may be obtained that represents an unstable filter. Fortunately, the problem can be eliminated through a well-known stabilization technique. In this technique, all the poles of the transfer function that are located outside the unit circle are replaced by their reciprocals and the transfer function is then multiplied by an appropriate multiplier constant which is equal to the reciprocal of the product of these poles (see p. 535 of [4]). For example, if

$$H(z) = \frac{N(z)}{D(z)} = \frac{N(z)}{D'(z)\prod_{i=1}^{k}(z - p_{u_i})} \tag{8.29}$$

is a transfer function with k poles p_{u_1}, p_{u_2}, ..., p_{u_k} that lie outside the unit circle, then a stable transfer function that yields the same amplitude response can be obtained as

$$H'(z) = H_0 \frac{N(z)}{D'(z)\prod_{i=1}^{k}(z - 1/p_{u_i})} = \frac{\sum_{i=0}^{N} a_i' z^i}{1 + \sum_{i=1}^{N} b_i' z^i} \tag{8.30a}$$

where

$$H_0 = \frac{1}{\prod_{i=1}^{k} p_{u_i}} \tag{8.30b}$$

In the design problem considered above, the poles of $H(z)$ were obtained as shown in column 2 of Table 8.1 by using command `roots` of MATLAB. Since $|p_i| > 1$ for $i = 1$ and 2, a complex-conjugate pair of poles are located outside the unit circle, which render the filter unstable. By applying the above stabilization technique, the poles in column 3 of Table 8.1 were obtained and multiplier constant H_0 was calculated as $H_0 = 0.54163196$.

Table 8.1 Poles of the IIR filters for Example 8.2 (*a*)

i	Poles of the unstable filter	Poles of the stabilized filter
1	$0.51495917 + 1.25741370j$	$0.27891834 + 0.68105544j$
2	$0.51495917 - 1.25741370j$	$0.27891834 - 0.68105544j$
3	$0.23514844 + 0.92879138j$	$0.23514844 + 0.92879138j$
4	$0.23514844 - 0.92879138j$	$0.23514844 - 0.92879138j$
5	$0.24539982 + 0.82867789j$	$0.24539982 + 0.82867789j$
6	$0.24539982 - 0.82867789j$	$0.24539982 - 0.82867789j$
7	$0.32452615 + 0.46022220j$	$0.32452615 + 0.46022220j$
8	$0.32452615 - 0.46022220j$	$0.32452615 - 0.46022220j$
9	$0.35906202 + 0.16438481j$	$0.35906202 + 0.16438481j$
10	$0.35906202 - 0.16438481j$	$0.35906202 - 0.16438481j$

220

By using Eq. (8.30a), coefficients $\mathbf{a}'$ and $\mathbf{b}'$ were obtained as

$$
\mathbf{a}' = \begin{bmatrix} 0.00398286 \\ 0.01467694 \\ 0.03683489 \\ 0.06538702 \\ 0.09111901 \\ 0.10113192 \\ 0.09071630 \\ 0.06481253 \\ 0.03631528 \\ 0.01440247 \\ 0.00386543 \end{bmatrix}, \quad
\mathbf{b}' = \begin{bmatrix} -2.88610955 \\ 5.98928394 \\ -8.20059471 \\ 8.75507027 \\ -7.05776764 \\ 4.44624218 \\ -2.10292453 \\ 0.72425530 \\ -0.16255342 \\ 0.01836567 \end{bmatrix}
$$

The largest magnitude of the poles of the modified transfer function is 0.9581, and thus the filter is stable.

The approximation error $|e(\mathbf{x}, \omega)|$ over the passband and stopband is plotted in Fig. 8.3 and the amplitude response of the filter is shown in Fig. 8.4.

(b) For part (b), the number of sampling points was set to 65, i.e., 29 and 36 in the frequency ranges $0 \leq \omega \leq \omega_p$ and $\omega_a \leq \omega \leq \pi$, respectively. The initial point and parameters ε_1 and ε_2 were the same as in part (a), and parameter l was set to 5. It took Algorithm 8.3 six iterations and 18.73 s of CPU time to converge to the solution point $\mathbf{x} = [\mathbf{a}^T \; \mathbf{b}^T]$ where

$$
\mathbf{a} = \begin{bmatrix} 0.00815296 \\ 0.03509437 \\ 0.09115541 \\ 0.16919427 \\ 0.24129855 \\ 0.27357739 \\ 0.24813555 \\ 0.17915173 \\ 0.09963780 \\ 0.03973358 \\ 0.00981327 \end{bmatrix}, \quad
\mathbf{b} = \begin{bmatrix} -2.02896582 \\ 3.98574025 \\ -3.65125139 \\ 2.56127374 \\ -0.11412527 \\ -1.16704564 \\ 1.36351210 \\ -0.77298905 \\ 0.25851314 \\ -0.03992105 \end{bmatrix}
$$

As can be verified, a complex-conjugate pair of poles of the transfer function obtained are located outside the unit circle. By applying the stabilization technique described in part (a), the coefficients of the modified transfer function

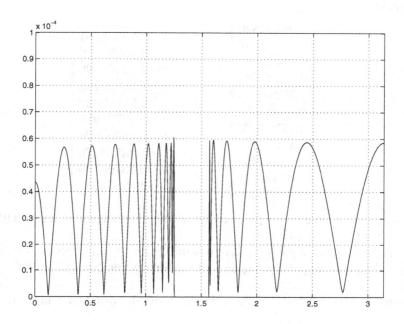

Figure 8.3. Error $|e(\mathbf{x}, \omega)|$ versus ω for Example 8.2(a).

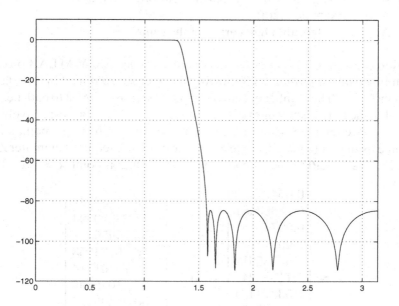

Figure 8.4. Amplitude response of the lowpass filter for Example 8.2(a).

were obtained as

$$\mathbf{a}' = \begin{bmatrix} 0.00584840 \\ 0.02517440 \\ 0.06538893 \\ 0.12136889 \\ 0.17309179 \\ 0.19624651 \\ 0.17799620 \\ 0.12851172 \\ 0.07147364 \\ 0.02850227 \\ 0.00703940 \end{bmatrix}, \quad \mathbf{b}' = \begin{bmatrix} -1.83238201 \\ 3.04688036 \\ -2.42167890 \\ 1.31022752 \\ 0.27609329 \\ -0.90732976 \\ 0.84795926 \\ -0.43579279 \\ 0.13706106 \\ -0.02054213 \end{bmatrix}$$

The largest magnitude of the poles of the modified transfer function is 0.9537 and thus the filter is stable.

The approximation error $|e(\mathbf{x}, \omega)|$ over the passband and stopband is plotted in Fig. 8.5 and the amplitude response of the filter is depicted in Fig. 8.6. ∎

The next example illustrates the application of Algorithms 8.2 and 8.4.

Example 8.3

(a) Applying Algorithm 8.2, design the 10th-order lowpass digital filter specified in Example 8.2(a).

(b) Applying Algorithm 8.4, carry out the same design.

Solution (a) The required design was obtained by using a MATLAB program that implements Algorithm 8.2 following the approach outlined in the solution of Example 8.2. The number of actual sample points, K, was set to 650, i.e., 289 and 361 in the frequency ranges $0 \leq \omega \leq \omega_p$ and $\omega_a \leq \omega \leq \pi$, respectively, and ε_1 and ε_2 were set to 3×10^{-9} and 10^{-15}, respectively. The initial point $\mathbf{x}_0$ was the same as in part (a) of Example 8.2. Algorithm 8.2 required eight iterations and 213.70 s of CPU time to converge to the solution point $\mathbf{x} = [\mathbf{a}^T \ \mathbf{b}^T]^T$ where

$$\mathbf{a} = \begin{bmatrix} 0.05487520 \\ 0.23393481 \\ 0.59719051 \\ 1.09174124 \\ 1.53685612 \\ 1.71358243 \\ 1.53374494 \\ 1.08715408 \\ 0.59319673 \\ 0.23174666 \\ 0.05398863 \end{bmatrix}, \quad \mathbf{b} = \begin{bmatrix} -5.21138732 \\ 18.28000994 \\ -39.14255091 \\ 66.45234153 \\ -78.76751214 \\ 76.41046395 \\ -50.05505315 \\ 25.84116347 \\ -6.76718946 \\ 0.68877840 \end{bmatrix}$$

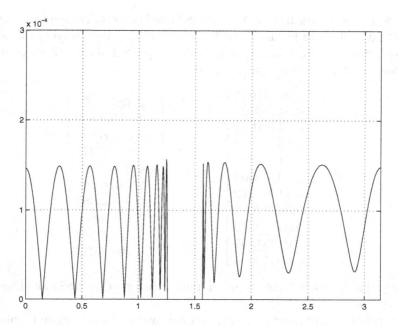

Figure 8.5. Error $|e(\mathbf{x}, \omega)|$ versus ω for Example 8.2(*b*).

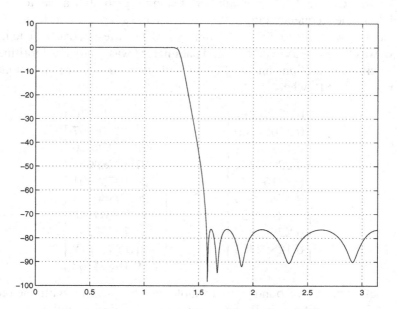

Figure 8.6. Amplitude response of the lowpass filter for Example 8.2(*b*).

As can be shown, the transfer function obtained has three complex-conjugate pairs of poles that are located outside the unit circle. By applying the stabilization technique described in part (a) of Example 8.2, the coefficients of the modified transfer function were obtained as

$$\mathbf{a}' = \begin{bmatrix} 0.00421864 \\ 0.01798421 \\ 0.04591022 \\ 0.08392980 \\ 0.11814890 \\ 0.13173509 \\ 0.11790972 \\ 0.08357715 \\ 0.04560319 \\ 0.01781599 \\ 0.00415048 \end{bmatrix}, \quad \mathbf{b}' = \begin{bmatrix} -2.49921097 \\ 4.87575840 \\ -6.01897510 \\ 5.92269310 \\ -4.27567184 \\ 2.41390695 \\ -0.98863984 \\ 0.28816806 \\ -0.05103514 \\ 0.00407073 \end{bmatrix}$$

The largest magnitude of the modified transfer function is 0.9532 and thus the filter is stable.

The approximation error $|e(\mathbf{x}, \omega)|$ over the passband and stopband is plotted in Fig. 8.7 and the amplitude response of the filter is depicted in Fig. 8.8.

(b) As in Example 8.2(b), the number of sampling points was set to 65, i.e., 29 and 36 in the frequency ranges $0 \le \omega \le \omega_p$ and $\omega_a \le \omega \le \pi$, respectively, and ε_1 and ε_2 were set to $\varepsilon = 10^{-9}$ and $\varepsilon_2 = 10^{-15}$, respectively. The initial point $\mathbf{x}_0$ was the same as in part (a) and parameter l was set to 4. Algorithm 8.4 required sixteen iterations and 48.38 s of CPU time to converge to a solution point $\mathbf{x} = [\mathbf{a}^T \ \mathbf{b}^T]^T$ where

$$\mathbf{a} = \begin{bmatrix} 0.01307687 \\ 0.05061800 \\ 0.12781582 \\ 0.22960471 \\ 0.32150671 \\ 0.35814899 \\ 0.32167525 \\ 0.22984873 \\ 0.12803465 \\ 0.05073663 \end{bmatrix}, \quad \mathbf{b} = \begin{bmatrix} -4.25811576 \\ 11.94976697 \\ -20.27972610 \\ 27.10889061 \\ -26.10756891 \\ 20.09430301 \\ -11.29104740 \\ 4.74405652 \\ -1.28479278 \\ 0.16834783 \end{bmatrix}$$

These coefficients correspond to an unstable transfer function with one pair of poles outside the unit circle. By applying the stabilization technique,

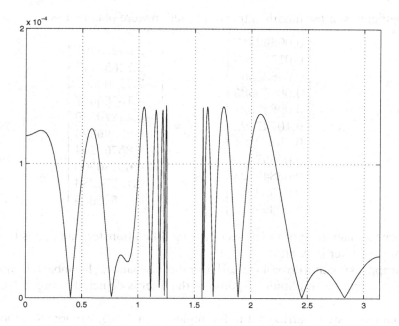

Figure 8.7.   Error $|e(\mathbf{x}, \omega)|$ versus ω for Example 8.3(a).

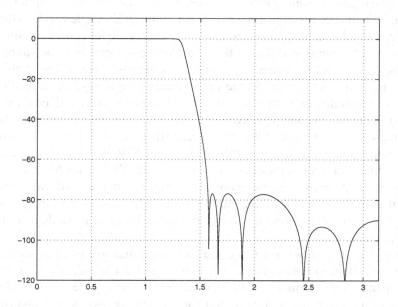

Figure 8.8.   Amplitude response of the lowpass filter for Example 8.3(a).

the coefficients of the modified transfer function were obtained as

$$\mathbf{a}' = \begin{bmatrix} 0.00392417 \\ 0.01518969 \\ 0.03835557 \\ 0.06890085 \\ 0.09647924 \\ 0.10747502 \\ 0.09652981 \\ 0.06897408 \\ 0.03842123 \\ 0.01522528 \\ 0.00393926 \end{bmatrix}, \quad \mathbf{b}' = \begin{bmatrix} -2.80254807 \\ 5.74653112 \\ -7.72509562 \\ 8.13565547 \\ -6.44870979 \\ 3.99996323 \\ -1.85761204 \\ 0.62780900 \\ -0.13776283 \\ 0.01515986 \end{bmatrix}$$

The largest magnitude of the poles of the modified transfer function is 0.9566 and thus the filter is stable.

The approximation error $|e(\mathbf{x}, \omega)|$ over the passband and stopband is plotted in Fig. 8.9 and the amplitude response of the filter is depicted in Fig. 8.10. ∎

From the designs carried out in Examples 8.2 and 8.3, we note that the use of the least-pth method with uniform sampling in Example 8.2(a) resulted in the lowest minimax error but a very large density of sample points was required to achieve a good design, which translates into a large amount of computation. Through the use of nonuniform variable sampling in Example 8.2(b), a design of practically the same quality was achieved with much less computation.

It should be mentioned that in the Charalambous algorithm, the value of ξ becomes progressively larger and approaches the minimum value of the objective function from below as the optimization progresses. As a result, the number of sample points that remain active is progressively reduced, i.e., the sizes of index sets I_1 and I_2 become progressively smaller. Consequently, by avoiding the computation of the partial derivatives of $e_i(\mathbf{x})$ for $i \in I_3$ through careful programming, the evaluation of gradient $\nabla\Psi$ (see Eq. (8.11)) can be carried out much more efficiently. In the above examples, we have not taken advantage of the above technique but our past experience has shown that when it is fully implemented, the Charalambous algorithm usually requires between 10 to 40% of the computation required by the least-pth method, depending on the application.

Finally, it should be mentioned that with optimization there is always an element of chance in obtaining a good design and, therefore, one would need to carry out a large number of different designs using a large set of randomly chosen initial points to be able to compare two alternative design algorithms such as Algorithms 8.1 and 8.2.

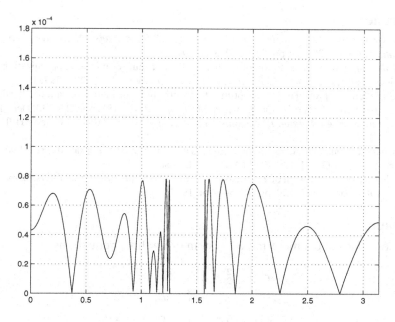

Figure 8.9.   Error $|e(\mathbf{x}, \omega)|$ versus ω for Example 8.3(b).

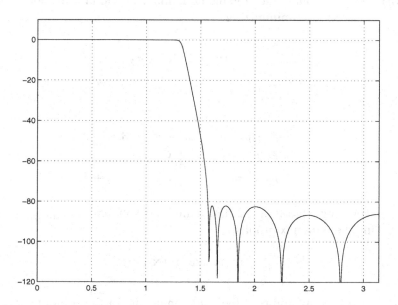

Figure 8.10.   Amplitude response of the lowpass filter for Example 8.3(b).

References

1 C. Charalambous, "A unified review of optimization," *IEEE Trans. Microwave Theory and Techniques*, vol. MTT-22, pp. 289–300, Mar. 1974.

2 C. Charalambous, "Acceleration of the least-*p*th algorithm for minimax optimization with engineering applications," *Mathematical Programming*, vol. 17, pp. 270–297, 1979.

3 A. Antoniou, "Improved minimax optimisation algorithms and their application in the design of recursive digital filters," *Proc. Inst. Elect. Eng.*, part G, vol. 138, pp. 724–730, Dec. 1991.

4 A. Antoniou, *Digital Signal Processing: Signals, Systems, and Filters*, McGraw-Hill, New York, 2005.

5 C. Charalambous and A. Antoniou, "Equalisation of recursive digital filters," *Proc. Inst. Elect. Eng.*, part G, vol. 127, pp. 219–225, Oct. 1980.

6 C. Charalambous, "Design of 2-dimensional circularly-symmetric digital filters," *Proc. Inst. Elect. Eng.*, part G, vol. 129, pp. 47–54, Apr. 1982.

Problems

8.1 Consider the overdetermined system of nonlinear equations

$$x_1^2 - x_2^2 - x_1 - 3x_2 = 2$$
$$x_1^3 - x_2^4 = -2$$
$$x_1^2 + x_2^3 + 2x_1 - x_2 = -1.1$$

(a) Using the Gauss-Newton method, find a solution for the above equations, $\mathbf{x}_{gn}$, by minimizing

$$F(\mathbf{x}) = \sum_{i=1}^{3} f_i^2(\mathbf{x})$$

where

$$f_1(\mathbf{x}) = x_1^2 - x_2^2 - x_1 - 3x_2 - 2$$
$$f_2(\mathbf{x}) = x_1^3 - x_2^4 + 2$$
$$f_3(\mathbf{x}) = x_1^2 + x_2^3 + 2x_1 - x_2 + 1.1$$

(b) Applying Algorithm 8.2, find a minimax solution, $\mathbf{x}_{minimax}$, by solving the minimax problem

$$\underset{\mathbf{x}}{\text{minimize}} \; \underset{1 \le i \le 3}{\max} |f_i(\mathbf{x})|$$

(c) Evaluate and compare the equation errors for the solutions $\mathbf{x}_{gn}$ and $\mathbf{x}_{minimax}$.

8.2 Verify the expression for the gradient $\nabla e_i(\mathbf{x})$ given in Eqs. (8.28b), (8.28c), and (8.28d).

8.3 Applying Algorithm 8.1, design a 12th-order highpass digital filter, assuming a desired amplitude response

$$M_0(\omega) = \begin{cases} 0 & \text{for } 0 \leq \omega \leq 0.45\pi \text{ rad/s} \\ 1 & \text{for } 0.5\pi \leq \omega \leq \pi \text{ rad/s} \end{cases}$$

The transfer function of the filter is of the form given by Eq. (8.2).

8.4 Repeat Problem 8.3 by applying Algorithm 8.2.

8.5 Repeat Problem 8.3 by applying Algorithm 8.3.

8.6 Repeat Problem 8.3 by applying Algorithm 8.4.

8.7 Applying Algorithm 8.1, design a 12th-order bandpass filter, assuming a desired amplitude response

$$M_0(\omega) = \begin{cases} 0 & \text{for } 0 \leq \omega \leq 0.3\pi \text{ rad/s} \\ 1 & \text{for } 0.375\pi \leq \omega \leq 0.625\pi \text{ rad/s} \\ 0 & \text{for } 0.7\pi \leq \omega \leq \pi \text{ rad/s} \end{cases}$$

The transfer function of the filter is of the form given by Eq. (8.2).

8.8 Repeat Problem 8.7 by applying Algorithm 8.2.

8.9 Repeat Problem 8.7 by applying Algorithm 8.3.

8.10 Repeat Problem 8.7 by applying Algorithm 8.4.

8.11 Applying Algorithm 8.1, design a 12th-order bandstop filter, assuming a desired amplitude response

$$M_0(\omega) = \begin{cases} 1 & \text{for } 0 \leq \omega \leq 0.35\pi \text{ rad/s} \\ 0 & \text{for } 0.425\pi \leq \omega \leq 0.575\pi \text{ rad/s} \\ 1 & \text{for } 0.65\pi \leq \omega \leq \pi \text{ rad/s} \end{cases}$$

The transfer function of the filter is of the form given by Eq. (8.2).

8.12 Repeat Problem 8.11 by applying Algorithm 8.2.

8.13 Repeat Problem 8.11 by applying Algorithm 8.3.

8.14 Repeat Problem 8.11 by applying Algorithm 8.4.

Chapter 9

APPLICATIONS OF UNCONSTRAINED OPTIMIZATION

9.1   Introduction

Optimization problems occur in many disciplines, for example, in engineering, physical sciences, social sciences, and commerce. In this chapter, we demonstrate the usefulness of the unconstrained optimization algorithms studied in this book by applying them to a number of problems in engineering. Applications of various constrained optimization algorithms will be presented in Chap. 16.

Optimization is particularly useful in the various branches of engineering like electrical, mechanical, chemical, and aeronautical engineering. The applications we consider here and in Chap. 16 are in the areas of digital signal processing, pattern recognition, automatic control, robotics, and telecommunications. For each selected application, sufficient background material is provided to assist the reader to understand the application. The steps involved are the problem formulation phase which converts the problem at hand into an unconstrained optimization problem, and the solution phase which involves selecting and applying an appropriate optimization algorithm.

In Sec. 9.2, we examine a problem of point-pattern matching in an unconstrained optimization framework. To this end, the concept of similarity transformation is introduced to quantify the meaning of 'best pattern matching'. In addition, it is shown that the optimal pattern from a database that best matches a given point pattern can be obtained by minimizing a convex quadratic function. In Sec. 9.3, we consider a problem known as the inverse kinematics of robotic manipulators which entails a system of nonlinear equations. The problem is first converted into an unconstrained minimization problem and then various methods studied earlier are applied and the results obtained are compared in terms of solution accuracy and computational efficiency. Throughout the discussion, the advantages of using an optimization-based solution method relative to a conventional closed-form method are stressed. In Sec. 9.4, we obtain weighted least-squares and minimax designs of finite-duration impulse-response (FIR) digital filters using unconstrained optimization.

9.2 Point-Pattern Matching

9.2.1 Motivation

A problem that arises in pattern recognition is the so-called *point-pattern matching problem*. In this problem, a pattern such as a printed or handwritten character, numeral, symbol, or even the outline of a manufactured part can be described by a set of points, say,

$$\mathcal{P} = \{\mathbf{p}_1, \ \mathbf{p}_2, \ \ldots, \ \mathbf{p}_n\} \tag{9.1}$$

where

$$\mathbf{p}_i = \begin{bmatrix} p_{i1} \\ p_{i2} \end{bmatrix}$$

is a vector in terms of the coordinates of the ith sample point. If the number of points in $\mathcal{P}$, n, is sufficiently large, then $\mathcal{P}$ in Eq. (9.1) describes the object accurately and $\mathcal{P}$ is referred to as a *point pattern* of the object. The same object viewed from a different distance and/or a different angle will obviously correspond to a different point pattern, $\tilde{\mathcal{P}}$, and it is of interest to examine whether or not two given patterns are matched to within a scaled rotation and a translation.

In a more general setting, we consider the following pattern-matching problem: We have a database that contains N standard point patterns $\{\mathcal{P}_1, \mathcal{P}_2, \ldots, \mathcal{P}_N\}$ where each $\mathcal{P}_i$ has the form of Eq. (9.1) and we need to find a pattern from the database that best matches a given point pattern $\mathcal{Q} = \{\mathbf{q}_1, \mathbf{q}_2, \ldots, \mathbf{q}_n\}$. In order to solve this problem, two issues need to be addressed. First, we need to establish a measure to quantify the meaning of 'best matching'. Second, we need to develop a solution method to find an optimal pattern $\mathcal{P}^*$ from the database that best matches pattern $\mathcal{Q}$ based on the chosen measure.

9.2.2 Similarity transformation

Two point patterns $\mathcal{P}$ and $\tilde{\mathcal{P}}$ are said to be *similar* if one pattern can be obtained by applying a scaled rotation plus a translation to the other. If pattern $\mathcal{P}$ is given by Eq. (9.1) and

$$\tilde{\mathcal{P}} = \{\tilde{\mathbf{p}}_1, \ \tilde{\mathbf{p}}_2, \ \ldots, \ \tilde{\mathbf{p}}_n\} \quad \text{with} \quad \tilde{\mathbf{p}}_i = [\tilde{p}_{i1} \ \tilde{p}_{i2}]^T$$

then $\mathcal{P}$ and $\tilde{\mathcal{P}}$ are similar if and only if there exist a rotation angle θ, a scaling factor η, and a translation vector $\mathbf{r} = [r_1 \ r_2]^T$ such that the relation

$$\tilde{\mathbf{p}}_i = \eta \begin{bmatrix} \cos\theta & -\sin\theta \\ \sin\theta & \cos\theta \end{bmatrix} \mathbf{p}_i + \begin{bmatrix} r_1 \\ r_2 \end{bmatrix} \tag{9.2}$$

holds for $i = 1, 2, \ldots, n$. A transformation that maps pattern $\mathcal{P}$ to pattern $\mathcal{Q}$ is said to be a *similarity transformation*. From Eq. (9.2), we see that a similarity transformation is characterized by the parameter column vector $[\eta \; \theta \; r_1 \; r_2]^T$. Note that the similarity transformation is a nonlinear function of parameters η and θ. This nonlinearity can lead to a considerable increase in the amount of computation required by the optimization process. This problem can be fixed by applying the variable substitution

$$a = \eta \cos\theta, \quad b = \eta \sin\theta$$

to Eq. (9.2) to obtain

$$\tilde{\mathbf{p}}_i = \begin{bmatrix} a & -b \\ b & a \end{bmatrix} \mathbf{p}_i + \begin{bmatrix} r_1 \\ r_2 \end{bmatrix} \tag{9.3}$$

Thus the parameter vector becomes $\mathbf{x} = [a \; b \; r_1 \; r_2]^T$. Evidently, the similarity transformation now depends *linearly* on the parameters.

9.2.3   Problem formulation

In a real-life problem, a perfect match between a given point pattern $\mathcal{Q}$ and a point pattern in the database is unlikely, and the best we can do is identify the closest pattern to $\mathcal{Q}$ to within a similarity transformation.

Let

$$\mathcal{Q} = \{\mathbf{q}_1, \mathbf{q}_2, \ldots, \mathbf{q}_n\}$$

be a given pattern and assume that

$$\tilde{\mathcal{P}}(\mathbf{x}) = \{\tilde{\mathbf{p}}_1, \tilde{\mathbf{p}}_2, \ldots, \tilde{\mathbf{p}}_n\}$$

is a transformed version of pattern

$$\mathcal{P} = \{\mathbf{p}_1, \mathbf{p}_2, \ldots, \mathbf{p}_n\}$$

Let these patterns be represented by the matrices

$$\mathbf{Q} = [\mathbf{q}_1 \; \mathbf{q}_2 \; \cdots \; \mathbf{q}_n], \; \tilde{\mathbf{P}}(\mathbf{x}) = [\tilde{\mathbf{p}}_1 \; \tilde{\mathbf{p}}_2 \; \cdots \; \tilde{\mathbf{p}}_n], \text{ and } \mathbf{P} = [\mathbf{p}_1 \; \mathbf{p}_2 \; \cdots \; \mathbf{p}_n]$$

respectively. A transformed pattern $\tilde{\mathcal{P}}$ that matches $\mathcal{Q}$ can be obtained by solving the unconstrained optimization problem

$$\underset{\mathbf{x}}{\text{minimize}} \; \|\tilde{\mathbf{P}}(\mathbf{x}) - \mathbf{Q}\|_F^2 \tag{9.4}$$

where $\| \cdot \|_F$ denotes the Frobenius norm (see Sec. A.8.2). The solution of the above minimization problem corresponds to finding the best transformation that would minimize the difference between patterns $\tilde{\mathcal{P}}$ and $\mathcal{Q}$ in the Frobenius sense. Since

$$\|\tilde{\mathbf{P}}(\mathbf{x}) - \mathbf{Q}\|_F^2 = \sum_{i=1}^{n} \|\tilde{\mathbf{p}}_i(x) - \mathbf{q}_i\|^2$$

the best transformation in the least-squares sense is obtained.

Now if $\mathbf{x}^*$ is the minimizer of the problem in Eq. (9.4), then the error

$$e(\tilde{\mathcal{P}}, \mathcal{Q}) = \|\tilde{\mathbf{P}}(\mathbf{x}^*) - \mathbf{Q}\|_F \tag{9.5}$$

is a measure of the dissimilarity between patterns $\tilde{\mathcal{P}}$ and $\mathcal{Q}$. Obviously, $e(\tilde{\mathcal{P}}, \mathcal{Q})$ should be as small as possible and a zero value would correspond to a perfect match.

9.2.4 Solution of the problem in Eq. (9.4)

On using Eq. (9.3), Eq. (9.5) gives

$$
\begin{aligned}
\|\tilde{\mathbf{P}}(\mathbf{x}) - \mathbf{Q}\|_F^2 &= \sum_{i=1}^{n} \|\tilde{\mathbf{p}}_i(x) - \mathbf{q}_i\|^2 \\
&= \sum_{i=1}^{n} \left\| \begin{bmatrix} ap_{i1} - bp_{i2} + r_1 \\ bp_{i1} + ap_{i2} + r_2 \end{bmatrix} - \mathbf{q}_i \right\|^2 \\
&= \sum_{i=1}^{n} \left\| \begin{bmatrix} p_{i1} & -p_{i2} & 1 & 0 \\ p_{i2} & p_{i1} & 0 & 1 \end{bmatrix} \mathbf{x} - \mathbf{q}_i \right\|^2 \\
&= \mathbf{x}^T \mathbf{H} \mathbf{x} - 2\mathbf{x}^T \mathbf{b} + \kappa
\end{aligned}
\tag{9.6a}
$$

where

$$
\mathbf{H} = \begin{bmatrix} \sum_{i=1}^{n} \mathbf{R}_i^T \mathbf{R}_i & \sum_{i=1}^{n} \mathbf{R}_i^T \\ \sum_{i=1}^{n} \mathbf{R}_i & n\mathbf{I}_2 \end{bmatrix}, \quad \mathbf{R}_i = \begin{bmatrix} p_{i1} & -p_{i2} \\ p_{i2} & p_{i1} \end{bmatrix} \tag{9.6b}
$$

$$
\mathbf{b} = \sum_{i=1}^{n} [\mathbf{R}_i \ \mathbf{I}_2]^T \mathbf{q}_i \tag{9.6c}
$$

$$
\kappa = \sum_{i=1}^{n} \|\mathbf{q}_i\|^2 \tag{9.6d}
$$

(see Prob. 9.1(a)). It can be readily verified that the Hessian $\mathbf{H}$ in Eq. (9.6b) is positive definite (see Prob. 9.1(b)) and hence it follows from Chap. 2 that the objective function in Eq. (9.4) is globally strictly convex and, therefore, has a unique global minimizer. Using Eq. (9.6a), the gradient of the objective function can be obtained as

$$\mathbf{g}(\mathbf{x}) = 2\mathbf{H}\mathbf{x} - 2\mathbf{b}$$

The unique global minimizer can be obtained in closed form by letting

$$\mathbf{g}(\mathbf{x}) = 2\mathbf{H}\mathbf{x} - 2\mathbf{b} = 0$$

and hence

$$\mathbf{x}^* = \mathbf{H}^{-1}\mathbf{b} \tag{9.7}$$

Since $\mathbf{H}$ is a positive definite matrix of size 4×4, its inverse exists and is easy to evaluate (see Prob. 9.1(c)).

9.2.5 Alternative measure of dissimilarity

As can be seen in Eq. (9.6a), the Frobenius norm of a matrix can be related to the L_2 norm of its column vectors. If we define two new vectors $\tilde{\mathbf{p}}(\mathbf{x})$ and $\mathbf{q}$ as

$$\tilde{\mathbf{p}}(\mathbf{x}) = \begin{bmatrix} \tilde{\mathbf{p}}_1(\mathbf{x}) \\ \tilde{\mathbf{p}}_2(\mathbf{x}) \\ \vdots \\ \tilde{\mathbf{p}}_n(\mathbf{x}) \end{bmatrix} \quad \text{and} \quad \mathbf{q} = \begin{bmatrix} \mathbf{q}_1 \\ \mathbf{q}_2 \\ \vdots \\ \mathbf{q}_n \end{bmatrix}$$

then Eq. (9.6) implies that

$$\|\tilde{\mathbf{P}}(\mathbf{x}) - \mathbf{Q}\|_F^2 = \|\tilde{\mathbf{p}}(\mathbf{x}) - \mathbf{q}\|^2$$

Hence the dissimilarity measure defined in Eq. (9.5) can be expressed as

$$e(\tilde{\mathcal{P}}, \mathcal{Q}) = \|\tilde{\mathbf{p}}(\mathbf{x}) - \mathbf{q}\|$$

An alternative of the above dissimilarity measure can be defined in terms of the L_{2p} norm

$$e_{2p}(\tilde{\mathcal{P}}, \mathcal{Q}) = \|\tilde{\mathbf{p}}(\mathbf{x}) - \mathbf{q}\|_{2p}$$

As p increases, $e_{2p}(\tilde{\mathcal{P}}, \mathcal{Q})$ approaches the L_{∞} norm of $\tilde{\mathbf{p}}(\mathbf{x}) - \mathbf{q}$ which is numerically equal to the maximum of the function. Therefore, solving the problem

$$\underset{\mathbf{x}}{\text{minimize}} \ e_{2p}(\tilde{\mathcal{P}}, \mathcal{Q}) = \|\tilde{\mathbf{p}}(\mathbf{x}) - \mathbf{q}\|_{2p} \tag{9.8}$$

with a sufficiently large p amounts to minimizing the maximum error between symbols $\tilde{\mathcal{P}}$ and $\mathcal{Q}$. If we let

$$\mathbf{r}_{i1} = [p_{i1} \ -p_{i2} \ 1 \ 0]^T$$
$$\mathbf{r}_{i2} = [p_{i2} \ p_{i1} \ 0 \ 1]^T$$
$$\mathbf{q}_i = \begin{bmatrix} q_{i1} \\ q_{i2} \end{bmatrix}$$

then the objective function in Eq. (9.8) can be expressed as

$$e_{2p}(\mathbf{x}) = \left\{ \sum_{i=1}^{n} [(\mathbf{r}_{i1}^T\mathbf{x} - q_{i1})^{2p} + (\mathbf{r}_{i2}^T\mathbf{x} - q_{i2})^{2p}] \right\}^{1/2p} \tag{9.9a}$$

The gradient and Hessian of $e_{2p}(\mathbf{x})$ can be evaluated as

$$\nabla e_{2p}(\mathbf{x}) = \frac{1}{e_{2p}^{2p-1}(\mathbf{x})} \sum_{i=1}^{n} [(\mathbf{r}_{i1}^T \mathbf{x} - q_{i1})^{2p-1} + (\mathbf{r}_{i2}^T \mathbf{x} - q_{i2})^{2p-1}] \qquad (9.9\text{b})$$

and

$$\nabla^2 e_{2p}(\mathbf{x}) = \frac{(2p-1)}{e_{2p}^{2p-1}(\mathbf{x})} \sum_{i=1}^{n} [(\mathbf{r}_{i1}^T \mathbf{x} - q_{i1})^{2p-2} \mathbf{r}_{i1} \mathbf{r}_{i1}^T + (\mathbf{r}_{i2}^T \mathbf{x} - q_{i2})^{2p-2} \mathbf{r}_{i2} \mathbf{r}_{i2}^T]$$
$$- \frac{(2p-1)}{e_{2p}(\mathbf{x})} \nabla e_{2p}(\mathbf{x}) \nabla^T e_{2p}(\mathbf{x}) \qquad (9.9\text{c})$$

respectively (see Prob. 9.3(*a*)). It can be shown that the Hessian $\nabla^2 e_{2p}(\mathbf{x})$ in Eq. (9.9c) is positive semidefinite for any $\mathbf{x} \in R^4$ and, therefore, the objective function $e_{2p}(\mathbf{x})$ is globally convex (see Prob. 9.3(*b*)).

Since the Hessian of $e_{2p}(\mathbf{x})$ is a 4×4 positive semidefinite matrix and is available in closed form, the Newton algorithm (Algorithm 5.3) with the Hessian matrix $\mathbf{H}_k$ modified according to Eq. (5.13) is an appropriate algorithm for the solution of the problem in Eq. (9.8). If the power $2p$ involved in the optimization problem is a power of 2, i.e., $2p = 2^K$, then the problem at hand can be solved by first solving the problem for the case $p = 1$ using Eq. (9.7). The minimizer so obtained can then be used as the initial point to minimize the objective function for $p = 2$. This procedure is then repeated for $p = 4, 8, 16, \ldots$ until two successive optimizations give the same maximum error to within a prescribed tolerance.

9.2.6   Handwritten character recognition

For illustration purposes, we consider the problem of recognizing a handwritten character using a database comprising the ten 'standard' characters shown in Fig. 9.1. Each character in the database can be represented by a point pattern of the form in Eq. (9.1) with $n = 196$, and the patterns for $a, c, e, \ldots$ can be denoted as $\mathcal{P}_a, \mathcal{P}_c, \mathcal{P}_e, \ldots$ where the subscript represents the associated character. Fig. 9.2 shows a set of sample points that form pattern $\mathcal{P}_a$ in the database. The character to be recognized is plotted in Fig. 9.3. It looks like a rotated e, it is of larger size relative to the corresponding character in the database, and it is largely located in the third quadrant. To apply the method discussed, the character in Fig. 9.3 is represented by a point pattern $\mathcal{Q}$ with $n = 196$.

The dissimilarity between each pattern $\mathcal{P}_{\text{character}}$ in the database and pattern $\mathcal{Q}$ is measured in terms of $e(\mathcal{P}_{\text{character}}, \mathcal{Q})$ in Eq. (9.5) and $e_{2p}(\mathcal{P}_{\text{character}}, \mathcal{Q})$ in Eq. (9.8) with $2p = 128$. Note that the minimization of $e(\mathcal{P}_{\text{character}}, \mathcal{Q})$ can be viewed as a special case of the problem in Eq. (9.8) with $p = 1$, and its solution can be obtained using Eq. (9.7). For the minimization of $e_{128}(\mathcal{P}_{\text{character}}, \mathcal{Q})$, a sequential implementation of the Newton method as described in Sec. 9.2.5 was used to obtain the solution. The results obtained are summarized in Ta-

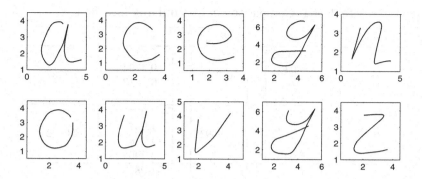

Figure 9.1. Ten standard characters in the database.

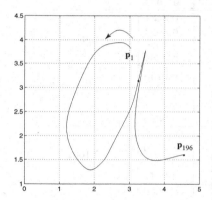

Figure 9.2. Sample points in pattern $\mathcal{P}_a$.   *Figure 9.3.* A character to be recognized.

ble 9.1 where $\mathbf{x}_2^*$ and $\mathbf{x}_{128}^*$ denote the minimizers of $e_2(\mathcal{P}_{\text{character}}, \mathcal{Q})$ and $e_{128}(\mathcal{P}_{\text{character}}, \mathcal{Q})$, respectively. From the table, it is evident that the character in Fig. 9.3 is most similar to character e.

See [1] for an in-depth investigation of dissimilarity and affine invariant distances between two-dimensional point patterns.

9.3   Inverse Kinematics for Robotic Manipulators

9.3.1   Position and orientation of a manipulator

Typically an industrial robot, also known as a robotic manipulator, comprises a chain of mechanical links with one end fixed relative to the ground and the other end, known as the *end-effector*, free to move. Motion is made possible in a manipulator by moving the joint of each link about its axis with an electric or hydraulic actuator.

Table 9.1 Comparison of dissimilarity measures

Character	$\mathbf{x}_2^*$	$e(\mathcal{P}, \mathcal{Q})$	$\mathbf{x}_{128}^*$	$e_{128}(\mathcal{P}, \mathcal{Q})$
a	$\begin{bmatrix} 0.8606 \\ 0.0401 \\ -8.9877 \\ -4.4466 \end{bmatrix}$	30.7391	$\begin{bmatrix} 0.4453 \\ 0.3764 \\ -6.8812 \\ -4.0345 \end{bmatrix}$	2.7287
c	$\begin{bmatrix} 0.8113 \\ 1.3432 \\ -5.5632 \\ -7.0455 \end{bmatrix}$	19.9092	$\begin{bmatrix} -0.0773 \\ 1.0372 \\ -4.4867 \\ -4.4968 \end{bmatrix}$	2.0072
e	$\begin{bmatrix} -1.1334 \\ 1.9610 \\ 0.6778 \\ -3.9186 \end{bmatrix}$	5.2524	$\begin{bmatrix} -1.0895 \\ 2.0307 \\ 0.6513 \\ -4.1631 \end{bmatrix}$	0.4541
g	$\begin{bmatrix} -0.2723 \\ 0.5526 \\ -3.5780 \\ -3.1246 \end{bmatrix}$	30.4058	$\begin{bmatrix} -0.0481 \\ 0.8923 \\ -2.7970 \\ -5.3467 \end{bmatrix}$	2.5690
n	$\begin{bmatrix} 0.0670 \\ 0.5845 \\ -5.6081 \\ -3.8721 \end{bmatrix}$	33.0044	$\begin{bmatrix} -0.0745 \\ 0.6606 \\ -5.2831 \\ -3.9995 \end{bmatrix}$	2.5260
o	$\begin{bmatrix} 1.0718 \\ 1.3542 \\ -6.0667 \\ -8.3572 \end{bmatrix}$	16.8900	$\begin{bmatrix} -0.2202 \\ 1.2786 \\ -3.1545 \\ -4.9915 \end{bmatrix}$	2.1602
u	$\begin{bmatrix} 0.3425 \\ 0.3289 \\ -8.5193 \\ -2.2115 \end{bmatrix}$	33.6184	$\begin{bmatrix} 0.0600 \\ 0.0410 \\ -6.8523 \\ -2.0225 \end{bmatrix}$	2.8700
v	$\begin{bmatrix} 1.7989 \\ -0.2632 \\ -12.0215 \\ -6.2948 \end{bmatrix}$	20.5439	$\begin{bmatrix} 1.1678 \\ 0.0574 \\ -9.6540 \\ -5.9841 \end{bmatrix}$	2.0183
y	$\begin{bmatrix} -0.1165 \\ 0.6660 \\ -3.8249 \\ -4.1959 \end{bmatrix}$	30.1985	$\begin{bmatrix} -0.0064 \\ 0.6129 \\ -4.2815 \\ -4.1598 \end{bmatrix}$	2.3597
z	$\begin{bmatrix} 0.1962 \\ 1.7153 \\ -3.2896 \\ -6.9094 \end{bmatrix}$	21.4815	$\begin{bmatrix} 0.0792 \\ 1.1726 \\ -4.4356 \\ -4.8665 \end{bmatrix}$	2.0220

One of the basic problems in robotics is the description of the position and orientation of the end-effector in terms of the joint variables. There are two types of joints: rotational joints for rotating the associated robot link, and translational joints for pushing and pulling the associated robot link along a

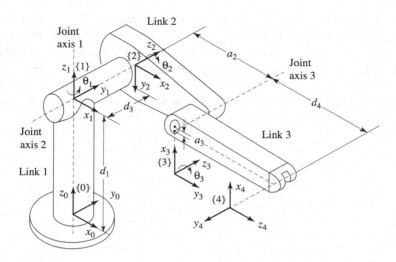

Figure 9.4. A three-link robotic manipulator.

straight line. However, joints in industrial robots are almost always rotational. Fig. 9.4 shows a three-joint industrial robot, where the three joints can be used to rotate links 1, 2, and 3. In this case, the end-effector is located at the end of link 3, whose position and orientation can be conveniently described relative to a fixed coordinate system which is often referred to as a *frame* in robotics. As shown in Fig. 9.4, frame $\{0\}$ is attached to the robot base and is fixed relative to the ground. Next, frames $\{1\}$, $\{2\}$, and $\{3\}$ are attached to joint axes 1, 2, and 3, respectively, and are subject to the following rules:

- The z axis of frame $\{i\}$ is along the joint axis i for $i = 1$, 2, 3.

- The x axis of frame $\{i\}$ is perpendicular to the z axes of frames $\{i\}$ and $\{i+1\}$ for $i = 1$, 2, 3.

- The y axis of frame $\{i\}$ is determined such that frame $\{i\}$ is a standard right-hand coordinate system.

- Frame $\{4\}$ is attached to the end of link 3 in such a way that the axes of frames $\{3\}$ and $\{4\}$ are in parallel and the distance between the z axes of these two frames is zero.

Having assigned the frames, the relation between two consecutive frames can be characterized by the so-called *Denavit-Hartenberg (D-H) parameters* [2] which are defined in the following table:

a_i: distance from the z_i axis to the z_{i+1} axis measured along the x_i axis
α_i: angle between the z_i axis and the z_{i+1} axis measured about the x_i axis
d_i: distance from the x_{i-1} axis to the x_i axis measured along the z_i axis
θ_i: angle between the x_{i-1} axis and the x_i axis measured about the z_i axis

As can be observed in Fig. 9.4, parameters d_1, a_2, and d_4 in this case represent the lengths of links 1, 2, and 3, respectively, d_3 represents the offset between link 1 and link 2, and a_3 represents the offset between link 2 and link 3. In addition, the above frame assignment also determines the angles $\alpha_0 = 0°$, $\alpha_1 = -90°$, $\alpha_2 = 0°$, and $\alpha_3 = -90°$. Table 9.2 summarizes the D-H parameters of the three-joint robot in Fig. 9.4 where the only variable parameters are θ_1, θ_2, and θ_3 which represent the rotation angles of joints 1, 2, and 3, respectively.

Table 9.2 D-H parameters of 3-link robot

i	α_{i-1}	a_{i-1}	d_i	θ_i
1	$0°$	0	d_1	θ_1
2	$-90°$	0	0	θ_2
3	$0°$	a_2	d_3	θ_3
4	$-90°$	a_3	d_4	$0°$

Since the D-H parameters a_{i-1}, α_{i-1}, d_i, and θ_i characterize the relation between frames $\{i-1\}$ and $\{i\}$, they can be used to describe the position and orientation of frame $\{i\}$ in relation to those of frame $\{i-1\}$. To this end, we define the so-called *homogeneous transformation* in terms of the 4×4 matrix

$$_{i}^{i-1}\mathbf{T} = \left[\begin{array}{c|c} _{i}^{i-1}\mathbf{R} & ^{i-1}\mathbf{p}_{i\text{ORG}} \\ \hline 0\,0\,0 & 1 \end{array} \right]_{4\times 4} \tag{9.10}$$

where vector $^{i-1}\mathbf{p}_{i\text{ORG}}$ denotes the position of the origin of frame $\{i\}$ with respect to frame $\{i-1\}$, and matrix $_{i}^{i-1}\mathbf{R}$ is an orthogonal matrix whose columns denote the x-, y-, and z-coordinate vectors of frame $\{i\}$ with respect to frame $\{i-1\}$. With the D-H parameters a_{i-1}, α_{i-1}, d_i, and θ_i known, the homogeneous transformation in Eq. (9.10) can be expressed as [2]

$$_{i}^{i-1}\mathbf{T} = \begin{bmatrix} c\theta_i & -s\theta_i & 0 & a_{i-1} \\ s\theta_i c\alpha_{i-1} & c\theta_i c\alpha_{i-1} & -\alpha_{i-1} & -s\alpha_{i-1}d_i \\ s\theta_i s\alpha_{i-1} & c\theta_i s\alpha_{i-1} & \alpha_{i-1} & \alpha_{i-1}d_i \\ 0 & 0 & 0 & 1 \end{bmatrix} \tag{9.11}$$

where $s\theta$ and $c\theta$ denote $\sin\theta$ and $\cos\theta$, respectively. The significance of the above formula is that it can be used to evaluate the position and orientation of the end-effector as

$$_{N}^{0}\mathbf{T} = {}_{1}^{0}\mathbf{T}\,{}_{2}^{1}\mathbf{T}\cdots{}_{N}^{N-1}\mathbf{T} \tag{9.12}$$

where each $_{i}^{i-1}\mathbf{T}$ on the right-hand side can be obtained using Eq. (9.11). The formula in Eq. (9.12) is often referred to as the *equation of forward kinematics*.

Example 9.1 Derive closed-form formulas for the position and orientation of the robot tip in Fig. 9.4 in terms of joint angles θ_1, θ_2, and θ_3.

Solution Using Table 9.2 and Eq. (9.11), the homogeneous transformations $_{i}^{i-1}\mathbf{T}$ for $i = 1$, 2, 3, and 4 are obtained as

$$
_{1}^{0}\mathbf{T} = \begin{bmatrix} c_1 & -s_1 & 0 & 0 \\ s_1 & c_1 & 0 & 0 \\ 0 & 0 & 1 & d_1 \\ 0 & 0 & 0 & 1 \end{bmatrix}, \quad _{2}^{1}\mathbf{T} = \begin{bmatrix} c_2 & -s_2 & 0 & 0 \\ 0 & 0 & 1 & 0 \\ -s_2 & -c_2 & 0 & 0 \\ 0 & 0 & 0 & 1 \end{bmatrix}
$$

$$
_{3}^{2}\mathbf{T} = \begin{bmatrix} c_3 & -s_3 & 0 & a_2 \\ s_3 & c_3 & 0 & 0 \\ 0 & 0 & 1 & d_3 \\ 0 & 0 & 0 & 1 \end{bmatrix}, \quad _{4}^{3}\mathbf{T} = \begin{bmatrix} 1 & 0 & 0 & a_3 \\ 0 & 0 & 1 & d_4 \\ 0 & -1 & 0 & 0 \\ 0 & 0 & 0 & 1 \end{bmatrix}
$$

With $N = 4$, Eq. (9.12) gives

$$
\begin{aligned}
_{4}^{0}\mathbf{T} &= {}_{1}^{0}\mathbf{T}{}_{2}^{1}\mathbf{T}{}_{3}^{2}\mathbf{T}{}_{4}^{3}\mathbf{T} \\
&= \begin{bmatrix} c_1c_{23} & s_1 & -c_1s_{23} & c_1(a_2c_2 + a_3c_{23} - d_4s_{23}) - d_3s_1 \\ s_1c_{23} & -c_1 & -s_1s_{23} & s_1(a_2c_2 + a_3c_{23} - d_4s_{23}) + d_3c_1 \\ -s_{23} & 0 & -c_{23} & d_1 - a_2s_2 - a_3s_{23} - d_4c_{23} \\ 0 & 0 & 0 & 1 \end{bmatrix}
\end{aligned}
$$

where $c_1 = \cos\theta_1$, $s_1 = \sin\theta_1$, $c_{23} = \cos(\theta_2 + \theta_3)$, and $s_{23} = \sin(\theta_2 + \theta_3)$. Therefore, the position of the robot tip with respect to frame $\{0\}$ is given by

$$
{}^{0}\mathbf{p}_{4\text{ORG}} = \begin{bmatrix} c_1(a_2c_2 + a_3c_{23} - d_4s_{23}) - d_3s_1 \\ s_1(a_2c_2 + a_3c_{23} - d_4s_{23}) + d_3c_1 \\ d_1 - a_2s_2 - a_3s_{23} - d_4c_{23} \end{bmatrix} \tag{9.13}
$$

and the orientation of the robot tip with respect to frame $\{0\}$ is characterized by the orthogonal matrix

$$
{}_{4}^{0}\mathbf{R} = \begin{bmatrix} c_1c_{23} & s_1 & -c_1s_{23} \\ s_1c_{23} & -c_1 & -s_1s_{23} \\ -s_{23} & 0 & -c_{23} \end{bmatrix} \tag{9.14}
$$

∎

9.3.2 Inverse kinematics problem

The joint angles of manipulator links are usually measured using sensors such as optical encoders that are attached to the link actuators. As discussed

242

in Sec. 9.3.1, when the joint angles θ_1, θ_2, $\ldots$, θ_n are known, the position and orientation of the end-effector can be evaluated using Eq. (9.12). A related and often more important problem is the *inverse kinematics problem* which is as follows: find the joint angles θ_i for $1 \leq i \leq n$ with which the manipulator's end-effector would achieve a *prescribed* position and orientation. The significance of the inverse kinematics lies in the fact that the tasks to be accomplished by a robot are usually in terms of trajectories in the Cartesian space that the robot's end-effector must follow. Under these circumstances, the position and orientation for the end-effector are known and the problem is to find the correct values of the joint angles that would move the robot's end-effector to the desired position and orientation.

Mathematically, the inverse kinematics problem can be described as the problem of finding the values θ_i for $1 \leq i \leq n$ that would satisfy Eq. (9.12) for a given $^0_N\mathbf{T}$. Since Eq. (9.12) is highly nonlinear, the problem of finding its solutions is not a trivial one [2]. For example, if a prescribed position of the end-effector for the three-link manipulator in Fig. 9.4 is given by $^0\mathbf{p}_{4\text{ORG}} = [p_x \; p_y \; p_z]^T$, then Eq. (9.13) gives

$$c_1(a_2c_2 + a_3c_{23} - d_4s_{23}) - d_3s_1 = p_x$$
$$s_1(a_2c_2 + a_3c_{23} - d_4s_{23}) + d_3c_1 = p_y \qquad (9.15)$$
$$d_1 - a_2s_2 - a_3s_{23} - d_4c_{23} = p_z$$

In the next section, we illustrate an optimization approach for the solution of the inverse kinematics problem on the basis of Eq. (9.15).

9.3.3 Solution of inverse kinematics problem

If we let

$$\mathbf{x} = [\theta_1 \; \theta_2 \; \theta_3]^T \qquad (9.16a)$$
$$f_1(\mathbf{x}) = c_1(a_2c_2 + a_3c_{23} - d_4s_{23}) - d_3s_1 - p_x \qquad (9.16b)$$
$$f_2(\mathbf{x}) = s_1(a_2c_2 + a_3c_{23} - d_4s_{23}) + d_3c_1 - p_y \qquad (9.16c)$$
$$f_3(\mathbf{x}) = d_1 - a_2s_2 - a_3s_{23} - d_4c_{23} - p_z \qquad (9.16d)$$

then Eq. (9.15) is equivalent to

$$f_1(\mathbf{x}) = 0 \qquad (9.17a)$$
$$f_2(\mathbf{x}) = 0 \qquad (9.17b)$$
$$f_3(\mathbf{x}) = 0 \qquad (9.17c)$$

To solve this system of nonlinear equations, we construct the objective function

$$F(\mathbf{x}) = f_1^2(\mathbf{x}) + f_2^2(\mathbf{x}) + f_3^2(\mathbf{x})$$

and notice that vector $\mathbf{x}^*$ solves Eq. (9.17) if and only if $F(\mathbf{x}^*) = 0$. Since function $F(\mathbf{x})$ is nonnegative, finding a solution point $\mathbf{x}$ for Eq. (9.17) amounts to finding a minimizer $\mathbf{x}^*$ at which $F(\mathbf{x}^*) = 0$. In other words, we can convert the inverse kinematics problem at hand into the unconstrained minimization problem

$$\text{minimize } F(\mathbf{x}) = \sum_{k=1}^{3} f_k^2(\mathbf{x}) \tag{9.18}$$

An advantage of this approach over conventional methods for inverse kinematics problems [2] is that when the desired position $[p_x\, p_y\, p_z]^T$ is *not* within the manipulator's reach, the conventional methods will fail to work and a conclusion that no solution exists will be drawn. With the optimization approach, however, minimizing function $F(\mathbf{x})$ will still yield a minimizer, say, $\mathbf{x}^* = [\theta_1^*\, \theta_2^*\, \theta_3^*]^T$, although the objective function $F(\mathbf{x})$ would not become zero at $\mathbf{x}^*$. In effect, an *approximate* solution of the problem would be obtained, which could be entirely satisfactory in most engineering applications. We shall illustrate this point further in Example 9.2 by means of computer simulations.

To apply the minimization algorithms studied earlier, we let

$$\mathbf{f}(\mathbf{x}) = \begin{bmatrix} f_1(\mathbf{x}) \\ f_2(\mathbf{x}) \\ f_3(\mathbf{x}) \end{bmatrix}$$

and compute the gradient of $F(\mathbf{x})$ as

$$\mathbf{g}(\mathbf{x}) = 2\mathbf{J}^T(\mathbf{x})\mathbf{f}(\mathbf{x}) \tag{9.19}$$

where the Jacobian matrix $\mathbf{J}(\mathbf{x})$ is given by

$$\begin{aligned} \mathbf{J}(\mathbf{x}) &= [\nabla f_1(\mathbf{x})\, \nabla f_2(\mathbf{x})\, \nabla f_3(\mathbf{x})]^T \\ &= \begin{bmatrix} -q_3 s_1 - d_3 c_1 & q_4 c_1 & q_2 c_1 \\ q_3 c_1 - d_3 s_1 & q_4 s_1 & q_2 s_1 \\ 0 & -q_3 & -q_1 \end{bmatrix} \end{aligned} \tag{9.20}$$

with $q_1 = a_2 c_{23} - d_4 s_{23}$, $q_2 = -a_3 s_{23} - d_4 c_{23}$, $q_3 = a_2 c_2 + q_1$, and $q_4 = -a_2 s_2 + q_2$. The Hessian of $F(\mathbf{x})$ is given by

$$\mathbf{H}(\mathbf{x}) = 2\mathbf{J}^T(\mathbf{x})\mathbf{J}(\mathbf{x}) + 2\sum_{k=1}^{3} f_k(\mathbf{x})\nabla^2 f_k(\mathbf{x}) \tag{9.21}$$

where $\nabla^2 f_k(\mathbf{x})$ is the Hessian of $f_k(\mathbf{x})$ (see Prob. 9.4).

Example 9.2 In the three-link manipulator depicted in Fig. 9.4, $d_1 = 66.04$ cm, $d_3 = 14.91$ cm, $d_4 = 43.31$ cm, $a_2 = 43.18$ cm, and $a_3 = 2.03$ cm. By applying a steepest-descent (SD), Newton (N), Gauss-Newton (GN), Fletcher-Reeves (FR) algorithm and then a quasi-Newton (QN) algorithm based on the Broyden-Fletcher-Goldfarb-Shanno updating formula in Eq. (7.57), determine the joint angles $\theta_i(t)$ for $i = 1,\ 2,\ 3$ and $-\pi \le t \le \pi$ such that the manipulator's end-effector tracks the desired trajectory $\mathbf{p}_d(t) = [p_x(t)\ p_y(t)\ p_z(t)]^T$ where

$$p_x(t) = 30\cos t, \quad p_y(t) = 100\sin t, \quad p_z(t) = 10t + 66.04$$

for $-\pi \le t \le \pi$ as illustrated in Fig. 9.5.

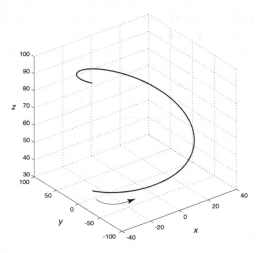

Figure 9.5. Desired Cartesian trajectory for Example 9.2.

Solution The problem was solved by applying Algorithms 5.1, 5.5, and 6.3 as the steepest-descent, Gauss-Newton, and Fletcher-Reeves algorithm, respectively, using the inexact line search in Steps 1 to 6 of Algorithm 7.3 in each case. The Newton algorithm used was essentially Algorithm 5.3 incorporating the Hessian-matrix modification in Eq. (5.13) as detailed below:

Algorithm 9.1 Newton algorithm
Step 1
Input $\mathbf{x}_0$ and initialize the tolerance ε.
Set $k = 0$.
Step 2
Compute $\mathbf{g}_k$ and $\mathbf{H}_k$.
Step 3
Compute the eigenvalues of $\mathbf{H}_k$ (see Sec. A.5).
Determine the smallest eigenvalue of $\mathbf{H}_k$, λ_{min}.
Modify matrix $\mathbf{H}_k$ to

$$\hat{\mathbf{H}}_k = \begin{cases} \mathbf{H}_k & \text{if } \lambda_{min} > 0 \\ \mathbf{H}_k + \gamma \mathbf{I}_n & \text{if } \lambda_{min} \le 0 \end{cases}$$

where

$$\gamma = -1.05\lambda_{min} + 0.1$$

Step 4
Compute $\hat{\mathbf{H}}_k^{-1}$ and $\mathbf{d}_k = -\hat{\mathbf{H}}_k^{-1}\mathbf{g}_k$
Step 5
Find α_k, the value of α that minimizes $f(\mathbf{x}_k + \alpha\mathbf{d}_k)$, using the inexact line search in Steps 1 to 6 of Algorithm 7.3.
Step 6
Set $\mathbf{x}_{k+1} = \mathbf{x}_k + \alpha_k\mathbf{d}_k$.
Compute $f_{k+1} = f(\mathbf{x}_{k+1})$.
Step 7
If $\|\alpha_k\mathbf{d}_k\| < \varepsilon$, then do:
    Output $\mathbf{x}^* = \mathbf{x}_{k+1}$ and $f(\mathbf{x}^*) = f_{k+1}$, and stop.
Otherwise, set $k = k + 1$ and repeat from Step 2.

The quasi-Newton algorithm used was essentially Algorithm 7.3 with a slightly modified version of Step 8b as follows:

Step 8b$'$
Compute $D = \delta_k^T \gamma_k$. If $D \le 0$, then set $\mathbf{S}_{k+1} = \mathbf{I}_n$, otherwise, compute $\mathbf{S}_{k+1}$ using Eq. (7.57).

At $t = t_k$, the desired trajectory can be described in terms of its Cartesian coordinates as

$$\mathbf{p}_d(t_k) = \begin{bmatrix} p_x(t_k) \\ p_y(t_k) \\ p_z(t_k) \end{bmatrix} = \begin{bmatrix} 30\cos t_k \\ 100\sin t_k \\ 10t_k + 66.04 \end{bmatrix}$$

where $-\pi \le t_k \le \pi$. Assuming 100 uniformly spaced sample points, the solution of the system of equations in Eq. (9.17) can obtained by solving the minimization problem in Eq. (9.18) for $k = 1, 2, \ldots, 100$, i.e., for $t_k = -\pi, \ldots, \pi$, using the specified D-H parameters. Since the gradient and Hessian of $F(\mathbf{x})$ are available (see Eqs. (9.19) and (9.21)), the problem can be solved using each of the five optimization algorithms specified in the description of the problem to obtain a minimizer $\mathbf{x}^*(t_k)$ in each case. If the objective function $F(\mathbf{x})$ turns out to be zero at $\mathbf{x}^*(t_k)$, then $\mathbf{x}^*(t_k)$ satisfies Eq. (9.17), and the joint angles specified by $\mathbf{x}^*(t_k)$ lead the manipulator's end-effector to the desired position precisely. On the other hand, if $F[\mathbf{x}^*(t_k)]$ is nonzero, then $\mathbf{x}^*(t_k)$ is taken as an approximate solution of the inverse kinematics problem at instant t_k.

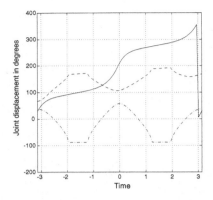

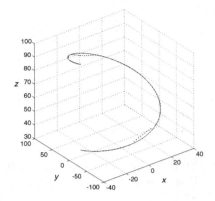

Figure 9.6. Optimal joint angles $\theta_1^*(t)$ (solid line), $\theta_2^*(t)$ (dashed line), and $\theta_3^*(t)$ (dot-dashed line).

Figure 9.7. End-effector's profile (dotted line) and the desired trajectory (solid line).

Once the minimizer $\mathbf{x}^*(t_k)$ is obtained, the above steps can be repeated at $t = t_{k+1}$ to obtain solution point $\mathbf{x}^*(t_{k+1})$. Since t_{k+1} differs from t_k only by a small amount and the profile of optimal joint angles is presumably continuous, $\mathbf{x}^*(t_{k+1})$ is expected to be in the vicinity of $\mathbf{x}^*(t_k)$. Therefore, the previous solution $\mathbf{x}^*(t_k)$ can be used as a reasonable initial point for the next optimization.[1]

The five optimization algorithms were applied to the problem at hand and were all found to work although with different performance in terms of solution accuracy and computational complexity. The solution obtained using the QN algorithm, $\mathbf{x}^*(t_k) = [\theta_1^*(t_k) \ \theta_2^*(t_k) \ \theta_3^*(t_k)]^T$ for $1 \le k \le 100$, is plotted in Fig. 9.6; the tracking profile of the end-effector is plotted as the dotted curve in Fig. 9.7 and is compared with the desired trajectory which is plotted as the solid curve. It turns out that the desired positions $\mathbf{p}_d(t_k)$ for $20 \le k \le 31$ and $70 \le k \le 81$ are beyond the manipulator's reach. As a result, we see in Fig. 9.7 that there are two small portions of the tracking profile that deviate from the desired trajectory, but even in this case, the corresponding $\mathbf{x}^*(t_k)$ still offers a reasonable approximate solution. The remaining part of the tracking profile coincides with the desired trajectory almost perfectly which simply means that for the desired positions within the manipulator's work space, $\mathbf{x}^*(t_k)$ offers a nearly exact solution.

The performance of the five algorithms in terms of the number of Kflops and iterations per sample point and the error at sample points within and outside the

---

[1] Choosing the initial point on the basis of *any* knowledge about the solution instead of a random initial point can lead to a large reduction in the amount of computation in most optimization problems.

work space is summarized in Table 9.3. The data supplied are in the form of averages with respect to 100 runs of the algorithms using random initializations. As can be seen, the average errors within the manipulator's work space for the solutions $\mathbf{x}^*(t_k)$ obtained using the steepest-descent and Fletcher-Reeves algorithms are much larger than those obtained using the Newton, Gauss-Newton, and QN algorithms, although the solutions obtained are still acceptable considering the relatively large size of the desired trajectory. The best results in terms of efficiency as well as accuracy are obtained by using the Newton and QN Algorithms.

Table 9.3 Performance comparisons for Example 9.2

Algorithm	Average number of Kflops per sample point	Average number of iterations per sample point	Average error within work space	Average error outside work space
SD	46.87	23.54	0.05	4.37
N	3.52	2.78	5.27×10^{-8}	4.37
GN	3.66	2.76	1.48×10^{-4}	7.77
FR	13.74	15.80	0.17	4.37
QN	6.07	3.40	2.84×10^{-5}	4.37

■

9.4   Design of Digital Filters

In this section, we will apply unconstrained optimization for the design of FIR digital filters. Different designs are possible depending on the type of FIR filter required and the formulation of the objective function. The theory and design principles of digital filters are quite extensive [3] and are beyond the scope of this book. To facilitate the understanding of the application of unconstrained optimization to the design of digital filters, we present a brief review of the highlights of the theory, properties, and characterization of digital filters in Appendix B, which should prove quite adequate in the present context.

The one design aspect of digital filters that can be handled quite efficiently with optimization is the approximation problem whereby the parameters of the filter have to be chosen to achieve a specified type of frequency response. Below, we examine two different designs (see Sec. B.9). In one design, we formulate a weighted least-squares objective function, i.e., one based on the square of the L_2 norm, for the design of linear-phase FIR filters and in another we obtain a minimax objective function, i.e., one based on the L_∞ norm.

The L_p norm of a vector where $p \geq 1$ is defined in Sec. A.8.1. Similarly, the L_p norm of a function $F(\omega)$ of a continuous variable ω can be defined with respect to the interval $[a, b]$ as

$$\|F(\omega)\|_p = \left(\int_a^b |F(\omega)|^p \, d\omega \right)^{1/p} \tag{9.22}$$

248

where $p \geq 1$ and if

$$\int_a^b |F(\omega)|^p \, d\omega \leq K < \infty$$

the L_p norm of $F(\omega)$ exists. If $F(\omega)$ is bounded with respect to the interval $[a, b]$, i.e., $|F(\omega)| \leq M$ for $\omega \in [a, b]$ where M is finite, then the L_∞ norm of $F(\omega)$ is defined as

$$\|F(\omega)\|_\infty = \max_{a \leq \omega \leq b} |F(\omega)| \qquad (9.23a)$$

and as in the case of the L_∞ norm of a vector, it can be verified that

$$\lim_{p \to \infty} \|F(\omega)\|_p = \|F(\omega)\|_\infty \qquad (9.23b)$$

(see Sec. B.9.1).

9.4.1   Weighted least-squares design of FIR filters

As shown in Sec. B.5.1, an FIR filter is completely specified by its transfer function which assumes the form

$$H(z) = \sum_{n=0}^N h_n z^{-n} \qquad (9.24)$$

where the coefficients h_n for $n = 0, 1, \ldots, n$ represent the impulse response of the filter.

9.4.1.1   Specified frequency response

Assuming a normalized sampling frequency of 2π, which corresponds to a normalized sampling period $T = 1$ s, the frequency response of an FIR filter is obtained as $H(e^{j\omega})$ by letting $z = e^{j\omega}$ in the transfer function (see Sec. B.8). In practice, the frequency response is required to approach some desired frequency response, $H_d(\omega)$, to within a specified error. Hence an FIR filter can be designed by formulating an objective function based on the difference between the actual and desired frequency responses (see Sec. B.9.3). Except in some highly specialized applications, the transfer function coefficients (or impulse response values) of a digital filters are real and, consequently, knowledge of the frequency response of the filter with respect to the positive half of the baseband fully characterizes the filter (see Sec. B.8). Under these circumstances, a weighted least-squares objective function that can be used to design FIR filters can be constructed as

$$e(\mathbf{x}) = \int_0^\pi W(\omega) |H(e^{j\omega}) - H_d(\omega)|^2 \, d\omega \qquad (9.25)$$

where $\mathbf{x} = [h_0 \ h_1 \ \cdots \ h_N]^T$ is an $N + 1$-dimensional variable vector representing the transfer function coefficients, ω is a normalized frequency variable which is assumed to be in the range 0 to π rad/s, and $W(\omega)$ is a predefined weighting function. The design is accomplished by finding the vector $\mathbf{x}^*$ that minimizes $e(\mathbf{x})$, and this can be efficiently done by means of unconstrained optimization.

Weighting is used to emphasize or deemphasize the objective function with respect to one or more ranges of ω. Without weighting, an optimization algorithm would tend to minimize the objective function uniformly with respect to ω. Thus if the objective function is multiplied by a weighting constant larger than unity for values of ω in a certain critical range but is left unchanged for all other frequencies, a reduced value of the objective function will be achieved with respect to the critical frequency range. This is due to the fact that the weighted objective function will tend to be minimized uniformly and thus the actual unweighted objective function will tend to be scaled down in proportion to the inverse of the weighting constant in the critical range of ω relative to its value at other frequencies. Similarly, if a weighting constant of value less than unity is used for a certain uncritical frequency range, an increased value of the objective will be the outcome with respect to the uncritical frequency range. Weighting is very important in practice because through the use of suitable scaling, the designer is often able to design a more economical filter for the required specifications. In the above example, the independent variable is frequency. In other applications, it could be time or some other independent parameter.

An important step in an optimization-based design is to express the objective function in terms of variable vector $\mathbf{x}$ *explicitly*. This facilitates the evaluation of the gradient and Hessian of the objective function. To this end, if we let

$$\mathbf{c}(\omega) = [1 \ \cos\omega \ \cdots \ \cos N\omega]^T \qquad (9.26a)$$

$$\mathbf{s}(\omega) = [0 \ \sin\omega \ \cdots \ \sin N\omega]^T \qquad (9.26b)$$

the frequency response of the filter can be expressed as

$$H(e^{j\omega}) = \sum_{n=0}^{N} h_n \cos n\omega - j \sum_{n=0}^{N} h_n \sin n\omega = \mathbf{x}^T \mathbf{c}(\omega) - j\mathbf{x}^T \mathbf{s}(\omega) \quad (9.27)$$

If we let

$$H_d(\omega) = H_r(\omega) - jH_i(\omega) \qquad (9.28)$$

where $H_r(\omega)$ and $-H_i(\omega)$ are the real and imaginary parts of $H_d(\omega)$, respectively, then Eqs. (9.27) and (9.28) give

$$\begin{aligned}
|H(e^{j\omega}) - H_d(\omega)|^2 &= [\mathbf{x}^T\mathbf{c}(\omega) - H_r(\omega)]^2 + [\mathbf{x}^T\mathbf{s}(\omega) - H_i(\omega)]^2 \\
&= \mathbf{x}^T[\mathbf{c}(\omega)\mathbf{c}^T(\omega) + \mathbf{s}(\omega)\mathbf{s}^T(\omega)]\mathbf{x} \\
&\quad - 2\mathbf{x}^T[\mathbf{c}(\omega)H_r(\omega) + \mathbf{s}(\omega)H_i(\omega)] + |H_d(\omega)|^2
\end{aligned}$$

Therefore, the objective function in Eq. (9.25) can be expressed as a quadratic function with respect to $\mathbf{x}$ of the form

$$e(\mathbf{x}) = \mathbf{x}^T \mathbf{Q} \mathbf{x} - 2\mathbf{x}^T \mathbf{b} + \kappa \tag{9.29}$$

where κ is a constant[2] and

$$\mathbf{Q} = \int_0^\pi W(\omega)[\mathbf{c}(\omega)\mathbf{c}^T(\omega) + \mathbf{s}(\omega)\mathbf{s}^T(\omega)]\, d\omega \tag{9.30}$$

$$\mathbf{b} = \int_0^\pi W(\omega)[H_r(\omega)\mathbf{c}(\omega) + H_i(\omega)\mathbf{s}(\omega)]\, d\omega \tag{9.31}$$

Matrix $\mathbf{Q}$ in Eq. (9.30) is positive definite (see Prob. 9.5). Hence the objective function $e(\mathbf{x})$ in Eq. (9.29) is globally strictly convex and has a unique global minimizer $\mathbf{x}^*$ given by

$$\mathbf{x}^* = \mathbf{Q}^{-1}\mathbf{b} \tag{9.32}$$

For the design of high-order FIR filters, the matrix $\mathbf{Q}$ in Eq. (9.30) is of a large size and the methods described in Sec. 6.4 can be used to find the minimizer without obtaining the inverse of matrix $\mathbf{Q}$.

9.4.1.2 Linear phase response

The frequency response of an FIR digital filter of order N (or length $N+1$) with linear phase response is given by

$$H(e^{j\omega}) = e^{-j\omega N/2} A(\omega) \tag{9.33}$$

Assuming an even-order filter, function $A(\omega)$ in Eq. (9.33) can be expressed as

$$A(\omega) = \sum_{n=0}^{N/2} a_n \cos n\omega \tag{9.34a}$$

$$a_n = \begin{cases} h_{N/2} & \text{for } n = 0 \\ 2h_{N/2-n} & \text{for } n \neq 0 \end{cases} \tag{9.34b}$$

(see Sec. B.9.2) and if the desired frequency response is assumed to be of the form

$$H_d(\omega) = e^{-j\omega N/2} A_d(\omega)$$

then the least-squares objective function

$$e_l(\mathbf{x}) = \int_0^\pi W(\omega)[A(\omega) - A_d(\omega)]^2\, d\omega \tag{9.35a}$$

---

[2]Symbol κ will be used to represent a constant throughout this chapter.

can be constructed where the variable vector is given by

$$\mathbf{x} = [a_0 \ a_1 \ \cdots \ a_{N/2}]^T \tag{9.35b}$$

If we now let

$$\mathbf{c}_l(\omega) = [1 \ \cos\omega \ \cdots \ \cos N\omega/2]^T \tag{9.36a}$$

$A(\omega)$ can be written in terms of the inner product $\mathbf{x}^T\mathbf{c}_l(\omega)$ and the objective function $e_l(\mathbf{x})$ in Eq. (9.35a) can be expressed as

$$e_l(\mathbf{x}) = \mathbf{x}^T\mathbf{Q}_l\mathbf{x} - 2\mathbf{x}^T\mathbf{b}_l + \kappa \tag{9.36b}$$

where κ is a constant, as before, with

$$\mathbf{Q}_l = \int_0^\pi W(\omega)\mathbf{c}_l(\omega)\mathbf{c}_l^T(\omega) \, d\omega \tag{9.37a}$$

$$\mathbf{b}_l = \int_0^\pi W(\omega)A_d(\omega)\mathbf{c}_l(\omega) \, d\omega \tag{9.37b}$$

Like matrix $\mathbf{Q}$ in Eq. (9.30), matrix $\mathbf{Q}_l$ in Eq. (9.37a) is positive definite; hence, like the objective function $e(\mathbf{x})$ in Eq. (9.29), $e_l(\mathbf{x})$ in Eq. (9.36b) is globally strictly convex and its unique global minimizer is given in closed form by

$$\mathbf{x}_l^* = \mathbf{Q}_l^{-1}\mathbf{b}_l \tag{9.38}$$

For filters of order less than 200, matrix $\mathbf{Q}_l$ in Eq. (9.38) is of size less than 100, and the formula in Eq. (9.38) requires a moderate amount of computation. For higher-order filters, the closed-form solution given in Eq. (9.38) becomes computationally very demanding and methods that do not require the computation of the inverse of matrix $\mathbf{Q}_l$ such as those studied in Sec. 6.4 would be preferred.

Example 9.3

    (a) Applying the above method, formulate the design of an even-order linear-phase lowpass FIR filter assuming the desired amplitude response

$$A_d(\omega) = \begin{cases} 1 & \text{for } 0 \le \omega \le \omega_p \\ 0 & \text{for } \omega_a \le \omega \le \pi \end{cases} \tag{9.39}$$

    where ω_p and ω_a are the passband and stopband edges, respectively (see Sec. B.9.1). Assume a normalized sampling frequency of 2π rad/s.

    (b) Using the formulation in part (a), design FIR filters with $\omega_p = 0.45\pi$ and $\omega_a = 0.5\pi$ for filter orders of 20, 40, 60, and 80.

Solution (*a*) A suitable weighting function $W(\omega)$ for this problem is

$$W(\omega) = \begin{cases} 1 & \text{for } 0 \le \omega \le \omega_p \\ \gamma & \text{for } \omega_a \le \omega \le \pi \\ 0 & \text{elsewhere} \end{cases} \qquad (9.40)$$

The value of γ can be chosen to emphasize or deemphasize the error function in the stopband relative to that in the passband. Since $W(\omega)$ is piecewise constant, the matrix $\mathbf{Q}_l$ in Eq. (9.37a) can be written as

$$\mathbf{Q}_l = \mathbf{Q}_{l1} + \mathbf{Q}_{l2}$$

where

$$\mathbf{Q}_{l1} = \int_0^{\omega_p} \mathbf{c}_l(\omega) \mathbf{c}_l^T(\omega) \, d\omega = \{q_{ij}^{(1)}\} \qquad \text{for } 1 \le i,\, j \le \frac{N+2}{2} \quad (9.41a)$$

and

$$\mathbf{Q}_{l2} = \gamma \int_{\omega_a}^{\pi} \mathbf{c}_l(\omega) \mathbf{c}_l^T(\omega) \, d\omega = \{q_{ij}^{(2)}\} \qquad \text{for } 1 \le i,\, j \le \frac{N+2}{2} \quad (9.41b)$$

with

$$q_{ij}^{(1)} = \begin{cases} \dfrac{\omega_p}{2} + \dfrac{\sin[2(i-1)\omega_p]}{4(i-1)} & \text{for } i = j \\[2ex] \dfrac{\sin[(i-j)\omega_p]}{2(i-j)} + \dfrac{\sin[(i+j-2)\omega_p]}{2(i+j-2)} & \text{for } i \ne j \end{cases} \qquad (9.42a)$$

and

$$q_{ij}^{(2)} = \begin{cases} \gamma\left[\dfrac{(\pi - \omega_a)}{2} - \dfrac{\sin[2(i-1)\omega_a]}{4(i-1)}\right] & \text{for } i = j \\[2ex] -\dfrac{\gamma}{2}\left[\dfrac{\sin[(i-j)\omega_a]}{(i-j)} + \dfrac{\sin[(i+j-2)\omega_a]}{(i+j-2)}\right] & \text{for } i \ne j \end{cases} \qquad (9.42b)$$

Note that for $i = j = 1$, the expressions in Eq. (9.42) are evaluated by taking the limit as $i \to 1$, which implies that

$$q_{11}^{(1)} = \omega_p \quad \text{and} \quad q_{11}^{(2)} = \gamma(\pi - \omega_a) \qquad (9.42c)$$

Vector $\mathbf{b}_l$ in Eq. (9.37b) is calculated as

$$\mathbf{b}_l = \int_0^{\omega_p} \mathbf{c}_l(\omega) \, d\omega = \{b_n\} \qquad (9.43a)$$

with

$$b_n = \frac{\sin[(n-1)\omega_p]}{(n-1)} \quad \text{for } 1 \le n \le \frac{N+2}{2} \quad (9.43b)$$

As before, for $n = 1$, the expression in Eq. (9.43b) is evaluated by taking the limit as $n \to 1$, which gives

$$b_1 = \omega_p \quad (9.43c)$$

(*b*) Optimal weighted least-squares designs for the various values of N were obtained by computing the minimizer $\mathbf{x}_l^*$ given by Eq. (9.38) and then evaluating the filter coefficients $\{h_i\}$ using Eq. (9.34b). The weighting constant γ was assumed to be 25. The amplitude responses of the FIR filters obtained are plotted in Fig. 9.8.

∎

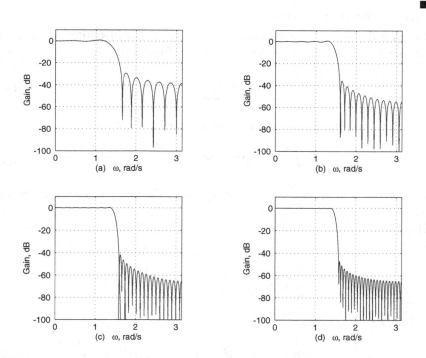

Figure 9.8. Amplitude responses of the filters in Example 9.3: (a) $N = 20$, (b) $N = 40$, (c) $N = 60$, (d) $N = 80$.

9.4.2 Minimax design of FIR filters

The Parks-McClellan algorithm and its variants have been the most efficient tools for the minimax design of FIR digital filters [3]–[5]. However, these algorithms apply only to the class of linear-phase FIR filters. The group delay introduced by these filters is constant and independent of frequency in the entire baseband (see Sec. B.8) but it can be quite large. In practice, a variable group delay in stopbands is of little concern and by allowing the phase response to be nonlinear in stopbands, FIR filters can be designed with constant group delay with respect to the passbands, which is significantly reduced relative to that achieved with filters that have a constant group delay throughout the entire baseband.

This section presents a least-pth approach to the design of low-delay FIR filters. For FIR filters, the weighted L_p error function with an even integer p can be shown to be globally convex.[3] This property, in conjunction with the availability of the gradient and Hessian of the objective function in closed form, enables us to develop an unconstrained optimization method for the design problem at hand.

9.4.2.1 Objective function

Given a desired frequency response $H_d(\omega)$ for an FIR filter, we want to determine the coefficients $\{h_n\}$ in the transfer function

$$H(z) = \sum_{n=0}^{N} h_n z^{-n} \tag{9.44}$$

such that the weighted L_{2p} approximation error

$$f(\mathbf{h}) = \left[\int_0^\pi W(\omega) |H(e^{j\omega}) - H_d(\omega)|^{2p} \, d\omega \right]^{1/2p} \tag{9.45}$$

is minimized, where $W(\omega) \geq 0$ is a weighting function, p is a positive integer, and $\mathbf{h} = [h_0 \ h_1 \ \cdots \ h_N]^T$.

If we let

$$
\begin{aligned}
H_d(\omega) &= H_{dr}(\omega) - j H_{di}(\omega) \\
\mathbf{c}(\omega) &= [1 \ \cos \omega \ \cdots \ \cos N\omega]^T \\
\mathbf{s}(\omega) &= [0 \ \sin \omega \ \cdots \ \sin N\omega]^T
\end{aligned}
$$

then Eq. (9.45) becomes

$$f(\mathbf{h}) = \left\{ \int_0^\pi W[(\mathbf{h}^T\mathbf{c} - H_{dr})^2 + (\mathbf{h}^T\mathbf{s} - H_{di})^2]^p \, d\omega \right\}^{1/2p} \tag{9.46}$$

where for simplicity the frequency dependence of W, $\mathbf{c}$, $\mathbf{s}$, H_{dr}, and H_{di} has been omitted. Now if we let

$$e_2(\omega) = [\mathbf{h}^T\mathbf{c}(\omega) - H_{dr}(\omega)]^2 + [\mathbf{h}^T\mathbf{s}(\omega) - H_{di}(\omega)]^2 \tag{9.47}$$

then the objective function can be expressed as

---

[3]Note that this property does not apply to infinite-duration impulse response (IIR) filters [3].

$$f(\mathbf{h}) = \left[\int_0^\pi W(\omega) e_2^p(\omega) \, d\omega \right]^{1/2p} \tag{9.48}$$

9.4.2.2 Gradient and Hessian of $f(\mathbf{h})$

Using Eq. (9.48), the gradient and Hessian of objective function $f(\mathbf{h})$ can be readily obtained as

$$\nabla f(\mathbf{h}) = f^{1-2p}(\mathbf{h}) \int_0^\pi W(\omega) e_2^{p-1}(\omega) \mathbf{q}(\omega) \, d\omega \tag{9.49a}$$

where

$$\mathbf{q}(\omega) = [\mathbf{h}^T \mathbf{c}(\omega) - H_{dr}(\omega)] \mathbf{c}(\omega) + [\mathbf{h}^T \mathbf{s}(\omega) - H_{di}(\omega)] \mathbf{s}(\omega) \tag{9.49b}$$

and

$$\nabla^2 f(\mathbf{h}) = \mathbf{H}_1 + \mathbf{H}_2 - \mathbf{H}_3 \tag{9.49c}$$

where

$$\mathbf{H}_1 = 2(p-1) f^{1-2p}(\mathbf{h}) \int_0^\pi W(\omega) e_2^{p-2}(\omega) \mathbf{q}(\omega) \mathbf{q}^T(\omega) \, d\omega \tag{9.49d}$$

$$\mathbf{H}_2 = f^{1-2p}(\mathbf{h}) \int_0^\pi W(\omega) e_2^{p-1}(\omega) [\mathbf{c}(\omega) \mathbf{c}^T(\omega) + \mathbf{s}(\omega) \mathbf{s}^T(\omega)] \, d\omega \tag{9.49e}$$

$$\mathbf{H}_3 = (2p-1) f^{-1}(\mathbf{h}) \nabla f(\mathbf{h}) \nabla^T f(\mathbf{h}) \tag{9.49f}$$

respectively.

Of central importance to the present algorithm is the property that for each and every positive integer p, the weighted L_{2p} objective function defined in Eq. (9.45) is convex in the entire parameter space $\mathcal{R}^{N+1}$. This property can be proved by showing that the Hessian $\nabla^2 f(\mathbf{h})$ is positive semidefinite for all $\mathbf{h} \in R^{N+1}$ (see Prob. 9.9).

9.4.2.3 Design algorithm

It is now quite clear that an FIR filter whose frequency response approximates a rather arbitrary frequency response $H_d(\omega)$ to within a given tolerance in the minimax sense can be obtained by minimizing $f(\mathbf{h})$ in Eq. (9.45) with a sufficiently large p. It follows from the above discussion that for a given p, $f(\mathbf{h})$ has a unique global minimizer. Therefore, any descent minimization algorithm, e.g., the steepest-descent, Newton, and quasi-Newton methods studied in previous chapters, can, in principle, be used to obtain the minimax design regardless of the initial design chosen. The amount of computation required to obtain the design is largely determined by the choice of optimization method as well as the initial point assumed.

A reasonable initial point can be deduced by using the L_2-optimal design obtained by minimizing $f(\mathbf{h})$ in Eq. (9.45) with $p = 1$. We can write

$$f(\mathbf{h}) = (\mathbf{h}^T \mathbf{Q} \mathbf{h} - 2 \mathbf{h}^T \mathbf{p} + \kappa)^{1/2} \tag{9.50a}$$

where

$$\mathbf{Q} = \int_0^\pi W(\omega)[\mathbf{c}(\omega)\mathbf{c}^T(\omega) + \mathbf{s}(\omega)\mathbf{s}^T(\omega)] \, d\omega \qquad (9.50\text{b})$$

$$\mathbf{p} = \int_0^\pi W(\omega)[H_{dr}(\omega)\mathbf{c}(\omega) + H_{di}(\omega)\mathbf{s}(\omega)] \, d\omega \qquad (9.50\text{c})$$

Since $\mathbf{Q}$ is positive definite, the global minimizer of $f(\mathbf{h})$ in Eq. (9.50a) can be obtained as the solution of the linear equation

$$\mathbf{Qh} = \mathbf{p} \qquad (9.51)$$

We note that $\mathbf{Q}$ in Eq. (9.51) is a symmetric Toeplitz matrix[4] for which fast algorithms are available to compute the solution of Eq. (9.51) [6].

The minimization of convex objective function $f(\mathbf{h})$ can be accomplished in a number of ways. Since the gradient and Hessian of $f(\mathbf{h})$ are available in closed-form and $\nabla^2 f(\mathbf{h})$ is positive semidefinite, the Newton method and the family of quasi-Newton methods are among the most appropriate.

From Eqs. (9.48) and (9.49), we note that $f(\mathbf{h})$, $\nabla f(\mathbf{h})$, and $\nabla^2 f(\mathbf{h})$ all involve integration which can be carried out using numerical methods. In computing $\nabla^2 f(\mathbf{h})$, the error introduced in the numerical integration can cause the Hessian to lose its positive definiteness but the problem can be easily fixed by modifying $\nabla^2 f(\mathbf{h})$ to $\nabla^2 f(\mathbf{h}) + \varepsilon \mathbf{I}$ where ε is a small positive scalar.

9.4.2.4 Direct and sequential optimizations

With a power p, weighting function $W(\omega)$, and an initial $\mathbf{h}$, say, $\mathbf{h}_0$, chosen, the design can be obtained directly or indirectly.

In a direct optimization, one of the unconstrained optimization methods is applied to minimize the L_{2p} objective function in Eq. (9.48) directly. Based on rather extensive trials, it was found that to achieve a near-minimax design, the value of p should be larger than 20 and for high-order FIR filters a value comparable to the filter order N should be used.

In sequential optimization, an L_{2p} optimization is first carried out with $p = 1$. The minimizer thus obtained, $\mathbf{h}^*$, is then used as the initial point in another optimization with $p = 2$. The same procedure is repeated for $p = 4, 8, 16, \ldots$ until the reduction in the objective function between two successive optimizations is less than a prescribed tolerance.

Example 9.4 Using the above direct and sequential approaches first with a Newton and then with a quasi-Newton algorithm, design a lowpass FIR filter of order $N = 54$ that would have approximately constant passband group delay of 23 s. Assume idealized passband and stopband gains of 1 and 0, respectively;

---

[4]A Toeplitz matrix is a matrix whose entries along each diagonal are constant [6].

a normalized sampling frequency $\omega_s = 2\pi$; passband edge $\omega_p = 0.45\pi$ and stopband edge $\omega_a = 0.55\pi$; $W(\omega) = 1$ in both the passband and stopband, and $W(\omega) = 0$ elsewhere.

Solution The design was carried out using the direct approach with $p = 128$ and the sequential approach with $p = 2, 4, 8, \ldots, 128$ by minimizing the objective function in Eq. (9.48) with the Newton algorithm and a quasi-Newton algorithm with the BFGS updating formula in Eq. (7.57). The Newton algorithm used was essentially the same as Algorithm 9.1 (see solution of Example 9.2) except that Step 3 was replaced by the following modified Step 3:

Step 3$'$
Modify matrix $\mathbf{H}_k$ to $\hat{\mathbf{H}}_k = \mathbf{H}_k + 0.1\mathbf{I}_n$

The quasi-Newton algorithm used was Algorithm 7.3 with the modifications described in the solution of Example 9.2.

A lowpass FIR filter that would satisfy the required specifications can be obtained by assuming a complex-valued idealized frequency response of the form

$$H_d(\omega) = \begin{cases} e^{-j23\omega} & \text{for } \omega \in [0,\ \omega_p] \\ 0 & \text{for } \omega \in [\omega_a,\ \omega_s/2] \end{cases}$$
$$= \begin{cases} e^{-j23\omega} & \text{for } \omega \in [0,\ 0.45\pi] \\ 0 & \text{for } \omega \in [0.55\pi,\ \pi] \end{cases}$$

(see Sec. B.9.2). The integrations in Eqs. (9.48), (9.49a), and (9.49c) can be carried out by using one of several available numerical methods for integration. A fairly simple and economical approach, which works well in optimization, is as follows: Given a continuous function $f(\omega)$ of ω, an approximate value of its integral over the interval $[a,\ b]$ can be obtained as

$$\int_a^b f(\omega)d\omega \approx \delta \sum_{i=1}^K f(\omega_i)$$

where $\delta = (b - a)/K$ and $\omega_1 = a + \delta/2$, $\omega_2 = a + 3\delta/2$, $\ldots$, $\omega_K = a + (2K - 1)\delta/2$. That is, we divide interval $[a,\ b]$ into K subintervals, add the values of the function at the midpoints of the K subintervals, and then multiply the sum obtained by δ.

The objective function in Eq. (9.48) was expressed as

$$f(\mathbf{h}) = \left[\int_0^{0.45\pi} e_2^p(\omega)d\omega \right]^{1/2p} + \left[\int_{0.55\pi}^{\pi} e_2^p(\omega)d\omega \right]^{1/2p}$$

and each integral was evaluated using the above approach with $K = 500$. The integrals in Eqs. (9.49a) and (9.49c) were evaluated in the same way.

258

The initial h was obtained by applying L_2 optimization to Eq. (9.50). All trials converged to the same near minimax design, and the sequential approach turned out to be more efficient than the direct approach. The Newton and quasi-Newton algorithms required 21.1 and 40.7 s of CPU time, respectively, on a PC with a Pentium 4, 3.2 GHz CPU. The amplitude response, passband error, and group delay characteristic of the filter obtained are plotted in Fig. 9.9a, b, and c, respectively. We note that an equiripple amplitude response was achieved in both the passband and stopband. The passband group delay varies between 22.9 and 23.1 but it is not equiripple. This is because the minimax optimization was carried out for the *complex-valued* frequency response $H_d(\omega)$, not the phase-response alone (see Eq. (9.45)).　　　　　∎

Example 9.5 Using the above direct and sequential approaches first with a Newton and then with a quasi-Newton algorithm, design a bandpass FIR filter of order $N = 160$ that would have approximately constant passband group delay of 65 s. Assume idealized passband and stopband gains of 1 and 0, respectively; normalized sampling frequency $= 2\pi$; passband edges $\omega_{p1} = 0.4\pi$ and $\omega_{p2} = 0.6\pi$; stopband edges $\omega_{a1} = 0.375\pi$ and $\omega_{a2} = 0.625\pi$; $W(\omega) = 1$ in the passband and $W(\omega) = 50$ in the stopbands, and $W(\omega) = 0$ elsewhere.

Solution The required design was carried out using the direct approach with $p = 128$ and the sequential approach with $p = 2, 4, 8, \ldots, 128$ by minimizing the objective function in Eq. (9.48) with the Newton and quasi-Newton algorithms described in Example 9.4.

A bandpass FIR filter that would satisfy the required specifications can be obtained by assuming a complex-valued idealized frequency response of the form

$$H_d(\omega) = \begin{cases} e^{-j65\omega} & \text{for } \omega \in [\omega_{p1}, \omega_{p2}] \\ 0 & \text{for } \omega \in [0, \omega_{a1}] \bigcup [\omega_{a2}, \omega_s/2] \end{cases}$$
$$= \begin{cases} e^{-j65\omega} & \text{for } \omega \in [0.4\pi, 0.6\pi] \\ 0 & \text{for } \omega \in [0, 0.375\pi] \bigcup [0.625\pi, \pi] \end{cases}$$

(see Sec. B.9.2). The objective function in Eq. (9.48) was expressed as

$$f(\mathbf{h}) = \left[\int_0^{0.375\pi} 50 e_2^p(\omega) d\omega \right]^{1/2p} + \left[\int_{0.4\pi}^{0.6\pi} e_2^p(\omega) d\omega \right]^{1/2p} + \left[\int_{0.625\pi}^{\pi} 50 e_2^p(\omega) d\omega \right]^{1/2p}$$

and the integrals at the right-hand side were evaluated using the numerical method in the solution of Example 9.4 with $K = 382, 236, 382$ respectively.

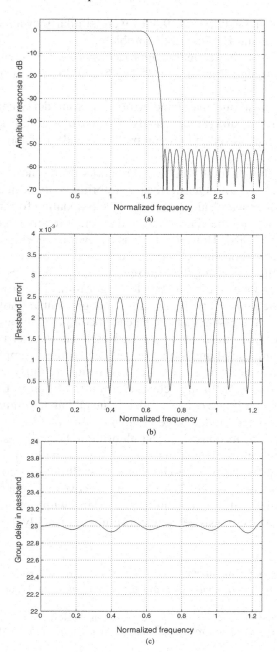

Figure 9.9. Minimax design of a lowpass filter with low passband group delay for Example 9.4: (a) Frequency response, (b) magnitude of the passband error, and (c) passband group delay.

260

The integrals in Eq. (9.49a) and (9.49c) were similarly evaluated in order to obtain the gradient and Hessian of the problem.

As in Example 9.4, the sequential approach was more efficient. The Newton and quasi-Newton algorithms required 173.5 and 201.8 s, respectively, on a Pentium 4 PC.

The amplitude response, passband error, and group delay characteristic are plotted in Fig. 9.10a, b, and c, respectively. We note that an equiripple amplitude response has been achieved in both the passband and stopband.

■

We conclude this chapter with some remarks on the numerical results of Examples 9.2, 9.4 and 9.5. Quasi-Newton algorithms, in particular algorithms using an inexact line-search along with the BFGS updating formula (e.g., Algorithm 7.3), are known to be very robust and efficient relative to other gradient-based algorithms [7]–[8]. However, the basic Newton algorithm used for these problems, namely, Algorithm 9.1, turned out to be more efficient than the quasi-Newton algorithm. This is largely due to certain unique features of the problems considered, which favor the basic Newton algorithm. The problem in Example 9.2 is a simple problem with only three independent variables and an well defined gradient and Hessian that can be easily computed through closed-form formulas. Furthermore, the inversion of the Hessian is almost a trivial task. The problems in Examples 9.4 and 9.5 are significantly more complex than that in Example 9.2; however, their gradients and Hessians are fairly easy to compute accurately and efficiently through closed-form formulas as in Example 9.2. In addition, these problems are convex with unique global minimums that are easy to locate. On the other hand, a large number of variables in the problem tends to be an impediment in quasi-Newton algorithms because, as was shown in Chap. 7, these algorithms would, in theory, require n iterations in an n-variable problem to compute the inverse-Hessian in a well defined convex quadratic problem (see proof of Theorem 7.3), more in noncovex nonquadratic problems. However, in multimodal[5] highly nonlinear problems with a moderate number of independent variables, quasi-Newton algorithms are usually the most efficient.

References

1 M. Werman and D. Weinshall, "Similarity and affine invariant distances between 2D point sets," *IEEE Trans. Pattern Analysis and Machine Intelligence*, vol. 17, pp. 810–814, August 1995.
2 J. J. Craig, *Introduction to Robotics*, 2nd ed., Addison-Wesley, 1989.
3 A. Antoniou, *Digital Signal Processing: Signals, Systems, and Filters,* McGraw-Hill, New York, 2005.
4 T. W. Parks and J. H. McClellan, "Chebyshev approximation for nonrecursive digital filters with linear phase," *IEEE Trans. Circuit Theory*, vol. 19, pp. 189-194, 1972.

[5]Problems with multiple minima.

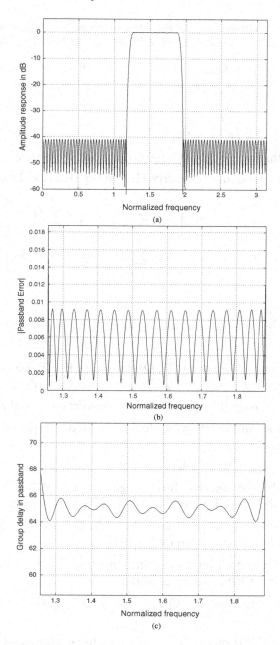

Figure 9.10. Minimax design of a bandpass filter with low passband group delay for Example 9.5: (a) Frequency response, (b) magnitude of passband error, (c) passband group delay.

5 T. W. Parks and C. S. Burrus, *Digital Filter Design*, Wiley, New York, 1987.
6 G. H. Golub and C. F. Van Loan, *Matrix Computations,* 2nd ed., The Johns Hopkins University Press, Baltimore, 1989.
7 R. Fletcher, *Practical Methods of Optimization*, vol. 1, Wiley, New York, 1980.
8 R. Fletcher, *Practical Methods of Optimization*, 2nd ed., Wiley, New York, 1987.

Problems

9.1 (a) Verify Eqs. (9.6a)–(9.6d).

(b) Show that matrix $\mathbf{H}$ in Eq. (9.6b) is positive definite.

(c) Show that the inverse matrix $\mathbf{H}^{-1}$ in Eq. (9.7) can be evaluated as

$$\mathbf{H}^{-1} = \begin{bmatrix} \gamma_4 \mathbf{I}_2 & -\frac{\gamma_4}{n}\sum_{i=1}^{n}\mathbf{R}_i^T \\ -\frac{\gamma_4}{n}\sum_{i=1}^{n}\mathbf{R}_i & \frac{1}{n}(1+\frac{\gamma_3\gamma_4}{n})\mathbf{I}_2 \end{bmatrix}$$

where

$$\gamma_1 = \sum_{i=1}^{n}p_{i1},\ \gamma_2 = \sum_{i=1}^{n}p_{i2},\ \gamma_3 = \gamma_1^2+\gamma_2^2,\ \gamma_4 = \left(\|\mathbf{P}\|_F^2 - \frac{\gamma_3}{n}\right)^{-1}$$

9.2 The dissimilarity measure $e(\tilde{\mathcal{P}},\mathcal{Q})$ defined in Eq. (9.5) is not symmetric, i.e., in general $e(\tilde{\mathcal{P}},\mathcal{Q}) \neq e(\mathcal{Q},\tilde{\mathcal{P}})$, which is obviously undesirable.

(a) Obtain a dissimilarity measure for two point patterns that is symmetric.

(b) Solve the minimization problem associated with the new dissimilarity measure.

9.3 (a) Verify Eqs. (9.9a)–(9.9c).

(b) Prove that the objective function given in Eq. (9.8) is globally convex. Hint: Show that for any $\mathbf{y} \in R^4$, $\mathbf{y}^T\nabla^2 e_{2p}(\mathbf{x})\mathbf{y} \geq 0$.

9.4 Derive formulas for the evaluation of $\nabla^2 f_k(\mathbf{x})$ for $k = 1,\ 2$, and 3 for the set of functions $f_k(\mathbf{x})$ given by Eq. (9.16).

9.5 Show that for a nontrivial weighting function $W(\omega) \geq 0$, the matrix $\mathbf{Q}$ given by Eq. (9.30) is positive definite.

9.6 Derive the expressions of $\mathbf{Q}_l$ and $\mathbf{b}_l$ given in Eqs. (9.41), (9.42), and (9.43).

9.7 Write a MATLAB program to implement the unconstrained optimization algorithm for the weighted least-squares design of linear-phase lowpass FIR digital filters studied in Sec. 9.4.1.2.

9.8 Develop an unconstrained optimization algorithm for the weighted least-squares design of linear-phase highpass digital filters.

9.9 Prove that the objective function given in Eq. (9.45) is globally convex. Hint: Show that for any $\mathbf{y} \in R^{N+1}$, $\mathbf{y}^T \nabla^2 f(\mathbf{h}) \mathbf{y} \geq 0$.

9.10 Develop a method based on unconstrained optimization for the design of FIR filters with low passband group delay allowing coefficients with complex values.

9.11 Consider the double inverted pendulum control system described in Example 1.2, where $\alpha = 16$, $\beta = 8$, $T_0 = 0.8$, $\Delta t = 0.02$, and $K = 40$. The initial state is set to $\mathbf{x}(0) = [\pi/6 \ 1 \ \pi/6 \ 1]^T$ and the constraints on the magnitude of control actions are $|u(i)| \leq m$ for $i = 0, 1, \ldots, K-1$ with $m = 112$.

(a) Use the singular-value decomposition technique (see Sec. A.9, especially Eqs. (A.43) and (A.44)) to eliminate the equality constraint $\mathbf{a}(\mathbf{u}) = \mathbf{0}$ in Eq. (1.9b).

(b) Convert the constrained problem obtained from part (a) to an unconstrained problem of the augmented objective function

$$F_\tau(\mathbf{u}) = \mathbf{u}^T \mathbf{u} - \tau \sum_{i=0}^{K-1} \ln[m - u(i)] - \tau \sum_{i=0}^{K-1} \ln[m + u(i)]$$

where the barrier parameter τ is fixed to a positive value in each round of minimization, which is then reduced to a smaller value at a fixed rate in the next round of minimization.

Note that in each round of minimization, a line search step should be carefully executed where the step-size α is limited to a *finite* interval $[0, \bar{\alpha}]$ that is determined by the constraints $|u(i)| \leq m$ for $0 \leq i \leq K - 1$.

Chapter 10

FUNDAMENTALS OF CONSTRAINED OPTIMIZATION

10.1 Introduction

The material presented so far dealt largely with principles, methods, and algorithms for unconstrained optimization. In this and the next five chapters, we build on the introductory principles of constrained optimization discussed in Secs. 1.4–1.6 and proceed to examine the underlying theory and structure of some very sophisticated and efficient constrained optimization algorithms.

The presence of constraints gives rise to a number of technical issues that are not encountered in unconstrained problems. For example, a search along the direction of the negative of the gradient of the objective function is a well justified technique for unconstrained minimization. However, in a constrained optimization problem points along such a direction may not satisfy the constraints and in such a case the search will not yield a solution of the problem. Consequently, new methods for determining feasible search directions have to be sought.

Many powerful techniques developed for constrained optimization problems are based on unconstrained optimization methods. If the constraints are simply given in terms of lower and/or upper limits on the parameters, the problem can be readily converted into an unconstrained problem. Furthermore, methods of transforming a constrained minimization problem into a sequence of unconstrained minimizations of an appropriate auxiliary function exist.

The purpose of this chapter is to lay a theoretical foundation for the development of various algorithms for constrained optimization. Equality and inequality constraints are discussed in general terms in Sec. 10.2. After a brief discussion on the classification of constrained optimization problems in Sec. 10.3, several variable transformation techniques for converting optimization problems with simple constraints into unconstrained problems are studied in Sec. 10.4. One of the most important concepts in constrained optimization, the concept of *Lagrange multipliers*, is introduced and a geometric interpretation of Lagrange multipliers is given in Sec. 10.5. The first-order necessary conditions for a point $\mathbf{x}^*$ to be a solution of a constrained problem, known as the *Karush-Kuhn-Tucker conditions*, are studied in Sec. 10.6 and the second-order conditions are discussed in Sec. 10.7. As in the unconstrained case, the concept of convexity plays an important role in the study of constrained optimization and it is discussed in Sec. 10.8. Finally, the concept of duality, which is of significant importance in the development and unification of optimization theory, is addressed in Sec. 10.9.

10.2   Constraints

10.2.1   Notation and basic assumptions

In its most general form, a constrained optimization problem is to find a vector $\mathbf{x}^*$ that solves the problem

$$\text{minimize } f(\mathbf{x}) \tag{10.1a}$$

$$\text{subject to:} \quad a_i(\mathbf{x}) = 0 \quad \text{for } i = 1, 2, \ldots, p \tag{10.1b}$$

$$c_j(\mathbf{x}) \geq 0 \quad \text{for } j = 1, 2, \ldots, q \tag{10.1c}$$

Throughout the chapter, we assume that the objective function $f(\mathbf{x})$ as well as the functions involved in the constraints in Eqs. (10.1b) and (10.1c), namely, $\{a_i(\mathbf{x})$ for $i = 1, 2, \ldots, p\}$ and $\{c_j(\mathbf{x})$ for $j = 1, 2, \ldots, q\}$, are continuous and have continuous second partial derivatives, i.e., $a_i(\mathbf{x}), c_j(\mathbf{x}) \in C^2$. Let $\mathcal{R}$ denote the feasible region for the problem in Eq. (10.1), which was defined in Sec. 1.5 as the set of points satisfying Eqs. (10.1b) and (10.1c), i.e.,

$$\mathcal{R} = \{\mathbf{x} : a_i(\mathbf{x}) = 0 \text{ for } i = 1, 2, \ldots, p, \ c_j(\mathbf{x}) \geq 0 \text{ for } j = 1, 2, \ldots, q\}$$

In this chapter as well as the rest of the book, we often need to compare two vectors or matrices entry by entry. For two matrices $\mathbf{A} = \{a_{ij}\}$ and $\mathbf{B} = \{b_{ij}\}$ of the same dimension, we use $\mathbf{A} \geq \mathbf{B}$ to denote $a_{ij} \geq b_{ij}$ for all i, j. Consequently, $\mathbf{A} \geq \mathbf{0}$ means $a_{ij} \geq 0$ for all i, j. We write $\mathbf{A} \succ \mathbf{0}$, $\mathbf{A} \succeq \mathbf{0}$, $\mathbf{A} \prec \mathbf{0}$, and $\mathbf{A} \preceq \mathbf{0}$ to denote that matrix $\mathbf{A}$ is positive definite, positive semidefinite, negative definite, and negative semidefinite, respectively.

10.2.2   Equality constraints

The set of equality constraints

$$a_1(\mathbf{x}) = 0$$
$$\vdots \tag{10.2}$$
$$a_p(\mathbf{x}) = 0$$

defines a hypersurface in R^n. Using vector notation, we can write

$$\mathbf{a}(\mathbf{x}) = [a_1(\mathbf{x})\, a_2(\mathbf{x}) \,\cdots\, a_p(\mathbf{x})]^T$$

and from Eq. (10.2), we have

$$\mathbf{a}(\mathbf{x}) = \mathbf{0} \tag{10.3}$$

Definition 10.1 A point $\mathbf{x}$ is called a *regular point* of the constraints in Eq. (10.2) if $\mathbf{x}$ satisfies Eq. (10.2) and column vectors $\nabla a_1(\mathbf{x})$, $\nabla a_2(\mathbf{x})$, ..., $\nabla a_p(\mathbf{x})$ are linearly independent.

∎

The definition states, in effect, that $\mathbf{x}$ is a regular point of the constraints, if it is a solution of Eq. (10.2) and the Jacobian $\mathbf{J}_e = [\nabla a_1(\mathbf{x}) \; \nabla a_2(\mathbf{x}) \; \cdots \; \nabla a_p(\mathbf{x})]^T$ has full row rank. The importance of a point $\mathbf{x}$ being regular for a given set of equality constraints lies in the fact that a tangent plane of the hypersurface determined by the constraints at a regular point $\mathbf{x}$ is well defined. Later in this chapter, the term 'tangent plane' will be used to express and describe important necessary as well as sufficient conditions for constrained optimization problems. Since $\mathbf{J}_e$ is a $p \times n$ matrix, it would not be possible for $\mathbf{x}$ to be a regular point of the constraints if $p > n$. This leads to an upper bound for the number of independent equality constraints, i.e., $p \leq n$. Furthermore, if $p = n$, in many cases the number of vectors $\mathbf{x}$ that satisfy Eq. (10.2) is finite and the optimization problem becomes a trivial one. For these reasons, we shall assume that $p < n$ throughout the rest of the book.

Example 10.1 Discuss and sketch the feasible region described by the equality constraints

$$-x_1 + x_3 - 1 = 0 \tag{10.4a}$$
$$x_1^2 + x_2^2 - 2x_1 = 0 \tag{10.4b}$$

Solution The Jacobian of the constraints is given by

$$\mathbf{J}_e(\mathbf{x}) = \begin{bmatrix} -1 & 0 & 1 \\ 2x_1 - 2 & 2x_2 & 0 \end{bmatrix}$$

which has rank 2 except at $\mathbf{x} = [1 \; 0 \; x_3]^T$. Since $\mathbf{x} = [1 \; 0 \; x_3]^T$ does not satisfy the constraint in Eq. (10.4b), any point $\mathbf{x}$ satisfying Eq. (10.4) is regular. The constraints in Eq. (10.4) describe a curve which is the intersection between the cylinder in Eq. (10.4b) and the plane in Eq. (10.4a). For the purpose of displaying the curve, we derive a parametric representation of the curve as follows. Eq. (10.4b) can be written as

$$(x_1 - 1)^2 + x_2^2 = 1$$

which suggests the parametric expressions

$$x_1 = 1 + \cos t \tag{10.5a}$$
$$x_2 = \sin t \tag{10.5b}$$

for x_1 and x_2. Now Eq. (10.5) in conjunction with Eq. (10.4a) gives

$$x_3 = 2 + \cos t \tag{10.5c}$$

With parameter t varying from 0 to 2π, Eq. (10.5) describes the curve shown in Fig. 10.1.

■

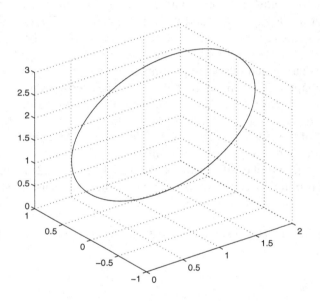

Figure 10.1. Constraints in Eq. (10.4) as a curve.

A particularly important class of equality constraints is the class of linear constraints where functions $a_i(\mathbf{x})$ are all linear. In this case, Eq. (10.2) becomes a system of linear equations which can be expressed as

$$\mathbf{A}\mathbf{x} = \mathbf{b} \tag{10.6}$$

where $\mathbf{A} \in R^{p \times n}$ is numerically equal to the Jacobian, i.e., $\mathbf{A} = \mathbf{J}_e$, and $\mathbf{b} \in R^{p \times 1}$. Since the Jacobian is a constant matrix, any solution point of Eq. (10.6) is a regular point if $\text{rank}(\mathbf{A}) = p$. If $\text{rank}(\mathbf{A}) = p' < p$, then there are two possibilities: either

$$\text{rank}([\mathbf{A}\ \mathbf{b}]) > \text{rank}(\mathbf{A}) \tag{10.7}$$

or

$$\text{rank}([\mathbf{A} \ \mathbf{b}]) = \text{rank}(\mathbf{A}) \tag{10.8}$$

If Eq. (10.7) is satisfied, then we conclude that contradictions exist in Eq. (10.6), and a careful examination of Eq. (10.6) is necessary to eliminate such contradictions. If Eq. (10.8) holds with $\text{rank}(\mathbf{A}) = p' < p$, then simple algebraic manipulations can be used to reduce Eq. (10.6) to an equivalent set of p' equality constraints

$$\hat{\mathbf{A}}\mathbf{x} = \hat{\mathbf{b}} \tag{10.9}$$

where $\hat{\mathbf{A}} \in R^{p' \times n}$ has rank p' and $\hat{\mathbf{b}} \in R^{p' \times 1}$. Further, linear equality constraints in the form of Eq. (10.9) with a full row rank $\hat{\mathbf{A}}$ can be eliminated so as to convert the problem to an unconstrained problem or to reduce the number of parameters involved. The reader is referred to Sec. 10.4.1.1 for the details.

When $\text{rank}(\mathbf{A}) = p' < p$, a numerically reliable way to reduce Eq. (10.6) to Eq. (10.9) is to apply the singular-value decomposition (SVD) to matrix $\mathbf{A}$. The basic theory pertaining to the SVD can be found in Sec. A.9. Applying the SVD to $\mathbf{A}$, we obtain

$$\mathbf{A} = \mathbf{U}\mathbf{\Sigma}\mathbf{V}^T \tag{10.10}$$

where $\mathbf{U} \in R^{p \times p}$ and $\mathbf{V} \in R^{n \times n}$ are orthogonal matrices and

$$\mathbf{\Sigma} = \left[\begin{array}{c|c} \mathbf{S} & \mathbf{0} \\ \hline \mathbf{0} & \mathbf{0} \end{array}\right]_{p \times n}$$

with $\mathbf{S} = \text{diag}\{\sigma_1, \sigma_2, \ldots, \sigma_{p'}\}$, and $\sigma_1 \geq \sigma_2 \geq \cdots \geq \sigma_{p'} > 0$. It follows that

$$\mathbf{A} = \mathbf{U}\left[\begin{array}{c} \hat{\mathbf{A}} \\ \mathbf{0} \end{array}\right]$$

with $\hat{\mathbf{A}} = \mathbf{S}[\mathbf{v}_1 \ \mathbf{v}_2 \ \cdots \ \mathbf{v}_{p'}]^T \in R^{p' \times n}$ where $\mathbf{v}_i$ denotes the ith column of $\mathbf{V}$, and Eq. (10.6) becomes

$$\left[\begin{array}{c} \hat{\mathbf{A}} \\ \mathbf{0} \end{array}\right]\mathbf{x} = \left[\begin{array}{c} \hat{\mathbf{b}} \\ \mathbf{0} \end{array}\right]$$

This leads to Eq. (10.9) where $\hat{\mathbf{b}}$ is formed by using the first p' entries of $\mathbf{U}^T\mathbf{b}$. Evidently, any solution point of Eq. (10.9) is a regular point.

In MATLAB, the SVD of a matrix $\mathbf{A}$ is performed by using command svd. The decomposition in Eq. (10.10) can be obtained by using

    [U, SIGMA, V]=svd(A);

The command svd can also be used to compute the rank of a matrix. We use svd(A) to compute the singular values of $\mathbf{A}$, and the number of the nonzero singular values of $\mathbf{A}$ is the rank of $\mathbf{A}$.[1]

---

[1] The rank of a matrix can also be found by using MATLAB command rank.

Example 10.2 Simplify the linear equality constraints

$$
\begin{aligned}
x_1 - 2x_2 + 3x_3 + 2x_4 &= 4 \\
2x_2 - x_3 &= 1 \\
2x_1 - 10x_2 + 9x_3 + 4x_4 &= 5
\end{aligned}
\tag{10.11}
$$

Solution It can be readily verified that rank$(\mathbf{A}) = $ rank$([\mathbf{A}\ \mathbf{b}]) = 2$. Hence the constraints in Eq. (10.11) can be reduced to a set of two equality constraints. The SVD of $\mathbf{A}$ yields

$$
\mathbf{U} = \begin{bmatrix}
0.2717 & -0.8003 & -0.5345 \\
-0.1365 & -0.5818 & 0.8018 \\
0.9527 & 0.1449 & 0.2673
\end{bmatrix}
$$

$$
\mathbf{\Sigma} = \begin{bmatrix}
14.8798 & 0 & 0 & 0 \\
0 & 1.6101 & 0 & 0 \\
0 & 0 & 0 & 0
\end{bmatrix}
$$

$$
\mathbf{V} = \begin{bmatrix}
0.1463 & -0.3171 & 0.6331 & -0.6908 \\
-0.6951 & -0.6284 & -0.3161 & -0.1485 \\
0.6402 & -0.3200 & -0.6322 & -0.2969 \\
0.2926 & -0.6342 & 0.3156 & 0.6423
\end{bmatrix}
$$

Therefore, the reduced set of equality constraints is given by

$$
2.1770x_1 - 10.3429x_2 + 9.5255x_3 + 4.3540x_4 = 5.7135 \tag{10.12a}
$$
$$
-0.5106x_1 - 1.0118x_2 - 0.5152x_3 - 1.0211x_4 = -3.0587 \tag{10.12b}
$$

∎

10.2.3 Inequality constraints

In this section, we discuss the class of inequality constraints. The discussion will be focused on their difference from as well as their relation to equality constraints. In addition, the convexity of a feasible region defined by linear inequalities will be addressed.

Consider the constraints

$$
\begin{aligned}
c_1(\mathbf{x}) &\geq 0 \\
c_2(\mathbf{x}) &\geq 0 \\
&\vdots \\
c_q(\mathbf{x}) &\geq 0
\end{aligned}
\tag{10.13}
$$

Unlike the number of equality constraints, the number of inequality constraints, q, is not required to be less than n. For example, if we consider the case

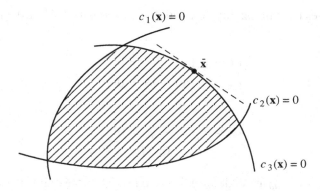

Figure 10.2.   Active and inactive constraints.

where all $c_j(\mathbf{x})$ for $1 \leq j \leq q$ are linear functions, then the constraints in Eq. (10.13) represent a polyhedron with q facets, and the number of facets in such a polyhedron is obviously unlimited.

The next two issues are concerned with the inequalities in Eq. (10.13). For a feasible point $\mathbf{x}$, these inequalities can be divided into two classes, the set of constraints with $c_i(\mathbf{x}) = 0$, which are called *active constraints*, and the set of constraints with $c_i(\mathbf{x}) > 0$, which are called *inactive constraints*. Since $c_i(\mathbf{x})$ are continuous functions, the constraints that are inactive at $\mathbf{x}$ will remain so in a sufficiently small neighborhood of $\mathbf{x}$. This means that the local properties of $\mathbf{x}$ will not be affected by the inactive constraints. On the other hand, when $c_i(\mathbf{x}) = 0$ the point $\mathbf{x}$ is on the boundary determined by the active constraints. Hence directions exist that would violate some of these constraints. In other words, active constraints restrict the feasible region of the neighborhoods of $\mathbf{x}$. For example, consider a constrained problem with the feasible region shown as the shaded area in Fig. 10.2. The problem involves three inequality constraints; constraints $c_1(\mathbf{x}) \geq 0$ and $c_2(\mathbf{x}) \geq 0$ are inactive while $c_3(\mathbf{x}) \geq 0$ is active at point $\mathbf{x} = \bar{\mathbf{x}}$ since $\bar{\mathbf{x}}$ is on the boundary characterized by $c_3(\mathbf{x}) = 0$. It can be observed that local searches in a neighborhood of $\bar{\mathbf{x}}$ will not be affected by the first two constraints but will be restricted to one side of the tangent line to the curve $c_3(\mathbf{x}) = 0$ at $\bar{\mathbf{x}}$. The concept of active constraints is an important one as it can be used to reduce the number of constraints that must be taken into account in a particular iteration and, therefore, often leads to improved computational efficiency.

Another approach to deal with inequality constraints is to convert them into equality constraints. For the sake of simplicity, we consider the problem

$$\text{minimize } f(\mathbf{x}) \qquad \mathbf{x} \in R^n \qquad (10.14a)$$

$$\text{subject to: } c_i(\mathbf{x}) \geq 0 \qquad \text{for } i = 1, 2, \ldots, q \qquad (10.14b)$$

which involves only inequality constraints. The constraints in Eq. (10.14b) are equivalent to

$$\hat{c}_1 = c_1(\mathbf{x}) - y_1 = 0$$
$$\hat{c}_2 = c_2(\mathbf{x}) - y_2 = 0$$
$$\vdots \qquad\qquad\qquad (10.15a)$$
$$\hat{c}_q = c_q(\mathbf{x}) - y_q = 0$$
$$y_i \geq 0 \qquad \text{for } 1 \leq i \leq q \qquad (10.15b)$$

where $y_1, y_2, \ldots, y_q$ are called *slack variables*. The constraints in Eq. (10.15b) can be eliminated by using the simple variable substitutions

$$y_i = \hat{y}_i^2 \qquad \text{for } 1 \leq i \leq q$$

If we let

$$\hat{\mathbf{x}} = [x_1 \ \cdots \ x_n \ \hat{y}_1 \ \cdots \ \hat{y}_q]^T$$

then the problem in Eq. (10.14) can be formulated as

$$\text{minimize } f(\hat{\mathbf{x}}) \qquad \hat{\mathbf{x}} \in E^{n+q} \qquad (10.16a)$$

$$\text{subject to: } \hat{c}_i(\hat{\mathbf{x}}) = 0 \qquad \text{for } i = 1, 2, \ldots, q \qquad (10.16b)$$

The idea of introducing slack variables to reformulate an optimization problem has been used successfully in the past, especially in linear programming, to transform a nonstandard problem into a standard problem (see Chap. 11 for the details).

We conclude this section by showing that there is a close relation between the linearity of inequality constraints to the convexity of the feasible region defined by the constraints. Although determining whether or not the region characterized by the inequality constraints in Eq. (10.13) is convex is not always easy, it can be readily shown that a feasible region defined by Eq. (10.13) with linear $c_i(\mathbf{x})$ is a *convex* polyhedron.

To demonstrate that this indeed is the case, we can write the linear inequality constraints as

$$\mathbf{C}\mathbf{x} \geq \mathbf{d} \qquad (10.17)$$

with $\mathbf{C} \in R^{q \times n}$, $\mathbf{d} \in R^{q \times 1}$. Let $\mathcal{R} = \{\mathbf{x} : \mathbf{C}\mathbf{x} \geq \mathbf{d}\}$ and assume that $\mathbf{x}_1$, $\mathbf{x}_2 \in \mathcal{R}$. For $\lambda \in [0, 1]$, the point $\mathbf{x} = \lambda\mathbf{x}_1 + (1 - \lambda)\mathbf{x}_2$ satisfies Eq. (10.17) because

$$\mathbf{C}\mathbf{x} = \lambda\mathbf{C}\mathbf{x}_1 + (1 - \lambda)\mathbf{C}\mathbf{x}_2$$
$$\geq \lambda\mathbf{d} + (1 - \lambda)\mathbf{d} = \mathbf{d}$$

Therefore, $\mathbf{Cx} \geq \mathbf{d}$ defines a convex set (see Sec. 2.7). In the literature, inequality constraints are sometimes given in the form

$$c_1(\mathbf{x}) \leq 0$$

$$\vdots \qquad\qquad\qquad (10.18)$$

$$c_q(\mathbf{x}) \leq 0$$

A similar argument can be used to show that if $c_i(\mathbf{x})$ for $1 \leq i \leq q$ in Eq. (10.18) are all linear functions, then the feasible region defined by Eq. (10.18) is convex.

10.3 Classification of Constrained Optimization Problems

In Sec. 1.6, we provided an introductory discussion on the various branches of mathematical programming. Here, we re-examine the classification issue paying particular attention to the structure of constrained optimization problems.

Constrained optimization problems can be classified according to the nature of the objective function and the constraints. For specific classes of problems, there often exist methods that are particularly suitable for obtaining solutions quickly and reliably. For example, for linear programming problems, the *simplex method* of Dantzig [1] and the *primal-dual interior-point methods* [2] have proven very efficient. For general convex programming problems, several *interior-point methods* that are particularly efficient have recently been developed [3][4].

Before discussing the classification, we formally describe the different types of minimizers of a general constrained optimization problem. In the following definitions, $\mathcal{R}$ denotes the feasible region of the problem in Eq. (10.1) and the set of points $\{\mathbf{x} : ||\mathbf{x} - \mathbf{x}^*|| \leq \delta\}$ with $\delta > 0$ is said to be a *ball* centered at $\mathbf{x}^*$.

Definition 10.2 Point $\mathbf{x}^*$ is a local constrained minimizer of the problem in Eq. (10.1) if there exists a ball $\mathcal{B}_{x^*} = \{\mathbf{x} : ||\mathbf{x} - \mathbf{x}^*|| \leq \delta\}$ with $\delta > 0$ such that $\mathcal{D}_{x^*} = \mathcal{B}_{x^*} \cap \mathcal{R}$ is nonempty and $f(\mathbf{x}^*) = \min\{f(\mathbf{x}) : \mathbf{x} \in \mathcal{D}_{x^*}\}$. ∎

Definition 10.3 Point $\mathbf{x}^*$ is a global constrained minimizer of the problem in Eq. (10.1) if $\mathbf{x}^* \in \mathcal{R}$ and $f(\mathbf{x}^*) = \min\{f(\mathbf{x}) : \mathbf{x} \in \mathcal{R}\}$ ∎

Definition 10.4 A constrained minimizer $\mathbf{x}^*$ is called a strong local minimizer if there exists a ball $\mathcal{B}_{x^*}$ such that $\mathcal{D}_{x^*} = \mathcal{B}_{x^*} \cap \mathcal{R}$ is nonempty and $\mathbf{x}^*$ is the only constrained minimizer in $\mathcal{D}_{x^*}$. ∎

10.3.1   Linear programming

The standard form of a linear programming (LP) problem can be stated as

$$\text{minimize } f(\mathbf{x}) = \mathbf{c}^T \mathbf{x} \qquad (10.19a)$$

$$\text{subject to:} \quad \mathbf{Ax} = \mathbf{b} \qquad (10.19b)$$

$$\mathbf{x} \geq \mathbf{0} \qquad (10.19c)$$

where $\mathbf{c} \in R^{n \times 1}$, $\mathbf{A} \in R^{p \times n}$, and $\mathbf{b} \in R^{p \times 1}$ are given. In words, we need to find a vector $\mathbf{x}^*$ that minimizes a linear objective function subject to the linear equality constraints in Eq. (10.19b) and the nonnegativity bounds in Eq. (10.19c).

LP problems may also be encountered in the nonstandard form

$$\text{minimize } \mathbf{c}^T \mathbf{x} \qquad (10.20a)$$

$$\text{subject to:} \quad \mathbf{Ax} \geq \mathbf{b} \qquad (10.20b)$$

By introducing slack variables in terms of vector $\mathbf{y}$ as

$$\mathbf{y} = \mathbf{Ax} - \mathbf{b}$$

Eq. (10.20b) can be expressed as

$$\mathbf{Ax} - \mathbf{y} = \mathbf{b} \qquad (10.21a)$$

and

$$\mathbf{y} \geq \mathbf{0} \qquad (10.21b)$$

If we express variable $\mathbf{x}$ as the difference of two nonnegative vectors $\mathbf{x}^+ \geq \mathbf{0}$ and $\mathbf{x}^- \geq \mathbf{0}$, i.e.,

$$\mathbf{x} = \mathbf{x}^+ - \mathbf{x}^-$$

and let

$$\hat{\mathbf{x}} = \begin{bmatrix} \mathbf{x}^+ \\ \mathbf{x}^- \\ \mathbf{y} \end{bmatrix}$$

then the objective function becomes

$$\hat{\mathbf{c}}^T \hat{\mathbf{x}} = [\mathbf{c}^T \ -\mathbf{c}^T \ \mathbf{0}] \hat{\mathbf{x}}$$

and the constraints in Eq. (10.21) can be written as

$$[\mathbf{A} \ -\mathbf{A} \ -\mathbf{I}] \, \hat{\mathbf{x}} = \mathbf{b}$$

and

$$\hat{\mathbf{x}} \geq \mathbf{0}$$

Therefore, the problem in Eq. (10.20) can be stated as the standard LP problem

$$\text{minimize } \hat{\mathbf{c}}^T \hat{\mathbf{x}} \tag{10.22a}$$

$$\text{subject to: } \hat{\mathbf{A}}\hat{\mathbf{x}} = \mathbf{b} \tag{10.22b}$$

$$\hat{\mathbf{x}} \geq \mathbf{0} \tag{10.22c}$$

where

$$\hat{\mathbf{c}} = \begin{bmatrix} \mathbf{c} \\ -\mathbf{c} \\ \mathbf{0} \end{bmatrix} \quad \text{and} \quad \hat{\mathbf{A}} = [\mathbf{A} \ -\mathbf{A} \ -\mathbf{I}]$$

The simplex and other methods that are very effective for LP problems will be studied in Chaps. 11 and 12.

10.3.2 Quadratic programming

The simplest, yet the most frequently encountered class of constrained nonlinear optimization problems, is the class of quadratic programming (QP) problems. In these problems, the objective function is quadratic and the constraints are linear, i.e.,

$$\text{minimize } f(\mathbf{x}) = \tfrac{1}{2}\mathbf{x}^T \mathbf{H}\mathbf{x} + \mathbf{x}^T \mathbf{p} + c \tag{10.23a}$$

$$\text{subject to: } \mathbf{A}\mathbf{x} = \mathbf{b} \tag{10.23b}$$

$$\mathbf{C}\mathbf{x} \geq \mathbf{d} \tag{10.23c}$$

In many applications, the Hessian of $f(\mathbf{x})$, $\mathbf{H}$, is positive semidefinite. This implies that $f(\mathbf{x})$ is a globally convex function. Since the feasible region determined by Eqs. (10.23b) and (10.23c) is always convex, QP problems with positive semidefinite $\mathbf{H}$ can be regarded as a special class of convex programming problems which will be further addressed in Sec. 10.3.3. Algorithms for solving QP problems will be studied in Chap. 13.

10.3.3 Convex programming

In a convex programming (CP) problem, a parameter vector is sought that minimizes a *convex* objective function subject to a set of constraints that define a *convex* feasible region for the problem [3][4]. Evidently, LP and QP problems with positive semidefinite Hessian matrices can be viewed as CP problems.

There are other types of CP problems that are of practical importance in engineering and science. As an example, consider the problem

$$\text{minimize } \ln(\det \mathbf{P}^{-1}) \tag{10.24a}$$

$$\text{subject to: } \mathbf{P} \succ \mathbf{0} \tag{10.24b}$$

$$\mathbf{v}_i^T \mathbf{P}\mathbf{v}_i \leq 1 \quad \text{for } i = 1, 2, \ldots, L \tag{10.24c}$$

276

where vectors $\mathbf{v}_i$ for $1 \leq i \leq L$ are given and the elements of matrix $\mathbf{P} = \mathbf{P}^T$ are the variables. It can be shown that if $\mathbf{P} \succ 0$ (i.e., $\mathbf{P}$ is positive definite), then $\ln(\det\mathbf{P}^{-1})$ is a convex function of $\mathbf{P}$ (see Prob. 10.6). In addition, if $\mathbf{p} = \mathbf{P}(:)$ denotes the vector obtained by lexicographically ordering the elements of matrix $\mathbf{P}$, then the set of vectors $\mathbf{p}$ satisfying the constraints in Eqs. (10.24b) and (10.24c) is convex and, therefore, Eq. (10.24) describes a CP problem. Algorithms for solving CP problems will be studied in Chap. 13.

10.3.4 General constrained optimization problem

The problem in Eq. (10.1) will be referred to as a *general constrained optimization* (GCO) *problem* if either $f(\mathbf{x})$ has a nonlinearity of higher order than second order and is not globally convex or at least one constraint is not convex.

Example 10.3 Classify the constrained problem (see [5]):

$$\text{minimize } f(\mathbf{x}) = \frac{1}{27\sqrt{3}}[(x_1 - 3)^2 - 9]x_2^3$$

$$\text{subject to:} \quad x_1/\sqrt{3} - x_2 \geq 0$$
$$x_1 + \sqrt{3}x_2 \geq 0$$
$$-x_1 - \sqrt{3}x_2 \geq -6$$
$$x_1 \geq 0$$
$$x_2 \geq 0$$

Solution The Hessian of $f(\mathbf{x})$ is given by

$$\mathbf{H}(\mathbf{x}) = \frac{2}{27\sqrt{3}} \begin{bmatrix} x_2^3 & 3(x_1 - 3)x_2^2 \\ 3(x_1 - 3)x_2^2 & 3[(x_1 - 3)^2 - 9]x_2 \end{bmatrix}$$

Note that $\mathbf{x} = [3 \ 1]^T$ satisfies all the constraints but $\mathbf{H}(\mathbf{x})$ is indefinite at point $\mathbf{x}$; hence $f(\mathbf{x})$ is not convex in the feasible region and the problem is a GCO problem. ∎

Very often GOP problems have multiple solutions that correspond to a number of distinct local minimizers. An effective way to obtain a good local solution in such a problem, especially when a reasonable initial point, say, $\mathbf{x}_0$, can be identified, is to tackle the problem by using a *sequential QP method*. In these methods, the highly nonlinear objective function is approximated in the neighborhood of point $\mathbf{x}_0$ in terms of a convex quadratic function while the nonlinear constraints are approximated in terms of linear constraints. In this way, the QP problem can be solved efficiently to obtain a solution, say, $\mathbf{x}_1$. The GCO problem is then approximated in the neighborhood of point $\mathbf{x}_1$ to yield a new

QP problem whose solution is $\mathbf{x}_2$. This process is continued until a certain convergence criterion, such as $||\mathbf{x}_k - \mathbf{x}_{k+1}||$ or $|f(\mathbf{x}_k) - f(\mathbf{x}_{k+1})| < \varepsilon$ where ε is a prescribed termination tolerance, is met. Sequential QP methods will be studied in detail in Chap. 15.

Another approach for the solution of a GCO problem is to reformulate the problem as a sequential unconstrained problem in which the objective function is modified taking the constraints into account. The *barrier function methods* are representatives of this class of approaches, and will be investigated in Chap. 15.

10.4   Simple Transformation Methods

A transformation method is a method that solves the problem in Eq. (10.1) by transforming the constrained optimization problem into an unconstrained optimization problem [6][7].

In this section, we shall study several simple transformation methods that can be applied when the equality constraints are linear equations or simple nonlinear equations, and when the inequality constraints are lower and/or upper bounds.

10.4.1   Variable elimination

10.4.1.1   Linear equality constraints

Consider the optimization problem

$$\text{minimize } f(\mathbf{x}) \tag{10.25a}$$

$$\text{subject to: } \mathbf{A}\mathbf{x} = \mathbf{b} \tag{10.25b}$$

$$c_i(\mathbf{x}) \geq 0 \qquad \text{for } 1 \leq i \leq q \tag{10.25c}$$

where $\mathbf{A} \in R^{p \times n}$ has full row rank, i.e., $\text{rank}(\mathbf{A}) = p$ with $p < n$. It can be shown that all solutions of Eq. (10.25b) are characterized by

$$\mathbf{x} = \mathbf{A}^+\mathbf{b} + [\mathbf{I}_n - \mathbf{A}^+\mathbf{A}]\hat{\boldsymbol{\phi}} \tag{10.26}$$

where $\mathbf{A}^+$ denotes the Moore-Penrose pseudo-inverse of $\mathbf{A}$ [8], $\mathbf{I}_n$ is the $n \times n$ identity matrix, and $\hat{\boldsymbol{\phi}}$ is an arbitrary n-dimensional parameter vector (see Prob. 10.7). The solutions expressed in Eq. (10.26) can be simplified considerably by using the SVD. As $\mathbf{A}$ has full row rank, the SVD of $\mathbf{A}$ gives

$$\mathbf{A} = \mathbf{U}\boldsymbol{\Sigma}\mathbf{V}^T$$

where $\mathbf{U} \in R^{p \times p}$ and $\mathbf{V} \in R^{n \times n}$ are orthogonal and $\boldsymbol{\Sigma} = [\mathbf{S} \ \mathbf{0}] \in R^{p \times n}$, $\mathbf{S} = \text{diag}\{\sigma_1, \sigma_2, \ldots, \sigma_p\}$, $\sigma_1 \geq \cdots \geq \sigma_p > 0$. Hence we have

$$\mathbf{A}^+ = \mathbf{A}^T(\mathbf{A}\mathbf{A}^T)^{-1} = \mathbf{V}\begin{bmatrix} \mathbf{S}^{-1} \\ \mathbf{0} \end{bmatrix}\mathbf{U}^T$$

and

$$I_n - A^+A = V \begin{bmatrix} 0 & 0 \\ 0 & I_{n-p} \end{bmatrix} V^T = V_r V_r^T$$

where $V_r = [v_{p+1} \ v_{p+2} \ \cdots \ v_n]$ contains the last $r = n - p$ columns of V. Therefore, Eq. (10.26) becomes

$$x = V_r \phi + A^+ b \qquad (10.27)$$

where $\phi \in R^{r \times 1}$ is an arbitrary r-dimensional vector. In words, Eq. (10.27) gives a complete characterization of all solutions that satisfy Eq. (10.25b). Substituting Eq. (10.27) into Eqs. (10.25a) and (10.25c), we obtain the equivalent optimization problem

$$\underset{\phi}{\text{minimize}} \ f(V_r \phi + A^+ b) \qquad (10.28a)$$

$$\text{subject to:} \quad c_i(V_r \phi + A^+ b) \geq 0 \qquad \text{for } 1 \leq i \leq q \qquad (10.28b)$$

in which the linear equality constraints are eliminated and the number of parameters is reduced from $n = \dim(x)$ to $r = \dim(\phi)$.

We note two features of the problem in Eq. (10.28). First, the size of the problem as compared with that of the problem in Eq. (10.25) is reduced from n to $r = n - p$. Once the problem in Eq. (10.28) is solved with a solution ϕ^*, Eq. (10.27) implies that x^* given by

$$x^* = V_r \phi^* + A^+ b \qquad (10.29)$$

is a solution of the problem in Eq. (10.25). Second, the linear relationship between x and ϕ as shown in Eq. (10.27) means that the degree of nonlinearity of the objective function $f(x)$ is preserved in the constrained problem of Eq. (10.28). If, for example, Eq. (10.25) is an LP or QP problem, then the problem in Eq. (10.28) is an LP or QP problem as well. Moreover, it can be shown that if the problem in Eq. (10.25) is a CP problem, then the reduced problem in Eq. (10.28) is also a CP problem.

A weak point of the above method is that performing the SVD of matrix A is computationally demanding, especially when the size of A is large. An alternative method that does not require the SVD is as follows. Assume that A has full row rank and let $P \in R^{n \times n}$ be a permutation matrix that would permute the columns of A such that

$$Ax = APP^T x = [A_1 \ A_2] \hat{x}$$

where $A_1 \in R^{p \times p}$ consists of p linearly independent columns of A and $\hat{x} = P^T x$ is simply a vector obtained by re-ordering the components of x accordingly. If we denote

$$\hat{x} = \begin{bmatrix} \tilde{x} \\ \psi \end{bmatrix} \qquad (10.30)$$

with $\tilde{\mathbf{x}} \in R^{p \times 1}$, $\psi \in R^{r \times 1}$, then Eq. (10.25b) becomes

$$\mathbf{A}_1 \tilde{\mathbf{x}} + \mathbf{A}_2 \psi = \mathbf{b}$$

i.e.,

$$\tilde{\mathbf{x}} = \mathbf{A}_1^{-1} \mathbf{b} - \mathbf{A}_1^{-1} \mathbf{A}_2 \psi$$

It follows that

$$\mathbf{x} = \mathbf{P} \hat{\mathbf{x}} = \mathbf{P} \begin{bmatrix} \tilde{\mathbf{x}} \\ \psi \end{bmatrix} = \mathbf{P} \begin{bmatrix} \mathbf{A}_1^{-1} \mathbf{b} - \mathbf{A}_1^{-1} \mathbf{A}_2 \psi \\ \psi \end{bmatrix}$$

$$\equiv \mathbf{W} \psi + \tilde{\mathbf{b}} \tag{10.31}$$

where

$$\mathbf{W} = \mathbf{P} \begin{bmatrix} -\mathbf{A}_1^{-1} \mathbf{A}_2 \\ \mathbf{I}_r \end{bmatrix} \in R^{n \times r}$$

$$\tilde{\mathbf{b}} = \mathbf{P} \begin{bmatrix} \mathbf{A}_1^{-1} \mathbf{b} \\ \mathbf{0} \end{bmatrix} \in R^{n \times 1}$$

The optimization problem in Eq. (10.25) is now reduced to

$$\underset{\psi}{\text{minimize}} \; f(\mathbf{W} \psi + \tilde{\mathbf{b}}) \tag{10.32a}$$

$$\text{subject to:} \quad c_i(\mathbf{W} \psi + \tilde{\mathbf{b}}) \geq 0 \quad \text{for } 1 \leq i \leq q \tag{10.32b}$$

Note that the new parameter vector ψ is actually a collection of r components from $\mathbf{x}$.

Example 10.4 Apply the above variable elimination method to minimize

$$f(\mathbf{x}) = \tfrac{1}{2} \mathbf{x}^T \mathbf{H} \mathbf{x} + \mathbf{x}^T \mathbf{p} + c \tag{10.33}$$

subject to the constraints in Eq. (10.11), where $\mathbf{x} = [x_1 \; x_2 \; x_3 \; x_4]^T$.

Solution Since $\text{rank}(\mathbf{A}) = \text{rank}([\mathbf{A} \; \mathbf{b}]) = 2$, the three constraints in Eq. (10.11) are consistent but redundant. It can be easily verified that the first two constraints in Eq. (10.11) are linearly independent; hence if we let

$$\mathbf{x} = \begin{bmatrix} \tilde{\mathbf{x}} \\ \psi \end{bmatrix} \quad \text{with} \quad \tilde{\mathbf{x}} = \begin{bmatrix} x_1 \\ x_2 \end{bmatrix} \quad \text{and} \quad \psi = \begin{bmatrix} x_3 \\ x_4 \end{bmatrix}$$

then Eq. (10.11) is equivalent to

$$\begin{bmatrix} 1 & -2 \\ 0 & 2 \end{bmatrix} \tilde{\mathbf{x}} + \begin{bmatrix} 3 & 2 \\ -1 & 0 \end{bmatrix} \psi = \begin{bmatrix} 4 \\ 1 \end{bmatrix}$$

i.e.,

$$\tilde{\mathbf{x}} = \begin{bmatrix} -2 & -2 \\ \frac{1}{2} & 0 \end{bmatrix} \boldsymbol{\psi} + \begin{bmatrix} 5 \\ \frac{1}{2} \end{bmatrix} \equiv \mathbf{W}\boldsymbol{\psi} + \tilde{\mathbf{b}} \qquad (10.34)$$

It follows that if we partition $\mathbf{H}$ and $\mathbf{p}$ in Eq. (10.33) as

$$\mathbf{H} = \begin{bmatrix} \mathbf{H}_{11} & \mathbf{H}_{12} \\ \mathbf{H}_{12}^T & \mathbf{H}_{22} \end{bmatrix} \quad \text{and} \quad \mathbf{p} = \begin{bmatrix} \mathbf{p}_1 \\ \mathbf{p}_2 \end{bmatrix}$$

with $\mathbf{H}_{11} \in R^{2\times2}$, $\mathbf{H}_{22} \in R^{2\times2}$, $\mathbf{p}_1 \in R^{2\times1}$, $\mathbf{p}_2 \in R^{2\times1}$, then Eq. (10.33) becomes

$$f(\boldsymbol{\psi}) = \tfrac{1}{2}\boldsymbol{\psi}^T \hat{\mathbf{H}} \boldsymbol{\psi} + \boldsymbol{\psi}^T \hat{\mathbf{p}} + \hat{c} \qquad (10.35)$$

where

$$\hat{\mathbf{H}} = \mathbf{W}^T \mathbf{H}_{11} \mathbf{W} + \mathbf{H}_{12}^T \mathbf{W} + \mathbf{W}^T \mathbf{H}_{12} + \mathbf{H}_{22}$$
$$\hat{\mathbf{p}} = \mathbf{H}_{12}^T \tilde{\mathbf{b}} + \mathbf{W}^T \mathbf{H}_{11} \tilde{\mathbf{b}} + \mathbf{p}_2 + \mathbf{W}^T \mathbf{p}_1$$
$$\hat{c} = \tfrac{1}{2}\tilde{\mathbf{b}}^T \mathbf{H}_{11} \tilde{\mathbf{b}} + \tilde{\mathbf{b}}^T \mathbf{p}_2 + c$$

The problem now reduces to minimizing $f(\boldsymbol{\psi})$ without constraints. By writing

$$\hat{\mathbf{H}} = [\mathbf{W}^T\ \mathbf{I}]\mathbf{H} \begin{bmatrix} \mathbf{W} \\ \mathbf{I} \end{bmatrix}$$

we note that $\hat{\mathbf{H}}$ is positive definite if $\mathbf{H}$ is positive definite. In such a case, the unique minimizer of the problem is given by

$$\mathbf{x}^* = \begin{bmatrix} \tilde{\mathbf{x}}^* \\ \boldsymbol{\psi}^* \end{bmatrix}$$

with

$$\boldsymbol{\psi}^* = -\hat{\mathbf{H}}^{-1}\hat{\mathbf{p}} \quad \text{and} \quad \tilde{\mathbf{x}}^* = \mathbf{W}\boldsymbol{\psi}^* + \tilde{\mathbf{b}}$$

∎

10.4.1.2 Nonlinear equality constraints

When the equality constraints are nonlinear, no general methods are available for variable elimination since solving a system of nonlinear equations is far more involved than solving a system of linear equations, if not impossible. However, in many cases the constraints can be appropriately manipulated to yield an equivalent constraint set in which some variables are expressed in terms of the rest of the variables so that the constraints can be partially or completely eliminated.

Example 10.5 Use nonlinear variable substitution to simplify the constrained problem

$$\text{minimize } f(\mathbf{x}) = -x_1^4 - 2x_2^4 - x_3^4 - x_1^2 x_2^2 - x_1^2 x_3^2 \quad (10.36)$$

$$\text{subject to: } a_1(\mathbf{x}) = x_1^4 + x_2^4 + x_3^4 - 25 = 0 \quad (10.37a)$$

$$a_2(\mathbf{x}) = 8x_1^2 + 14x_2^2 + 7x_3^2 - 56 = 0 \quad (10.37b)$$

Solution By writing Eq. (10.37b) as

$$x_3^2 = -\tfrac{8}{7}x_1^2 - 2x_2^2 + 8$$

the constraint in Eq. (10.37b) as well as variable x_3 in Eqs. (10.36) and (10.37a) can be eliminated, and an equivalent minimization problem can be formulated as

$$\text{minimize } f(\mathbf{x}) = -\tfrac{57}{49}x_1^4 - 6x_2^4 - \tfrac{25}{7}x_1^2 x_2^2 + \tfrac{72}{7}x_1^2 + 32x_2^2 \quad (10.38)$$

$$\text{subject to: } a_1(\mathbf{x}) = \tfrac{113}{49}x_1^4 + 5x_2^4 + \tfrac{32}{7}x_1^2 x_2^2 - \tfrac{128}{7}x_1^2 - 32x_2^2 + 39 = 0 \quad (10.39)$$

To eliminate Eq. (10.39), we write the equation as

$$5x_2^4 + (\tfrac{32}{7}x_1^2 - 32)x_2^2 + (\tfrac{113}{49}x_1^4 - \tfrac{128}{7}x_1^2 + 39) = 0$$

and treat it as a quadratic equation of x_2^2. In this way

$$x_2^2 = -(\tfrac{16}{35}x_1^2 - \tfrac{16}{5}) \pm \frac{1}{10}\sqrt{\left(-\tfrac{212}{49}x_1^4 + \tfrac{512}{7}x_1^2 + 244\right)} \quad (10.40)$$

By substituting Eq. (10.40) into Eq. (10.38), we obtain a minimization problem with only one variable.

The plus and minus signs in Eq. (10.40) mean that we have to deal with two separate cases, and the minimizer can be determined by comparing the results for the two cases. It should be noted that the polynomial under the square root in Eq. (10.40) assumes a negative value for large x_1; therefore, the one-dimensional minimization problem must be solved on an interval where the square root yields real values.

■

10.4.2   Variable transformations
10.4.2.1   Nonnegativity bounds

The nonnegativity bound

$$x_i \geq 0$$

can be eliminated by using the variable transformation [7]

$$x_i = y_i^2 \quad (10.41)$$

Similarly, the constraint $x_i \geq d$ can be eliminated by using the transformation

$$x_i = d + y_i^2 \tag{10.42}$$

and one can readily verify that $x_i \leq d$ can be eliminated by using the transformation

$$x_i = d - y_i^2 \tag{10.43}$$

Although these transformations are simple and easy to use, these bounds are eliminated at the cost of increasing the degree of nonlinearity of the objective function as well as the remaining constraints, which may, in turn, reduce the efficiency of the optimization process.

Example 10.6 Apply a variable transformation to simplify the constrained problem

$$\text{minimize } f(\mathbf{x}) = -x_1^2 - 2x_2^2 - x_3^2 - x_1 x_2 - x_1 x_3 \tag{10.44}$$

$$\text{subject to: } b_1(\mathbf{x}) = x_1^2 + x_2^2 + x_3^2 - 25 = 0 \tag{10.45a}$$

$$b_2(\mathbf{x}) = 8x_1 + 14x_2 + 7x_3 - 56 = 0 \tag{10.45b}$$

$$x_i \geq 0 \qquad i = 1,\ 2,\ 3,\ 4 \tag{10.45c}$$

Solution The nonnegativity bounds in the problem can be eliminated by using the transformation in Eq. (10.41). While eliminating Eq. (10.45c), the transformation changes Eqs. (10.44), (10.45a), and (10.45b) to Eqs. (10.36), (10.37a), and (10.37b), respectively, where the y_i's have been renamed as x_i's.

∎

10.4.2.2 Interval-type constraints

The hyperbolic tangent function defined by

$$y = \tanh(z) = \frac{e^z - e^{-z}}{e^z + e^{-z}} \tag{10.46}$$

is a differentiable monotonically increasing function that maps the entire 1-D space $-\infty < z < \infty$ onto the interval $-1 < y < 1$ as can be seen in Fig. 10.3. This in conjunction with the linear transformation

$$x = \frac{(b-a)}{2}y + \frac{b+a}{2} \tag{10.47}$$

transforms the infinite interval $(-\infty,\ \infty)$ into the open interval $(a,\ b)$. By writing $\tanh(z)$ as

$$\tanh(z) = \frac{e^{2z} - 1}{e^{2z} + 1}$$

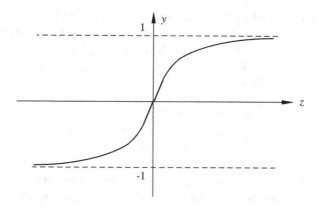

Figure 10.3. The hyperbolic tangent function.

we note that evaluating $\tanh(z)$ has about the same numerical complexity as the exponential function.

An alternative transformation for Eq. (10.46) is one that uses the inverse tangent function

$$y = \frac{2}{\pi} \tan^{-1} z \qquad (10.48)$$

which is also differentiable and monotonically increasing. As the transformations in Eqs. (10.46) and (10.48) are nonlinear, applying them to eliminate interval-type constraints will in general increase the nonlinearity of the objective function as well as the remaining constraints.

Example 10.7 In certain engineering problems, an nth-order polynomial

$$p(z) = z^n + d_{n-1} z^{n-1} + \cdots + d_1 z + d_0$$

is required to have zeros inside the unit circle of the z plane, for example, the denominator of the transfer function in discrete-time systems and digital filters [9]. Such polynomials are sometimes called *Schur polynomials*.

Find a suitable transformation for coefficients d_0 and d_1 which would ensure that the second-order polynomial

$$p(z) = z^2 + d_1 z + d_0$$

is always a Schur polynomial.

Solution The zeros of $p(z)$ are located inside the unit circle if and only if [9]

$$\begin{aligned} d_0 &< 1 \\ d_1 - d_0 &< 1 \\ d_1 + d_0 &> -1 \end{aligned} \qquad (10.49)$$

284

The region described by the constraints in Eq. (10.49) is the triangle shown in Fig. 10.4. For a fixed $d_0 \in (-1, 1)$, the line segment inside the triangle shown as a dashed line is characterized by d_1 varying from $-(1+d_0)$ to $1+d_0$. As d_0 varies from -1 to 1, the line segment will cover the entire triangle. This observation suggests the transformation

$$d_0 = \tanh(b_0)$$
$$d_1 = [1 + \tanh(b_0)] \tanh(b_1) \tag{10.50}$$

which provides a one-to-one correspondence between points in the triangle in the (d_0, d_1) space and points in the entire (b_0, b_1) space. In other words, $p(z)$ is transformed into the polynomial

$$p(z) = z^2 + [1 + \tanh(b_0)] \tanh(b_1) z + \tanh(b_0) \tag{10.51}$$

which is always a Schur polynomial for *any* finite values of b_0 and b_1.

This characterization of second-order Schur polynomials has been found to be useful in the design of stable recursive digital filters [10]. ∎

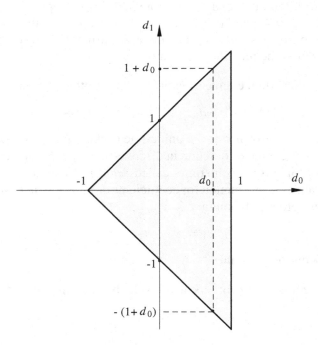

Figure 10.4. Region of the d_1 versus d_0 plane for which $p(z)$ is a Schur polynomial.

10.5   Lagrange Multipliers

Lagrange multipliers play a crucial role in the study of constrained optimization. On the one hand, the conditions imposed on the Lagrange multipliers are always an integral part of various necessary and sufficient conditions and, on the other, they provide a natural connection between constrained and corresponding unconstrained optimization problems; each individual Lagrange multiplier can be interpreted as the rate of change in the objective function with respect to changes in the associated constraint function [7]. In simple terms, if $\mathbf{x}^*$ is a local minimizer of a constrained minimization problem, then in addition to $\mathbf{x}^*$ being a feasible point, the gradient of the objective function at $\mathbf{x}^*$ has to be a linear combination of the gradients of the constraint functions, and the Lagrange multipliers are the coefficients in that linear combination. Moreover, the Lagrange multipliers associated with inequality constraints have to be nonnegative and the multipliers associated with inactive inequality constraints have to be zero. Collectively, these conditions are known as the *Karush-Kuhn-Tucker conditions* (KKT).

In what follows, we introduce the concept of Lagrange multipliers through a simple example and then develop the KKT conditions for an arbitrary problem with equality constraints.

10.5.1   An example

Let us consider the minimization of the objective function $f(x_1, x_2, x_3, x_4)$ subject to the equality constraints

$$a_1(x_1, x_2, x_3, x_4) = 0 \tag{10.52a}$$
$$a_2(x_1, x_2, x_3, x_4) = 0 \tag{10.52b}$$

If these constraints can be expressed as

$$x_3 = h_1(x_1, x_2) \tag{10.53a}$$
$$x_4 = h_2(x_1, x_2) \tag{10.53b}$$

then they can be eliminated by substituting Eq. (10.53) into the objective function which will assume the form $f[x_1, x_2, h_1(x_1, x_2), h_2(x_1, x_2)]$. If $\mathbf{x}^* = [x_1^* \; x_2^* \; x_3^* \; x_4^*]^T$ is a local minimizer of the original constrained optimization problem, then $\hat{\mathbf{x}}^* = [x_1^* \; x_2^*]^T$ is a local minimizer of the problem

$$\text{minimize } f[x_1, x_2, h_1(x_1, x_2), h_2(x_1, x_2)]$$

It, therefore, follows that at $\hat{\mathbf{x}}^*$ we have

$$\nabla f = \begin{bmatrix} \frac{\partial f}{\partial x_1} \\ \frac{\partial f}{\partial x_2} \end{bmatrix} = 0$$

Since variables x_3 and x_4 in the constraints of Eq. (10.53) are related to variables x_1 and x_2, the use of the chain rule for the partial derivatives in ∇f gives

$$\frac{\partial f}{\partial x_1} + \frac{\partial f}{\partial x_3}\frac{\partial h_1}{\partial x_1} + \frac{\partial f}{\partial x_4}\frac{\partial h_2}{\partial x_1} = 0$$

$$\frac{\partial f}{\partial x_2} + \frac{\partial f}{\partial x_3}\frac{\partial h_1}{\partial x_2} + \frac{\partial f}{\partial x_4}\frac{\partial h_2}{\partial x_2} = 0$$

From Eqs. (10.52) and (10.53), we have

$$\frac{\partial a_1}{\partial x_1} + \frac{\partial a_1}{\partial x_3}\frac{\partial h_1}{\partial x_1} + \frac{\partial a_1}{\partial x_4}\frac{\partial h_2}{\partial x_1} = 0$$

$$\frac{\partial a_1}{\partial x_2} + \frac{\partial a_1}{\partial x_3}\frac{\partial h_1}{\partial x_2} + \frac{\partial a_1}{\partial x_4}\frac{\partial h_2}{\partial x_2} = 0$$

$$\frac{\partial a_2}{\partial x_1} + \frac{\partial a_2}{\partial x_3}\frac{\partial h_1}{\partial x_1} + \frac{\partial a_2}{\partial x_4}\frac{\partial h_2}{\partial x_1} = 0$$

$$\frac{\partial a_2}{\partial x_2} + \frac{\partial a_2}{\partial x_3}\frac{\partial h_1}{\partial x_2} + \frac{\partial a_2}{\partial x_4}\frac{\partial h_2}{\partial x_2} = 0$$

The above six equations can now be expressed as

$$\begin{bmatrix} \nabla^T f(\mathbf{x}) \\ \nabla^T a_1(\mathbf{x}) \\ \nabla^T a_2(\mathbf{x}) \end{bmatrix} \begin{bmatrix} 1 & 0 \\ 0 & 1 \\ \frac{\partial h_1}{\partial x_1} & \frac{\partial h_1}{\partial x_2} \\ \frac{\partial h_2}{\partial x_1} & \frac{\partial h_2}{\partial x_2} \end{bmatrix} = \mathbf{0} \tag{10.54}$$

This equation implies that $\nabla f(\mathbf{x}^*)$, $\nabla a_1(\mathbf{x}^*)$, and $\nabla a_2(\mathbf{x}^*)$ are linearly dependent (see Prob. 10.9). Hence there exist constants α, β, γ which are not all zero such that

$$\alpha\nabla f(\mathbf{x}^*) + \beta\nabla a_1(\mathbf{x}^*) + \gamma\nabla a_2(\mathbf{x}^*) = \mathbf{0} \tag{10.55}$$

If we assume that $\mathbf{x}^*$ is a regular point of the constraints, then α in Eq. (10.55) cannot be zero and Eq. (10.55) can be simplified to

$$\nabla f(\mathbf{x}^*) - \lambda_1^*\nabla a_1(\mathbf{x}^*) - \lambda_2^*\nabla a_2(\mathbf{x}^*) = \mathbf{0} \tag{10.56}$$

and, therefore

$$\nabla f(\mathbf{x}^*) = \lambda_1^*\nabla a_1(\mathbf{x}^*) + \lambda_2^*\nabla a_2(\mathbf{x}^*)$$

where $\lambda_1^* = -\beta/\alpha$, $\lambda_2^* = -\gamma/\alpha$. In words, we conclude that *at a local minimizer of the constrained optimization problem, the gradient of the objective function is a linear combination of the gradients of the constraints.* Constants λ_1^* and λ_2^* in Eq. (10.56) are called the *Lagrange multipliers* of the constrained problem. In the rest of this section, we examine the concept of Lagrange multipliers from a different perspective.

10.5.2   Equality constraints

We now consider the constrained optimization problem

$$\text{minimize } f(\mathbf{x}) \tag{10.57a}$$

$$\text{subject to:} \quad a_i(\mathbf{x}) = 0 \quad \text{for } i = 1, 2, \ldots, p \tag{10.57b}$$

following an approach used by Fletcher in [7, Chap. 9]. Let $\mathbf{x}^*$ be a local minimizer of the problem in Eq. (10.57). By using the Taylor series of constraint function $a_i(\mathbf{x})$ at $\mathbf{x}^*$, we can write

$$\begin{aligned} a_i(\mathbf{x}^* + \mathbf{s}) &= a_i(\mathbf{x}^*) + \mathbf{s}^T \nabla a_i(\mathbf{x}^*) + o(||\mathbf{s}||) \\ &= \mathbf{s}^T \nabla a_i(\mathbf{x}^*) + o(||\mathbf{s}||) \end{aligned} \tag{10.58}$$

since $a_i(\mathbf{x}^*) = 0$. If $\mathbf{s}$ is a feasible vector at $\mathbf{x}^*$, then $a_i(\mathbf{x}^* + \mathbf{s}) = 0$ and hence Eq. (10.58) implies that

$$\mathbf{s}^T \nabla a_i(\mathbf{x}^*) = 0 \quad \text{for } i = 1, 2, \ldots, p \tag{10.59}$$

In other words, $\mathbf{s}$ is feasible if it is orthogonal to the gradients of the constraint functions. Now we project the gradient $\nabla f(\mathbf{x}^*)$ orthogonally onto the space spanned by $\{\nabla a_1(\mathbf{x}^*), \nabla a_2(\mathbf{x}^*), \ldots, \nabla a_p(\mathbf{x}^*)\}$. If we denote the projection as

$$\sum_{i=1}^{p} \lambda_i^* \nabla a_i(\mathbf{x}^*)$$

then $\nabla f(\mathbf{x}^*)$ can be expressed as

$$\nabla f(\mathbf{x}^*) = \sum_{i=1}^{p} \lambda_i^* \nabla a_i(\mathbf{x}^*) + \mathbf{d} \tag{10.60}$$

where $\mathbf{d}$ is orthogonal to $\nabla a_i(\mathbf{x}^*)$ for $i = 1, 2, \ldots, p$.

In what follows, we show that if $\mathbf{x}^*$ is a local minimizer then $\mathbf{d}$ must be zero. The proof is accomplished by contradiction. Assume that $\mathbf{d} \neq \mathbf{0}$ and let $\mathbf{s} = -\mathbf{d}$. Since $\mathbf{s}$ is orthogonal to $\nabla a_i(\mathbf{x}^*)$ by virtue of Eq. (10.59), $\mathbf{s}$ is feasible at $\mathbf{x}^*$. Now we use Eq. (10.60) to obtain

$$\mathbf{s}^T \nabla f(\mathbf{x}^*) = \mathbf{s}^T \left(\sum_{i=1}^{p} \lambda_i^* \nabla a_i(\mathbf{x}^*) + \mathbf{d} \right) = -||\mathbf{d}||^2 < 0$$

This means that $\mathbf{s}$ is a descent direction at $\mathbf{x}^*$ which contradicts the fact that $\mathbf{x}^*$ is a minimizer. Therefore, $\mathbf{d} = \mathbf{0}$ and Eq. (10.60) becomes

$$\nabla f(\mathbf{x}^*) = \sum_{i=1}^{p} \lambda_i^* \nabla a_i(\mathbf{x}^*) \tag{10.61}$$

In effect, for an arbitrary constrained problem with equality constraints, the gradient of the objective function at a local minimizer *is equal to the linear combination of the gradients of the equality constraint functions with the Lagrange multipliers as the coefficients.*

For the problem in Eq. (10.1) with both equality and inequality constraints, Eq. (10.61) needs to be modified to include those inequality constraints that are *active* at $\mathbf{x}^*$. This more general case is treated in Sec. 10.6.

Example 10.8 Determine the Lagrange multipliers for the optimization problem

$$\text{minimize } f(\mathbf{x})$$
$$\text{subject to:} \quad \mathbf{Ax} = \mathbf{b}$$

where $\mathbf{A} \in R^{p \times n}$ is assumed to have full row rank. Also discuss the case where the constraints are nonlinear.

Solution Eq. (10.61) in this case becomes

$$\mathbf{g}^* = \mathbf{A}^T \boldsymbol{\lambda}^* \tag{10.62}$$

where $\boldsymbol{\lambda}^* = [\lambda_1^* \ \lambda_2^* \ \cdots \ \lambda_p^*]^T$ and $\mathbf{g}^* = \nabla f(\mathbf{x}^*)$. By virtue of Eq. (10.62), the Lagrange multipliers are uniquely determined as

$$\boldsymbol{\lambda}^* = (\mathbf{A}\mathbf{A}^T)^{-1} \mathbf{A} \mathbf{g}^* = (\mathbf{A}^T)^+ \mathbf{g}^* \tag{10.63}$$

where $(\mathbf{A}^T)^+$ denotes the Moore-Penrose pseudo-inverse of $\mathbf{A}^T$.

For the case of nonlinear equality constraints, a similar conclusion can be reached in terms of the Jacobian of the constraints in Eq. (10.57b). If we let

$$\mathbf{J}_e(\mathbf{x}) = [\nabla a_1(\mathbf{x}) \ \nabla a_2(\mathbf{x}) \ \cdots \ \nabla a_p(\mathbf{x})]^T \tag{10.64}$$

then the Lagrange multipliers λ_i^* for $1 \le i \le p$ in Eq. (10.61) are uniquely determined as

$$\boldsymbol{\lambda}^* = [\mathbf{J}_e^T(\mathbf{x}^*)]^+ \mathbf{g}^* \tag{10.65}$$

provided that $\mathbf{J}_e(\mathbf{x})$ has full row rank at $\mathbf{x}^*$. ∎

The concept of Lagrange multipliers can also be explained from a different perspective. If we introduce the function

$$L(\mathbf{x}, \ \boldsymbol{\lambda}) = f(\mathbf{x}) - \sum_{i=1}^{p} \lambda_i a_i(\mathbf{x}) \tag{10.66}$$

as the *Lagrangian* of the optimization problem, then the condition in Eq. (10.61) and the constraints in Eq. (10.57b) can be written as

$$\nabla_x L(\mathbf{x}, \ \boldsymbol{\lambda}) = \mathbf{0} \quad \text{for } \{\mathbf{x}, \boldsymbol{\lambda}\} = \{\mathbf{x}^*, \ \boldsymbol{\lambda}^*\} \tag{10.67a}$$

and

$$\nabla_\lambda L(\mathbf{x}, \ \lambda) = 0 \qquad \text{for } \{\mathbf{x}, \lambda\} = \{\mathbf{x}^*, \ \lambda^*\} \qquad (10.67\text{b})$$

respectively. The numbers of equations in Eqs. (10.67a) and (10.67b) are n and p, respectively, and the total number of equations is consistent with the number of parameters in $\mathbf{x}$ and λ, i.e., $n + p$. Now if we define the gradient operator ∇ as

$$\nabla = \begin{bmatrix} \nabla_x \\ \nabla_\lambda \end{bmatrix}$$

then Eqs. (10.67a) and (10.67b) can be expressed as

$$\nabla L(\mathbf{x}, \ \lambda) = 0 \qquad \text{for } \{\mathbf{x}, \ \lambda\} = \{\mathbf{x}^*, \lambda^*\} \qquad (10.68)$$

From the above analysis, we see that the Lagrangian incorporates the constraints into a modified objective function in such a way that a constrained minimizer $\mathbf{x}^*$ is connected to an *unconstrained* minimizer $\{\mathbf{x}^*, \ \lambda^*\}$ for the augmented objective function $L(\mathbf{x}, \ \lambda)$ where the augmentation is achieved with the p Lagrange multipliers.

Example 10.9 Solve the problem

$$\text{minimize } f(\mathbf{x}) = \tfrac{1}{2}\mathbf{x}^T \mathbf{H} \mathbf{x} + \mathbf{x}^T \mathbf{p}$$

$$\text{subject to:} \quad \mathbf{A}\mathbf{x} = \mathbf{b}$$

where $\mathbf{H} \succ 0$ and $\mathbf{A} \in R^{p \times n}$ has full row rank.

Solution In Example 10.4 we solved a similar problem by eliminating the equality constraints. Here, we define the Lagrangian

$$L(\mathbf{x}, \ \lambda) = \tfrac{1}{2}\mathbf{x}^T \mathbf{H} \mathbf{x} + \mathbf{x}^T \mathbf{p} - \lambda^T(\mathbf{A}\mathbf{x} - \mathbf{b})$$

and apply the condition in Eq. (10.68) to obtain

$$\nabla L(\mathbf{x}, \ \lambda) = \begin{bmatrix} \mathbf{H}\mathbf{x} + \mathbf{p} - \mathbf{A}^T \lambda \\ -\mathbf{A}\mathbf{x} + \mathbf{b} \end{bmatrix}$$

$$= \begin{bmatrix} \mathbf{H} & -\mathbf{A}^T \\ -\mathbf{A} & 0 \end{bmatrix} \begin{bmatrix} \mathbf{x} \\ \lambda \end{bmatrix} + \begin{bmatrix} \mathbf{p} \\ \mathbf{b} \end{bmatrix} = 0 \qquad (10.69)$$

Since $\mathbf{H} \succ 0$ and $\text{rank}(\mathbf{A}) = p$, we can show that the matrix

$$\begin{bmatrix} \mathbf{H} & -\mathbf{A}^T \\ -\mathbf{A} & 0 \end{bmatrix}$$

is nonsingular (see [13, Chap. 14]) and, therefore, Eq. (10.69) has the unique
solution

$$\begin{bmatrix} \mathbf{x}^* \\ \boldsymbol{\lambda}^* \end{bmatrix} = - \begin{bmatrix} \mathbf{H} & -\mathbf{A}^T \\ -\mathbf{A} & 0 \end{bmatrix}^{-1} \begin{bmatrix} \mathbf{p} \\ \mathbf{b} \end{bmatrix}$$

It follows that

$$\mathbf{x}^* = \mathbf{H}^{-1}(\mathbf{A}^T\boldsymbol{\lambda}^* - \mathbf{p}) \tag{10.70a}$$

where

$$\boldsymbol{\lambda}^* = (\mathbf{A}\mathbf{H}^{-1}\mathbf{A}^T)^{-1}(\mathbf{A}\mathbf{H}^{-1}\mathbf{p} + \mathbf{b}) \tag{10.70b}$$

In Sec. 10.8, it will be shown that $\mathbf{x}^*$ given by Eq. (10.70a) with $\boldsymbol{\lambda}^*$ determined using Eq. (10.70b) is the unique, global minimizer of the constrained minimization problem.

■

10.5.3 Tangent plane and normal plane

The first derivative of a smooth function of one variable indicates the direction along which the function increases. Similarly, the gradient of a smooth multivariable function indicates the direction along which the function increases at the greatest rate. This fact can be verified by using the first-order approximation of the Taylor series of the function, namely,

$$f(\mathbf{x}^* + \boldsymbol{\delta}) = f(\mathbf{x}^*) + \boldsymbol{\delta}^T\nabla f(\mathbf{x}^*) + o(||\boldsymbol{\delta}||)$$

If $||\boldsymbol{\delta}||$ is small, then the value of the function increases by $\boldsymbol{\delta}^T\nabla f(\mathbf{x}^*)$ which reaches the maximum when the direction of $\boldsymbol{\delta}$ coincides with that of $\nabla f(\mathbf{x}^*)$.

Two interrelated concepts that are closely related to the gradients of the objective function and the constraints of the optimization problem in Eq. (10.57) are the *tangent plane* and *normal plane*.

The tangent plane of a smooth function $f(\mathbf{x})$ at a given point $\mathbf{x}^*$ can be defined in two ways as follows. If $C_{\mathbf{x}^*}$ is the contour surface of $f(\mathbf{x})$ that passes through point $\mathbf{x}^*$, then we can think of the tangent plane as a hyperplane in R^n that touches $C_{\mathbf{x}^*}$ at and only at point $\mathbf{x}^*$. Alternatively, the tangent plane can be defined as a hyperplane that passes through point $\mathbf{x}^*$ with $\nabla f(\mathbf{x}^*)$ as the normal. For example, for $n = 2$ the contours, tangent plane, and gradient of a smooth function are related to each other as illustrated in Fig. 10.5.

Following the above discussion, the tangent plane at point $\mathbf{x}^*$ can be defined analytically as the set

$$\mathcal{T}_{\mathbf{x}^*} = \{\mathbf{x} : \nabla f(\mathbf{x}^*)^T(\mathbf{x} - \mathbf{x}^*) = 0\}$$

In other words, *a point* $\mathbf{x}$ *lies on the tangent plane if the vector that connects* $\mathbf{x}^*$ *to* $\mathbf{x}$ *is orthogonal to the gradient* $\nabla f(\mathbf{x}^*)$, as can be seen in Fig. 10.5.

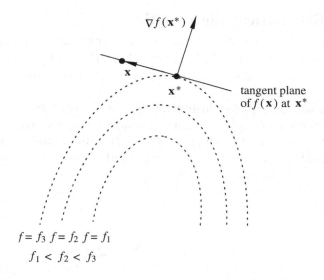

Figure 10.5. Relation of tangent plane to contours and gradient.

Proceeding in the same way, a tangent plane can be defined for a surface that is characterized by *several* equations. Let S be the surface defined by the equations

$$a_i(\mathbf{x}) = 0 \qquad \text{for } i = 1, 2, \ldots, p$$

and assume that $\mathbf{x}^*$ is a point satisfying these constraints, i.e., $\mathbf{x}^* \in S$. The tangent plane of S at $\mathbf{x}^*$ is given by

$$\mathcal{T}_{\mathbf{x}^*} = \{\mathbf{x} : \mathbf{J}_e(\mathbf{x}^*)(\mathbf{x} - \mathbf{x}^*) = \mathbf{0}\} \qquad (10.71)$$

where $\mathbf{J}_e$ is the Jacobian defined by Eq. (10.64). From Eq. (10.71), we conclude that the tangent plane of S is actually an $(n - p)$-dimensional hyperplane in space R^n. For example, in the case of Fig. 10.5 we have $n = 2$ and $p = 1$ and hence the tangent plane degenerates into a straight line.

The normal plane can similarly be defined. Given a set of equations $a_i(\mathbf{x}) = 0$ for $1 \leq i \leq p$ and a point $\mathbf{x}^* \in S$, the normal plane at $\mathbf{x}^*$ is given by

$$\mathcal{N}_{\mathbf{x}^*} = \{\mathbf{x} : \mathbf{x} - \mathbf{x}^* = \sum_{i=1}^{p} \alpha_i \nabla a_i(\mathbf{x}^*) \text{ for } \alpha_i \in R\} \qquad (10.72)$$

It follows that $\{\mathcal{N}_{\mathbf{x}^*} - \mathbf{x}^*\}$ is the range of matrix $\mathbf{J}_e^T(\mathbf{x}^*)$, and hence it is a p-dimensional subspace in R^n. More importantly, $\mathcal{T}_{\mathbf{x}^*}$ and $\mathcal{N}_{\mathbf{x}^*}$ are orthogonal to each other.

10.5.4   Geometrical interpretation

On the basis of the preceding definitions, a geometrical interpretation of the necessary condition in Eq. (10.61) is possible [7][11] as follows: *If $\mathbf{x}^*$ is a constrained local minimizer, then the vector $\nabla f(\mathbf{x}^*)$ must lie in the normal plane $\mathcal{N}_{\mathbf{x}^*}$.*

A two-variable example is illustrated in Fig. 10.6 where several contours of the objective function $f(x_1, x_2)$ and the only equality constraint $a_1(x_1, x_2) = 0$ are depicted. Note that at feasible point $\tilde{\mathbf{x}}$, $\nabla f(\tilde{\mathbf{x}})$ lies exactly in the normal plane generated by $\nabla a_1(\tilde{\mathbf{x}})$ only when $\tilde{\mathbf{x}}$ coincides with $\mathbf{x}^*$, the minimizer of $f(\mathbf{x})$ subject to constraint $a_1(\mathbf{x}) = 0$.

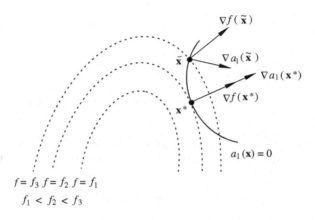

Figure 10.6.   Geometrical interpretation of Eq. (10.61): $\nabla f(\tilde{\mathbf{x}})$ lies in $\mathcal{N}_{\mathbf{x}^*}$ if $\tilde{\mathbf{x}} = \mathbf{x}^*$ where $\mathbf{x}^*$ is a minimizer.

Eq. (10.61) may also hold when $\mathbf{x}^*$ is a minimizer as illustrated in Fig. 10.7a and b, or a maximizer as shown in Fig. 10.7c, or $\mathbf{x}^*$ is neither a minimizer nor a maximizer. In addition, for a local minimizer, the Lagrange multipliers can be either positive as in Fig. 10.7a or negative as in Fig. 10.7b.

Example 10.10 Construct the geometrical interpretation of Eq. (10.61) for the three-variable problem

$$\text{minimize } f(\mathbf{x}) = x_1^2 + x_2^2 + \tfrac{1}{4}x_3^2$$
$$\text{subject to:} \quad a_1(\mathbf{x}) = -x_1 + x_3 - 1 = 0$$
$$a_2(\mathbf{x}) = x_1^2 + x_2^2 - 2x_1 = 0$$

Solution As was discussed in Example 10.1, the above constraints describe the curve obtained as the intersection of the cylinder $a_2(\mathbf{x}) = 0$ with the plane

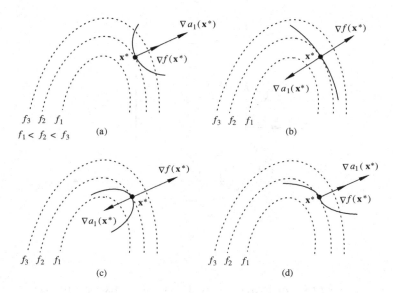

Figure 10.7. Geometrical interpretation of Eq. (10.61): (a) $\mathbf{x}^*$ is a minimizer with $\lambda^* > 0$; (b) $\mathbf{x}^*$ is a minimizer with $\lambda^* < 0$; (c) $\mathbf{x}^*$ is a maximizer; (d) $\mathbf{x}^*$ is neither a minimizer nor a maximizer.

$a_1(\mathbf{x}) = 0$. Fig. 10.8 shows that the constrained problem has a global minimizer $\mathbf{x}^* = [0\ 0\ 1]^T$. At $\mathbf{x}^*$, the tangent plane in Eq. (10.71) becomes a line that passes through $\mathbf{x}^*$ and is parallel with the x_2 axis while the normal plane $\mathcal{N}_{\mathbf{x}^*}$ is the plane spanned by

$$\nabla a_1(\mathbf{x}^*) = \begin{bmatrix} -1 \\ 0 \\ 1 \end{bmatrix} \quad \text{and} \quad \nabla a_2(\mathbf{x}^*) = \begin{bmatrix} -2 \\ 0 \\ 0 \end{bmatrix}$$

which is identical to plane $x_2 = 0$. Note that at $\mathbf{x}^*$

$$\nabla f(\mathbf{x}^*) = \begin{bmatrix} 0 \\ 0 \\ \frac{1}{2} \end{bmatrix}$$

As is expected, $\nabla f(\mathbf{x}^*)$ lies in the normal plane $\mathcal{N}_{\mathbf{x}^*}$ (see Fig. 10.8) and can be expressed as

$$\nabla f(\mathbf{x}^*) = \lambda_1^* \nabla a_1(\mathbf{x}^*) + \lambda_2^* \nabla a_2(\mathbf{x}^*)$$

where $\lambda_1^* = \frac{1}{2}$ and $\lambda_2^* = -\frac{1}{4}$ are the Lagrange multipliers. ■

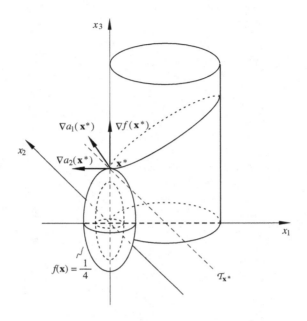

Figure 10.8. An interpretation of Eq. (10.61) for Example 10.10.

10.6   First-Order Necessary Conditions

The necessary conditions for a point $\mathbf{x}^*$ to be a local minimizer are useful in two situations: (a) They can be used to exclude those points that do not satisfy at least one of the necessary conditions from the candidate points; (b) they become sufficient conditions when the objective function in question is convex (see Sec. 10.8 for details).

10.6.1   Equality constraints

Based on the discussion in Sec. 10.5, the first-order necessary conditions for a minimum for the problem in Eq. (10.57) can be summarized in terms of the following theorem.

Theorem 10.1 *First-order necessary conditions for a minimum, equality constraints If* $\mathbf{x}^*$ *is a constrained local minimizer of the problem in Eq. (10.57) and is a regular point of the constraints in Eq. (10.57b), then*

(a) $a_i(\mathbf{x}^*) = 0$ *for* $i = 1, 2, \ldots, p$, *and*   (10.73) (*b*) *there exist Lagrange multipliers* λ_i^* *for* $i = 1, 2, \ldots, p$ *such that*

$$\nabla f(\mathbf{x}^*) = \sum_{i=1}^{p} \lambda_i^* \nabla a_i(\mathbf{x}^*) \qquad (10.74)$$

∎

Eq. (10.74) can be expressed in terms of the Jacobian $\mathbf{J}_e(\mathbf{x})$ (see Eq. (10.64)) as

$$\mathbf{g}(\mathbf{x}^*) - \mathbf{J}_e^T(\mathbf{x}^*)\boldsymbol{\lambda}^* = \mathbf{0}$$

where $\mathbf{g}(\mathbf{x}) = \nabla f(\mathbf{x})$. In other words, *if* $\mathbf{x}^*$ *is a local minimizer* of the problem in Eq. (10.57), *then there exists a vector* $\boldsymbol{\lambda}^* \in R^p$ *such that the* $(n + p)$-*dimensional vector* $[\mathbf{x}^{*T} \; \boldsymbol{\lambda}^{*T}]^T$ *satisfies the* $n + p$ *nonlinear equations*

$$\begin{bmatrix} \mathbf{g}(\mathbf{x}^*) - \mathbf{J}_e^T(\mathbf{x}^*)\boldsymbol{\lambda}^* \\ \mathbf{a}(\mathbf{x}^*) \end{bmatrix} = \mathbf{0} \qquad (10.75)$$

Theorem 10.1 can be related to the first-order necessary conditions for a minimum for the case of unconstrained minimization in Theorem 2.1 (see Sec. 2.5) as follows. If function $f(\mathbf{x})$ is minimized without constraints, we can consider the problem as the special case of the problem in Eq. (10.57) where the number of constraints is reduced to zero. In such a case, condition (*a*) of Theorem 10.1 is satisfied automatically and condition (*a*) of Theorem 2.1 must hold. On the other hand, condition (*b*) becomes

$$\nabla f(\mathbf{x}^*) = \mathbf{0}$$

which is condition (*b*) of Theorem 2.1.

If $\mathbf{x}^*$ is a local minimizer and $\boldsymbol{\lambda}^*$ is the associated vector of Lagrange multipliers, the set $\{\mathbf{x}^*, \; \boldsymbol{\lambda}^*\}$ may be referred to as the *minimizer set* or *minimizer* for short.

Example 10.11 Find the points that satisfy the necessary conditions for a minimum for the problem in Example 10.10.

Solution We have

$$\mathbf{g}(\mathbf{x}) = \begin{bmatrix} 2x_1 \\ 2x_2 \\ \frac{1}{2}x_3 \end{bmatrix}, \quad \mathbf{J}_e^T(\mathbf{x}) = \begin{bmatrix} -1 & 2x_1 - 2 \\ 0 & 2x_2 \\ 1 & 0 \end{bmatrix}$$

Hence Eq. (10.75) becomes

$$2x_1 + \lambda_1 - \lambda_2(2x_1 - 2) = 0$$
$$2x_2 - 2\lambda_2 x_2 = 0$$
$$x_3 - 2\lambda_1 = 0$$
$$-x_1 + x_3 - 1 = 0$$
$$x_1^2 + x_2^2 - 2x_1 = 0$$

296

Solving the above system of equations, we obtain two solutions, i.e.,

$$\mathbf{x}_1^* = \begin{bmatrix} 0 \\ 0 \\ 1 \end{bmatrix} \quad \text{and} \quad \boldsymbol{\lambda}_1^* = \begin{bmatrix} \frac{1}{2} \\ -\frac{1}{4} \end{bmatrix}$$

and

$$\mathbf{x}_2^* = \begin{bmatrix} 2 \\ 0 \\ 3 \end{bmatrix} \quad \text{and} \quad \boldsymbol{\lambda}_2^* = \begin{bmatrix} \frac{3}{2} \\ \frac{11}{4} \end{bmatrix}$$

The first solution, $\{\mathbf{x}_1^*, \boldsymbol{\lambda}_1^*\}$, is the global minimizer set as can be observed in Fig. 10.8. Later on in Sec. 10.7, we will show that $\{\mathbf{x}_2^*, \boldsymbol{\lambda}_2^*\}$ is not a minimizer set.

■

10.6.2 Inequality constraints

Consider now the general constrained optimization problem in Eq. (10.1) and let $\mathbf{x}^*$ be a local minimizer. The set $\mathcal{J}(\mathbf{x}^*) \subseteq \{1, 2, \ldots, q\}$ is the set of indices j for which the constraints $c_j(\mathbf{x}) \geq 0$ are *active* at $\mathbf{x}^*$, i.e., $c_j(\mathbf{x}^*) = 0$. At point $\mathbf{x}^*$, the feasible directions are characterized only by the equality constraints *and* those inequality constraints $c_j(\mathbf{x})$ with $j \in \mathcal{J}(\mathbf{x}^*)$, and are not influenced by the inequality constraints that are *inactive*. As a matter of fact, for an inactive constraint $c_j(\mathbf{x}) \geq 0$, the feasibility of $\mathbf{x}^*$ implies that

$$c_j(\mathbf{x}^*) > 0$$

This leads to

$$c_j(\mathbf{x}^* + \boldsymbol{\delta}) > 0$$

for any $\boldsymbol{\delta}$ with a sufficiently small $||\boldsymbol{\delta}||$.

If there are K active inequality constraints at $\mathbf{x}^*$ and

$$\mathcal{J}(\mathbf{x}^*) = \{j_1, j_2, \ldots, j_K\} \tag{10.76}$$

then Eq. (10.61) needs to be modified to

$$\nabla f(\mathbf{x}^*) = \sum_{i=1}^{p} \lambda_i^* \nabla a_i(\mathbf{x}^*) + \sum_{k=1}^{K} \mu_{j_k}^* \nabla c_{j_k}(\mathbf{x}^*) \tag{10.77}$$

In words, Eq. (10.77) states that *the gradient at* $\mathbf{x}^*$, $\nabla f(\mathbf{x}^*)$, *is a linear combination of the gradients of all the constraint functions that are active at* $\mathbf{x}^*$.

An argument similar to that used in Sec. 10.5.2 to explain why Eq. (10.77) must hold for a local minimum of the problem in Eq. (10.1) is as follows [7]. We start by assuming that $\mathbf{x}^*$ is a regular point for the constraints that are active at $\mathbf{x}^*$. Let j_k be one of the indices from $\mathcal{J}(\mathbf{x}^*)$ and assume that $\mathbf{s}$ is a feasible vector at $\mathbf{x}^*$. Using the Taylor series of $c_{j_k}(\mathbf{x})$, we can write

$$
\begin{aligned}
c_{j_k}(\mathbf{x}^* + \mathbf{s}) &= c_{j_k}(\mathbf{x}^*) + \mathbf{s}^T \nabla c_{j_k}(\mathbf{x}^*) + o(||\mathbf{s}||) \\
&= \mathbf{s}^T \nabla c_{j_k}(\mathbf{x}^*) + o(||\mathbf{s}||)
\end{aligned}
$$

Since $\mathbf{s}$ is feasible, $c_{j_k}(\mathbf{x}^* + \mathbf{s}) \geq 0$ which leads to

$$
\mathbf{s}^T \nabla c_{j_k}(\mathbf{x}^*) \geq 0 \tag{10.78}
$$

Now we orthogonally project $\nabla f(\mathbf{x}^*)$ onto the space spanned by $\mathcal{S} = \{\nabla a_i(\mathbf{x}^*)$ for $1 \leq i \leq p$ and $\nabla c_{j_k}(\mathbf{x}^*)$ for $1 \leq k \leq K\}$. Since the projection is on $\mathcal{S}$, it can be expressed as a linear combination of vectors $\{\nabla a_i(\mathbf{x}^*)$ for $1 \leq i \leq p$ and $\nabla c_{j_k}(\mathbf{x}^*)$ for $1 \leq k \leq K\}$, i.e.,

$$
\sum_{i=1}^{p} \lambda_i^* \nabla a_i(\mathbf{x}^*) + \sum_{k=1}^{K} \mu_{j_k}^* \nabla c_{j_k}(\mathbf{x}^*)
$$

for some λ_i^*'s and $\mu_{j_k}^*$'s. If we denote the difference between $\nabla f(\mathbf{x}^*)$ and this projection by $\mathbf{d}$, then we can write

$$
\nabla f(\mathbf{x}^*) = \sum_{i=1}^{p} \lambda_i^* \nabla a_i(\mathbf{x}^*) + \sum_{k=1}^{K} \mu_{j_k}^* \nabla c_{j_k}(\mathbf{x}^*) + \mathbf{d} \tag{10.79}
$$

Since $\mathbf{d}$ is orthogonal to $\mathcal{S}$, $\mathbf{d}$ is orthogonal to $\nabla a_i(\mathbf{x}^*)$ and $\nabla c_{j_k}(\mathbf{x}^*)$; hence $\mathbf{s} = -\mathbf{d}$ is a feasible direction (see Eqs. (10.59) and (10.78)); however, Eq. (10.79) gives

$$
\mathbf{s}^T \nabla f(\mathbf{x}^*) = -||\mathbf{d}||^2 < 0
$$

meaning that $\mathbf{s}$ would be a descent direction at $\mathbf{x}^*$. This contradicts the fact that $\mathbf{x}^*$ is a local minimizer. Therefore, $\mathbf{d} = \mathbf{0}$ and Eq. (10.77) holds. Constants λ_i^* and $\mu_{j_k}^*$ in Eq. (10.77) are the Lagrange multipliers for equality and inequality constraints, respectively.

Unlike the Lagrange multipliers associated with equality constraints, which can be either positive or negative, those associated with *active* inequality constraints must be nonnegative, i.e.,

$$
\mu_{j_k}^* \geq 0 \qquad \text{for } 1 \leq k \leq K \tag{10.80}
$$

We demonstrate the validity of Eq. (10.80) by contradiction. Suppose that $\mu_{j_{k^*}}^* < 0$ for some j_{k^*}. Since the gradients in $\mathcal{S}$ are linearly independent, the system

$$
\begin{bmatrix}
\nabla^T a_1(\mathbf{x}^*) \\
\vdots \\
\nabla^T a_p(\mathbf{x}^*) \\
\nabla^T c_{j_1}(\mathbf{x}^*) \\
\vdots \\
\nabla^T c_{j_K}(\mathbf{x}^*)
\end{bmatrix}
\mathbf{s} =
\begin{bmatrix}
0 \\
\vdots \\
0 \\
1 \\
0 \\
\vdots \\
0
\end{bmatrix}
$$

has a solution for **s**, where the vector on the right-hand side of the above equation has only one nonzero entry corresponding to $\nabla^T c_{j_{k*}}(\mathbf{x}^*)$. Hence we have a vector **s** satisfying the equations

$$
\mathbf{s}^T \nabla a_i(\mathbf{x}^*) = 0 \quad \text{for } 1 \le i \le p
$$

$$
\mathbf{s}^T \nabla c_{j_k} =
\begin{cases}
1 & \text{for } k = k^* \\
0 & \text{otherwise}
\end{cases}
$$

It follows from Eqs. (10.59) and (10.78) that **s** is feasible. By virtue of Eq. (10.77), we obtain

$$
\mathbf{s}^T \nabla f(\mathbf{x}^*) = \mu^*_{j_{k*}} < 0
$$

Hence **s** is a descent direction at $\mathbf{x}^*$ which contradicts the fact that $\mathbf{x}^*$ is a local minimizer. This proves Eq. (10.80). The following theorem, known as the KKT conditions [12], summarizes the above discussion.

Theorem 10.2 *Karush-Kuhn-Tucker conditions* *If $\mathbf{x}^*$ is a local minimizer of the problem in Eq. (10.1) and is regular for the constraints that are active at $\mathbf{x}^*$, then*

(a) $a_i(\mathbf{x}^*) = 0$ *for* $1 \le i \le p$,

(b) $c_j(\mathbf{x}^*) \ge 0$ *for* $1 \le j \le q$,

(c) *there exist Lagrange multipliers λ^*_i for $1 \le i \le p$ and μ^*_j for $1 \le j \le q$ such that*

$$
\nabla f(\mathbf{x}^*) = \sum_{i=1}^{p} \lambda^*_i \nabla a_i(\mathbf{x}^*) + \sum_{j=1}^{q} \mu^*_j \nabla c_j(\mathbf{x}^*) \tag{10.81}
$$

(d) $\lambda^*_i a_i(\mathbf{x}^*) = 0$ *for* $1 \le i \le p$, $\qquad\qquad$ (10.82a)

$\mu^*_j c_j(\mathbf{x}^*) = 0$ *for* $1 \le j \le q$, *and* $\qquad\qquad$ (10.82b)

(e) $\mu^*_j \ge 0$ *for* $1 \le j \le q$. $\qquad\qquad$ (10.83)

■

Some remarks on the KKT conditions stated in Theorem 10.2 are in order. Conditions (a) and (b) simply mean that $\mathbf{x}^*$ must be a feasible point. The

$p+q$ equations in Eq. (10.82) are often referred to as the *complementarity KKT conditions*. They state that λ_i^* and $a_i(\mathbf{x}^*)$ cannot be nonzero simultaneously, and μ_j^* and $c_j(\mathbf{x}^*)$ cannot be nonzero simultaneously. Note that condition (a) implies the condition in Eq. (10.82a) regardless of whether λ_i^* is zero or not. For the equality conditions in Eq. (10.82b), we need to distinguish those constraints that are active at $\mathbf{x}^*$, i.e.,

$$c_j(\mathbf{x}^*) = 0 \qquad \text{for } j \in \mathcal{J}(\mathbf{x}^*) = \{j_1, j_2, \ldots, j_K\}$$

from those that are inactive at $\mathbf{x}^*$, i.e.,

$$c_j(\mathbf{x}^*) > 0 \qquad \text{for } j \in \{1, 2, \ldots, q\} \backslash \mathcal{J}(\mathbf{x}^*)$$

where $\mathcal{I} \backslash \mathcal{J}$ denotes the system indices in $\mathcal{I}$, that are not in $\mathcal{J}$. From Eq. (10.82b),

$$\mu_j^* = 0 \qquad \text{for } j \in \{1, 2, \ldots, q\} \backslash \mathcal{J}(\mathbf{x}^*)$$

which reduces Eq. (10.81) to Eq. (10.77); however, μ_j may be nonzero for $j \in \mathcal{J}(\mathbf{x}^*)$. Condition ($e$) states that

$$\mu_j^* \geq 0 \qquad \text{for } j \in \mathcal{J}(\mathbf{x}^*) \tag{10.84}$$

The nonnegativity of the Lagrange multipliers associated with inequality constraints can be explained using Fig. 10.9. For the sake of simplicity, let us assume that $p = 0$ and $q = 1$ in which case the optimization problem would involve only one inequality constraint, namely,

$$c_1(\mathbf{x}) \geq 0 \tag{10.85}$$

If the minimizer $\mathbf{x}^*$ happens to be inside the feasible region $\mathcal{R}$ defined by the constraint in Eq. (10.85) (see Fig. 10.9a), then $\nabla f(\mathbf{x}^*) = \mathbf{0}$ and $\mu_1^* = 0$. If $\mathbf{x}^*$ is on the boundary of $\mathcal{R}$ (see Fig. 10.9b), then Eq. (10.81) implies that

$$\nabla f(\mathbf{x}^*) = \mu_1^* \nabla c_1(\mathbf{x}^*)$$

As can be seen in Fig. 10.9b, $\nabla c_1(\mathbf{x}^*)$ is a vector pointing towards the interior of the feasible region, since $c_1(\mathbf{x}^*) = 0$ and $c(\mathbf{x}) > 0$ inside $\mathcal{R}$, and similarly $\nabla f(\mathbf{x}^*)$ is a vector pointing towards the interior of $\mathcal{R}$. This in conjunction with the above equation implies that $\nabla f(\mathbf{x}^*)$ and $\nabla c_1(\mathbf{x}^*)$ must be in the same direction and hence $\mu_1^* > 0$. It should be stressed that the nonnegativity of the Lagrange multipliers holds only for those multipliers associated with *inequality* constraints. As was illustrated in Fig. 10.7a and b, nonzero Lagrange multipliers associated with equality constraints can be either positive or negative.

There are a total of p (equality) $+K$ (inequality) Lagrange multipliers that may be nonzero, and there are n entries in parameter vector $\mathbf{x}$. It is interesting

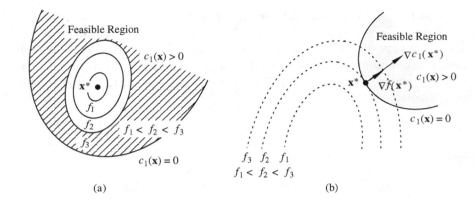

Figure 10.9. Nonnegativity of Lagrange multipliers: (a) $\mathbf{x}^*$ is a minimizer in the interior of the feasible region; (b) $\mathbf{x}^*$ is a minimizer on the boundary of the feasible region.

to note that the KKT conditions involve the same number of equations, i.e.,

$$\mathbf{g}(\mathbf{x}^*) - \mathbf{J}_e^T(\mathbf{x}^*)\boldsymbol{\lambda}^* - \hat{\mathbf{J}}_{ie}^T(\mathbf{x}^*)\hat{\boldsymbol{\mu}}^* = \mathbf{0} \qquad (10.86a)$$
$$\mathbf{a}(\mathbf{x}^*) = \mathbf{0} \qquad (10.86b)$$
$$\hat{\mathbf{c}}(\mathbf{x}^*) = \mathbf{0} \qquad (10.86c)$$

where

$$\hat{\boldsymbol{\mu}}^* = [\mu_{j1}^* \ \mu_{j2}^* \ \cdots \ \mu_{jK}^*]^T \qquad (10.87a)$$
$$\hat{\mathbf{J}}_{ie}(\mathbf{x}) = [\nabla c_{j1}(\mathbf{x}) \ \nabla c_{j2}(\mathbf{x}) \ \cdots \ \nabla c_{jK}(\mathbf{x})]^T \qquad (10.87b)$$
$$\hat{\mathbf{c}}(\mathbf{x}) = [c_{j1}(\mathbf{x}) \ c_{j2}(\mathbf{x}) \ \cdots \ c_{jK}(\mathbf{x})]^T \qquad (10.87c)$$

Example 10.12 Solve the constrained minimization problem

$$\text{minimize } f(\mathbf{x}) = x_1^2 + x_2^2 - 14x_1 - 6x_2$$
$$\text{subject to: } \quad c_1(\mathbf{x}) = 2 - x_1 - x_2 \geq 0$$
$$c_2(\mathbf{x}) = 3 - x_1 - 2x_2 \geq 0$$

by applying the KKT conditions.

Solution The KKT conditions imply that

$$2x_1 - 14 + \mu_1 + \mu_2 = 0$$
$$2x_2 - 6 + \mu_1 + 2\mu_2 = 0$$
$$\mu_1(2 - x_1 - x_2) = 0$$
$$\mu_2(3 - x_1 - 2x_2) = 0$$
$$\mu_1 \geq 0$$
$$\mu_2 \geq 0$$

One way to find the solution in this simple case is to consider all possible cases with regard to active constraints and verify the nonnegativity of the μ_i's obtained [13].

Case 1 No active constraints

If there are no active constraints, we have $\mu_1^* = \mu_2^* = 0$, which leads to

$$\mathbf{x}^* = \begin{bmatrix} 7 \\ 3 \end{bmatrix}$$

Obviously, this $\mathbf{x}^*$ violates both constraints and it is not a solution.

Case 2 One constraint active

If only the first constraint is active, then we have $\mu_2^* = 0$, and

$$2x_1 - 14 + \mu_1 = 0$$
$$2x_2 - 6 + \mu_1 = 0$$
$$2 - x_1 - x_2 = 0$$

Solving this system of equations, we obtain

$$\mathbf{x}^* = \begin{bmatrix} 3 \\ -1 \end{bmatrix} \quad \text{and} \quad \mu_1^* = 8$$

Since $\mathbf{x}^*$ also satisfies the second constraint, $\mathbf{x}^* = [3 \ -1]^T$ and $\boldsymbol{\mu}^* = [8 \ 0]^T$ satisfy the KKT conditions.

If only the second constraint is active, then $\mu_1^* = 0$ and the KKT conditions become

$$2x_1 - 14 + \mu_2 = 0$$
$$2x_2 - 6 + 2\mu_2 = 0$$
$$3 - x_1 - x_2 = 0$$

The solution of this system of equations is given by

$$\mathbf{x}^* = \begin{bmatrix} \frac{14}{3} \\ -\frac{5}{3} \end{bmatrix} \quad \text{and} \quad \mu_2^* = \frac{14}{3}$$

As $\mathbf{x}^*$ violates the first constraint, the above $\mathbf{x}^*$ and $\boldsymbol{\mu}^*$ do not satisfy the KKT conditions.

Case 3 Both constraints active

If both constraints are active, we have

$$2x_1 - 14 + \mu_1 + \mu_2 = 0$$
$$2x_2 - 6 + \mu_1 + 2\mu_2 = 0$$
$$2 - x_1 - x_2 = 0$$
$$3 - x_1 - 2x_2 = 0$$

The solution to this system of equations is given by

$$\mathbf{x}^* = \begin{bmatrix} 1 \\ 1 \end{bmatrix} \quad \text{and} \quad \boldsymbol{\mu}^* = \begin{bmatrix} 20 \\ -8 \end{bmatrix}$$

Since $\mu_2^* < 0$, this is not a solution of the optimization problem.

Therefore, the only candidate for a minimizer of the problem is

$$\mathbf{x}^* = \begin{bmatrix} 3 \\ -1 \end{bmatrix}, \quad \boldsymbol{\mu}^* = \begin{bmatrix} 8 \\ 0 \end{bmatrix}$$

As can be observed in Fig. 10.10, the above point is actually the global minimizer.

$\blacksquare$

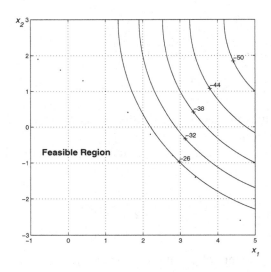

Figure 10.10. Contours of $f(\mathbf{x})$ and the two constraints for Example 10.12.

10.7   Second-Order Conditions

As in the unconstrained case, there are second-order conditions

(*a*) that must be satisfied for a point to be a local minimizer (i.e., necessary conditions), and

(*b*) that will assure that a point is a local minimizer (i.e., sufficient conditions).

The conditions in the constrained case are more complicated than their unconstrained counterparts due to the involvement of the various constraints, as may be expected.

10.7.1 Second-order necessary conditions

Suppose $\mathbf{x}^*$ is a local minimizer for the equality-constrained problem in Eq. (10.57) and is a regular point of the constraints in Eq. (10.57b). A second-order condition can be derived by examining the behavior of $f(\mathbf{x})$ in a neighborhood of $\mathbf{x}^*$. If $\mathbf{s}$ is a feasible direction at $\mathbf{x}^*$, then $a_i(\mathbf{x}^* + \mathbf{s}) = 0$ for $1 \le i \le p$, which in conjunction with Eq. (10.66) implies that

$$f(\mathbf{x}^* + \mathbf{s}) = L(\mathbf{x}^* + \mathbf{s}, \; \boldsymbol{\lambda}^*) \tag{10.88}$$

where $\boldsymbol{\lambda}^*$ satisfies Eq. (10.74). By using the Taylor expansion of $L(\mathbf{x}^* + \mathbf{s}, \; \boldsymbol{\lambda}^*)$ at $\{\mathbf{x}^*, \; \boldsymbol{\lambda}^*\}$ and Theorem 10.1, we obtain

$$
\begin{aligned}
f(\mathbf{x}^* + \mathbf{s}) &= L(\mathbf{x}^*, \; \boldsymbol{\lambda}^*) + \mathbf{s}^T \nabla_x L(\mathbf{x}^*, \; \boldsymbol{\lambda}^*) + \tfrac{1}{2}\mathbf{s}^T \nabla_x^2 L(\mathbf{x}^*, \; \boldsymbol{\lambda}^*)\mathbf{s} + o(||\mathbf{s}||^2) \\
&= f(\mathbf{x}^*) + \tfrac{1}{2}\mathbf{s}^T \nabla_x^2 L(\mathbf{x}^*, \; \boldsymbol{\lambda}^*)\mathbf{s} + o(||\mathbf{s}||^2) \tag{10.89}
\end{aligned}
$$

Using an argument similar to that used in the proof of Theorem 2.2, it can be shown that by virtue of $\mathbf{x}^*$ being a local minimizer, we have

$$\mathbf{s}^T \nabla_x^2 L(\mathbf{x}^*, \; \boldsymbol{\lambda}^*)\mathbf{s} \ge 0 \tag{10.90}$$

From Eqs. (10.59) and (10.64), it is clear that $\mathbf{s}$ is feasible at $\mathbf{x}^*$ if

$$\mathbf{J}_e(\mathbf{x}^*)\mathbf{s} = \mathbf{0}$$

i.e., $\mathbf{s} \in \mathcal{N}[\mathbf{J}_e(\mathbf{x}^*)]$, which is the null space of $\mathbf{J}_e(\mathbf{x}^*)$. Since this null space can be characterized by a basis of the space, Eq. (10.90) is equivalent to the positive semidefiniteness of $\mathbf{N}^T(\mathbf{x}^*)\nabla_x^2 L(\mathbf{x}^*, \; \boldsymbol{\lambda}^*)\mathbf{N}(\mathbf{x}^*)$ where $\mathbf{N}(\mathbf{x}^*)$ is a matrix whose columns form a basis of $\mathcal{N}[\mathbf{J}_e(\mathbf{x}^*)]$. These results can be summarized in terms of Theorem 10.3.

Theorem 10.3 *Second-order necessary conditions for a minimum, equality constraints If $\mathbf{x}^*$ is a constrained local minimizer of the problem in Eq. (10.57) and is a regular point of the constraints in Eq. (10.57b), then*
   (a) $a_i(\mathbf{x}^*) = 0$ for $i = 1, 2, \ldots, p$,
   (b) there exist λ_i^* for $i = 1, 2, \ldots, p$ such that

$$\nabla f(\mathbf{x}^*) = \sum_{i=1}^{p} \lambda_i^* \nabla a_i(\mathbf{x}^*)$$

   (c) $\mathbf{N}^T(\mathbf{x}^*)\nabla_x^2 L(\mathbf{x}^*, \; \boldsymbol{\lambda}^*)\mathbf{N}(\mathbf{x}^*) \succeq \mathbf{0}$. $\qquad\qquad$ (10.91)
$\blacksquare$

Example 10.13 In Example 10.11 it was found that

$$\mathbf{x}_2^* = \begin{bmatrix} 2 \\ 0 \\ 3 \end{bmatrix} \quad \text{and} \quad \boldsymbol{\lambda}_2^* = \begin{bmatrix} \frac{3}{2} \\ \frac{11}{4} \end{bmatrix}$$

satisfy the first-order necessary conditions for a minimum for the problem of Example 10.10. Check whether the second-order necessary conditions for a minimum are satisfied.

Solution We can write

$$\nabla_x^2 L(\mathbf{x}_2^*, \boldsymbol{\lambda}_2^*) = \begin{bmatrix} -\frac{7}{2} & 0 & 0 \\ 0 & -\frac{7}{2} & 0 \\ 0 & 0 & \frac{1}{2} \end{bmatrix}$$

and

$$\mathbf{J}_e(\mathbf{x}_2^*) = \begin{bmatrix} -1 & 0 & 1 \\ 2 & 0 & 0 \end{bmatrix}$$

It can be readily verified that the null space of $\mathbf{J}_e(\mathbf{x}_2^*)$ is the one-dimensional space spanned by $\mathbf{N}(x_2^*) = [0 \ 1 \ 0]^T$. Since

$$\mathbf{N}^T(\mathbf{x}_2^*)\nabla_x^2(\mathbf{x}_2^*, \boldsymbol{\lambda}_2^*)\mathbf{N}(\mathbf{x}_2^*) = -\frac{7}{2} < 0 \qquad (10.92)$$

we conclude that $\{\mathbf{x}_2^*, \boldsymbol{\lambda}_2^*\}$ does not satisfy the second-order necessary conditions.

∎

For the general constrained optimization problem in Eq. (10.1), a second-order condition similar to Eq. (10.91) can be derived as follows. Let $\mathbf{x}^*$ be a local minimizer of the problem in Eq. (10.1) and $\mathcal{J}(\mathbf{x}^*)$ be the index set for the inequality constraints that are active at $\mathbf{x}^*$ (see Eq. (10.76)). A direction s is said to be feasible at $\mathbf{x}^*$ if

$$a_i(\mathbf{x}^* + \mathbf{s}) = 0 \qquad \text{for } 1 \le i \le p \qquad (10.93a)$$
$$c_j(\mathbf{x}^* + \mathbf{s}) = 0 \qquad \text{for } j \in \mathcal{J}(\mathbf{x}^*) \qquad (10.93b)$$

Recall that the Lagrangian for the problem in Eq. (10.1) is defined by

$$L(\mathbf{x}, \boldsymbol{\lambda}, \boldsymbol{\mu}) = f(\mathbf{x}) - \sum_{i=1}^{p} \lambda_i a_i(\mathbf{x}) - \sum_{j=1}^{q} \mu_j c_j(\mathbf{x}) \qquad (10.94)$$

If $\boldsymbol{\lambda}^*$ and $\boldsymbol{\mu}^*$ are the Lagrange multipliers described in Theorem 10.2, then the constraints in Eqs. (10.1b) and (10.1c) and the complementarity condition in Eq. (10.82) imply that

$$f(\mathbf{x}^*) = L(\mathbf{x}^*, \boldsymbol{\lambda}^*, \boldsymbol{\mu}^*) \qquad (10.95)$$

From Eqs. (10.81), (10.93), and (10.95), we have

$$\begin{aligned} f(\mathbf{x}^* + \mathbf{s}) &= L(\mathbf{x}^* + \mathbf{s}, \boldsymbol{\lambda}^*, \boldsymbol{\mu}^*) \\ &= L(\mathbf{x}^*, \boldsymbol{\lambda}^*, \boldsymbol{\mu}^*) + \mathbf{s}^T \nabla_x L(\mathbf{x}^*, \boldsymbol{\lambda}^*, \boldsymbol{\mu}^*) \\ &\quad + \frac{1}{2}\mathbf{s}^T \nabla_x^2 L(\mathbf{x}^*, \boldsymbol{\lambda}^*, \boldsymbol{\mu}^*)\mathbf{s} + o(||\mathbf{s}||^2) \\ &= f(\mathbf{x}^*) + \frac{1}{2}\mathbf{s}^T \nabla_x^2 L(\mathbf{x}^*, \boldsymbol{\lambda}^*, \boldsymbol{\mu}^*)\mathbf{s} + o(||\mathbf{s}||^2) \end{aligned}$$

This in conjunction with the fact that $f(\mathbf{x}^*) \leq f(\mathbf{x}^* + \mathbf{s})$ implies that

$$\mathbf{s}^T \nabla_x^2 L(\mathbf{x}^*, \boldsymbol{\lambda}^*, \boldsymbol{\mu}^*)\mathbf{s} \geq 0 \tag{10.96}$$

for any $\mathbf{s}$ feasible at $\mathbf{x}^*$.

From Eq. (10.93), the feasible directions at $\mathbf{x}^*$ are those directions that are orthogonal to the gradients of the constraints that are active at $\mathbf{x}^*$, namely,

$$\mathbf{J}(\mathbf{x}^*)\mathbf{s} = \begin{bmatrix} \mathbf{J}_e(\mathbf{x}^*) \\ \hat{\mathbf{J}}_{ie}(\mathbf{x}^*) \end{bmatrix} \mathbf{s} = \mathbf{0} \tag{10.97}$$

where $\hat{\mathbf{J}}_{ie}(\mathbf{x})$ is given by Eq. (10.87b). Hence the feasible directions at $\mathbf{x}^*$ are characterized by the null space of $\mathbf{J}(\mathbf{x}^*)$, denoted as $\mathcal{N}[\mathbf{J}(\mathbf{x}^*)]$, and the condition in Eq. (10.96) assures the positive semidefiniteness of $\mathbf{N}^T(\mathbf{x}^*)\nabla_x^2 L(\mathbf{x}^*, \boldsymbol{\lambda}^*, \boldsymbol{\mu}^*)$ $\mathbf{N}(\mathbf{x}^*)$ where $\mathbf{N}(\mathbf{x}^*)$ is a matrix whose columns form a basis of $\mathcal{N}[\mathbf{J}(\mathbf{x}^*)]$. A set of necessary conditions for the general constrained optimization problem in Eq. (10.1) can now be summarized in terms of the following theorem.

Theorem 10.4 *Second-order necessary conditions for a minimum, general constrained problem* If $\mathbf{x}^*$ *is a constrained local minimizer of the problem in Eq. (10.1) and is a regular point of the constraints in Eqs. (10.1b) and (10.1c), then*

(a) $a_i(\mathbf{x}^*) = 0$ for $1 \leq i \leq p$,
(b) $c_j(\mathbf{x}^*) \geq 0$ for $1 \leq j \leq q$,
(c) there exist Lagrange multipliers λ_i^*'s and μ_j^*'s such that

$$\nabla f(\mathbf{x}^*) = \sum_{i=1}^{p} \lambda_i^* \nabla a_i(\mathbf{x}^*) + \sum_{j=1}^{q} \mu_j^* \nabla c_j(\mathbf{x}^*)$$

(d) $\lambda_i^* a_i(\mathbf{x}^*) = 0$ for $1 \leq i \leq p$ and $\mu_j^* c_j(\mathbf{x}^*) = 0$ for $1 \leq j \leq q$,
(e) $\mu_j^* \geq 0$ for $1 \leq j \leq q$, and
(f) $\mathbf{N}^T(\mathbf{x}^*)\nabla_x^2 L(\mathbf{x}^*, \boldsymbol{\lambda}^*, \boldsymbol{\mu}^*)\mathbf{N}(\mathbf{x}^*) \succeq \mathbf{0}$. $\tag{10.98}$

∎

10.7.2 Second-order sufficient conditions

For the constrained problem in Eq. (10.57), second-order sufficient conditions for a point $\mathbf{x}^*$ to be a local minimizer can be readily obtained from Eq. (10.89), where $\{\mathbf{x}^*, \boldsymbol{\lambda}^*\}$ is assumed to satisfy the first-order necessary conditions described in Theorem 10.1. Using an argument similar to that used in the proof of Theorem 2.4, we can show that a point $\mathbf{x}^*$ that satisfies the conditions in Theorem 10.1 is a local minimizer if the matrix

$$\mathbf{N}^T(\mathbf{x}^*)\nabla_x^2 L(\mathbf{x}^*, \boldsymbol{\lambda}^*)\mathbf{N}(\mathbf{x}^*)$$

is positive definite.

Theorem 10.5 *Second-order sufficient conditions for a minimum, equality constraints* *If* $\mathbf{x}^*$ *is a regular point of the constraints in Eq. (10.57b), then it is a strong local minimizer of Eq. (10.57) if*
(a) $a_i(\mathbf{x}^*) = 0$ *for* $1 \le i \le p$,
(b) *there exist Lagrange multipliers* λ_i^* *for* $i = 1, 2, \ldots, p$ *such that*

$$\nabla f(\mathbf{x}^*) = \sum_{i=1}^{p} \lambda_i^* \nabla a_i(\mathbf{x}^*)$$

(c) $\mathbf{N}^T(\mathbf{x}^*) \nabla_x^2 L(\mathbf{x}^*, \boldsymbol{\lambda}^*) \mathbf{N}(\mathbf{x}) \succ \mathbf{0}$, $\qquad\qquad\qquad\qquad\qquad$ (10.99)
i.e., $\nabla_x^2 L(\mathbf{x}^*, \boldsymbol{\lambda}^*)$ *is positive definite in the null space* $\mathcal{N}[\mathbf{J}(\mathbf{x}^*)]$. ∎

Example 10.14 Check whether the second-order sufficient conditions for a minimum are satisfied in the minimization problem of Example 10.10.

Solution We compute

$$\nabla_x^2 L(\mathbf{x}_1^*, \boldsymbol{\lambda}_1^*) = \begin{bmatrix} \frac{5}{2} & 0 & 0 \\ 0 & \frac{5}{2} & 0 \\ 0 & 0 & \frac{1}{2} \end{bmatrix}$$

which is positive definite in the entire E^3. Hence Theorem 10.5 implies that $\mathbf{x}_1^*$ is a strong, local minimizer. ∎

A set of sufficient conditions for point $\mathbf{x}^*$ to be a local minimizer for the general constrained problem in Eq. (10.1) is given by the following theorem.

Theorem 10.6 *Second-order sufficient conditions for a minimum, general constrained problem* *A point* $\mathbf{x}^* \in R^n$ *is a strong local minimizer of the problem in Eq. (10.1) if*
(a) $a_i(\mathbf{x}^*) = 0$ *for* $1 \le i \le p$,
(b) $c_j(\mathbf{x}^*) \ge 0$ *for* $1 \le j \le q$,
(c) $\mathbf{x}^*$ *is a regular point of the constraints that are active at* $\mathbf{x}^*$,
(d) *there exist* λ_i^*'s *and* μ_j^*'s *such that*

$$\nabla f(\mathbf{x}^*) = \sum_{i=1}^{p} \lambda_i^* \nabla a_i(\mathbf{x}^*) + \sum_{j=1}^{q} \mu_j^* \nabla c_j(\mathbf{x}^*)$$

(e) $\lambda_i^* a_i(\mathbf{x}^*) = 0$ *for* $1 \le i \le p$ *and* $\mu_j^* c_j(\mathbf{x}^*) = 0$ *for* $1 \le j \le q$,
(f) $\mu_j^* \ge 0$ *for* $1 \le j \le q$, *and*

(g) $\mathbf{N}^T(\mathbf{x}^*)\nabla_x^2 L(\mathbf{x}^*, \boldsymbol{\lambda}^*, \boldsymbol{\mu}^*)\mathbf{N}(\mathbf{x}^*) \succ \mathbf{0}$ (10.100)
where $\mathbf{N}(\mathbf{x}^*)$ *is a matrix whose columns form a basis of the null space of* $\tilde{\mathbf{J}}(\mathbf{x}^*)$ *defined by*

$$\tilde{\mathbf{J}}(\mathbf{x}^*) = \begin{bmatrix} \mathbf{J}_e(\mathbf{x}^*) \\ \tilde{\mathbf{J}}_{ie}(\mathbf{x}^*) \end{bmatrix} \tag{10.101}$$

The Jacobian $\tilde{\mathbf{J}}_{ie}(\mathbf{x}^*)$ *is the matrix whose rows are composed of those gradients of inequality constraints that are active at* $\mathbf{x}^*$, *i.e.,* $\nabla^T c_j(\mathbf{x}^*)$, *with* $c_j(\mathbf{x}^*) = 0$ *and* $\mu_j^* > 0$.

Proof Let us suppose that $\mathbf{x}^*$ satisfies conditions (a) to (g) but is not a strong local minimizer. Under these circumstances there would exist a sequence of feasible points $\mathbf{x}_k \to \mathbf{x}^*$ such that $f(\mathbf{x}_k) \leq f(\mathbf{x}^*)$. If we write $\mathbf{x}_k = \mathbf{x}^* + \delta_k \mathbf{s}_k$ with $\|\mathbf{s}_k\| = 1$ for all k, then we may assume that $\delta_k > 0$, and $\mathbf{s}_k \to \mathbf{s}^*$ for some vector $\mathbf{s}^*$ with $\|\mathbf{s}^*\| = 1$ and

$$f(\mathbf{x}^* + \delta_k \mathbf{s}_k) - f(\mathbf{x}^*) \leq 0$$

which leads to

$$\nabla^T f(\mathbf{x}^*)\mathbf{s}^* \leq 0 \tag{10.102}$$

Since $\mathbf{s}^*$ is feasible at $\mathbf{x}^*$, we have

$$\nabla^T a_i(\mathbf{x}^*)\mathbf{s}^* = 0 \tag{10.103}$$

If $\mathcal{J}_p(\mathbf{x}^*)$ is the index set for inequality constraints that are active at $\mathbf{x}^*$ and are associated with strictly positive Lagrange multipliers, then

$$c_j(\mathbf{x}_k) - c_j(\mathbf{x}^*) = c_j(\mathbf{x}_k) \geq 0 \qquad \text{for } j \in \mathcal{J}_p(\mathbf{x}^*)$$

i.e.,

$$c_j(\mathbf{x}^* + \delta_k \mathbf{s}_k) - c_j(\mathbf{x}^*) \geq 0$$

which leads to

$$\nabla^T c_j(\mathbf{x}^*)\mathbf{s}^* \geq 0 \qquad \text{for } j \in \mathcal{J}_p(\mathbf{x}^*) \tag{10.104}$$

Now the inequality in Eq. (10.104) cannot occur since otherwise conditions (d), (e), (f) in conjunction with Eqs. (10.102), (10.103) would imply that

$$0 \geq \nabla^T f(\mathbf{x}^*)\mathbf{s}^* = \sum_{i=1}^{p} \lambda_i^* \nabla^T a_i(\mathbf{x}^*)\mathbf{s}^* + \sum_{j=1}^{q} \mu_j^* \nabla^T c_j(\mathbf{x}^*)\mathbf{s}^* > 0$$

i.e., $0 > 0$ which is a contradiction. Hence

$$\nabla^T c_j(\mathbf{x}^*)\mathbf{s}^* = 0 \qquad \text{for } j \in \mathcal{J}_p(\mathbf{x}^*) \tag{10.105}$$

From Eqs. (10.103) and (10.104), it follows that $\mathbf{s}^*$ belongs to the null space of $\tilde{\mathbf{J}}(\mathbf{x}^*)$ and so condition (g) implies that $\mathbf{s}^{*T}\nabla_x^2 L(\mathbf{x}^*, \boldsymbol{\lambda}^*, \boldsymbol{\mu}^*)\mathbf{s}^* > 0$. Since $\mathbf{x}_k \rightarrow \mathbf{x}^*$, we have $\mathbf{s}_k^T \nabla_x^2 L(\mathbf{x}^*, \boldsymbol{\lambda}^*, \boldsymbol{\mu}^*)\mathbf{s}_k > 0$ for a sufficiently large k. Using the condition in (d), the Taylor expansion of $L(\mathbf{x}_k, \boldsymbol{\lambda}^*, \boldsymbol{\mu}^*)$ at $\mathbf{x}^*$ gives

$$
\begin{aligned}
L(\mathbf{x}_k^*, \boldsymbol{\lambda}^*, \boldsymbol{\mu}^*) &= L(\mathbf{x}^*, \boldsymbol{\lambda}^*, \boldsymbol{\mu}^*) + \delta_k \mathbf{s}_k^T \nabla_x L(\mathbf{x}^*, \boldsymbol{\lambda}^*, \boldsymbol{\mu}^*) \\
&\quad + \tfrac{1}{2}\delta_k^2 \mathbf{s}_k^T \nabla_x^2 L(\mathbf{x}^*, \boldsymbol{\lambda}^*, \boldsymbol{\mu}^*)\mathbf{s}_k + o(\delta_k^2) \\
&= f(\mathbf{x}^*) + \tfrac{1}{2}\delta_k^2 \mathbf{s}_k^T \nabla_x^2 L(\mathbf{x}^*, \boldsymbol{\lambda}^*, \boldsymbol{\mu}^*)\mathbf{s}_k + o(\delta_k^2)
\end{aligned}
$$

This in conjunction with the inequalities $f(\mathbf{x}_k) \geq L(\mathbf{x}_k, \boldsymbol{\lambda}^*, \boldsymbol{\mu}^*)$ and $f(\mathbf{x}_k) \leq f(\mathbf{x}^*)$ leads to

$$
0 \geq f(\mathbf{x}_k) - f(\mathbf{x}^*) \geq \tfrac{1}{2}\delta_k^2 \mathbf{s}_k^T \nabla_x^2 L(\mathbf{x}^*, \boldsymbol{\lambda}^*, \boldsymbol{\mu}^*)\mathbf{s}_k + o(\delta_k^2) \qquad (10.106)
$$

So, for a sufficiently large k the right-hand side of Eq. (10.106) becomes strictly positive, which leads to the contradiction $0 > 0$. This completes the proof.

∎

Example 10.15 Use Theorem 10.6 to check the solution of the minimization problem discussed in Example 10.12.

Solution The candidate for a local minimizer was found to be

$$
\mathbf{x}^* = \begin{bmatrix} 3 \\ -1 \end{bmatrix}, \quad \boldsymbol{\mu}^* = \begin{bmatrix} 8 \\ 0 \end{bmatrix}
$$

Since the constraints are linear,

$$
\nabla_x^2 L(\mathbf{x}^*, \boldsymbol{\lambda}^*, \boldsymbol{\mu}^*) = \nabla^2 f(\mathbf{x}^*) = \begin{bmatrix} 2 & 0 \\ 0 & 2 \end{bmatrix}
$$

which is positive definite in the entire E^2. Therefore, $\{\mathbf{x}^*, \boldsymbol{\mu}^*\}$ satisfies all the conditions of Theorem 10.6 and hence $\mathbf{x}^*$ is a strong local minimizer. As was observed in Fig. 10.10, $\mathbf{x}^*$ is actually the global minimizer of the problem.

∎

10.8 Convexity

Convex functions and their basic properties were studied in Sec. 2.7 and the unconstrained optimization of convex functions was discussed in Sec. 2.8. The concept of convexity is also important in constrained optimization. In unconstrained optimization, the properties of convex functions are of interest when these functions are defined over a convex set. In a constrained optimization, the objective function is minimized with respect to the feasible region which is characterized by the constraints imposed. As may be expected, the concept of convexity can be fully used to achieve useful optimization results when both the objective function and the feasible region are convex. In Sec. 10.2, these problems were referred to as CP problems. A typical problem of this class can be formulated as

$$\text{minimize } f(\mathbf{x}) \tag{10.107a}$$
$$\text{subject to} \quad a_i(\mathbf{x}) = \mathbf{a}_i^T \mathbf{x} - b_i \qquad \text{for } 1 \leq i \leq p \tag{10.107b}$$
$$c_j(\mathbf{x}) \geq 0 \qquad \text{for } 1 \leq j \leq q \tag{10.107c}$$

where $f(\mathbf{x})$ and $-c_j(\mathbf{x})$ for $1 \leq j \leq q$ are convex functions. The main results, which are analogous to those in Sec. 2.8, are described by the next two theorems.

Theorem 10.7 *Globalness and convexity of minimizers in CP problems*
  (a) *If $\mathbf{x}^*$ is a local minimizer of a CP problem, then $\mathbf{x}^*$ is also a global minimizer.*
  (b) *The set of minimizers of a CP problem, denoted as S, is convex.*
  (c) *If the objective function $f(\mathbf{x})$ is strictly convex on the feasible region $\mathcal{R}$, then the global minimizer is unique.*

Proof
  (a) If $\mathbf{x}^*$ is a local minimizer that is not a global minimizer, then there is a feasible $\hat{\mathbf{x}}$ such that $f(\hat{\mathbf{x}}) < f(\mathbf{x}^*)$. If we let $\mathbf{x}_\tau = \tau\hat{\mathbf{x}} + (1 - \tau)\mathbf{x}^*$ for $0 < \tau < 1$, then the convexity of $f(\mathbf{x})$ implies that

$$f(\mathbf{x}_\tau) \leq \tau f(\hat{\mathbf{x}}) + (1 - \tau)f(\mathbf{x}^*) < f(\mathbf{x}^*)$$

  no matter how close $\mathbf{x}_\tau$ is to $\mathbf{x}^*$. This contradicts the assumption that $\mathbf{x}^*$ is a local minimizer since $f(\mathbf{x}^*)$ is supposed to assume the smallest value in a sufficiently small neighborhood of $\mathbf{x}^*$. Hence $\mathbf{x}^*$ is a global minimizer.
  (b) Let $\mathbf{x}_a,\ \mathbf{x}_b \in S$. From part (a), it follows that $\mathbf{x}_a$ and $\mathbf{x}_b$ are global minimizers. If $\mathbf{x}_\tau = \tau\mathbf{x}_a + (1 - \tau)\mathbf{x}_b$ for $0 \leq \tau \leq 1$, then the convexity of $f(\mathbf{x})$ leads to

$$f(\mathbf{x}_\tau) \leq \tau f(\mathbf{x}_a) + (1 - \tau)f(\mathbf{x}_b) = f(\mathbf{x}_a)$$

  Since $\mathbf{x}_a$ is a global minimizer, $f(\mathbf{x}_\tau) \geq f(\mathbf{x}_a)$. Hence $f(\mathbf{x}_\tau) = f(\mathbf{x}_a)$, i.e., $\mathbf{x}_\tau \in S$ for each τ, thus S is convex.
  (c) Suppose that the solution set S contains two distinct points $\mathbf{x}_a$ and $\mathbf{x}_b$ and $\mathbf{x}_\tau$ is defined as in part (b) with $0 < \tau < 1$. Since $\mathbf{x}_a \neq \mathbf{x}_b$ and $\tau \in (0, 1)$, we have $\mathbf{x}_\tau \neq \mathbf{x}_a$. By using the strict convexity of $f(\mathbf{x})$, we would conclude that $f(\mathbf{x}_\tau) < f(\mathbf{x}_a)$ which contradicts the assumption that $\mathbf{x}_a \in S$. Therefore, the global minimizer is unique.
∎

It turns out that in a CP problem, the KKT conditions become sufficient for $\mathbf{x}^*$ to be a global minimizer as stated in the following theorem.

Theorem 10.8 *Sufficiency of KKT conditions in CP problems* If $\mathbf{x}^*$ is a regular point of the constraints in Eqs. (10.107b) and (10.107c), and satisfies the KKT conditions stated in Theorem 10.2, where $f(\mathbf{x})$ is convex and $a_i(\mathbf{x})$ and $c_j(\mathbf{x})$ are given by Eqs. (10.107b) and (10.107c), respectively, then it is a global minimizer.

Proof For a feasible point $\hat{\mathbf{x}}$ with $\hat{\mathbf{x}} \neq \mathbf{x}^*$, we have $a_i(\hat{\mathbf{x}}) = 0$ for $1 \leq i \leq p$ and $c_j(\hat{\mathbf{x}}) \geq 0$ for $1 \leq j \leq q$. In terms of the notation used in Theorem 10.2, we can write

$$f(\hat{\mathbf{x}}) \geq f(\hat{\mathbf{x}}) - \sum_{j=1}^{q} \mu_j^* c_j(\hat{\mathbf{x}})$$

Since $f(\mathbf{x})$ and $-c_j(\mathbf{x})$ are convex, then from Theorem 2.12, we have

$$f(\hat{\mathbf{x}}) \geq f(\mathbf{x}^*) + \nabla^T f(\mathbf{x}^*)(\hat{\mathbf{x}} - \mathbf{x}^*)$$

and

$$-c_j(\hat{\mathbf{x}}) \geq -c_j(\mathbf{x}^*) - \nabla^T c_j(\mathbf{x}^*)(\hat{\mathbf{x}} - \mathbf{x}^*)$$

It follows that

$$f(\hat{\mathbf{x}}) \geq f(\mathbf{x}^*) + \nabla^T f(\mathbf{x}^*)(\hat{\mathbf{x}} - \mathbf{x}^*) - \sum_{j=1}^{q} \mu_j^* \nabla^T c_j(\mathbf{x}^*)(\hat{\mathbf{x}} - \mathbf{x}^*) - \sum_{j=1}^{q} \mu_j^* c_j(\mathbf{x}^*)$$

In the light of the complementarity conditions in Eq. (10.82b), the last term in the above inequality is zero and hence we have

$$f(\hat{\mathbf{x}}) \geq f(\mathbf{x}^*) + [\nabla f(\mathbf{x}^*) - \sum_{j=1}^{q} \mu_j^* \nabla c_j(\mathbf{x}^*)]^T (\hat{\mathbf{x}} - \mathbf{x}^*) \qquad (10.108)$$

Since $a_i(\hat{\mathbf{x}}) = a_i(\mathbf{x}^*) = 0$, we get

$$0 = a_i(\hat{\mathbf{x}}) - a_i(\mathbf{x}^*) = \mathbf{a}_i^T (\hat{\mathbf{x}} - \mathbf{x}^*) = \nabla^T a_i(\mathbf{x}^*)(\hat{\mathbf{x}} - \mathbf{x}^*)$$

Multiplying the above equality by $-\lambda_i^*$ and then adding it to the inequality in Eq. (10.108) for $1 \leq i \leq p$, we obtain

$$f(\hat{\mathbf{x}}) \geq f(\mathbf{x}^*) + [\nabla f(\mathbf{x}^*) - \sum_{i=1}^{p} \lambda_i^* \nabla a_i(\mathbf{x}^*) - \sum_{j=1}^{q} \mu_j^* \nabla c_j(\mathbf{x}^*)]^T (\hat{\mathbf{x}} - \mathbf{x}^*)$$

From Eq (10.81), the last term in the above inequality is zero, which leads to $f(\hat{\mathbf{x}}) \geq f(\mathbf{x}^*)$. This shows that $f(\mathbf{x}^*)$ is a global minimum.

∎

10.9   Duality

The concept of duality as applied to optimization is essentially a problem transformation that leads to an indirect but sometimes more efficient solution method. In a duality-based method the original problem, which is referred to as the *primal* problem, is transformed into a problem in which the parameters are the Lagrange multipliers of the primal. The transformed problem is called the *dual* problem. In the case where the number of inequality constraints is much greater than the dimension of $\mathbf{x}$, solving the dual problem to find the Lagrange multipliers and then finding $\mathbf{x}^*$ for the primal problem becomes an attractive alternative. For LP problems, a duality theory has been developed to serve as the foundation of modern primal-dual interior-point methods, (see Sec. 11.4 for the details).

A popular duality-based method is the *Wolfe dual* [14], which is concerned with the CP problem in Eq. (10.107). The main results of the Wolfe dual are described in terms of the following theorem.

Theorem 10.9 *Duality in convex programming* *Let $\mathbf{x}^*$ be a minimizer, and $\boldsymbol{\lambda}^*$, $\boldsymbol{\mu}^*$ be the associated Lagrange multipliers of the problem in Eq. (10.107). If $\mathbf{x}^*$ is a regular point of the constraints, then $\mathbf{x}^*$, $\boldsymbol{\lambda}^*$, and $\boldsymbol{\mu}^*$ solve the dual problem*

$$\underset{\mathbf{x},\,\boldsymbol{\lambda},\,\boldsymbol{\mu}}{maximize}\ L(\mathbf{x},\ \boldsymbol{\lambda},\ \boldsymbol{\mu}) \tag{10.109a}$$

$$\text{subject to}:\ \ \nabla_x L(\mathbf{x},\ \boldsymbol{\lambda},\ \boldsymbol{\mu}) = \mathbf{0} \tag{10.109b}$$

$$\boldsymbol{\mu} \geq \mathbf{0} \tag{10.109c}$$

In addition, $f(\mathbf{x}^*) = L(\mathbf{x}^*,\ \boldsymbol{\lambda}^*,\ \boldsymbol{\mu}^*)$.

Proof By virtue of Theorem 10.2, $f(\mathbf{x}^*) = L(\mathbf{x}^*,\ \boldsymbol{\lambda}^*,\ \boldsymbol{\mu}^*)$ and $\boldsymbol{\mu}^* \geq \mathbf{0}$. For a set $\{\mathbf{x},\ \boldsymbol{\lambda},\ \boldsymbol{\mu}\}$ that is feasible for the problem in Eq. (10.109), we have $\boldsymbol{\mu} \geq \mathbf{0}$ and $\nabla_x L(\mathbf{x},\ \boldsymbol{\lambda},\ \boldsymbol{\mu}) = \mathbf{0}$. Hence

$$L(\mathbf{x}^*,\ \boldsymbol{\lambda}^*,\ \boldsymbol{\mu}^*) = f(\mathbf{x}^*)$$

$$\geq f(\mathbf{x}^*) - \sum_{i=1}^{p} \lambda_i a_i(\mathbf{x}^*) - \sum_{j=1}^{q} \mu_j c_j(\mathbf{x}^*) = L(\mathbf{x}^*,\ \boldsymbol{\lambda},\ \boldsymbol{\mu})$$

With $\boldsymbol{\mu} \geq \mathbf{0}$, the Lagrangian $L(\mathbf{x},\ \boldsymbol{\lambda},\ \boldsymbol{\mu})$ is convex and, therefore,

$$L(\mathbf{x}^*,\ \boldsymbol{\lambda},\ \boldsymbol{\mu}) \geq L(\mathbf{x},\ \boldsymbol{\lambda},\ \boldsymbol{\mu}) + (\mathbf{x}^* - \mathbf{x})^T \nabla_x L(\mathbf{x},\ \boldsymbol{\lambda},\ \boldsymbol{\mu}) = L(\mathbf{x},\ \boldsymbol{\lambda},\ \boldsymbol{\mu})$$

Hence $L(\mathbf{x}^*,\ \boldsymbol{\lambda}^*,\ \boldsymbol{\mu}^*) \geq L(\mathbf{x},\ \boldsymbol{\lambda},\ \boldsymbol{\mu})$, i.e., set $\{\mathbf{x}^*,\ \boldsymbol{\lambda}^*,\ \boldsymbol{\mu}^*\}$ solves the problem in Eq. (10.109). ∎

312

Example 10.16 Find the Wolfe dual of the standard-form LP problem

$$\text{minimize } \mathbf{c}^T\mathbf{x} \tag{10.110a}$$

$$\text{subject to: } \mathbf{Ax} = \mathbf{b} \quad \mathbf{A} \in R^{p \times n} \tag{10.110b}$$

$$\mathbf{x} \geq \mathbf{0} \tag{10.110c}$$

Solution The Lagrangian is given by

$$L(\mathbf{x}, \boldsymbol{\lambda}, \boldsymbol{\mu}) = \mathbf{c}^T\mathbf{x} - (\mathbf{Ax} - \mathbf{b})^T\boldsymbol{\lambda} - \mathbf{x}^T\boldsymbol{\mu}$$

From Theorem 10.9, the Wolfe dual of the problem in Eq. (10.110) is the maximization problem

$$\underset{\mathbf{x}, \boldsymbol{\lambda}, \boldsymbol{\mu}}{\text{maximize }} \mathbf{x}^T(\mathbf{c} - \mathbf{A}^T\boldsymbol{\lambda} - \boldsymbol{\mu}) + \mathbf{b}^T\boldsymbol{\lambda} \tag{10.111a}$$

$$\text{subject to: } \mathbf{c} - \mathbf{A}^T\boldsymbol{\lambda} - \boldsymbol{\mu} = \mathbf{0} \tag{10.111b}$$

$$\boldsymbol{\mu} \geq \mathbf{0} \tag{10.111c}$$

Using Eq. (10.111b), the objective function in Eq. (10.111a) can be simplified and the dual problem can be stated as

$$\underset{\boldsymbol{\lambda}, \boldsymbol{\mu}}{\text{maximize }} \mathbf{b}^T\boldsymbol{\lambda} \tag{10.112a}$$

$$\text{subject to: } \mathbf{c} - \mathbf{A}^T\boldsymbol{\lambda} - \boldsymbol{\mu} = \mathbf{0} \tag{10.112b}$$

$$\boldsymbol{\mu} \geq \mathbf{0} \tag{10.112c}$$

∎

References

1 G. B. Dantzig, *Linear Programming and Extensions*, Princeton University Press, Princeton, NJ., 1963.
2 S. J. Wright, *Primal-Dual Interior-Point Methods*, SIAM, Philadelphia, 1997.
3 S. Boyd, L. El Ghaoui, E. Feron, and V. Balakrishnan, *Linear Matrix Inequalities in System and Control Theory*, SIAM, Philadelphia, 1994.
4 Y. Nesterov and A. Nemirovskii, *Interior-Point Polynomial Algorithms in Convex Programming*, SIAM, Philadelphia, 1994.
5 J. T. Betts, "An accelerated multiplier method for nonlinear programming," *JOTA*, vol. 21, no. 2, pp. 137–174, 1977.
6 G. Van der Hoek, *Reduction Methods in Nonlinear Programming*, Mathematical Centre Tracts, vol. 126, Mathematisch Centrum, Amsterdam, 1980.
7 R. Fletcher, *Practical Methods of Optimization*, 2nd ed., Wiley, New York, 1987.
8 G. H. Golub and C. F. Van Loan, *Matrix Computations*, 2nd ed., Baltimore, Johns Hopkins University Press, Baltimore, 1989.

9 A. Antoniou, *Digital Signal Processing: Signals, Systems, and Filters*, McGraw-Hill, New York, 2005.

10 W.-S. Lu, "A parameterization method for the design of IIR digital filters with prescribed stability margin," in *Proc. Int. Symp. Circuits Syst.*, pp. 2381–2384, June 1997.

11 E. K. P. Chong and S. H. Żak, *An Introduction to Optimization*, Wiley, New York, 1996.

12 H. W. Kuhn and A. W. Tucker, "Nonlinear programming," in *Proc. 2nd Berkeley Symp.*, pp. 481–492, Berkeley, CA, 1951.

13 D. G. Luenberger, *Linear and Nonlinear Programming*, 2nd ed., Addison-Wesley, Reading, MA, 1984.

14 P. Wolfe, "A duality theorem for nonlinear programming," *Quar. Appl. Math*, vol. 19, pp. 239–244, 1961.

Problems

10.1 A trigonometric polynomial is given by

$$A(\omega) = \sum_{k=0}^{n} a_k \cos k\omega \qquad (P10.1)$$

and Ω_p, Ω_a are sets given by

$$\Omega_p = \{\omega_{p0}, \omega_{p1}, \ldots, \omega_{pN}\} \subseteq [0, \omega_p]$$
$$\Omega_a = \{\omega_{a0}, \omega_{a1}, \ldots, \omega_{aM}\} \subseteq [\omega_a, \pi]$$

with $\omega_p \leq \omega_a$. Coefficients a_k for $k = 0, 1, \ldots, n$ are required in (P10.1) such that the upper bound δ in

$$|A(\omega) - 1| \leq \delta \qquad \text{for } \omega \in \Omega_p \qquad (P10.2)$$

and

$$|A(\omega)| \leq \delta \qquad \text{for } \omega \in \Omega_a \qquad (P10.3)$$

is minimized. Formulate the above problem as a constrained minimization problem.

10.2 Consider the trigonometric polynomial $A(\omega)$ given in Prob. P10.1. Suppose we need to find a_k, for $k = 0, 1, \ldots, n$ such that

$$J = \int_0^{\omega_p} [A(\omega) - 1]^2 d\omega + \int_{\omega_a}^{\pi} W(\omega) A^2(\omega) d\omega \qquad (P10.4)$$

is minimized subject to constraints in Eqs. (P10.2) and (P10.3), where $W(\omega) \geq 0$ is a weighting function, and δ is treated as a known positive scalar. Formulate the above problem as a constrained optimization.

10.3 (*a*) Write a MATLAB function to examine whether the equality constraints in $\mathbf{Ax} = \mathbf{b}$ are (i) inconsistent, or (ii) consistent but redundant, or (iii) consistent without redundancy.

314

(b) Modify the MATLAB function obtained from part (a) so that if $\mathbf{Ax} = \mathbf{b}$ is found to be consistent but redundant, the constraints are reduced to $\hat{\mathbf{A}}\mathbf{x} = \hat{\mathbf{b}}$ such that (i) $\hat{\mathbf{A}}\mathbf{x} = \hat{\mathbf{b}}$ describes the same feasible region and (ii) the constraints in $\hat{\mathbf{A}}\mathbf{x} = \hat{\mathbf{b}}$ are not redundant.

10.4 In Sec. 10.3.1, it was shown that the LP problem in Eq. (10.20) can be converted into the standard-form LP problem of Eq. (10.19). Show that the standard-form LP problem in Eq. (10.19) can be converted into the problem in Eq. (10.20). Hint: Use Eq. (10.27).

10.5 (a) Apply the result of Prob. 10.4 to convert the LP problem

$$\text{minimize } f(\mathbf{x}) = x_1 + 2x_2 + 11x_3 + 2x_4$$

$$\text{subject to: } \quad a_1(\mathbf{x}) = x_1 + x_2 + x_3 + 2x_4 = 3$$
$$a_2(\mathbf{x}) = x_2 + 2x_3 + 4x_4 = 3$$
$$a_3(\mathbf{x}) = 2x_3 + x_4 = 2$$
$$c_i(\mathbf{x}) = x_i \geq 0 \qquad \text{for } i = 1,\ 2,\ 3,\ 4$$

into the problem in Eq. (10.20).

(b) Solve the LP problem obtained in part (a).

(c) Use the result of part (b) to solve the standard-form LP problem in part (a).

10.6 (a) Prove that if $\mathbf{P}$ is positive definite, then $\ln(\det \mathbf{P}^{-1})$ is a convex function of $\mathbf{P}$.

(b) Prove that if $\mathbf{p} = \mathbf{P}(:)$ denotes the vector obtained by lexicographically ordering matrix $\mathbf{P}$, then the set of vectors satisfying the constraints in Eqs. (10.24b) and (10.24c) is convex.

10.7 Prove that all solutions of $\mathbf{Ax} = \mathbf{b}$ are characterized by Eq. (10.26). To simplify the proof, assume that $\mathbf{A} \in R^{p \times n}$ has full row rank. In this case the pseudo-inverse of $\mathbf{A}^+$ is given by

$$\mathbf{A}^+ = \mathbf{A}^T(\mathbf{AA}^T)^{-1}$$

10.8 The feasible region shown in Fig. P10.8 can be described by

$$\mathcal{R}: \begin{cases} c < x_1 < 400 \\ 1 < x_2 < 61 \\ x_2 < x_1/c \end{cases}$$

where $c > 0$ is a constant. Find variable transformations $x_1 = T_1(t_1, t_2)$ and $x_2 = T_2(t_1, t_2)$ such that $-\infty < t_1, t_2 < \infty$ describe the same feasible region.

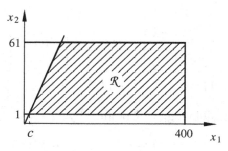

Figure P10.8.

10.9 Show that $\nabla f(\mathbf{x})$, $\nabla a_1(\mathbf{x})$, and $\nabla a_2(\mathbf{x})$ that satisfy Eq. (10.54) are linearly dependent.

Hint: Apply the singular-value decomposition to

$$
\begin{bmatrix}
1 & 0 \\
0 & 1 \\
\frac{\partial h_1}{\partial x_1} & \frac{\partial h_1}{\partial x_2} \\
\frac{\partial h_2}{\partial x_1} & \frac{\partial h_2}{\partial x_2}
\end{bmatrix}
$$

10.10 (*a*) Provide an example to demonstrate that $\mathbf{Ax} \geq \mathbf{b}$ does not imply $\mathbf{MAx} \geq \mathbf{Mb}$ in general, even if $\mathbf{M}$ is positive definite.

(*b*) Which condition on $\mathbf{MAx} \geq \mathbf{b}$ implies $\mathbf{MAx} \geq \mathbf{Mb}$?

10.11 Use two methods, namely, Eq. (10.27) and the Lagrange multiplier method, to solve the problem

$$\text{minimize } f(\mathbf{x}) = \tfrac{1}{2}\mathbf{x}^T\mathbf{Hx} + \mathbf{x}^T\mathbf{p}$$

$$\text{subject to: } \mathbf{Ax} = \mathbf{b}$$

where

$$
\mathbf{H} =
\begin{bmatrix}
\mathbf{H}_1 & \mathbf{H}_2 & \mathbf{H}_3 & \mathbf{H}_4 \\
\mathbf{H}_2 & \mathbf{H}_1 & \mathbf{H}_2 & \mathbf{H}_3 \\
\mathbf{H}_3 & \mathbf{H}_2 & \mathbf{H}_1 & \mathbf{H}_2 \\
\mathbf{H}_4 & \mathbf{H}_3 & \mathbf{H}_2 & \mathbf{H}_1
\end{bmatrix}
$$

with

$$
\mathbf{H}_1 =
\begin{bmatrix}
10 & 8 & 7 & 6 \\
8 & 10 & 8 & 7 \\
7 & 8 & 10 & 8 \\
6 & 7 & 8 & 10
\end{bmatrix}
$$

$$
\mathbf{H}_2 =
\begin{bmatrix}
3 & 2 & 1 & 0 \\
2 & 3 & 2 & 1 \\
1 & 2 & 3 & 2 \\
0 & 1 & 2 & 3
\end{bmatrix},\quad
\mathbf{H}_3 =
\begin{bmatrix}
2 & 1 & 0 & 0 \\
1 & 2 & 1 & 0 \\
0 & 1 & 2 & 1 \\
0 & 0 & 1 & 2
\end{bmatrix},\quad
\mathbf{H}_4 = \mathbf{I}_4
$$

$$\mathbf{p} = \begin{bmatrix} \mathbf{p}_1 \\ \mathbf{p}_2 \\ \mathbf{p}_3 \\ \mathbf{p}_4 \end{bmatrix}, \; \mathbf{p}_1 = \begin{bmatrix} 1 \\ -1 \\ 2 \\ -2 \end{bmatrix}, \; \mathbf{p}_2 = \begin{bmatrix} 0 \\ 0 \\ 0 \\ 0 \end{bmatrix}, \; \mathbf{p}_3 = \begin{bmatrix} 2 \\ 2 \\ -4 \\ 4 \end{bmatrix}, \; \mathbf{p}_4 = \begin{bmatrix} 0 \\ 0 \\ 0 \\ 0 \end{bmatrix}$$

$$\mathbf{A} = [\mathbf{H}_3 \; \mathbf{H}_2 \; \mathbf{H}_1 \; \mathbf{H}_4], \; \mathbf{b} = [1 \; 14 \; 3 \; -4]^T$$

10.12 Consider the feasible region $\mathcal{R}$ defined by

$$\mathcal{R}: \quad a_i(\mathbf{x}) = 0 \qquad \text{for } i = 1, 2, \ldots, p$$
$$c_j(\mathbf{x}) \geq 0 \qquad \text{for } j = 1, 2, \ldots, q$$

At a feasible point $\mathbf{x}$, let $\mathcal{J}(\mathbf{x})$ be the active index set for the inequality constraints at $\mathbf{x}$, and define the sets $\mathcal{F}(\mathbf{x})$ and $F(\mathbf{x})$ as

$$\mathcal{F}(\mathbf{x}) = \{\mathbf{s} : \mathbf{s} \text{ is feasible at } \mathbf{x}\}$$

and

$$F(\mathbf{x}) = \{\mathbf{s} : \quad \mathbf{s}^T \nabla a_i(\mathbf{x}) = 0 \qquad \text{for } i = 1, 2, \ldots, p$$
$$\text{and} \quad \mathbf{s}^T \nabla c_j(\mathbf{x}) \geq 0 \qquad \text{for } j \in \mathcal{J}(\mathbf{x})\}$$

respectively. Prove that $\mathcal{F}(\mathbf{x}) \subseteq F(\mathbf{x})$, i.e., set $F(\mathbf{x})$ contains set $\mathcal{F}(\mathbf{x})$.

10.13 Prove that if at a feasible $\mathbf{x}$ one of the following conditions is satisfied, then $\mathcal{F}(\mathbf{x}) = F(\mathbf{x})$:

(i) The constraints that are active at $\mathbf{x}$ are all linear.

(ii) Vectors $\nabla a_i(\mathbf{x})$ for $i = 1, 2, \ldots, p$ and $\nabla c_j(\mathbf{x})$ for those $c_j(\mathbf{x})$ that are active at $\mathbf{x}$ are linearly independent.

10.14 In the literature, the assumption that $\mathcal{F}(\mathbf{x}) = F(\mathbf{x})$ is known as the *constraint qualification* of $\mathbf{x}$. Verify that the constraint qualification assumption does not hold at $\mathbf{x} = \mathbf{0}$ when the constraints are given by

$$c_1(\mathbf{x}) = x_1^3 - x_2$$
$$c_2(\mathbf{x}) = x_2$$

Hint: Check the vector $\mathbf{s} = [-1 \; 0]^T$.

10.15 Consider the constrained minimization problem (see [12])

$$\text{minimize } f(\mathbf{x}) = (x_1 - 2)^2 + x_2^2$$

$$\text{subject to: } c_1(\mathbf{x}) = x_1 \geq 0$$
$$c_2(\mathbf{x}) = x_2 \geq 0$$
$$c_3(\mathbf{x}) = (1 - x_1)^3 - x_2 \geq 0$$

(a) Using a graphical solution, show that $\mathbf{x}^* = [1\ 0]^T$ is the global minimizer.

(b) Verify that $\mathbf{x}^*$ is not a regular point.

(c) Show that there exist no $\mu_2 \geq 0$ and $\mu_3 \geq 0$ such that

$$\nabla f(\mathbf{x}^*) = \mu_2 \nabla c_2(\mathbf{x}^*) + \mu_3 \nabla c_3(\mathbf{x}^*)$$

10.16 Given column vectors $\nu_1, \nu_2, \ldots, \nu_q$, define the polyhedral cone C as

$$C = \{\nu : \nu = \sum_{i=1}^{q} \mu_i \nu_i, \ \mu_i \geq 0\}$$

Prove that C is closed and convex.

10.17 Let $\mathbf{g}$ be a vector that does not belong to set C in Prob. 10.16. Prove that there exists a hyperplane $\mathbf{s}^T\mathbf{x} = 0$ that separates C and $\mathbf{g}$.

10.18 Given column vectors $\nu_1, \nu_2, \ldots, \nu_q$ and $\mathbf{g}$, show that the set

$$S = \{\mathbf{s} : \mathbf{s}^T\mathbf{g} < 0 \text{ and } \mathbf{s}^T\nu_i \geq 0, \text{ for } i = 1, 2, \ldots, q\}$$

is empty if and only if there exist $\mu_i \geq 0$ such that

$$\mathbf{g} = \sum_{i=1}^{q} \mu_i \nu_i$$

(This is known as Farkas' lemma.)

Hint: Use the results of Probs. 10.16 and 10.17.

10.19 Let $\mathcal{J}(\mathbf{x}^*) = \{j_1, j_2, \ldots, j_K\}$ be the active index set at $\mathbf{x}^*$ for the constraints in Eq. (10.1c). Show that the set

$$S = \{\mathbf{s} : \mathbf{s}^T\nabla f(\mathbf{x}^*) < 0, \ \mathbf{s}^T\nabla a_i(\mathbf{x}^*) = 0 \text{ for } i = 1, 2, \ldots, p,$$
$$\text{and } \mathbf{s}^T\nabla c_j(\mathbf{x}^*) \geq 0 \text{ for } j \in \mathcal{J}(\mathbf{x}^*)\}$$

is empty if and only if there exist multipliers λ_i^* for $1 \leq i \leq p$ and $\mu_j^* \geq 0$, such that

$$\nabla f(\mathbf{x}^*) = \sum_{i=1}^{p} \lambda_i^* \nabla a_i(\mathbf{x}^*) + \sum_{j \in \mathcal{J}(\mathbf{x}^*)} \mu_j^* \nabla c_j(\mathbf{x}^*)$$

(This is known as the Extension of Farkas' lemma.)

10.20 Using the KKT conditions, find solution candidates for the following CP problem

$$\text{minimize } x_1^2 + x_2^2 - 2x_1 - 4x_2 + 9$$

$$
\begin{aligned}
\text{subject to:} \qquad x_1 &\geq 1 \\
x_2 &\geq 0 \\
-\tfrac{1}{2}x_1 - x_2 + \tfrac{3}{2} &\geq 0
\end{aligned}
$$

10.21 Consider the constrained minimization problem

$$\text{minimize } f(\mathbf{x}) = -x_1^3 + x_2^3 - 2x_1 x_3^2$$

$$
\begin{aligned}
\text{subject to:} \quad 2x_1 + x_2^2 + x_3 - 5 &= 0 \\
5x_1^2 - x_2^2 - x_3 &\geq 2 \\
x_1 \geq 0, \ x_2 \geq 0, \ x_3 &\geq 0
\end{aligned}
$$

(a) Write the KKT conditions for the solution points of the problem.

(b) Vector $\mathbf{x}^* = [1 \ 0 \ 3]^T$ is known to be a local minimizer. At $\mathbf{x}^*$, find λ_1^* and μ_i^* for $1 \leq i \leq 4$, and verify that $\mu_i^* \geq 0$ for $1 \leq i \leq 4$.

(c) Examine the second-order conditions for set $(\mathbf{x}^*, \, \boldsymbol{\lambda}^*, \, \boldsymbol{\mu}^*)$.

10.22 Consider the QP problem

$$\text{minimize } f(\mathbf{x}) = \tfrac{1}{2}\mathbf{x}^T \mathbf{H}\mathbf{x} + \mathbf{x}^T \mathbf{p}$$

$$
\begin{aligned}
\text{subject to:} \quad \mathbf{Ax} &= \mathbf{b} \\
\mathbf{x} &\geq \mathbf{0}
\end{aligned}
$$

(a) Write the KKT conditions for the solution points of the problem.

(b) Derive the Wolfe dual of the problem.

(c) Let set $(\mathbf{x}, \, \boldsymbol{\lambda}, \, \boldsymbol{\mu})$ be feasible for the primal and dual problems, and denote their objective functions as $f(\mathbf{x})$ and $h(\mathbf{x}, \, \boldsymbol{\lambda}, \, \boldsymbol{\mu})$, respectively. Evaluate the duality gap defined by

$$\delta(\mathbf{x}, \, \boldsymbol{\lambda}, \, \boldsymbol{\mu}) = f(\mathbf{x}) - h(\mathbf{x}, \, \boldsymbol{\lambda}, \, \boldsymbol{\mu})$$

and show that $\delta(\mathbf{x}, \, \boldsymbol{\lambda}, \, \boldsymbol{\mu})$ is always nonnegative for a feasible $(\mathbf{x}, \, \boldsymbol{\lambda}, \, \boldsymbol{\mu})$.

10.23 Consider the minimization problem

$$\text{minimize } f(\mathbf{x}) = \mathbf{c}^T \mathbf{x}$$

$$
\begin{aligned}
\text{subject to:} \quad \mathbf{Ax} &= \mathbf{0} \\
\|\mathbf{x}\| &\leq 1
\end{aligned}
$$

where $\|\mathbf{x}\|$ denotes the Euclidean norm of $\mathbf{x}$.

(*a*) Show that this is a CP problem.

(*b*) Derive the KKT conditions for the solution points of the problem.

(*c*) Show that if $\mathbf{c} - \mathbf{A}^T\boldsymbol{\lambda} \neq \mathbf{0}$ where $\boldsymbol{\lambda}$ satisfies

$$\mathbf{A}\mathbf{A}^T\boldsymbol{\lambda} = \mathbf{A}\mathbf{c}$$

then the minimizer is given by

$$\mathbf{x}^* = -\frac{\mathbf{c} - \mathbf{A}^T\boldsymbol{\lambda}}{\|\mathbf{c} - \mathbf{A}^T\boldsymbol{\lambda}\|}$$

Otherwise, any feasible $\mathbf{x}$ is a solution.

10.24 Consider the minimization problem

$$\text{minimize } f(\mathbf{x}) = \mathbf{c}^T\mathbf{x}$$

$$\text{subject to: } \|\mathbf{A}\mathbf{x}\| \leq 1$$

(*a*) Show that this is a CP problem.

(*b*) Derive the KKT conditions for the solution points of the problem.

(*c*) Show that if the solution of the equation

$$\mathbf{A}^T\mathbf{A}\mathbf{y} = \mathbf{c}$$

is nonzero, then the minimizer is given by

$$\mathbf{x}^* = -\frac{\mathbf{y}}{\|\mathbf{A}\mathbf{y}\|}$$

Otherwise, any feasible $\mathbf{x}$ is a solution.

Chapter 11

LINEAR PROGRAMMING
PART I: THE SIMPLEX METHOD

11.1   Introduction

Linear programming (LP) problems occur in a diverse range of real-life applications in economic analysis and planning, operations research, computer science, medicine, and engineering. In such problems, it is known that any minima occur at the vertices of the feasible region and can be determined through a 'brute-force' or exhaustive approach by evaluating the objective function at all the vertices of the feasible region. However, the number of variables involved in a practical LP problem is often very large and an exhaustive approach would entail a considerable amount of computation. In 1947, Dantzig developed a method for the solution of LP problems known as the *simplex method* [1][2]. Although in the worst case, the simplex method is known to require an exponential number of iterations, for typical standard-form problems the number of iterations required is just a small multiple of the problem dimension [3]. For this reason, the simplex method has been the primary method for solving LP problems since its introduction.

In Sec. 11.2, the general theory of constrained optimization developed in Chap. 10 is applied to derive optimality conditions for LP problems. The geometrical features of LP problems are discussed and connected to the several issues that are essential in the development of the simplex method. In Sec. 11.3, the simplex method is presented for alternative-form LP problems as well as for standard-form LP problems from a linear-algebraic perspective.

11.2   General Properties
11.2.1   Formulation of LP problems

In Sec. 10.3.1, the standard-form LP problem was stated as

$$\text{minimize } f(\mathbf{x}) = \mathbf{c}^T\mathbf{x} \tag{11.1a}$$

$$\text{subject to: } \mathbf{A}\mathbf{x} = \mathbf{b} \tag{11.1b}$$

$$\mathbf{x} \geq \mathbf{0} \tag{11.1c}$$

where $\mathbf{c} \in R^{n \times 1}$ with $\mathbf{c} \neq \mathbf{0}$, $\mathbf{A} \in R^{p \times n}$, and $\mathbf{b} \in R^{p \times 1}$ are given. Throughout this chapter, we assume that $\mathbf{A}$ is of full row rank, i.e., $\text{rank}(\mathbf{A}) = p$. For the standard-form LP problem in Eq. (11.1) to be a meaningful LP problem, full row rank in $\mathbf{A}$ implies that $p < n$.

For a fixed scalar β, the equation $\mathbf{c}^T\mathbf{x} = \beta$ describes an affine manifold in the n-dimensional Euclidean space E^n (see Sec. A.15). For example, with $n = 2$, $\mathbf{c}^T\mathbf{x} = \beta$ represents a line and $\mathbf{c}^T\mathbf{x} = \beta$ for $\beta = \beta_1, \beta_2, \ldots$ represents a family of parallel lines. The normal of these lines is $\mathbf{c}$, and for this reason vector $\mathbf{c}$ is often referred to as the *normal vector* of the objective function.

Another LP problem, which is often encountered in practice, involves minimizing a linear function subject to inequality constraints, i.e.,

$$\text{minimize } f(\mathbf{x}) = \mathbf{c}^T\mathbf{x} \tag{11.2a}$$

$$\text{subject to: } \mathbf{A}\mathbf{x} \geq \mathbf{b} \tag{11.2b}$$

where $\mathbf{c} \in R^{n \times 1}$ with $\mathbf{c} \neq \mathbf{0}$, $\mathbf{A} \in R^{p \times n}$, and $\mathbf{b} \in R^{p \times 1}$ are given. This will be referred to as the *alternative-form* LP problem hereafter. If we let

$$\mathbf{A} = \begin{bmatrix} \mathbf{a}_1^T \\ \mathbf{a}_2^T \\ \vdots \\ \mathbf{a}_p^T \end{bmatrix}, \quad \mathbf{b} = \begin{bmatrix} b_1 \\ b_2 \\ \vdots \\ b_p \end{bmatrix}$$

then the p constraints in Eq. (11.2b) can be written as

$$\mathbf{a}_i^T\mathbf{x} \geq b_i \qquad \text{for } i = 1, 2, \ldots, p$$

where vector $\mathbf{a}_i$ is the normal of the ith inequality constraint, and $\mathbf{A}$ is usually referred to as the *constraint matrix*.

By introducing a p-dimensional slack vector variable $\mathbf{y}$, Eq. (11.2b) can be reformulated as

$$\mathbf{A}\mathbf{x} - \mathbf{y} = \mathbf{b} \qquad \text{for } \mathbf{y} \geq \mathbf{0}$$

Furthermore, vector variable $\mathbf{x}$ can be decomposed as

$$\mathbf{x} = \mathbf{x}^+ - \mathbf{x}^- \quad \text{with} \quad \mathbf{x}^+ \geq \mathbf{0} \quad \text{and} \quad \mathbf{x}^- \geq \mathbf{0}$$

Hence if we let

$$\hat{\mathbf{x}} = \begin{bmatrix} \mathbf{x}^+ \\ \mathbf{x}^- \\ \mathbf{y} \end{bmatrix}, \quad \hat{\mathbf{c}} = \begin{bmatrix} \mathbf{c} \\ -\mathbf{c} \\ \mathbf{0} \end{bmatrix}, \quad \hat{\mathbf{A}} = [\mathbf{A} \; -\mathbf{A} \; -\mathbf{I}_p]$$

then Eq. (11.2) can be expressed as a standard-form LP problem, i.e.,

$$\text{minimize } f(\mathbf{x}) = \hat{\mathbf{c}}^T \hat{\mathbf{x}} \qquad (11.3\text{a})$$
$$\text{subject to: } \quad \hat{\mathbf{A}} \hat{\mathbf{x}} = \mathbf{b} \qquad (11.3\text{b})$$
$$\hat{\mathbf{x}} \geq \mathbf{0} \qquad (11.3\text{c})$$

Likewise, the most general LP problem with both equality and inequality constraints, i.e.,

$$\text{minimize } f(\mathbf{x}) = \mathbf{c}^T \mathbf{x} \qquad (11.4\text{a})$$
$$\text{subject to: } \quad \mathbf{A}\mathbf{x} = \mathbf{b} \qquad (11.4\text{b})$$
$$\mathbf{C}\mathbf{x} \geq \mathbf{d} \qquad (11.4\text{c})$$

can be expressed as a standard-form LP problem with respect to an augmented variable $\hat{\mathbf{x}}$. It is primarily for these reasons that the standard-form LP problem in Eq. (11.1) has been employed most often as the prototype for the description and implementation of various LP algorithms. Nonstandard LP problems, particularly the problem in Eq. (11.2), may be encountered directly in a variety of applications. Although the problem in Eq. (11.2) can be reformulated as a standard-form LP problem, the increase in problem size leads to reduced computational efficiency which can sometimes be a serious problem particularly when the number of inequality constraints is large. In what follows, the underlying principles pertaining to the LP problems in Eqs. (11.1) and (11.2) will be described separately to enable us to solve each of these problems directly without the need of converting the one form into the other.

11.2.2 Optimality conditions

Since linear functions are convex (or concave), an LP problem can be viewed as a convex programming problem. By applying Theorems 10.8 and 10.2 to the problem in Eq. (11.1), the following theorem can be deduced.

Theorem 11.1 *Karush-Kuhn-Tucker conditions for standard-form LP problem* If $\mathbf{x}^*$ *is regular for the constraints that are active at* $\mathbf{x}^*$, *then it is a global solution of the LP problem in Eq. (11.1) if and only if*
(a) $\mathbf{A}\mathbf{x}^* = \mathbf{b}$, $\qquad (11.5\text{a})$
(b) $\mathbf{x}^* \geq \mathbf{0}$, $\qquad (11.5\text{b})$
(c) there exist Lagrange multipliers $\boldsymbol{\lambda}^* \in R^{p \times 1}$ *and* $\boldsymbol{\mu}^* \in R^{n \times 1}$ *such that* $\boldsymbol{\mu}^* \geq \mathbf{0}$ *and*

$$\mathbf{c} = \mathbf{A}^T \boldsymbol{\lambda}^* + \boldsymbol{\mu}^* \qquad (11.5\text{c})$$

(d) $\mu_i^* x_i^* = 0$ *for* $1 \le i \le n$. (11.5d)

∎

The first two conditions in Eq. (11.5) simply say that solution $\mathbf{x}^*$ must be a feasible point. In Eq. (11.5c), constraint matrix $\mathbf{A}$ and vector $\mathbf{c}$ are related through the Lagrange multipliers $\boldsymbol{\lambda}^*$ and $\boldsymbol{\mu}^*$.

An immediate observation on the basis of Eqs. (11.5a)–(11.5d) is that in most cases solution $\mathbf{x}^*$ cannot be strictly feasible. Here we take the term 'strictly feasible points' to mean those points that satisfy the equality constraints in Eq. (11.5a) with $x_i^* > 0$ for $1 \le i \le n$. From Eq. (11.5d), $\boldsymbol{\mu}^*$ must be a zero vector for a strictly feasible point $\mathbf{x}^*$ to be a solution. Hence Eq. (11.5c) becomes

$$\mathbf{c} = \mathbf{A}^T \boldsymbol{\lambda}^* \tag{11.6}$$

In other words, for a strictly feasible point to be a minimizer of the standard-form LP problem in Eq. (11.1), the n-dimensional vector $\mathbf{c}$ must lie in the p-dimensional subspace spanned by the p columns of $\mathbf{A}^T$. Since $p < n$, the probability that Eq. (11.6) is satisfied is very small. Therefore, any solutions of the problem are very likely to be located on the *boundary* of the feasible region.

Example 11.1 Solve the LP problem

$$\text{minimize } f(\mathbf{x}) = x_1 + 4x_2 \tag{11.7a}$$

$$\text{subject to: } \quad x_1 + x_2 = 1 \tag{11.7b}$$

$$\mathbf{x} \ge \mathbf{0} \tag{11.7c}$$

Solution As shown in Fig. 11.1, the feasible region of the above problem is the segment of the line $x_1 + x_2 = 1$ in the first quadrant, the dashed lines are contours of the form $f(\mathbf{x}) = \text{constant}$, and the arrow points to the steepest descent direction of $f(\mathbf{x})$. We have

$$\mathbf{c} = \begin{bmatrix} 1 \\ 4 \end{bmatrix} \quad \text{and} \quad \mathbf{A}^T = \begin{bmatrix} 1 \\ 1 \end{bmatrix}$$

Since $\mathbf{c}$ and $\mathbf{A}^T$ are linearly independent, Eq. (11.6) cannot be satisfied and, therefore, no interior feasible point can be a solution. This leaves two end points to verify. From Fig. 11.1 it is evident that the unique minimizer is $\mathbf{x}^* = \begin{bmatrix} 1 & 0 \end{bmatrix}^T$.

At $\mathbf{x}^*$ the constraint in Eq. (11.7b) and the second constraint in Eq. (11.7c) are active, and since the Jacobian of these constraints, namely,

$$\begin{bmatrix} 1 & 1 \\ 0 & 1 \end{bmatrix}$$

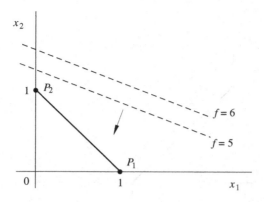

Figure 11.1. LP problem in Example 11.1.

is nonsingular, $\mathbf{x}^*$ is a regular point. Now Eq. (11.5d) gives $\mu_1^* = 0$, which leads to Eq. (11.5c) with

$$\lambda^* = 1 \quad \text{and} \quad \mu_2^* = 3$$

This confirms that $\mathbf{x}^* = [1\ 0]^T$ is indeed a global solution.

Note that if the objective function is changed to

$$f(\mathbf{x}) = \mathbf{c}^T \mathbf{x} = 4x_1 + 4x_2$$

then Eq. (11.6) is satisfied with $\lambda^* = 4$ and any feasible point becomes a global solution. In fact, the objective function remains constant in the feasible region, i.e.,

$$f(\mathbf{x}) = 4(x_1 + x_2) = 4 \qquad \text{for } \mathbf{x} \in \mathcal{R}$$

A graphical interpretation of this situation is shown in Fig. 11.2 ∎

Note that the conditions in Theorems 10.2 and 10.8 are also applicable to the alternative-form LP problem in Eq. (11.2) since the problem is, in effect, a convex programming (CP) problem. These conditions can be summarized in terms of the following theorem.

Theorem 11.2 *Necessary and sufficient conditions for a minimum in alternative-form LP problem If* $\mathbf{x}^*$ *is regular for the constraints in Eq. (11.2b) that are active at* $\mathbf{x}^*$, *then it is a global solution of the problem in Eq. (11.2) if and only if*

(a) $\mathbf{A}\mathbf{x}^* \geq \mathbf{b}$,   (11.8a)
(b) there exists a $\boldsymbol{\mu}^* \in R^{p \times 1}$ *such that* $\boldsymbol{\mu}^* \geq \mathbf{0}$ *and*

$$\mathbf{c} = \mathbf{A}^T \boldsymbol{\mu}^* \qquad\qquad (11.8b)$$

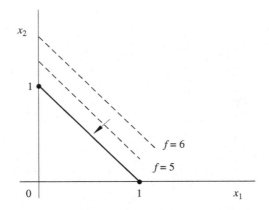

Figure 11.2. LP problem in Example 11.1 with $f(\mathbf{x}) = 4x_1 + 4x_2$.

(c) $\mu_i^*(\mathbf{a}_i^T \mathbf{x}^* - b_i) = 0$ *for* $1 \leq i \leq p$ (11.8c)
where $\mathbf{a}_i^T$ *is the ith row of* $\mathbf{A}$.

               ∎

The observation made with regard to Theorem 11.1, namely, that the solutions of the problem are very likely to be located on the boundary of the feasible region, also applies to Theorem 11.2. As a matter of fact, if $\mathbf{x}^*$ is a strictly feasible point satisfying Eq. (11.8c), then $\mathbf{A}\mathbf{x}^* > \mathbf{b}$ and the complementarity condition in Eq. (11.8c) implies that $\boldsymbol{\mu}^* = \mathbf{0}$. Hence Eq. (11.8b) cannot be satisfied unless $\mathbf{c} = \mathbf{0}$, which would lead to a meaningless LP problem. In other words, any solutions of Eq. (11.8) can only occur on the boundary of the feasible region defined by Eq. (11.2b).

Example 11.2 Solve the LP problem

$$\text{minimize } f(\mathbf{x}) = -x_1 - 4x_2$$

$$\text{subject to:} \quad x_1 \geq 0$$

$$-x_1 \geq -2$$

$$x_2 \geq 0$$

$$-x_1 - x_2 + 3.5 \geq 0$$

$$-x_1 - 2x_2 + 6 \geq 0$$

Solution The five constraints can be expressed as $\mathbf{Ax} \geq \mathbf{b}$ with

$$\mathbf{A} = \begin{bmatrix} 1 & 0 \\ -1 & 0 \\ 0 & 1 \\ -1 & -1 \\ -1 & -2 \end{bmatrix}, \quad \mathbf{b} = \begin{bmatrix} 0 \\ -2 \\ 0 \\ -3.5 \\ -6 \end{bmatrix}$$

The feasible region is the polygon shown in Fig. 11.3.

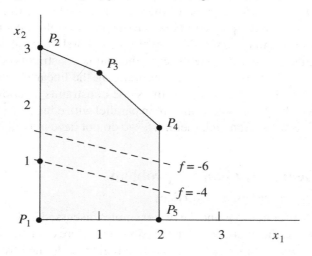

Figure 11.3. Feasible region in Example 11.2.

Since the solution cannot be inside the polygon, we consider the five edges of the polygon. We note that at any point $\mathbf{x}$ on an edge other than the five vertices P_i for $1 \leq i \leq 5$ only *one* constraint is active. This means that only one of the five μ_i's is nonzero. At such an $\mathbf{x}$, Eq. (11.8b) becomes

$$\mathbf{c} = \begin{bmatrix} -1 \\ -4 \end{bmatrix} = \mu_i \mathbf{a}_i \qquad (11.9)$$

where $\mathbf{a}_i$ is the transpose of the ith row in $\mathbf{A}$. Since each $\mathbf{a}_i$ is linearly independent of $\mathbf{c}$, no μ_i exists that satisfies Eq. (11.9). This then leaves the five vertices for verification. At point $P_1 = [0\ 0]^T$, both the first and third constraints are active and Eq. (11.8b) becomes

$$\begin{bmatrix} -1 \\ -4 \end{bmatrix} = \begin{bmatrix} 1 & 0 \\ 0 & 1 \end{bmatrix} \begin{bmatrix} \mu_1 \\ \mu_3 \end{bmatrix}$$

which gives $\mu_1 = -1$ and $\mu_3 = -4$. Since condition (*b*) of Theorem 11.2 is violated, P_1 is not a solution. At point $P_2 = [0\ 3]^T$, both the first and fifth constraints are active, and Eq. (11.8b) becomes

$$\begin{bmatrix} -1 \\ -4 \end{bmatrix} = \begin{bmatrix} 1 & -1 \\ 0 & -2 \end{bmatrix} \begin{bmatrix} \mu_1 \\ \mu_5 \end{bmatrix}$$

which gives $\mu_1 = 1$ and $\mu_5 = 2$. Since the rest of the μ_i's are all zero, conditions (a)–(c) of Theorem 11.2 are satisfied with $\boldsymbol{\mu} \equiv \boldsymbol{\mu}^* = [1\ 0\ 0\ 0\ 2]^T$ and $P_2 = [0\ 3]^T$ is a minimizer, i.e., $\mathbf{x} \equiv \mathbf{x}^* = P_2$. One can go on to check the rest of the vertices to confirm that point P_2 is the unique solution to the problem. However, the uniqueness of the solution is obvious from Fig. 11.3.

We conclude the example with two remarks on the solution's uniqueness. Later on, we will see that the solution can also be verified by using the positivity of those μ_i's that are associated with active inequality constraints (see Theorem 11.7 in Sec. 11.2.4.2). If we consider minimizing the linear function $f(\mathbf{x}) = \mathbf{c}^T\mathbf{x}$ with $\mathbf{c} = [-1\ -2]^T$ subject to the same constraints as above, then the contours defined by $f(\mathbf{x}) = $ constant are in parallel with edge $\overline{P_2 P_3}$. Hence any point on $\overline{P_2 P_3}$ is a solution and, therefore, we do not have a unique solution. ∎

11.2.3 Geometry of an LP problem

11.2.3.1 Facets, edges, and vertices

The optimality conditions and the two examples discussed in Sec. 11.2.2 indicate that points on the boundary of the feasible region are of critical importance in LP. For the two-variable case, the feasible region $\mathcal{R}$ defined by Eq. (11.2b) is a polygon, and the facets and edges of $\mathcal{R}$ are the same. For problems with $n > 2$, they represent different geometrical structures which are increasingly difficult to visualize and formal definitions for these structures are, therefore, necessary.

In general, the feasible region defined by $\mathcal{R} = \{\mathbf{x}: \mathbf{A}\mathbf{x} \geq \mathbf{b}\}$ is a convex polyhedron. A set of points, $\mathcal{F}$, in the n-dimensional space E^n is said to be a *face* of a convex polyhedron $\mathcal{R}$ if the condition $\mathbf{p}_1, \mathbf{p}_2 \in \mathcal{F}$ implies that $(\mathbf{p}_1 + \mathbf{p}_2)/2 \in \mathcal{F}$. The dimension of a face is defined as the dimension of $\mathcal{F}$. Depending on its dimension, a face can be a facet, an edge, or a vertex. If l is the dimension of a face $\mathcal{F}$, then a *facet* of $\mathcal{F}$ is an $(l-1)$-dimensional face, an *edge* of $\mathcal{F}$ is a one-dimensional face, and a *vertex* of $\mathcal{F}$ is a zero-dimensional face [4]. As an example, Fig. 11.4 shows the convex polyhedron defined by the constraints

$$x_1 + x_2 + x_3 \leq 1$$

$$x_1 \geq 0, \quad x_2 \geq 0, \quad x_3 \geq 0$$

i.e.,

$$\mathbf{A}\mathbf{x} \geq \mathbf{b} \tag{11.10}$$

with

$$A = \begin{bmatrix} -1 & -1 & -1 \\ 1 & 0 & 0 \\ 0 & 1 & 0 \\ 0 & 0 & 1 \end{bmatrix}, \quad b = \begin{bmatrix} -1 \\ 0 \\ 0 \\ 0 \end{bmatrix}$$

The polyhedron is a three-dimensional face which has four facets, six edges, and four vertices.

In the case where $n = 2$, a feasible region defined by $Ax \geq b$ becomes a polygon and facets become edges. As can be seen in Fig. 11.3, the vertices of a polygon are the points where *two* inequality constraints become active. In the case where $n = 3$, Fig. 11.4 suggests that vertices are the points where *three* inequality constraints become active. In general, we define a vertex point as follows [3].

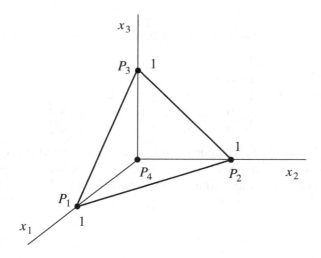

Figure 11.4. Polyhedron defined by Eq. (11.10) and its facets, edges, and vertices.

Definition 11.1 A vertex is a feasible point P at which there exist at least n active constraints which contain n linearly independent constraints where n is the dimension of x. Vertex P is said to be *nondegenerate* if exactly n constraints are active at P or *degenerate* if more than n constraints are active at P. ∎

Definition 11.1 covers the general case where both equality and inequality constraints are present. Linearly independent active constraints are the constraints that are active at P and the matrix whose rows are the vectors associated with the active constraints is of full row rank. At point P_1 in Fig. 11.1, for example, the equality constraint in Eq. (11.7b) and one of the inequality constraints, i.e., $x_2 \geq 0$, are active. This in conjunction with the nonsingularity of the associated matrix

$$\begin{bmatrix} 1 & 1 \\ 0 & 1 \end{bmatrix}$$

implies that P_1 is a nondegenerate vertex. It can be readily verified that point P_2 in Fig. 11.1, points P_i for $i = 1, 2, \ldots, 5$ in Fig. 11.3, and points P_i for $i = 1, 2, \ldots, 4$ in Fig. 11.4 are also nondegenerate vertices.

As another example, the feasible region characterized by the constraints

$$x_1 + x_2 + x_3 \leq 1$$
$$0.5x_1 + 2x_2 + x_3 \leq 1$$
$$x_1 \geq 0, \quad x_2 \geq 0, \quad x_3 \geq 0$$

i.e.,

$$\mathbf{Ax} \geq \mathbf{b} \tag{11.11}$$

with

$$\mathbf{A} = \begin{bmatrix} -1 & -1 & -1 \\ -0.5 & -2 & -1 \\ 1 & 0 & 0 \\ 0 & 1 & 0 \\ 0 & 0 & 1 \end{bmatrix}, \quad \mathbf{b} = \begin{bmatrix} -1 \\ -1 \\ 0 \\ 0 \\ 0 \end{bmatrix}$$

is illustrated in Fig. 11.5. The convex polyhedron has five facets, eight edges, and five vertices. At vertex P_5 four constraints are active but since $n = 3$, P_5 is degenerate. The other four vertices, namely, P_1, P_2, P_3 and P_4, are nondegenerate.

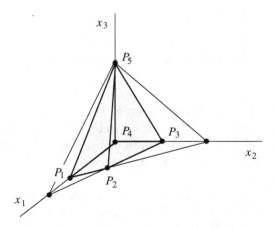

Figure 11.5. A feasible region with a degenerate vertex.

11.2.3.2   Feasible descent directions

A vector $\mathbf{d} \in R^{n\times 1}$ is said to be a *feasible descent direction* at a feasible point $\mathbf{x} \in R^{n\times 1}$ if $\mathbf{d}$ is a feasible direction as defined by Def. 2.4 and the linear objective function strictly decreases along $\mathbf{d}$, i.e., $f(\mathbf{x}+\alpha\mathbf{d}) < f(\mathbf{x})$ for $\alpha > 0$, where $f(\mathbf{x}) = \mathbf{c}^T\mathbf{x}$. Evidently, this implies that

$$\frac{1}{\alpha}[f(\mathbf{x} + \alpha\mathbf{d}) - f(\mathbf{x})] = \mathbf{c}^T\mathbf{d} < 0 \tag{11.12}$$

For the problem in Eq. (11.2), we denote as $\mathbf{A}_a$ the matrix whose rows are the rows of $\mathbf{A}$ that are associated with the constraints which are active at $\mathbf{x}$. We call $\mathbf{A}_a$ the *active constraint matrix* at $\mathbf{x}$. If $\mathcal{J} = \{j_1, j_2, \ldots, j_K\}$ is the set of indices that identify active constraints at $\mathbf{x}$, then

$$\mathbf{A}_a = \begin{bmatrix} \mathbf{a}_{j1}^T \\ \mathbf{a}_{j2}^T \\ \vdots \\ \mathbf{a}_{jK}^T \end{bmatrix} \tag{11.13}$$

satisfies the system of equations

$$\mathbf{a}_j^T\mathbf{x} = b_j \quad \text{for } j \in \mathcal{J}$$

For $\mathbf{d}$ to be a feasible direction, we must have

$$\mathbf{A}_a(\mathbf{x} + \alpha\mathbf{d}) \geq \mathbf{b}_a$$

where $\mathbf{b}_a = [b_{j1}\ b_{j2}\ \cdots\ b_{jK}]^T$. It follows that

$$\mathbf{A}_a\mathbf{d} \geq \mathbf{0}$$

which in conjunction with Eq. (11.12) characterizes a feasible descent direction $\mathbf{d}$ such that

$$\mathbf{A}_a\mathbf{d} \geq \mathbf{0} \quad \text{and} \quad \mathbf{c}^T\mathbf{d} < 0 \tag{11.14}$$

Since $\mathbf{x}^*$ is a solution of the problem in Eq. (11.2) if and only if no feasible descent directions exist at $\mathbf{x}^*$, we can state the following theorem.

Theorem 11.3 *Necessary and sufficient conditions for a minimum in alternative-form LP problem* Point $\mathbf{x}^*$ *is a solution of the problem in Eq. (11.2) if and only if it is feasible and*

$$\mathbf{c}^T\mathbf{d} \geq 0 \quad \text{for all } \mathbf{d} \text{ with } \mathbf{A}_{a*}\mathbf{d} \geq \mathbf{0} \tag{11.15}$$

where $\mathbf{A}_{a*}$ is the active constraint matrix at $\mathbf{x}^*$.

For the standard-form LP problem in Eq. (11.1), a feasible descent direction $\mathbf{d}$ at a feasible point $\mathbf{x}^*$ satisfies the constraints

$$\mathbf{Ad} = \mathbf{0}$$
$$d_j \geq 0 \qquad \text{for } j \in \mathcal{J}_*$$

and

$$\mathbf{c}^T \mathbf{d} \leq 0$$

where $\mathcal{J}_* = \{j_1, j_2, \ldots, j_K\}$ is the set of indices for the constraints in Eq. (11.1c) that are active at $\mathbf{x}^*$. This leads to the following theorem.

Theorem 11.4 *Necessary and sufficient conditions for a minimum in standard-form LP problem Point $\mathbf{x}^*$ is a solution of the LP problem in Eq. (11.1) if and only if it is a feasible point and*

$$\mathbf{c}^T \mathbf{d} \geq 0 \text{ for all } \mathbf{d} \text{ with } \mathbf{d} \in \mathcal{N}(\mathbf{A}) \text{ and } d_j \geq 0 \text{ for } j \in \mathcal{J}_* \qquad (11.16)$$

where $\mathcal{N}(\mathbf{A})$ denotes the null space of $\mathbf{A}$.

∎

11.2.3.3 Finding a vertex

Examples 11.1 and 11.2 discussed in Sec. 11.2.2 indicate that any solutions of the LP problems in Eqs. (11.1) and (11.2) can occur at vertex points. In Sec. 11.2.3.4, it will be shown that under some reasonable conditions, a vertex minimizer always exists. In what follows, we describe an iterative strategy that can be used to find a minimizer vertex for the LP problem in Eq. (11.2) starting with a feasible point $\mathbf{x}_0$.

In the kth iteration, if the active constraint matrix at $\mathbf{x}_k$, $\mathbf{A}_{a_k}$, has rank n, then $\mathbf{x}_k$ itself is already a vertex. So let us assume that $\text{rank}(\mathbf{A}_{a_k}) < n$. From a linear algebra perspective, the basic idea here is to generate a feasible point $\mathbf{x}_{k+1}$ such that the active constraint matrix at $\mathbf{x}_{k+1}$, $\mathbf{A}_{a_{k+1}}$, is an *augmented* version of $\mathbf{A}_{a_k}$ with $\text{rank}(\mathbf{A}_{a_{k+1}})$ increased by one. In other words, $\mathbf{x}_{k+1}$ is a point such that (a) it is feasible, (b) all the constraints that are active at $\mathbf{x}_k$ remain active at $\mathbf{x}_{k+1}$, and (c) there is a new active constraint at $\mathbf{x}_{k+1}$, which was inactive at $\mathbf{x}_k$. In this way, a vertex can be identified in a finite number of steps.

Let

$$\mathbf{x}_{k+1} = \mathbf{x}_k + \alpha_k \mathbf{d}_k \qquad (11.17)$$

To assure that all active constraints at $\mathbf{x}_k$ remain active at $\mathbf{x}_{k+1}$, we must have

$$\mathbf{A}_{a_k} \mathbf{x}_{k+1} = \mathbf{b}_{a_k}$$

where $\mathbf{b}_{a_k}$ is composed of the entries of $\mathbf{b}$ that are associated with the constraints which are active at $\mathbf{x}_k$. Since $\mathbf{A}_{a_k} \mathbf{x}_k = \mathbf{b}_{a_k}$, it follows that

$$\mathbf{A}_{a_k} \mathbf{d}_k = \mathbf{0} \qquad (11.18)$$

Since $\text{rank}(\mathbf{A}_{a_k}) < n$, the solutions of Eq. (11.18) form the null space of $\mathbf{A}_{a_k}$ of dimension $n - \text{rank}(\mathbf{A}_{a_k})$. Now for a fixed $\mathbf{x}_k$ and $\mathbf{d}_k \in \mathcal{N}(\mathbf{A}_{a_k})$, we call an inactive constraint $\mathbf{a}_i^T \mathbf{x}_k - b_i > 0$ *decreasing* with respect to $\mathbf{d}_k$ if $\mathbf{a}_i^T \mathbf{d}_k < 0$. If the ith constraint is a decreasing constraint with respect to $\mathbf{d}_k$, then moving from $\mathbf{x}_k$ to $\mathbf{x}_{k+1}$ along $\mathbf{d}_k$, the constraint becomes

$$
\begin{aligned}
\mathbf{a}_i^T \mathbf{x}_{k+1} - b_i &= \mathbf{a}_i^T (\mathbf{x}_k + \alpha_k \mathbf{d}_k) - b_i \\
&= (\mathbf{a}_i^T \mathbf{x}_k - b_i) + \alpha_k \mathbf{a}_i^T \mathbf{d}_k
\end{aligned}
$$

with $\mathbf{a}_i^T \mathbf{x}_k - b_i > 0$ and $\mathbf{a}_i^T \mathbf{d}_k < 0$. A positive α_k that makes the ith constraint active at point $\mathbf{x}_{k+1}$ can be identified as

$$
\alpha_k = \frac{\mathbf{a}_i^T \mathbf{x}_k - b_i}{-\mathbf{a}_i^T \mathbf{d}_k} \tag{11.19}
$$

It should be stressed, however, that moving the point along $\mathbf{d}_k$ also affects other inactive constraints and care must be taken to ensure that the value of α_k used does not lead to an infeasible $\mathbf{x}_{k+1}$. From the above discussion, we note two problems that need to be addressed, namely, how to find a direction $\mathbf{d}_k$ in the null space $\mathcal{N}(\mathbf{A}_{a_k})$ such that there is at least one decreasing constraint with respect to $\mathbf{d}_k$ and, if such a $\mathbf{d}_k$ is found, how to determine the step size α_k in Eq. (11.17).

Given $\mathbf{x}_k$ and $\mathbf{A}_{a_k}$, we can find an inactive constraint whose normal $\mathbf{a}_i^T$ is linearly independent of the rows of $\mathbf{A}_{a_k}$. It follows that the system of equations

$$
\begin{bmatrix} \mathbf{A}_{a_k} \\ \mathbf{a}_i^T \end{bmatrix} \mathbf{d}_k = \begin{bmatrix} \mathbf{0} \\ -1 \end{bmatrix} \tag{11.20}
$$

has a solution $\mathbf{d}_k$ with $\mathbf{d}_k \in \mathcal{N}(\mathbf{A}_{a_k})$ and $\mathbf{a}_i^T \mathbf{d}_k < 0$. Having determined $\mathbf{d}_k$, the set of indices corresponding to decreasing constraints with respect to $\mathbf{d}_k$ can be defined as

$$
\mathcal{I}_k = \{i : \mathbf{a}_i^T \mathbf{x}_k - b_i > 0, \ \mathbf{a}_i^T \mathbf{d}_k < 0\}
$$

The value of α_k can be determined as the value for which $\mathbf{x}_k + \alpha_k \mathbf{d}_k$ intersects the nearest new constraint. Hence, α_k can be calculated as

$$
\alpha_k = \min_{i \in \mathcal{I}_k} \left(\frac{\mathbf{a}_i^T \mathbf{x}_k - b_i}{-\mathbf{a}_i^T \mathbf{d}_k} \right) \tag{11.21}
$$

If $i = i^*$ is an index in $\mathcal{I}_k$ that yields the α_k in Eq. (11.21), then it is quite clear that at point $\mathbf{x}_{k+1} = \mathbf{x}_k + \alpha_k \mathbf{d}_k$ the active constraint matrix becomes

$$
\mathbf{A}_{a_{k+1}} = \begin{bmatrix} \mathbf{A}_{a_k} \\ \mathbf{a}_{i*}^T \end{bmatrix} \tag{11.22}
$$

where $\text{rank}(\mathbf{A}_{a_{k+1}}) = \text{rank}(\mathbf{A}_{a_k}) + 1$. By repeating the above steps, a feasible point $\mathbf{x}_K$ with $\text{rank}(\mathbf{A}_{a_K}) = n$ will eventually be reached, and point $\mathbf{x}_K$ is then deemed to be a vertex.

Example 11.3 Starting from point $\mathbf{x}_0 = [1\ 1]^T$, apply the iterative procedure described above to find a vertex for the LP problem in Example 11.2.

Solution Since the components of the residual vector at $\mathbf{x}_0$, namely,

$$\mathbf{r}_0 = \mathbf{A}\mathbf{x}_0 - \mathbf{b} = \begin{bmatrix} 1 \\ 1 \\ 1 \\ 1.5 \\ 3 \end{bmatrix}$$

are all positive, there are no active constraints at $\mathbf{x}_0$. If the first constraint (whose residual is the smallest) is chosen to form equation Eq. (11.20), we have

$$[1\ 0]\mathbf{d}_0 = -1$$

which has a (nonunique) solution $\mathbf{d}_0 = [-1\ 0]^T$. The set $\mathcal{I}_0$ in this case contains only one index, i.e.,

$$\mathcal{I}_0 = \{1\}$$

Using Eq. (11.21), we obtain $\alpha_0 = 1$ with $i^* = 1$. Hence

$$\mathbf{x}_1 = \mathbf{x}_0 + \alpha_0 \mathbf{d}_0 = \begin{bmatrix} 1 \\ 1 \end{bmatrix} + \begin{bmatrix} -1 \\ 0 \end{bmatrix} = \begin{bmatrix} 0 \\ 1 \end{bmatrix}$$

with

$$\mathbf{A}_{a_1} = [1\ 0]$$

At point $\mathbf{x}_1$, the residual vector is given by

$$\mathbf{r}_1 = \mathbf{A}\mathbf{x}_1 - \mathbf{b} = \begin{bmatrix} 0 \\ 2 \\ 1 \\ 2.5 \\ 4 \end{bmatrix}$$

Now if the third constraint (whose residual is the smallest) is chosen to form

$$\begin{bmatrix} \mathbf{A}_{a_1} \\ \mathbf{a}_3^T \end{bmatrix} \mathbf{d}_1 = \begin{bmatrix} 1 & 0 \\ 0 & 1 \end{bmatrix} \mathbf{d}_1 = \begin{bmatrix} 0 \\ -1 \end{bmatrix}$$

we obtain $\mathbf{d}_1 = [0\ -1]^T$. It follows that

$$\mathcal{I}_1 = \{3\}$$

From Eq. (11.21), $\alpha_1 = 1$ with $i^* = 3$ and, therefore,

$$\mathbf{x}_2 = \mathbf{x}_1 + \alpha_1 \mathbf{d}_1 = \begin{bmatrix} 0 \\ 0 \end{bmatrix}$$

with

$$\mathbf{A}_{a_2} = \begin{bmatrix} 1 & 0 \\ 0 & 1 \end{bmatrix}$$

Since $\text{rank}(\mathbf{A}_{a_2}) = 2 = n$, $\mathbf{x}_2$ is a vertex. A graphical illustration of this solution procedure is shown in Fig. 11.6.

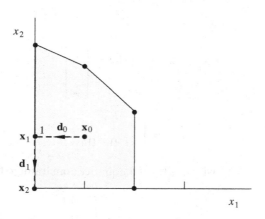

Figure 11.6. Search path for a vertex starting from point $\mathbf{x}_0$.

The iterative strategy described above can also be applied to the standard-form LP problem in Eq. (11.1). Note that the presence of the equality constraints in Eq. (11.1b) means that at any feasible point $\mathbf{x}_k$, the active constraint matrix $\mathbf{A}_{a_k}$ always contains $\mathbf{A}$ as a submatrix.

Example 11.4 Find a vertex for the convex polygon

$$x_1 + x_2 + x_3 = 1$$

such that

$$\mathbf{x} \geq \mathbf{0}$$

starting with $\mathbf{x}_0 = [\frac{1}{3} \ \frac{1}{3} \ \frac{1}{3}]^T$.

Solution At $\mathbf{x}_0$, matrix $\mathbf{A}_{a_0}$ is given by $\mathbf{A}_{a_0} = [1 \ 1 \ 1]$. Note that the residual vector at $\mathbf{x}_k$ for the standard-form LP problem is always given by $\mathbf{r}_k = \mathbf{x}_k$.

Hence $\mathbf{r}_0 = \begin{bmatrix} \frac{1}{3} & \frac{1}{3} & \frac{1}{3} \end{bmatrix}^T$. If the first inequality constraint is chosen to form Eq. (11.20), then we have

$$\begin{bmatrix} 1 & 1 & 1 \\ 1 & 0 & 0 \end{bmatrix} \mathbf{d}_0 = \begin{bmatrix} 0 \\ -1 \end{bmatrix}$$

which has a (nonunique) solution $\mathbf{d}_0 = [-1\ 1\ 0]^T$. It follows that

$$\mathcal{I}_0 = \{1\}$$

and

$$\alpha_0 = \frac{1}{3} \quad \text{with} \quad i^* = 1$$

Hence

$$\mathbf{x}_1 = \mathbf{x}_0 + \alpha_0 \mathbf{d}_0 = \begin{bmatrix} 0 \\ \frac{2}{3} \\ \frac{1}{3} \end{bmatrix}$$

At $\mathbf{x}_1$,

$$\mathbf{A}_{a_1} = \begin{bmatrix} 1 & 1 & 1 \\ 1 & 0 & 0 \end{bmatrix}$$

and $\mathbf{r}_1 = \begin{bmatrix} 0 & \frac{2}{3} & \frac{1}{3} \end{bmatrix}^T$. Choosing the third inequality constraint to form Eq. (11.20), we have

$$\begin{bmatrix} 1 & 1 & 1 \\ 1 & 0 & 0 \\ 0 & 0 & 1 \end{bmatrix} \mathbf{d}_1 = \begin{bmatrix} 0 \\ 0 \\ -1 \end{bmatrix}$$

which leads to $\mathbf{d}_1 = [0\ 1\ -1]^T$. Consequently,

$$\mathcal{I}_1 = \{3\}$$

and

$$\alpha_1 = \frac{1}{3} \quad \text{with} \quad i^* = 3$$

Therefore,

$$\mathbf{x}_2 = \mathbf{x}_1 + \alpha_1 \mathbf{d}_1 = \begin{bmatrix} 0 \\ 1 \\ 0 \end{bmatrix}$$

At $\mathbf{x}_2$,

$$\mathbf{A}_{a_2} = \begin{bmatrix} 1 & 1 & 1 \\ 1 & 0 & 0 \\ 0 & 0 & 1 \end{bmatrix}$$

Hence rank$(\mathbf{A}_{a_2}) = 3$, indicating that $\mathbf{x}_2$ is a vertex. The search path that leads to vertex $\mathbf{x}_2$ is illustrated in Fig. 11.7.

∎

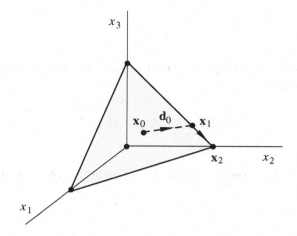

Figure 11.7. Search path for a vertex for Example 11.4.

11.2.3.4 Two implementation issues

There are two issues that have to be dealt with when using the method in Sec. 11.2.3.3 to find a vertex. First, as in the case of any iterative optimization method, we need to identify a feasible point. As will be shown shortly, this problem can itself be treated as an LP problem, which is often referred to as a *phase-1 LP problem*. Second, in order to move from point $\mathbf{x}_k$ to point $\mathbf{x}_{k+1}$, we need to identify a constraint, say, the ith constraint, which is inactive at $\mathbf{x}_k$ such that $\mathbf{a}_i^T$ is linearly independent of the rows of $\mathbf{A}_{a_k}$. Obviously, this is a rank determination problem. Later on in this subsection, we will describe a method for rank determination based on the QR decomposition of matrix $\mathbf{A}_{a_k}$.

I. Finding a feasible point. Finding a feasible point for the LP problem in Eq. (11.2) amounts to finding a vector $\mathbf{x}_0 \in R^{n \times 1}$ such that

$$\mathbf{A}\mathbf{x}_0 \geq \mathbf{b}$$

To this end, we consider the modified constraints

$$\mathbf{A}\mathbf{x} + \phi\mathbf{e} \geq \mathbf{b} \qquad (11.23)$$

where ϕ is an auxiliary scalar variable and $\mathbf{e} = [1 \ 1 \ \cdots \ 1]^T$. Evidently, if $\mathbf{x} = \mathbf{0}$ and $\phi = \phi_0 = \max(0, \ b_1, \ b_2, \ \ldots, \ b_p)$ in Eq. (11.23) where b_i is the ith component of $\mathbf{b}$ in Eq. (11.23), then $\phi \geq 0$ and Eq. (11.23) is satisfied because

$$\mathbf{A}\mathbf{0} + \phi_0\mathbf{e} \geq \mathbf{b}$$

In other words, if we define the augmented vector $\hat{\mathbf{x}}$ as

$$\hat{\mathbf{x}} = \begin{bmatrix} \mathbf{x} \\ \phi \end{bmatrix}$$

then the initial value

$$\hat{x}_0 = \begin{bmatrix} 0 \\ \phi_0 \end{bmatrix} \tag{11.24}$$

satisfies the constraints in Eq. (11.23). This suggests that a phase-1 LP problem can be formulated as

$$\text{minimize } \phi \tag{11.25a}$$

$$\text{subject to: } \mathbf{Ax} + \phi\mathbf{e} \geq \mathbf{b} \tag{11.25b}$$

$$\phi \geq 0 \tag{11.25c}$$

A feasible initial point for this problem is given by Eq. (11.24). If the solution is assumed to be

$$\hat{x}^* = \begin{bmatrix} \mathbf{x}^* \\ \phi^* \end{bmatrix} = \begin{bmatrix} \mathbf{x}^* \\ 0 \end{bmatrix}$$

then at $\hat{x}^*$ the constraints in Eq. (11.25b) become $\mathbf{Ax}^* \geq \mathbf{b}$ and hence $\mathbf{x}^*$ is a feasible point for the original LP problem. If $\phi^* > 0$, we conclude that no feasible point exists for constraints $\mathbf{Ax} \geq \mathbf{b}$ and ϕ^* then represents a single perturbation of the constraint in Eq. (11.2b) with minimum L_∞ norm to ensure feasibility. In effect, point $\mathbf{x}^*$ would become feasible if the constraints were modified to

$$\mathbf{Ax} \geq \tilde{\mathbf{b}} \quad \text{with } \tilde{\mathbf{b}} = \mathbf{b} - \phi^*\mathbf{e} \tag{11.26}$$

II. Finding a linearly independent $\mathbf{a}_i^T$. Assume that at $\mathbf{x}_k$, $\text{rank}(\mathbf{A}_{a_k}) = r_k$ with $r_k < n$. Finding a normal vector $\mathbf{a}_i^T$ associated with an inactive constraint at $\mathbf{x}_k$ such that $\mathbf{a}_i^T$ is linearly independent of the rows of $\mathbf{A}_{a_k}$ is equivalent to finding an $\mathbf{a}_i^T$ such that $\text{rank}(\hat{\mathbf{A}}_{a_k}) = r_k + 1$ where

$$\hat{\mathbf{A}}_{a_k} = \begin{bmatrix} \mathbf{A}_{a_k} \\ \mathbf{a}_i^T \end{bmatrix} \tag{11.27}$$

An effective way of finding the rank of a matrix obtained through finite-precision computations is to perform QR decomposition with column pivoting, which can be done through the use of the Householder QR decomposition described in Sec. A.12.2 (see also [5, Chap. 5]). On applying this procedure to a matrix $\mathbf{M} \in R^{n \times m}$ with $m \leq n$, after r steps of the procedure we obtain

$$\mathbf{MP}^{(r)} = \mathbf{Q}^{(r)}\mathbf{R}^{(r)}$$

where $\mathbf{Q}^{(r)} \in R^{n \times n}$ is an orthogonal matrix, $\mathbf{P}^{(r)} \in R^{m \times m}$ is a permutation matrix, and

$$\mathbf{R}^{(r)} = \begin{bmatrix} \mathbf{R}_{11}^{(r)} & \mathbf{R}_{12}^{(r)} \\ 0 & \mathbf{R}_{22}^{(r)} \end{bmatrix}$$

where $\mathbf{R}_{11}^{(r)} \in R^{r \times r}$ is nonsingular and upper-triangular. If $||\mathbf{R}_{22}^{(r)}||_2$ is negligible, then the numerical rank of $\mathbf{M}$ is deemed to be r. A reasonable condition for terminating the QR decomposition is

$$||\mathbf{R}_{22}^{(r)}||_2 \le \varepsilon ||\mathbf{M}||_2 \qquad (11.28)$$

where ε is some small machine-dependent parameter. When Eq. (11.28) is satisfied, block $\mathbf{R}_{22}^{(r)}$ is set to zero and the QR decomposition of $\mathbf{M}$ becomes

$$\mathbf{MP} = \mathbf{QR}$$

where $\mathbf{P} = \mathbf{P}^{(r)}$, $\mathbf{Q} = \mathbf{Q}^{(r)}$, and

$$\mathbf{R} = \begin{bmatrix} \mathbf{R}_{11}^{(r)} & \mathbf{R}_{12}^{(r)} \\ \mathbf{0} & \mathbf{0} \end{bmatrix} \qquad (11.29)$$

For matrix $\mathbf{A}_{a_k}$ in Eq. (11.27), the above QR decomposition can be applied to $\mathbf{A}_{a_k}^T \in R^{n \times r}$ with $r < n$, i.e.,

$$\mathbf{A}_{a_k}^T \mathbf{P} = \mathbf{QR} \qquad (11.30)$$

where $\mathbf{R}$ has the form of Eq. (11.29), and the size of $\mathbf{R}_{11}^{(r)}$ gives the rank of $\mathbf{A}_{a_k}$. A nice feature of the QR decomposition method is that if matrix $\mathbf{A}_{a_k}$ is altered in some way, for example, by adding a rank-one matrix or appending a row (or column) to it or deleting a row (or column) from it, the QR decomposition of the altered matrix can be obtained based on the QR decomposition of matrix $\mathbf{A}_{a_k}$ with a computationally simple updating procedure (see Sec. A.12 and [5, Chap. 12]). In the present case, we are interested in the QR decomposition of $\hat{\mathbf{A}}_{a_k}^T$ in Eq. (11.27), which is obtained from $\mathbf{A}_{a_k}^T$ by appending $\mathbf{a}_i$ as the last column. If we let

$$\hat{\mathbf{P}} = \begin{bmatrix} \mathbf{P} & \mathbf{0} \\ \mathbf{0} & 1 \end{bmatrix}$$

it follows from Eq. (11.30) that

$$\begin{aligned} \mathbf{Q}^T \hat{\mathbf{A}}_{a_k}^T \hat{\mathbf{P}} &= \mathbf{Q}^T [\mathbf{A}_{a_k}^T \; \mathbf{a}_i] \hat{\mathbf{P}} \\ &= [\mathbf{Q}^T \mathbf{A}_{a_k}^T \mathbf{P} \; \mathbf{Q}^T \mathbf{a}_i] = [\mathbf{R} \; \mathbf{w}_i] \\ &= \begin{bmatrix} \mathbf{R}_{11}^{(r)} & \mathbf{R}_{12}^{(r)} & \\ & & \mathbf{w}_i \\ \mathbf{0} & \mathbf{0} & \end{bmatrix} \end{aligned} \qquad (11.31)$$

where $\mathbf{w}_i = \mathbf{Q}^T \mathbf{a}_i$ is a column vector with n entries. Note that if we apply $n - r + 1$ Givens rotations (see Sec. A.11.2 and [5, Chap. 5]) $\mathbf{J}_l^T$ for $1 \le l \le$

$n - r - 1$ to $\mathbf{w}_i$ successively so that

$$\mathbf{J}_{n-r-1}^T \cdots \mathbf{J}_2^T \cdot \mathbf{J}_1^T \mathbf{w}_i = \begin{bmatrix} \psi_1 \\ \vdots \\ \psi_r \\ \psi_{r+1} \\ 0 \\ \vdots \\ 0 \end{bmatrix} \Bigg\} (n - r - 1) \text{ zeros} \tag{11.32}$$

then the structure of $\mathbf{R}$ is not changed. Now by defining

$$\mathbf{J} = \mathbf{J}_1 \; \mathbf{J}_2 \; \cdots \; \mathbf{J}_{n-r-1} \quad \text{and} \quad \hat{\mathbf{Q}} = \mathbf{Q}\mathbf{J}$$

Eqs. (11.31) and (11.32) yield

$$\hat{\mathbf{Q}}^T \hat{\mathbf{A}}_{a_k}^T \hat{\mathbf{P}} = \mathbf{J}[\mathbf{R} \; \mathbf{w}_i] = \begin{bmatrix} \hat{\mathbf{R}}_{11}^{(r)} & \hat{\mathbf{R}}_{12}^{(r)} & \begin{array}{c} \psi_1 \\ \vdots \\ \psi_r \\ \hline \psi_{r+1} \\ 0 \end{array} \\ \mathbf{0} & \mathbf{0} & \begin{array}{c} \vdots \\ 0 \end{array} \end{bmatrix} \tag{11.33}$$

where $\hat{\mathbf{R}}_{11}^{(r)}$ is an $r \times r$ nonsingular upper triangular matrix. If ψ_{r+1} is negligible, then $\text{rank}(\hat{\mathbf{A}}_{a_k})$ may be deemed to be r; hence $\mathbf{a}_i^T$ is not linearly independent of the rows of $\mathbf{A}_{a_k}$. However, if ψ_{r+1} is not negligible, then Eq. (11.33) shows that $\text{rank}(\hat{\mathbf{A}}_{a_k}) = r + 1$ so that $\mathbf{a}_i^T$ is a desirable vector for Eq. (11.20). By applying a permutation matrix $\mathbf{P}_a$ to Eq. (11.33) to interchange the $(r + 1)$th column with the last column for the matrix on the right-hand side, the updated QR decomposition of $\hat{\mathbf{A}}_{a_k}^T$ is obtained as

$$\mathbf{A}_{a_k}^T \tilde{\mathbf{P}} = \hat{\mathbf{Q}}\hat{\mathbf{R}}$$

where $\tilde{\mathbf{P}} = \hat{\mathbf{P}}\mathbf{P}_a$ is a permutation matrix and $\hat{\mathbf{R}}$ is given by

$$\mathbf{R} = \begin{bmatrix} \hat{\mathbf{R}}_{11}^{(r+1)} & \tilde{\mathbf{R}}_{12}^{(r+1)} \\ \mathbf{0} & \mathbf{0} \end{bmatrix}$$

with

$$\hat{\mathbf{R}}_{11}^{(r+1)} = \begin{bmatrix} \hat{\mathbf{R}}_{11}^{(r)} & \psi_1 \\ & \vdots \\ \mathbf{0} & \psi_{r+1} \end{bmatrix}$$

11.2.4   Vertex minimizers

11.2.4.1   Finding a vertex minimizer

The iterative method for finding a vertex described in Sec. 11.2.3.3 does not involve the objective function $f(\mathbf{x}) = \mathbf{c}^T\mathbf{x}$. Consequently, the vertex obtained may not be a minimizer (see Example 11.3). However, as will be shown in the next theorem, if we start the iterative method at a minimizer, a vertex would eventually be reached without increasing the objective function which would, therefore, be a vertex minimizer.

Theorem 11.5 *Existence of a vertex minimizer in alternative-form LP problem* *If the minimum of $f(\mathbf{x})$ in the alternative-form LP problem of Eq. (11.2) is finite, then there is a vertex minimizer.*

Proof If $\mathbf{x}_0$ is a minimizer, then $\mathbf{x}_0$ is finite and satisfies the conditions stated in Theorem 11.2. Hence there exists a $\boldsymbol{\mu}^* \geq \mathbf{0}$ such that

$$\mathbf{c} = \mathbf{A}^T\boldsymbol{\mu}^* \tag{11.34}$$

By virtue of the complementarity condition in Eq. (11.8c), Eq. (11.34) can be written as

$$\mathbf{c} = \mathbf{A}_{a_0}^T\boldsymbol{\mu}_a^* \tag{11.35}$$

where $\mathbf{A}_{a_0}$ is the active constraint matrix at $\mathbf{x}_0$ and $\boldsymbol{\mu}_a^*$ is composed of the entries of $\boldsymbol{\mu}^*$ that correspond to the active constraints. If $\mathbf{x}_0$ is not a vertex, the method described in Sec. 11.2.3.3 can be applied to yield a point $\mathbf{x}_1 = \mathbf{x}_0 + \alpha_0\mathbf{d}_0$ which is closer to a vertex, where $\mathbf{d}_0$ is a feasible direction that satisfies the condition $\mathbf{A}_{a_0}\mathbf{d}_0 = \mathbf{0}$ (see Eq. (11.18)). It follows that at $\mathbf{x}_1$ the objective function remains the same as at $\mathbf{x}_0$, i.e.,

$$f(\mathbf{x}_1) = \mathbf{c}^T\mathbf{x}_1 = \mathbf{c}^T\mathbf{x}_0 + \alpha_0\boldsymbol{\mu}_a^{*T}\mathbf{A}_{a_0}\mathbf{d}_0 = \mathbf{c}^T\mathbf{x}_0 = f(\mathbf{x}_0)$$

which means that $\mathbf{x}_1$ is a minimizer. If $\mathbf{x}_1$ is not yet a vertex, then the process is continued to generate minimizers $\mathbf{x}_2$, $\mathbf{x}_3$, ... until a vertex minimizer is reached.

∎

Theorem 11.5 also applies to the standard-form LP problem in Eq. (11.1). To prove this, let $\mathbf{x}_0$ be a finite minimizer of Eq. (11.1). It follows from Eq. (11.5c) that

$$\mathbf{c} = \mathbf{A}^T\boldsymbol{\lambda}^* + \boldsymbol{\mu}^* \tag{11.36}$$

The complementarity condition implies that Eq. (11.36) can be written as

$$\mathbf{c} = \mathbf{A}^T\boldsymbol{\lambda}^* + \mathbf{I}_0^T\boldsymbol{\mu}_a^* \tag{11.37}$$

where $\mathbf{I}_0$ consists of the rows of the $n \times n$ identity matrix that are associated with the inequality constraints in Eq. (11.3c) that are active at $\mathbf{x}_0$, and $\boldsymbol{\mu}_a^*$ is composed of the entries of $\boldsymbol{\mu}^*$ that correspond to the active (inequality) constraints. At $\mathbf{x}_0$, the active constraint matrix $\mathbf{A}_{a_0}$ is given by

$$\mathbf{A}_{a_0} = \begin{bmatrix} \mathbf{A} \\ \mathbf{I}_0 \end{bmatrix} \tag{11.38}$$

Hence Eq. (11.37) becomes

$$\mathbf{c} = \mathbf{A}_{a_0}^T \boldsymbol{\eta}_a^* \quad \text{with} \quad \boldsymbol{\eta}_a^* = \begin{bmatrix} \boldsymbol{\lambda}^* \\ \boldsymbol{\mu}_a^* \end{bmatrix} \tag{11.39}$$

which is the counterpart of Eq. (11.35) for the problem in Eq. (11.1). The rest of the proof is identical with that of Theorem 11.5. We can, therefore, state the following theorem.

Theorem 11.6 *Existence of a vertex minimizer in standard-form LP problem* *If the minimum of $f(\mathbf{x})$ in the LP problem of Eq. (11.1) is finite, then a vertex minimizer exists.*

■

11.2.4.2 Uniqueness

A key feature in the proofs of Theorems 11.5 and 11.6 is the connection of vector $\mathbf{c}$ to the active constraints as described by Eqs. (11.35) and (11.39) through the Lagrange multipliers $\boldsymbol{\mu}^*$ and $\boldsymbol{\lambda}^*$. As will be shown in the next theorem, the Lagrange multipliers also play a critical role in the uniqueness of a vertex minimizer.

Theorem 11.7 *Uniqueness of minimizer of alternative-form LP problem* Let $\mathbf{x}^*$ be a vertex minimizer of the LP problem in Eq. (11.2) at which

$$\mathbf{c}^T = \mathbf{A}_{a^*}^T \boldsymbol{\mu}_a^*$$

where $\boldsymbol{\mu}_a^* \geq \mathbf{0}$ is defined in the proof of Theorem 11.5. If $\boldsymbol{\mu}_a^* > \mathbf{0}$, then $\mathbf{x}^*$ is the unique vertex minimizer of Eq. (11.2).

Proof Let us suppose that there is another vertex minimizer $\tilde{\mathbf{x}} \neq \mathbf{x}^*$. We can write

$$\tilde{\mathbf{x}} = \mathbf{x}^* + \mathbf{d}$$

with $\mathbf{d} = \tilde{\mathbf{x}} - \mathbf{x}^* \neq \mathbf{0}$. Since both $\mathbf{x}^*$ and $\tilde{\mathbf{x}}$ are feasible, $\mathbf{d}$ is a feasible direction which implies that $\mathbf{A}_{a^*} \mathbf{d} \geq \mathbf{0}$. Since $\mathbf{x}^*$ is a vertex, $\mathbf{A}_{a^*}$ is nonsingular; hence

$\mathbf{A}_{a^*}\mathbf{d} \geq \mathbf{0}$ together with $\mathbf{d} \neq \mathbf{0}$ implies that at least one component of $\mathbf{A}_{a^*}\mathbf{d}$, say, $(\mathbf{A}_{a^*}\mathbf{d})_i$, is strictly positive. We then have

$$0 = f(\tilde{\mathbf{x}}) - f(\mathbf{x}^*) = \mathbf{c}^T\tilde{\mathbf{x}} - \mathbf{c}^T\mathbf{x}^* = \mathbf{c}^T\mathbf{d}$$
$$= \boldsymbol{\mu}_a^{*T}\mathbf{A}_{a^*}\mathbf{d} \geq (\boldsymbol{\mu}_a^*)_i \cdot (\mathbf{A}_{a^*}\mathbf{d})_i > 0$$

The above contradiction implies that another minimizer $\tilde{\mathbf{x}}$ cannot exist.

∎

For the standard-form LP problem in Eq. (11.1), the following theorem applies.

Theorem 11.8 *Uniqueness of minimizer of standard-form LP problem* Consider the LP problem in Eq. (11.1) and let $\mathbf{x}^*$ be a vertex minimizer at which

$$\mathbf{c}^T = \mathbf{A}_{a^*}^T\boldsymbol{\eta}_a^*$$

with

$$\mathbf{A}_{a^*} = \begin{bmatrix} \mathbf{A} \\ \mathbf{I}_* \end{bmatrix}, \quad \boldsymbol{\eta}_a^* = \begin{bmatrix} \boldsymbol{\lambda}^* \\ \boldsymbol{\mu}_a^* \end{bmatrix}$$

where $\mathbf{I}_*$ consists of the rows of the $n \times n$ identity matrix that are associated with the inequality constraints in Eq. (11.1c) that are active at $\mathbf{x}^*$, $\boldsymbol{\lambda}^*$ and $\boldsymbol{\mu}^*$ are the Lagrange multipliers in Eq. (11.5c), and $\boldsymbol{\mu}_a^*$ consists of the entries of $\boldsymbol{\mu}^*$ associated with active (inequality) constraints. If $\boldsymbol{\mu}_a^* > \mathbf{0}$, then $\mathbf{x}^*$ is the unique vertex minimizer of the problem in Eq. (11.1).

∎

Theorem 11.8 can be proved by assuming that there is another minimizer $\tilde{\mathbf{x}}$ and then using an argument similar to that in the proof of Theorem 11.7 with some minor modifications. Direction $\mathbf{c}$ being feasible implies that

$$\mathbf{A}_{a^*}\mathbf{d} = \begin{bmatrix} \mathbf{A}\mathbf{d} \\ \mathbf{I}_*\mathbf{d} \end{bmatrix} = \begin{bmatrix} \mathbf{0} \\ \mathbf{I}_*\mathbf{d} \end{bmatrix} \geq \mathbf{0} \tag{11.40}$$

where $\mathbf{I}_*\mathbf{d}$ consists of the components of $\mathbf{d}$ that are associated with the active (inequality) constraints at $\mathbf{x}^*$. Since $\mathbf{A}_{a^*}$ is nonsingular, Eq. (11.40) in conjunction with $\mathbf{d} \neq \mathbf{0}$ implies that at least one component of $\mathbf{I}_*\mathbf{d}$, say, $(\mathbf{I}_*\mathbf{d})_i$, is strictly positive. This yields the contradiction

$$0 = f(\tilde{\mathbf{x}}) - f(\mathbf{x}^*) = \mathbf{c}^T\mathbf{d}$$
$$= [\boldsymbol{\lambda}^{*T}\boldsymbol{\mu}_a^{*T}]\mathbf{A}_a\mathbf{d} = \boldsymbol{\mu}_a^{*T}\mathbf{I}_*\mathbf{d} \geq (\boldsymbol{\mu}_a^*)_i \cdot (\mathbf{I}_*\mathbf{d})_i$$
$$> 0$$

The strict positiveness of the Lagrange multiplier $\boldsymbol{\mu}_a^*$ is critical for the uniqueness of the solution. As a matter of fact, if the vertex minimizer $\mathbf{x}^*$ is *nondegenerate* (see Def. 11.1 in Sec. 11.2.3.1), then any zero entries in $\boldsymbol{\mu}_a^*$ imply the nonuniqueness of the solution. The reader is referred to [3, Sec. 7.7] for the details.

11.3   Simplex Method

11.3.1   Simplex method for alternative-form LP problem

In this section, we consider a general method for the solution of the LP problem in Eq. (11.2) known as the *simplex method*. It was shown in Sec. 11.2.4 that if the minimum value of the objective function in the feasible region is finite, then a vertex minimizer exists. Let x_0 be a vertex and assume that it is not a minimizer. The simplex method generates an adjacent vertex x_1 with $f(x_1) < f(x_0)$ and continues doing so until a vertex minimizer is reached.

11.3.1.1   Nondegenerate case

To simplify our discussion, we assume that all vertices are nondegenerate, i.e., at a vertex there are exactly n active constraints. This assumption is often referred to as the *nondegeneracy assumption* [3] in the literature.

Given a vertex x_k, a vertex x_{k+1} is said to be *adjacent* to x_k if $A_{a_{k+1}}$ differs from A_{a_k} by *one* row. In terms of the notation used in Sec. 11.2.3.2, we denote A_{a_k} as

$$A_{a_k} = \begin{bmatrix} a_{j_1}^T \\ a_{j_2}^T \\ \vdots \\ a_{j_n}^T \end{bmatrix}$$

where a_{j_l} is the normal of the j_lth constraint in Eq. (11.2b). Associated with A_{a_k} is the index set

$$\mathcal{J}_k = \{j_1, \ j_2, \ \ldots, \ j_n\}$$

Obviously, if $\mathcal{J}_k$ and $\mathcal{J}_{k+1}$ have exactly $(n-1)$ members, vertices x_k and x_{k+1} are adjacent. At vertex x_k, the simplex method verifies whether x_k is a vertex minimizer, and if it is not, it finds an adjacent vertex x_{k+1} that yields a *reduced* value of the objective function. Since a vertex minimizer exists and there is only a finite number of vertices, the simplex method will find a solution after a finite number of iterations.

Under the nondegeneracy assumption, A_{a_k} is square and nonsingular. Hence there exists a $\mu_k \in R^{n \times 1}$ such that

$$c = A_{a_k}^T \mu_k \tag{11.41}$$

Since x_k is a feasible point, by virtue of Theorem 11.2 we conclude that x_k is a vertex minimizer if and only if

$$\mu_k \geq 0 \tag{11.42}$$

In other words, x_k is not a vertex minimizer if and only if at least one component of μ_k, say, $(\mu_k)_l$, is negative.

Assume that $\mathbf{x}_k$ is not a vertex minimizer and let

$$(\boldsymbol{\mu}_k)_l < 0 \tag{11.43}$$

The simplex method finds an edge as a feasible descent direction $\mathbf{d}_k$ that points from $\mathbf{x}_k$ to an adjacent vertex $\mathbf{x}_{k+1}$ given by

$$\mathbf{x}_{k+1} = \mathbf{x}_k + \alpha_k \mathbf{d}_k \tag{11.44}$$

It was shown in Sec. 11.2.3.2 that a feasible descent direction $\mathbf{d}_k$ is characterized by

$$\mathbf{A}_{a_k} \mathbf{d}_k \geq \mathbf{0} \quad \text{and} \quad \mathbf{c}^T \mathbf{d}_k < 0 \tag{11.45}$$

To find an edge that satisfies Eq. (11.45), we denote the lth coordinate vector (i.e., the lth column of the $n \times n$ identity matrix) as $\mathbf{e}_l$ and examine vector $\mathbf{d}_k$ that solves the equation

$$\mathbf{A}_{a_k} \mathbf{d}_k = \mathbf{e}_l \tag{11.46}$$

From Eq. (11.46), we note that $\mathbf{A}_{a_k} \mathbf{d}_k \geq \mathbf{0}$. From Eqs. (11.41), (11.43), and (11.46), we have

$$\mathbf{c}^T \mathbf{d}_k = \boldsymbol{\mu}_k^T \mathbf{A}_{a_k} \mathbf{d}_k = \boldsymbol{\mu}_k^T \mathbf{e}_l = (\boldsymbol{\mu}_k)_i < 0$$

and hence $\mathbf{d}_k$ satisfies Eq. (11.45) and, therefore, it is a feasible descent direction. Moreover, for $i \neq l$ Eq. (11.46) implies that

$$\mathbf{a}_{j_i}^T (\mathbf{x}_k + \alpha \mathbf{d}_k) = \mathbf{a}_{j_i}^T \mathbf{x}_k + \alpha \mathbf{a}_{j_i}^T \mathbf{d}_k = b_{j_i}$$

Therefore, there are exactly $n - 1$ constraints that are active at $\mathbf{x}_k$ and remain active at $\mathbf{x}_k + \alpha \mathbf{d}_k$. This means that $\mathbf{x}_k + \alpha \mathbf{d}_k$ with $\alpha > 0$ is an edge that connects $\mathbf{x}_k$ to an adjacent vertex $\mathbf{x}_{k+1}$ with $f(\mathbf{x}_{k+1}) < f(\mathbf{x}_k)$. By using an argument similar to that in Sec. 11.2.3.3, the right step size α_k can be identified as

$$\alpha_k = \min_{i \in \mathcal{I}_k} \left(\frac{\mathbf{a}_i^T \mathbf{x}_k - b_i}{-\mathbf{a}_i^T \mathbf{d}_k} \right) \tag{11.47}$$

where $\mathcal{I}_k$ contains the indices of the constraints that are inactive at $\mathbf{x}_k$ with $\mathbf{a}_i^T \mathbf{d}_k < 0$, i.e.,

$$\mathcal{I}_k = \{i : \mathbf{a}_i^T \mathbf{x}_k - b_i > 0 \text{ and } \mathbf{a}_i^T \mathbf{d}_k < 0\} \tag{11.48}$$

Once α_k is calculated, the next vertex $\mathbf{x}_{k+1}$ is determined by using Eq. (11.44).

Now if $i^* \in \mathcal{I}_k$ is the index that achieves the minimum in Eq. (11.47), i.e.,

$$\alpha_k = \frac{\mathbf{a}_{i^*}^T \mathbf{x}_k - b_{i^*}}{-\mathbf{a}_{i^*}^T \mathbf{d}_k}$$

346

then at $\mathbf{x}_{k+1}$ the i^*th constraint becomes active. With the j_lth constraint leaving $\mathbf{A}_{a_k}$ and the i^*th constraint entering $\mathbf{A}_{a_{k+1}}$, there are exactly n active constraints at $\mathbf{x}_{k+1}$ and $\mathbf{A}_{a_{k+1}}$ given by

$$\mathbf{A}_{a_{k+1}} = \begin{bmatrix} \mathbf{a}_{j_1}^T \\ \vdots \\ \mathbf{a}_{j_{l-1}}^T \\ \mathbf{a}_{i^*}^T \\ \mathbf{a}_{j_{l+1}}^T \\ \vdots \\ \mathbf{a}_{j_n}^T \end{bmatrix} \qquad (11.49)$$

and the index set is given by

$$\mathcal{J}_{k+1} = \{j_1, \ldots, j_{l-1}, i^*, j_{l+1}, \ldots, j_n\} \qquad (11.50)$$

A couple of remarks on the method described are in order. First, when the Lagrange multiplier vector $\boldsymbol{\mu}_k$ determined by using Eq. (11.41) contains more than one negative component, a 'textbook rule' is to select the index l in Eq. (11.46) that corresponds to the most negative component in $\boldsymbol{\mu}_k$ [3]. Second, Eq. (11.47) can be modified to deal with the LP problem in Eq. (11.2) with an unbounded minimum. If the LP problem at hand does not have a bounded minimum, then at some iteration k the index set $\mathcal{I}_k$ will become empty which signifies an unbounded solution of the LP problem. Below, we summarize an algorithm that implements the simplex method and use two examples to illustrate its application.

Algorithm 11.1 Simplex algorithm for the alternative-form LP problem in Eq. (11.2), nondegenerate vertices
Step 1
Input vertex $\mathbf{x}_0$, and form $\mathbf{A}_{a_0}$ and $\mathcal{J}_0$.
Set $k = 0$.
Step 2
Solve

$$\mathbf{A}_{a_k}^T \boldsymbol{\mu}_k = \mathbf{c} \qquad (11.51)$$

for $\boldsymbol{\mu}_k$.
If $\boldsymbol{\mu}_k \geq \mathbf{0}$, stop ($\mathbf{x}_k$ is a vertex minimizer); otherwise, select the index l that corresponds to the most negative component in $\boldsymbol{\mu}_k$.
Step 3
Solve

$$\mathbf{A}_{a_k} \mathbf{d}_k = \mathbf{e}_l \qquad (11.52)$$

for $\mathbf{d}_k$.

Step 4

Compute the residual vector

$$\mathbf{r}_k = \mathbf{A}\mathbf{x}_k - \mathbf{b} = (r_i)_{i=1}^p \tag{11.53a}$$

If the index set

$$\mathcal{I}_k = \{i : r_i > 0 \text{ and } \mathbf{a}_i^T \mathbf{d}_k < 0\} \tag{11.53b}$$

is empty, stop (the objective function tends to $-\infty$ in the feasible region); otherwise, compute

$$\alpha_k = \min_{i \in \mathcal{I}_k} \left(\frac{r_i}{-\mathbf{a}_i^T \mathbf{d}_k} \right) \tag{11.53c}$$

and record the index i^* with $\alpha_k = r_{i^*}/(-\mathbf{a}_{i^*}^T \mathbf{d}_k)$.

Step 5

Set

$$\mathbf{x}_{k+1} = \mathbf{x}_k + \alpha_k \mathbf{d}_k \tag{11.54}$$

Update $\mathbf{A}_{a_{k+1}}$ and $\mathcal{J}_{k+1}$ using Eqs. (11.49) and (11.50), respectively. Set $k = k + 1$ and repeat from Step 2.

Example 11.5 Solve the LP problem in Example 11.2 with initial vertex $\mathbf{x}_0 = [2\ 1.5]^T$ using the simplex method.

Solution From Example 11.2 and Fig. 11.3, the objective function is given by

$$f(\mathbf{x}) = \mathbf{c}^T \mathbf{x} = -x_1 - 4x_2$$

and the constraints are given by $\mathbf{A}\mathbf{x} \geq \mathbf{b}$ with

$$\mathbf{A} = \begin{bmatrix} 1 & 0 \\ -1 & 0 \\ 0 & 1 \\ -1 & -1 \\ -1 & -2 \end{bmatrix} \quad \text{and} \quad \mathbf{b} = \begin{bmatrix} 0 \\ -2 \\ 0 \\ -3.5 \\ -6 \end{bmatrix}$$

We note that at vertex $\mathbf{x}_0$, the second and fourth constraints are active and hence

$$\mathbf{A}_{a_0} = \begin{bmatrix} -1 & 0 \\ -1 & -1 \end{bmatrix}, \quad \mathcal{J}_0 = \{2,\ 4\}$$

Solving $\mathbf{A}_{a_0}^T \boldsymbol{\mu}_0 = \mathbf{c}$ for $\boldsymbol{\mu}_0$ where $\mathbf{c} = [-1\ -4]^T$, we obtain $\boldsymbol{\mu}_0 = [-3\ 4]^T$. This shows that $\mathbf{x}_0$ is not a minimizer and $l = 1$. Next we solve

$$\mathbf{A}_{a_0} \mathbf{d}_0 = \mathbf{e}_1$$

for d_0 to obtain $d_0 = [-1 \ 1]^T$. From Fig. 11.3, it is evident that d_0 is a feasible descent direction at x_0. The residual vector at x_0 is given by

$$r_0 = Ax_0 - b = \begin{bmatrix} 2 \\ 0 \\ 1.5 \\ 0 \\ 1 \end{bmatrix}$$

which shows that the first, third, and fifth constraints are inactive at x_0. Furthermore,

$$\begin{bmatrix} a_1^T \\ a_3^T \\ a_5^T \end{bmatrix} d_0 = \begin{bmatrix} 1 & 0 \\ 0 & 1 \\ -1 & -2 \end{bmatrix} \begin{bmatrix} -1 \\ 1 \end{bmatrix} = \begin{bmatrix} -1 \\ 1 \\ -1 \end{bmatrix}$$

Hence

$$\mathcal{I}_0 = \{1, 5\}$$

and

$$\alpha_0 = \min \left(\frac{r_1}{-a_1^T d_0}, \frac{r_5}{-a_5^T d_0} \right) = 1$$

The next vertex is obtained as

$$x_1 = x_0 + \alpha_0 d_0 = \begin{bmatrix} 1 \\ 2.5 \end{bmatrix}$$

with

$$A_{a_1} = \begin{bmatrix} -1 & -2 \\ -1 & -1 \end{bmatrix} \quad \text{and} \quad \mathcal{J}_1 = \{5, 4\}$$

This completes the first iteration.

The second iteration starts by solving $A_{a_1}^T \mu_1 = c$ for μ_1. It is found that $\mu_1 = [3 \ -2]^T$. Hence x_1 is not a minimizer and $l = 2$. By solving

$$A_{a_1} d_1 = e_2$$

we obtain the feasible descent direction $d_1 = [-2 \ 1]^T$. Next we compute the residual vector at x_1 as

$$r_1 = Ax_1 - b = \begin{bmatrix} 1 \\ 1 \\ 2.5 \\ 0 \\ 0 \end{bmatrix}$$

which indicates that the first three constraints are inactive at x_1. By evaluating

$$\begin{bmatrix} a_1^T \\ a_2^T \\ a_3^T \end{bmatrix} d_1 = \begin{bmatrix} 1 & 0 \\ -1 & 0 \\ 0 & 1 \end{bmatrix} \begin{bmatrix} -2 \\ 1 \end{bmatrix} = \begin{bmatrix} -2 \\ 2 \\ 1 \end{bmatrix}$$

we obtain

$$\mathcal{I}_1 = \{1\}$$

and

$$\alpha_1 = \frac{r_1}{-\mathbf{a}_1^T \mathbf{d}_1} = \tfrac{1}{2}$$

This leads to

$$\mathbf{x}_2 = \mathbf{x}_1 + \alpha_1 \mathbf{d}_1 = \begin{bmatrix} 0 \\ 3 \end{bmatrix}$$

with

$$\mathbf{A}_{a_2} = \begin{bmatrix} -1 & -2 \\ 1 & 0 \end{bmatrix} \quad \text{and} \quad \mathcal{J}_2 = \{5, \, 1\}$$

which completes the second iteration.

Vertex $\mathbf{x}_2$ is confirmed to be a minimizer at the beginning of the third iteration since the equation

$$\mathbf{A}_{a_2}^T \boldsymbol{\mu}_2 = \mathbf{c}$$

yields nonnegative Lagrange multipliers $\boldsymbol{\mu}_2 = [2 \; 3]^T$.

∎

Example 11.6 Solve the LP problem

$$\text{minimize } f(\mathbf{x}) = x_1 + x_2$$

$$\text{subject to:} \qquad -x_1 \geq -2$$
$$-x_2 \geq -2$$
$$2x_1 - x_2 \geq -2$$
$$-2x_1 - x_2 \geq -4$$

Solution The constraints can be written as $\mathbf{A}\mathbf{x} \geq \mathbf{b}$ with

$$\mathbf{A} = \begin{bmatrix} -1 & 0 \\ 0 & -1 \\ 2 & -1 \\ -2 & -1 \end{bmatrix} \quad \text{and} \quad \mathbf{b} = \begin{bmatrix} -2 \\ -2 \\ -2 \\ -4 \end{bmatrix}$$

The feasible region defined by the constraints is illustrated in Fig. 11.8. Note that the feasible region is unbounded.

Assume that we are given vertex $\mathbf{x}_0 = [1 \; 2]^T$ to start Algorithm 11.1. At $\mathbf{x}_0$, the second and fourth constraints are active and so

$$\mathbf{A}_{a_0} = \begin{bmatrix} 0 & -1 \\ -2 & -1 \end{bmatrix} \quad \text{and} \quad \mathcal{J}_0 = \{2, \, 4\}$$

350

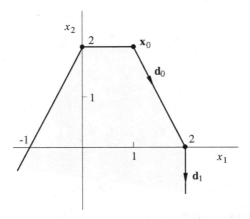

Figure 11.8. Feasible region for Example 11.6.

Equation $\mathbf{A}_{a_0}^T \boldsymbol{\mu}_0 = \mathbf{c}$ yields $\boldsymbol{\mu}_0 = [-\frac{1}{2} \ -\frac{1}{2}]^T$ and hence $\mathbf{x}_0$ is not a minimizer. Since both components of $\boldsymbol{\mu}_0$ are negative, we can choose index l to be either 1 or 2. Choosing $l = 1$, Eq. (11.46) becomes $\mathbf{A}_{a_0} \mathbf{d}_0 = \mathbf{e}_1$ which gives $\mathbf{d}_0 = [\frac{1}{2} \ -1]^T$. The residual vector at $\mathbf{x}_0$ is given by

$$
\mathbf{r}_0 = \mathbf{A}\mathbf{x}_0 - \mathbf{b} = \begin{bmatrix} 1 \\ 0 \\ 2 \\ 0 \end{bmatrix}
$$

Hence the first and third constraints are inactive at $\mathbf{x}_0$. We now compute

$$
\begin{bmatrix} \mathbf{a}_1^T \\ \mathbf{a}_3^T \end{bmatrix} \mathbf{d}_0 = \begin{bmatrix} -1 & 0 \\ 2 & -1 \end{bmatrix} \begin{bmatrix} \frac{1}{2} \\ -1 \end{bmatrix} = \begin{bmatrix} -\frac{1}{2} \\ 2 \end{bmatrix}
$$

to identify index set $\mathcal{I}_0 = \{1\}$. Hence

$$
\alpha_0 = \frac{r_1}{-\mathbf{a}_1^T \mathbf{d}_0} = 2
$$

and the next vertex is given by

$$
\mathbf{x}_1 = \mathbf{x}_0 + \alpha_0 \mathbf{d}_0 = \begin{bmatrix} 2 \\ 0 \end{bmatrix}
$$

with

$$
\mathbf{A}_{a_1} = \begin{bmatrix} -1 & 0 \\ -2 & -1 \end{bmatrix} \quad \text{and} \quad \mathcal{J}_1 = \{1, 4\}
$$

Next we examine whether or not $\mathbf{x}_1$ is a minimizer by solving $\mathbf{A}_{a_1}^T \boldsymbol{\mu}_1 = \mathbf{c}$. This gives $\boldsymbol{\mu}_1 = [3 \ -2]^T$ indicating that $\mathbf{x}_1$ is not a minimizer and $l = 2$. Solving

$\mathbf{A}_{a_1}\mathbf{d}_1 = \mathbf{e}_2$ for $\mathbf{d}_1$, we obtain $\mathbf{d}_1 = [0 \ -1]^T$. At $\mathbf{x}_1$ the residual vector is given by

$$\mathbf{r}_1 = \mathbf{A}\mathbf{x}_1 - \mathbf{b} = \begin{bmatrix} 0 \\ 2 \\ 6 \\ 0 \end{bmatrix}$$

Hence the second and third constraints are inactive. Next we evaluate

$$\begin{bmatrix} \mathbf{a}_2^T \\ \mathbf{a}_3^T \end{bmatrix} \mathbf{d}_1 = \begin{bmatrix} 0 & -1 \\ 2 & -1 \end{bmatrix} \begin{bmatrix} 0 \\ -1 \end{bmatrix} = \begin{bmatrix} 1 \\ 1 \end{bmatrix}$$

Since $\mathcal{I}_1$ is empty, we conclude that the solution of this LP problem is unbounded.

∎

11.3.1.2 Degenerate case

When some of the vertices associated with the problem are degenerate, Algorithm 11.1 needs several minor modifications. At a degenerate vertex, say, $\mathbf{x}_k$, the number of active constraints is larger than n minus the dimension of variable vector $\mathbf{x}$. Consequently, the number of rows in matrix $\mathbf{A}_{a_k}$ is larger than n and matrix $\mathbf{A}_{a_k}$ should be replaced in Steps 2 and 3 of Algorithm 11.1 by a matrix $\hat{\mathbf{A}}_{a_k}$ that is composed of n *linearly independent* rows of $\mathbf{A}_{a_k}$. Likewise, $\mathbf{A}_{a_0}$ in Step 1 and $\mathbf{A}_{a_{k+1}}$ in Step 5 should be replaced by $\hat{\mathbf{A}}_{a_0}$ and $\hat{\mathbf{A}}_{a_{k+1}}$, respectively.

The set of constraints corresponding to the rows in $\hat{\mathbf{A}}_{a_k}$ is called a *working set* of active constraints and in the literature $\hat{\mathbf{A}}_{a_k}$ is often referred to as a *working-set matrix*.

Associated with $\hat{\mathbf{A}}_{a_k}$ is the *working index set* denoted as

$$\mathcal{W}_k = \{w_1, w_2, \ldots, w_n\}$$

which contains the indices of the rows of $\hat{\mathbf{A}}_{a_k}$ as they appear in matrix $\mathbf{A}$. Some additional modifications of the algorithm in terms of the notation just introduced are to replace $\mathcal{J}_0$ in Step 1 and $\mathcal{J}_{k+1}$ in Step 5 by $\mathcal{W}_0$ and $\mathcal{W}_{k+1}$, respectively, and to redefine the index set $\mathcal{I}_k$ in Eq. (11.48) as

$$\mathcal{I}_k = \{i : i \notin \mathcal{W}_k \text{ and } \mathbf{a}_i^T\mathbf{d}_k < 0\} \tag{11.55}$$

Relative to $\mathcal{I}_k$ in Eq. (11.48), the modified $\mathcal{I}_k$ in Eq. (11.55) also includes the indices of the constraints that are active at $\mathbf{x}_k$ but are excluded from $\hat{\mathbf{A}}_{a_k}$ and which satisfy the inequality $\mathbf{a}_i^T\mathbf{d}_k < 0$.

Obviously, for a nondegenerate vertex $\mathbf{x}_k$, $\hat{\mathbf{A}}_{a_k} = \mathbf{A}_{a_k}$ and there is only *one* working set of active constraints that includes all the active constraints at

352

$\mathbf{x}_k$ and $\mathcal{I}_k$ does not contain indices of any active constraints. For a degenerate vertex $\mathbf{x}_k$, however, $\hat{\mathbf{A}}_{a_k}$ is not unique and, as Eq. (11.55) indicates, $\mathcal{I}_k$ may contain indices of active constraints. When $\mathcal{I}_k$ does include the index of an active constraint, the associated residual is zero. Consequently, the step size α_k computed using Eq. (11.53c) is also zero, which implies that $\mathbf{x}_{k+1} = \mathbf{x}_k$. Although under such circumstances the working index set $\mathcal{W}_{k+1}$ will differ from $\mathcal{W}_k$, the possibility of generating an infinite sequence of working index sets without moving from a given vertex does exist. For an example where such 'cycling' occurs, see [3, Sec. 8.3.2].

Cycling can be avoided by using an approach proposed by Bland [6]. The approach is known as *Bland's least-index rule* for deleting and adding constraints and is as follows:

1. In Step 2 of Algorithm 11.1, if the Lagrange multiplier μ_k has more than one negative components, then index l is selected as the smallest index in the working index set $\mathcal{W}_k$ corresponding to a negative component of μ_k, i.e.,

$$l = \min_{w_i \in \mathcal{W}_k,\ (\mu_k)_i < 0} (w_i) \qquad (11.56)$$

2. In Step 4, if there are more than one indices that yield the optimum α_k in Eq. (11.53c), then the associated constraints are called *blocking constraints*, and i^* is determined as the smallest index of a blocking constraint.

The steps of the modified simplex algorithm are as follows.

Algorithm 11.2 Simplex algorithm for the alternative-form LP problem in Eq. (11.2), degenerate vertices
Step 1
Input vertex $\mathbf{x}_0$ and form a working-set matrix $\hat{\mathbf{A}}_{a_0}$ and a working-index set $\mathcal{W}_0$.
Set $k = 0$.
Step 2
Solve

$$\hat{\mathbf{A}}_{a_k}^T \mu_k = \mathbf{c} \qquad (11.57)$$

for μ_k.
If $\mu_k \geq 0$, stop (vertex $\mathbf{x}_k$ is a minimizer); otherwise, select index l using Eq. (11.56).
Step 3
Solve

$$\hat{\mathbf{A}}_{a_k} \mathbf{d}_k = \mathbf{e}_l \qquad (11.58)$$

for $\mathbf{d}_k$.

Step 4
Form index set $\mathcal{I}_k$ using Eq. (11.55).
If $\mathcal{I}_k$ is empty, stop (the objective function tends to $-\infty$ in the feasible region).
Step 5
Compute the residual vector

$$\mathbf{r}_k = \mathbf{A}\mathbf{x}_k - \mathbf{b} = (r_i)_{i=1}^p$$

parameter

$$\delta_i = \frac{r_i}{-\mathbf{a}_i^T \mathbf{d}_k} \qquad \text{for } i \in \mathcal{I}_k \qquad (11.59a)$$

and

$$\alpha_k = \min_{i \in \mathcal{I}_k} (\delta_i) \qquad (11.59b)$$

Record index i^* as

$$i^* = \min_{\delta_i = \alpha_k} (i) \qquad (11.59c)$$

Step 6
Set $\mathbf{x}_{k+1} = \mathbf{x}_k + \alpha_k \mathbf{d}_k$.
Update $\hat{\mathbf{A}}_{a_{k+1}}$ by deleting row $\mathbf{a}_l^T$ and adding row $\mathbf{a}_{i^*}^T$ and update index set $\mathcal{W}_{k+1}$ accordingly.
Set $k = k + 1$ and repeat from Step 2.

Example 11.7 Solve the LP problem

$$\text{minimize } f(\mathbf{x}) = -2x_1 - 3x_2 + x_3 + 12x_4$$

$$\text{subject to: } \quad x_1 \ge 0, \; x_2 \ge 0, \; x_3 \ge 0, \; x_4 \ge 0$$

$$2x_1 + 9x_2 - x_3 - 9x_4 \ge 0$$

$$-\frac{1}{3}x_1 - x_2 + \tfrac{1}{3}x_3 + 2x_4 \ge 0$$

(See [3, p. 351].)

Solution We start with $\mathbf{x}_0 = [0 \; 0 \; 0 \; 0]^T$ which is obviously a degenerate vertex. Applying Algorithm 11.2, the first iteration results in the following computations:

$$\hat{\mathbf{A}}_{a_0} = \begin{bmatrix} 1 & 0 & 0 & 0 \\ 0 & 1 & 0 & 0 \\ 0 & 0 & 1 & 0 \\ 0 & 0 & 0 & 1 \end{bmatrix}$$

$$\mathcal{W} = \{1, 2, 3, 4\}$$

$$\boldsymbol{\mu}_0 = [-2 \ -3 \ 1 \ 12]^T$$
$$l = 1$$
$$\mathbf{d}_0 = [1 \ 0 \ 0 \ 0]^T$$
$$\mathbf{r}_0 = [0 \ 0 \ 0 \ 0 \ 0 \ 0]^T$$
$$\mathcal{I}_0 = \{6\}$$
$$\alpha_0 = 0$$
$$i^* = 6$$
$$\mathbf{x}_1 = \mathbf{x}_0 = [0 \ 0 \ 0 \ 0]^T$$
$$\hat{\mathbf{A}}_{a_1} = \begin{bmatrix} -\frac{1}{3} & -1 & \frac{1}{3} & 2 \\ 0 & 1 & 0 & 0 \\ 0 & 0 & 1 & 0 \\ 0 & 0 & 0 & 1 \end{bmatrix}$$
$$\mathcal{W}_1 = \{6, \ 2, \ 3, \ 4\}$$

Note that although $\mathbf{x}_1 = \mathbf{x}_0$, $\hat{\mathbf{A}}_{a_1}$ differs from $\hat{\mathbf{A}}_{a_0}$. Repeating from Step 2, the second iteration $(k = 1)$ gives

$$\boldsymbol{\mu}_1 = [6 \ 3 \ -1 \ 0]^T$$
$$l = 3$$
$$\mathbf{d}_1 = [1 \ 0 \ 1 \ 0]^T$$
$$\mathbf{r}_1 = [0 \ 0 \ 0 \ 0]^T$$
$$\mathcal{I}_1 = \{\phi\}$$

where $\mathcal{I}_1$ is an empty set. Therefore, in the feasible region the objective function tends to $-\infty$.

As a matter of fact, all the points along the feasible descent direction $\mathbf{d}_1$ are feasible, i.e.,

$$\mathbf{x} = \mathbf{x}_1 + \alpha \mathbf{d}_1 = [\alpha \ 0 \ \alpha \ 0]^T \qquad \text{for } \alpha > 0$$

where $f(\mathbf{x}) = -\alpha$ approaches $-\infty$ as $\alpha \to +\infty$. ∎

11.3.2   Simplex method for standard-form LP problems

11.3.2.1   Basic and nonbasic variables

For a standard-form LP problem of the type given in Eq. (11.1) with a matrix $\mathbf{A}$ of full row rank, the p equality constraints in Eq. (11.1b) are always treated as active constraints. As was discussed in Sec. 10.4.1, these constraints reduce the number of 'free' variables from n to $n - p$. In other words, the p equality constraints can be used to express p dependent variables in terms of $n - p$ independent variables. Let $\mathbf{B}$ be the matrix that consists of p linearly independent columns of $\mathbf{A}$. If the variable vector $\mathbf{x}$ is partitioned accordingly, then we can write the equality constraint in Eq. (11.2b) as

$$\mathbf{A}\mathbf{x} = [\mathbf{B} \ \mathbf{N}] \begin{bmatrix} \mathbf{x}_B \\ \mathbf{x}_N \end{bmatrix} = \mathbf{B}\mathbf{x}_B + \mathbf{N}\mathbf{x}_N = \mathbf{b} \qquad (11.60)$$

The variables contained in $\mathbf{x}_B$ and $\mathbf{x}_N$ are called *basic* and *nonbasic* variables, respectively. Since $\mathbf{B}$ is nonsingular, the basic variables can be expressed in terms of the nonbasic variables as

$$\mathbf{x}_B = \mathbf{B}^{-1}\mathbf{b} - \mathbf{B}^{-1}\mathbf{N}\mathbf{x}_N \qquad (11.61)$$

At vertex $\mathbf{x}_k$, there are at least n active constraints. Hence in addition to the p equality constraints, there are at least $n - p$ inequality constraints that become active at $\mathbf{x}_k$. Therefore, for a standard-form LP problem a vertex contains at least $n - p$ zero components. The next theorem describes an interesting property of $\mathbf{A}$.

Theorem 11.9 *Linear independence of columns in matrix* $\mathbf{A}$ *The columns of* $\mathbf{A}$ *corresponding to strictly positive components of a vertex* $\mathbf{x}_k$ *are linearly independent.*

Proof We adopt the proof used in [3]. Let $\hat{\mathbf{B}}$ be formed by the columns of $\mathbf{A}$ that correspond to strictly positive components of $\mathbf{x}_k$, and let $\hat{\mathbf{x}}_k$ be the collection of the positive components of $\mathbf{x}_k$. If $\hat{\mathbf{B}}\hat{\mathbf{w}} = \mathbf{0}$ for some nonzero $\hat{\mathbf{w}}$, then it follows that

$$\mathbf{A}\mathbf{x}_k = \hat{\mathbf{B}}\hat{\mathbf{x}}_k = \hat{\mathbf{B}}(\hat{\mathbf{x}}_k + \alpha\hat{\mathbf{w}}) = \mathbf{b}$$

for any scalar α. Since $\hat{\mathbf{x}}_k > \mathbf{0}$, there exists a sufficiently small $\alpha_+ > 0$ such that

$$\hat{\mathbf{y}}_k = \hat{\mathbf{x}}_k + \alpha\hat{\mathbf{w}} > \mathbf{0} \qquad \text{for} \ -\alpha_+ \le \alpha \le \alpha_+$$

Now let $\mathbf{y}_k \in R^{n \times 1}$ be such that the components of $\mathbf{y}_k$ corresponding to $\hat{\mathbf{x}}_k$ are equal to the components of $\hat{\mathbf{y}}_k$ and the remaining components of $\mathbf{y}_k$ are zero. Evidently, we have

$$\mathbf{A}\mathbf{y}_k = \hat{\mathbf{B}}\hat{\mathbf{y}}_k = \mathbf{b}$$

and

$$\mathbf{y}_k \ge \mathbf{0} \qquad \text{for} \ -\alpha_+ \le \alpha \le \alpha_+$$

Note that with $\alpha = 0$, $\mathbf{y}_k = \mathbf{x}_k$ is a vertex, and when α varies from $-\alpha_+$ to α_+, vertex $\mathbf{x}_k$ would lie between two feasible points on a straight line, which is a contradiction. Hence $\hat{\mathbf{w}}$ must be zero and the columns of $\hat{\mathbf{B}}$ are linearly independent.

∎

By virtue of Theorem 11.9, we can use the columns of $\hat{\mathbf{B}}$ as a set of core basis vectors to construct a nonsingular square matrix $\mathbf{B}$. If $\hat{\mathbf{B}}$ already contains p columns, we assume that $\mathbf{B} = \hat{\mathbf{B}}$; otherwise, we augment $\hat{\mathbf{B}}$ with additional columns of $\mathbf{A}$ to obtain a square nonsingular $\mathbf{B}$. Let the index set associated with $\mathbf{B}$ at $\mathbf{x}_k$ be denoted as $\mathcal{I}_\beta = \{\beta_1, \beta_2, \ldots, \beta_p\}$. With matrix $\mathbf{B}$ so formed, matrix $\mathbf{N}$ in Eq. (11.60) can be constructed with those $n - p$ columns of $\mathbf{A}$ that are not in $\mathbf{B}$. Let $\mathcal{I}_N = \{\nu_1, \nu_2, \ldots, \nu_{n-p}\}$ be the index set for the columns of $\mathbf{N}$ and let $\mathbf{I}_N$ be the $(n - p) \times n$ matrix composed of rows $\nu_1, \nu_2, \ldots, \nu_{n-p}$ of the $n \times n$ identity matrix. With this notation, it is clear that at vertex $\mathbf{x}_k$ the active constraint matrix $\mathbf{A}_{a_k}$ contains the working-set matrix

$$\hat{\mathbf{A}}_{a_k} = \begin{bmatrix} \mathbf{A} \\ \mathbf{I}_N \end{bmatrix} \tag{11.62}$$

as an $n \times n$ submatrix. It can be shown that matrix $\hat{\mathbf{A}}_{a_k}$ in Eq. (11.62) is nonsingular. In fact if $\hat{\mathbf{A}}_{a_k}\mathbf{x} = \mathbf{0}$ for some $\mathbf{x}$, then we have

$$\mathbf{B}\mathbf{x}_B + \mathbf{N}\mathbf{x}_N = \mathbf{0} \quad \text{and} \quad \mathbf{x}_N = \mathbf{0}$$

It follows that

$$\mathbf{x}_B = -\mathbf{B}^{-1}\mathbf{N}\mathbf{x}_N = \mathbf{0}$$

and hence $\mathbf{x} = \mathbf{0}$. Therefore, $\hat{\mathbf{A}}_{a_k}$ is nonsingular. In summary, at a vertex $\mathbf{x}_k$ a working set of active constraints for the application of the simplex method can be obtained with three simple steps as follows:

(a) Select the columns in matrix $\mathbf{A}$ that correspond to the strictly positive components of $\mathbf{x}_k$ to form matrix $\hat{\mathbf{B}}$.

(b) If the number of columns in $\hat{\mathbf{B}}$ is equal to p, take $\mathbf{B} = \hat{\mathbf{B}}$; otherwise, $\hat{\mathbf{B}}$ is augmented with additional columns of $\mathbf{A}$ to form a square nonsingular matrix $\mathbf{B}$.

(c) Determine the index set $\mathcal{I}_N$ and form matrix $\mathbf{I}_N$.

Example 11.8 Identify working sets of active constraints at vertex $\mathbf{x} = [3\ 0\ 0\ 0]^T$ for the LP problem

$$\text{minimize } f(\mathbf{x}) = x_1 - 2x_2 - x_4$$

$$\text{subject to:} \qquad 3x_1 + 4x_2 + x_3 = 9$$
$$2x_1 + x_2 + x_4 = 6$$
$$x_1 \geq 0,\ x_2 \geq 0,\ x_3 \geq 0,\ x_4 \geq 0$$

Solution It is easy to verify that point $\mathbf{x} = [3\ 0\ 0\ 0]^T$ is a degenerate vertex at which there are five active constraints. Since x_1 is the only strictly positive component, $\hat{\mathbf{B}}$ contains only the first column of $\mathbf{A}$, i.e.,

$$\hat{\mathbf{B}} = \begin{bmatrix} 3 \\ 2 \end{bmatrix}$$

Matrix $\hat{\mathbf{B}}$ can be augmented, for example, by using the second column of $\mathbf{A}$ to generate a nonsingular $\mathbf{B}$ as

$$\mathbf{B} = \begin{bmatrix} 3 & 4 \\ 2 & 1 \end{bmatrix}$$

This leads to

$$\mathcal{I}_N = \{3, 4\} \quad \text{and} \quad \hat{\mathbf{A}}_a = \begin{bmatrix} 3 & 4 & 1 & 0 \\ 2 & 1 & 0 & 1 \\ 0 & 0 & 1 & 0 \\ 0 & 0 & 0 & 1 \end{bmatrix}$$

Since vertex $\mathbf{x}$ is degenerate, matrix $\hat{\mathbf{A}}_a$ is not unique. As a reflection of this nonuniqueness, there are two possibilities for augmenting $\hat{\mathbf{B}}$. Using the third column of $\mathbf{A}$ for the augmentation, we have

$$\mathbf{B} = \begin{bmatrix} 3 & 1 \\ 2 & 0 \end{bmatrix}$$

which gives

$$\mathcal{I}_N = \{2, 4\}$$

and

$$\hat{\mathbf{A}}_a = \begin{bmatrix} 3 & 4 & 1 & 0 \\ 2 & 1 & 0 & 1 \\ 0 & 1 & 0 & 0 \\ 0 & 0 & 0 & 1 \end{bmatrix}$$

Alternatively, augmenting $\hat{\mathbf{B}}$ with the fourth column of $\mathbf{A}$ yields

$$\mathbf{B} = \begin{bmatrix} 3 & 0 \\ 2 & 1 \end{bmatrix}$$

which gives

$$\mathcal{I}_N = \{2, 3\}$$

and

$$\hat{\mathbf{A}}_a = \begin{bmatrix} 3 & 4 & 1 & 0 \\ 2 & 1 & 0 & 1 \\ 0 & 1 & 0 & 0 \\ 0 & 0 & 1 & 0 \end{bmatrix}$$

It can be easily verified that all three $\hat{\mathbf{A}}_a$'s are nonsingular.

■

11.3.2.2 Algorithm for standard-form LP problem

Like Algorithms 11.1 and 11.2, an algorithm for the standard-form LP problem based on the simplex method can start with a vertex, and the steps of Algorithm 11.2 can serve as a framework for the implementation. A major difference from Algorithms 11.1 and 11.2 is that the special structure of the working-set matrix $\hat{\mathbf{A}}_{a_k}$ in Eq. (11.62) can be utilized in Steps 2 and 3, which would result in reduced computational complexity.

At a vertex $\mathbf{x}_k$, the nonsingularity of the working-set matrix $\hat{\mathbf{A}}_{a_k}$ given by Eq. (11.62) implies that there exist $\boldsymbol{\lambda}_k \in R^{p \times 1}$ and $\hat{\boldsymbol{\mu}}_k \in R^{(n-p) \times 1}$ such that

$$\mathbf{c} = \hat{\mathbf{A}}_{a_k}^T \begin{bmatrix} \boldsymbol{\lambda}_k \\ \hat{\boldsymbol{\mu}}_k \end{bmatrix} = \mathbf{A}^T \boldsymbol{\lambda}_k + \mathbf{I}_N^T \hat{\boldsymbol{\mu}}_k \tag{11.63}$$

If $\boldsymbol{\mu}_k \in R^{n \times 1}$ is the vector with zero basic variables and the components of $\hat{\boldsymbol{\mu}}_k$ as its nonbasic variables, then Eq. (11.63) can be expressed as

$$\mathbf{c} = \mathbf{A}^T \boldsymbol{\lambda}_k + \boldsymbol{\mu}_k \tag{11.64}$$

By virtue of Theorem 11.1, vertex $\mathbf{x}_k$ is a minimizer if and only if $\hat{\boldsymbol{\mu}}_k \geq \mathbf{0}$. If we use a permutation matrix, $\mathbf{P}$, to rearrange the components of $\mathbf{c}$ in accordance with the partition of $\mathbf{x}_k$ into basic and nonbasic variables as in Eq. (11.60), then Eq. (11.63) gives

$$\mathbf{Pc} = \begin{bmatrix} \mathbf{c}_B \\ \mathbf{c}_N \end{bmatrix} = \mathbf{PA}^T \boldsymbol{\lambda}_k + \mathbf{PI}_N^T \hat{\boldsymbol{\mu}}_k$$

$$= \begin{bmatrix} \mathbf{B}^T \\ \mathbf{N}^T \end{bmatrix} \boldsymbol{\lambda}_k + \begin{bmatrix} \mathbf{0} \\ \hat{\boldsymbol{\mu}}_k \end{bmatrix}$$

It follows that

$$\mathbf{B}^T \boldsymbol{\lambda}_k = \mathbf{c}_B \tag{11.65}$$

and

$$\hat{\boldsymbol{\mu}}_k = \mathbf{c}_N - \mathbf{N}^T \boldsymbol{\lambda}_k \tag{11.66}$$

Since $\mathbf{B}$ is nonsingular, $\boldsymbol{\lambda}_k$ and $\hat{\boldsymbol{\mu}}_k$ can be computed using Eqs. (11.65) and (11.66), respectively. Note that the system of equations that need to be solved is of size $p \times p$ rather than $n \times n$ as in Step 2 of Algorithms 11.1 and 11.2.

If some entry in $\hat{\boldsymbol{\mu}}_k$ is negative, then $\mathbf{x}_k$ is not a minimizer and a search direction $\mathbf{d}_k$ needs to be determined. Note that the Lagrange multipliers $\hat{\boldsymbol{\mu}}_k$ are *not* related to the equality constraints in Eq. (11.1b) but are related to those bound constraints in Eq. (11.1c) that are active and are associated with the nonbasic variables. If the search direction $\mathbf{d}_k$ is partitioned according to the basic and nonbasic variables, $\mathbf{x}_B$ and $\mathbf{x}_N$, into $\mathbf{d}_k^{(B)}$ and $\mathbf{d}_k^{(N)}$, respectively, and if $(\hat{\boldsymbol{\mu}}_k)_l < 0$, then assigning

$$\mathbf{d}_k^{(N)} = \mathbf{e}_l \tag{11.67}$$

where $\mathbf{e}_l$ is the lth column of the $(n - p) \times (n - p)$ identity matrix, yields a search direction $\mathbf{d}_k$ that makes the ν_lth constraint inactive without affecting other bound constraints that are associated with the nonbasic variables. In order to assure the feasibility of $\mathbf{d}_k$, it is also required that $\mathbf{A}\mathbf{d}_k = \mathbf{0}$ (see Theorem 11.4). This requirement can be described as

$$\mathbf{A}\mathbf{d}_k = \mathbf{B}\mathbf{d}_k^{(B)} + \mathbf{N}\mathbf{d}_k^{(N)} = \mathbf{B}\mathbf{d}_k^{(B)} + \mathbf{N}\mathbf{e}_l = \mathbf{0} \qquad (11.68)$$

where $\mathbf{N}\mathbf{e}_l$ is actually the ν_lth column of $\mathbf{A}$. Hence $\mathbf{d}_k^{(B)}$ can be determined by solving the system of equations

$$\mathbf{B}\mathbf{d}_k^{(B)} = -\mathbf{a}_{\nu_l} \qquad (11.69a)$$

where

$$\mathbf{a}_{\nu_l} = \mathbf{N}\mathbf{e}_l \qquad (11.69b)$$

Together, Eqs. (11.67) and (11.69) determine the search direction $\mathbf{d}_k$. From Eqs. (11.63), (11.67), and (11.68), it follows that

$$\mathbf{c}^T\mathbf{d}_k = \lambda_k^T\mathbf{A}\mathbf{d}_k + \hat{\boldsymbol{\mu}}_k^T\mathbf{I}_N\mathbf{d}_k = \hat{\boldsymbol{\mu}}_k^T\mathbf{d}_k^{(N)} = \hat{\boldsymbol{\mu}}_k^T\mathbf{e}_l$$
$$= (\hat{\boldsymbol{\mu}}_k)_l < 0$$

Therefore, $\mathbf{d}_k$ is a feasible descent direction. From Eqs. (11.67) and (11.69), it is observed that unlike the cases of Algorithms 11.1 and 11.2 where finding a feasible descent search direction requires the solution of a system of n equations (see Eqs. (11.52) and (11.58)), the present algorithm involves the solution of a system of p equations.

Considering the determination of step size α_k, we note that a point $\mathbf{x}_k + \alpha\mathbf{d}_k$ with any α satisfies the constraints in Eq. (11.1b), i.e.,

$$\mathbf{A}(\mathbf{x}_k + \alpha\mathbf{d}_k) = \mathbf{A}\mathbf{x}_k + \alpha\mathbf{A}\mathbf{d}_k = \mathbf{b}$$

Furthermore, Eq. (11.67) indicates that with any positive α, $\mathbf{x}_k + \alpha\mathbf{d}_k$ does not violate the constraints in Eq. (11.1c) that are associated with the nonbasic variables. Therefore, the only constraints that are sensitive to step size α_k are those that are associated with the basic variables and are decreasing along direction $\mathbf{d}_k$. When limited to the basic variables, $\mathbf{d}_k$ becomes $\mathbf{d}_k^{(B)}$. Since the normals of the constraints in Eq. (11.1c) are simply coordinate vectors, a bound constraint associated with a basic variable is decreasing along $\mathbf{d}_k$ if the associated component in $\mathbf{d}_k^{(B)}$ is negative. In addition, the special structure of the inequality constraints in Eq. (11.1c) also implies that the residual vector, when limited to basic variables in $\mathbf{x}_B$, is $\mathbf{x}_B$ itself.

The above analysis leads to a simple step that can be used to determine the index set

$$\mathcal{I}_k = \{i : (\mathbf{d}_k^{(B)})_i < 0\} \qquad (11.70)$$

and, if $\mathcal{I}_k$ is not empty, to determine α_k as

$$\alpha_k = \min_{i \in \mathcal{I}_k} \left[\frac{(\mathbf{x}_k^{(B)})_i}{(-\mathbf{d}_k^{(B)})_i} \right] \qquad (11.71)$$

where $\mathbf{x}_k^{(B)}$ denotes the vector for the basic variables of $\mathbf{x}_k$. If i^* is the index in $\mathcal{I}_k$ that achieves α_k, then the i^*th component of $\mathbf{x}_k^{(B)} + \alpha_k \mathbf{d}_k^{(B)}$ is zero. This zero component is then interchanged with the lth component of $\mathbf{x}_k^{(N)}$ which is now not zero but α_k. The vector $\mathbf{x}_k^{(B)} + \alpha \mathbf{d}_k^{(B)}$ after this updating becomes $\mathbf{x}_{k+1}^{(B)}$ and, of course, $\mathbf{x}_{k+1}^{(N)}$ remains a zero vector. Matrices $\mathbf{B}$ and $\mathbf{N}$ as well as the associated index sets $\mathcal{I}_B$ and $\mathcal{I}_N$ also need to be updated accordingly. An algorithm based on the above principles is as follows.

Algorithm 11.3 Simplex algorithm for the standard-form LP problem of Eq. (11.1)

Step 1
Input vertex $\mathbf{x}_0$, set $k = 0$, and form $\mathbf{B}$, $\mathbf{N}$, $\mathbf{x}_0^{(B)}$, $\mathcal{I}_B = \{\beta_1^{(0)}, \beta_2^{(0)}, \ldots, \beta_p^{(0)}\}$, and $\mathcal{I}_N = \{\nu_1^{(0)}, \nu_2^{(0)}, \ldots, \nu_{n-p}^{(0)}\}$.

Step 2
Partition vector $\mathbf{c}$ into $\mathbf{c}_B$ and $\mathbf{c}_N$.
Solve Eq. (11.65) for $\boldsymbol{\lambda}_k$ and compute $\hat{\boldsymbol{\mu}}_k$ using Eq. (11.66).
If $\hat{\boldsymbol{\mu}}_k \geq 0$, stop ($\mathbf{x}_k$ is a vertex minimizer); otherwise, select the index l that corresponds to the most negative component in $\hat{\boldsymbol{\mu}}_k$.

Step 3
Solve Eq. (11.69a) for $\mathbf{d}_k^{(B)}$ where $\mathbf{a}_{\nu_l}$ is the $\nu_l^{(k)}$th column of $\mathbf{A}$.

Step 4
Form index set $\mathcal{I}_k$ in Eq. (11.70).
If $\mathcal{I}_k$ is empty then stop (the objective function tends to $-\infty$ in the feasible region); otherwise, compute α_k using Eq. (11.71) and record the index i^* with $\alpha_k = (\mathbf{x}_k^{(B)})_{i^*}/(-\mathbf{d}^{(B)})_{i^*}$.

Step 5
Compute $\mathbf{x}_{k+1}^{(B)} = \mathbf{x}_k^{(B)} + \alpha_k \mathbf{d}_k^{(B)}$ and replace its i^*th zero component by α_k.
Set $\mathbf{x}_{k+1}^{(N)} = \mathbf{0}$.
Update $\mathbf{B}$ and $\mathbf{N}$ by interchanging the lth column of $\mathbf{N}$ with the i^*th column of $\mathbf{B}$.

Step 6
Update $\mathcal{I}_B$ and $\mathcal{I}_N$ by interchanging index $\nu_l^{(k)}$ of $\mathcal{I}_N$ with index $\beta_{i^*}^{(B)}$ of $\mathcal{I}_B$.

Use the $\mathbf{x}_{k+1}^{(B)}$ and $\mathbf{x}_{k+1}^{(N)}$ obtained in Step 5 in conjunction with $\mathcal{I}_B$ and $\mathcal{I}_N$ to form $\mathbf{x}_{k+1}$.

Set $k = k + 1$ and repeat from Step 2.

Example 11.9 Solve the standard-form LP problem

$$\text{minimize } f(\mathbf{x}) = 2x_1 + 9x_2 + 3x_3$$

subject to:
$$-2x_1 + 2x_2 + x_3 - x_4 = 1$$
$$x_1 + 4x_2 - x_3 - x_5 = 1$$
$$x_1 \geq 0, \ x_2 \geq 0, \ x_3 \geq 0, \ x_4 \geq 0, \ x_5 \geq 0$$

Solution From Eq. (11.1)

$$\mathbf{A} = \begin{bmatrix} -2 & 2 & 1 & -1 & 0 \\ 1 & 4 & -1 & 0 & -1 \end{bmatrix}, \quad \mathbf{b} = \begin{bmatrix} 1 \\ 1 \end{bmatrix}$$

and

$$\mathbf{c} = [2\ 9\ 3\ 0\ 0]^T$$

To identify a vertex, we set $x_1 = x_3 = x_4 = 0$ and solve the system

$$\begin{bmatrix} 2 & 0 \\ 4 & -1 \end{bmatrix} \begin{bmatrix} x_2 \\ x_5 \end{bmatrix} = \begin{bmatrix} 1 \\ 1 \end{bmatrix}$$

for x_2 and x_5. This leads to $x_2 = 1/2$ and $x_5 = 1$; hence

$$\mathbf{x}_0 = [0\ \tfrac{1}{2}\ 0\ 0\ 1]^T$$

is a vertex. Associated with $\mathbf{x}_0$ are $\mathcal{I}_B = \{2, 5\}, \mathcal{I}_N = \{1, 3, 4\}$

$$\mathbf{B} = \begin{bmatrix} 2 & 0 \\ 4 & -1 \end{bmatrix}, \quad \mathbf{N} = \begin{bmatrix} -2 & 1 & -1 \\ 1 & -1 & 0 \end{bmatrix}, \quad \text{and} \quad \mathbf{x}_0^{(B)} = \begin{bmatrix} \tfrac{1}{2} \\ 1 \end{bmatrix}$$

Partitioning $\mathbf{c}$ into

$$\mathbf{c}_B = [9\ 0]^T \quad \text{and} \quad \mathbf{c}_N = [2\ 3\ 0]^T$$

and solving Eq. (11.65) for $\boldsymbol{\lambda}_0$, we obtain $\boldsymbol{\lambda}_0 = [\tfrac{9}{2}\ 0]^T$. Hence Eq. (11.66) gives

$$\hat{\boldsymbol{\mu}}_0 = \begin{bmatrix} 2 \\ 3 \\ 0 \end{bmatrix} - \begin{bmatrix} -2 & 1 \\ 1 & -1 \\ -1 & 0 \end{bmatrix} \begin{bmatrix} \tfrac{9}{2} \\ 0 \end{bmatrix} = \begin{bmatrix} 11 \\ -\tfrac{2}{3} \\ \tfrac{9}{2} \end{bmatrix}$$

Since $(\hat{\boldsymbol{\mu}}_0)_2 < 0$, $\mathbf{x}_0$ is not a minimizer, and $l = 2$. Next, we solve Eq. (11.69a) for $\mathbf{d}_0^{(B)}$ with $\nu_2^{(0)} = 3$ and $\mathbf{a}_3 = [1\ -1]^T$, which yields

$$\mathbf{d}_0^{(B)} = \begin{bmatrix} -\tfrac{1}{2} \\ -3 \end{bmatrix} \quad \text{and} \quad \mathcal{I}_0 = \{1, 2\}$$

Hence

$$\alpha_0 = \min\left(1, \tfrac{1}{3}\right) = \tfrac{1}{3} \quad \text{and} \quad i^* = 2$$

To find $\mathbf{x}_1^{(B)}$, we compute

$$\mathbf{x}_0^{(B)} + \alpha_0 \mathbf{d}_0^{(B)} = \begin{bmatrix} \tfrac{1}{3} \\ 0 \end{bmatrix}$$

and replace its i^*th component by α_0, i.e.,

$$\mathbf{x}_1^{(B)} = \begin{bmatrix} \tfrac{1}{3} \\ \tfrac{1}{3} \\ \tfrac{1}{3} \end{bmatrix} \quad \text{with} \quad \mathbf{x}_1^{(N)} = \begin{bmatrix} 0 \\ 0 \end{bmatrix}$$

Now we update $\mathbf{B}$ and $\mathbf{N}$ as

$$\mathbf{B} = \begin{bmatrix} 2 & 1 \\ 4 & -1 \end{bmatrix} \quad \text{and} \quad \mathbf{N} = \begin{bmatrix} -2 & 0 & -1 \\ 1 & -1 & 0 \end{bmatrix}$$

and update $\mathcal{I}_B$ and $\mathcal{I}_N$ as $\mathcal{I}_B = \{2, 3\}$ and $\mathcal{I}_N = \{1, 5, 4\}$. The vertex obtained is

$$\mathbf{x}_1 = \begin{bmatrix} 0 & \tfrac{1}{3} & \tfrac{1}{3} & 0 & 0 \end{bmatrix}^T$$

to complete the first iteration.

The second iteration starts with the partitioning of $\mathbf{c}$ into

$$\mathbf{c}_B = \begin{bmatrix} 9 \\ 3 \end{bmatrix} \quad \text{and} \quad \mathbf{c}_N = \begin{bmatrix} 2 \\ 0 \\ 0 \end{bmatrix}$$

Solving Eq. (11.65) for $\boldsymbol{\lambda}_1$, we obtain $\boldsymbol{\lambda}_1 = [\tfrac{7}{2} \ \tfrac{1}{2}]^T$ which leads to

$$\hat{\boldsymbol{\mu}}_1 = \begin{bmatrix} 2 \\ 0 \\ 0 \end{bmatrix} - \begin{bmatrix} -2 & 1 \\ 0 & -1 \\ -1 & 0 \end{bmatrix} \begin{bmatrix} \tfrac{7}{2} \\ \tfrac{1}{2} \end{bmatrix} = \begin{bmatrix} \tfrac{17}{2} \\ \tfrac{1}{2} \\ \tfrac{7}{2} \end{bmatrix}$$

Since $\hat{\boldsymbol{\mu}}_1 > 0$, $\mathbf{x}_1$ is the unique vertex minimizer.

■

We conclude this section with a remark on the degenerate case. For a standard-form LP problem, a vertex $\mathbf{x}_k$ is degenerate if it has more than $n - p$ zero components. With the notation used in Sec. 11.3.2.1, the matrix $\hat{\mathbf{B}}$ associated with a degenerate vertex contains less than p columns and hence the index set $\mathcal{I}_B$ contains at least one index that corresponds to a *zero* component of $\mathbf{x}_k$. Consequently, the index set $\mathcal{I}_k$ defined by Eq. (11.70) may contain an index corresponding to a zero component of $\mathbf{x}_k$. If this happens, then obviously the step size determined using Eq. (11.71) is $\alpha_k = 0$, which would lead to $\mathbf{x}_{k+1} = \mathbf{x}_k$ and from this point on, cycling would occur. In order to prevent cycling, modifications should be made in Steps 2 and 4 of Algorithm 11.3, for example, using Bland's least-index rule.

11.3.3   Tabular form of the simplex method

For LP problems of very small size, the simplex method can be applied in terms of a *tabular form* in which the input data such as $\mathbf{A}$, $\mathbf{b}$, and $\mathbf{c}$ are used to form a table which evolves in a more explicit manner as simplex iterations proceed.

Consider the standard-form LP problem in Eq. (11.1) and assume that at vertex $\mathbf{x}_k$ the equality constraints are expressed as

$$\mathbf{x}_k^{(B)} + \mathbf{B}^{-1}\mathbf{N}\mathbf{x}_k^{(N)} = \mathbf{B}^{-1}\mathbf{b} \qquad (11.72)$$

From Eq. (11.64), the objective function is given by

$$\begin{aligned} \mathbf{c}^T\mathbf{x}_k &= \boldsymbol{\mu}_k^T\mathbf{x}_k + \boldsymbol{\lambda}_k^T\mathbf{A}\mathbf{x}_k \\ &= \mathbf{O}^T\mathbf{x}_k^{(B)} + \hat{\boldsymbol{\mu}}_k^T\mathbf{x}_k^{(N)} + \boldsymbol{\lambda}_k^T\mathbf{b} \qquad (11.73) \end{aligned}$$

So the important data at the kth iteration can be put together in a tabular form as shown in Table 11.1 from which we observe the following:

(a) If $\hat{\boldsymbol{\mu}}_k \geq 0$, $\mathbf{x}_k$ is a minimizer.
(b) Otherwise, an appropriate rule can be used to choose a negative component in $\hat{\boldsymbol{\mu}}_k$, say, $(\hat{\boldsymbol{\mu}}_k)_l < 0$. As can be seen in Eq. (11.69), the column in $\mathbf{B}^{-1}\mathbf{N}$ that is right above $(\hat{\boldsymbol{\mu}}_k)_l$ gives $-\mathbf{d}_k^{(B)}$. In the discussion that follows, this column will be referred to as the pivot column. In addition, the variable in $\mathbf{x}_N^T$ that corresponds to $(\hat{\boldsymbol{\mu}}_k)_l$ is the variable chosen as a *basic* variable.
(c) Since $\mathbf{x}_k^{(N)} = \mathbf{0}$, Eq. (11.72) implies that $\mathbf{x}_k^{(B)} = \mathbf{B}^{-1}\mathbf{b}$. Therefore, the far-right p-dimensional vector gives $\mathbf{x}_k^{(B)}$.
(d) Since $\mathbf{x}_k^{(N)} = \mathbf{0}$, Eq. (11.73) implies that the number in the lower-right corner of Table 11.1 is equal to $-f(\mathbf{x}_k)$.

Table 11.1 Simplex method, kth iteration

$\mathbf{x}_B^T$	$\mathbf{x}_N^T$	
$\mathbf{I}$	$\mathbf{B}^{-1}\mathbf{N}$	$\mathbf{B}^{-1}\mathbf{b}$
$\mathbf{O}^T$	$\hat{\boldsymbol{\mu}}_k^T$	$-\boldsymbol{\lambda}_k^T\mathbf{b}$

Taking the LP problem discussed in Example 11.8 as an example, at $\mathbf{x}_0$ the table assumes the form shown in Table 11.2. Since $(\hat{\boldsymbol{\mu}}_0)_2 < 0$, $\mathbf{x}_0$ is not a minimizer. As was shown above, $(\hat{\boldsymbol{\mu}}_0)_2 < 0$ also suggests that x_3 is the variable in $\mathbf{x}_0^{(N)}$ that will become a basic variable, and the vector above $(\hat{\boldsymbol{\mu}}_0)_2$, $\left[\frac{1}{2} \ 3\right]^T$, is the pivot column $-\mathbf{d}_0^{(B)}$. It follows from Eqs. (11.70) and (11.71) that only

Table 11.2 Simplex method, Example 11.8

Basic Variables		Nonbasic Variables				
x_2	x_5	x_1	x_3	x_4	$\mathbf{B}^{-1}\mathbf{b}$	$-\boldsymbol{\lambda}_k^T\mathbf{b}$
1	0	-1	$\frac{1}{2}$	$-\frac{1}{2}$	$\frac{1}{2}$	
0	1	-5	3	-2	1	
0	0	11	$-\frac{3}{2}$	$\frac{9}{2}$		$-\frac{9}{2}$

the *positive* components of the pivot column should be used to compute the ratio $(\mathbf{x}_0^{(B)})_i/(-\mathbf{d}_0^{(B)})_i$ where $\mathbf{x}_0^{(B)}$ is the far-right column in the table. The index that yields the minimum ratio is $i^* = 2$. This suggests that the second basic variable, x_5, should be exchanged with x_3 to become a nonbasic variable. To transform x_3 into the second basic variable, we use elementary row operations to transform the pivot column into the i^*th coordinate vector. In the present case, we add $-1/6$ times the second row to the first row, and then multiply the second row by $1/3$. The table assumes the form in Table 11.3.

Table 11.3 Simplex method, Example 11.8 continued

Basic Variables		Nonbasic Variables				
x_2	x_5	x_1	x_3	x_4	$\mathbf{B}^{-1}\mathbf{b}$	$-\boldsymbol{\lambda}_k^T\mathbf{b}$
1	$-\frac{1}{6}$	$-\frac{1}{6}$	0	$-\frac{1}{6}$	$\frac{1}{3}$	
0	$\frac{1}{3}$	$-\frac{5}{3}$	1	$-\frac{2}{3}$	$\frac{1}{3}$	
0	0	11	$-\frac{3}{2}$	$\frac{9}{2}$		$-\frac{9}{2}$

Next we interchange the columns associated with variables x_3 and x_5 to form the updated basic and nonbasic variables, and then add $3/2$ times the second row to the last row to eliminate the nonzero Lagrange multiplier associated with variable x_3. This leads to the table shown as Table 11.4.

Table 11.4 Simplex method, Example 11.8 continued

Basic Variables		Nonbasic Variables				
x_2	x_3	x_1	x_5	x_4	$\mathbf{B}^{-1}\mathbf{b}$	$-\lambda_k^T\mathbf{b}$
1	0	$-\frac{1}{6}$	$-\frac{1}{6}$	$-\frac{1}{6}$	$\frac{1}{3}$	
0	1	$-\frac{5}{3}$	$\frac{1}{3}$	$-\frac{2}{3}$	$\frac{1}{3}$	
0	0	$\frac{17}{2}$	$\frac{1}{2}$	$\frac{7}{2}$		-4

The Lagrange multipliers $\hat{\mu}_1$ in the last row of Table 11.4 are all positive and hence $\mathbf{x}_1$ is the unique minimizer. Vector $\mathbf{x}_1$ is specified by $\mathbf{x}_1^{(B)} = \begin{bmatrix} \frac{1}{3} & \frac{1}{3} \end{bmatrix}^T$ in the far-right column and $\mathbf{x}_1^{(N)} = [0\ 0\ 0]^T$. In conjunction with the composition of the basic and nonbasic variables, $\mathbf{x}_1^{(B)}$ and $\mathbf{x}_1^{(N)}$ yield

$$\mathbf{x}_1 = \begin{bmatrix} 0 & \frac{1}{3} & \frac{1}{3} & 0 & 0 \end{bmatrix}^T$$

At $\mathbf{x}_1$, the lower-right corner of Table 11.4 gives the minimum of the objective function as $f(\mathbf{x}_1) = 4$.

11.3.4 Computational complexity

As in any iterative algorithm, the computational complexity of a simplex algorithm depends on both the number of iterations it requires to converge and the amount of computation in each iteration.

11.3.4.1 Computations per iteration

For an LP problem of the type given in Eq. (11.2) with nondegenerate vertices, the major computational effort in each iteration is to solve two transposed $n \times n$ linear systems, i.e.,

$$\mathbf{A}_{a_k}^T \boldsymbol{\mu}_k = \mathbf{c} \quad \text{and} \quad \mathbf{A}_{a_k} \mathbf{d}_k = \mathbf{e}_l \tag{11.74}$$

(see Steps 2 and 3 of Algorithm 11.1). For the degenerate case, matrix $\mathbf{A}_{a_k}$ in Eq. (11.74) is replaced by working-set matrix $\hat{\mathbf{A}}_{a_k}$ which has the same size as $\mathbf{A}_{a_k}$. For the problem in Eq. (11.1), the computational complexity in each iteration is largely related to solving two transposed $p \times p$ linear systems, namely,

$$\mathbf{B}^T \boldsymbol{\lambda}_k = \mathbf{c}_B \quad \text{and} \quad \mathbf{B}\mathbf{d}_k^{(B)} = -\mathbf{a}_{\nu_l} \tag{11.75}$$

(see Steps 2 and 3 of Algorithm 11.3). Noticing the similarity between the systems in Eqs. (11.74) and (11.75), we conclude that the computational efficiency in each iteration depends critically on how efficiently two transposed linear systems of a given size are solved. A reliable and efficient approach to solve a linear system of equations in which the number of unknowns is equal to the number of equations (often called a *square system*) with a nonsingular asymmetric system matrix such as $\mathbf{A}_{a_k}$ in Eq. (11.74) and $\mathbf{B}$ in Eq. (11.75) is to use one of several matrix factorization-based methods. These include the LU factorization with pivoting and the Householder orthogonalization-based QR factorization [3][5]. The number of floating-point operations (flops) required to solve an n-variable square system using the LU factorization and QR factorization methods are $2n^3/3$ and $4n^3/3$, respectively, (see Sec. A.12). It should be stressed that although the QR factorization requires more flops, it is comparable with the LU factorization in efficiency when memory traffic and vectorization overhead are taken into account [5, Chap. 5]. Another desirable feature of the QR factorization method is the guaranteed numerical stability, particularly when the system is ill-conditioned.

For the systems in Eqs. (11.74) and (11.75), there are two important features that can lead to further reduction in the amount of computation. First, each of the two systems involves a pair of matrices that are the *transposes* of each other. So when matrix factorization is performed for the first system, the transposed version of the factorization can be utilized to solve the second system. Second, in each iteration, the matrix is obtained from the matrix used in the preceding iteration through a *rank-one modification*. Specifically, Step 5 of Algorithms 11.1 updates $\mathbf{A}_{a_k}$ by replacing one of its rows with the normal vector of the constraint that just becomes active, while Step 6 of Algorithm 11.3 updates $\mathbf{B}$ by replacing one of its columns with the column in $\mathbf{N}$ that corresponds to the new basic variable. Let

$$
\mathbf{A}_{a_k} = \begin{bmatrix} \mathbf{a}_{j_1}^T \\ \mathbf{a}_{j_2}^T \\ \vdots \\ \mathbf{a}_{j_n}^T \end{bmatrix}
$$

and assume that $\mathbf{a}_{i*}^T$ is used to replace $\mathbf{a}_{j_l}^T$ in the updating of $\mathbf{A}_{a_k}$ to $\mathbf{A}_{a_{k+1}}$. Under these circumstances

$$
\mathbf{A}_{a_{k+1}} = \mathbf{A}_{a_k} + \mathbf{\Delta}_a \tag{11.76a}
$$

where $\mathbf{\Delta}_a$ is the rank-one matrix

$$
\mathbf{\Delta}_a = \mathbf{e}_{j_l}(\mathbf{a}_{i*}^T - \mathbf{a}_{j_l}^T) \tag{11.76b}
$$

with $\mathbf{e}_{j_l}$ being the j_lth coordinate vector. Similarly, if we denote matrix $\mathbf{B}$ in the kth and $(k+1)$th iterations as $\mathbf{B}_k$ and $\mathbf{B}_{k+1}$, respectively, then

$$
\mathbf{B}_{k+1} = \mathbf{B}_k + \mathbf{\Delta}_b \tag{11.77a}
$$

$$
\mathbf{\Delta}_b = (\mathbf{b}_{i*}^{(k+1)} - \mathbf{b}_{i*}^{(k)})\mathbf{e}_{i*}^T \tag{11.77b}
$$

where $\mathbf{b}_{i*}^{(k+1)}$ and $\mathbf{b}_{i*}^{(k)}$ are the $i*$th columns in $\mathbf{B}_{k+1}$ and $\mathbf{B}_k$, respectively. Efficient algorithms for updating the LU and QR factorizations of a matrix with a rank-one modification, which require only $O(n^2)$ flops, are available in the literature. The reader is referred to [3, Chap. 4], [5, Chap. 12], [7, Chap. 3], and Sec. A.12 for the details.

As a final remark on the matter, LP problems encountered in practice often involve a large number of parameters and the associated large-size system matrix $\mathbf{A}_{a_k}$ or $\mathbf{B}$ is often very *sparse*.[1] Sparse linear systems can be solved using specially designed algorithms that take full advantage of either particular patterns of sparsity that the system matrix exhibits or the general sparse nature of the matrix. Using these algorithms, reduction in the number of flops as well as the required storage space can be significant. (See Sec. 2.7 of [8] for an introduction to several useful methods and further references on the subject.)

11.3.4.2 Performance in terms of number of iterations

The number of iterations required for a given LP problem to converge depends on the data that specify the problem and on the initial point, and is difficult to predict accurately [3]. As far as the simplex method is concerned, there is a worse-case analysis on the computational complexity of the method on the one hand, and observations on the algorithm's practical performance on the other hand.

Considering the alternative-form LP problem in Eq. (11.2), in the worst case, the simplex method entails examining *every* vertex to find the minimizer. Consequently, the number of iterations would grow exponentially with the problem size. In 1972, Klee and Minty [9] described the following well-known LP problem

$$\text{maximize} \sum_{j=1}^{n} 10^{n-j} x_j \tag{11.78a}$$

$$\text{subject to:} \quad x_i + 2 \sum_{j=1}^{i-1} 10^{i-j} x_j \le 100^{i-1} \quad \text{for } i = 1, 2, \ldots, n \tag{11.78b}$$

$$x_j \ge 0 \quad \text{for } j = 1, 2, \ldots, n \tag{11.78c}$$

For each n, the LP problem involves $2n$ inequality constraints. By introducing n slack variables $s_1, s_2, \ldots, s_n$ and adding them to the constraints in Eq. (11.78b) to convert the constraints into equalities, it was shown that if we start with the initial point $s_i = 100^{i-1}$ and $x_i = 0$ for $i = 1, 2, \ldots, n$, then the simplex method has to perform $2^n - 1$ iterations to obtain the solution. However, the chances of encountering the worst case scenario in a real-life LP problem are extremely small. In fact, the simplex method is usually very efficient,

---

[1] A matrix is said to be sparse if only a relatively small number of its elements are nonzero.

368

and consistently requires a number of iterations that is a small multiple of the problem dimension [10], typically, 2 or 3 times.

References

1 G. B. Dantzig, "Programming in a linear structure," *Comptroller*, USAF, Washington, D.C., Feb. 1948.
2 G. B. Dantzig, *Linear Programming and Extensions*, Princeton University Press, Princeton, NJ, 1963.
3 P. E. Gill, W. Murray, and M. H. Wright, *Numerical Linear Algebra and Optimization*, vol. I, Addison-Wesley, Reading, 1991.
4 R. Saigal, *LP problem: A Modern Integrated Analysis*, Kluwer Academic, Norwell, 1995.
5 G. H. Golub and C. F. Van Loan, *Matrix Computation*, 2nd ed., The Johns Hopkins University Press, Baltimore, 1989.
6 R. G. Bland, "New finite pivoting rules for the simplex method," *Math. Operations Research*, vol. 2, pp. 103–108, May 1977.
7 J. E. Dennis, Jr. and R. B. Schnabel, *Numerical Methods for Unconstrained Optimization and Nonlinear Equations*, SIAM, Philadelphia, 1996.
8 W. H. Press, S. A. Teukolsky, W. T. Vetterling, and B. P. Flannery, *Numerical Recipes in C*, 2nd ed., Cambridge University Press, Cambridge, UK, 1992.
9 V. Klee and G. Minty, "How good is the simplex method?" in *Inequalities*, O. Shisha ed., pp. 159–175, Academic Press, New York, 1972.
10 M. H. Wright, "Interior methods for constrained optimization," *Acta Numerica*, vol. 1, pp. 341–407, 1992.

Problems

11.1 (a) Develop a MATLAB function to generate the data matrices $\mathbf{A}$, $\mathbf{b}$, and $\mathbf{c}$ for the LP problem formulated in Prob. 10.1. Inputs of the function should include the order of polynomial $A(\omega)$, n, passband edge ω_p, stopband edge ω_a, number of grid points in the passband, N, and number of grid points in the stopband, M.

(b) Applying the MATLAB function obtained in part (a) with $n = 30$, $\omega_p = 0.45\pi$, $\omega_a = 0.55\pi$, and $M = N = 30$, obtain matrices $\mathbf{A}$, $\mathbf{b}$, and $\mathbf{c}$ for Prob. 10.1.

11.2 (a) Develop a MATLAB function that would find a vertex of the feasible region defined by

$$\mathbf{A}\mathbf{x} \geq \mathbf{b} \qquad (P11.1)$$

The function may look like x=find_v(A,b,x0) and should accept a general pair $(\mathbf{A},\ \mathbf{b})$ that defines a nonempty feasible region through (P11.1), and a feasible initial point $\mathbf{x}_0$.

(b) Test the MATLAB function obtained by applying it to the LP problem in Example 11.2 using several different initial points.

(c) Develop a MATLAB function that would find a vertex of the feasible region defined by $\mathbf{A}\mathbf{x} = \mathbf{b}$ and $\mathbf{x} \geq 0$.

11.3 (*a*) Develop a MATLAB function that would implement Algorithm 11.1. The function may look like x=lp_nd1(A,b,c,x0) where x_0 is a feasible initial point.

(*b*) Apply the MATLAB function obtained to the LP problems in Examples 11.2 and 11.6.

11.4 (*a*) Develop a MATLAB function that would implement Algorithm 11.1 without requiring a feasible initial point. The code can be developed by implementing the technique described in the first part of Sec. 11.2.3.4 using the code obtained from Prob. 11.3(*a*).

(*b*) Apply the MATLAB function obtained to the LP problems in Examples 11.2 and 11.6.

11.5 In connection with the LP problem in Eq. (11.2), use Farkas' Lemma (see Prob. 10.18) to show that if **x** is a feasible point but not a minimizer, then at **x** there always exists a feasible descent direction.

11.6 (*a*) Using a graphical approach, describe the feasible region $\mathcal{R}$ defined by

$$x_1 \geq 0$$
$$x_2 \geq 0$$
$$x_1 + x_2 - 1 \geq 0$$
$$x_1 - 2x_2 + 4 \geq 0$$
$$x_1 - x_2 + 1 \geq 0$$
$$-5x_1 + 2x_2 + 15 \geq 0$$
$$-5x_1 + 6x_2 + 5 \geq 0$$
$$-x_1 - 4x_2 + 14 \geq 0$$

(*b*) Identify the degenerate vertices of $\mathcal{R}$.

11.7 (*a*) By modifying the MATLAB function obtained in Prob. 11.3(*a*), implement Algorithm 11.2. The function may look like x=lp_d1(A,b,c, x0) where x_0 is a feasible initial point.

(*b*) Apply the MATLAB function obtained to the LP problem

$$\text{minimize } f(\mathbf{x}) = x_1$$

$$\text{subject to: } \mathbf{x} \in \mathcal{R}$$

where $\mathcal{R}$ is the polygon described in Prob. 11.6(a).

11.8 Consider the LP problem

$$\text{minimize } f(\mathbf{x}) = -2x_1 - 3x_2 + x_3 + 12x_4$$

$$\text{subject to:} \quad 2x_1 + 9x_2 - x_3 - 9x_4 \geq 0$$
$$-x_1/3 - x_2 + x_3/3 + 2x_4 \geq 0$$
$$x_i \geq 0 \qquad \text{for } i = 1, \, 2, \, 3, \, 4$$

(See [3, p. 351].)

(a) Show that this LP problem does not have finite minimizers.
   Hint: Any points of the form $[r \; 0 \; r \; 0]^T$ with $r \geq 0$ are feasible.

(b) Apply Algorithm 11.1 to the LP problem using $\mathbf{x}_0 = \mathbf{0}$ as a starting point, and observe the results.

(c) Apply Algorithm 11.2 to the LP problem using $\mathbf{x}_0 = \mathbf{0}$ as a starting point.

11.9 Applying an appropriate LP algorithm, solve the problem

$$\text{minimize } f(\mathbf{x}) = -4x_1 - 8x_3$$

$$\text{subject to:} \quad 16x_1 - x_2 + 5x_3 \leq 1$$
$$2x_1 + 4x_3 \leq 1$$
$$10x_1 + x_2 \leq 1$$
$$x_i \leq 1 \qquad \text{for } i = 1, \, 2, \, 3$$

11.10 Applying Algorithm 11.1, solve the LP problem

$$\text{minimize } f(\mathbf{x}) = x_1 - 4x_2$$

$$\text{subject to:} \quad -x_1 + x_2 + 2 \geq 0$$
$$-x_1 - x_2 + 6 \geq 0$$
$$x_i \geq 0 \qquad \text{for } i = 1, \, 2$$

Draw the path of the simplex steps using $\mathbf{x}_0 = [2 \; 0]^T$ as a starting point.

11.11 Applying Algorithm 11.2, solve the LP problem

$$\text{minimize } f(\mathbf{x}) = 2x_1 - 6x_2 - x_3$$

$$\text{subject to:} \quad -3x_1 + x_2 - 2x_3 + 7 \geq 0$$
$$2x_1 - 4x_2 + 12 \geq 0$$
$$4x_1 - 3x_2 - 3x_3 + 14 \geq 0$$
$$x_i \geq 0 \qquad \text{for } i = 1, \, 2, \, 3$$

11.12 Applying Algorithm 11.2, solve the LP problem described in Prob. 10.1 with $n = 30$, $\omega_p = 0.45\pi$, $\omega_a = 0.55\pi$, and $M = N = 30$. Note that the matrices $\mathbf{A}$, $\mathbf{b}$, and $\mathbf{c}$ of the problem can be generated using the MATLAB function developed in Prob. 11.1.

11.13 (*a*) Develop a MATLAB function that would implement Algorithm 11.3.

(*b*) Apply the MATLAB function obtained in part (*a*) to the LP problem in Example 11.8.

11.14 (*a*) Convert the LP problem in Prob. 11.10 to a standard-form LP problem by introducing slack variables.

(*b*) Apply Algorithm 11.3 to the LP problem obtained in part (*a*) and compare the results with those obtained in Prob. 11.10.

11.15 (*a*) Convert the LP problem in Prob. 11.11 to a standard-form LP problem by introducing slack variables.

(*b*) Apply Algorithm 11.3 to the LP problem obtained in part (*a*) and compare the results with those of Prob. 11.11.

11.16 Applying Algorithm 11.3, solve the LP problem

$$\text{minimize } f(\mathbf{x}) = x_1 + 1.5x_2 + x_3 + x_4$$

$$\text{subject to:} \quad x_1 + 2x_2 + x_3 + 2x_4 = 3$$
$$x_1 + x_2 + 2x_3 + 4x_4 = 5$$
$$x_i \geq 0 \qquad \text{for } i = 1,\ 2,\ 3,\ 4$$

11.17 Applying Algorithm 11.3, solve the LP problem

$$\text{minimize } f(\mathbf{x}) = x_1 + 0.5x_2 + 2x_3$$

$$\text{subject to:} \quad x_1 + x_2 + 2x_3 = 3$$
$$2x_1 + x_2 + 3x_3 = 5$$
$$x_i \geq 0 \qquad \text{for } i = 1,\ 2,\ 3$$

11.18 Based on the remarks given at the end of Sec. 11.3.2, develop a step-by-step description of an algorithm that extends Algorithm 11.3 to the degenerate case.

11.19 Develop a MATLAB function to implement the algorithm developed in Prob. 11.18.

11.20 (*a*) Convert the LP problem in Prob. 11.8 to a standard-form LP problem. Note that only *two* slack variables need to be introduced.

(*b*) Apply Algorithm 11.3 to the problem formulated in part (*a*) using an initial point $\mathbf{x}_0 = \mathbf{0}$, and observe the results.

(*c*) Applying the algorithm developed in Prob. 11.18, solve the problem formulated in part (*a*) using an initial point $\mathbf{x}_0 = \mathbf{0}$.

372

11.21 Consider the nonlinear minimization problem

$$\text{minimize } f(\mathbf{x}) = -2x_1 - 2.5x_2$$
$$\text{subject to: } 1 - x_1^2 - x_2^2 \geq 0$$
$$x_1 \geq 0, \ x_2 \geq 0$$

(a) Find an approximate solution of this problem by solving the LP problem with the same linear objective function subject to $\mathbf{x} \in \mathcal{P}$ where $\mathcal{P}$ is a polygon in the first quadrant of the $(x_1, \ x_2)$ plane that contains the feasible region described above.

(b) Improve the approximate solution obtained in part (a) by using a polygon with an increased number of edges.

Chapter 12

LINEAR PROGRAMMING
PART II: INTERIOR-POINT METHODS

12.1   Introduction

A paper by Karmarkar in 1984 [1] and substantial progress made since that time have led to the field of modern *interior-point methods* for linear programming (LP). Unlike the family of simplex methods considered in Chap. 11, which approach the solution through a sequence of iterates that move from vertex to vertex along the edges on the boundary of the feasible polyhedron, the iterates generated by interior-point algorithms approach the solution from the *interior* of a polyhedron. Although the claims about the efficiency of the algorithm in [1] have not been substantiated in general, extensive computational testing has shown that a number of interior-point algorithms are much more efficient than simplex methods for large-scale LP problems [2].

In this chapter, we study several representative interior-point methods. Our focus will be on algorithmic development rather than theoretical analysis of the methods. *Duality* is a concept of central importance in modern interior-point methods. In Sec. 12.2, we discuss several basic concepts of a duality theory for linear programming. These include primal-dual solutions and central path. Two important primal interior-point methods, namely, the primal affine-scaling method and the primal Newton barrier method will be studied in Secs. 12.3 and 12.4, respectively. In Sec. 12.5, we present two primal-dual path-following methods. One of these methods, namely, Mehrotra's predictor-corrector algorithm [3], has been the basis of most interior-point software for LP developed since 1990.

12.2 Primal-Dual Solutions and Central Path

12.2.1 Primal-dual solutions

The concept of duality was first introduced in Sec. 10.9 for the general convex programming problem (10.107) and the main results of the Wolfe dual, namely, the results of Theorem 10.9 as applied to LP problems were briefly discussed in Example 10.16. In this section, we present several additional results concerning duality, which are of importance for the development of modern interior-point methods.

Consider the standard-form LP problem

$$\text{minimize } f(\mathbf{x}) = \mathbf{c}^T \mathbf{x} \tag{12.1a}$$

$$\text{subject to:} \quad \mathbf{Ax} = \mathbf{b} \tag{12.1b}$$

$$\mathbf{x} \geq \mathbf{0} \tag{12.1c}$$

where matrix $\mathbf{A} \in R^{p \times n}$ is of full row rank as the primal problem (see Sec. 10.9). By applying Theorem 10.9 to Eq. (12.1), we obtain the *dual* problem

$$\text{maximize} \quad h(\boldsymbol{\lambda}) = \mathbf{b}^T \boldsymbol{\lambda} \tag{12.2a}$$

$$\text{subject to:} \quad \mathbf{A}^T \boldsymbol{\lambda} + \boldsymbol{\mu} = \mathbf{c} \tag{12.2b}$$

$$\boldsymbol{\mu} \geq \mathbf{0} \tag{12.2c}$$

(see Example 10.16).

Two basic questions concerning the LP problems in Eqs. (12.1) and (12.2) are:

(a) Under what conditions will the solutions of these problems exist?

(b) How are the feasible points and solutions of the primal and dual related?

An LP problem is said to be *feasible* if its feasible region is not empty. The problem in Eq. (12.1) is said to be *strictly feasible* if there exists an $\mathbf{x}$ that satisfies Eq. (12.1b) with $\mathbf{x} > \mathbf{0}$. Likewise, the LP problem in Eq. (12.2) is said to be strictly feasible if there exist $\boldsymbol{\lambda}$ and $\boldsymbol{\mu}$ that satisfy Eq. (12.2b) with $\boldsymbol{\mu} > \mathbf{0}$. It is known that $\mathbf{x}^*$ is a minimizer of the problem in Eq. (12.1) if and only if there exist $\boldsymbol{\lambda}^*$ and $\boldsymbol{\mu}^* \geq \mathbf{0}$ such that

$$\mathbf{A}^T \boldsymbol{\lambda}^* + \boldsymbol{\mu}^* = \mathbf{c} \tag{12.3a}$$

$$\mathbf{Ax}^* = \mathbf{b} \tag{12.3b}$$

$$x_i^* \mu_i^* = 0 \quad \text{for } 1 \leq i \leq n \tag{12.3c}$$

$$\mathbf{x}^* \geq \mathbf{0}, \quad \boldsymbol{\mu}^* \geq \mathbf{0} \tag{12.3d}$$

For the primal problem, $\boldsymbol{\lambda}^*$ and $\boldsymbol{\mu}^*$ in Eq. (12.3) are the Lagrange multipliers. It can be readily verified that a set of vectors $\{\boldsymbol{\lambda}^*, \boldsymbol{\mu}^*\}$ satisfying Eq. (12.3) is a maximizer for the dual problem in Eq. (12.2), and x^* in Eq. (12.3) may be interpreted as the Lagrange multipliers for the dual problem. A set $\{x^*, \lambda^*, \mu^*\}$ satisfying Eq. (12.3) is called a *primal-dual solution*. It follows that $\{x^*, \lambda^*, \mu^*\}$ is a primal-dual solution if and only if x^* solves the primal and $\{\lambda^*, \mu^*\}$ solves the dual [3]. The next two theorems address the existence and boundedness of primal-dual solutions.

Theorem 12.1 *Existence of a primal-dual solution* A *primal-dual solution exists if the primal and dual problems are both feasible.*

Proof If point x is feasible for the LP problem in Eq. (12.1) and $\{\lambda, \mu\}$ is feasible for the LP problem in Eq. (12.2), then set

$$\lambda^T b \leq \lambda^T b + \mu^T x = \lambda^T A x + \mu^T x$$
$$= (A^T \lambda + \mu)^T x = c^T x \qquad (12.4)$$

Since $f(x) = c^T x$ has a finite lower bound in the feasible region, there exists a set $\{x^*, \lambda^*, \mu^*\}$ that satisfies Eq. (12.3). Evidently, this x^* solves the problem in Eq. (12.1). From Eq. (12.4), $h(\lambda)$ has a finite upper bound and $\{\lambda^*, \mu^*\}$ solves the problem in Eq. (12.2). Consequently, the set $\{x^*, \lambda^*, \mu^*\}$ is a primal-dual solution. ∎

Theorem 12.2 *Strict feasibility of primal-dual solutions* *If the primal and dual problems are both feasible, then*

(a) *solutions of the primal problem are bounded if the dual is strictly feasible;*
(b) *solutions of the dual problem are bounded if the primal is strictly feasible;*
(c) *primal-dual solutions are bounded if the primal and dual are both strictly feasible.*

Proof The statement in (c) is an immediate consequence of (a) and (b). To prove (a), we first note that by virtue of Theorem 12.1 a solution of the primal exists. Below we follow [3] to show the boundedness. Let $\{\lambda, \mu\}$ be strictly feasible for the dual, x be feasible for the primal, and x^* be a solution of the primal. It follows that

$$\mu^T x^* = (c - A^T \lambda)^T x^*$$
$$= c^T x^* - \lambda^T A x^* = c^T x^* - \lambda^T b$$
$$\leq c^T x - \lambda^T b = \mu^T x$$

Since $x^* \geq 0$ and $\mu > 0$, we conclude that

$$\mu_i^* x_i^* \leq \mu^T x^* \leq \mu^T x$$

Hence

$$x_i^* \leq \frac{1}{\mu_i^*} \boldsymbol{\mu}^T \mathbf{x} \leq \max_{1 \leq i \leq n} \left(\frac{1}{\mu_i^*} \right) \cdot \boldsymbol{\mu}^T \mathbf{x}$$

and $\mathbf{x}^*$ is bounded.

Part (b) can be proved in a similar manner.

∎

From Eq. (12.3), we observe that

$$\mathbf{c}^T \mathbf{x}^* = [(\boldsymbol{\mu}^*)^T + (\boldsymbol{\lambda}^*)^T \mathbf{A}] \mathbf{x}^* = (\boldsymbol{\lambda}^*)^T \mathbf{A} \mathbf{x}^* = (\boldsymbol{\lambda}^*)^T \mathbf{b} \qquad (12.5)$$

i.e.,

$$f(\mathbf{x}^*) = h(\boldsymbol{\lambda}^*)$$

If we define the *duality gap* as

$$\delta(\mathbf{x}, \boldsymbol{\lambda}) = \mathbf{c}^T \mathbf{x} - \mathbf{b}^T \boldsymbol{\lambda} \qquad (12.6)$$

then Eq. (12.4) and Eq. (12.5) imply that $\delta(\mathbf{x}, \boldsymbol{\lambda})$ is always nonnegative with $\delta(\mathbf{x}^*, \boldsymbol{\lambda}^*) = 0$. Moreover, for any feasible $\mathbf{x}$ and $\boldsymbol{\lambda}$, we have

$$\mathbf{c}^T \mathbf{x} \geq \mathbf{c}^T \mathbf{x}^* = \mathbf{b}^T \boldsymbol{\lambda}^* \geq \mathbf{b}^T \boldsymbol{\lambda}$$

Hence

$$0 \leq \mathbf{c}^T \mathbf{x} - \mathbf{c}^T \mathbf{x}^* \leq \delta(\mathbf{x}, \boldsymbol{\lambda}) \qquad (12.7)$$

Eq. (12.7) indicates that the duality gap can serve as a bound on the closeness of $f(\mathbf{x})$ to $f(\mathbf{x}^*)$ [2].

12.2.2 Central path

Another important concept related to primal-dual solutions is central path. By virtue of Eq. (12.3), set $\{\mathbf{x}, \boldsymbol{\lambda}, \boldsymbol{\mu}\}$ with $\mathbf{x} \in R^n$, $\boldsymbol{\lambda} \in R^p$, and $\boldsymbol{\mu} \in R^n$ is a primal-dual solution if it satisfies the conditions

$$\mathbf{A}\mathbf{x} = \mathbf{b} \qquad \text{with } \mathbf{x} \geq \mathbf{0} \qquad (12.8a)$$
$$\mathbf{A}^T \boldsymbol{\lambda} + \boldsymbol{\mu} = \mathbf{c} \qquad \text{with } \boldsymbol{\mu} \geq \mathbf{0} \qquad (12.8b)$$
$$\mathbf{X}\boldsymbol{\mu} = \mathbf{0} \qquad (12.8c)$$

where $\mathbf{X} = \text{diag}\{x_1, x_2, \ldots, x_n\}$. The central path for a standard-form LP problem is defined as a set of vectors $\{\mathbf{x}(\tau), \boldsymbol{\lambda}(\tau), \boldsymbol{\mu}(\tau)\}$ that satisfy the conditions

$$\mathbf{A}\mathbf{x} = \mathbf{b} \qquad \text{with } \mathbf{x} > \mathbf{0} \qquad (12.9a)$$
$$\mathbf{A}^T \boldsymbol{\lambda} + \boldsymbol{\mu} = \mathbf{c} \qquad \text{with } \boldsymbol{\mu} > \mathbf{0} \qquad (12.9b)$$
$$\mathbf{X}\boldsymbol{\mu} = \tau \mathbf{e} \qquad (12.9c)$$

where τ is a strictly positive scalar parameter, and $\mathbf{e} = [1 \ 1 \ \cdots \ 1]^T$. For each fixed $\tau > 0$, the vectors in the set $\{\mathbf{x}(\tau), \boldsymbol{\lambda}(\tau), \boldsymbol{\mu}(\tau)\}$ satisfying Eq. (12.9) can be viewed as sets of points in R^n, R^p, and R^n, respectively, and when τ varies, the corresponding points form a set of trajectories called the *central path*. On comparing Eq. (12.9) with Eq. (12.8), it is obvious that the central path is closely related to the primal-dual solutions. From Eqs. (12.9a) and (12.9b), every point on the central path is strictly feasible. Hence the central path lies in the interior of the feasible regions of the problems in Eqs. (12.1) and (12.2), and it approaches a primal-dual solution as $\tau \to 0$.

A more explicit relation of the central path with the primal-dual solution can be observed using the duality gap defined in Eq. (12.6). Given $\tau > 0$, let $\{\mathbf{x}(\tau), \boldsymbol{\lambda}(\tau), \boldsymbol{\mu}(\tau)\}$ be on the central path. From Eq. (12.9), the duality gap $\delta[\mathbf{x}(\tau), \boldsymbol{\lambda}(\tau)]$ is given by

$$
\begin{aligned}
\delta[\mathbf{x}(\tau), \boldsymbol{\lambda}(\tau)] &= \mathbf{c}^T \mathbf{x}(\tau) - \mathbf{b}^T \boldsymbol{\lambda}(\tau) \\
&= [\boldsymbol{\lambda}^T(\tau)\mathbf{A} + \boldsymbol{\mu}^T(\tau)]\mathbf{x}(\tau) - \mathbf{b}^T \boldsymbol{\lambda}(\tau) \\
&= \boldsymbol{\mu}^T(\tau)\mathbf{x}(\tau) = n\tau
\end{aligned}
\tag{12.10}
$$

Hence the duality gap along the central path converges linearly to zero as τ approaches zero. Consequently, as $\tau \to 0$ the objective function of the primal problem, $\mathbf{c}^T \mathbf{x}(\tau)$, and the objective function of the dual problem, $\mathbf{b}^T \boldsymbol{\lambda}(\tau)$, approach the same optimal value.

Example 12.1 Sketch the central path of the LP problem

$$
\begin{aligned}
\text{minimize } & f(\mathbf{x}) = -2x_1 + x_2 - 3x_3 \\
\text{subject to: } & x_1 + x_2 + x_3 = 1 \\
& x_1 \geq 0, \ x_2 \geq 0, \ x_3 \geq 0
\end{aligned}
$$

Solution With $\mathbf{c} = [-2 \ 1 \ -3]^T$, $\mathbf{A} = [1 \ 1 \ 1]$, and $\mathbf{b} = 1$, Eq. (12.9) becomes

$$x_1 + x_2 + x_3 = 1 \tag{12.11a}$$
$$\lambda + \mu_1 = -2 \tag{12.11b}$$
$$\lambda + \mu_2 = 1 \tag{12.11c}$$
$$\lambda + \mu_3 = -3 \tag{12.11d}$$
$$x_1 \mu_1 = \tau \tag{12.11e}$$
$$x_2 \mu_2 = \tau \tag{12.11f}$$
$$x_3 \mu_3 = \tau \tag{12.11g}$$

where $x_i > 0$ and $\mu_i > 0$ for $i = 1, 2, 3$. From Eqs. (12.11b) – (12.11d), we have

$$\mu_1 = -2 - \lambda \tag{12.12a}$$

$$\mu_2 = 1 - \lambda \qquad (12.12b)$$

$$\mu_3 = -3 - \lambda \qquad (12.12c)$$

Hence $\mu_i > 0$ for $1 \le i \le 3$ if

$$\lambda < -3 \qquad (12.13)$$

If we assume that λ satisfies Eq. (12.13), then Eqs. (12.11e) – (12.11g) and (12.11a) yield

$$-\frac{1}{2+\lambda} + \frac{1}{1-\lambda} - \frac{1}{3+\lambda} = \frac{1}{\tau}$$

i.e.,

$$\frac{1}{\tau}\lambda^3 + \left(\frac{4}{\tau} + 3\right)\lambda^2 + \left(\frac{1}{\tau} + 8\right)\lambda + \left(1 - \frac{6}{\tau}\right) = 0 \qquad (12.14)$$

The central path can now be constructed by finding a root of Eq. (12.14), $\hat{\lambda}$, that satisfies Eq. (12.13), by computing μ_i for $1 \le i \le 3$ using Eq. (12.12) with $\lambda = \hat{\lambda}$, and then evaluating x_i for $1 \le i \le 3$ using Eq. (12.11) with $\lambda = \hat{\lambda}$. Fig. 12.1 shows the $\mathbf{x}(\tau)$ component of the central path for $\tau_0 = 5$ and $\tau_f = 10^{-4}$. Note that the entire trajectory lies inside the triangle which is the feasible region of the problem, and approaches vertex $[0\ 0\ 1]^T$ which is the unique minimizer of the LP problem.

∎

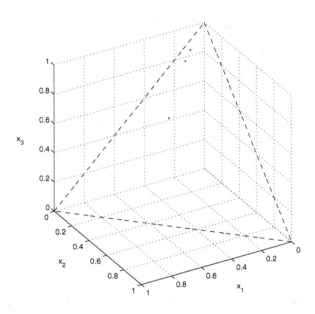

Figure 12.1. Trajectory of $\mathbf{x}(\tau)$ in Example 12.1.

12.3   Primal Affine-Scaling Method

The primal affine-scaling (PAS) method was proposed in [4][5] as a modification of Karmarkar's interior-point algorithm [1] for LP problems. Though conceptually simple, the method has been found effective, particularly for large-scale problems.

Consider the standard-form LP problem in Eq. (12.1) and let x_k be a strictly feasible point. The two major steps of the PAS method involve moving x_k along a *projected steepest-descent direction* and *scaling* the resulting point to center it in the feasible region in a transformed space.

For a linear objective function $f(x) = c^T x$, the steepest-descent direction is $-c$. At a feasible point x_k, moving along $-c$ does not guarantee the feasibility of the next iterate since Ac is most likely nonzero. The PAS method moves x_k along the direction that is the orthogonal projection of $-c$ onto the null space of A. This direction d_k is given by

$$d_k = -Pc \tag{12.15}$$

where P is the projection matrix given by

$$P = I - A^T (AA^T)^{-1} A \tag{12.16}$$

It can be readily verified that $AP = 0$. Hence if the next point is denoted as

$$x_{k+1} = x_k + \alpha_k d_k$$

then x_{k+1} satisfies Eq. (12.1b), i.e.,

$$\begin{aligned} Ax_{k+1} &= Ax_k + \alpha_k Ad_k \\ &= b - \alpha_k APc = b \end{aligned}$$

If matrix A is expressed in terms of its singular-value decomposition (SVD) as

$$A = U[\Sigma \ 0]V^T$$

where $U \in R^{p \times p}$ and $V \in R^{n \times n}$ are orthogonal and Σ is positive definite and diagonal, then the projection matrix becomes

$$P = V \begin{bmatrix} 0 & 0 \\ 0 & I_{n-p} \end{bmatrix} V^T$$

This gives

$$c^T Pc = \sum_{j=p+1}^{n} [(V^T c)_j]^2 \tag{12.17}$$

which is always nonnegative and strictly positive as long as one of the last $n-p$ components in $\mathbf{V}^T\mathbf{c}$ is nonzero. It follows that for any $\alpha_k > 0$

$$f(\mathbf{x}_{k+1}) = \mathbf{c}^T\mathbf{x}_{k+1} = \mathbf{c}^T\mathbf{x}_k - \alpha_k\mathbf{c}^T\mathbf{Pc}$$
$$\leq \mathbf{c}^T\mathbf{x}_k = f(\mathbf{x}_k)$$

and $f(\mathbf{x}_{k+1})$ will be strictly less than $f(\mathbf{x}_k)$ if at least one of the last $n-p$ components in $\mathbf{V}^T\mathbf{c}$ is nonzero.

The search direction, $\mathbf{d}_k$, determined by using Eqs. (12.15) and (12.16) is *independent* of the current point $\mathbf{x}_k$ and the progress that can be made along such a constant direction may become insignificant particularly when $\mathbf{x}_k$ is close to the boundary of the feasible region. A crucial step in the PAS method that overcomes this difficulty is to transform the original LP problem at the kth iteration from that in Eq. (12.1) to an equivalent LP problem in which point $\mathbf{x}_k$ is at a more 'central' position so as to achieve significant reduction in $f(\mathbf{x})$ along the projected steepest-descent direction.

For the standard-form LP problem in Eq. (12.1), the nonnegativity bounds in Eq. (12.1c) suggest that the point $\mathbf{e} = [1\ 1\ \cdots\ 1]^T$, which is situated at an equal distance from each x_i axis for $1 \leq i \leq n$, can be considered as a central point. The *affine scaling* transformation defined by

$$\bar{\mathbf{x}} = \mathbf{X}^{-1}\mathbf{x} \tag{12.18a}$$

with

$$\mathbf{X} = \text{diag}\{(\mathbf{x}_k)_1,\ (\mathbf{x}_k)_2,\ \ldots,\ (\mathbf{x}_k)_n\} \tag{12.18b}$$

maps point $\mathbf{x}_k$ to $\mathbf{e}$, and the equivalent LP problem given by this transformation is

$$\text{minimize } \bar{f}(\bar{\mathbf{x}}) = \bar{\mathbf{c}}^T\bar{\mathbf{x}} \tag{12.19a}$$
$$\text{subject to:}\quad \bar{\mathbf{A}}\bar{\mathbf{x}} = \mathbf{b} \tag{12.19b}$$
$$\bar{\mathbf{x}} \geq \mathbf{0} \tag{12.19c}$$

where $\bar{\mathbf{c}} = \mathbf{Xc}$ and $\bar{\mathbf{A}} = \mathbf{AX}$. If the next point is generated along the projected steepest-descent direction from $\mathbf{x}_k$, then

$$\bar{\mathbf{x}}_{k+1} = \bar{\mathbf{x}}_k + \alpha_k\bar{\mathbf{d}}_k = \mathbf{e} + \alpha_k\bar{\mathbf{d}}_k \tag{12.20}$$

where

$$\bar{\mathbf{d}}_k = -\bar{\mathbf{P}}\bar{\mathbf{c}} = -[\mathbf{I} - \bar{\mathbf{A}}^T(\bar{\mathbf{A}}\bar{\mathbf{A}}^T)^{-1}\bar{\mathbf{A}}]\bar{\mathbf{c}}$$
$$= -[\mathbf{I} - \mathbf{XA}^T(\mathbf{AX}^2\mathbf{A}^T)^{-1}\mathbf{AX}]\mathbf{Xc} \tag{12.21}$$

Equation (12.20) can be written in terms of the original variables as

$$\mathbf{x}_{k+1} = \mathbf{x}_k + \alpha_k\mathbf{d}_k \tag{12.22a}$$

with

$$\mathbf{d}_k = \mathbf{X}\bar{\mathbf{d}}_k = -[\mathbf{X}^2 - \mathbf{X}^2\mathbf{A}^T(\mathbf{A}\mathbf{X}^2\mathbf{A}^T)^{-1}\mathbf{A}\mathbf{X}^2]\mathbf{c} \quad (12.22b)$$

which is called the *primal affine-scaling direction* [6]. In order to compare the two search directions given by Eqs. (12.15) and (12.22b), we write vector $\mathbf{d}_k$ in Eq. (12.22b) as

$$\mathbf{d}_k = -\mathbf{X}\bar{\mathbf{P}}\mathbf{X}\mathbf{c} \quad (12.23)$$

with

$$\bar{\mathbf{P}} = \mathbf{I} - \mathbf{X}\mathbf{A}^T(\mathbf{A}\mathbf{X}^2\mathbf{A}^T)^{-1}\mathbf{A}\mathbf{X} \quad (12.24)$$

Note that matrix $\mathbf{P}$ in Eq. (12.16) is the projection matrix for $\mathbf{A}$ while matrix $\bar{\mathbf{P}}$ in Eq. (12.24) is the projection matrix for $\mathbf{A}\mathbf{X}$, which depends on both $\mathbf{A}$ and the present point $\mathbf{x}_k$. Consequently, $\mathbf{A}\mathbf{X}\bar{\mathbf{P}} = \mathbf{0}$, which in conjunction with Eqs. (12.22a) and (12.23) implies that if $\mathbf{x}_k$ is strictly feasible, then $\mathbf{x}_{k+1}$ satisfies Eq. (12.1b), i.e.,

$$\mathbf{A}\mathbf{x}_{k+1} = \mathbf{A}\mathbf{x}_k + \alpha_k\mathbf{A}\mathbf{d}_k = \mathbf{b} - \alpha_k\mathbf{A}\mathbf{X}\bar{\mathbf{P}}\mathbf{X}\mathbf{c} = \mathbf{b}$$

It can also be shown that for any $\alpha_k > 0$ in Eq. (12.22a),

$$f(\mathbf{x}_{k+1}) \le f(\mathbf{x}_k)$$

(see Prob. 12.2) and if at least one of the last $n - p$ components of $\mathbf{V}_k^T\mathbf{X}\mathbf{c}$ is nonzero, the above inequality becomes strict. Here matrix $\mathbf{V}_k$ is the $n \times n$ orthogonal matrix obtained from the SVD of $\mathbf{A}\mathbf{X}$, i.e.,

$$\mathbf{A}\mathbf{X} = \mathbf{U}_k[\boldsymbol{\Sigma}_k\ \mathbf{0}]\mathbf{V}_k^T$$

Having calculated the search direction $\mathbf{d}_k$ using Eq. (12.22b), the step size α_k in Eq. (12.22a) can be chosen such that $\mathbf{x}_{k+1} > \mathbf{0}$. In practice, α_k is chosen as [6]

$$\alpha_k = \gamma\alpha_{\max} \quad (12.25a)$$

where $0 < \gamma < 1$ is a constant, usually close to unity, and

$$\alpha_{\max} = \min_{i \text{ with } (\mathbf{d}_k)_i < 0} \left[-\frac{(\mathbf{x}_k)_i}{(\mathbf{d}_k)_i}\right] \quad (12.25b)$$

The PAS algorithm can now be summarized as follows.

382

Algorithm 12.1 Primal affine-scaling algorithm for the standard-form LP problem

Step 1
Input $\mathbf{A}$, $\mathbf{c}$, and a strictly feasible initial point $\mathbf{x}_0$.
Set $k = 0$ and initialize the tolerance ε.
Evaluate $f(\mathbf{x}_k) = \mathbf{c}^T\mathbf{x}_k$.
Step 2
Form $\mathbf{X}$ at $\mathbf{x}_k$ and compute $\mathbf{d}_k$ using Eq. (12.22b).
Step 3
Calculate the step size α_k using Eq. (12.25).
Step 4
Set $\mathbf{x}_{k+1} = \mathbf{x}_k + \alpha_k\mathbf{d}_k$ and evaluate $f(\mathbf{x}_{k+1}) = \mathbf{c}^T\mathbf{x}_{k+1}$.
Step 5
If
$$\frac{|f(\mathbf{x}_k) - f(\mathbf{x}_{k+1})|}{\max(1, |f(\mathbf{x}_k)|)} < \varepsilon$$
output $\mathbf{x}^* = \mathbf{x}_{k+1}$ and stop; otherwise, set $k = k + 1$ and repeat from Step 2.

Example 12.2 Solve the standard-form LP problem in Example 11.9 using the PAS algorithm.

Solution A strictly feasible initial point is $\mathbf{x}_0 = [0.2\ 0.7\ 1\ 1\ 1]^T$. With $\gamma = 0.9999$ and $\varepsilon = 10^{-4}$, Algorithm 12.1 converged to the solution

$$\mathbf{x}^* = \begin{bmatrix} 0.000008 \\ 0.333339 \\ 0.333336 \\ 0.000000 \\ 0.000029 \end{bmatrix}$$

after 4 iterations. The sequence of the iterates obtained is given in Table 12.1.

Table 12.1 Sequence of points $\{\mathbf{x}_k$ for $k = 0, 1, \ldots, 4\}$ in Example 12.2

$\mathbf{x}_0$	$\mathbf{x}_1$	$\mathbf{x}_2$	$\mathbf{x}_3$	$\mathbf{x}_4$
0.200000	0.099438	0.000010	0.000010	0.000008
0.700000	0.454077	0.383410	0.333357	0.333339
1.000000	0.290822	0.233301	0.333406	0.333336
1.000000	0.000100	0.000100	0.000100	0.000000
1.000000	0.624922	0.300348	0.000030	0.000029

12.4 Primal Newton Barrier Method

12.4.1 Basic idea

In the primal Newton barrier (PNB) method [2][7], the inequality constraints in Eq. (12.1c) are incorporated in the objective function by adding a *logarithmic barrier function*. The subproblem obtained has the form

$$\text{minimize } f_\tau(\mathbf{x}) = \mathbf{c}^T \mathbf{x} - \tau \sum_{i=1}^{n} \ln x_i \qquad (12.26a)$$

$$\text{subject to: } \quad \mathbf{A}\mathbf{x} = \mathbf{b} \qquad (12.26b)$$

where τ is a strictly positive scalar. The term $-\tau \sum_{i=1}^{n} \ln x_i$ in Eq. (12.26a) is called a *'barrier' function* for the reason that if we start with an initial $\mathbf{x}_0$ which is strictly inside the feasible region, then the term is well defined and acts like a barrier that prevents any component x_i from becoming zero. The scalar τ is known as the *barrier parameter*. The effect of the barrier function on the original LP problem depends largely on the magnitude of τ. If we start with an interior point, $\mathbf{x}_0$, then under certain conditions to be examined below for a given $\tau > 0$, a unique solution of the subproblem in Eq. (12.26) exists. Thus, if we solve the subproblem in Eq. (12.26) for a series of values of τ, a series of solutions are obtained that converge to the solution of the original LP problem as $\tau \rightarrow 0$. In effect, the PNB method solves the LP problem through the solution of a sequence of optimization problems [8] as in the minimax optimization methods of Chap. 8.

In a typical sequential optimization method, there are three issues that need to be addressed. These are:

(*a*) For each fixed $\tau > 0$ does a minimizer of the subproblem in Eq. (12.26) exist?

(*b*) If $\mathbf{x}_\tau^*$ is a minimizer of the problem in Eq. (12.26) and $\mathbf{x}^*$ is a minimizer of the problem in Eq. (12.1), how close is $\mathbf{x}_\tau^*$ to $\mathbf{x}^*$ as $\tau \rightarrow 0$?

(*c*) For each fixed $\tau > 0$, how do we compute or estimate $\mathbf{x}_\tau^*$?

12.4.2 Minimizers of subproblem

Throughout the rest of the section, we assume that the primal in Eq. (12.1) and the dual in Eq. (12.2) are both strictly feasible. Let $\tau > 0$ be fixed and $\mathbf{x}_0$ be a strictly feasible point for the problem in Eq. (12.1). At $\mathbf{x}_0$ the objective function of Eq. (12.26a), $f_\tau(\mathbf{x}_0)$, is well defined. By virtue of Theorem 12.2, the above assumption implies that solutions of the primal exist and are bounded. Under these circumstances, it can be shown that for a given $\varepsilon > 0$ the set

$$\mathcal{S}_0 = \{\mathbf{x} : \mathbf{x} \text{ is strictly feasible for problem (12.1)}; f_\tau(\mathbf{x}) \leq f_\tau(\mathbf{x}_0) + \varepsilon\}$$

is compact for all $\tau > 0$ (see Theorem 4 in [2]). This implies that $f_\tau(\mathbf{x})$ has a local minimizer $\mathbf{x}_\tau^*$ at an interior point of $\mathcal{S}_0$. We can compute the gradient and Hessian of $f_\tau(\mathbf{x})$ as

$$\nabla f_\tau(\mathbf{x}) = \mathbf{c} - \tau \mathbf{X}^{-1}\mathbf{e} \tag{12.27a}$$

$$\nabla^2 f_\tau(\mathbf{x}) = \tau \mathbf{X}^{-2} \tag{12.27b}$$

with $\mathbf{X} = \operatorname{diag}\{x_1, x_2, \ldots, x_n\}$ and $\mathbf{e} = [1 \; 1 \; \cdots \; 1]^T$. Since $f_\tau(\mathbf{x})$ is convex, $\mathbf{x}_\tau^*$ in $\mathcal{S}_0$ is a global minimizer of the problem in Eq. (12.26).

12.4.3   A convergence issue

Let $\{\tau_k\}$ be a sequence of barrier parameters that are monotonically decreasing to zero and $\mathbf{x}_k^*$ be the minimizer of the problem in Eq. (12.26) with $\tau = \tau_k$. It follows that

$$\mathbf{c}^T\mathbf{x}_k^* - \tau_k \sum_{i=1}^n \ln(\mathbf{x}_k^*)_i \leq \mathbf{c}^T\mathbf{x}_{k+1}^* - \tau_k \sum_{i=1}^n \ln(\mathbf{x}_{k+1}^*)_i$$

and

$$\mathbf{c}^T\mathbf{x}_{k+1}^* - \tau_{k+1} \sum_{i=1}^n \ln(\mathbf{x}_{k+1}^*)_i \leq \mathbf{c}^T\mathbf{x}_k^* - \tau_{k+1} \sum_{i=1}^n \ln(\mathbf{x}_k^*)_i$$

These equations yield (see Prob. 12.11(a))

$$f(\mathbf{x}_{k+1}^*) = \mathbf{c}^T\mathbf{x}_{k+1}^* \leq \mathbf{c}^T\mathbf{x}_k^* = f(\mathbf{x}_k^*) \tag{12.28}$$

i.e., the objective function of the original LP problem in Eq. (12.1) is a monotonically decreasing function of sequence $\{\mathbf{x}_k^*$ for $k = 0, 1, \ldots\}$. An immediate consequence of Eq. (12.28) is that all the minimizers, $\mathbf{x}_k^*$, are contained in the compact set

$$\mathcal{S} = \{\mathbf{x} : \mathbf{x} \text{ is feasible for the problem in Eq. (12.1) and } f(\mathbf{x}) \leq f(\mathbf{x}_0)\}$$

Therefore, sequence $\{\mathbf{x}_k^*\}$ contains at least one convergent subsequence, which for the sake of simplicity is denoted again as $\{\mathbf{x}_k^*\}$, namely,

$$\lim_{k\to\infty} \mathbf{x}_k^* = \mathbf{x}^* \tag{12.29}$$

It can be shown that the limit vector $\mathbf{x}^*$ in Eq. (12.29) is a minimizer of the primal problem in Eq. (12.1) [2][8]. Moreover, the closeness of $\mathbf{x}_k^*$ to $\mathbf{x}^*$ can be related to the magnitude of the barrier parameter τ_k as follows. Problem in Eq. (12.1) is said to be *nondegenerate* if there are exactly p strictly positive components in $\mathbf{x}^*$ and is said to be *degenerate* otherwise. In [9] and [10], it was shown that

$$\|\mathbf{x}_k^* - \mathbf{x}^*\| = O(\tau_k) \quad \text{if the problem in Eq. (12.1) is nondegenerate}$$

and

$$||\mathbf{x}_k^* - \mathbf{x}^*|| = O(\tau_k^{1/2}) \quad \text{if the problem in Eq. (12.1) is degenerate}$$

The sequence of minimizers for the subproblem in Eq. (12.26) can also be related to the central path of the problems in Eqs. (12.1) and (12.2). To see this, we write the Karush-Kuhn-Tucker (KKT) condition in Eq. (10.74) for the subproblem in Eq. (12.26) at $\mathbf{x}_k^*$ as

$$\mathbf{A}^T \boldsymbol{\lambda}_k + \tau_k \mathbf{X}^{-1}\mathbf{e} = \mathbf{c} \tag{12.30}$$

where $\mathbf{X} = \text{diag}\{(\mathbf{x}_k^*)_1, (\mathbf{x}_k^*)_2, \ldots, (\mathbf{x}_k^*)_n\}$. If we let

$$\boldsymbol{\mu}_k = \tau_k \mathbf{X}^{-1}\mathbf{e} \tag{12.31}$$

then with $\mathbf{x}_k^*$ being a strictly feasible point, Eqs. (12.30) and (12.31) lead to

$$\mathbf{A}\mathbf{x}_k^* = \mathbf{b} \quad \text{with } \mathbf{x}_k^* > \mathbf{0} \tag{12.32a}$$
$$\mathbf{A}^T \boldsymbol{\lambda}_k + \boldsymbol{\mu}_k = \mathbf{c} \quad \text{with } \boldsymbol{\mu}_k > \mathbf{0} \tag{12.32b}$$
$$\mathbf{X}\boldsymbol{\mu}_k = \tau_k\mathbf{e} \tag{12.32c}$$

On comparing Eq. (12.32) with Eq. (12.9), we conclude that the sequences of points $\{\mathbf{x}_k^*, \boldsymbol{\lambda}_k, \boldsymbol{\mu}_k\}$ are on the central path for the problems in Eqs. (12.1) and Eq. (12.2). Further, since $\mathbf{x}^*$ is a minimizer of the problem in Eq. (12.1), there exist $\boldsymbol{\lambda}^*$ and $\boldsymbol{\mu}^* \geq \mathbf{0}$ such that

$$\mathbf{A}\mathbf{x}^* = \mathbf{b} \quad \text{with } \mathbf{x}^* \geq \mathbf{0} \tag{12.33a}$$
$$\mathbf{A}^T \boldsymbol{\lambda}^* + \boldsymbol{\mu}^* = \mathbf{c} \quad \text{with } \boldsymbol{\mu}^* \geq \mathbf{0} \tag{12.33b}$$
$$\mathbf{X}^* \boldsymbol{\mu}^* = \mathbf{0} \tag{12.33c}$$

where $\mathbf{X}^* = \text{diag}\{(\mathbf{x}^*)_1, (\mathbf{x}^*)_2, \ldots, (\mathbf{x}^*)_n\}$. By virtue of Eq. (12.29) and $\tau_k \to 0$, Eqs. (12.32c) and (12.33c) imply that $\boldsymbol{\mu}_k \to \boldsymbol{\mu}^*$. From Eqs. (12.32b) and (12.33b), we have

$$\lim_{k \to \infty} \mathbf{A}^T(\boldsymbol{\lambda}_k - \boldsymbol{\lambda}^*) = \mathbf{0} \tag{12.34}$$

Since $\mathbf{A}^T$ has full column rank, Eq. (12.34) implies that $\boldsymbol{\lambda}_k \to \boldsymbol{\lambda}^*$. Therefore, by letting $k \to \infty$ in Eq. (12.32), we obtain Eq. (12.33). In other words, as $k \to \infty$ the sequences of points $\{\mathbf{x}_k^*, \boldsymbol{\lambda}_k^*, \boldsymbol{\mu}_k^*\}$ converge to a primal-dual solution $\{\mathbf{x}^*, \boldsymbol{\lambda}^*, \boldsymbol{\mu}^*\}$ of the problems in Eqs. (12.1) and (12.2).

12.4.4 Computing a minimizer of the problem in Eq. (12.26)

For a fixed $\tau > 0$, the PNB method starts with a strictly feasible point $\mathbf{x}_0$ and proceeds iteratively to find points $\mathbf{x}_k$ and

$$\mathbf{x}_{k+1} = \mathbf{x}_k + \alpha_k\mathbf{d}_k$$

such that the search direction satisfies the equality

$$\mathbf{A}\mathbf{d}_k = \mathbf{0} \tag{12.35}$$

The constraints in Eq. (12.35) ensure that if $\mathbf{x}_k$ satisfies Eq. (12.26b), then so does $\mathbf{x}_{k+1}$, i.e.,

$$\mathbf{A}\mathbf{x}_{k+1} = \mathbf{A}\mathbf{x}_k + \alpha_k \mathbf{A}\mathbf{d}_k = \mathbf{b}$$

To find a descent direction, a second-order approximation of the problem in Eq. (12.26) is employed using the gradient and Hessian of $f_\tau(\mathbf{x})$ in Eq. (12.27), namely,

$$\text{minimize } \tfrac{1}{2}\tau \mathbf{d}^T \mathbf{X}^{-2}\mathbf{d} + \mathbf{d}^T(\mathbf{c} - \tau \mathbf{X}^{-1}\mathbf{e}) \tag{12.36a}$$

$$\text{subject to: } \mathbf{A}\mathbf{d} = \mathbf{0} \tag{12.36b}$$

For a strictly feasible $\mathbf{x}_k$, $\mathbf{X}^{-2}$ is positive definite. Hence Eq. (12.36) is a convex programming problem whose solution $\mathbf{d}_k$ satisfies the KKT conditions

$$\tau \mathbf{X}^{-2}\mathbf{d}_k + \mathbf{c} - \tau \mathbf{X}^{-1}\mathbf{e} = \mathbf{A}^T\boldsymbol{\lambda} \tag{12.37a}$$

$$\mathbf{A}\mathbf{d}_k = \mathbf{0} \tag{12.37b}$$

From Eq. (12.37), we obtain

$$\mathbf{d}_k = \mathbf{x}_k + \frac{1}{\tau}\mathbf{X}^2(\mathbf{A}^T\boldsymbol{\lambda} - \mathbf{c}) \tag{12.38a}$$

and

$$\mathbf{A}\mathbf{X}^2\mathbf{A}^T\boldsymbol{\lambda} = \tau \mathbf{A}\mathbf{d}_k + \mathbf{A}\mathbf{X}^2\mathbf{c} - \tau \mathbf{A}\mathbf{x}_k$$
$$= \mathbf{A}(\mathbf{X}^2\mathbf{c} - \tau\mathbf{x}_k) \tag{12.38b}$$

We see that the search direction $\mathbf{d}_k$ in the PNB method is determined by using Eq. (12.38a) with a $\boldsymbol{\lambda}$ obtained by solving the $p \times p$ symmetric positive-definite system in Eq. (12.38b).

Having determined $\mathbf{d}_k$, a line search along $\mathbf{d}_k$ can be carried out to determine a scalar $\alpha_k > 0$ such that $\mathbf{x}_k + \alpha\mathbf{d}_k$ remains strictly feasible and $\mathbf{f}_\tau(\mathbf{x}_k + \alpha\mathbf{d}_k)$ is minimized with respect to the range $0 \le \alpha \le \bar{\alpha}_k$ where $\bar{\alpha}_k$ is the largest possible scalar for $\mathbf{x}_k + \alpha\mathbf{d}_k$ to be strictly feasible. If we let

$$\mathbf{x}_k = \begin{bmatrix} x_1 \\ x_2 \\ \vdots \\ x_n \end{bmatrix} \quad \text{and} \quad \mathbf{d}_k = \begin{bmatrix} d_1 \\ d_2 \\ \vdots \\ d_n \end{bmatrix}$$

the strict feasibility of $\mathbf{x}_k + \alpha\mathbf{d}_k$ can be assured, i.e., $x_i + \alpha d_i > 0$ for $1 \le i \le n$, if $\alpha < x_i/(-d_i)$ for all $1 \le i \le n$. Hence point $\mathbf{x}_k + \alpha\mathbf{d}_k$ will remain strictly feasible if α satisfies the condition

$$\alpha < \min_{i \text{ with } d_i < 0} \left[\frac{x_i}{(-d_i)} \right]$$

In practice, the upper bound of α, namely,

$$\bar{\alpha}_k = 0.99 \times \min_{i \text{ with } d_i < 0} \left[\frac{x_i}{(-d_i)} \right] \tag{12.39}$$

gives satisfactory results. At $\mathbf{x}_k$, the line search for function $\mathbf{f}_\tau(\mathbf{x}_k + \alpha\mathbf{d}_k)$ is carried out on the closed interval $[0, \bar{\alpha}_k]$. Since

$$\frac{d^2 f_\tau(\mathbf{x}_k + \alpha\mathbf{d}_k)}{d\alpha^2} = \tau \sum_{i=1}^{n} \frac{d_i}{(x_i + \alpha d_i)^2} > 0$$

$f_\tau(\mathbf{x}_k + \alpha\mathbf{d}_k)$ is strictly convex on $[0, \bar{\alpha}_k]$ and has a unique minimum. One of the search methods discussed in Chapter 4 can be used to find the minimizer, α_k, and the new point is obtained as $\mathbf{x}_{k+1} = \mathbf{x}_k + \alpha_k\mathbf{d}_k$.

The PNB algorithm can be summarized as follows.

Algorithm 12.2 Primal Newton barrier algorithm for the standard-form LP problem
Step 1
Input $\mathbf{A}$, $\mathbf{c}$, and a strictly feasible initial point $\mathbf{x}_0$.
Set $l = 0$, initialize the barrier parameter such that $\tau_0 > 0$, and input the outer-loop tolerance $\varepsilon_{\text{outer}}$.
Step 2
Set $k = 0$ and $\mathbf{x}_0^{(l)} = \mathbf{x}_l$, and input the inner-loop tolerance $\varepsilon_{\text{inner}}$.
    Step 3.1
    Use Eq. (12.38) with $\tau = \tau_l$ to calculate $\mathbf{d}_k^{(l)}$ at $\mathbf{x}_k^{(l)}$.
    Step 3.2
    Use Eq. (12.39) to calculate $\bar{\alpha}_k$ where $\mathbf{x}_k = \mathbf{x}_k^{(l)}$ and $\mathbf{d}_k = \mathbf{d}_k^{(l)}$.
    Step 3.3
    Use a line search (e.g., a line search based on the golden-section method) to determine $\alpha_k^{(l)}$.
    Step 3.4
    Set $\mathbf{x}_{k+1}^{(l)} = \mathbf{x}_k^{(l)} + \alpha_k^{(l)}\mathbf{d}_k^{(l)}$.
    Step 3.5
    If $||\alpha_k^{(l)}\mathbf{d}_k^{(l)}|| < \varepsilon_{\text{inner}}$, set $\mathbf{x}_{l+1} = \mathbf{x}_{k+1}^{(l)}$ and go to Step 4; otherwise, set $k = k + 1$ and repeat from Step 3.1.
Step 4
If $||\mathbf{x}_l - \mathbf{x}_{l+1}|| < \varepsilon_{\text{outer}}$, output $\mathbf{x}^* = \mathbf{x}_{l+1}$, and stop; otherwise, choose $\tau_{l+1} < \tau_l$, set $l = l + 1$, and repeat from Step 2.

Example 12.3 Apply the PNB algorithm to the LP problem in Example 12.1.

Solution We start with $\mathbf{c} = [-2 \ 1 \ -3]^T$, $\mathbf{A} = [1 \ 1 \ 1]$, $\mathbf{b} = 1$, and $\mathbf{x}_0 = \left[\frac{1}{3} \ \frac{1}{3} \ \frac{1}{3}\right]^T$ which is strictly feasible, and employ the golden-section method (see Sec. 4.4) to perform the line search in Step 3.3. Parameter τ_i is chosen as $\tau_{l+1} = \sigma\tau_l$ with $\sigma = 0.1$.

With $\tau_0 = 0.1$ and $\varepsilon_{\text{outer}} = 10^{-4}$, Algorithm 12.2 took six iterations to converge to the solution

$$\mathbf{x}^* = \begin{bmatrix} 0.000007 \\ 0.000001 \\ 0.999992 \end{bmatrix}$$

The number of flops required was 5.194K. The path of the sequence $\{\mathbf{x}_l$ for $l = 0, 1, \ldots, 6\}$ is shown in Fig. 12.2. ∎

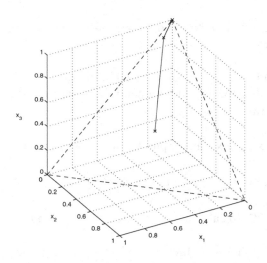

Figure 12.2. Iteration path in Example 12.3

12.5 Primal-Dual Interior-Point Methods

The methods studied in Secs. 12.3 and 12.4 are primal interior-point methods in which the dual is not explicitly involved. Primal-dual methods, on the other hand, solve the primal and dual LP problems simultaneously, and have emerged as the most efficient interior-point methods for LP problems. In this section, we examine two important primal-dual interior-point methods, namely, a primal-dual path-following and a nonfeasible-initialization primal-dual path-following methods.

12.5.1 Primal-dual path-following method

The path-following method to be discussed here is based on the work reported in [11]–[13]. Consider the standard-form LP problem in Eq. (12.1) and its dual in Eq. (12.2) and let $\mathbf{w}_k = \{\mathbf{x}_k, \boldsymbol{\lambda}_k, \boldsymbol{\mu}_k\}$ where $\mathbf{x}_k$ is strictly feasible for the primal and $\{\boldsymbol{\lambda}_k, \boldsymbol{\mu}_k\}$ is strictly feasible for the dual. We need to find an increment vector $\boldsymbol{\delta}_w = \{\boldsymbol{\delta}_x, \boldsymbol{\delta}_\lambda, \boldsymbol{\delta}_\mu\}$ such that the next iterate $\mathbf{w}_{k+1} = \{\mathbf{x}_{k+1}, \boldsymbol{\lambda}_{k+1}, \boldsymbol{\mu}_{k+1}\} = \{\mathbf{x}_k + \boldsymbol{\delta}_x, \boldsymbol{\lambda}_k + \boldsymbol{\delta}_\lambda, \boldsymbol{\mu}_k + \boldsymbol{\delta}_\mu\}$ remains strictly feasible and approaches the central path defined by Eq. (12.9) with $\tau = \tau_{k+1} > 0$. In the path-following method, a suitable $\boldsymbol{\delta}_w$ is obtained as a first-order approximate solution of Eq. (12.9). If $\mathbf{w}_{k+1}$ satisfies Eq. (12.9) with $\tau = \tau_{k+1}$, then

$$\mathbf{A}\boldsymbol{\delta}_x = \mathbf{0} \tag{12.40a}$$

$$\mathbf{A}^T \boldsymbol{\delta}_\lambda + \boldsymbol{\delta}_\mu = \mathbf{0} \tag{12.40b}$$

$$\Delta\mathbf{X}\boldsymbol{\mu}_k + \mathbf{X}\boldsymbol{\delta}_\mu + \Delta\mathbf{X}\boldsymbol{\delta}_\mu = \tau_{k+1}\mathbf{e} - \mathbf{X}\boldsymbol{\mu}_k \tag{12.40c}$$

where

$$\Delta\mathbf{X} = \text{diag}\{(\boldsymbol{\delta}_x)_1, (\boldsymbol{\delta}_x)_2, \ldots, (\boldsymbol{\delta}_x)_n\} \tag{12.41}$$

If the only second-order term in Eq. (12.40c), namely, $\Delta\mathbf{X}\boldsymbol{\delta}_\mu$ is neglected, then Eq. (12.40) is approximated by the system of linear equations

$$\mathbf{A}\boldsymbol{\delta}_x = \mathbf{0} \tag{12.42a}$$

$$\mathbf{A}^T \boldsymbol{\delta}_\lambda + \boldsymbol{\delta}_\mu = \mathbf{0} \tag{12.42b}$$

$$\mathbf{M}\boldsymbol{\delta}_x + \mathbf{X}\boldsymbol{\delta}_\mu = \tau_{k+1}\mathbf{e} - \mathbf{X}\boldsymbol{\mu}_k \tag{12.42c}$$

where term $\Delta\mathbf{X}\boldsymbol{\mu}_k$ in Eq. (12.40c) has been replaced by $\mathbf{M}\boldsymbol{\delta}_x$ with

$$\mathbf{M} = \text{diag}\{(\boldsymbol{\mu}_k)_1, (\boldsymbol{\mu}_k)_2, \ldots, (\boldsymbol{\mu}_k)_n\} \tag{12.43}$$

Solving Eq. (12.42) for $\boldsymbol{\delta}_w$, we obtain

$$\boldsymbol{\delta}_\lambda = \mathbf{Y}\mathbf{A}\mathbf{y} \tag{12.44a}$$

$$\boldsymbol{\delta}_\mu = -\mathbf{A}^T \boldsymbol{\delta}_\lambda \tag{12.44b}$$

$$\boldsymbol{\delta}_x = -\mathbf{y} - \mathbf{D}\boldsymbol{\delta}_\mu \tag{12.44c}$$

where

$$\mathbf{D} = \mathbf{M}^{-1}\mathbf{X} \tag{12.44d}$$

$$\mathbf{Y} = (\mathbf{A}\mathbf{D}\mathbf{A}^T)^{-1} \tag{12.44e}$$

and

$$\mathbf{y} = \mathbf{x}_k - \tau_{k+1}\mathbf{M}^{-1}\mathbf{e} \tag{12.44f}$$

Since $\mathbf{x}_k > \mathbf{0}$, $\boldsymbol{\mu}_k > \mathbf{0}$, and $\mathbf{A}$ has full row rank, matrix $\mathbf{Y}$ in Eq. (12.44e) is the inverse of a $p \times p$ positive definite matrix, and calculating $\mathbf{Y}$ is the major computation effort in the evaluation of $\boldsymbol{\delta}_w$ using Eq. (12.44).

From Eqs. (12.42a) and (12.42b), the new iterate $\mathbf{w}_{k+1}$ satisfies Eqs. (12.9a) and (12.9b) but not necessarily Eq. (12.9c) because Eq. (12.42c) is merely a linear approximation of Eq. (12.40c). If we define vector $\mathbf{f}(\mathbf{w}_k) = [f_1(\mathbf{w}_k) \ f_2(\mathbf{w}_k) \cdots f_n(\mathbf{w}_k)]^T$ with

$$f_i(\mathbf{w}_k) = (\boldsymbol{\mu}_k)_i \cdot (\mathbf{x}_k)_i \qquad \text{for } 1 \le i \le n$$

then Eqs. (12.9c) and (12.10) suggest that the L_2 norm $\|\mathbf{f}(\mathbf{w}_k) - \tau_k \mathbf{e}\|$ can be viewed as a measure of the closeness of $\mathbf{w}_k$ to the central path. In [13], it was shown that if an initial point $\mathbf{w}_0 = \{\mathbf{x}_0, \ \boldsymbol{\lambda}_0, \ \boldsymbol{\mu}_0\}$ is chosen such that (a) $\mathbf{x}_0$ is strictly feasible for the primal and $\{\boldsymbol{\lambda}_0, \ \boldsymbol{\mu}_0\}$ is strictly feasible for the dual and (b)

$$\|\mathbf{f}(\mathbf{w}_0) - \tau_0 \mathbf{e}\| \le \theta \tau_0 \tag{12.45a}$$

where $\tau_0 = (\boldsymbol{\mu}_0^T \mathbf{x}_0)/n$ and θ satisfies the conditions

$$0 \le \theta \le \frac{1}{2} \tag{12.45b}$$

$$\frac{\theta^2 + \delta^2}{2(1 - \theta)} \le \left(1 - \frac{\delta}{\sqrt{n}}\right)\theta \tag{12.45c}$$

for some $\delta \in (0, \ \sqrt{n})$, then the iterate

$$\mathbf{w}_{k+1} = \mathbf{w}_k + \boldsymbol{\delta}_w \tag{12.46}$$

where $\boldsymbol{\delta}_w = \{\boldsymbol{\delta}_x, \ \boldsymbol{\delta}_\lambda, \ \boldsymbol{\delta}_\mu\}$ is given by Eq. (12.44) with

$$\tau_{k+1} = \left(1 - \frac{\delta}{\sqrt{n}}\right)\tau_k \tag{12.47}$$

will remain strictly feasible and satisfy the conditions

$$\|\mathbf{f}(\mathbf{w}_{k+1}) - \tau_{k+1}\mathbf{e}\| \le \theta \tau_{k+1} \tag{12.48}$$

and

$$\boldsymbol{\mu}_{k+1}^T \mathbf{x}_{k+1} = n\tau_{k+1} \tag{12.49}$$

Since $0 < \delta/\sqrt{n} < 1$, it follows from Eq. (12.47) that $\tau_k = (1 - \delta/n)^k \tau_0 \to 0$ as $k \to \infty$. From Eqs. (12.49) and (12.10), the duality gap tends to zero, i.e., $\delta(\mathbf{x}_k, \boldsymbol{\lambda}_k) \to 0$, as $k \to \infty$. In other words, $\mathbf{w}_k$ converges to a primal-dual solution as $k \to \infty$. The above method can be implemented in terms of the following algorithm [13].

Algorithm 12.3 Primal-dual path-following algorithm for the standard-form LP problem
Step 1
Input $\mathbf{A}$ and a strictly feasible $\mathbf{w}_0 = \{\mathbf{x}_0, \boldsymbol{\lambda}_0, \boldsymbol{\mu}_0\}$ that satisfies Eq. (12.45). Set $k = 0$ and initialize the tolerance ε for the duality gap.
Step 2
If $\boldsymbol{\mu}_k^T \mathbf{x}_k \leq \varepsilon$, output solution $\mathbf{w}^* = \mathbf{w}_k$ and stop; otherwise, continue with Step 3.
Step 3
Set τ_{k+1} using Eq. (12.47) and compute $\boldsymbol{\delta}_w = \{\boldsymbol{\delta}_x, \boldsymbol{\delta}_\lambda, \boldsymbol{\delta}_\mu\}$ using Eq. (12.44).
Step 4
Set $\mathbf{w}_{k+1}$ using Eq. (12.46). Set $k = k + 1$ and repeat from Step 2.

A couple of remarks concerning Step 1 of the algorithm are in order. First, values of θ and δ that satisfy Eqs. (12.45b) and (12.45c) exist. For example, it can be readily verified that $\theta = 0.4$ and $\delta = 0.4$ meet Eqs. (12.45b) and (12.45c) for any $n \geq 2$. Second, in order to find an initial $\mathbf{w}_0$ that satisfies Eq. (12.45a), we can introduce an augmented pair of primal-dual LP problems such that (a) a strictly feasible initial point can be easily identified for the augmented problem and (b) a solution of the augmented problem will yield a solution of the original problem [13]. A more general remedy for dealing with this initialization problem is to develop a 'nonfeasible-initialization algorithm' so that a point $\mathbf{w}_0$ that satisfies $\mathbf{x}_0 > 0$ and $\boldsymbol{\mu}_0 > 0$ but not necessarily Eq. (12.9) can be used as the initial point. Such a primal-dual path-following algorithm will be studied in Sec. 12.5.2.

It is important to stress that even for problems of moderate size, the choice $\delta = 0.4$ yields a factor $(1 - \delta/\sqrt{n})$ which is close to unity and, therefore, parameter τ_{k+1} determined using Eq. (12.47) converges to zero slowly and a large number of iterations are required to reach a primal-dual solution. In the literature, interior-point algorithms of this type are referred to as *short-step* path-following algorithms [3]. In practice, Algorithm 12.3 is modified to allow larger changes in parameter τ so as to accelerate the convergence [6][14]. It was proposed in [14] that τ_{k+1} be chosen as

$$\tau_{k+1} = \frac{\boldsymbol{\mu}_k^T \mathbf{x}_k}{n + \rho} \tag{12.50}$$

with $\rho > \sqrt{n}$. In order to assume the strict feasibility of the next iterate, the modified path-following algorithm assigns

$$\mathbf{w}_{k+1} = \mathbf{w}_k + \alpha_k \boldsymbol{\delta}_w \tag{12.51}$$

where $\delta_w = \{\delta_x, \delta_\lambda, \delta_\mu\}$ is calculated using Eq. (12.44), and

$$\alpha_k = (1 - 10^{-6})\alpha_{\max} \tag{12.52a}$$

with $\alpha_{\max}$ being determined as

$$\alpha_{\max} = \min(\alpha_p, \alpha_d) \tag{12.52b}$$

where

$$\alpha_p = \min_{i \text{ with } (\delta_x)_i < 0} \left[-\frac{(\mathbf{x}_k)_i}{(\delta_x)_i} \right] \tag{12.52c}$$

$$\alpha_d = \min_{i \text{ with } (\delta_\mu)_i < 0} \left[-\frac{(\mu_k)_i}{(\delta_\mu)_i} \right] \tag{12.52d}$$

The modified algorithm assumes the following form.

Algorithm 12.4 Modified version of Algorithm 12.3
Step 1
Input $\mathbf{A}$ and a strictly feasible $\mathbf{w}_0 = \{\mathbf{x}_0, \lambda_0, \mu_0\}$.
Set $k = 0$ and $\rho > \sqrt{n}$, and initialize the tolerance ε for the duality gap.
Step 2
If $\mu_k^T \mathbf{x}_k \leq \varepsilon$, output solution $\mathbf{w}^* = \mathbf{w}_k$ and stop; otherwise, continue with Step 3.
Step 3
Set τ_{k+1} using Eq. (12.50) and compute $\delta_w = \{\delta_x, \delta_\lambda, \delta_\mu\}$ using Eq. (12.44).
Step 4
Compute step size α_k using Eq. (12.52) and set $\mathbf{w}_{k+1}$ using Eq. (12.51). Set $k = k + 1$ and repeat from Step 2.

Example 12.4 Apply Algorithm 12.4 to the LP problems in
(a) Example 12.3
(b) Example 12.2

Solution (a) In order to apply the algorithm to the LP problem in Example 12.3, we have used the method described in Example 12.1 to find an initial $\mathbf{w}_0$ on the central path with $\tau_0 = 5$. The vector $\mathbf{w}_0$ obtained is $\{\mathbf{x}_0, \lambda_0, \mu_0\}$ with

$$\mathbf{x}_0 = \begin{bmatrix} 0.344506 \\ 0.285494 \\ 0.370000 \end{bmatrix}, \quad \lambda_0 = -16.513519, \quad \mu_0 = \begin{bmatrix} 14.513519 \\ 17.513519 \\ 13.513519 \end{bmatrix}$$

With $\rho = 7\sqrt{n}$ and $\varepsilon = 10^{-6}$, Algorithm 12.4 converges after eight iterations to the solution

$$\mathbf{x}^* = \begin{bmatrix} 0.000000 \\ 0.000000 \\ 1.000000 \end{bmatrix}$$

The number of flops required was 858.

(b) In order to apply the algorithm to the LP problem in Example 12.2, we have to find a strictly feasible initial point $\mathbf{w}_0$ first. By using the method described in Sec. 10.4.1, a vector $\mathbf{x}$ that satisfies Eq. (12.9a) can be obtained as

$$\mathbf{x} = \mathbf{V}_r \boldsymbol{\phi} + \mathbf{A}^+ \mathbf{b}$$

where $\mathbf{V}_r$ is composed of the last $n - p$ columns of matrix $\mathbf{V}$ from the SVD of matrix $\mathbf{A}$ and $\boldsymbol{\phi} \in R^{(n-p) \times 1}$ is a free parameter vector. From Eq. (10.27), we have

$$\mathbf{x} = \begin{bmatrix} 0.5980 & 0.0000 & 0.0000 \\ 0.0608 & 0.1366 & 0.1794 \\ 0.6385 & 0.5504 & -0.2302 \\ -0.4358 & 0.8236 & 0.1285 \\ 0.2027 & -0.0039 & 0.9478 \end{bmatrix} \begin{bmatrix} \phi_1 \\ \phi_2 \\ \phi_3 \end{bmatrix} + \begin{bmatrix} -0.1394 \\ 0.2909 \\ 0.0545 \\ -0.0848 \\ -0.0303 \end{bmatrix} \quad (12.53)$$

The requirement $x_1 > 0$ is met if

$$\phi_1 > 0.2331 \quad (12.54a)$$

If we assume that $\phi_2 = \phi_3 > 0$, then $x_2 > 0$ and $x_3 > 0$. To satisfy the inequalities $x_4 > 0$ and $x_5 > 0$, we require

$$-0.4358\phi_1 + 0.9572\phi_2 > 0.0848 \quad (12.54b)$$

and

$$0.2027\phi_1 + 0.9439\phi_2 > 0.0303 \quad (12.54c)$$

Obviously, $\phi_1 = 0.5$ and $\phi_2 = 0.5$ satisfy Eq. (12.54) and lead to a strictly feasible initial point

$$\mathbf{x}_0 = \begin{bmatrix} 0.1596 \\ 0.4793 \\ 0.5339 \\ 0.1733 \\ 0.5430 \end{bmatrix}$$

Next we can write Eq. (12.9b) as

$$\boldsymbol{\mu} = \mathbf{c} - \mathbf{A}^T \boldsymbol{\lambda} = \begin{bmatrix} 2 + 2\lambda_1 - \lambda_2 \\ 9 - 2\lambda_1 - 4\lambda_2 \\ 3 - \lambda_1 + \lambda_2 \\ \lambda_1 \\ \lambda_2 \end{bmatrix}$$

from which it is easy to verify that $\lambda_0 = [1\ 1]^T$ leads to $\mu_0 = [3\ 3\ 3\ 1\ 1]^T > 0$ and $\{\lambda_0,\ \mu_0\}$ satisfies Eq. (12.9b).

The application of Algorithm 12.4 using the above w_0, $\rho = 12\sqrt{n}$, and $\varepsilon = 10^{-5}$ led to the solution

$$\mathbf{x}^* = \begin{bmatrix} 0.000000 \\ 0.333333 \\ 0.333333 \\ 0.000000 \\ 0.000000 \end{bmatrix}$$

in seven iterations. The number of flops required was 2.48K.

■

12.5.2 A nonfeasible-initialization primal-dual path-following method

Both Algorithms 12.3 and 12.4 require an initial $w_0 = \{x_0,\ \lambda_0,\ \mu_0\}$ with x_0 being strictly feasible for the primal and $\{\lambda_0,\ \mu_0\}$ being strictly feasible for the dual. As can be observed from Example 12.4, finding such an initial point is not straightforward, even for problems of small size, and it would certainly be highly desirable to start with an initial point w_0 that is not necessarily feasible. In the literature, interior-point algorithms that accept nonfeasible initial points are often referred to as *nonfeasible-initialization* or *nonfeasible-start* algorithms. As described in [6], if w_k is nonfeasible in the sense that it does not satisfy Eqs. (12.1b) and (12.2b), then a reasonable way to generate the next point is to find a set of vector increments $\delta_w = \{\delta_x,\ \delta_\lambda,\ \delta_\mu\}$ such that $w_k + \delta_w$ satisfies Eqs. (12.1b) and (12.2b). Based on this approach, the basic idea presented in Sec. 12.5.1 can be used to construct a *nonfeasible-initialization primal-dual path-following* algorithm [15].

Let $w_k = \{x_k,\ \lambda_k,\ \mu_k\}$ be such that only the conditions $x_k > 0$ and $\mu_k > 0$ are assumed. We need to obtain the next iterate

$$w_{k+1} = w_k + \alpha_k \delta_w$$

such that $x_{k+1} > 0$ and $\mu_{k+1} > 0$, and that $\delta_w = \{\delta_x,\ \delta_\lambda,\ \delta_\mu\}$ satisfies the conditions

$$A(x_k + \delta_x) = b \tag{12.55a}$$
$$A^T(\lambda_k + \delta_\lambda) + (\mu_k + \delta_\mu) = c \tag{12.55b}$$
$$M\delta_x + X\delta_\mu = \tau_{k+1}e - X\mu_k \tag{12.55c}$$

Note that Eq. (12.55c) is the same as Eq. (12.42c) which is a linear approximation of Eq. (12.40c) but Eqs. (12.55a) and (12.55b) differ from Eqs. (12.42a) and (12.42b) since in the present case the feasibility of w_k is not assumed. At the kth iteration, $\mathbf{w}_k$ is known; hence Eq. (12.55) is a system of linear equations for $\{\boldsymbol{\delta}_x, \boldsymbol{\delta}_\lambda, \boldsymbol{\delta}_\mu\}$, which can be written as

$$\mathbf{A}\boldsymbol{\delta}_x = \mathbf{r}_p \qquad (12.56a)$$

$$\mathbf{A}^T\boldsymbol{\delta}_\lambda + \boldsymbol{\delta}_\mu = \mathbf{r}_d \qquad (12.56b)$$

$$\mathbf{M}\boldsymbol{\delta}_x + \mathbf{X}\boldsymbol{\delta}_\mu = \tau_{k+1}\mathbf{e} - \mathbf{X}\boldsymbol{\mu}_k \qquad (12.56c)$$

where $\mathbf{r}_p = \mathbf{b} - \mathbf{A}\mathbf{x}_k$ and $\mathbf{r}_d = \mathbf{c} - \mathbf{A}^T\boldsymbol{\lambda}_k - \boldsymbol{\mu}_k$ are the residuals for the primal and dual constraints, respectively. Solving Eq. (12.56) for $\boldsymbol{\delta}_w$, we obtain

$$\boldsymbol{\delta}_\lambda = \mathbf{Y}(\mathbf{A}\mathbf{y} + \mathbf{A}\mathbf{D}\mathbf{r}_d + \mathbf{r}_p) \qquad (12.57a)$$

$$\boldsymbol{\delta}_\mu = -\mathbf{A}^T\boldsymbol{\delta}_\lambda + \mathbf{r}_d \qquad (12.57b)$$

$$\boldsymbol{\delta}_x = -\mathbf{y} - \mathbf{D}\boldsymbol{\delta}_\mu \qquad (12.57c)$$

where $\mathbf{D}$, $\mathbf{Y}$, and $\mathbf{y}$ are defined by Eqs. (12.44d) – (12.44f), respectively. It should be stressed that if the new iterate $\mathbf{w}_{k+1}$ is set as in Eq. (12.51) with α_k determined using Eq. (12.52), then $\mathbf{x}_{k+1}$ and $\boldsymbol{\mu}_{k+1}$ remain strictly positive but $\mathbf{w}_{k+1}$ is not necessarily strictly feasible unless α_k happens to be unity. As the iterations proceed, the new iterates generated get closer and closer to the central path and approach to a primal-dual solution. The nonfeasible-initialization interior-point algorithm is summarized as follows.

Algorithm 12.5 Nonfeasible-initialization primal-dual path-following algorithm for the standard-form LP problem
Step 1
Input $\mathbf{A}$, $\mathbf{b}$, $\mathbf{c}$, and $\mathbf{w}_0 = \{\mathbf{x}_0, \boldsymbol{\lambda}_0, \boldsymbol{\mu}_0\}$ with $\mathbf{x}_0 > 0$ and $\boldsymbol{\mu}_0 > 0$.
Set $k = 0$ and $\rho > \sqrt{n}$, and initialize the tolerance ε for the duality gap.
Step 2
If $\boldsymbol{\mu}_k^T\mathbf{x}_k \le \varepsilon$, output solution $\mathbf{w}^* = \mathbf{w}_k$ and stop; otherwise, continue with Step 3.
Step 3
Set τ_{k+1} using Eq. (12.50) and compute $\boldsymbol{\delta}_w = (\boldsymbol{\delta}_x, \boldsymbol{\delta}_\lambda, \boldsymbol{\delta}_\mu)$ using Eq. (12.57).
Step 4
Compute step size α_k using Eq. (12.52) and set $\mathbf{w}_{k+1}$ using Eq. (12.51).
Set $k = k + 1$ and repeat from Step 2.

Example 12.5 Apply Algorithm 12.5 to the LP problems in
(a) Example 12.3
(b) Example 12.2
with nonfeasible initial points.

Solution (a) In order to apply the algorithm to the LP problem in Example 12.3, we can use $\mathbf{w}_0 = \{\mathbf{x}_0, \lambda_0, \boldsymbol{\mu}_0\}$ with

$$\mathbf{x}_0 = \begin{bmatrix} 0.4 \\ 0.3 \\ 0.4 \end{bmatrix}, \quad \lambda_0 = 0.5, \quad \text{and} \quad \boldsymbol{\mu}_0 = \begin{bmatrix} 1.0 \\ 0.5 \\ 1.0 \end{bmatrix}$$

So $\mathbf{x}_0 > \mathbf{0}$ and $\boldsymbol{\mu}_0 > \mathbf{0}$ but $\mathbf{w}_0$ is not feasible. With $\varepsilon = 10^{-6}$ and $\rho = 7\sqrt{n}$, Algorithm 12.5 took eight iterations to converge to the solution

$$\mathbf{x}^* = \begin{bmatrix} 0.000000 \\ 0.000000 \\ 1.000000 \end{bmatrix}$$

The number of flops required was 1.21K. Fig. 12.3 shows point $\mathbf{x}_0$ and the first three iterates, i.e., $\mathbf{x}_k$ for $k = 0, 1, 2, 3$, as compared to the central path which is shown as a dotted curve.

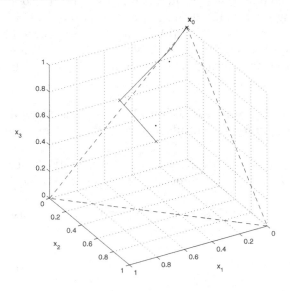

Figure 12.3. Iteration path in Example 12.5(a) as compared to the central path.

(b) For the LP problem in Example 12.2, we can use $\mathbf{w}_0 = \{\mathbf{x}_0, \lambda_0, \boldsymbol{\mu}_0\}$ with

$$\mathbf{x}_0 = \begin{bmatrix} 1.0 \\ 0.1 \\ 0.1 \\ 2.0 \\ 5.0 \end{bmatrix}, \quad \lambda_0 = \begin{bmatrix} -1 \\ 1 \end{bmatrix}, \quad \text{and} \quad \boldsymbol{\mu}_0 = \begin{bmatrix} 1.0 \\ 0.1 \\ 0.2 \\ 1.0 \\ 10.0 \end{bmatrix}$$

With $\varepsilon = 10^{-8}$ and $\rho = 12\sqrt{n}$, the algorithm took 13 iterations to converge to the solution

$$\mathbf{x}^* = \begin{bmatrix} 0.000000 \\ 0.333333 \\ 0.333333 \\ 0.000000 \\ 0.000000 \end{bmatrix}$$

The number of flops required was 6.96K.

∎

12.5.3 Predictor-corrector method

The predictor-corrector method (PCM) proposed by Mehrotra [16] can be viewed as an important improved primal-dual path-following algorithm relative to the algorithms studied in Secs. 12.5.1 and 12.5.2. As a matter of fact, most interior-point software available since 1990 is based on Mehrotra's PCM algorithm [3]. Briefly speaking, improvement is achieved by including the effect of the second-order term $\Delta\mathbf{X}\delta_\mu$ in Eq. (12.40c) using a prediction-correction strategy rather than simply neglecting it. In addition, in this method the parameter τ in Eq. (12.9c) is assigned a value according to the relation

$$\tau = \sigma\hat{\tau}$$

where $\hat{\tau} = (\boldsymbol{\mu}^T\mathbf{x})/n$ and $0 < \sigma < 1$. The scalar σ, which is referred to as *centering* parameter, is determined *adaptively* in each iteration based on whether good progress has been made in the prediction phase.

At the kth iteration, there are three steps in the PCM algorithm that produce the next iterate $\mathbf{w}_{k+1} = \{\mathbf{x}_{k+1}, \boldsymbol{\lambda}_{k+1}, \boldsymbol{\mu}_{k+1}\}$ with $\mathbf{x}_{k+1} > 0$ and $\boldsymbol{\mu}_{k+1} > 0$ as described below (see Chap. 10 in [3]).

1. *Generate an affine-scaling 'predictor' direction δ_w^{aff} using a linear approximation of the KKT conditions in Eq. (12.8).*
   Let $\mathbf{w}_k = \{\mathbf{x}_k, \boldsymbol{\lambda}_k, \boldsymbol{\mu}_k\}$ with $\mathbf{x}_k > 0$ and $\boldsymbol{\mu}_k > 0$, and consider an increment $\delta_w^{\text{aff}} = \{\delta_x^{\text{aff}}, \delta_\lambda^{\text{aff}}, \delta_\mu^{\text{aff}}\}$ such that $\mathbf{w}_k + \delta_w^{\text{aff}}$ linearly approximates the KKT conditions in Eq. (12.8). Under these circumstances, δ_w^{aff} satisfies the equations

$$\mathbf{A}\delta_x^{\text{aff}} = \mathbf{r}_p \tag{12.58a}$$

$$\mathbf{A}^T\delta_\lambda^{\text{aff}} + \delta_\mu^{\text{aff}} = \mathbf{r}_d \tag{12.58b}$$

$$\mathbf{M}\delta_x^{\text{aff}} + \mathbf{X}\delta_\mu^{\text{aff}} = -\mathbf{X}\mathbf{M}\mathbf{e} \tag{12.58c}$$

where

$$\mathbf{r}_p = \mathbf{b} - \mathbf{A}\mathbf{x}_k \tag{12.58d}$$

$$\mathbf{r}_d = \mathbf{c} - \mathbf{A}^T\boldsymbol{\lambda}_k - \boldsymbol{\mu}_k \tag{12.58e}$$

Solving Eq. (12.58) for δ_w^{aff}, we obtain

$$\delta_\lambda^{\text{aff}} = \mathbf{Y}(\mathbf{b} + \mathbf{A}\mathbf{D}\mathbf{r}_d) \tag{12.59a}$$

$$\delta_\mu^{\text{aff}} = \mathbf{r}_d - \mathbf{A}^T \delta_\lambda^{\text{aff}} \tag{12.59b}$$

$$\delta_x^{\text{aff}} = -\mathbf{x}_k - \mathbf{D}\delta_\mu^{\text{aff}} \tag{12.59c}$$

where

$$\mathbf{D} = \mathbf{M}^{-1}\mathbf{X} \tag{12.59d}$$

$$\mathbf{Y} = (\mathbf{A}\mathbf{D}\mathbf{A}^T)^{-1} \tag{12.59e}$$

Along the directions δ_x^{aff} and δ_μ^{aff}, two scalars α_p^{aff} and α_d^{aff} are determined as

$$\alpha_p^{\text{aff}} = \max_{0 \le \alpha \le 1, \ \mathbf{x}_k + \alpha \delta_x^{\text{aff}} \ge 0} (\alpha) \tag{12.60a}$$

and

$$\alpha_d^{\text{aff}} = \max_{0 \le \alpha \le 1, \ \boldsymbol{\mu}_k + \alpha \delta_\mu^{\text{aff}} \ge 0} (\alpha) \tag{12.60b}$$

A hypothetical value of τ_{k+1}, denoted as τ_{aff}, is then determined as

$$\tau_{\text{aff}} = \frac{1}{n}[(\boldsymbol{\mu}_k + \alpha_d^{\text{aff}} \delta_\mu^{\text{aff}})^T (\mathbf{x}_k + \alpha_p^{\text{aff}} \delta_x^{\text{aff}})] \tag{12.61}$$

2. *Determine the centering parameter σ_k.*
   A heuristic choice of σ_k, namely,

$$\sigma_k = \left(\frac{\tau_{\text{aff}}}{\hat{\tau}_k}\right)^3 \tag{12.62}$$

with

$$\hat{\tau}_k = \frac{1}{n}(\boldsymbol{\mu}_k^T \mathbf{x}_k) \tag{12.63}$$

was suggested in [16] and was found effective in extensive computational testing. Intuitively, if $\tau_{\text{aff}} \ll \hat{\tau}_k$, then the predictor direction δ_w^{aff} given by Eq. (12.59) is good and we should use a small centering parameter σ_k to substantially reduce the magnitude of parameter $\tau_{k+1} = \sigma_k \hat{\tau}_k$. If τ_{aff} is close to $\hat{\tau}_k$, then we should choose σ_k close to unity so as to move the next iterate $\mathbf{w}_{k+1}$ closer to the central path.

3. *Generate a 'corrector' direction to compensate for the nonlinearity in the affine-scaling direction.*
   The corrector direction $\delta_w^c = \{\delta_x^c, \ \delta_\lambda^c, \ \delta_\mu^c\}$ is determined using Eq. (12.40) with the term $\mathbf{X}\boldsymbol{\mu}_k$ in Eq. (12.40c) neglected and the second-order term $\Delta\mathbf{X}\delta_\mu$ in Eq. (12.40c) replaced by $\Delta\mathbf{X}^{\text{aff}}\delta_\mu^{\text{aff}}$ where $\Delta\mathbf{X}^{\text{aff}} = \text{diag}\{(\delta_x^{\text{aff}})_1,$

$(\delta_x^{\text{aff}})_2, \ldots, (\delta_x^{\text{aff}})_n\}$. The reason that term $\mathbf{X}\boldsymbol{\mu}_k$ is neglected is because it has been included in Eq. (12.58c) where $\mathbf{X}\mathbf{M}\mathbf{e} = \mathbf{X}\boldsymbol{\mu}_k$. In the primal-dual path-following algorithms studied in Secs. 12.5.1 and 12.5.2, the second-order term $\Delta\mathbf{X}\boldsymbol{\delta}_\mu$ was dropped to obtain the *linear* systems in Eqs. (12.42) and (12.56). The PCM method approximates this second-order term with the increment vectors $\boldsymbol{\delta}_x$ and $\boldsymbol{\delta}_\mu$ obtained from the predictor direction. Having made the above modifications, the equations to be used to compute $\boldsymbol{\delta}_w^c$ become

$$\mathbf{A}\boldsymbol{\delta}_x^c = \mathbf{0} \tag{12.64a}$$
$$\mathbf{A}^T\boldsymbol{\delta}_\lambda^c + \boldsymbol{\delta}_\mu^c = \mathbf{0} \tag{12.64b}$$
$$\mathbf{M}\boldsymbol{\delta}_x^c + \mathbf{X}\boldsymbol{\delta}_\mu^c = \tau_{k+1}\mathbf{e} - \Delta\mathbf{X}^{\text{aff}}\boldsymbol{\delta}_\mu^{\text{aff}} \tag{12.64c}$$

where

$$\tau_{k+1} = \sigma_k\hat{\tau}_k \tag{12.64d}$$

with σ_k and $\hat{\tau}_k$ given by Eqs. (12.62) and (12.63), respectively. Solving Eq. (12.64) for $\boldsymbol{\delta}_w^c$, we obtain

$$\boldsymbol{\delta}_\lambda^c = \mathbf{Y}\mathbf{A}\mathbf{y} \tag{12.65a}$$
$$\boldsymbol{\delta}_\mu^c = -\mathbf{A}^T\boldsymbol{\delta}_\lambda^c \tag{12.65b}$$
$$\boldsymbol{\delta}_x^c = -\mathbf{y} - \mathbf{D}\boldsymbol{\delta}_\mu^c \tag{12.65c}$$

where $\mathbf{D}$ and $\mathbf{Y}$ are given by Eqs. (12.59d) and (12.59e), respectively, and

$$\mathbf{y} = \mathbf{M}^{-1}(\Delta\mathbf{X}^{\text{aff}}\boldsymbol{\delta}_\mu^{\text{aff}} - \tau_{k+1}\mathbf{e}) \tag{12.65d}$$

The predictor and corrector directions are now combined to obtain the search direction $\{\boldsymbol{\delta}_x, \boldsymbol{\delta}_\lambda, \boldsymbol{\delta}_\mu\}$ where

$$\boldsymbol{\delta}_x = \boldsymbol{\delta}_x^{\text{aff}} + \boldsymbol{\delta}_x^c \tag{12.66a}$$
$$\boldsymbol{\delta}_\lambda = \boldsymbol{\delta}_\lambda^{\text{aff}} + \boldsymbol{\delta}_\lambda^c \tag{12.66b}$$
$$\boldsymbol{\delta}_\mu = \boldsymbol{\delta}_\mu^{\text{aff}} + \boldsymbol{\delta}_\mu^c \tag{12.66c}$$

and the new iterate is given by

$$\mathbf{w}_{k+1} = \mathbf{w}_k + \{\alpha_{k,p}\boldsymbol{\delta}_x, \; \alpha_{k,d}\boldsymbol{\delta}_\lambda, \; \alpha_{k,d}\boldsymbol{\delta}_\mu\} \tag{12.67}$$

where the step sizes for $\boldsymbol{\delta}_x$ and $(\boldsymbol{\delta}_\lambda, \boldsymbol{\delta}_\mu)$ are determined separately as

$$\alpha_{k,p} = \min(0.99\,\alpha_{\max}^{(p)}, 1) \tag{12.68a}$$
$$\alpha_{\max}^{(p)} = \max_{\alpha\geq 0,\, \mathbf{x}_k+\alpha\boldsymbol{\delta}_x\geq\mathbf{0}} (\alpha) \tag{12.68b}$$
$$\alpha_{k,d} = \min(0.99\,\alpha_{\max}^{(d)}, 1) \tag{12.68c}$$
$$\alpha_{\max}^{(d)} = \max_{\alpha\geq 0,\, \boldsymbol{\mu}_k+\alpha\boldsymbol{\delta}_\mu\geq\mathbf{0}} (\alpha) \tag{12.68d}$$

400

A note on the computational complexity of the method is in order. From Eq. (12.66), we see that the search direction is obtained by computing δ_w^{aff} and δ_w^c; hence the two linear systems in Eqs. (12.58) and (12.64) have to be solved. However, the system matrices for Eqs. (12.58) and (12.64) are *identical* and, consequently, the computational effort required by the PCM algorithm is increased only slightly relative to that required by the primal-dual path-following algorithms discussed in the preceding sections. This can also be observed from the fact that matrices $\mathbf{Y}$ and $\mathbf{D}$ used to solve Eq. (12.58) can also be used to solve Eq. (12.64). A step-by-step summary of the PCM algorithm is given below.

Algorithm 12.6 Mehrotra's predictor-corrector algorithm for the standard-form LP problem
Step 1
Input $\mathbf{A}$, $\mathbf{b}$, $\mathbf{c}$, and $\mathbf{w}_0 = \{\mathbf{x}_0, \lambda_0, \mu_0\}$ with $\mathbf{x}_0 > 0$ and $\mu_0 > 0$.
Set $k = 0$ and $\hat{\tau}_0 = (\mu_0^T \mathbf{x}_0)/n$, and initialize the tolerance ε for the duality gap.
Step 2
If $\mu_k^T \mathbf{x}_k \le \varepsilon$, output solution $\mathbf{w}^* = \mathbf{w}_k$ and stop; otherwise, go to Step 3.
Step 3
Compute predictor direction $\{\delta_x^{\text{aff}}, \delta_\lambda^{\text{aff}}, \delta_\mu^{\text{aff}}\}$ using Eq. (12.59).
Step 4
Compute τ_{aff} using Eqs. (12.60) and (12.61) and determine τ_{k+1} as

$$\tau_{k+1} = \sigma_k \hat{\tau}_k$$

where σ_k and $\hat{\tau}_k$ are evaluated using Eqs. (12.62) and (12.63).
Step 5
Compute corrector direction $\{\delta_x^c, \delta_\lambda^c, \delta_\mu^c\}$ using Eq. (12.65).
Step 6
Obtain search direction $\{\delta_x, \delta_\lambda, \delta_\mu\}$ using Eq. (12.66) and evaluate step sizes $\alpha_{k,p}$ and $\alpha_{k,d}$ using Eq. (12.68).
Step 7
Set $\mathbf{w}_{k+1}$ using Eq. (12.67).
Set $k = k + 1$ and repeat from Step 2.

Example 12.6 Apply Algorithm 12.6 to the LP problems in
(a) Example 12.3
(b) Example 12.2
with nonfeasible initial points.

Solution

(a) We can use the same $\mathbf{w}_0$ and ε as in Example 12.5(a) to start Algorithm 12.6. It took six iterations for the algorithm to converge to the solution

$$\mathbf{x}^* = \begin{bmatrix} 0.000000 \\ 0.000000 \\ 1.000000 \end{bmatrix}$$

The number of flops required was 1.268K, which entails a slight increase as compared to that in Example 12.5(a) but the solution $\mathbf{x}^*$ is more accurate. Fig. 12.4 shows point $\mathbf{x}_0$ and the first three iterates, i.e., $\mathbf{x}_k$ for $k = 0, 1, 2, 3$ as compared to the central path which is plotted as the dotted curve.

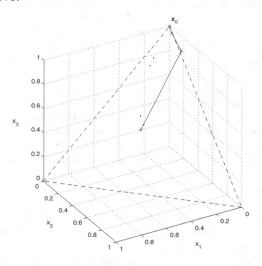

Figure 12.4.   Iteration path in Example 12.6(a) as compared to the central path.

(b) The same $\mathbf{w}_0$ and ε as in Example 12.5(b) were used here. The algorithm took 11 iterations to converge to the solution

$$\mathbf{x}^* = \begin{bmatrix} 0.000000 \\ 0.333333 \\ 0.333333 \\ 0.000000 \\ 0.000000 \end{bmatrix}$$

The number of flops required was 7.564K. This is slightly larger than the number of flops in Example 12.5(b) but some improvement in the accuracy of the solution has been achieved.

■

References

1 N. K. Karmarkar, "A new polynomial time algorithm for linear programming," *Combinatorica*, vol. 4, pp. 373–395, 1984.

2 M. H. Wright, "Interior methods for constrained optimization," *Acta Numerica*, pp. 341–407, Cambridge Univ. Press, Cambridge, UK, 1992.

3 S. J. Wright, *Primal-Dual Interior-Point Methods*, SIAM, Philadelphia, 1997.

4 E. R. Barnes, "A variation on Karmarkar's algorithm for solving linear programming problems," *Math. Programming*, vol. 36, pp. 174–182, 1986.

5 R. J. Vanderbei, M. S. Meketon, and B. A. Freedman, "A modification of Karmarkar's linear programming algorithm," *Algorithmica*, vol. 1, pp. 395–407, 1986.

6 S. G. Nash and A. Sofer, *Linear and Nonlinear Programming*, McGraw-Hill, New York, 1996.

7 P. E. Gill, W. Murray, M. A. Saunders, J. A. Tomlin, and M. H. Wright, "On projected Newton barrier methods for linear programming and an equivalence to Karmarkar's projective method," *Math. Programming*, vol. 36, pp. 183–209, 1986.

8 A. V. Fiacco and G. P. McCormick, *Nonlinear Programming: Sequential Unconstrained Minimization Techniques*, Wiley, New York, 1968 (Republished by SIAM, 1990).

9 K. Jittorntrum, *Sequential Algorithms in Nonlinear Programming*, Ph.D. thesis, Australian National University, 1978.

10 K. Jittorntrum and M. R. Osborne, "Trajectory analysis and extrapolation in barrier function methods," *J. Australian Math. Soc.*, Series B, vol. 20, pp. 352–369, 1978.

11 M. Kojima, S. Mizuno, and A. Yoshise, "A primal-dual interior point algorithm for linear programming," *Progress in Mathematical Programming: Interior Point and Related Methods*, N. Megiddo ed., pp. 29–47, Springer Verlag, New York, 1989.

12 N. Megiddo, "Pathways to the optimal set in linear programming," *Progress in Mathematical Programming: Interior Point and Related Methods*, N. Megiddo ed., pp. 131–158, Springer Verlag, New York, 1989.

13 R. D. C. Monteiro and I. Adler, "Interior path following primal-dual algorithms, Part I: Linear programming," *Math. Programming*, vol. 44, pp. 27–41, 1989.

14 Y. Ye, *Interior-Point Algorithm: Theory and Analysis*, Wiley, New York, 1997.

15 I. J. Lustig, R. E. Marsten, and D. F. Shanno, "Computational experience with a primal-dual interior point method for linear programming," *Linear Algebra and Its Applications*, vol. 152, pp. 191–222, 1991.

16 S. Mehrotra, "On the implementation of a primal-dual interior point method," *SIAM J. Optimization*, vol. 2, pp. 575–601, 1992.

Problems

12.1 This problem concerns the central path of the LP problem described in Example 12.1.

(a) For a sample number of values τ ranging from 500 to 10^{-3}, use MATLAB command roots to evaluate the roots λ of Eq. (12.14) with $\lambda < -3$.

(b) Generate a trajectory (x_1, x_2, x_3) similar to that in Fig. 12.1.

(c) Change the range of τ from $[10^{-3}, 500]$ to $[10^{-2}, 200]$ and then to $[2.5 \times 10^{-2}, 20]$ and observe the trajectories (x_1, x_2, x_3) obtained.

12.2 Consider the LP problem Eq. (12.1) and let $\mathbf{x}_{k+1}$ be determined using Eq. (12.22). Show that if $\alpha_k > 0$, then $f(\mathbf{x}_{k+1}) \leq f(\mathbf{x}_k)$ and this inequality holds strictly if at least one of the last $n - p$ components of $\mathbf{V}_k^T \mathbf{X} \mathbf{c}$ is nonzero, where $\mathbf{V}_k$ is the $n \times n$ orthogonal matrix obtained from the SVD of $\mathbf{A}\mathbf{X}$: $\mathbf{A}\mathbf{X} = \mathbf{U}_k[\mathbf{\Sigma}_k\ \mathbf{0}]\mathbf{V}_k^T$.

12.3 (a) Apply the PAS algorithm to solve the LP problem in Prob. 11.16. Compare the results with those obtained in Prob. 11.16.

(b) Apply the PAS algorithm to solve the LP problem in Prob. 11.17. Compare the results with those obtained in Prob. 11.17.

12.4 (a) Derive the KKT conditions for the minimizer of the problem in Eq. (12.26).

(b) Relate the KKT conditions obtained in part (a) to the central path of the original LP problem in Eq. (12.9).

12.5 (a) Apply the PNB algorithm to solve the LP problem in Prob. 11.16. Compare the results with those obtained in Prob. 12.3(a).

(b) Apply the PNB algorithm to solve the LP problem in Prob. 11.17. Compare the results with those obtained in Prob. 12.3(b).

12.6 Develop a PNB algorithm that is directly applicable to the LP problem in Eq. (11.2).

Hint: Denote $\mathbf{A}$ and $\mathbf{b}$ in Eq. (11.2b) as

$$\mathbf{A} = \begin{bmatrix} \mathbf{a}_1^T \\ \mathbf{a}_2^T \\ \vdots \\ \mathbf{a}_p^T \end{bmatrix} \quad \text{and} \quad \mathbf{b} = \begin{bmatrix} b_1 \\ b_2 \\ \vdots \\ b_p \end{bmatrix}$$

and consider the logarithmic barrier function

$$f_\tau(\mathbf{x}) = \mathbf{c}^T\mathbf{x} - \tau \sum_{i=1}^{p} \ln(\mathbf{a}_i^T\mathbf{x} - b_i)$$

where $\tau > 0$ is a barrier parameter.

12.7 Using the initial point $[x_1\ x_2\ x_3\ s_1\ s_2\ s_3] = [0.5\ 1\ 10\ 0.5\ 98\ 9870]^T$, solve the LP problem

$$\text{minimize } 100x_1 + 10x_2 + x_3$$

$$\text{subject to:} \quad s_1 + x_1 = 1$$
$$s_2 + 2x_1 + x_2 = 100$$
$$s_3 + 200x_1 + 20x_2 + x_3 = 10000$$
$$x_i \geq 0,\ s_i \geq 0 \text{ for } i = 1, 2, 3$$

404

by using

(a) the PAS algorithm.

(b) the PNB algorithm.

12.8 The primal Newton barrier method discussed in Sec. 12.4 is related to the primal LP problem in Eq. (12.1). It is possible to develop a dual Newton barrier (DNB) method in terms of the following steps:

(a) Define a dual subproblem similar to that in Eq. (12.26) for the LP problem in Eq. (12.2).

(b) Derive the first-order optimality conditions for the subproblem obtained in part (a).

(c) Show that the points satisfying these first-order conditions are on the primal-dual central path.

(d) Develop a DNB algorithm for solving the dual problem in Eq. (12.2).

12.9 Consider the standard-form LP problem in Eq. (12.1). A strictly feasible point $x^* > 0$ is said to be the *analytic center* of the feasible region if x^* is the farthest away from all the boundaries of the feasible region in the sense that x^* solves the problem

$$\text{minimize} \; - \sum_{i=1}^{n} \ln x_i \tag{P12.1a}$$

$$\text{subject to:} \quad Ax = b \tag{P12.1b}$$

(a) Derive the KKT conditions for the minimizer of the problem in Eq. (P12.1).

(b) Are the KKT conditions necessary and sufficient conditions?

(c) Use the KKT conditions obtained to find the analytic center for the LP problem in Example 12.1.

12.10 Generalize the concept of analytic center discussed in Prob. 12.9 to the feasible region given by $Ax \geq b$, where $A \in R^{p \times n}$ with $p > n$, and $\text{rank}(A) = n$.

12.11 (a) Prove the inequality in Eq. (12.28).

(b) Drive the formulas in Eqs. (12.38a) and (12.38b).

12.12 Develop a primal path-following interior-point algorithm for the primal LP problem in Eq. (12.1) in several steps as described below.

(a) Formulate a subproblem by adding a logarithmic barrier function to the objective function, i.e.,

$$\text{minimize } f_\tau(\mathbf{x}) = \mathbf{c}^T \mathbf{x} - \tau \sum_{i=1}^{n} \ln x_i$$

$$\text{subject to:} \quad \mathbf{A}\mathbf{x} = \mathbf{b}$$

where $\tau > 0$ is a barrier parameter.

(b) Show that the KKT conditions for the minimizer of the above sub-problem can be expressed as

$$\mathbf{A}\mathbf{x} = \mathbf{b}$$
$$\mathbf{c} - \mathbf{A}^T \boldsymbol{\lambda} - \tau \mathbf{X}^{-1} \mathbf{e} = \mathbf{0}$$

where $\mathbf{X} = \text{diag}(\mathbf{x})$ and $\mathbf{e} = [1 \ 1 \ \cdots \ 1]^T$.

(c) At the kth iteration, let $\mathbf{x}_{k+1} = \mathbf{x}_k + \mathbf{d}$ such that $\mathbf{x}_{k+1}$ would better approximate the above KKT conditions. Show that up to first-order approximation, we would require that $\mathbf{d}$ satisfy the equations

$$\mathbf{A}\mathbf{d} = \mathbf{0} \qquad \text{(P12.2a)}$$
$$\tau \mathbf{X}^{-2} \mathbf{d} + \mathbf{c} - \tau \mathbf{X}^{-1} \mathbf{e} - \mathbf{A}^T \boldsymbol{\lambda}_k = \mathbf{0} \qquad \text{(P12.2b)}$$

where $\mathbf{X} = \text{diag}\{\mathbf{x}_k\}$.

(d) Show that the search direction $\mathbf{d}$ in Eq. (P12.2) can be obtained as

$$\mathbf{d} = \mathbf{x}_k - \frac{1}{\tau}\mathbf{X}^2 \boldsymbol{\mu}_k \qquad \text{(P12.3a)}$$

where

$$\boldsymbol{\mu}_k = \mathbf{c} - \mathbf{A}^T \boldsymbol{\lambda}_k \qquad \text{(P12.3b)}$$
$$\boldsymbol{\lambda}_k = (\mathbf{A}\mathbf{X}^2\mathbf{A}^T)^{-1}\mathbf{A}\mathbf{X}^2(\mathbf{c} - \tau \mathbf{X}^{-1}\mathbf{e}) \qquad \text{(P12.3c)}$$

(e) Based on the results obtained in parts (a)-(d), describe a primal path-following interior-point algorithm.

12.13 (a) Apply the algorithm developed in Prob. 12.12 to the LP problem in Prob. 11.16.

(b) Compare the results obtained in part (a) with those of Prob. 12.3(a) and Prob. 12.5(a).

12.14 (a) Apply the algorithm developed in Prob. 12.12 to the LP problem in Prob. 11.17.

(b) Compare the results obtained in part (a) with those of Prob. 12.3(b) and Prob. 12.5(b).

12.15 Show that the search direction determined by Eq. (P12.3) can be expressed as

$$\mathbf{d} = -\frac{1}{\tau}\mathbf{X}\bar{\mathbf{P}}\mathbf{x}\mathbf{c} + \mathbf{X}\bar{\mathbf{P}}\mathbf{e} \qquad \text{(P12.4)}$$

where $\bar{P} = I - XA^T(AX^2A^T)^{-1}AX$ is the projection matrix given by Eq. (12.24).

12.16 In the literature, the two terms on the right-hand side of Eq. (P12.4) are called the *primal affine-scaling direction* and *centering direction*, respectively. Justify the use of this terminology.
Hint: Use the results of Sec. 12.3 and Prob. 12.9.

12.17 (a) Derive the formulas in Eq. (12.44) using Eq. (12.42).
(b) Derive the formulas in Eq. (12.57) using Eq. (12.56).

12.18 (a) Apply Algorithm 12.4 to the LP problem in Prob. 11.16. Compare the results obtained with those of Probs. 12.3(a), 12.5(a) and 12.13(a).
(b) Apply Algorithm 12.4 to the LP problem 11.17. Compare the results obtained with those of Probs. 12.3(b), 12.5(b), and 12.14(a).

12.19 (a) Apply Algorithm 12.5 to the LP problem in Prob. 11.16 with a nonfeasible initial point $\{x_0, \lambda_0, \mu_0\}$ with $x_0 > 0$ and $\mu_0 > 0$. Compare the results obtained with those of Prob. 12.18(a).
(b) Apply Algorithm 12.5 to the LP problem in Prob. 11.17 with a nonfeasible initial point $\{x_0, \lambda_0, \mu_0\}$ with $x_0 > 0$ and $\mu_0 > 0$. Compare the results obtained with those of Prob. 12.18(b).

12.20 (a) Derive the formulas in Eq. (12.59) using Eq. (12.58).
(b) Derive the formulas in Eq. (12.65) using Eq. (12.64).

12.21 (a) Apply Algorithm 12.6 to the LP problem in Prob. 11.16 with the same nonfeasible initial point used in Prob. 12.19(a). Compare the results obtained with those of Prob. 12.19(a).
(b) Apply Algorithm 12.6 to the LP problem in Prob. 11.17 with the same nonfeasible initial point used in Prob. 12.19(b). Compare the results obtained with those of Prob. 12.19(b).

12.22 Consider the nonstandard-form LP problem

$$\text{minimize } c^T x$$

$$\text{subject to: } Ax \geq b$$

where $c \in R^{n \times 1}$, $A \in R^{p \times n}$, and $b \in R^{p \times 1}$ with $p > n$. Show that its solution x^* can be obtained by solving the standard-form LP problem

$$\text{minimize } -b^T x$$

$$\text{subject to: } A^T x = c$$

$$x \geq 0$$

using a primal-dual algorithm and then taking the optimal Lagrange multiplier vector λ^* as x^*.

Chapter 13

QUADRATIC AND CONVEX PROGRAMMING

13.1 Introduction

Quadratic programming (QP) is a family of methods, techniques, and algorithms that can be used to minimize quadratic objective functions subject to linear constraints. On the one hand, QP shares many combinatorial features with linear programming (LP) and, on the other, it is often used as the basis of constrained nonlinear programming. In fact, the computational efficiency of a nonlinear programming algorithm is often heavily dependent on the efficiency of the QP algorithm involved.

An important branch of QP is convex QP where the objective function is a convex quadratic function. A generalization of convex QP is convex programming (CP) where the objective function is convex but not necessarily quadratic and the feasible region is convex.

In this chapter, we will first study convex QP problems with equality constraints and describe a QR-decomposition-based solution method. Next, two active set methods for strictly convex QP problems are discussed in detail. These methods can be viewed as direct extensions of the simplex method discussed in Chap. 11. In Sec. 13.4, the concepts of central path and duality gap are extended to QP and two primal-dual path-following methods are studied. In addition, the concept of complementarity for convex QP is examined and its relation to that in LP is discussed. In Secs. 13.5 and 13.6, certain important classes of CP algorithms known as cutting-plane and ellipsoid algorithms are introduced.

Two special branches of CP known as *semidefinite programming* (SDP) and *second-order cone programming* (SOCP) have been the subject of intensive research during the past several years. The major algorithms for SDP and SOCP and related concepts will be studied in Chap. 14.

13.2 Convex QP Problems with Equality Constraints

The problem we consider in this section is

$$\text{minimize } f(\mathbf{x}) = \tfrac{1}{2}\mathbf{x}^T\mathbf{H}\mathbf{x} + \mathbf{x}^T\mathbf{p} \tag{13.1a}$$

$$\text{subject to: } \mathbf{A}\mathbf{x} = \mathbf{b} \tag{13.1b}$$

where $\mathbf{A} \in R^{p \times n}$. We assume in the rest of this section that the Hessian $\mathbf{H}$ is symmetric and positive semidefinite, $\mathbf{A}$ has full row rank, and $p < n$. From Sec. 10.4.1, the solutions of the problem in Eq. (13.1b) assume the form

$$\mathbf{x} = \mathbf{V}_r\boldsymbol{\phi} + \mathbf{A}^+\mathbf{b} \tag{13.2}$$

where $\mathbf{V}_r$ is composed of the last $n - p$ columns of $\mathbf{V}$ and $\mathbf{V}$ is obtained from the singular-value decomposition (SVD) of $\mathbf{A}$, namely, $\mathbf{U}\boldsymbol{\Sigma}\mathbf{V}^T$. By using Eq. (13.2), the constraints in Eq. (13.1b) can be eliminated to yield the unconstrained minimization problem

$$\text{minimize } \hat{f}(\boldsymbol{\phi}) = \tfrac{1}{2}\boldsymbol{\phi}^T\hat{\mathbf{H}}\boldsymbol{\phi} + \boldsymbol{\phi}^T\hat{\mathbf{p}} \tag{13.3a}$$

where

$$\hat{\mathbf{H}} = \mathbf{V}_r^T\mathbf{H}\mathbf{V}_r \tag{13.3b}$$

and

$$\hat{\mathbf{p}} = \mathbf{V}_r^T(\mathbf{H}\mathbf{A}^+\mathbf{b} + \mathbf{p}) \tag{13.3c}$$

If $\mathbf{H}$ in Eq. (13.3b) is positive definite, then $\hat{\mathbf{H}}$ is also positive definite and the unique global minimizer of the problem in Eq. (13.1) is given by

$$\mathbf{x}^* = \mathbf{V}_r\boldsymbol{\phi}^* + \mathbf{A}^+\mathbf{b} \tag{13.4a}$$

where $\boldsymbol{\phi}^*$ is a solution of the linear system of equations

$$\hat{\mathbf{H}}\boldsymbol{\phi} = -\hat{\mathbf{p}} \tag{13.4b}$$

If $\mathbf{H}$ is positive semidefinite, then $\hat{\mathbf{H}}$ in Eq. (13.3b) may be either positive definite or positive semidefinite. If $\hat{\mathbf{H}}$ is positive definite, then $\mathbf{x}^*$ given by Eq. (13.4a) is the unique global minimizer of the problem in Eq. (13.1). If $\hat{\mathbf{H}}$ is positive semidefinite, then there are two possibilities: (a) If $\hat{\mathbf{p}}$ can be expressed as a linear combination of the columns of $\hat{\mathbf{H}}$, then $\hat{f}(\boldsymbol{\phi})$ has infinitely many global minimizers and so does $f(\mathbf{x})$; (b) if $\hat{\mathbf{p}}$ is not a linear combination of the columns of $\hat{\mathbf{H}}$, then $\hat{f}(\boldsymbol{\phi})$, and therefore $f(\mathbf{x})$, has no minimizers.

An alternative and often more economical approach to obtain Eq. (13.2) is to use the QR decomposition of $\mathbf{A}^T$, i.e.,

$$\mathbf{A}^T = \mathbf{Q}\begin{bmatrix} \mathbf{R} \\ \mathbf{0} \end{bmatrix} \tag{13.5}$$

where $\mathbf{Q}$ is an $n \times n$ orthogonal and $\mathbf{R}$ is a $p \times p$ upper triangular matrix (see Sec. A.12 and [1]). Using Eq. (13.5), the constraints in Eq. (13.1b) can be expressed as

$$\mathbf{R}^T \hat{\mathbf{x}}_1 = \mathbf{b}$$

where $\hat{\mathbf{x}}_1$ is the vector composed of the first p elements of $\hat{\mathbf{x}}$ with

$$\hat{\mathbf{x}} = \mathbf{Q}^T \mathbf{x}$$

If we denote

$$\hat{\mathbf{x}} = \begin{bmatrix} \hat{\mathbf{x}}_1 \\ \boldsymbol{\phi} \end{bmatrix} \quad \text{and} \quad \mathbf{Q} = [\mathbf{Q}_1 \ \mathbf{Q}_2]$$

with $\boldsymbol{\phi} \in R^{(n-p) \times 1}$, $\mathbf{Q}_1 \in R^{n \times p}$, and $\mathbf{Q}_2 \in R^{n \times (n-p)}$, then

$$\mathbf{x} = \mathbf{Q}\hat{\mathbf{x}} = \mathbf{Q}_2 \boldsymbol{\phi} + \mathbf{Q}_1 \hat{\mathbf{x}}_1 = \mathbf{Q}_2 \boldsymbol{\phi} + \mathbf{Q}_1 \mathbf{R}^{-T} \mathbf{b}$$

i.e.,

$$\mathbf{x} = \mathbf{Q}_2 \boldsymbol{\phi} + \mathbf{Q}_1 \mathbf{R}^{-T} \mathbf{b} \tag{13.6a}$$

which is equivalent to Eq. (13.2). The parameterized solutions in Eq. (13.6a) can be used to convert the problem in Eq. (13.1) to the reduced-size unconstrained problem in Eq. (13.3) where $\hat{\mathbf{H}}$ and $\hat{\mathbf{p}}$ are given by

$$\hat{\mathbf{H}} = \mathbf{Q}_2^T \mathbf{H} \mathbf{Q}_2 \tag{13.6b}$$

and

$$\hat{\mathbf{p}} = \mathbf{Q}_2^T (\mathbf{H} \mathbf{Q}_1 \mathbf{R}^{-T} \mathbf{b} + \mathbf{p})$$

respectively. If $\mathbf{H}$ is positive definite, the unique global minimizer of the problem in Eq. (13.1) can be determined as

$$\mathbf{x}^* = \mathbf{Q}_2 \boldsymbol{\phi}^* + \mathbf{Q}_1 \mathbf{R}^{-T} \mathbf{b} \tag{13.7}$$

where $\boldsymbol{\phi}^*$ is a solution of Eq. (13.4b) with $\hat{\mathbf{H}}$ given by Eq. (13.6b).

In both approaches discussed above, $\hat{\mathbf{H}}$ is positive definite and the system in Eq. (13.4b) can be solved efficiently through the LDL^T (see Chap. 5) or Cholesky decomposition (see Sec. A.13).

Example 13.1 Solve the QP problem

$$\text{minimize } f(\mathbf{x}) = \tfrac{1}{2}(x_1^2 + x_2^2) + 2x_1 + x_2 - x_3 \tag{13.8a}$$

$$\text{subject to: } \mathbf{A}\mathbf{x} = \mathbf{b} \tag{13.8b}$$

where

$$\mathbf{A} = [0 \ 1 \ 1], \quad \mathbf{b} = 1$$

Solution Since matrix $\mathbf{H}$ is positive semidefinite in this case, the SVD of $\mathbf{A}$ leads to

$$\mathbf{V}_r = \begin{bmatrix} 1 & 0 \\ 0 & \frac{1}{\sqrt{2}} \\ 0 & -\frac{1}{\sqrt{2}} \end{bmatrix} \quad \text{and} \quad \mathbf{A}^+ = \begin{bmatrix} 0 \\ \frac{1}{2} \\ \frac{1}{2} \end{bmatrix}$$

Since

$$\hat{\mathbf{H}} = \mathbf{V}_r^T \mathbf{H} \mathbf{V}_r = \begin{bmatrix} 1 & 0 \\ 0 & \frac{1}{\sqrt{2}} \end{bmatrix}$$

is positive definite, the use of Eq. (13.4a) yields the unique global minimizer as

$$\begin{aligned} \mathbf{x}^* &= \mathbf{V}_r \boldsymbol{\phi}^* + \mathbf{A}^+ \mathbf{b} \\ &= \begin{bmatrix} 1 & 0 \\ 0 & \frac{1}{\sqrt{2}} \\ 0 & -\frac{1}{\sqrt{2}} \end{bmatrix} \begin{bmatrix} -2.0000 \\ -3.5355 \end{bmatrix} + \begin{bmatrix} 0 \\ \frac{1}{2} \\ \frac{1}{2} \end{bmatrix} = \begin{bmatrix} -2 \\ -2 \\ 3 \end{bmatrix} \end{aligned}$$

Alternatively, the problem can be solved by using the QR decomposition of $\mathbf{A}^T$. From Eq. (13.5), we have

$$\mathbf{Q} = \begin{bmatrix} 0 & \frac{\sqrt{2}}{2} & \frac{\sqrt{2}}{2} \\ \frac{\sqrt{2}}{2} & -0.5 & 0.5 \\ \frac{\sqrt{2}}{2} & 0.5 & -0.5 \end{bmatrix}, \quad \mathbf{R} = \sqrt{2}$$

which leads to

$$\mathbf{Q}_1 = \begin{bmatrix} 0 \\ \frac{\sqrt{2}}{2} \\ \frac{\sqrt{2}}{2} \end{bmatrix}, \quad \mathbf{Q}_2 = \begin{bmatrix} \frac{\sqrt{2}}{2} & \frac{\sqrt{2}}{2} \\ -0.5 & 0.5 \\ 0.5 & -0.5 \end{bmatrix}$$

and

$$\hat{\mathbf{H}} = \begin{bmatrix} 0.75 & 0.25 \\ 0.25 & 0.75 \end{bmatrix}, \quad \hat{\mathbf{p}} = \begin{bmatrix} 0.1642 \\ 2.6642 \end{bmatrix}$$

Hence

$$\boldsymbol{\phi}^* = \begin{bmatrix} 1.0858 \\ -3.9142 \end{bmatrix}$$

The same solution, i.e., $\mathbf{x}^* = [-2 \ -2 \ 3]^T$, can be obtained by using Eq. (13.7). Note that if the constraint matrix $\mathbf{A}$ is changed to

$$\mathbf{A} = [1 \ 0 \ 0] \tag{13.9}$$

then

$$\mathbf{V}_r = \begin{bmatrix} 0 & 0 \\ 1 & 0 \\ 0 & 1 \end{bmatrix}, \quad \hat{\mathbf{H}} = \begin{bmatrix} 1 & 0 \\ 0 & 0 \end{bmatrix}$$

and

$$\hat{\mathbf{p}} = \begin{bmatrix} 1 \\ -1 \end{bmatrix}$$

Obviously, $\hat{\mathbf{p}}$ cannot be expressed as a linear combination of the columns of $\hat{\mathbf{H}}$ in this case and hence the problem in Eq. (13.8) with $\mathbf{A}$ given by Eq. (13.9) does not have a finite solution.

If the objective function is modified to

$$f(\mathbf{x}) = \tfrac{1}{2}(x_1^2 + x_2^2) + 2x_1 + x_2$$

then with $\mathbf{A}$ given by Eq. (13.9), we have

$$\hat{\mathbf{p}} = \begin{bmatrix} 1 \\ 0 \end{bmatrix}$$

In this case, $\hat{\mathbf{p}}$ is a linear combination of the columns of $\hat{\mathbf{H}}$ and hence there are infinitely many solutions. As a matter of fact, it can be readily verified that any $\mathbf{x}^* = [1 \ -1 \ x_3]^T$ with an arbitrary x_3 is a global minimizer of the problem. ∎

The problem in Eq. (13.1) can also be solved by using the first-order necessary conditions described in Theorem 10.1, which are given by

$$\mathbf{H}\mathbf{x}^* + \mathbf{p} - \mathbf{A}^T\boldsymbol{\lambda}^* = 0$$
$$-\mathbf{A}\mathbf{x}^* + \mathbf{b} = 0$$

i.e.,

$$\begin{bmatrix} \mathbf{H} & -\mathbf{A}^T \\ -\mathbf{A} & 0 \end{bmatrix} \begin{bmatrix} \mathbf{x}^* \\ \boldsymbol{\lambda}^* \end{bmatrix} = - \begin{bmatrix} \mathbf{p} \\ \mathbf{b} \end{bmatrix} \tag{13.10}$$

If $\mathbf{H}$ is positive definite and $\mathbf{A}$ has full row rank, then the system matrix in Eq. (13.10) is nonsingular (see Eq. 10.69) and the solution $\mathbf{x}^*$ from Eq. (13.10) is the unique global minimizer of the problem in Eq. (13.1). Hence the solution $\mathbf{x}^*$ and Lagrange multipliers $\boldsymbol{\lambda}^*$ can be expressed as

$$\boldsymbol{\lambda}^* = (\mathbf{A}\mathbf{H}^{-1}\mathbf{A}^T)^{-1}(\mathbf{A}\mathbf{H}^{-1}\mathbf{p} + \mathbf{b}) \tag{13.11a}$$
$$\mathbf{x}^* = \mathbf{H}^{-1}(\mathbf{A}\boldsymbol{\lambda}^* - \mathbf{p}) \tag{13.11b}$$

The solution of the symmetric system in Eq. (13.10) can be obtained using numerical methods that are often more reliable and efficient than the formulas in Eq. (13.11) (see Chap. 10 of [1] for the details).

13.3   Active-Set Methods for Strictly Convex QP Problems

The general form of a QP problem is to minimize a quadratic function subject to a set of linear equality and a set of linear inequality constraints. Using Eq. (13.2) or Eq. (13.6a), the equality constraints can be eliminated and without loss of generality the problem can be reduced to a QP problem subject to only linear inequality constraints as

$$\text{minimize } f(\mathbf{x}) = \tfrac{1}{2}\mathbf{x}^T \mathbf{H}\mathbf{x} + \mathbf{x}^T \mathbf{p} \tag{13.12a}$$

$$\text{subject to:} \quad \mathbf{A}\mathbf{x} \geq \mathbf{b} \tag{13.12b}$$

where $\mathbf{A} \in R^{p \times n}$. The Karush-Kuhn-Tucker (KKT) conditions of the problem at a minimizer $\mathbf{x}$ are given by

$$\mathbf{H}\mathbf{x} + \mathbf{p} - \mathbf{A}^T \boldsymbol{\mu} = \mathbf{0} \tag{13.13a}$$

$$(\mathbf{a}_i^T \mathbf{x} - b_i)\mu_i = 0 \quad \text{for } i = 1, 2, \ldots, p \tag{13.13b}$$

$$\mu_i \geq 0 \quad \text{for } i = 1, 2, \ldots, p \tag{13.13c}$$

$$\mathbf{A}\mathbf{x} \geq \mathbf{b} \tag{13.13d}$$

To focus our attention on the major issues, we assume in the rest of this section that $\mathbf{H}$ is positive definite and all vertices of the feasible region are nondegenerate. First, we consider the possibility of having a solution $\{\mathbf{x}^*, \ \boldsymbol{\mu}^*\}$ for Eq. (13.13) with $\mathbf{x}^*$ in the interior of the feasible region $\mathcal{R}$. If this is the case, then $\mathbf{A}\mathbf{x}^* > \mathbf{b}$ and Eq. (13.13b) implies that $\boldsymbol{\mu}^* = \mathbf{0}$, and Eq. (13.13a) gives

$$\mathbf{x}^* = -\mathbf{H}^{-1}\mathbf{p} \tag{13.14}$$

which is the unique global minimizer of $f(\mathbf{x})$ if there are no constraints. Therefore, we conclude that solutions of the problem in Eq. (13.12) are on the boundary of the feasible region $\mathcal{R}$ unless the unconstrained minimizer in Eq. (13.14) is an interior point of $\mathcal{R}$. In any given iteration, the search direction in an active set method is determined by treating the constraints that are active at the iterate as a set of equality constraints while neglecting the rest of the constraints. In what follows, we describe first a primal active set method [2][3] and then a dual active set method [4] for the problem in Eq. (13.12).

13.3.1 Primal active-set method

Let $\mathbf{x}_k$ be a feasible iterate obtained in the kth iteration and assume that $\mathcal{J}_k$ is the index set of the active constraints, which is often referred to as the *active set*, at $\mathbf{x}_k$. The next iterate is given by

$$\mathbf{x}_{k+1} = \mathbf{x}_k + \alpha_k \mathbf{d}_k \tag{13.15}$$

The constraints that are active at $\mathbf{x}_k$ will remain active if

$$\mathbf{a}_j^T \mathbf{x}_{k+1} - b_j = 0 \quad \text{for } j \in \mathcal{J}_k$$

which leads to

$$\mathbf{a}_j^T \mathbf{d}_k = 0 \quad \text{for } j \in \mathcal{J}_k$$

The objective function at $\mathbf{x}_k + \mathbf{d}$ becomes

$$f_k(\mathbf{d}) = \tfrac{1}{2}\mathbf{d}^T \mathbf{H}\mathbf{d} + \mathbf{d}^T \mathbf{g}_k + c_k$$

where

$$\mathbf{g}_k = \mathbf{p} + \mathbf{H}\mathbf{x}_k \qquad (13.16)$$

and c_k is a constant. A major step in the active set method is to solve the QP subproblem

$$\text{minimize } \hat{f}(\mathbf{d}) = \tfrac{1}{2}\mathbf{d}^T\mathbf{H}\mathbf{d} + \mathbf{d}^T\mathbf{g}_k \qquad (13.17a)$$

$$\text{subject to: } \mathbf{a}_j^T\mathbf{d} = 0 \qquad \text{for } j \in \mathcal{J}_k \qquad (13.17b)$$

and this can be accomplished by using one of the methods described in the preceding section.

If the solution of the problem in Eq. (13.17) is denoted as $\mathbf{d}_k$, then there are two possibilities: either $\mathbf{d}_k = \mathbf{0}$ or $\mathbf{d}_k \neq \mathbf{0}$.

If $\mathbf{d}_k = \mathbf{0}$, then the first-order necessary conditions imply that there exist μ_j for $j \in \mathcal{J}_k$ such that

$$\mathbf{H}\mathbf{x}_k + \mathbf{p} - \sum_{j \in \mathcal{J}_k} \mu_j \mathbf{a}_j = \mathbf{0} \qquad (13.18)$$

i.e.,

$$\mathbf{H}\mathbf{x}_k + \mathbf{p} - \mathbf{A}_{a_k}^T \hat{\boldsymbol{\mu}} = \mathbf{0} \qquad (13.19)$$

where $\mathbf{A}_{a_k}$ is the matrix composed of those rows of $\mathbf{A}$ that are associated with the constraints that are active at $\mathbf{x}_k$ and $\hat{\boldsymbol{\mu}}$ is the vector composed of the μ_i's in Eq. (13.18). If we augment vector $\hat{\boldsymbol{\mu}}$ to n-dimensional vector $\boldsymbol{\mu}$ by padding zeros at the places corresponding to those rows of $\mathbf{A}$ that are inactive at $\mathbf{x}_k$, then Eq. (13.19) can be written as

$$\mathbf{H}\mathbf{x}_k + \mathbf{p} - \mathbf{A}^T\boldsymbol{\mu} = \mathbf{0}$$

which is the same as Eq. (13.13a). Since $\mathbf{x}_k$ is a feasible point, it satisfies Eq. (13.13d). Moreover, because of the way vector $\boldsymbol{\mu}$ is constructed, the complementarity condition in Eq. (13.13b) is also satisfied. So the first-order necessary conditions in Eq. (13.13), which are also sufficient conditions since the present problem is a convex QP problem, will be satisfied at $\mathbf{x}_k$ if $\hat{\boldsymbol{\mu}} \geq \mathbf{0}$. In such a case, $\mathbf{x}_k$ can be deemed to be the unique global solution and the iteration can be terminated. On the other hand, if one of the components of $\hat{\boldsymbol{\mu}}$, say, μ_i, is negative, then if point $\mathbf{x}$ moves along a feasible direction at $\mathbf{x}_k$, say, $\tilde{\mathbf{d}}$, where the ith constraint becomes inactive while all the other constraints that were active at $\mathbf{x}_k$ remain active, then the objective function will *decrease*. As a matter of fact, at $\mathbf{x}_k$ we have $\mathbf{a}_j^T\tilde{\mathbf{d}} = 0$ for $j \in \mathcal{J}_k$, $j \neq i$, and $\mathbf{a}_i^T\tilde{\mathbf{d}} > 0$. From Eq. (13.19), we have

$$\nabla^T f(\mathbf{x}_k)\tilde{\mathbf{d}} = (\mathbf{H}\mathbf{x}_k + \mathbf{p})^T\tilde{\mathbf{d}} = \hat{\boldsymbol{\mu}}^T\mathbf{A}_{a_k}^T\tilde{\mathbf{d}} = \sum_{j \in \mathcal{J}_k} \mu_j \mathbf{a}_j^T\tilde{\mathbf{d}}$$

$$= \mu_i(\mathbf{a}_i^T\tilde{\mathbf{d}}) < 0$$

Consequently, active set $\mathcal{J}_k$ can be updated by removing index i from $\mathcal{J}_k$. For the sake of simplicity, the updated index set is again denoted as $\mathcal{J}_k$. If there are more than one negative Lagrange multipliers, then the index associated with the most negative Lagrange multiplier is removed.

It should be stressed that in an implementation of the method described, verifying whether or not $\mathbf{d}_k$ is zero can be carried out without solving the problem in Eq. (13.17). At point $\mathbf{x}_k$, we can write Eq. (13.19) as

$$\mathbf{A}_{a_k}^T \hat{\boldsymbol{\mu}} = \mathbf{g}_k \tag{13.20}$$

where $\mathbf{g}_k$ is given by Eq. (13.16). It is well known that a solution $\hat{\boldsymbol{\mu}}$ exists if and only if

$$\text{rank}[\mathbf{A}_{a_k}^T \ \mathbf{g}_k] = \text{rank}(\mathbf{A}_{a_k}^T) \tag{13.21}$$

SVD- and QR-decomposition-based methods are available for checking the condition in Eq. (13.21) [1][5]. If the condition in Eq. (13.21) is met, the components of $\hat{\boldsymbol{\mu}}$ are examined to determine whether $\mathbf{x}_k$ is the solution or $\mathcal{J}_k$ needs to be updated. Otherwise, the subproblem in Eq. (13.17) is solved.

If $\mathbf{d}_k \neq \mathbf{0}$, then parameter α_k in Eq. (13.15) needs to be determined to assure the feasibility of $\mathbf{x}_{k+1}$. Using Eq. (13.17b), the optimal α_k can be determined as

$$\alpha_k = \min \left\{ 1, \ \min_{\substack{i \notin \mathcal{J}_k \\ \mathbf{a}_i^T \mathbf{d}_k < 0}} \frac{\mathbf{a}_i^T \mathbf{x}_k - b_i}{-\mathbf{a}_i^T \mathbf{d}_k} \right\} \tag{13.22}$$

If $\alpha_k < 1$, then a new constraint becomes active at $\mathbf{x}_{k+1}$. The active set $\mathcal{J}_{k+1}$ at $\mathbf{x}_{k+1}$ is obtained by adding the index of the new active constraint, j_k, to $\mathcal{J}_k$.

The active-set method can be implemented in terms of the following algorithm.

Algorithm 13.1 Primal active-set algorithm for QP problems with inequality constraints
Step 1
Input a feasible point, $\mathbf{x}_0$, identify the active set $\mathcal{J}_0$, form matrix $\mathbf{A}_{a_0}$, and set $k = 0$.
Step 2
Compute $\mathbf{g}_k$ using Eq. (13.16).
Check the rank condition in Eq. (13.21); if Eq. (13.21) does not hold, go to Step 4.
Step 3
Solve Eq. (13.20) for $\hat{\boldsymbol{\mu}}$. If $\hat{\boldsymbol{\mu}} \geq \mathbf{0}$, output $\mathbf{x}_k$ as the solution and stop; otherwise, remove the index that is associated with the most negative Lagrange multiplier from $\mathcal{J}_k$.
Step 4
Solve the problem in Eq. (13.17) for $\mathbf{d}_k$.

Step 5
Find α_k using Eq. (13.22) and set $\mathbf{x}_{k+1} = \mathbf{x}_k + \alpha_k \mathbf{d}_k$.
Step 6
If $\alpha_k < 1$, construct $\mathcal{J}_{k+1}$ by adding the index that yields the minimum in Eq. (13.22) to $\mathcal{J}_k$; otherwise, let $\mathcal{J}_{k+1} = \mathcal{J}_k$.
Step 7
Set $k = k + 1$ and repeat from Step 2.

Algorithm 13.1 requires a feasible initial point $\mathbf{x}_0$ that satisfies the constraints $\mathbf{Ax}_0 \geq \mathbf{b}$. Such a point can be identified by using, for example, the method described in Sec. 11.2.3.4. The method involves solving an LP problem of size $n + 1$ for which a feasible initial point can be easily identified.

Example 13.2 Find the shortest distance between triangles $\mathcal{R}$ and $\mathcal{S}$ shown in Fig. 13.1 and the points $\mathbf{r}^* \in \mathcal{R}$ and $\mathbf{s}^* \in \mathcal{S}$ that yield the minimum distance.

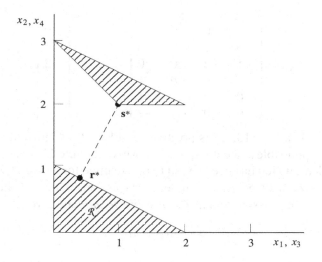

Figure 13.1. Triangles $\mathcal{R}$ and $\mathcal{S}$ in Example 13.2.

Solution Let $\mathbf{r} = [x_1 \ x_2]^T \in \mathcal{R}$ and $\mathbf{s} = [x_3 \ x_4]^T \in \mathcal{S}$. The square of the distance between $\mathbf{r}$ and $\mathbf{s}$ is given by

$$(x_1 - x_3)^2 + (x_2 - x_4)^2 = \mathbf{x}^T \mathbf{H} \mathbf{x}$$

where

$$\mathbf{H} = \begin{bmatrix} 1 & 0 & -1 & 0 \\ 0 & 1 & 0 & -1 \\ -1 & 0 & 1 & 0 \\ 0 & -1 & 0 & 1 \end{bmatrix} \tag{13.23}$$

and $\mathbf{x} = [x_1 \ x_2 \ x_3 \ x_4]^T$ is constrained to satisfy the inequalities

$$x_1 \geq 0$$
$$x_2 \geq 0$$
$$x_1 + 2x_2 \leq 2$$
$$x_4 \geq 2$$
$$x_3 + x_4 \geq 3$$
$$x_3 + 2x_4 \leq 6$$

The problem can be formulated as the QP problem

$$\text{minimize } f(\mathbf{x}) = \tfrac{1}{2}\mathbf{x}^T \mathbf{H} \mathbf{x} \tag{13.24a}$$

$$\text{subject to:} \quad \mathbf{A}\mathbf{x} \geq \mathbf{b} \tag{13.24b}$$

where

$$\mathbf{A} = \begin{bmatrix} 1 & 0 & 0 & 0 \\ 0 & 1 & 0 & 0 \\ -1 & -2 & 0 & 0 \\ 0 & 0 & 0 & 1 \\ 0 & 0 & 1 & 1 \\ 0 & 0 & -1 & -2 \end{bmatrix}, \quad \mathbf{b} = \begin{bmatrix} 0 \\ 0 \\ -2 \\ 2 \\ 3 \\ -6 \end{bmatrix}$$

Since matrix $\mathbf{H}$ in Eq. (13.23) is positive semidefinite, Algorithm 13.1 is not immediately applicable since it requires a positive definite $\mathbf{H}$. One approach to fix this problem, which has been found to be effective for convex QP problems of moderate size with positive semidefinite Hessian, is to introduce a small perturbation to the Hessian to make it *positive definite*, i.e., we let

$$\tilde{\mathbf{H}} = \mathbf{H} + \delta \mathbf{I} \tag{13.25}$$

where $\mathbf{I}$ is the identity matrix and δ is a small positive scalar. Modifying matrix $\mathbf{H}$ in Eq. (13.23) to $\tilde{\mathbf{H}}$ as in Eq. (13.25) with $\delta = 10^{-9}$ and then applying Algorithm 13.1 to the modified QP problem with an initial point $\mathbf{x}_0 = [2 \ 0 \ 0 \ 3]^T$, the minimizer was obtained in 5 iterations as $\mathbf{x}^* = [0.4 \ 0.8 \ 1.0 \ 2.0]^T$. Hence $\mathbf{r}^* = [0.4 \ 0.8]^T$ and $\mathbf{s}^* = [1 \ 2]^T$ and the shortest distance between these points is 1.341641 (see Fig. 13.1).
∎

13.3.2 Dual active-set method

A dual active-set method for the QP problem in Eq. (13.12) with $\mathbf{H}$ positive definite was proposed by Goldfarb and Idnani [4]. The method is essentially the active-set method described in the preceding section but applied to the dual of the problem in Eq. (13.12). According to Theorem 10.9, the dual is the maximization problem

$$\text{maximize } [-\tfrac{1}{2}\mu^T \mathbf{AH}^{-1}\mathbf{A}^T\mu + \mu^T(\mathbf{AH}^{-1}\mathbf{p} + \mathbf{b})]$$

$$\text{subject to: } \quad \mu \geq 0$$

which is equivalent to

$$\text{minimize } h(\mu) = \tfrac{1}{2}\mu^T \mathbf{AH}^{-1}\mathbf{A}^T\mu - \mu^T(\mathbf{AH}^{-1}\mathbf{p} + \mathbf{b}) \qquad (13.26a)$$

$$\text{subject to: } \quad \mu \geq 0 \qquad (13.26b)$$

Once the minimizer of the problem in Eq. (13.26), μ^*, is determined, the minimizer of the primal is obtained from one of the KKT conditions, i.e.,

$$\mathbf{Hx} + \mathbf{p} - \mathbf{A}^T\mu = 0$$

which gives

$$\mathbf{x}^* = \mathbf{H}^{-1}(\mathbf{A}^T\mu^* - \mathbf{p}) \qquad (13.27)$$

The advantages of the dual problem in Eq. (13.26) include:

(a) A feasible initial point can be easily identified as any vector with non-negative entries, e.g., $\mu_0 = 0$.

(b) The constraint matrix in Eq. (13.26b) is the $p \times p$ identity matrix. Consequently, the dual problem always satisfies the nondegeneracy assumption.

(c) The dual problem only involves bound-type inequality constraints which considerably simplify the computations required in Algorithm 13.1. For example, checking the rank condition in Eq. (13.21) for the dual problem entails examining whether the components of $\mathbf{g}_k$ that correspond to those indices not in the active set are all zero.

As in the primal active-set method discussed in Sec. 13.3.1, a major step in the dual active-set method is to solve the QP subproblem which is the dual of the QP problem in Eq. (13.17). This QP subproblem can be reduced to the unconstrained optimization problem

$$\text{minimize } \tfrac{1}{2}\tilde{\mathbf{d}}^T \tilde{\mathbf{H}}\tilde{\mathbf{d}} + \tilde{\mathbf{d}}^T \tilde{\mathbf{g}}_k$$

where $\tilde{\mathbf{d}}$ is the column vector obtained by deleting the components of $\mathbf{d}$ whose indices are in $\mathcal{J}_k$, $\tilde{\mathbf{H}}$ is the principal submatrix of $\mathbf{H}$ obtained by deleting the columns and rows associated with index set $\mathcal{J}_k$ and $\tilde{\mathbf{g}}_k$ is obtained by deleting the components of $\mathbf{g}_k$ whose indices are in $\mathcal{J}_k$.

13.4 Interior-Point Methods for Convex QP Problems

In this section, we discuss several interior-point methods for convex QP problems that can be viewed as natural extensions of the interior-point methods discussed in Chap. 12 for LP problems.

13.4.1   Dual QP problem, duality gap, and central path

By introducing slack variables and splitting free variables into positive and negative parts, we can reformulate the QP problem in Eq. (13.12) as

$$\text{minimize } f(\mathbf{x}) = \tfrac{1}{2}\mathbf{x}^T\mathbf{H}\mathbf{x} + \mathbf{x}^T\mathbf{p} \tag{13.28a}$$

$$\text{subject to: } \quad \mathbf{A}\mathbf{x} = \mathbf{b} \tag{13.28b}$$

$$\mathbf{x} \geq 0 \tag{13.28c}$$

where $\mathbf{H} \in R^{n \times n}$ is positive semidefinite and $\mathbf{A} \in R^{p \times n}$ has full row rank. By applying Theorem 10.9 to Eq. (13.28), the dual problem can be obtained as

$$\text{maximize; } h(\mathbf{x}, \boldsymbol{\lambda}, \boldsymbol{\mu}) = -\tfrac{1}{2}\mathbf{x}^T\mathbf{H}\mathbf{x} + \boldsymbol{\lambda}^T\mathbf{b} \tag{13.29a}$$

$$\text{subject to: } \quad \mathbf{A}^T\boldsymbol{\lambda} + \boldsymbol{\mu} - \mathbf{H}\mathbf{x} = \mathbf{p} \tag{13.29b}$$

$$\boldsymbol{\mu} \geq 0 \tag{13.29c}$$

The necessary and sufficient conditions for vector $\mathbf{x}$ to be the global minimizer of the problem in Eq. (13.28) are the KKT conditions which are given by

$$\mathbf{A}\mathbf{x} - \mathbf{b} = 0 \quad \text{for } \mathbf{x} \geq 0 \tag{13.30a}$$

$$\mathbf{A}^T\boldsymbol{\lambda} + \boldsymbol{\mu} - \mathbf{H}\mathbf{x} - \mathbf{p} = 0 \quad \text{for } \boldsymbol{\mu} \geq 0 \tag{13.30b}$$

$$\mathbf{X}\boldsymbol{\mu} = 0 \tag{13.30c}$$

where $\mathbf{X} = \text{diag}\{x_1, x_2, \ldots, x_n\}$.

Let set $\{\mathbf{x}, \boldsymbol{\lambda}, \boldsymbol{\mu}\}$ be feasible for the problems in Eqs. (13.28) and (13.29). The duality gap, which was defined in Eq. (12.6) for LP problems, can be obtained for $\{\mathbf{x}, \boldsymbol{\lambda}, \boldsymbol{\mu}\}$ as

$$\begin{aligned}\delta(\mathbf{x}, \boldsymbol{\lambda}, \boldsymbol{\mu}) &= f(\mathbf{x}) - h(\mathbf{x}, \boldsymbol{\lambda}, \boldsymbol{\mu}) = \mathbf{x}^T\mathbf{H}\mathbf{x} + \mathbf{x}^T\mathbf{p} - \boldsymbol{\lambda}^T\mathbf{b} \\ &= \mathbf{x}^T(\mathbf{A}^T\boldsymbol{\lambda} + \boldsymbol{\mu}) - \boldsymbol{\lambda}^T\mathbf{b} = \mathbf{x}^T\boldsymbol{\mu}\end{aligned} \tag{13.31}$$

which is always nonnegative and is equal to zero at solution $\{\mathbf{x}^*, \boldsymbol{\lambda}^*, \boldsymbol{\mu}^*\}$ because of the complementarity condition in Eq. (13.30c).

Based on Eq. (13.30), the concept of central path, which was initially introduced for LP problems in Sec. 12.2.2, can be readily extended to the problems in Eqs. (13.28) and (13.29) as the parameterized set $\mathbf{w}(\tau) = \{\mathbf{x}(\tau), \boldsymbol{\lambda}(\tau), \boldsymbol{\mu}(\tau)\}$ that satisfies the conditions

$$\mathbf{A}\mathbf{x} - \mathbf{b} = 0 \quad \text{for } \mathbf{x} > 0 \tag{13.32a}$$

$$\mathbf{A}^T\boldsymbol{\lambda} + \boldsymbol{\mu} - \mathbf{H}\mathbf{x} - \mathbf{p} = 0 \quad \text{for } \boldsymbol{\mu} > 0 \tag{13.32b}$$

$$\mathbf{X}\boldsymbol{\mu} = \tau\mathbf{e} \tag{13.32c}$$

where τ is a strictly positive scalar parameter and $e = [1\ 1\ \cdots\ 1]^T \in R^n$. It follows that every point on the central path is strictly feasible and the entire central path lies in the interior of the feasible regions described by Eqs. (13.28b), (13.28c), (13.29b), and (13.29c). On comparing Eq. (13.32) with Eq. (13.30), we see that as $\tau \to 0$ the central path approaches set $w^* = \{x^*,\ \lambda^*,\ \mu^*\}$ which solves the problems in Eqs. (13.28) and (13.29) simultaneously. This can also be seen by computing the duality gap on the central path, i.e.,

$$\delta[\mathbf{x}(\tau),\ \boldsymbol{\lambda}(\tau),\ \boldsymbol{\mu}(\tau)] = \mathbf{x}^T(\tau)\boldsymbol{\mu}(\tau) = n\tau \tag{13.33}$$

Hence the duality gap approaches zero linearly as $\tau \to 0$.

As in the LP case, the equations in Eq. (13.32) that define the central path for the problem in Eq. (13.28) and its dual can be interpreted as the KKT conditions for the modified minimization problem

$$\text{minimize } \hat{f}(\mathbf{x}) = \tfrac{1}{2}\mathbf{x}^T\mathbf{H}\mathbf{x} + \mathbf{x}^T\mathbf{p} - \tau \sum_{i=1}^{n} \ln x_i \tag{13.34a}$$

$$\text{subject to:} \quad \mathbf{A}\mathbf{x} = \mathbf{b} \tag{13.34b}$$

where $\tau > 0$ is the barrier parameter (see Sec. 12.4). In order to ensure that $\hat{f}(\mathbf{x})$ in Eq. (13.34a) is well defined, it is required that

$$\mathbf{x} > 0 \tag{13.34c}$$

The KKT conditions for the problem in Eq. (13.34) are given by

$$\mathbf{A}\mathbf{x} - \mathbf{b} = 0 \quad \text{for } \mathbf{x} > 0 \tag{13.35a}$$
$$\mathbf{A}^T\boldsymbol{\lambda} - \tau\mathbf{X}^{-1}\mathbf{e} - \mathbf{H}\mathbf{x} - \mathbf{p} = 0 \tag{13.35b}$$

If we let $\boldsymbol{\mu} = \tau\mathbf{X}^{-1}\mathbf{e}$, then $\mathbf{x} > 0$ implies that $\boldsymbol{\mu} > 0$ and Eq. (13.35b) can be written as

$$\mathbf{A}^T\boldsymbol{\lambda} + \boldsymbol{\mu} - \mathbf{H}\mathbf{x} - \mathbf{p} = 0 \quad \text{for } \boldsymbol{\mu} > 0 \tag{13.36a}$$

and

$$\mathbf{X}\boldsymbol{\mu} = \tau\mathbf{e} \tag{13.36b}$$

Consequently, Eqs. (13.35a), (13.36a), and (13.36b) are identical with Eqs. (13.32a), (13.32b), and (13.32c), respectively.

In what follows, we describe a primal-dual path-following method similar to the one proposed by Monteiro and Adler [6] which is an extension of their work on LP [7] described in Sec. 12.5. We then discuss the class of monotone linear complementarity problems (LCP's) and its variant known as the class of mixed LCP's, and recast convex QP problems as mixed LCP's (see Chap. 8 in [8]).

420

13.4.2 A primal-dual path-following method for convex QP problems

Consider the convex QP problem in Eq. (13.28) and let $\mathbf{w}_k = \{\mathbf{x}_k, \boldsymbol{\lambda}_k, \boldsymbol{\mu}_k\}$ be such that $\mathbf{x}_k$ is strictly feasible for the primal problem in Eq. (13.28) and $\mathbf{w}_k = \{\mathbf{x}_k, \boldsymbol{\lambda}_k, \boldsymbol{\mu}_k\}$ is strictly feasible for the dual problem in Eq. (13.29). We require an increment set $\delta_w = \{\delta_x, \delta_\lambda, \delta_\mu\}$ such that the next iterate $\mathbf{w}_{k+1} = \{\mathbf{x}_{k+1}, \boldsymbol{\lambda}_{k+1}, \boldsymbol{\mu}_{k+1}\} = \mathbf{w}_k + \delta_w$ remains strictly feasible and approaches the central path defined by Eq. (13.32). If $\mathbf{w}_k$ were to satisfy Eq. (13.32) with $\tau = \tau_{k+1}$, we would have

$$-\mathbf{H}\delta_x + \mathbf{A}^T\delta_\lambda + \delta_\mu = 0 \tag{13.37a}$$
$$\mathbf{A}\delta_x = 0 \tag{13.37b}$$
$$\Delta\mathbf{X}\boldsymbol{\mu}_k + \mathbf{X}\delta_\mu + \Delta\mathbf{X}\delta_\mu = \tau_{k+1}\mathbf{e} - \mathbf{X}\boldsymbol{\mu}_k \tag{13.37c}$$

where $\Delta\mathbf{X} = \text{diag}\{(\delta_x)_1, (\delta_x)_2, \ldots, (\delta_x)_n\}$. If the second-order term in Eq. (13.37c), namely, $\Delta\mathbf{X}\delta_\mu$, is neglected, then Eq. (13.37) becomes the system of *linear* equations

$$-\mathbf{H}\delta_x + \mathbf{A}^T\delta_\lambda + \delta_\mu = 0 \tag{13.38a}$$
$$\mathbf{A}\delta_x = 0 \tag{13.38b}$$
$$\mathbf{M}\delta_x + \mathbf{X}\delta_\mu = \tau_{k+1}\mathbf{e} - \mathbf{X}\boldsymbol{\mu}_k \tag{13.38c}$$

where $\mathbf{M} = \text{diag}\{(\boldsymbol{\mu}_k)_1, (\boldsymbol{\mu}_k)_2, \ldots, (\boldsymbol{\mu}_k)_n\}$. These equations can be expressed in matrix form as

$$\begin{bmatrix} -\mathbf{H} & \mathbf{A}^T & \mathbf{I} \\ \mathbf{A} & 0 & 0 \\ \mathbf{M} & 0 & \mathbf{X} \end{bmatrix} \delta_w = \begin{bmatrix} 0 \\ 0 \\ \tau_{k+1}\mathbf{e} - \mathbf{X}\boldsymbol{\mu}_k \end{bmatrix} \tag{13.39}$$

A good choice of parameter τ_{k+1} in Eqs. (13.38) and (13.39) is

$$\tau_{k+1} = \frac{\mathbf{x}_k^T\boldsymbol{\mu}_k}{n+\rho} \quad \text{with } \rho \geq \sqrt{n} \tag{13.40}$$

It can be shown that for a given tolerance ε for the duality gap, this choice of τ_{k+1} will reduce the primal-dual potential function which is defined as

$$\psi_{n+\rho}(\mathbf{x}, \boldsymbol{\mu}) = (n+\rho)\ln(\mathbf{x}^T\boldsymbol{\mu}) - \sum_{i=1}^{n}\ln(x_i\mu_i) \tag{13.41}$$

to a small but constant amount. This would lead to an iteration complexity of $O(\rho\ln(1/\varepsilon))$ (see Chap. 4 in [9]).

The solution of Eq. (13.38) can be obtained as

$$\boldsymbol{\delta}_\lambda = \mathbf{Y}\mathbf{y} \tag{13.42a}$$

$$\boldsymbol{\delta}_x = \boldsymbol{\Gamma}\mathbf{X}\mathbf{A}^T\boldsymbol{\delta}_\lambda - \mathbf{y} \tag{13.42b}$$

$$\boldsymbol{\delta}_\mu = \mathbf{H}\boldsymbol{\delta}_x - \mathbf{A}^T\boldsymbol{\delta}_\lambda \tag{13.42c}$$

where

$$\boldsymbol{\Gamma} = (\mathbf{M} + \mathbf{X}\mathbf{H})^{-1} \tag{13.42d}$$

$$\mathbf{Y} = (\mathbf{A}\boldsymbol{\Gamma}\mathbf{X}\mathbf{A}^T)^{-1}\mathbf{A} \tag{13.42e}$$

and

$$\mathbf{y} = \boldsymbol{\Gamma}(\mathbf{X}\boldsymbol{\mu}_k - \tau_{k+1}\mathbf{e}) \tag{13.42f}$$

Since $\mathbf{x}_k > \mathbf{0}$ and $\boldsymbol{\mu}_k > \mathbf{0}$, matrices $\mathbf{X}$ and $\mathbf{M}$ are positive definite. Therefore, $\mathbf{X}^{-1}\mathbf{M} + \mathbf{H}$ is also positive definite and the inverse of the matrix

$$\mathbf{M} + \mathbf{X}\mathbf{H} = \mathbf{X}(\mathbf{X}^{-1}\mathbf{M} + \mathbf{H})$$

exists. Moreover, since $\mathbf{A}$ has full row rank, the matrix

$$\mathbf{A}\boldsymbol{\Gamma}\mathbf{X}\mathbf{A}^T = \mathbf{A}(\mathbf{X}^{-1}\mathbf{M} + \mathbf{H})^{-1}\mathbf{A}^T$$

is also positive definite and hence nonsingular. Therefore, matrices $\boldsymbol{\Gamma}$ and $\mathbf{Y}$ in Eq. (13.42) are well defined.

Once $\boldsymbol{\delta}_w$ is calculated, an appropriate α_k needs to be determined such that

$$\mathbf{w}_{k+1} = \mathbf{w}_k + \alpha_k\boldsymbol{\delta}_w \tag{13.43}$$

remains strictly feasible. Such an α_k can be chosen in the same way as in the primal-dual interior-point algorithm discussed in Sec. 12.5.1, i.e.,

$$\alpha_k = (1 - 10^{-6})\alpha_{\max} \tag{13.44a}$$

where

$$\alpha_{\max} = \min(\alpha_p, \alpha_d) \tag{13.44b}$$

with

$$\alpha_p = \min_{i \text{ with } (\boldsymbol{\delta}_x)_i < 0} \left[-\frac{(\mathbf{x}_k)_i}{(\boldsymbol{\delta}_x)_i} \right] \tag{13.44c}$$

$$\alpha_d = \min_{i \text{ with } (\boldsymbol{\delta}_\mu)_i < 0} \left[-\frac{(\boldsymbol{\mu}_k)_i}{(\boldsymbol{\delta}_\mu)_i} \right] \tag{13.44d}$$

The method described can be implemented in terms of the following algorithm.

Algorithm 13.2 Primal-dual path-following algorithm for convex QP problems
Step 1
Input a strictly feasible $\mathbf{w}_0 = \{\mathbf{x}_0, \boldsymbol{\lambda}_0, \boldsymbol{\mu}_0\}$.
Set $k = 1$ and $\rho \geq \sqrt{n}$, and initialize the tolerance ε for duality gap.
Step 2
If $\mathbf{x}_k^T \boldsymbol{\mu}_k \leq \varepsilon$, output solution $\mathbf{w}^* = \mathbf{w}_k$, and stop; otherwise, continue with Step 3.
Step 3
Set τ_{k+1} using Eq. (13.40) and compute $\boldsymbol{\delta}_w = \{\boldsymbol{\delta}_x, \boldsymbol{\delta}_\lambda, \boldsymbol{\delta}_\mu\}$ using Eqs. (13.42a) to (13.42c).
Step 4
Compute α_k using Eq. (13.44) and update $\mathbf{w}_{k+1}$ using Eq. (13.43).
Set $k = k + 1$ and repeat from Step 2.

13.4.3 Nonfeasible-initialization primal-dual path-following method for convex QP problems

Algorithm 13.2 requires a strictly feasible $\mathbf{w}_0$ which might be difficult to obtain particularly for large-scale problems. The idea described in Sec. 12.5.2 can be used to develop a nonfeasible-initialization primal-dual path-following algorithm for convex QP problems. Let $\mathbf{w}_k = \{\mathbf{x}_k, \boldsymbol{\lambda}_k, \boldsymbol{\mu}_k\}$ be such that $\mathbf{x}_k > 0$ and $\boldsymbol{\mu}_k > 0$ but which may not satisfy Eqs. (13.32a) and (13.32b). We need to find the next iterate

$$\mathbf{w}_{k+1} = \mathbf{w}_k + \alpha_k \boldsymbol{\delta}_w$$

such that $\mathbf{x}_{k+1} > 0$ and $\boldsymbol{\mu}_{k+1} > 0$, and that $\boldsymbol{\delta}_w = \{\boldsymbol{\delta}_x, \boldsymbol{\delta}_\lambda, \boldsymbol{\delta}_\mu\}$ satisfies the equations

$$-\mathbf{H}(\mathbf{x}_k + \boldsymbol{\delta}_x) - \mathbf{p} + \mathbf{A}^T(\boldsymbol{\lambda}_k + \boldsymbol{\delta}_\lambda) + (\boldsymbol{\mu}_k + \boldsymbol{\delta}_\mu) = 0$$
$$\mathbf{A}(\mathbf{x}_k + \boldsymbol{\delta}_x) = \mathbf{b}$$
$$\mathbf{M}\boldsymbol{\delta}_x + \mathbf{X}\boldsymbol{\delta}_\mu = \tau_{k+1}\mathbf{e} - \mathbf{X}\boldsymbol{\mu}_k$$

i.e.,

$$-\mathbf{H}\boldsymbol{\delta}_x + \mathbf{A}^T\boldsymbol{\delta}_\lambda + \boldsymbol{\delta}_\mu = \mathbf{r}_d$$
$$\mathbf{A}\boldsymbol{\delta}_x = \mathbf{r}_p$$
$$\mathbf{M}\boldsymbol{\delta}_x + \mathbf{X}\boldsymbol{\delta}_\mu = \tau_{k+1}\mathbf{e} - \mathbf{X}\boldsymbol{\mu}_k$$

where

$$\mathbf{r}_d = \mathbf{H}\mathbf{x}_k + \mathbf{p} - \mathbf{A}^T\boldsymbol{\lambda}_k - \boldsymbol{\mu}_k \qquad (13.45\text{a})$$
$$\mathbf{r}_p = \mathbf{b} - \mathbf{A}\mathbf{x}_k \qquad (13.45\text{b})$$

The above system of linear equations can be expressed as

$$\begin{bmatrix} -\mathbf{H} & \mathbf{A}^T & \mathbf{I} \\ \mathbf{A} & 0 & 0 \\ \mathbf{M} & 0 & \mathbf{X} \end{bmatrix} \delta_w = \begin{bmatrix} \mathbf{r}_d \\ \mathbf{r}_p \\ \tau_{k+1}\mathbf{e} - \mathbf{X}\boldsymbol{\mu}_k \end{bmatrix} \qquad (13.46)$$

On comparing Eq. (13.46) with Eq. (13.39), we see that δ_w becomes the search direction determined by using Eq. (13.39) when the residual vectors $\mathbf{r}_p$ and $\mathbf{r}_d$ are reduced to zero. Note that in general the elimination of $\mathbf{r}_p$ and $\mathbf{r}_d$ cannot be accomplished in a single iteration because the next iterate also depends on α_k which may not be unity. The solution of Eq. (13.46) can be obtained as

$$\delta_\lambda = \mathbf{Y}_0(\mathbf{A}\mathbf{y}_d + \mathbf{r}_p) \qquad (13.47a)$$

$$\delta_x = \boldsymbol{\Gamma}\mathbf{X}\mathbf{A}^T\delta_\lambda - \mathbf{y}_d \qquad (13.47b)$$

$$\delta_\mu = \mathbf{H}\delta_x - \mathbf{A}^T\delta_\lambda + \mathbf{r}_d \qquad (13.47c)$$

where

$$\boldsymbol{\Gamma} = (\mathbf{M} + \mathbf{X}\mathbf{H})^{-1} \qquad (13.47d)$$

$$\mathbf{Y}_0 = (\mathbf{A}\boldsymbol{\Gamma}\mathbf{X}\mathbf{A}^T)^{-1} \qquad (13.47e)$$

$$\mathbf{y}_d = \boldsymbol{\Gamma}[\mathbf{X}(\boldsymbol{\mu}_k + \mathbf{r}_d) - \tau_{k+1}\mathbf{e}] \qquad (13.47f)$$

$$\tau_{k+1} = \frac{\mathbf{x}_k^T\boldsymbol{\mu}_k}{n + \rho} \quad \text{with } \rho \geq \sqrt{n} \qquad (13.47g)$$

Obviously, if residual vectors $\mathbf{r}_p$ and $\mathbf{r}_d$ are reduced to zero, the vector $\delta_w = \{\delta_x, \delta_\lambda, \delta_\mu\}$ determined by using Eq. (13.47) is identical with that obtained using Eq. (13.42). Once δ_w is determined, α_k can be calculated using Eq. (13.44). The above principles lead to the following algorithm.

Algorithm 13.3 Nonfeasible-initialization primal-dual path-following algorithm for convex QP problems
Step 1
Input a set $\mathbf{w}_0 = \{\mathbf{x}_0, \boldsymbol{\lambda}_0, \boldsymbol{\mu}_0\}$ with $\mathbf{x}_0 > 0$ and $\boldsymbol{\mu}_0 > 0$.
Set $k = 0$ and $\rho \geq \sqrt{n}$, and initialize the tolerance ε for the duality gap.
Step 2
If $\mathbf{x}_k^T\boldsymbol{\mu}_k \leq \varepsilon$, output solution $\mathbf{w}^* = \mathbf{w}_k$ and stop; otherwise, continue with Step 3.
Step 3
Compute τ_{k+1} using Eq. (13.47g) and determine $\delta_w = \{\delta_x, \delta_\lambda, \delta_\mu\}$ using Eq. (13.47).
Step 4
Compute α_k using Eq. (13.44) and update $\mathbf{w}_{k+1}$ using Eq. (13.43).
Set $k = k + 1$ and repeat from Step 2.

Example 13.3 Solve the convex QP problem

$$\text{minimize } f(\mathbf{x}) = \tfrac{1}{2}\mathbf{x}^T \begin{bmatrix} 4 & 0 & 0 \\ 0 & 1 & -1 \\ 0 & -1 & 1 \end{bmatrix} \mathbf{x} + \mathbf{x}^T \begin{bmatrix} -8 \\ -6 \\ -6 \end{bmatrix} \qquad (13.48\text{a})$$

$$\text{subject to: } x_1 + x_2 + x_3 = 3 \qquad (13.48\text{b})$$

$$\mathbf{x} \geq 0 \qquad (13.48\text{c})$$

Solution The problem can be solved by using either Algorithm 13.2 or Algorithm 13.3. Using a strictly feasible point $\mathbf{x}_0 = [1\ 1\ 1]^T$ and assigning $\lambda_0 = -7$ and $\boldsymbol{\mu}_0 = [3\ 1\ 1]^T$, it took Algorithm 13.2 11 iterations and 3681 flops to converge to the solution

$$\mathbf{x}^* = \begin{bmatrix} 0.500000 \\ 1.250000 \\ 1.250000 \end{bmatrix}$$

On the other hand, using a nonfeasible initial point $\mathbf{x}_0 = [1\ 2\ 2]^T$ and assigning $\lambda_0 = -1$, $\boldsymbol{\mu}_0 = [0.2\ 0.2\ 0.2]^T$, $\rho = n + 2\sqrt{n}$, and $\varepsilon = 10^{-5}$, Algorithm 13.3 took 13 iterations and 4918 flops to converge to the solution

$$\mathbf{x}^* = \begin{bmatrix} 0.500001 \\ 1.249995 \\ 1.249995 \end{bmatrix}$$

∎

Example 13.4 Solve the shortest-distance problem described in Example 13.2 by using Algorithm 13.3.

Solution By letting $\mathbf{x} = \mathbf{x}^+ - \mathbf{x}^-$ where $\mathbf{x}^+ \geq 0$ and $\mathbf{x}^- \geq 0$, and then introducing slack vector $\boldsymbol{\eta} \geq 0$, the problem in Eq. (13.12) can be converted into a QP problem of the type given in Eq. (13.28), i.e.,

$$\text{minimize } \tfrac{1}{2}\hat{\mathbf{x}}^T \hat{\mathbf{H}}\hat{\mathbf{x}} + \hat{\mathbf{x}}^T \hat{\mathbf{p}} \qquad (13.49\text{a})$$

$$\text{subject to: } \hat{\mathbf{A}}\hat{\mathbf{x}} = \mathbf{b} \qquad (13.49\text{b})$$

$$\hat{\mathbf{x}} \geq 0 \qquad (13.49\text{c})$$

where

$$\hat{\mathbf{H}} = \begin{bmatrix} \mathbf{H} & -\mathbf{H} & 0 \\ -\mathbf{H} & \mathbf{H} & 0 \\ 0 & 0 & 0 \end{bmatrix}, \quad \hat{\mathbf{p}} = \begin{bmatrix} \mathbf{p} \\ -\mathbf{p} \\ 0 \end{bmatrix}, \quad \hat{\mathbf{x}} = \begin{bmatrix} \mathbf{x}^+ \\ \mathbf{x}^- \\ \boldsymbol{\eta} \end{bmatrix}$$

$$\hat{\mathbf{A}} = [\mathbf{A}\ -\mathbf{A}\ -\mathbf{I}_p]$$

and $n = 14$, $p = 6$. We note that $\hat{\mathbf{H}}$ is positive semidefinite if $\mathbf{H}$ is positive semidefinite. Since a strictly feasible initial $\mathbf{w}_0$ is difficult to find in this example, Algorithm 13.3 was used with $\mathbf{x}_0 = \text{ones}\{14, 1\}$, $\boldsymbol{\lambda}_0 = -\text{ones}\{6, 1\}$, $\boldsymbol{\mu}_0 = \text{ones}\{14, 1\}$, where ones$\{m, 1\}$ represents a column vector of dimension m whose elements are all equal to one. Assigning $\varepsilon = 10^{-5}$ and $\rho = n + 20\sqrt{n}$, the algorithm took 11 iterations and 215 Kflops to converge to $\hat{\mathbf{x}}^*$ whose first 8 elements were then used to obtain

$$\mathbf{x}^* = \begin{bmatrix} 0.400002 \\ 0.799999 \\ 1.000001 \\ 2.000003 \end{bmatrix}$$

The shortest distance can be obtained as 1.341644.

Note that we do not need to introduce a small perturbation to matrix $\mathbf{H}$ to make it positive definite in this example as was the case in Example 13.2. ∎

13.4.4 Linear complementarity problems

The linear complementarity problem (LCP) is to find a vector pair $\{\mathbf{x}, \boldsymbol{\mu}\}$ in R^n that satisfies the relations

$$\mathbf{K}\mathbf{x} + \mathbf{q} = \boldsymbol{\mu} \tag{13.50a}$$
$$\mathbf{x} \geq \mathbf{0} \quad \text{for } \boldsymbol{\mu} \geq \mathbf{0} \tag{13.50b}$$
$$\mathbf{x}^T \boldsymbol{\mu} = 0 \tag{13.50c}$$

where $\mathbf{K} \in R^{n \times n}$ and $\mathbf{q} \in R^n$ are given, and $\mathbf{K}$ is positive semidefinite. Although the problem described in Eq. (13.50) is not an optimization problem, its solution can be related to the minimization problem

$$\text{minimize } f(\hat{\mathbf{x}}) = \hat{\mathbf{x}}_1^T \hat{\mathbf{x}}_2 \tag{13.51a}$$
$$\text{subject to:} \quad \mathbf{A}\hat{\mathbf{x}} = \mathbf{b} \tag{13.51b}$$
$$\hat{\mathbf{x}} \geq \mathbf{0} \tag{13.51c}$$

where

$$\hat{\mathbf{x}} = \begin{bmatrix} \hat{\mathbf{x}}_1 \\ \hat{\mathbf{x}}_2 \end{bmatrix} = \begin{bmatrix} \mathbf{x} \\ \boldsymbol{\mu} \end{bmatrix}, \quad \mathbf{A} = [\mathbf{K} \ -\mathbf{I}_n], \quad \text{and} \quad \mathbf{b} = -\mathbf{q}$$

Note that the objective function $\{f(\hat{\mathbf{x}})\}$ in Eq. (13.51a) can be expressed as

$$f(\hat{\mathbf{x}}) = \tfrac{1}{2}\hat{\mathbf{x}}^T \begin{bmatrix} \mathbf{0} & \mathbf{I}_n \\ \mathbf{I}_n & \mathbf{0} \end{bmatrix} \hat{\mathbf{x}}$$

Hence the problem in Eq. (13.51) is a QP problem with an *indefinite* Hessian.

A variant of the LCP which is well connected to *convex* QP is the *mixed* LCP which entails finding a vector pair $\{\mathbf{x}, \boldsymbol{\mu}\}$ in R^n and vector $\boldsymbol{\lambda} \in R^p$ such that

$$\begin{bmatrix} \mathbf{K}_{11} & \mathbf{K}_{12} \\ \mathbf{K}_{21} & \mathbf{K}_{22} \end{bmatrix} \begin{bmatrix} \mathbf{x} \\ \boldsymbol{\lambda} \end{bmatrix} + \begin{bmatrix} \mathbf{q}_1 \\ \mathbf{q}_2 \end{bmatrix} = \begin{bmatrix} \boldsymbol{\mu} \\ 0 \end{bmatrix} \tag{13.52a}$$

$$\mathbf{x} \geq 0, \; \boldsymbol{\mu} \geq 0 \tag{13.52b}$$

$$\mathbf{x}^T \boldsymbol{\mu} = 0 \tag{13.52c}$$

where matrix $\mathbf{K} \in R^{(n+p) \times (n+p)}$ given by

$$\begin{bmatrix} \mathbf{K}_{11} & \mathbf{K}_{12} \\ \mathbf{K}_{21} & \mathbf{K}_{22} \end{bmatrix}$$

is not necessarily symmetric but is positive semidefinite in the sense that

$$\mathbf{y}^T \mathbf{K} \mathbf{y} \geq 0 \qquad \text{for any } \mathbf{y} \in R^{n+p} \tag{13.53}$$

The LCP described by Eq. (13.50) can be viewed as a special mixed LCP where dimension p is 0. Again, the mixed LCP as stated in Eq. (13.52) is not an optimization problem. However, it is closely related to the standard-form LP problem in Eq. (11.1) as well as the convex QP problem in Eq. (13.28). In order to see the relation of Eq. (13.52) to the LP problem in Eq. (11.1), note that the conditions in Eqs. (13.52b) and (13.52c) imply that

$$x_i \mu_i = 0 \qquad \text{for } i = 1, 2, \ldots, n$$

which is the complementarity condition in Eq. (11.5d). Hence the KKT conditions in Eq. (11.5) can be restated as

$$\begin{bmatrix} 0 & -\mathbf{A}^T \\ \mathbf{A} & 0 \end{bmatrix} \begin{bmatrix} \mathbf{x} \\ \boldsymbol{\lambda} \end{bmatrix} + \begin{bmatrix} \mathbf{c} \\ -\mathbf{b} \end{bmatrix} = \begin{bmatrix} \boldsymbol{\mu} \\ 0 \end{bmatrix} \tag{13.54a}$$

$$\mathbf{x} \geq 0, \; \boldsymbol{\mu} \geq 0 \tag{13.54b}$$

$$\mathbf{x}^T \boldsymbol{\mu} = 0 \tag{13.54c}$$

Since matrix

$$\mathbf{K} = \begin{bmatrix} 0 & -\mathbf{A}^T \\ \mathbf{A} & 0 \end{bmatrix}$$

is positive semidefinite in the sense of Eq. (13.53) (see Prob. 13.10(a)), we note that standard-form LP problems can be formulated as mixed LCP's.

For the convex QP problem in Eq. (13.28), the KKT conditions given in Eq. (13.30) can be written as

$$\begin{bmatrix} \mathbf{H} & -\mathbf{A}^T \\ \mathbf{A} & 0 \end{bmatrix} \begin{bmatrix} \mathbf{x} \\ \boldsymbol{\lambda} \end{bmatrix} + \begin{bmatrix} \mathbf{p} \\ -\mathbf{b} \end{bmatrix} = \begin{bmatrix} \boldsymbol{\mu} \\ 0 \end{bmatrix} \tag{13.55a}$$

$$\mathbf{x} \geq 0, \quad \boldsymbol{\mu} \geq 0 \tag{13.55b}$$

$$\mathbf{x}^T \boldsymbol{\mu} = 0 \tag{13.55c}$$

where

$$\mathbf{K} = \begin{bmatrix} \mathbf{H} & -\mathbf{A}^T \\ \mathbf{A} & \mathbf{0} \end{bmatrix}$$

is positive semidefinite if $\mathbf{H}$ is positive semidefinite (see Prob. 13.10(b)). From the above analysis, we see that the class of mixed LCP covers standard-form LP problems, convex QP problems, and LCPs.

Let $\mathbf{w}_k = \{\mathbf{x}_k, \boldsymbol{\lambda}_k, \boldsymbol{\mu}_k\}$ be the kth iterate with $\mathbf{x}_k > 0$, and let $\boldsymbol{\mu}_k > 0$ and the $(k+1)$th iterate be

$$\mathbf{w}_{k+1} = \mathbf{w}_k + \alpha_k \boldsymbol{\delta}_w \tag{13.56}$$

where the search direction $\boldsymbol{\delta}_w = \{\boldsymbol{\delta}_x, \boldsymbol{\delta}_\lambda, \boldsymbol{\delta}_\mu\}$ is chosen to satisfy the relations

$$\begin{bmatrix} \mathbf{K}_{11} & \mathbf{K}_{12} \\ \mathbf{K}_{21} & \mathbf{K}_{22} \end{bmatrix} \begin{bmatrix} \mathbf{x}_k + \boldsymbol{\delta}_x \\ \boldsymbol{\lambda}_k + \boldsymbol{\delta}_\lambda \end{bmatrix} + \begin{bmatrix} \mathbf{q}_1 \\ \mathbf{q}_2 \end{bmatrix} = \begin{bmatrix} \boldsymbol{\mu}_k + \boldsymbol{\delta}_\mu \\ \mathbf{0} \end{bmatrix}$$

$$(\mathbf{x}_k + \boldsymbol{\delta}_x)^T (\boldsymbol{\mu}_k + \boldsymbol{\delta}_\mu) \approx \mathbf{x}_k^T \boldsymbol{\mu}_k + \boldsymbol{\delta}_x^T \boldsymbol{\mu}_k + \mathbf{x}_k^T \boldsymbol{\delta}_\mu = \tau_{k+1} \mathbf{e}$$

These equations can be expressed as

$$\begin{bmatrix} -\mathbf{K}_{11} & -\mathbf{K}_{12} & \mathbf{I} \\ \mathbf{K}_{21} & \mathbf{K}_{22} & \mathbf{0} \\ \mathbf{M} & \mathbf{0} & \mathbf{X} \end{bmatrix} \begin{bmatrix} \boldsymbol{\delta}_x \\ \boldsymbol{\delta}_\lambda \\ \boldsymbol{\delta}_\mu \end{bmatrix} = \begin{bmatrix} \mathbf{r}_1 \\ \mathbf{r}_2 \\ \tau_{k+1}\mathbf{e} - \mathbf{X}\boldsymbol{\mu}_k \end{bmatrix} \tag{13.57a}$$

where $\mathbf{M} = \text{diag}\{(\boldsymbol{\mu}_k)_1, (\boldsymbol{\mu}_k)_2, \ldots, (\boldsymbol{\mu}_k)_n\}$, $\mathbf{X} = \text{diag}\{(\mathbf{x}_k)_1, (\mathbf{x}_k)_2, \ldots, (\mathbf{x}_k)_n\}$, and

$$\mathbf{r}_1 = \mathbf{K}_{11}\mathbf{x}_k + \mathbf{K}_{12}\boldsymbol{\lambda}_k - \boldsymbol{\mu}_k + \mathbf{q}_1 \tag{13.57b}$$

$$\mathbf{r}_2 = -\mathbf{K}_{21}\mathbf{x}_k - \mathbf{K}_{22}\boldsymbol{\lambda}_k - \mathbf{q}_2 \tag{13.57c}$$

$$\tau_{k+1} = \frac{\mathbf{x}_k^T \boldsymbol{\mu}_k}{n + \rho}, \qquad \text{with } \rho \geq \sqrt{n} \tag{13.57d}$$

It can be readily verified that with $\mathbf{K}_{11} = \mathbf{K}_{22} = \mathbf{0}$, $\mathbf{K}_{21} = -\mathbf{K}_{12}^T = \mathbf{A}$, $\mathbf{q}_1 = \mathbf{c}$, and $\mathbf{q}_2 = -\mathbf{b}$, Eq. (13.57a) becomes Eq. (12.56) which determines the search direction for the nonfeasible-initialization primal-dual path-following algorithm in Sec. 12.5.2. Likewise, with $\mathbf{K}_{11} = \mathbf{H}$, $\mathbf{K}_{21} = -\mathbf{K}_{12}^T = \mathbf{A}$, $\mathbf{K}_{22} = \mathbf{0}$, $\mathbf{q}_1 = \mathbf{p}$, and $\mathbf{q}_2 = -\mathbf{b}$, Eqs. (13.57a) to (13.57d) become Eqs. (13.46), (13.45a), (13.45b), and (13.47g) which determine the search direction for the nonfeasible-initialization primal-dual path-following algorithm for the convex QP in Sec. 13.4.3. Once $\boldsymbol{\delta}_w$ is determined by solving Eq. (13.57), α_k can be calculated using Eq. (13.44). The above method can be implemented in terms of the following algorithm.

428

Algorithm 13.4 Nonfeasible-initialization interior-point algorithm for mixed LCP problems

Step 1
Input an initial point $\mathbf{w}_0 = \{\mathbf{x}_0, \boldsymbol{\lambda}_0, \boldsymbol{\mu}_0\}$ with $\mathbf{x}_0 > 0$ and $\boldsymbol{\mu}_0 > 0$.
Set $k = 0$ and $\rho > \sqrt{n}$, and initialize the tolerance ε for $\mathbf{x}_k^T \boldsymbol{\mu}_k$.

Step 2
If $\mathbf{x}_k^T \boldsymbol{\mu}_k \leq \varepsilon$, output solution $\mathbf{w}^* = \mathbf{w}_k$ and stop; otherwise, continue with Step 3.

Step 3
Compute τ_{k+1} using Eq. (13.57d) and determine $\boldsymbol{\delta}_w = (\boldsymbol{\delta}_x, \boldsymbol{\delta}_\lambda, \boldsymbol{\delta}_\mu)$ by solving Eq. (13.57a).

Step 4
Compute α_k using Eq. (13.44) and set $\mathbf{w}_{k+1} = \mathbf{w}_k + \alpha_k \boldsymbol{\delta}_w$.
Set $k = k + 1$ and repeat from Step 2.

13.5 Cutting-Plane Methods for CP Problems

Cutting-plane methods for CP problems are of importance as they make good use of the convexity of the problems at hand. Unlike many descent methods for convex problems, cutting-plane methods entail easy-to-apply termination criteria that assure the solution's optimality to a prescribed accuracy.

An important concept associated with CP is the concept of *subgradient*. In what follows, we adopt the approach described in [10] to introduce this concept and then move on to describe a cutting-plane algorithm proposed by Kelley [11].

13.5.1 Subgradients

The concept of subgradient is a natural generalization of the concept of gradient. If a function $f(\mathbf{x})$ is convex and differentiable, then it is known from Theorem 2.12 that at point $\mathbf{x}$, we have

$$f(\hat{\mathbf{x}}) \geq f(\mathbf{x}) + \nabla f(\mathbf{x})^T (\hat{\mathbf{x}} - \mathbf{x}) \qquad \text{for all } \hat{\mathbf{x}} \qquad (13.58)$$

This equation states, in effect, that the tangent to the surface defined by $f(\mathbf{x})$ at point $\mathbf{x}$ always lies below the surface, as shown in Fig. 2.8.

Definition 13.1 If $f(\mathbf{x})$ is convex but not necessarily differentiable, then vector $\mathbf{g} \in R^n$ is said to be a *subgradient* of $f(\mathbf{x})$ at $\mathbf{x}$ if

$$f(\hat{\mathbf{x}}) \geq f(\mathbf{x}) + \mathbf{g}^T (\hat{\mathbf{x}} - \mathbf{x}) \qquad \text{for all } \hat{\mathbf{x}} \qquad (13.59)$$

∎

On comparing Eq. (13.59) with Eq. (13.58), we note that the gradient of a differentiable convex function is a subgradient. For this reason, the commonly used notation $\mathbf{g}$ for gradient will also be adopted to represent a subgradient.

An important property in connection with subgradients is that *a convex function has at least one subgradient at every point* [12]. The right-hand side of the inequality in Eq. (13.59) may be viewed as a linear approximation of $f(\mathbf{x})$, and this linear function is a lower bound of $f(\mathbf{x})$ which is tight at point $\mathbf{x}$ meaning that the lower bound becomes an equality at $\mathbf{x}$. Geometrically, the subgradients at a point $\mathbf{x}$ for the case where the convex function $f(\mathbf{x})$ is not differentiable correspond to different tangent lines at $\mathbf{x}$. This is illustrated in Fig. 13.2, where the two subgradients of $f(x)$ at x^* are given by $g_1 = \tan\theta_1$ and $g_2 = \tan\theta_2$.

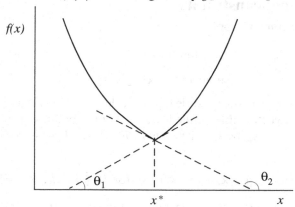

Figure 13.2. Two subgradients of $f(x)$ at x^* for a case where $f(x)$ is not differentiable at x^*.

From Eq. (13.59), it follows that $f(\hat{\mathbf{x}}) \geq f(\mathbf{x})$ as long as $\mathbf{g}^T(\hat{\mathbf{x}} - \mathbf{x}) \geq 0$. Note that for a fixed point, $\mathbf{g}^T(\hat{\mathbf{x}} - \mathbf{x}) = 0$ defines a hyperplane which passes through point $\mathbf{x}$ with $\mathbf{g}$ as its normal. This hyperplane divides the entire R^n space into two parts. On the one side of the hyperplane where each point $\hat{\mathbf{x}}$ satisfies $\mathbf{g}^T(\hat{\mathbf{x}} - \mathbf{x}) \geq 0$, no minimizers can exist since $f(\hat{\mathbf{x}}) \geq f(\mathbf{x})$. Consequently, a minimizer of $f(\mathbf{x})$ can only be found on the other side of the plane, which is characterized by the set of points $\{\hat{\mathbf{x}} : \mathbf{g}^T(\hat{\mathbf{x}} - \mathbf{x}) \leq 0\}$. From this discussion, we see that in an optimization context the concept of the subgradient is useful as it facilitates the definition of a 'cutting plane' in the parameter space, which can be used to reduce the region of search for a minimizer.

There are several important special cases in which the computation of a subgradient of a convex $f(\mathbf{x})$ can be readily carried out (see Prob. 13.12) as follows:

(a) If $f(\mathbf{x})$ is differentiable, then the gradient of $f(\mathbf{x})$ is a subgradient;
(b) If $\alpha > 0$, then a subgradient of $\alpha f(\mathbf{x})$ is given by $\alpha \mathbf{g}$ where $\mathbf{g}$ is a subgradient of $f(\mathbf{x})$;
(c) If $f(\mathbf{x}) = f_1(\mathbf{x}) + f_2(\mathbf{x}) + \cdots + f_r(\mathbf{x})$ where $f_i(\mathbf{x})$ for $1 \leq i \leq r$ are convex, then $\mathbf{g} = \mathbf{g}_1 + \mathbf{g}_2 + \cdots + \mathbf{g}_r$ is a subgradient of $f(\mathbf{x})$ where $\mathbf{g}_i$ is a subgradient of $f_i(\mathbf{x})$;

(d) If

$$f(\mathbf{x}) = \max[f_1(\mathbf{x}), \ f_2(\mathbf{x}), \ \ldots, \ f_r(\mathbf{x})]$$

where $f_i(\mathbf{x})$ for $1 \leq i \leq r$ are convex, then at point $\mathbf{x}$ there is at least one index i^* with $1 \leq i^* \leq r$ such that $f(\mathbf{x}) = f_{i^*}(\mathbf{x})$. In this case a subgradient of $f_{i^*}(\mathbf{x})$, $\mathbf{g}_{i^*}$, is a subgradient of $f(\mathbf{x})$.

13.5.2 Kelley's cutting-plane method for CP problems with bound constraints

Consider the convex problem

$$\text{minimize } f(\mathbf{x}) \tag{13.60a}$$

$$\text{subject to:} \quad \mathbf{x}_l \leq \mathbf{x} \leq \mathbf{x}_u \tag{13.60b}$$

where $f(\mathbf{x})$ is convex in the feasible region $\mathcal{R}$ described by Eq. (13.60b), and $\mathbf{x}_l$ and $\mathbf{x}_u$ are given vectors that define lower and upper bounds of $\mathbf{x}$, respectively.

Let $\mathbf{x}_0, \ \mathbf{x}_1, \ \ldots, \ \mathbf{x}_k$ be $k + 1$ points in $\mathcal{R}$. Since $f(\mathbf{x})$ is convex in $\mathcal{R}$, we have

$$f(\mathbf{x}) \geq f(\mathbf{x}_i) + \mathbf{g}^T(\mathbf{x}_i)(\mathbf{x} - \mathbf{x}_i) \quad \text{for } 0 \leq i \leq k, \ \mathbf{x} \in \mathcal{R} \tag{13.61}$$

where $\mathbf{g}(\mathbf{x}_i)$ is a subgradient of $f(\mathbf{x})$ at $\mathbf{x}_i$. Hence $f(\mathbf{x})$ has a lower bound

$$f(\mathbf{x}) \geq f_{l, \, k}(\mathbf{x}) \tag{13.62a}$$

where $f_{l, \, k}(\mathbf{x})$ is the piecewise linear convex function

$$f_{l, \, k}(\mathbf{x}) = \max_{0 \leq i \leq k} \left[f(\mathbf{x}_i) + \mathbf{g}^T(\mathbf{x}_i)(\mathbf{x} - \mathbf{x}_i) \right] \tag{13.62b}$$

Eq. (13.62a) is illustrated in Fig. 13.3 for the one-dimensional case with $k = 2$. As can be seen, the objective function $f(\mathbf{x})$ is bounded from below by the globally convex, piecewise linear function $f_{l, \, k}(\mathbf{x})$.

Three observations can be made based on Eq. (13.62) and Fig. 13.3. First, the lower bound $f_{l, \, k}(\mathbf{x})$ is tight at points $\mathbf{x}_0, \ \mathbf{x}_1, \ \ldots, \ \mathbf{x}_k$. Second, if $\mathbf{x}^*$ is a minimizer of $f(\mathbf{x})$ in $\mathcal{R}$, then $f^* = f(\mathbf{x}^*)$ is bounded from below by $L_k = \min_{\mathbf{x} \in \mathcal{R}}[f_{l, \, k}(\mathbf{x})]$. If we let

$$U_k = \min_{0 \leq i \leq k} [f(\mathbf{x}_i)] \tag{13.63}$$

then we have

$$L_k \leq f^* \leq U_k$$

Therefore, when k increases both the lower and upper bounds become tighter, i.e.,

$$L_k \leq L_{k+1} \leq f^* \leq U_{k+1} \leq U_k \tag{13.64}$$

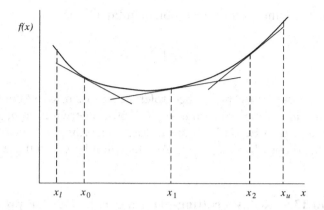

Figure 13.3.   A single-variable interpretation of functions $f(\mathbf{x})$ and $f_{l,\,2}(\mathbf{x})$.

Third, as k increases, the minimizer of the lower-bound function $f_{l,\,k}(\mathbf{x})$ can serve as an approximate solution of the problem in Eq. (13.60).

Note that minimizing $f_{l,k}(\mathbf{x})$ subject to $\mathbf{x} \in \mathcal{R}$ is an LP problem which is equivalent to

$$\text{minimize } L \tag{13.65a}$$

$$\text{subject to: } \quad f_{l,\,k}(\mathbf{x}) \le L \tag{13.65b}$$

$$\mathbf{x}_l \le \mathbf{x} \le \mathbf{x}_u \tag{13.65c}$$

If we let

$$\mathbf{z} = \begin{bmatrix} \mathbf{x} \\ L \end{bmatrix}, \quad \mathbf{c} = \begin{bmatrix} \mathbf{0} \\ 1 \end{bmatrix} \tag{13.66a}$$

$$\mathbf{A}_k = \begin{bmatrix} -\mathbf{g}^T(\mathbf{x}_0) & 1 \\ \vdots & \\ -\mathbf{g}^T(\mathbf{x}_k) & 1 \\ \mathbf{I} & 0 \\ -\mathbf{I} & 0 \end{bmatrix}, \quad \mathbf{b}_k = \begin{bmatrix} f(\mathbf{x}_0) - \mathbf{g}^T(\mathbf{x}_0)\mathbf{x}_0 \\ \vdots \\ f(\mathbf{x}_k) - \mathbf{g}^T(\mathbf{x}_k)\mathbf{x}_k \\ \mathbf{x}_l \\ -\mathbf{x}_u \end{bmatrix} \tag{13.66a}$$

where $\mathbf{I}$ denotes the $n \times n$ identity matrix, then the problem in Eq. (13.65) can be stated as the LP problem

$$\text{minimize } \mathbf{c}^T \mathbf{z} \tag{13.67a}$$

$$\text{subject to: } \quad \mathbf{A}\mathbf{z} \ge \mathbf{b} \tag{13.67b}$$

Let us denote the minimizer of the problem in Eq. (13.65) as

$$\mathbf{z}^* = \begin{bmatrix} \mathbf{x}_k^* \\ L_k \end{bmatrix}$$

If $U_k - L_k$ is less than a prescribed tolerance ε, then $\mathbf{x}_k^*$ is considered an acceptable solution of the problem in Eq. (13.60); otherwise, point $\mathbf{x}_{k+1}$ is set to $\mathbf{x}^*$ and $\mathbf{A}$ and $\mathbf{b}$ in Eq. (13.67) are updated accordingly. The above steps are then repeated until $U_k - L_k \leq \varepsilon$. An algorithm based on these ideas is as follows.

Algorithm 13.5 Kelley's cutting-plane algorithm for CP problems with bound constraints
Step 1
Input an initial feasible point $\mathbf{x}_0$.
Set $k = 0$ and initialize the tolerance ε.
Step 2
Evaluate $\mathbf{A}_k$ and $\mathbf{b}_k$ by using Eq. (13.66) and solve the LP problem in Eq. (13.67) to obtain minimizer $\mathbf{x}_k^*$.
Step 3
Compute L_k and U_k.
If $U_k - L_k \leq \varepsilon$, output $\mathbf{x}^* = \mathbf{x}_k^*$, and stop; otherwise, set $k = k + 1$, $\mathbf{x}_{k+1} = \mathbf{x}_k^*$, and repeat from Step 2.

It follows from Eq. (13.64) that with $U_k - L_k \leq \varepsilon$ the solution $\mathbf{x}_k^*$ obtained with Kelley's algorithm ensures that $|f(\mathbf{x}_k^*) - f(\mathbf{x}^*)| \leq \varepsilon$. Moreover, it can be shown [10] that $U_k - L_k$ approaches zero as k increases and, therefore, the algorithm always terminates.

A problem with Kelley's algorithm is that the number of constraints in Eq. (13.66) grows with the number of iterations performed and so the computational complexity of each iteration will increase accordingly. However, if each LP subproblem starts with a good initial point, it can converge to the minimizer in a small number of iterations and the algorithm becomes practical. The minimizer $\mathbf{x}_k^*$ can serve as the initial point for the $(k + 1)$th iteration. In effect, as the minimizer $\mathbf{x}_k^*$ satisfies Eq. (13.67b) where $\mathbf{A} = \mathbf{A}_k$ and $\mathbf{b} = \mathbf{b}_k$ with $\mathbf{A}_k$, $\mathbf{b}_k$ defined by Eq. (13.66) and $\mathbf{x}_{k+1} = \mathbf{x}_k^*$, the newly added constraint in the $(k + 1)$th iteration, i.e.,

$$f(\mathbf{x}) \geq f(\mathbf{x}_{k+1}) + \mathbf{g}^T(\mathbf{x}_{k+1})(\mathbf{x} - \mathbf{x}_{k+1})$$

is tightly satisfied at $\mathbf{x}_k^*$ and hence $\mathbf{x}_k^*$ is a feasible point. Moreover, as can be seen in Fig. 13.4, $\mathbf{x}_k^*$ is a good initial point for iteration $k + 1$.

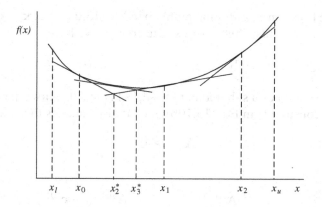

Figure 13.4. Point $\mathbf{x}_2^*$ serves as a good initial point in the 3rd iteration.

13.5.3 Kelley's cutting-plane method for CP problems with general inequality constraints

The general convex problem

$$\text{minimize } f(\mathbf{x}) \tag{13.68a}$$

$$\text{subject to: } c_j(\mathbf{x}) \geq 0 \quad \text{for } j = 1, 2, \ldots, q \tag{13.68b}$$

where $f(\mathbf{x})$ and $-c_j(\mathbf{x})$ for $j = 1, 2, \ldots, q$ are convex functions, can be converted to

$$\text{minimize } L \tag{13.69a}$$

$$\text{subject to: } f(\mathbf{x}) \leq L \tag{13.69b}$$

$$c_j(\mathbf{x}) \geq 0 \quad \text{for } j = 1, 2, \ldots, q \tag{13.69c}$$

With $\mathbf{z} = [\mathbf{x}^T \ L]^T$ and $\mathbf{c} = [0 \ \cdots \ 0 \ 1]^T$, the problem in Eq. (13.69) can be formulated as

$$\text{minimize } \mathbf{c}^T\mathbf{z}$$

$$\text{subject to: } \hat{c}_j(\mathbf{z}) \geq 0 \quad \text{for } j = 0, 1, \ldots, q$$

where $\hat{c}_0(\mathbf{z}) = L - f(\mathbf{x})$ and $\hat{c}_j(\mathbf{z}) = c_j(\mathbf{x})$ for $j = 1, 2, \ldots, q$. Obviously, functions $-\hat{c}_j(\mathbf{z})$ are all convex in $\mathbf{z}$. Therefore, without loss of generality, we can consider the CP problem

$$\text{minimize } f(\mathbf{x}) = \mathbf{c}^T\mathbf{x} \tag{13.70a}$$

$$\text{subject to: } c_j(\mathbf{x}) \geq 0 \quad \text{for } j = 1, 2, \ldots, q \tag{13.70b}$$

where functions $-c_j(\mathbf{x})$ are differentiable and convex.

The convexity of the constraint functions in Eq. (13.70b) can be utilized to generate piecewise linear lower-bound functions in a way similar to that used for objective function $f(\mathbf{x})$ for the problem in Eq. (13.60). Let $\mathbf{x}_0, \mathbf{x}_1, \ldots, \mathbf{x}_k$ be $k + 1$ distinct points. Since $-c_j(\mathbf{x})$ are convex, we have

$$-c_j(\mathbf{x}) \geq -c_j(\mathbf{x}_i) + \mathbf{h}_j^T(\mathbf{x}_i)(\mathbf{x} - \mathbf{x}_i) \qquad \text{for } 0 \leq i \leq k, \, 1 \leq j \leq q$$

where $\mathbf{h}_j^T(\mathbf{x}_i)$ denotes a subgradient of $-c_j(\mathbf{x})$ at $\mathbf{x}_i$. It follows that if point $\mathbf{x}$ satisfies the constraints in Eq. (13.70b), then it also satisfies the constraint

$$\mathbf{A}_k \mathbf{x} \geq \mathbf{b}_k \tag{13.71}$$

where

$$\mathbf{A}_k = \begin{bmatrix} \mathbf{A}^{(0)} \\ \vdots \\ \mathbf{A}^{(k)} \end{bmatrix}, \qquad \mathbf{b}_k = \begin{bmatrix} \mathbf{A}^{(0)}\mathbf{x}_0 - \mathbf{c}^{(0)} \\ \vdots \\ \mathbf{A}^{(k)}\mathbf{x}_k - \mathbf{c}^{(k)} \end{bmatrix}$$

$$\mathbf{A}^{(i)} = \begin{bmatrix} -\mathbf{h}_1^T(\mathbf{x}_i) \\ \vdots \\ -\mathbf{h}_q^T(\mathbf{x}_i) \end{bmatrix}, \qquad \mathbf{c}^{(i)} = \begin{bmatrix} c_1(\mathbf{x}_i) \\ \vdots \\ c_q(\mathbf{x}_i) \end{bmatrix}$$

At the kth iteration, the cutting-plane algorithm solves the LP problem

$$\text{minimize } f(\mathbf{x}) = \mathbf{c}^T \mathbf{x} \tag{13.72a}$$

$$\text{subject to: } \mathbf{A}_k \mathbf{x} \geq \mathbf{b}_k \tag{13.72b}$$

Since the feasible region $\mathcal{R}_{k-1}$ described by Eq. (13.72b) *contains* the feasible region described by Eq. (13.70b), the minimizer of the problem in Eq. (13.72), $\mathbf{x}_{k-1}^*$, might violate some of the constraints in Eq. (13.70b). Let us denote $\mathbf{x}_k^*$ as $\mathbf{x}_{k+1}$. If $\mathbf{x}_{k+1}$ satisfies Eq. (13.70b), then obviously $\mathbf{x}_{k+1}$ is the solution of the problem in Eq. (13.70) and the algorithm terminates. Otherwise, if j^* is the index for the most negative $c_j(\mathbf{x}_{k+1})$, then the constraints in Eq. (13.72b) are updated by including the linear constraint

$$c_{j^*}(\mathbf{x}_{k+1}) - \mathbf{h}_{j^*}^T(\mathbf{x}_{k+1})(\mathbf{x} - \mathbf{x}_{k+1}) \geq 0 \tag{13.73}$$

In other words, the feasible region of the problem in Eq. (13.72) is reduced to the intersection of $\mathcal{R}_{k-1}$ and the half-plane defined by Eq. (13.73). The updated constraints can be expressed as

$$\mathbf{A}_{k+1} \mathbf{x} \geq \mathbf{b}_{k+1} \tag{13.74}$$

where

$$\mathbf{A}_{k+1} = \begin{bmatrix} \mathbf{A}_k \\ -\mathbf{h}_{j^*}^T(\mathbf{x}_{k+1}) \end{bmatrix}, \quad \mathbf{b}_{k+1} = \begin{bmatrix} \mathbf{b}_k \\ -\mathbf{h}_{j^*}^T(\mathbf{x}_{k+1})\mathbf{x}_{k+1} - c_{j^*}(\mathbf{x}_{k+1}) \end{bmatrix}$$

The iterations continue until the LP subproblem reaches a solution $\mathbf{x}^*$ at which the most negative constraint function $c_j(\mathbf{x}^*)$ in Eq. (13.70b) is no less than $-\varepsilon$, where ε is a prescribed tolerance for nonfeasibility.

An algorithm for the problem in Eq. (13.70) based on Kelley's method is as follows.

Algorithm 13.6 Kelley's cutting-plane algorithm for CP problems with inequality constraints
Step 1
Input an initial point $\mathbf{x}_0$.
Set $k = 0$ and initialize the tolerance ε.
Step 2
Evaluate $\mathbf{A}_k$ and $\mathbf{b}_k$ in Eq. (13.71).
Step 3
Solve the LP problem in Eq. (13.72) to obtain minimizer $\mathbf{x}_k^*$.
Step 4
If $\min\{c_j(\mathbf{x}_k^*), 1 \le j \le q\} \ge -\varepsilon$, output $\mathbf{x}^* = \mathbf{x}_k^*$ and stop; otherwise, set $k = k + 1$, $\mathbf{x}_{k+1} = \mathbf{x}_k^*$, update $\mathbf{A}_k$ and $\mathbf{b}_k$ in Eq. (13.72b) by using Eq. (13.74), and repeat from Step 3.

Example 13.5 The two ellipses in Fig. 13.5 are described by

$$c_1(\mathbf{x}) = -[x_1 \ x_2] \begin{bmatrix} \frac{1}{4} & 0 \\ 0 & 1 \end{bmatrix} \begin{bmatrix} x_1 \\ x_2 \end{bmatrix} + [x_1 \ x_2] \begin{bmatrix} \frac{1}{2} \\ 0 \end{bmatrix} + \frac{3}{4} \ge 0$$

$$c_2(\mathbf{x}) = -\frac{1}{8}[x_3 \ x_4] \begin{bmatrix} 5 & 3 \\ 3 & 5 \end{bmatrix} \begin{bmatrix} x_3 \\ x_4 \end{bmatrix} + [x_3 \ x_4] \begin{bmatrix} \frac{11}{2} \\ \frac{13}{2} \end{bmatrix} - \frac{35}{2} \ge 0$$

where $\mathbf{x} = [x_1 \ x_2 \ x_3 \ x_4]^T$. Find the shortest distance between the two ellipses using Algorithm 13.6.

Solution The problem can be formulated as the constrained minimization problem

$$\text{minimize } f(\mathbf{x}) = \tfrac{1}{2}[(x_1 - x_3)^2 + (x_2 - x_4)^2]$$

$$\text{subject to:} \quad c_1(\mathbf{x}) \ge 0 \quad \text{and} \quad c_2(\mathbf{x}) \ge 0$$

The quadratic objective function has a positive-definite constant Hessian, and obviously the quadratic constraint functions $-c_1(\mathbf{x})$ and $-c_2(\mathbf{x})$ are convex functions. Hence this is a CP problem. In order to apply Algorithm 13.6, we convert the problem at hand into

$$\text{minimize } \mathbf{c}^T \mathbf{z}$$

$$\text{subject to:} \quad \hat{c}_i(\mathbf{z}) \ge 0 \quad \text{for } i = 0, 1, 2$$

436

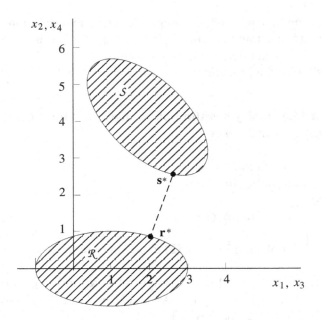

Figure 13.5. Distance between two ellipses (Example 13.5).

where $\mathbf{z} = [x_1 \ x_2 \ x_3 \ x_4 \ L]^T$, $\mathbf{c} = [0 \ 0 \ 0 \ 0 \ 1]^T$, $\hat{c}_0(\mathbf{z}) = L - f(\mathbf{x})$, $\hat{c}_1(\mathbf{z}) = c_1(\mathbf{x})$, and $\hat{c}_2(\mathbf{z}) = c_2(\mathbf{x})$.

With

$$\mathbf{x}_0 = \begin{bmatrix} 1.5 \\ 0.5 \\ 2.5 \\ 4.0 \end{bmatrix}, \quad L_0 = 1, \quad \text{and} \quad \varepsilon = 10^{-7},$$

the algorithm took 186 iterations and 10.75 Mflops to converge to the solution

$$\mathbf{x}^* = \begin{bmatrix} 1.992222 \\ 0.868259 \\ 2.577907 \\ 2.475862 \end{bmatrix}$$

which corresponds to the solution points $\mathbf{r}^* \in \mathcal{R}$ and $\mathbf{s}^* \in \mathcal{S}$ given by

$$\mathbf{r}^* = \begin{bmatrix} 1.992222 \\ 0.868259 \end{bmatrix} \quad \text{and} \quad \mathbf{s}^* = \begin{bmatrix} 2.577907 \\ 2.475862 \end{bmatrix}$$

These points give the shortest distance between $\mathcal{R}$ and $\mathcal{S}$ as $||\mathbf{r}^* - \mathbf{s}^*|| = 1.710969$.

∎

13.6   Ellipsoid Methods

Another class of cutting-plane algorithms, known as *ellipsoid algorithms*, was developed by Shor, Nemirovski, and Yudin during the 70's and was used by Khachiyan [13] to demonstrate the existence of polynomial-time algorithms for LP. Starting from an initial ellipsoid E_0 which contains a minimizer, an ellipsoid algorithm generates a hyperplane that passes through the center of the ellipsoid to divide it into two parts, one of which, denoted as E_{0h}, contains the minimizer. The algorithm then continues by generating another ellipsoid E_1 that entails minimum volume which contains E_{0h}. Next, a hyperplane that passes through the center of E_1 is generated to cut E_1 in half, where the half containing the minimizer is denoted as E_{1h}. The algorithm goes on to generate a sequence of progressively smaller ellipsoids, each of which containing the minimizer. After a sufficiently large number of iterations, the volume of the ellipsoid shrinks to zero and the minimizer is localized. Below we describe a basic ellipsoid method reported in [10].

13.6.1   Basic ellipsoid method for unconstrained CP problems

Consider minimizing a convex objective function $f(\mathbf{x})$ whose subgradient is denoted as $\mathbf{g}(\mathbf{x})$. Assume that $f(\mathbf{x})$ has a minimizer and that an ellipsoid E_0 that contains the minimizer as an interior point has been identified. At the kth iteration of the algorithm, an ellipsoid E_k in the n-dimensional Euclidean space E^n is described as

$$E_k = \{\mathbf{x} : (\mathbf{x} - \mathbf{x}_k)^T \mathbf{A}_k^{-1}(\mathbf{x} - \mathbf{x}_k) \le 1\}$$

where $\mathbf{x}_k$ is the center of the ellipsoid and $\mathbf{A}_k$ is a symmetric and positive-definite matrix. The lengths of the semi-axes of E_k are the square roots of the eigenvalues of $\mathbf{A}_k$ and the volume of E_k is given by [10] as

$$\text{vol}(E_k) = \beta_n \sqrt{\det(\mathbf{A}_k)}$$

where β_n is the volume of the unit ball in E^n given by

$$\beta_n = \frac{\pi^{n/2}}{\Gamma(\frac{n}{2} + 1)}$$

and $\Gamma(x)$ is the gamma function whose value at $n/2 + 1$ can be evaluated as

$$\Gamma\left(\frac{n}{2} + 1\right) = \begin{cases} \left(\frac{n}{2}\right)! & \text{for } n \text{ even} \\ \frac{\sqrt{\pi}}{2^{(n+1)/2}} \displaystyle\prod_{k=1}^{(n+1)/2} (2k - 1) & \text{for } n \text{ odd} \end{cases}$$

The hyperplane $P_k = \{\mathbf{x} : \mathbf{g}_k^T(\mathbf{x} - \mathbf{x}_k) = 0\}$, where $\mathbf{g}_k$ denotes a subgradient of $f(\mathbf{x})$ at $\mathbf{x}_k$, passes through the center of the ellipsoid and cuts the ellipsoid in half. Since $f(\mathbf{x})$ is convex, we have

$$f(\mathbf{x}) \geq f(\mathbf{x}_k) + \mathbf{g}_k^T(\mathbf{x} - \mathbf{x}_k) \tag{13.75}$$

Hence only the half of ellipsoid E_{kh} obtained by the intersection

$$E_{kh} = E_k \cap \{\mathbf{x} : \mathbf{g}_k^T(\mathbf{x} - \mathbf{x}_k) \leq 0\}$$

contains the minimizer. The next ellipsoid that contains E_{kh} with minimum volume is given by [13] as

$$E_{k+1} = \{\mathbf{x} : (\mathbf{x} - \mathbf{x}_{k+1})^T \mathbf{A}_{k+1}^{-1}(\mathbf{x} - \mathbf{x}_{k+1}) \leq 1\}$$

where

$$\mathbf{x}_{k+1} = \mathbf{x}_k - \frac{\mathbf{A}_k \tilde{\mathbf{g}}_k}{n+1} \tag{13.76a}$$

$$\mathbf{A}_{k+1} = \frac{n^2}{n^2 - 1}\left(\mathbf{A}_k - \frac{2}{n+1}\mathbf{A}_k \tilde{\mathbf{g}}_k \tilde{\mathbf{g}}_k^T \mathbf{A}_k\right) \tag{13.76b}$$

$$\tilde{\mathbf{g}}_k = \frac{\mathbf{g}_k}{(\mathbf{g}_k^T \mathbf{A}_k \mathbf{g}_k)^{1/2}} \tag{13.76c}$$

and has a minimum volume

$$\text{vol}(E_{k+1}) = \left(\frac{n}{n+1}\right)^{(n+1)/2}\left(\frac{n}{n-1}\right)^{(n-1)/2}\text{vol}(E_k)$$
$$< e^{-1/2n}\text{vol}(E_k) \tag{13.77}$$

Note that the volume-reduction rate depends only on the dimension of the parameter space. In the case of $n = 2$, for example, the reduction rate is 0.7698. Moreover, Eq. (13.77) implies that

$$\text{vol}(E_k) < e^{-k/2n}\text{vol}(E_0)$$

and hence vol (E_k) approaches zero as $k \to \infty$. This in conjunction with the fact that E_k for any k contains a minimizer proves the convergence of the algorithm.

An easy-to-use criterion for terminating the algorithm can be derived as follows by using the convexity property of $f(\mathbf{x})$. If $\mathbf{x}^*$ is the minimizer contained in E_k, then Eq. (13.75) implies that

$$f(\mathbf{x}^*) \geq f(\mathbf{x}_k) + \mathbf{g}_k^T(\mathbf{x}^* - \mathbf{x}_k)$$

Hence

$$f(\mathbf{x}_k) - f(\mathbf{x}^*) \leq -\mathbf{g}_k^T(\mathbf{x}^* - \mathbf{x}_k) \leq \max_{\mathbf{x} \in E_k}[-\mathbf{g}_k^T(\mathbf{x} - \mathbf{x}_k)]$$

It can be shown (see Prob. 13.14) that

$$\max_{\mathbf{x}\in E_k}[-\mathbf{g}_k^T(\mathbf{x}-\mathbf{x}_k)] = (\mathbf{g}_k^T\mathbf{A}_k\mathbf{g}_k)^{1/2} \tag{13.78}$$

Therefore, we can terminate the algorithm if

$$(\mathbf{g}_k^T\mathbf{A}_k\mathbf{g}_k)^{1/2} < \varepsilon$$

where ε is a prescribed tolerance. The method leads to the following algorithm.

Algorithm 13.7 Ellipsoid algorithm for unconstrained CP problems
Step 1
Input an initial ellipsoid defined by a positive-definite matrix $\mathbf{A}_0$ with center $\mathbf{x}_0$ that contains at least one minimizer.
Set $k = 0$ and initialize the tolerance ε.
Step 2
Evaluate a subgradient $\mathbf{g}_k$ and compute $\gamma_k = (\mathbf{g}_k^T\mathbf{A}_k\mathbf{g}_k)^{1/2}$.
Step 3
If $\gamma_k < \varepsilon$, then output $\mathbf{x}^* = \mathbf{x}_k$ and stop; otherwise, continue with Step 4.
Step 4
Compute $\tilde{\mathbf{g}}_k$, $\mathbf{x}_{k+1}$, and $\mathbf{A}_{k+1}$ using Eq. (13.76).
Set $k = k + 1$ and repeat from Step 2.

Example 13.6 It is known that the function

$$f(\mathbf{x}) = (x_1 - 5x_2 + 4)^2 + (7x_1 + 11x_2 - 18)^4$$

is globally convex and has a unique minimum at $\mathbf{x}^* = [1\ 1]^T$. Find the minimum point by applying Algorithm 13.7 with three different initial ellipses.

Solution Three possible initial ellipses are

$$E_0^{(1)} : \quad \frac{(x_1 - 4)^2}{6^2} + \frac{(x_2 + 1)^2}{3^2} \leq 1$$

$$E_0^{(2)} : \quad \frac{(x_1 - 31)^2}{40^2} + \frac{(x_2 - 11)^2}{16^2} \leq 1$$

$$E_0^{(3)} : \quad \frac{(x_1 - 61)^2}{80^2} + \frac{(x_2 + 29)^2}{48^2} \leq 1$$

and each can be shown to contain the minimizer. The three ellipses can be represented by

$$E_0^{(i)} = \{\mathbf{x} : (\mathbf{x} - \mathbf{x}_0^{(i)})^T\mathbf{A}_0^{(i)^{-1}}(\mathbf{x} - \mathbf{x}_0^{(i)}) \leq 1\}$$

where

$$\mathbf{x}_0^{(1)} = \begin{bmatrix} 4 \\ -1 \end{bmatrix}, \quad \mathbf{A}_0^{(1)} = \begin{bmatrix} 36 & 0 \\ 0 & 9 \end{bmatrix}$$

$$\mathbf{x}_0^{(2)} = \begin{bmatrix} 31 \\ 11 \end{bmatrix}, \quad \mathbf{A}_0^{(2)} = \begin{bmatrix} 1600 & 0 \\ 0 & 256 \end{bmatrix}$$

$$\mathbf{x}_0^{(3)} = \begin{bmatrix} 61 \\ -29 \end{bmatrix}, \quad \mathbf{A}_0^{(3)} = \begin{bmatrix} 6400 & 0 \\ 0 & 2304 \end{bmatrix}$$

With $\varepsilon = 10^{-7}$, Algorithm 13.7 quickly converged to the solution. The numerical results obtained from the three trials are summarized in Table 13.1.

Table 13.1 Numerical results for Example 13.6

$\mathbf{x}_0$	$\mathbf{x}^*$	Iterations	Kflops
$[4 \ -1]^T$	$[0.999975 \ 0.999984]^T$	59	4.344
$[31 \ 11]^T$	$[0.999989 \ 1.000013]^T$	68	5.001
$[61 \ -29]^T$	$[0.999560 \ 0.999911]^T$	73	5.374

∎

13.6.2 Ellipsoid method for constrained CP problems

The ellipsoid algorithm studied in Sec. 13.6.1 can be extended to deal with the convex problem

$$\text{minimize } f(\mathbf{x}) \tag{13.79a}$$

$$\text{subject to: } c_j(\mathbf{x}) \geq 0 \quad \text{for } j = 1, 2, \ldots, q \tag{13.79b}$$

where $f(\mathbf{x})$ and $-c_j(\mathbf{x})$ for $j = 1, 2, \ldots, q$ are convex functions. At the kth iteration of the algorithm, we examine the center of ellipsoid E_k, $\mathbf{x}_k$, to see whether or not it is a feasible point. If $\mathbf{x}_k$ is feasible, then we perform the iteration in the same way as in the unconstrained case. Since the iteration yields a new point $\mathbf{x}_{k+1}$ at which the objective function $f(\mathbf{x})$ is reduced as in the case where Algorithm 13.7 is used, such an iteration is referred to as an *objective iteration*.

If $\mathbf{x}_k$ is not feasible, then at least one constraint is violated at $\mathbf{x}_k$. Let j^* be the index for the most negative $c_{j^*}(\mathbf{x}_k)$. Since $-c_{j^*}(\mathbf{x})$ is convex, we have

$$-c_{j^*}(\mathbf{x}) \geq -c_{j^*}(\mathbf{x}_k) + \mathbf{h}_k^T(\mathbf{x} - \mathbf{x}_k) \quad \text{for all } \mathbf{x} \tag{13.80}$$

where $\mathbf{h}_k$ is a subgradient of $-c_{j^*}(\mathbf{x})$ at $\mathbf{x}_k$. It follows that at any point $\mathbf{x}$ with $\mathbf{h}_k^T(\mathbf{x} - \mathbf{x}_k) \geq 0$, we have $c_{j^*}(\mathbf{x}) < 0$, i.e., $\mathbf{x}$ is nonfeasible. Hence the hyperplane defined by $\mathbf{h}_k^T(\mathbf{x} - \mathbf{x}_k) = 0$ divides ellipsoid E_k into two parts, one of which is a region where every point is nonfeasible and, therefore, can be excluded in the subsequent iterations. Under these circumstances, the part of E_k that should be kept is obtained by the intersection

$$E_{kh} = E_k \cap \{\mathbf{x} : \mathbf{h}_k^T(\mathbf{x} - \mathbf{x}_k) \leq 0\}$$

Using an argument similar to that used in Sec. 13.6.1, we conclude that the next ellipsoid that contains E_{kh} with minimum volume is given by

$$E_{k+1} = \{\mathbf{x} : (\mathbf{x} - \mathbf{x}_{k+1})^T \mathbf{A}_{k+1}^{-1}(\mathbf{x} - \mathbf{x}_{k+1}) \leq 1\} \qquad (13.81a)$$

where

$$\mathbf{x}_{k+1} = \mathbf{x}_k - \frac{\mathbf{A}_k \tilde{\mathbf{g}}_k}{n+1} \qquad (13.81b)$$

$$\mathbf{A}_{k+1} = \frac{n^2}{n^2 - 1}\left(\mathbf{A}_k - \frac{2}{n+1}\mathbf{A}_k \tilde{\mathbf{g}}_k \tilde{\mathbf{g}}_k^T \mathbf{A}_k\right) \qquad (13.81c)$$

$$\tilde{\mathbf{g}}_k = \frac{\mathbf{h}_k}{(\mathbf{h}_k^T \mathbf{A}_k \mathbf{h}_k)^{1/2}} \qquad (13.81d)$$

It follows from the above analysis that Eq. (13.81) generates a new ellipsoid with a center $\mathbf{x}_{k+1}$ that is more likely to be feasible but does not necessarily reduce the objective function. For this reason the iteration associated with Eq. (13.81) is referred to as a *constraint iteration*. If $\mathbf{x}_{k+1}$ is indeed feasible, then the next iterate $\mathbf{x}_{k+2}$ can be generated using an objective iteration but if $\mathbf{x}_{k+1}$ is still nonfeasible, then another constraint iteration must be carried out to generate point $\mathbf{x}_{k+2}$. The iterations continue in this manner until a point, say, $\mathbf{x}_K$, is reached that satisfies all the constraints in Eq. (13.79b) and $(\mathbf{g}_K^T \mathbf{A}_K \mathbf{g}_K)^{1/2} < \varepsilon$, where $\mathbf{g}_K$ is a subgradient of $f(\mathbf{x})$ at $\mathbf{x}_K$.

Note that in a constraint iteration, the convexity of $-c_{j*}(\mathbf{x}_k)$ leads to

$$-c_{j*}(\mathbf{x}_k) + c_{j*}(\hat{\mathbf{x}}) \leq -\mathbf{h}_k^T(\hat{\mathbf{x}} - \mathbf{x}_k)$$
$$\leq \max_{\mathbf{x} \in E_k} -\mathbf{h}_k^T(\mathbf{x} - \mathbf{x}_k) = (\mathbf{h}_k^T \mathbf{A}_k \mathbf{h}_k)^{1/2}$$

where $\hat{\mathbf{x}}$ denotes the point at which $c_{j*}(\mathbf{x})$ reaches its maximum in E_k. It follows that

$$c_{j*}(\hat{\mathbf{x}}) \leq c_{j*}(\mathbf{x}_k) + (\mathbf{h}_k^T \mathbf{A}_k \mathbf{h}_k)^{1/2}$$

and hence $c_{j*}(\hat{\mathbf{x}}) < 0$ if

$$c_{j*}(\mathbf{x}_k) + (\mathbf{h}_k^T \mathbf{A}_k \mathbf{h}_k)^{1/2} < 0 \qquad (13.82)$$

Since $\hat{\mathbf{x}}$ is a maximizer of $c_{j*}(\mathbf{x})$ in ellipsoid E_k, the condition in Eq. (13.82) implies that $c_{j*}(\mathbf{x}) < 0$ in the entire E_k. In effect, no feasible points exist in E_k in such a case. Therefore, Eq. (13.82) can serve as a criterion as to whether the iteration should be terminated or not. An algorithm based on the approach is as follows.

Algorithm 13.8 Ellipsoid method for CP constrained problems
Step 1
Input an initial ellipsoid defined by a positive definite $\mathbf{A}_0$ with center $\mathbf{x}_0$ that contains at least one minimizer.
Set $k = 0$ and initialize the tolerance ε.
Step 2
If $\mathbf{x}_k$ is feasible continue with Step 2a, otherwise, go to Step 2b.
    Step 2a
    Evaluate a subgradient of $f(\mathbf{x})$ at $\mathbf{x}_k$ denoted as $\mathbf{g}_k$.
    Compute $\tilde{\mathbf{g}}_k = \mathbf{g}_k/(\mathbf{g}_k^T \mathbf{A}_k \mathbf{g}_k)^{1/2}$ and $\gamma_k = (\mathbf{g}_k^T \mathbf{A}_k \mathbf{g}_k)^{1/2}$.
    Go to Step 4.
    Step 2b
    Let $c_{j*}(\mathbf{x})$ be a constraint function such that

$$c_{j*}(\mathbf{x}_k) = \min_{1 \leq j \leq q} [c_j(\mathbf{x}_k)] < 0$$

    Evaluate a subgradient of $-c_{j*}(\mathbf{x})$ at $\mathbf{x}_k$ and denote it as $\mathbf{h}_k$.
    Compute $\tilde{\mathbf{g}}_k = \mathbf{h}_k/(\mathbf{h}_k^T \mathbf{A}_k \mathbf{h}_k)^{1/2}$.
    Continue with Step 3.
Step 3
If the condition in Eq. (13.82) holds, terminate the iteration; otherwise, go to Step 5.
Step 4
If $\gamma_k < \varepsilon$, output $\mathbf{x}^* = \mathbf{x}_k$ and stop; otherwise, continue with Step 5.
Step 5
Compute

$$\mathbf{x}_{k+1} = \mathbf{x}_k - \frac{\mathbf{A}_k \tilde{\mathbf{g}}_k}{n+1}$$

$$\mathbf{A}_{k+1} = \frac{n^2}{n^2 - 1}\left(\mathbf{A}_k - \frac{2}{n+1}\mathbf{A}_k \tilde{\mathbf{g}}_k \tilde{\mathbf{g}}_k^T \mathbf{A}_k\right)$$

Set $k = k + 1$ and repeat from Step 2.

Example 13.7 By applying Algorithm 13.8 find the minimizer of the convex problem with the objective function given in Example 13.6 and the constraints given by

$$c_1(\mathbf{x}) = 2x_1 - x_2 - 6 \geq 0$$
$$c_2(\mathbf{x}) = -2x_1 - x_2 + 10 \geq 0$$
$$c_3(\mathbf{x}) = -3x_1 + x_2 + 15 \geq 0$$
$$c_4(\mathbf{x}) = 3x_1 + x_2 - 9 \geq 0$$

Solution From Example 13.6, we know that the minimizer of the unconstrained counterpart of this problem is $\mathbf{x}_u^* = [1 \; 1]^T$ at which $c_1(\mathbf{x}_u^*) < 0$ and $c_4(\mathbf{x}_u^*) < 0$. Hence $\mathbf{x}_u^*$ is *not* a feasible point for the present constrained CP problem. We start the algorithm with ellipse $E_0^{(1)}$ defined in Example 13.6 for which

$$\mathbf{x}_0 = \begin{bmatrix} 4 \\ -1 \end{bmatrix} \quad \text{and} \quad \mathbf{A}_0 = \begin{bmatrix} 36 & 0 \\ 0 & 9 \end{bmatrix}$$

Ellipse $E_0^{(1)}$ is large enough to contain the entire feasible region. Therefore, it also contains the minimizer of the problem. Notice that $c_j(\mathbf{x}_0) > 0$ for $1 \le j \le 4$ and hence $\mathbf{x}_0$ is a feasible initial point. With $\varepsilon = 10^{-7}$, it took the algorithm 76 iterations to converge to the minimizer

$$\mathbf{x}^* = \begin{bmatrix} 3.063142 \\ -0.189377 \end{bmatrix}$$

which yields $f(\mathbf{x}^*) = 67.570003$. The number of flops used was 7627. ∎

References

1 G. H. Golub and C. F. Van Loan, *Matrix Computation*, 2nd ed., Baltimore, Johns Hopkins University Press, MD, 1989.

2 R. Fletcher, *Practical Methods of Optimization*, 2nd ed., Wiley, NY, 1987.

3 P. E. Gill, W. Murray, and M. H. Wright, *Practical Optimization*, Academic Press, NY, 1981.

4 D. Goldfarb and A. Idnani, "A numerically stable dual method for solving strictly convex quadratic programs," *Math. Prog.*, vol. 27, pp. 1–33, 1983.

5 C. L. Lawson and R. J. Hanson, *Solving Least Squares Problems*, Prentice Hall, Englewood Cliffs, NJ, 1974.

6 R. D. C. Monteiro and I. Adler, "Interior path following primal-dual algorithms, Part II: Convex quadratic programming," *Math. Programming*, vol. 44, pp. 45–66, 1989.

7 R. D. C. Monteiro and I. Adler, "Interior path following primal-dual algorithms, Part I: Linear programming," *Math. Programming*, vol. 44, pp. 27–41, 1989.

8 S. J. Wright, *Primal-Dual Interior-Point Methods*, SIAM, Philadelphia, PA, 1997.

9 Y. Ye, *Interior Point Algorithms: Theory and Analysis*, Wiley, NY, 1997.

10 S. P. Boyd and C. H. Barratt, *Linear Controller Design: Limits of Performance*, Prentice Hall, Englewood Cliffs, NJ, 1991.

11 J. E. Kelley, "The cutting-plane method for solving convex programs," *J. SIAM*, vol. 8, pp. 703–712, Dec. 1960.

12 R. T. Rockafellar, *Convex Analysis*, 2nd ed., Princeton University Press, Princeton, NJ, 1970.

13 L. G. Khachiyan, "A polynomial algorithm in linear programming," *Soviet Math. Doklady*, vol. 20, pp. 191–194, 1979.

Problems

13.1 Let $H(\omega) = \sum_{i=0}^{N} a_i \cos i\omega$ and $\mathbf{x} = [a_0 \ a_1 \ \cdots \ a_N]^T$. Show that the constrained optimization problem

$$\text{minimize } f(\mathbf{x}) = \int_0^{\pi} W(\omega)|H(\omega) - H_d(\omega)|^2 d\omega$$

$$\text{subject to: } \quad |H(\omega_k) - H_d(\omega_k)| \leq \delta_k \qquad \text{for } k = 1, 2, \ldots, K$$

is a convex QP problem. In this problem, $H_d(\omega)$ and $W(\omega)$ are given real-valued functions, $W(\omega) \geq 0$ is a weighting function, $\{\omega : \omega_k, \ k = 1, 2, \ldots, K\}$ is a set of grid points on $[0, \ \pi]$, and $\delta_k > 0$ for $1 \leq k \leq K$ are constants.

13.2 Solve the QP problems
   (a)

$$\text{minimize } f(\mathbf{x}) = 2x_1^2 + x_2^2 + x_1 x_2 - x_1 - x_2$$

$$\text{subject to: } \quad x_1 + x_2 = 1$$

   (b)

$$\text{minimize } f(\mathbf{x}) = 1.5x_1^2 - x_1 x_2 + x_2^2 - x_2 x_3 + 0.5x_3^2 + x_1 + x_2 + x_3$$

$$\text{subject to: } \quad x_1 + 2x_2 + x_3 = 4$$

   by using each of the following three methods: the SVD, QR decomposition, and the Lagrange-multiplier methods.

13.3 By applying Algorithm 13.1, solve the following QP problems:
   (a)

$$\text{minimize } f(\mathbf{x}) = 3x_1^2 + 3x_2^2 - 10x_1 - 24x_2$$

$$\text{subject to: } \quad -2x_1 - x_2 \geq -4$$
$$\mathbf{x} \geq \mathbf{0}$$

   with $\mathbf{x}_0 = [0 \ 0]^T$.
   (b)

$$\text{minimize } f(\mathbf{x}) = x_1^2 - x_1 x_2 + x_2^2 - 3x_1$$

$$\text{subject to: } \quad -x_1 - x_2 \geq -2$$
$$\mathbf{x} \geq \mathbf{0}$$

   with $\mathbf{x}_0 = [0 \ 0]^T$.

(c)

$$\text{minimize } f(\mathbf{x}) = x_1^2 + 0.5x_2^2 - x_1x_2 - 3x_1 - x_2$$
$$\text{subject to: } \quad -x_1 - x_2 \geq -2$$
$$-2x_1 + x_2 \geq -2$$
$$\mathbf{x} \geq \mathbf{0}$$

with $\mathbf{x}_0 = [0\ 0]^T$.

(d)

$$\text{minimize } f(\mathbf{x}) = x_1^2 + x_2^2 + 0.5x_3^2 + x_1x_2 + x_1x_3 - 4x_1 - 3x_2 - 2x_3$$
$$\text{subject to: } \quad -x_1 - x_2 - x_3 \geq -3$$
$$\mathbf{x} \geq \mathbf{0}$$

with $\mathbf{x}_0 = [0\ 0\ 0]^T$.

13.4 Verify that the solution of Eq. (13.39) is given by Eq. (13.42).

13.5 (a) Convert the QP problems in Prob. 13.3 into the form in Eq. (13.28).

(b) Solve the QP problems obtained in part (a) by applying Algorithm 13.2.

13.6 Verify that the solution of Eq. (13.46) is given by Eq. (13.47).

13.7 (a) Solve the QP problems obtained in Prob. 13.5(a) by applying Algorithm 13.3.

(b) Compare the results obtained in part (a) with those obtained in Prob. 13.5(b).

13.8 Show that if $\mathbf{H}$ is positive definite, $\mathbf{A}$ is of full row rank, $\mu_k > 0$, and $\mathbf{x}_k > 0$, then Eq. (13.39) has a unique solution for δ_w.

13.9 (a) By applying Algorithm 13.2 solve the following QP problem:

$$\text{minimize } \tfrac{1}{2}\mathbf{x}^T(\mathbf{h}\mathbf{h}^T)\mathbf{x} + \mathbf{x}^T\mathbf{p}$$
$$\text{subject to: } \quad \mathbf{A}\mathbf{x} = \mathbf{b}$$
$$\mathbf{x} \geq \mathbf{0}$$

where

$$\mathbf{h} = \begin{bmatrix} 1 \\ -4 \\ 2 \\ 1 \end{bmatrix}, \quad \mathbf{A} = [1\ 1\ 1\ 1], \quad b = 4, \quad \mathbf{p} = \begin{bmatrix} -1 \\ 0 \\ 7 \\ 4 \end{bmatrix}$$

with $\mathbf{x}_0 = [1\ 1\ 1\ 1]^T$, $\lambda_0 = -2$, and $\mu_0 = [1\ 2\ 9\ 6]^T$.

(b) By applying Algorithm 13.3, solve the QP problem in part (a) with $\mathbf{x}_0 = [3\ 3\ 3\ 3]^T$, $\lambda_0 = 1$, and $\mu_0 = [1\ 1\ 1\ 1]^T$. Compare the solution obtained with that of part (a).

13.10 Show that

(a)

$$\mathbf{K} = \begin{bmatrix} \mathbf{0} & -\mathbf{A}^T \\ \mathbf{A} & \mathbf{0} \end{bmatrix}$$

is positive semidefinite in the sense of Eq. (13.53).

(b) Show that if $\mathbf{H}$ is positive semidefinite, then

$$\mathbf{K} = \begin{bmatrix} \mathbf{H} & -\mathbf{A}^T \\ \mathbf{A} & \mathbf{0} \end{bmatrix}$$

is positive definite in the sense of Eq. (13.53).

13.11 (a) Convert the QP problem in Prob. 13.9(a) using the initial values for $\mathbf{x}_0$, λ_0, and μ_0 given in Prob. 13.9(b) to a mixed LCP problem.

(b) Solve the LCP problem obtained in part (a) by applying Algorithm 13.4.

(c) Compare the solutions obtained with those obtained in Prob. 13.9(b).

13.12 Demonstrate the validity of the following:

(a) If $f(\mathbf{x})$ is differentiable, then the gradient of $f(\mathbf{x})$ is a subgradient.

(b) If $\alpha > 0$, then a subgradient of $\alpha f(\mathbf{x})$ is given by $\alpha \mathbf{g}$ where $\mathbf{g}$ is a subgradient of $f(\mathbf{x})$.

(c) If $f(\mathbf{x}) = f_1(\mathbf{x}) + f_2(\mathbf{x}) + \cdots + f_r(\mathbf{x})$ where function $f_i(\mathbf{x})$ for $1 \le i \le r$ are convex, then $\mathbf{g} = \mathbf{g}_1 + \mathbf{g}_2 + \cdots + \mathbf{g}_r$ is a subgradient of $f(\mathbf{x})$ where $\mathbf{g}_i$ is a subgradient of $f_i(\mathbf{x})$.

(d) If

$$f(\mathbf{x}) = \max[f_1(\mathbf{x}),\ f_2(\mathbf{x}),\ \ldots,\ f_r(\mathbf{x})]$$

where $f_i(\mathbf{x})$ for $1 \le i \le r$ are convex, then at point $\mathbf{x}$ there is at least one index i^* with $1 \le i^* \le r$ such that $f(\mathbf{x}) = f_{i^*}(\mathbf{x})$. In this case a subgradient of $f_{i^*}(\mathbf{x})$, $\mathbf{g}_{i^*}$, is a subgradient of $f(\mathbf{x})$.

13.13 Consider the problem of finding the shortest distance between the circular and elliptic disks shown in Fig. P13.13.

(a) Using the following 'sequential QP' approach, obtain an approximate solution of the problem: (i) Replace the disks by polygons with, say, k edges that approximate the circle and ellipse from either inside or outside; (ii) formulate the problem of finding the shortest distance between the two polygons as a QP problem; (iii) apply one of the

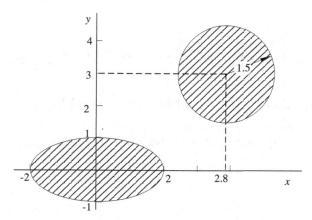

Figure P13.13.

algorithms described in Sec. 13.4 to find a solution of the QP problem; (iv) improve the solution accuracy by increasing the number of edges, k, of each of the two polygons.

(b) Formulate the shortest distance problem as a CP problem and solve it by using Algorithm 13.6.

(c) Solve the CP problem formulated in part (b) by applying Algorithm 13.8 and compare the results with those obtained in part (b).

13.14 Let $\mathbf{A} \in R^{n \times n}$ be positive definite and $\mathbf{g} \in R^{n \times 1}, \mathbf{x} \in R^{n \times 1}$ be arbitrarily given vectors. Show that

$$\max_{\mathbf{z} \in E_x}[-\mathbf{g}^T(\mathbf{z} - \mathbf{x})] = (\mathbf{g}^T \mathbf{A} \mathbf{g})^{1/2}$$

where

$$E_x = \{\mathbf{z} : (\mathbf{z} - \mathbf{x})^T \mathbf{A}^{-1}(\mathbf{z} - \mathbf{x}) \leq 1\}$$

Hint: Assume that $\mathbf{A} = \text{diag}\{\sigma_1, \sigma_2, \ldots, \sigma_n\}$ with $\sigma_i > 0$ for $1 \leq i \leq n$ without loss of generality.

13.15 Consider the least-squares minimization problem with quadratic inequality (LSQI) constraints which arises in cases where the solution to the ordinary least-squares problem needs to be regularized [1], namely,

$$\text{minimize } ||\mathbf{A}\mathbf{x} - \mathbf{b}||$$

$$\text{subject to: } ||\mathbf{B}\mathbf{x}|| \leq \delta$$

where $\mathbf{A} \in R^{m \times n}$, $\mathbf{b} \in R^{m \times 1}$, $\mathbf{B} \in R^{n \times n}$ with $\mathbf{B}$ nonsingular and $\delta \geq 0$.

(a) Convert the LSQI problem to the standard CP problem in Eq. (13.68).

(b) With

$$\mathbf{A} = \begin{bmatrix} 2 & 0 \\ 0 & 1 \\ 0 & 0 \end{bmatrix}, \quad \mathbf{b} = \begin{bmatrix} 4 \\ 2 \\ 3 \end{bmatrix}, \quad \mathbf{B} = \begin{bmatrix} 1 & 0 \\ 0 & 1 \end{bmatrix}, \quad \text{and} \quad \delta = 0.1$$

apply Algorithms 13.6 and 13.8 to solve the LSQI problem.

(c) Apply the algorithms used in part (b) to the case where δ is increased to 1. Compare the solution obtained with those in part (b).

Chapter 14

SEMIDEFINITE AND SECOND-ORDER CONE PROGRAMMING

14.1   Introduction

Semidefinite programming (SDP) is a branch of convex programming (CP) that has been a subject of intensive research since the early 1990's [1]–[9]. The continued interest in SDP has been motivated mainly by two reasons. First, many important classes of optimization problems such as linear-programming (LP) and convex quadratic-programming (QP) problems can be viewed as SDP problems, and many CP problems of practical usefulness that are neither LP nor QP problems can also be formulated as SDP problems. Second, several interior-point methods that have proven efficient for LP and convex QP problems have been extended to SDP in recent years.

Another important branch of convex programming is *second-order cone programming* (SOCP). Although quite specialized, this branch of optimization can deal effectively with many analysis and design problems in various disciplines. Furthermore, as for SDP, efficient interior-point methods are available for the solution of SOCP problems.

This chapter starts with the formulation of the primal and dual SDP problems. It then demonstrates that several useful CP problems can be formulated in an SDP setting. After an introduction of several basic properties of the primal-dual solutions of an SDP problem, a detailed account on several efficient interior-point SDP methods is provided. The methods considered include the primal-dual interior-point methods studied in [5]–[9] and the projective method proposed in [4][14]. The last two sections of the chapter are devoted to the primal and dual SOCP formulations and their relations to corresponding LP, QP, and SDP formulations; they also include an interior-point algorithm as well as several examples that illustrate how several important CP problems can be formulated as SOCP problems.

14.2   Primal and Dual SDP Problems

14.2.1   Notation and definitions

Let $\mathcal{S}^n$ be the space of real symmetric $n \times n$ matrices. The standard inner product on $\mathcal{S}^n$ is defined by

$$\mathbf{A} \cdot \mathbf{B} = \text{trace}(\mathbf{AB}) = \sum_{i=1}^{n} \sum_{j=1}^{n} a_{ij} b_{ij}$$

where $\mathbf{A} = \{a_{ij}\}$ and $\mathbf{B} = \{b_{ij}\}$ are two members of $\mathcal{S}^n$.

The primal SDP problem is defined as

$$\text{minimize } \mathbf{C} \cdot \mathbf{X} \tag{14.1a}$$

$$\text{subject to:} \quad \mathbf{A}_i \cdot \mathbf{X} = b_i \quad \text{for } i = 1, 2, \ldots, p \tag{14.1b}$$

$$\mathbf{X} \succeq \mathbf{0} \tag{14.1c}$$

where $\mathbf{C}$, $\mathbf{X}$, and $\mathbf{A}_i$ for $1 \leq i \leq p$ are members of $\mathcal{S}^n$ and the notation in Eq. (14.1c) denotes that $\mathbf{X}$ is positive semidefinite (see Sec. 10.2). It can be readily verified that the problem formulated in Eq. (14.1) is a CP problem (see Prob. 14.1). An important feature of the problem is that the variable involved is a *matrix* rather than a vector. Despite this distinction, SDP is closely related to several important classes of optimization problems. For example, if matrices $\mathbf{C}$ and $\mathbf{A}_i$ for $1 \leq i \leq p$ are all diagonal matrices, i.e.,

$$\mathbf{C} = \text{diag}\{\mathbf{c}\}, \quad \mathbf{A}_i = \text{diag}\{\mathbf{a}_i\}$$

with $\mathbf{c} \in R^{n \times 1}$ and $\mathbf{a}_i \in R^{n \times 1}$ for $1 \leq i \leq p$, then the problem in Eq. (14.1) is reduced to the standard-form LP problem

$$\text{minimize } \mathbf{c}^T \mathbf{x} \tag{14.2a}$$

$$\text{subject to:} \quad \mathbf{Ax} = \mathbf{b} \tag{14.2b}$$

$$\mathbf{x} \geq \mathbf{0} \tag{14.2c}$$

where $\mathbf{A} \in R^{p \times n}$ is a matrix with $\mathbf{a}_i^T$ as its ith row, $\mathbf{b} = [b_1 \ b_2 \ \cdots \ b_p]^T$, and vector $\mathbf{x} \in R^{n \times 1}$ is the diagonal of $\mathbf{X}$. The similarity between Eqs. (14.1a) and (14.2a) and between Eqs. (14.1b) and (14.2b) is quite evident. To see the similarity between Eq. (14.1c) and (14.2c), we need the concept of *convex cone*.

Definition 14.1 A convex cone $\mathcal{K}$ is a convex set such that $\mathbf{x} \in \mathcal{K}$ implies that $\alpha \mathbf{x} \in \mathcal{K}$ for any scalar $\alpha \geq 0$.

■

It can be readily verified that both sets $\{\mathbf{X} : \mathbf{X} \in R^{n \times n}, \mathbf{X} \succeq \mathbf{0}\}$ and $\{\mathbf{x} : \mathbf{x} \in R^{n \times 1}, \mathbf{x} \geq \mathbf{0}\}$ are convex cones (see Prob. 14.2). We now recall that the dual of the LP problem in Eq. (14.2) is given by

$$\text{maximize } \mathbf{b}^T \mathbf{y} \tag{14.3a}$$

$$\text{subject to: } \mathbf{A}^T \mathbf{y} + \mathbf{s} = \mathbf{c} \tag{14.3b}$$

$$\mathbf{s} \geq \mathbf{0} \tag{14.3c}$$

(see Chap. 12) and, therefore, the *dual* SDP problem with respect to the primal SDP problem in Eq. (14.1) can be obtained as

$$\text{maximize } \mathbf{b}^T \mathbf{y} \tag{14.4a}$$

$$\text{subject to: } \sum_{i=1}^{p} y_i \mathbf{A}_i + \mathbf{S} = \mathbf{C} \tag{14.4b}$$

$$\mathbf{S} \succeq \mathbf{0} \tag{14.4c}$$

when $\mathbf{S}$ is a slack variable that can be regarded as a matrix counterpart of the slack vector $\mathbf{s}$ in Eq. (14.3). To justify the maximization problem in Eq. (14.4) as a dual of the problem in Eq. (14.1), we assume that there exist $\mathbf{X} \in \mathcal{S}^n$, $\mathbf{y} \in R^p$, and $\mathbf{S} \in \mathcal{S}^n$ with $\mathbf{X} \succeq \mathbf{0}$ and $\mathbf{S} \succeq \mathbf{0}$ such that $\mathbf{X}$ is feasible for the primal and $\{\mathbf{y}, \mathbf{S}\}$ is feasible for the dual, and evaluate

$$\mathbf{C} \cdot \mathbf{X} - \mathbf{b}^T \mathbf{y} = \left(\sum_{i=1}^{p} y_i \mathbf{A}_i + \mathbf{S} \right) \cdot \mathbf{X} - \mathbf{b}^T \mathbf{y}$$

$$= \mathbf{S} \cdot \mathbf{X} \geq 0 \tag{14.5}$$

where the first and second equalities follow from Eq. (14.4b) and the inequality is a consequence of the fact that both $\mathbf{S}$ and $\mathbf{X}$ are positive semidefinite (see Prob. 14.3). Later in Sec. 14.3, it will be shown that if $\mathbf{X}^*$ is a solution of the primal and $\mathbf{y}^*$ is a solution of the dual, then

$$\mathbf{S}^* \cdot \mathbf{X}^* = 0 \tag{14.6}$$

where $\mathbf{S}^*$ is determined from Eq. (14.4b), i.e.,

$$\mathbf{S}^* = \mathbf{C} - \sum_{i=1}^{p} y_i^* \mathbf{A}_i$$

From Eqs. (14.5) and (14.6), it follows that

$$\mathbf{C} \cdot \mathbf{X}^* - \mathbf{b}^T \mathbf{y}^* = 0 \tag{14.7}$$

Eqs. (14.5) and (14.7) suggest that a duality gap similar to that in Eq. (12.6) can be defined for the problems in Eqs. (14.1) and (14.4) as

$$\delta(\mathbf{X}, \mathbf{y}) = \mathbf{C} \cdot \mathbf{X} - \mathbf{b}^T \mathbf{y} \tag{14.8}$$

452

for $\mathbf{X} \in \mathcal{F}_p$ and $\{\mathbf{y}, \mathbf{S}\} \in \mathcal{F}_d$ where $\mathcal{F}_p$ and $\mathcal{F}_d$ are the feasible sets for the primal and dual defined by

$$\mathcal{F}_p = \{\mathbf{X} : \mathbf{X} \succeq 0, \ \mathbf{A}_i \cdot \mathbf{X} = b_i \quad \text{for } 1 \le i \le p\}$$

$$\mathcal{F}_d = \left\{ \{\mathbf{y}, \mathbf{S}\} : \sum_{i=1}^{p} y_i \mathbf{A}_i + \mathbf{S} = \mathbf{C}, \ \mathbf{S} \succeq 0 \right\}$$

respectively. From Eqs. (14.5) and (14.7), it follows that for any $\mathbf{X} \in \mathcal{F}_p$ and $\{\mathbf{y}, \mathbf{S}\} \in \mathcal{F}_d$ the duality gap $\delta(\mathbf{X}, \mathbf{y})$ is nonnegative and the gap is reduced to zero at the solutions $\mathbf{X}^*$ and $\mathbf{S}^*$ of the primal and dual problems.

If we combine the constraints in Eqs. (14.4b) and (14.4c) into one inequality constraint, the dual SDP problem becomes

$$\text{maximize } \mathbf{b}^T \mathbf{y}$$
$$\text{subject to:} \quad \mathbf{C} - \sum_{i=1}^{p} y_i \mathbf{A}_i \succeq 0$$

This is obviously equivalent to the following minimization problem

$$\text{minimize } \mathbf{c}^T \mathbf{x} \tag{14.9a}$$
$$\text{subject to:} \quad \mathbf{F}(\mathbf{x}) \succeq 0 \tag{14.9b}$$

where $\mathbf{c} \in R^{p \times 1}$, $\mathbf{x} \in R^{p \times 1}$, and

$$\mathbf{F}(\mathbf{x}) = \mathbf{F}_0 + \sum_{i=1}^{p} x_i \mathbf{F}_i$$

with $\mathbf{F}_i \in \mathcal{S}^n$ for $0 \le i \le p$. Notice that the positive semidefinite constraint on matrix $\mathbf{F}(\mathbf{x})$ in Eq. (14.9b) is dependent on vector $\mathbf{x}$ in an *affine* manner. In the literature, the type of problems described by Eq. (14.9) are often referred to as *convex optimization* problems with *linear matrix inequality* (LMI) constraints, and have found many applications in science and engineering [3][10]. Since the minimization problem in Eq. (14.9) is equivalent to a dual SDP problem, the problem itself is often referred to as an SDP problem.

14.2.2 Examples

(i) *LP Problems* As we have seen in Sec. 14.2.1, standard-form LP problems can be viewed as a special class of SDP problems where the matrices $\mathbf{C}$ and $\mathbf{A}_i$ for $1 \le i \le p$ in Eq. (14.1) are all diagonal.

The alternative-form LP problem

$$\text{minimize } \mathbf{c}^T \mathbf{x} \tag{14.10a}$$
$$\text{subject to:} \quad \mathbf{A}\mathbf{x} \ge \mathbf{b}, \ \mathbf{A} \in R^{p \times n} \tag{14.10b}$$

which was studied extensively in Chap. 11 (see Eq. (11.2)), can be viewed as a linear minimization problem with LMI constraints. This can be demonstrated by expressing matrices $\mathbf{F}_i$ for $0 \leq i \leq n$ in Eq. (14.9b) as

$$\mathbf{F}_0 = -\mathrm{diag}\{\mathbf{b}\}, \quad \mathbf{F}_i = \mathrm{diag}\{\mathbf{a}_i\} \qquad \text{for } i = 1, 2, \ldots, n \qquad (14.11)$$

where $\mathbf{a}_i$ denotes the ith column of $\mathbf{A}$.

(ii) *Convex QP Problems* The general convex QP problem

$$\text{minimize } \mathbf{x}^T \mathbf{H} \mathbf{x} + \mathbf{p}^T \mathbf{x} \qquad \text{with } \mathbf{H} \succeq \mathbf{0} \qquad (14.12a)$$

$$\text{subject to: } \mathbf{A}\mathbf{x} \geq \mathbf{b} \qquad (14.12b)$$

which was studied in Chap. 13 (see Eq. (13.1)), can be formulated as

$$\text{minimize } \delta \qquad (14.13a)$$

$$\text{subject to: } \mathbf{x}^T \mathbf{H} \mathbf{x} + \mathbf{p}^T \mathbf{x} \leq \delta \qquad (14.13b)$$

$$\mathbf{A}\mathbf{x} \geq \mathbf{b} \qquad (14.13c)$$

where δ is an auxiliary scalar variable.

Since $\mathbf{H}$ is positive semidefinite, we can find a matrix $\hat{\mathbf{H}}$ such that $\mathbf{H} = \hat{\mathbf{H}}^T \hat{\mathbf{H}}$ (see proof of Theorem 7.2); hence the constraint in Eq. (14.13b) can be expressed as

$$\delta - \mathbf{p}^T \mathbf{x} - (\hat{\mathbf{H}}\mathbf{x})^T (\hat{\mathbf{H}}\mathbf{x}) \geq 0 \qquad (14.14)$$

It can be shown (see Prob. 14.4(a)) that the inequality in Eq. (14.14) holds if and only if

$$\mathbf{G}(\delta, \mathbf{x}) = \begin{bmatrix} \mathbf{I}_n & \hat{\mathbf{H}}\mathbf{x} \\ (\hat{\mathbf{H}}\mathbf{x})^T & \delta - \mathbf{p}^T \mathbf{x} \end{bmatrix} \succeq \mathbf{0} \qquad (14.15)$$

where $\mathbf{I}_n$ is the $n \times n$ identity matrix. Note that matrix $\mathbf{G}(\delta, \mathbf{x})$ is *affine* with respect to variables $\mathbf{x}$ and δ. In addition, the linear constraints in Eq. (14.13c) can be expressed as

$$\mathbf{F}(\mathbf{x}) = \mathbf{F}_0 + \sum_{i=1}^{n} x_i \mathbf{F}_i \succeq \mathbf{0} \qquad (14.16)$$

where the $\mathbf{F}_i$ for $0 \leq i \leq n$ are given by Eq. (14.11). Therefore, by defining an augmented vector

$$\hat{\mathbf{x}} = \begin{bmatrix} \delta \\ \mathbf{x} \end{bmatrix} \qquad (14.17)$$

the convex QP problem in Eq. (14.12) can be reformulated as the SDP problem

$$\text{minimize } \hat{\mathbf{c}}^T \hat{\mathbf{x}} \qquad (14.18a)$$

$$\text{subject to: } \mathbf{E}(\hat{\mathbf{x}}) \succeq \mathbf{0} \qquad (14.18b)$$

where $\hat{\mathbf{c}} \in R^{n+1}$ with

$$\hat{\mathbf{c}} = [1 \; 0 \; \cdots \; 0]^T \tag{14.19}$$

and

$$\mathbf{E}(\hat{\mathbf{x}}) = \mathrm{diag}\{\mathbf{G}(\delta, \; \mathbf{x}), \; \mathbf{F}(\mathbf{x})\}$$

(iii) *Convex QP Problems with Quadratic Constraints* Now let us consider the CP problem

$$\text{minimize } \mathbf{x}^T \mathbf{H} \mathbf{x} + \mathbf{p}^T \mathbf{x} \tag{14.20a}$$

$$\text{subject to:} \quad \mathbf{x}^T \mathbf{Q}_i \mathbf{x} + \mathbf{q}_i^T \mathbf{x} + r_i \leq 0 \qquad \text{for } i = 1, 2, \ldots, p \tag{14.20b}$$

where $\mathbf{H} \succeq \mathbf{0}$ and $\mathbf{Q}_i \succeq \mathbf{0}$ for $1 \leq i \leq p$. The class of problems represented by Eq. (14.20) covers the conventional convex QP problems represented by Eq. (14.12) as a subclass if $\mathbf{Q}_i = \mathbf{0}$ for all i. Again, by introducing an auxiliary scalar variable δ, the problem in Eq. (14.20) can be converted to

$$\text{minimize } \delta \tag{14.21a}$$

$$\text{subject to:} \quad \mathbf{x}^T \mathbf{H} \mathbf{x} + \mathbf{p}^T \mathbf{x} \leq \delta \tag{14.21b}$$

$$\mathbf{x}^T \mathbf{Q}_i \mathbf{x} + \mathbf{q}_i^T \mathbf{x} + r_i \leq 0 \qquad \text{for } 1 \leq i \leq p \tag{14.21c}$$

As in the convex QP case, the constraint in Eq. (14.21b) is equivalent to the constraint in Eq. (14.15) and the constraints in Eq. (14.21c) are equivalent to

$$\mathbf{F}_i(\mathbf{x}) = \begin{bmatrix} \mathbf{I}_n & \hat{\mathbf{Q}}_i \mathbf{x} \\ (\hat{\mathbf{Q}}_i \mathbf{x})^T & -\mathbf{q}_i^T \mathbf{x} - r_i \end{bmatrix} \succeq \mathbf{0} \qquad \text{for } 1 \leq i \leq p \tag{14.22}$$

where $\hat{\mathbf{Q}}_i$ is related to $\mathbf{Q}_i$ by the equation $\mathbf{Q}_i = \hat{\mathbf{Q}}_i^T \hat{\mathbf{Q}}_i$. Consequently, the quadratically constrained convex QP problem in Eq. (14.20) can be formulated as the SDP problem

$$\text{minimize } \hat{\mathbf{c}}^T \hat{\mathbf{x}} \tag{14.23a}$$

$$\text{subject to:} \quad \mathbf{E}(\hat{\mathbf{x}}) \succeq \mathbf{0} \tag{14.23b}$$

where $\hat{\mathbf{x}}$ and $\hat{\mathbf{c}}$ are given by Eqs. (14.17) and (14.19), respectively, and

$$\mathbf{E}(\hat{\mathbf{x}}) = \mathrm{diag}\{\mathbf{G}(\delta, \; \mathbf{x}), \; \mathbf{F}_1(\mathbf{x}), \; \mathbf{F}_2(\mathbf{x}), \; \ldots, \; \mathbf{F}_p(\mathbf{x})\}$$

where $\mathbf{G}(\delta, \; \mathbf{x})$ and $\mathbf{F}_i(\mathbf{x})$ are given by Eqs. (14.15) and (14.22), respectively.

There are many other types of CP problems that can be recast as SDP problems. One of them is the problem of minimizing the maximum eigenvalue of an affine matrix that can arise in structure optimization, control theory, and other areas [1][11]. This problem can be formulated as an SDP problem of the form in Eq. (14.4) (see Prob. 14.5). The reader is referred to [3][10] for more examples.

14.3 Basic Properties of SDP Problems
14.3.1 Basic assumptions

The feasible sets $\mathcal{F}_p$ and $\mathcal{F}_a$ for the primal and dual problems were defined in Sec. 14.2.1. A matrix $\mathbf{X}$ is said to be *strictly feasible* for the primal problem in Eq. (14.1) if it satisfies Eq. (14.1b) and $\mathbf{X} \succ 0$. Such a matrix $\mathbf{X}$ can be viewed as an *interior point* of $\mathcal{F}_p$. If we let

$$\mathcal{F}_p^o = \{\mathbf{X} : \mathbf{X} \succ 0, \ \mathbf{A}_i \cdot \mathbf{X} = b_i \text{ for } 1 \leq i \leq p\}$$

then $\mathcal{F}_p^o$ is the set of all interior points of $\mathcal{F}_p$ and $\mathbf{X}$ is strictly feasible for the primal if $\mathbf{X} \in \mathcal{F}_p^o$. Similarly, we can define the set of all interior points of $\mathcal{F}_d$ as

$$\mathcal{F}_d^o = \left\{ \{\mathbf{y}, \mathbf{S}\} : \sum_{i=1}^{p} y_i \mathbf{A}_i + \mathbf{S} = \mathbf{C}, \ \mathbf{S} \succ 0 \right\}$$

and a pair $\{\mathbf{y}, \mathbf{S}\}$ is said to be *strictly feasible* for the dual problem in Eq. (14.4) if $\{\mathbf{y}, \mathbf{S}\} \in \mathcal{F}_d^o$.

Unless otherwise stated, the following assumptions will be made in the rest of the chapter:

1. There exists a strictly feasible point $\mathbf{X}$ for the primal problem in Eq. (14.1) and a strictly feasible pair $\{\mathbf{y}, \mathbf{S}\}$ for the dual problem in Eq. (14.4). In other words, both $\mathcal{F}_p^o$ and $\mathcal{F}_d^o$ are nonempty.
2. Matrices $\mathbf{A}_i$ for $i = 1, 2, \ldots, p$ in Eq. (14.1b) are linearly independent, i.e., they span a p-dimensional linear space in $\mathcal{S}^n$.

The first assumption ensures that the optimization problem at hand can be tackled by using an *interior-point* approach. The second assumption, on the other hand, can be viewed as a matrix counterpart of the assumption made for the LP problem in Eq. (14.2) that the row vectors in matrix $\mathbf{A}$ in Eq. (14.2b) are linearly independent.

14.3.2 Karush-Kuhn-Tucker conditions

The Karush-Kuhn-Tucker (KKT) conditions for the SDP problem in Eq. (14.1) can be stated as follows: Matrix $\mathbf{X}^*$ is a minimizer of the problem in Eq. (14.1) if and only if there exist a matrix $\mathbf{S}^* \in \mathcal{S}^n$ and a vector $\mathbf{y}^* \in R^p$ such that

$$\sum_{i=1}^{p} y_i^* \mathbf{A}_i + \mathbf{S}^* = \mathbf{C} \tag{14.24a}$$

$$\mathbf{A}_i \cdot \mathbf{X}^* = b_i \qquad \text{for } 1 \leq i \leq p \tag{14.24b}$$

$$\mathbf{S}^* \mathbf{X}^* = 0 \tag{14.24c}$$

$$\mathbf{X}^* \succeq 0, \ \mathbf{S}^* \succeq 0 \tag{14.24d}$$

As noted in Sec. 14.2.1, if $\mathbf{A}_i = \text{diag}\{\mathbf{a}_i\}$ and $\mathbf{C} = \text{diag}\{\mathbf{c}\}$ with $\mathbf{a}_i \in R^n$ and $\mathbf{c} \in R^n$ for $1 \leq i \leq p$, the problem in Eq. (14.1) becomes a standard-form LP problem. In such a case, matrix $\mathbf{X}^*$ in Eq. (14.24) is also diagonal and the conditions in Eq. (14.24) become identical with those in Eq. (12.3), which are the KKT conditions for the LP problem in Eq. (14.2).

While the equations in (14.24a) and (14.24b) are linear, the complementarity constraint in Eq. (14.24c) is a nonlinear matrix equation. It can be shown that under the assumptions made in Sec. 14.3.1, the solution of Eq. (14.24) exists (see Theorem 3.1 of [3]). Furthermore, if we denote a solution of Eq. (14.24) as $\{\mathbf{X}^*, \mathbf{y}^*, \mathbf{S}^*\}$, then it can be readily verified that $\{\mathbf{y}^*, \mathbf{S}^*\}$ is a maximizer for the dual problem in Eq. (14.4). For these reasons, a set $\{\mathbf{X}^*, \mathbf{y}^*, \mathbf{S}^*\}$ satisfying Eq. (14.24) is called a *primal-dual solution*. It follows that $\{\mathbf{X}^*, \mathbf{y}^*, \mathbf{S}^*\}$ is a primal-dual solution if and only if $\mathbf{X}^*$ solves the primal problem in Eq. (14.1) and $\{\mathbf{y}^*, \mathbf{S}^*\}$ solves the dual problem in Eq. (14.4).

14.3.3 Central path

As we have seen in Chaps. 12 and 13, the concept of central path plays an important role in the development of interior-point algorithms for LP and QP problems. For the SDP problems in Eqs. (14.1) and (14.4), the central path consists of set $\{\mathbf{X}(\tau), \mathbf{y}(\tau), \mathbf{S}(\tau)\}$ such that for each $\tau > 0$ the equations

$$\sum_{i=1}^{p} y_i(\tau)\mathbf{A}_i + \mathbf{S}(\tau) = \mathbf{C} \tag{14.25a}$$

$$\mathbf{A}_i \cdot \mathbf{X}(\tau) = b_i \qquad \text{for } 1 \leq i \leq p \tag{14.25b}$$

$$\mathbf{X}(\tau)\mathbf{S}(\tau) = \tau\mathbf{I} \tag{14.25c}$$

$$\mathbf{S}(\tau) \succeq \mathbf{0}, \ \mathbf{X}(\tau) \succeq \mathbf{0} \tag{14.25d}$$

are satisfied.

Using Eqs. (14.8) and (14.25), the duality gap on the central path can be evaluated as

$$\delta[\mathbf{X}(\tau), \mathbf{y}(\tau)] = \mathbf{C} \cdot \mathbf{X}(\tau) - \mathbf{b}^T \mathbf{y}(\tau)$$

$$= \left[\sum_{i=1}^{p} y_i(\tau)\mathbf{A}_i + \mathbf{S}(\tau)\right] \cdot \mathbf{X}(\tau) - \mathbf{b}^T \mathbf{y}(\tau)$$

$$= \mathbf{S}(\tau) \cdot \mathbf{X}(\tau) = \text{trace}[\mathbf{S}(\tau)\mathbf{X}(\tau)]$$

$$= \text{trace}(\tau\mathbf{I}) = n\tau \tag{14.26}$$

which implies that

$$\lim_{\tau \to 0} \delta[\mathbf{X}(\tau), \mathbf{y}(\tau)] = 0$$

Therefore, the limiting set $\{\mathbf{X}^*, \mathbf{y}^*, \mathbf{S}^*\}$ obtained from $\mathbf{X}(\tau) \to \mathbf{X}^*$, $\mathbf{y}(\tau) \to \mathbf{y}^*$, and $\mathbf{S}(\tau) \to \mathbf{S}^*$ as $\tau \to 0$ is a primal-dual solution. This claim can also be confirmed by examining Eqs. (14.25a)–(14.25d) which, as τ approaches zero, become the KKT conditions in Eq. (14.24). In other words, as $\tau \to 0$, the central path approaches a primal-dual solution. In the subsequent sections, several algorithms will be developed to generate iterates that converge to a primal-dual solution by following the central path of the problem. Since $\mathbf{X}(\tau)$ and $\mathbf{S}(\tau)$ are positive semidefinite and satisfy Eqs. (14.24a) and (14.24b), respectively, $\mathbf{X}(\tau) \in \mathcal{F}_p$ and $\{\mathbf{y}(\tau), \mathbf{S}(\tau)\} \in \mathcal{F}_d$. Furthermore, the relaxed complementarity condition in Eq. (14.24c) implies that for each $\tau > 0$ both $\mathbf{X}(\tau)$ and $\mathbf{S}(\tau)$ are nonsingular; hence $\mathbf{X}(\tau) \succ \mathbf{0}$ and $\mathbf{S}(\tau) \succ \mathbf{0}$, which imply that $\mathbf{X}(\tau) \in \mathcal{F}_p^{\circ}$ and $\{\mathbf{y}(\tau), \mathbf{S}(\tau)\} \in \mathcal{F}_d^{\circ}$. In other words, for each $\tau > 0$, $\mathbf{X}(\tau)$ and $\{\mathbf{y}(\tau), \mathbf{S}(\tau)\}$ are in the *interior* of the feasible regions for the problems in Eqs. (14.1) and (14.4), respectively. Therefore, a path-following algorithm that generates iterates that follow the central path is intrinsically an interior-point algorithm.

14.3.4   Centering condition

On comparing Eqs. (14.24) and (14.25), we see that the only difference between the two systems of equations is that the complementarity condition in Eq. (14.24c) is relaxed in Eq. (14.25c). This equation is often referred to as the *centering condition* since the central path is parameterized by introducing variable τ in Eq. (14.25c). Obviously, if $\mathbf{X}(\tau) = \mathrm{diag}\{x_1(\tau), x_2(\tau), \ldots, x_n(\tau)\}$ and $\mathbf{S}(\tau) = \mathrm{diag}\{s_1(\tau), s_2(\tau), \ldots, s_n(\tau)\}$ as in LP problems, the centering condition is reduced to n scalar equations, i.e.,

$$x_i(\tau)s_i(\tau) = \tau \qquad \text{for } 1 \leq i \leq n \tag{14.27}$$

In general, the centering condition in Eq. (14.25c) involves n^2 nonlinear equations and, consequently, it is far more complicated than the condition in Eq. (14.27). In what follows, we describe a linear algebraic analysis to reveal the similarity between the general centering condition and the condition in Eq. (14.27) [8]. Since $\mathbf{X}(\tau)$ and $\mathbf{S}(\tau)$ are positive definite, their eigenvalues are strictly positive. Let $\delta_1(\tau) \geq \delta_2(\tau) \geq \cdots \geq \delta_n(\tau) > 0$ and $0 < \gamma_1(\tau) \leq \gamma_2(\tau) \leq \cdots \leq \gamma_n(\tau)$ be the eigenvalues of $\mathbf{X}(\tau)$ and $\mathbf{S}(\tau)$, respectively. There exists an orthogonal matrix $\mathbf{Q}(\tau)$ such that

$$\mathbf{X}(\tau) = \mathbf{Q}(\tau) \, \mathrm{diag}\{\delta_1(\tau), \delta_2(\tau), \ldots, \delta_n(\tau)\} \mathbf{Q}^T(\tau)$$

From Eq. (14.25c), it follows that

$$\mathbf{S}(\tau) = \tau \mathbf{X}^{-1}(\tau)$$
$$= \mathbf{Q}(\tau) \, \mathrm{diag}\left\{\frac{\tau}{\delta_1(\tau)}, \frac{\tau}{\delta_2(\tau)}, \ldots, \frac{\tau}{\delta_n(\tau)}\right\} \mathbf{Q}^T(\tau)$$
$$= \mathbf{Q}(\tau) \, \mathrm{diag}\{\gamma_1(\tau), \gamma_2(\tau), \ldots, \gamma_n(\tau)\} \mathbf{Q}^T(\tau)$$

458

which leads to

$$\delta_i(\tau)\gamma_i(\tau) = \tau \qquad \text{for } 1 \le i \le n \qquad (14.28)$$

As $\tau \to 0$, we have $\delta_i(\tau) \to \delta_i^*$ and $\gamma_i(\tau) \to \gamma_i^*$ where $\delta_1^* \ge \delta_2^* \ge \cdots \ge \delta_n^* > 0$ and $0 \le \gamma_1^* \le \gamma_2^* \le \cdots \le \gamma_n^*$ are the eigenvalues of $\mathbf{X}^*$ and $\mathbf{S}^*$, respectively, and Eq. (14.28) becomes

$$\delta_i^* \gamma_i^* = 0 \qquad \text{for } 1 \le i \le n \qquad (14.29)$$

We note that the relations between the eigenvalues of $\mathbf{X}(\tau)$ and $\mathbf{S}(\tau)$ as specified by Eq. (14.28) resemble the scalar centering conditions in Eq. (14.27). In addition, there is an interesting similarity between Eq. (14.29) and the complementarity conditions in LP problems (see Eq. (12.3c)).

14.4   Primal-Dual Path-Following Method
14.4.1   Reformulation of centering condition

A primal-dual path-following algorithm for SDP usually generates iterates by obtaining approximate solutions of Eq. (14.25) for a sequence of decreasing $\tau_i > 0$ for $k = 0, 1, \ldots$. If we let

$$\mathbf{G}(\mathbf{X}, \mathbf{y}, \mathbf{S}) = \begin{bmatrix} \sum_{i=1}^p y_i \mathbf{A}_i + \mathbf{S} - \mathbf{C} \\ \mathbf{A}_1 \cdot \mathbf{X} - b_1 \\ \vdots \\ \mathbf{A}_p \cdot \mathbf{X} - b_p \\ \mathbf{XS} - \tau \mathbf{I} \end{bmatrix} \qquad (14.30)$$

then Eqs. (14.25a) to (14.25c) can be expressed as

$$\mathbf{G}(\mathbf{X}, \mathbf{y}, \mathbf{S}) = \mathbf{0} \qquad (14.31)$$

We note that the domain of function $\mathbf{G}$ is in $\mathcal{S}^n \times R^p \times \mathcal{S}^n$ while the range of $\mathbf{G}$ is in $\mathcal{S}^n \times R^p \times R^{n \times n}$ simply because matrix $\mathbf{XS} - \tau \mathbf{I}$ is not symmetric in general although both $\mathbf{X}$ and $\mathbf{S}$ are symmetric. This domain inconsistency would cause difficulties if, for example, the Newton method were to be applied to Eq. (14.31) to obtain an approximate solution. Several approaches that deal with this nonsymmetrical problem are available, see, for example, [5]–[8]. In [8], Eq. (14.25c) is rewritten in symmetric form as

$$\mathbf{XS} + \mathbf{SX} = 2\tau \mathbf{I} \qquad (14.32)$$

Accordingly, function $\mathbf{G}$ in Eq. (14.30) is modified as

$$\mathbf{G}(\mathbf{X},\ \mathbf{y},\ \mathbf{S}) = \begin{bmatrix} \sum_{i=1}^{p} y_i \mathbf{A}_i + \mathbf{S} - \mathbf{C} \\ \mathbf{A}_1 \cdot \mathbf{X} - b_1 \\ \vdots \\ \mathbf{A}_p \cdot \mathbf{X} - b_p \\ \mathbf{XS} + \mathbf{SX} - 2\tau\mathbf{I} \end{bmatrix} \tag{14.33}$$

and its range is now in $\mathcal{S}^n \times R^p \times \mathcal{S}^n$. It can be shown that if $\mathbf{X} \succeq \mathbf{0}$ or $\mathbf{S} \succeq \mathbf{0}$, then Eqs. (14.25c) and (14.32) are equivalent (see Prob. 14.6).

In the Newton method, we start with a given set $\{\mathbf{X},\ \mathbf{y},\ \mathbf{S}\}$ and find increments $\Delta\mathbf{X}$, $\Delta\mathbf{y}$, and $\Delta\mathbf{S}$ with $\Delta\mathbf{X}$ and $\Delta\mathbf{S}$ symmetric such that set $\{\Delta\mathbf{X},\ \Delta\mathbf{y},\ \Delta\mathbf{S}\}$ satisfies the linearized equations

$$\sum_{i=1}^{p} \Delta y_i \mathbf{A}_i + \Delta\mathbf{S} = \mathbf{C} - \mathbf{S} - \sum_{i=1}^{p} y_i \mathbf{A}_i \tag{14.34a}$$

$$\mathbf{A}_i \cdot \Delta\mathbf{X} = b_i - \mathbf{A}_i \cdot \mathbf{X} \quad \text{for } 1 \leq i \leq p \tag{14.34b}$$

$$\mathbf{X}\Delta\mathbf{S} + \Delta\mathbf{S}\mathbf{X} + \Delta\mathbf{X}\mathbf{S} + \mathbf{S}\Delta\mathbf{X} = 2\tau\mathbf{I} - \mathbf{XS} - \mathbf{SX} \tag{14.34c}$$

Eq. (14.34) contains matrix equations with matrix variables $\Delta\mathbf{X}$ and $\Delta\mathbf{S}$. A mathematical operation known as *symmetric Kronecker product* [8] (see also Sec. A.14) turns out to be effective in dealing with this type of linear equations.

14.4.2   Symmetric Kronecker product

Given matrices $\mathbf{K}$, $\mathbf{M}$, and $\mathbf{N}$ in $R^{n \times n}$, the general asymmetric Kronecker product $\mathbf{M} \otimes \mathbf{N}$ with $\mathbf{M} = \{m_{ij}\}$ is defined as

$$\mathbf{M} \otimes \mathbf{N} = \begin{bmatrix} m_{11}\mathbf{N} & \cdots & m_{1n}\mathbf{N} \\ \vdots & & \vdots \\ m_{n1}\mathbf{N} & \cdots & m_{nn}\mathbf{N} \end{bmatrix}$$

(see Sec. A.14). To deal with matrix variables, it is sometimes desirable to represent a matrix $\mathbf{K}$ as a vector, denoted as $\text{nvec}(\mathbf{K})$, which stacks the columns of $\mathbf{K}$. It can be readily verified that

$$(\mathbf{M} \otimes \mathbf{N})\text{nvec}(\mathbf{K}) = \text{nvec}(\mathbf{NKM}^T) \tag{14.35}$$

The usefulness of Eq. (14.35) is that if a matrix equation involves terms like $\mathbf{NKM}^T$, where $\mathbf{K}$ is a matrix variable, then Eq. (14.35) can be used to convert $\mathbf{NKM}^T$ into a vector variable multiplied by a known matrix.

If a matrix equation contains a symmetric term given by $(\mathbf{NKM}^T + \mathbf{MKN}^T)/2$ where $\mathbf{K} \in \mathcal{S}^n$ is a matrix variable, then the term can be readily handled using the *symmetric Kronecker product* of $\mathbf{M}$ and $\mathbf{N}$, denoted as $\mathbf{M} \odot \mathbf{N}$, which is defined by the identity

$$(\mathbf{M} \odot \mathbf{N})\mathrm{svec}(\mathbf{K}) = \mathrm{svec}[\tfrac{1}{2}(\mathbf{NKM}^T + \mathbf{MKN}^T)] \tag{14.36}$$

where $\mathrm{sevc}(\mathbf{K})$ converts symmetric matrix $\mathbf{K} = \{k_{ij}\}$ into a vector of dimension $n(n+1)/2$ as

$$\mathrm{svec}(\mathbf{K}) = [k_{11} \ \sqrt{2}k_{12} \ \cdots \ \sqrt{2}k_{1n} \ k_{22} \ \sqrt{2}k_{23} \ \cdots \ \sqrt{2}k_{2n} \ \cdots \ k_{nn}]^T \tag{14.37}$$

Note that the standard inner product of $\mathbf{A}$ and $\mathbf{B}$ in $\mathcal{S}^n$ can be expressed as the standard inner product of vectors $\mathrm{svec}(\mathbf{A})$ and $\mathrm{svec}(\mathbf{B})$, i.e.,

$$\mathbf{A} \cdot \mathbf{B} = \mathrm{svec}(\mathbf{A})^T \mathrm{svec}(\mathbf{B}) \tag{14.38}$$

If we use a matrix $\mathbf{K} = \{k_{ij}\}$ with only one nonzero element k_{ij} for $1 \le i \le j \le n$, then Eq. (14.36) can be used to obtain each column of $\mathbf{M} \odot \mathbf{N}$. Based on this observation, a simple algorithm can be developed to obtain the $n(n+1)/2$-dimensional matrix $\mathbf{M} \odot \mathbf{N}$ (see Prob. 14.8). The following lemma describes an explicit relation between the eigenvalues and eigenvectors of $\mathbf{M} \odot \mathbf{N}$ and the eigenvalues and eigenvectors of $\mathbf{M}$ and $\mathbf{N}$ (see Prob. 14.9).

Lemma 14.1 *If* $\mathbf{M}$ *and* $\mathbf{N}$ *are symmetric matrices satisfying the relation* $\mathbf{MN} = \mathbf{NM}$, *then the* $n(n+1)/2$ *eigenvalues of* $\mathbf{M} \odot \mathbf{N}$ *are given by*

$$\tfrac{1}{2}(\alpha_i\beta_j + \beta_i\alpha_j) \qquad \text{for } 1 \le i \le j \le n$$

and the corresponding orthonormal eigenvectors are given by

$$\begin{aligned} \mathrm{svec}(\mathbf{v}_i\mathbf{v}_i^T) & \qquad \text{if } i = j \\ \tfrac{1}{\sqrt{2}}\mathrm{svec}(\mathbf{v}_i\mathbf{v}_j^T + \mathbf{v}_j\mathbf{v}_i^T) & \qquad \text{if } i < j \end{aligned}$$

where α_i *for* $1 \le i \le n$ *and* β_j *for* $1 \le j \le n$ *are the eigenvalues of* $\mathbf{M}$ *and* $\mathbf{N}$, *respectively, and* $\mathbf{v}_i$ *for* $1 \le i \le n$ *is a common basis of orthonormal eigenvectors of* $\mathbf{M}$ *and* $\mathbf{N}$.

14.4.3 Reformulation of Eq. (14.34)

Eq. (14.34c) can be expressed in terms of the symmetric Kronecker product as

$$(\mathbf{X} \odot \mathbf{I})\mathrm{svec}(\mathbf{\Delta S}) + (\mathbf{S} \odot \mathbf{I})\mathrm{svec}(\mathbf{\Delta X}) = \mathrm{svec}[\tau\mathbf{I} - \tfrac{1}{2}(\mathbf{XS} + \mathbf{SX})] \tag{14.39}$$

For the sake of simplicity, we denote

$$\text{svec}(\mathbf{\Delta X}) = \mathbf{\Delta x} \tag{14.40a}$$

$$\text{svec}(\mathbf{\Delta S}) = \mathbf{\Delta s} \tag{14.40b}$$

$$\mathbf{S} \odot \mathbf{I} = \mathbf{E} \tag{14.40c}$$

$$\mathbf{X} \odot \mathbf{I} = \mathbf{F} \tag{14.40d}$$

$$\text{svec}[\tau \mathbf{I} - \tfrac{1}{2}(\mathbf{XS} + \mathbf{SX})] = \mathbf{r}_c \tag{14.40e}$$

With this notation, Eq. (14.39) becomes

$$\mathbf{E}\mathbf{\Delta x} + \mathbf{F}\mathbf{\Delta s} = \mathbf{r}_c$$

To simplify Eqs. (14.34a) and (14.34b), we let

$$\mathbf{A} = \begin{bmatrix} [\text{svec}(\mathbf{A}_1)]^T \\ [\text{svec}(\mathbf{A}_2)]^T \\ \vdots \\ [\text{svec}(\mathbf{A}_p)]^T \end{bmatrix} \tag{14.41a}$$

$$\mathbf{x} = \text{svec}(\mathbf{X}) \tag{14.41b}$$

$$\mathbf{y} = [y_1 \ y_2 \ \cdots \ y_p]^T \tag{14.41c}$$

$$\mathbf{\Delta y} = [\Delta y_1 \ \Delta y_2 \ \cdots \ \Delta y_p]^T \tag{14.41d}$$

$$\mathbf{r}_p = \mathbf{b} - \mathbf{A}\mathbf{x} \tag{14.41e}$$

$$\mathbf{r}_d = \text{svec}[\mathbf{C} - \mathbf{S} - \text{mat}(\mathbf{A}^T\mathbf{y})] \tag{14.41f}$$

where $\text{mat}(\cdot)$ is the inverse of $\text{svec}(\cdot)$. With the use of Eqs. (14.41a)–(14.41f), Eqs. (14.34a) and (14.34b) can now be written as

$$\mathbf{A}^T\mathbf{\Delta y} + \mathbf{\Delta s} = \mathbf{r}_d$$

$$\mathbf{A}\mathbf{\Delta x} = \mathbf{r}_p$$

and, therefore, Eq. (14.34) can be reformulated as

$$\mathbf{J} \begin{bmatrix} \mathbf{\Delta x} \\ \mathbf{\Delta y} \\ \mathbf{\Delta s} \end{bmatrix} = \begin{bmatrix} \mathbf{r}_d \\ \mathbf{r}_p \\ \mathbf{r}_c \end{bmatrix} \tag{14.42}$$

where

$$\mathbf{J} = \begin{bmatrix} 0 & \mathbf{A}^T & \mathbf{I} \\ \mathbf{A} & 0 & 0 \\ \mathbf{E} & 0 & \mathbf{F} \end{bmatrix}$$

It can be readily verified that the solution of Eq. (14.42) is given by

$$\mathbf{\Delta x} = -\mathbf{E}^{-1}[\mathbf{F}(\mathbf{r}_d - \mathbf{A}^T\mathbf{\Delta y}) - \mathbf{r}_c] \tag{14.43a}$$

$$\mathbf{\Delta s} = \mathbf{r}_d - \mathbf{A}^T\mathbf{\Delta y} \tag{14.43b}$$

$$\mathbf{M}\mathbf{\Delta y} = \mathbf{r}_p + \mathbf{A}\mathbf{E}^{-1}(\mathbf{F}\mathbf{r}_d - \mathbf{r}_c) \tag{14.43c}$$

where matrix $\mathbf{M}$, which is known as the *Schur complement* matrix, is given by

$$\mathbf{M} = \mathbf{A}\mathbf{E}^{-1}\mathbf{F}\mathbf{A}^{T}$$

From Eq. (14.43), we see that solving the system of linear equations in Eq. (14.42) involves evaluating $\mathbf{E}^{-1}$ and computing $\mathbf{\Delta y}$ from the linear system in Eq. (14.43c). Hence the computational complexity is mainly determined by the computations required to solve the system in Eq. (14.43c) [8]. Matrix $\mathbf{J}$ in Eq. (14.42) is actually the Jacobian matrix of function $\mathbf{G}$ defined by Eq. (14.33). From Eq. (14.43), it can be shown that $\mathbf{J}$ is nonsingular (i.e., Eq. (14.42) has a unique solution) if and only if $\mathbf{M}$ is nonsingular. It can also be shown that if $\mathbf{XS} + \mathbf{SX} \succ \mathbf{0}$ then $\mathbf{M}$ is nonsingular [12]. Therefore, $\mathbf{XS} + \mathbf{SX} \succ \mathbf{0}$ is a sufficient condition for Eq. (14.43) to have a unique solution set $\{\mathbf{\Delta x}, \mathbf{\Delta y}, \mathbf{\Delta s}\}$.

14.4.4 Primal-dual path-following algorithm

The above analysis leads to the following algorithm.

Algorithm 14.1 Primal-dual path-following algorithm for SDP problems
Step 1
Input $\mathbf{A}_i$ for $1 \leq i \leq p$, $\mathbf{b} \in R^p$, $\mathbf{C} \in R^{n \times n}$, and a strictly feasible set $\{\mathbf{X}_p, \mathbf{y}_0, \mathbf{S}_0\}$ that satisfies Eqs. (14.1b) and (14.4b) with $\mathbf{X}_0 \succ \mathbf{0}$ and $\mathbf{S}_0 \succ \mathbf{0}$.
Choose a scalar σ in the range $0 \leq \sigma < 1$.
Set $k = 0$ and initialize the tolerance ε for the duality gap δ_k.
Step 2
Compute

$$\delta_k = \frac{\mathbf{X}_k \cdot \mathbf{S}_k}{n}$$

Step 3
If $\delta_k \leq \varepsilon$, output solution $\{\mathbf{X}_k, \mathbf{y}_k, \mathbf{S}_k\}$ and stop; otherwise, set

$$\tau_k = \sigma \frac{\mathbf{X}_k \cdot \mathbf{S}_k}{n} \qquad (14.44)$$

and continue with Step 4.
Step 4
Solve Eq. (14.42) using Eqs. (14.43a)–(14.43c) where $\mathbf{X} = \mathbf{X}_k, \mathbf{y} = \mathbf{y}_k$, $\mathbf{S} = \mathbf{S}_k$, and $\tau = \tau_k$.
Convert the solution $\{\mathbf{\Delta x}, \mathbf{\Delta y}, \mathbf{\Delta s}\}$ into $\{\mathbf{\Delta X}, \mathbf{\Delta y}, \mathbf{\Delta S}\}$ with $\mathbf{\Delta X} = \text{mat}(\mathbf{\Delta x})$ and $\mathbf{\Delta S} = \text{mat}(\mathbf{\Delta s})$.
Step 5
Choose a parameter γ in the range $0 < \gamma < 1$ and determine parameters α and β as

$$\alpha = \min(1, \ \gamma\hat{\alpha}) \tag{14.45a}$$

$$\beta = \min(1, \ \gamma\hat{\beta}) \tag{14.45b}$$

where

$$\hat{\alpha} = \max_{\mathbf{X}_k + \bar{\alpha}\mathbf{\Delta X} \succeq \mathbf{0}} (\bar{\alpha}) \quad \text{and} \quad \hat{\beta} = \max_{\mathbf{S}_k + \bar{\beta}\mathbf{\Delta S} \succeq \mathbf{0}} (\bar{\beta})$$

Step 6 Set

$$\mathbf{X}_{k+1} = \mathbf{X}_k + \alpha\mathbf{\Delta X} \tag{14.46a}$$

$$\mathbf{y}_{k+1} = \mathbf{y}_k + \beta\mathbf{\Delta y} \tag{14.46b}$$

$$\mathbf{S}_{k+1} = \mathbf{S}_k + \beta\mathbf{\Delta S} \tag{14.46c}$$

Set $k = k + 1$ and repeat from Step 2.

A couple of remarks on Step 5 of the algorithm are in order. First, it follows from Eq. (14.45) that if the increments $\mathbf{\Delta X}$, $\mathbf{\Delta y}$, and $\mathbf{\Delta S}$ obtained in Step 4 are such that $\mathbf{X}_k + \mathbf{\Delta X} \in \mathcal{F}_p^o$ and $\{\mathbf{y}_k + \mathbf{\Delta y}, \ \mathbf{S}_k + \mathbf{\Delta S}\} \in \mathcal{F}_d^o$, then we should use $\alpha = 1$ and $\beta = 1$. Otherwise, we should use $\alpha = \gamma\hat{\alpha}$ and $\beta = \gamma\hat{\beta}$ where $0 < \gamma < 1$ to ensure that $\mathbf{X}_{k+1} \in \mathcal{F}_p^o$ and $\{\mathbf{y}_{k+1}, \ \mathbf{S}_{k+1}\} \in \mathcal{F}_d^o$. Typically, a γ in the range $0.9 \leq \gamma \leq 0.99$ works well in practice. Second, the numerical values of $\hat{\alpha}$ and $\hat{\beta}$ can be determined using the eigendecomposition of symmetric matrices as follows. Since $\mathbf{X}_k \succ \mathbf{0}$, the Cholesky decomposition (see Sec. A.13) of $\mathbf{X}_k$ gives

$$\mathbf{X}_k = \hat{\mathbf{X}}_k^T \hat{\mathbf{X}}_k$$

Now if we perform an eigendecomposition of the symmetric matrix

$$(\hat{\mathbf{X}}_k^T)^{-1}\mathbf{\Delta X}\hat{\mathbf{X}}_k^{-1}$$

as

$$(\hat{\mathbf{X}}_k^T)^{-1}\mathbf{\Delta X}\hat{\mathbf{X}}_k^{-1} = \mathbf{U}^T\mathbf{\Lambda}\mathbf{U}$$

where $\mathbf{U}$ is orthogonal and $\mathbf{\Lambda} = \text{diag}\{\lambda_1, \ \lambda_2, \ \ldots, \ \lambda_n\}$, we get

$$\begin{aligned}
\mathbf{X}_k + \bar{\alpha}\mathbf{\Delta X} &= \hat{\mathbf{X}}_k^T[\mathbf{I} + \bar{\alpha}(\hat{\mathbf{X}}_k^T)^{-1}\mathbf{\Delta X}\hat{\mathbf{X}}_k^{-1}]\hat{\mathbf{X}}_k \\
&= \hat{\mathbf{X}}_k^T(\mathbf{I} + \bar{\alpha}\mathbf{U}^T\mathbf{\Lambda}\mathbf{U})\hat{\mathbf{X}}_k \\
&= (\mathbf{U}\hat{\mathbf{X}}_k)^T(\mathbf{I} + \bar{\alpha}\mathbf{\Lambda})(\mathbf{U}\hat{\mathbf{X}}_k)
\end{aligned}$$

Hence $\mathbf{X}_k\bar{\alpha}\mathbf{\Delta X} \succeq \mathbf{0}$ if and only if $\mathbf{I} + \bar{\alpha}\mathbf{\Lambda} = \text{diag}\{1 + \bar{\alpha}\lambda_1, 1 + \bar{\alpha}\lambda_2, \ \ldots, \ 1 + \bar{\alpha}\lambda_n\} \succeq \mathbf{0}$. If $\min\{\lambda_i\} \geq 0$, then $\mathbf{I} + \bar{\alpha}\mathbf{\Lambda} \succeq \mathbf{0}$ holds for any $\bar{\alpha} \geq 0$; otherwise, the largest $\bar{\alpha}$ to assure the positive definiteness of $\mathbf{I} + \bar{\alpha}\mathbf{\Lambda}$ is given by

$$\hat{\alpha} = \frac{1}{\max_i(-\lambda_i)} \tag{14.47}$$

Therefore, the numerical value of α in Eq. (14.45a) can be obtained as

$$\alpha = \begin{cases} 1 & \text{if all } \lambda_i \geq 0 \\ \min(1, \, \gamma\hat{\alpha}) & \text{otherwise} \end{cases} \tag{14.48}$$

where $\hat{\alpha}$ is determined using Eq. (14.47). Similarly, the numerical value of β in Eq. (14.45b) can be obtained as

$$\beta = \begin{cases} 1 & \text{if all } \mu_i \geq 0 \\ \min(1, \, \gamma\hat{\beta}) & \text{otherwise} \end{cases} \tag{14.49}$$

where the μ_i's are the eigenvalues of $(\hat{\mathbf{S}}_k^T)^{-1}\mathbf{\Delta}\mathbf{S}\hat{\mathbf{S}}_k^{-1}$ with $\mathbf{S}_k = \hat{\mathbf{S}}_k^T\hat{\mathbf{S}}_k$ and

$$\hat{\beta} = \frac{1}{\max_i(-\mu_i)}$$

The numerical value of the centering parameter σ should be in the range of $[0, 1)$. For small-scale applications, the choice

$$\sigma = \frac{n}{15\sqrt{n} + n} \tag{14.50}$$

is usually satisfactory.

Example 14.1 Find scalars y_1, y_2, and y_3 such that the maximum eigenvalue of $\mathbf{F} = \mathbf{A}_0 + y_1\mathbf{A}_1 + y_2\mathbf{A}_2 + y_3\mathbf{A}_3$ with

$$\mathbf{A}_0 = \begin{bmatrix} 2 & -0.5 & -0.6 \\ -0.5 & 2 & 0.4 \\ -0.6 & 0.4 & 3 \end{bmatrix}, \quad \mathbf{A}_1 = \begin{bmatrix} 0 & 1 & 0 \\ 1 & 0 & 0 \\ 0 & 0 & 0 \end{bmatrix}$$

$$\mathbf{A}_2 = \begin{bmatrix} 0 & 0 & 1 \\ 0 & 0 & 0 \\ 1 & 0 & 0 \end{bmatrix}, \quad \mathbf{A}_3 = \begin{bmatrix} 0 & 0 & 0 \\ 0 & 0 & 1 \\ 0 & 1 & 0 \end{bmatrix}$$

is minimized.

Solution This problem can be formulated as the SDP problem

$$\text{maximize } \mathbf{b}^T \mathbf{y} \tag{14.51a}$$

$$\text{subject to:} \quad \sum_{i=1}^{4} y_i \mathbf{A}_i + \mathbf{S} = \mathbf{C} \tag{14.51b}$$

$$\mathbf{S} \succeq \mathbf{0} \tag{14.51c}$$

where $\mathbf{b} = [0\ 0\ 0\ 1]^T$, $\mathbf{y} = [y_1\ y_2\ y_3\ y_4]^T$, $\mathbf{C} = -\mathbf{A}_0$, $\mathbf{A}_4 = \mathbf{I}$, and $-y_4$ is the maximum eigenvalue of matrix $\mathbf{F}$ (see Prob. 14.5). We observe that the optimization problem in Eq. (14.51) is of the type described by Eq. (14.4) with $n = 3$ and $p = 4$.

It is easy to verify that the set $\{\mathbf{X}_0,\ \mathbf{y}_0,\ \mathbf{S}_0\}$ with

$$\mathbf{X}_0 = \tfrac{1}{3}\mathbf{I}, \ \mathbf{y}_0 = [0.2\ 0.2\ 0.2\ -4]^T \quad \text{and} \quad \mathbf{S}_0 = \begin{bmatrix} 2 & 0.3 & 0.4 \\ 0.3 & 2 & -0.6 \\ 0.4 & -0.6 & 1 \end{bmatrix}$$

is strictly feasible for the associated primal-dual problems. The matrix $\mathbf{A}$ in Eq. (14.41a) is in this case a 4×6 matrix given by

$$\mathbf{A} = \begin{bmatrix} 0 & \sqrt{2} & 0 & 0 & 0 & 0 \\ 0 & 0 & \sqrt{2} & 0 & 0 & 0 \\ 0 & 0 & 0 & 0 & \sqrt{2} & 0 \\ 1 & 0 & 0 & 1 & 0 & 1 \end{bmatrix}$$

At the initial point, the maximum eigenvalue of $\mathbf{F}$ is 3.447265. With $\sigma = n/(15\sqrt{n} + n) = 0.1035$, $\gamma = 0.9$, and $\varepsilon = 10^{-3}$, it took Algorithm 14.1 four iterations and 26 Kflops to converge to the solution set $\{\mathbf{X}^*,\ \mathbf{y}^*,\ \mathbf{S}^*\}$ where

$$\mathbf{y}^* = \begin{bmatrix} 0.392921 \\ 0.599995 \\ -0.399992 \\ -3.000469 \end{bmatrix}$$

By using the first three components of $\mathbf{y}^*$, i.e., y_1, y_2, and y_3, the maximum eigenvalue of $\mathbf{F} = \mathbf{A}_0 + y_1 \mathbf{A}_1 + y_2 \mathbf{A}_2 + y_3 \mathbf{A}_3$ is found to be 3. ∎

14.5 Predictor-Corrector Method

The algorithm studied in Sec. 14.4 can be improved by incorporating a predictor-corrector rule proposed by Mehrotra [13] for LP problems (see Sec. 12.5.3).

As in the LP case, there are two steps in each iteration of a predictor-corrector method. Let us assume that we are now in the kth iteration of the algorithm. In the first step, a *predictor direction* $\{\boldsymbol{\Delta}\mathbf{X}^{(p)},\ \boldsymbol{\Delta}\mathbf{y}^{(p)},\ \boldsymbol{\Delta}\mathbf{S}^{(p)}\}$ is first identified by using a linear approximation of the KKT conditions. This set $\{\mathbf{\Delta X}^{(p)},\ \mathbf{\Delta y}^{(p)},\ \mathbf{\Delta S}^{(p)}\}$ can be obtained by setting $\tau = 0$ in Eq. (14.40e) to obtain

$$\mathbf{r}_c = \text{svec}\left[-\tfrac{1}{2}(\mathbf{X}_k\mathbf{S}_k + \mathbf{S}_k\mathbf{X}_k)\right] \qquad (14.52)$$

and then using Eq. (14.43). Next, the numerical values of α_p and β_p can be determined as

$$\alpha_p = \min(1,\ \gamma\hat{\alpha}) \qquad (14.53\text{a})$$
$$\beta_p = \min(1,\ \gamma\hat{\beta}) \qquad (14.53\text{b})$$

where

$$\hat{\alpha} = \max_{\mathbf{X}_k + \bar{\alpha}\mathbf{\Delta X}^{(p)} \succeq \mathbf{0}} (\bar{\alpha})$$
$$\hat{\beta} = \max_{\mathbf{S}_k + \bar{\beta}\mathbf{\Delta S}^{(p)} \succeq \mathbf{0}} (\bar{\beta})$$

in a way similar to that described in Eqs. (14.48) and (14.49). The centering parameter σ_k is then computed as

$$\sigma_k = \left[\frac{(\mathbf{X}_k + \alpha_p\mathbf{\Delta X}^{(p)}) \cdot (\mathbf{S}_k + \beta_p\mathbf{\Delta S}^{(p)})}{\mathbf{X}_k \cdot \mathbf{S}_k}\right]^3 \qquad (14.54)$$

and is used to determine the value of τ_k in Eq. (14.44), i.e.,

$$\tau_k = \sigma_k \frac{\mathbf{X}_k \cdot \mathbf{S}_k}{n} \qquad (14.55)$$

In the second step, the parameter τ_k in Eq. (14.55) is utilized to compute

$$\mathbf{r}_c = \tau_k\mathbf{I} - \tfrac{1}{2}(\mathbf{X}_k\mathbf{S}_k + \mathbf{S}_k\mathbf{X}_k + \mathbf{\Delta X}^{(p)}\mathbf{\Delta S}^{(p)} + \mathbf{\Delta S}^{(p)}\mathbf{\Delta X}^{(p)}) \qquad (14.56)$$

and the vector $\mathbf{r}_c$ in Eq. (14.56) is then used in Eq. (14.43) to obtain the *corrector direction* $\{\mathbf{\Delta X}^{(c)},\ \mathbf{\Delta y}^{(c)},\ \mathbf{\Delta S}^{(c)}\}$. The set $\{\mathbf{X}_k,\ \mathbf{y}_k,\ \mathbf{S}_k\}$ is then updated as

$$\mathbf{X}_{k+1} = \mathbf{X}_k + \alpha_c\mathbf{\Delta X}^{(c)} \qquad (14.57\text{a})$$
$$\mathbf{y}_{k+1} = \mathbf{y}_k + \beta_c\mathbf{\Delta y}^{(c)} \qquad (14.57\text{b})$$
$$\mathbf{S}_{k+1} = \mathbf{S}_k + \beta_c\mathbf{\Delta S}^{(c)} \qquad (14.57\text{c})$$

where α_c and β_c are given by

$$\alpha_c = \min(1,\ \gamma\hat{\alpha}) \qquad (14.58\text{a})$$
$$\beta_c = \min(1,\ \gamma\hat{\beta}) \qquad (14.58\text{b})$$

where

$$\hat{\alpha} = \max_{\mathbf{X}_k + \bar{\alpha}\boldsymbol{\Delta}\mathbf{X}^{(c)} \succeq 0} (\bar{\alpha})$$

$$\hat{\beta} = \max_{\mathbf{S}_k + \bar{\beta}\boldsymbol{\Delta}\mathbf{S}^{(c)} \succeq 0} (\bar{\beta})$$

The above approach can be implemented in terms of the following algorithm.

Algorithm 14.2 Predictor-corrector algorithm for SDP problems
Step 1
Input $\mathbf{A}_i$ for $1 \le i \le p$, $\mathbf{b} \in R^p$, and $\mathbf{C} \in R^{n \times n}$, and a strictly feasible set $\{\mathbf{X}_0, \mathbf{y}_0, \mathbf{S}_0\}$ that satisfies Eqs. (14.1b) and (14.4b) with $\mathbf{X}_0 \succ \mathbf{0}$ and $\mathbf{S}_0 \succ \mathbf{0}$. Set $k = 0$ and initialize the tolerance ε for the duality gap δ_k.
Step 2
Compute

$$\delta_k = \frac{\mathbf{X}_k \cdot \mathbf{S}_k}{n}$$

Step 3
If $\delta_k \le \varepsilon$, output solution $\{\mathbf{X}_k, \mathbf{y}_k, \mathbf{S}_k\}$ and stop; otherwise, continue with Step 4.
Step 4
Compute $\{\boldsymbol{\Delta}\mathbf{X}^{(p)}, \boldsymbol{\Delta}\mathbf{y}^{(p)}, \boldsymbol{\Delta}\mathbf{S}^{(p)}\}$ using Eq. (14.43) with $\mathbf{X} = \mathbf{X}_k$, $\mathbf{y} = \mathbf{y}_k$, $\mathbf{S} = \mathbf{S}_k$, and $\mathbf{r}_c$ given by Eq (14.52).
Choose a parameter γ in the range $0 < \gamma \le 1$ and compute α_p and β_p using Eq. (14.53) and evaluate σ_k using Eq. (14.54).
Compute τ_k using Eq. (14.55).
Step 5
Compute $\{\boldsymbol{\Delta}\mathbf{X}^{(c)}, \boldsymbol{\Delta}\mathbf{y}^{(c)}, \boldsymbol{\Delta}\mathbf{S}^{(c)}\}$ using Eq. (14.43) with $\mathbf{X} = \mathbf{X}_k$, $\mathbf{y} = \mathbf{y}_k$, $\mathbf{S} = \mathbf{S}_k$, and $\mathbf{r}_c$ given by Eq. (14.56).
Step 6
Compute α_c and β_c using Eq. (14.58).
Step 7
Obtain set $\{\mathbf{X}_{k+1}, \mathbf{y}_{k+1}, \mathbf{S}_{k+1}\}$ using Eq. (14.57).
Set $k = k + 1$ and repeat from Step 2.

Example 14.2 Apply Algorithm 14.2 to the shortest distance problem in Example 13.5.

Solution From Sec. 14.2.2, we can first formulate the problem as a CP problem of the form given by Eq. (14.21), i.e.,

$$\text{minimize } \delta$$
$$\text{subject to:} \quad \mathbf{x}^T \mathbf{H} \mathbf{x} \leq \delta$$
$$\mathbf{x}^T \mathbf{Q}_1 \mathbf{x} + \mathbf{q}_1^T \mathbf{x} + r_1 \leq 0$$
$$\mathbf{x}^T \mathbf{Q}_2 \mathbf{x} + \mathbf{q}_2^T \mathbf{x} + r_2 \leq 0$$

with

$$\mathbf{x} = \begin{bmatrix} x_1 \\ x_2 \\ x_3 \\ x_4 \end{bmatrix}, \quad \mathbf{H} = \begin{bmatrix} 1 & 0 & -1 & 0 \\ 0 & 1 & 0 & -1 \\ -1 & 0 & 1 & 0 \\ 0 & -1 & 0 & 1 \end{bmatrix}$$

$$\mathbf{Q}_1 = \begin{bmatrix} \frac{1}{4} & 0 \\ 0 & 1 \end{bmatrix}, \quad \mathbf{Q}_2 = \begin{bmatrix} 5 & 3 \\ 3 & 5 \end{bmatrix}, \quad \mathbf{q}_1 = \begin{bmatrix} -\frac{1}{2} \\ 0 \end{bmatrix}, \quad \mathbf{q}_2 = \begin{bmatrix} -44 \\ -52 \end{bmatrix}$$

$$r_1 = -\frac{3}{4}, \quad \text{and} \quad r_2 = 140$$

The above CP problem can be converted into the SDP problem in Eq. (14.23) with

$$\hat{\mathbf{c}} = \begin{bmatrix} 0 \\ 0 \\ 0 \\ 0 \\ 1 \end{bmatrix}, \quad \hat{\mathbf{x}} = \begin{bmatrix} x_1 \\ x_2 \\ x_3 \\ x_4 \\ \delta \end{bmatrix}$$

and

$$\mathbf{E}(\hat{\mathbf{x}}) = \text{diag}\{\mathbf{G}(\delta, \ \mathbf{x}), \ \mathbf{F}_1(\mathbf{x}), \ \mathbf{F}_2(\mathbf{x})\}$$

with

$$\mathbf{G}(\delta, \ \mathbf{x}) = \begin{bmatrix} \mathbf{I}_4 & \hat{\mathbf{H}}\mathbf{x} \\ (\hat{\mathbf{H}}\mathbf{x})^T & \delta \end{bmatrix}, \quad \hat{\mathbf{H}} = \begin{bmatrix} 1 & 0 & -1 & 0 \\ 0 & -1 & 0 & 1 \\ 0 & 0 & 0 & 0 \\ 0 & 0 & 0 & 0 \end{bmatrix}$$

$$\mathbf{F}_1(\mathbf{x}) = \begin{bmatrix} \mathbf{I}_2 & \hat{\mathbf{Q}}_1 \mathbf{x}_a \\ (\hat{\mathbf{Q}}_1 \mathbf{x}_a)^T & -\mathbf{q}_1^T \mathbf{x}_a - r_1 \end{bmatrix}, \quad \hat{\mathbf{Q}}_1 = \begin{bmatrix} \frac{1}{2} & 0 \\ 0 & 1 \end{bmatrix}, \quad \mathbf{x}_a = \begin{bmatrix} x_1 \\ x_2 \end{bmatrix}$$

$$\mathbf{F}_2(\mathbf{x}) = \begin{bmatrix} \mathbf{I}_2 & \hat{\mathbf{Q}}_2 \mathbf{x}_b \\ (\hat{\mathbf{Q}}_2 \mathbf{x}_b)^T & -\mathbf{q}_2^T \mathbf{x}_b - r_2 \end{bmatrix}, \quad \hat{\mathbf{Q}}_2 = \begin{bmatrix} 2 & 2 \\ -1 & 1 \end{bmatrix}, \quad \mathbf{x}_b = \begin{bmatrix} x_3 \\ x_4 \end{bmatrix}$$

The SDP problem in Eq. (14.23) is equivalent to the standard SDP problem in Eq. (14.4) with $p = 5$, $n = 11$,

$$\mathbf{y} = \begin{bmatrix} x_1 \\ x_2 \\ x_3 \\ x_4 \\ \delta \end{bmatrix}, \quad \mathbf{b} = \begin{bmatrix} 0 \\ 0 \\ 0 \\ 0 \\ -1 \end{bmatrix}$$

Matrices $\mathbf{A}_i$ for $1 \leq i \leq 5$ and $\mathbf{C}$ are given by

$$\mathbf{A}_1 = -\text{diag}\left\{\begin{bmatrix} \mathbf{0}_4 & \mathbf{h}_1 \\ \mathbf{h}_1^T & 0 \end{bmatrix}, \begin{bmatrix} \mathbf{0}_2 & \mathbf{q}_{11} \\ \mathbf{q}_{11}^T & -\mathbf{q}_1(1) \end{bmatrix}, \mathbf{0}_3\right\}$$

$$\mathbf{A}_2 = -\text{diag}\left\{\begin{bmatrix} \mathbf{0}_4 & \mathbf{h}_2 \\ \mathbf{h}_2^T & 0 \end{bmatrix}, \begin{bmatrix} \mathbf{0}_2 & \mathbf{q}_{12} \\ \mathbf{q}_{12}^T & -\mathbf{q}_1(2) \end{bmatrix}, \mathbf{0}_3\right\}$$

$$\mathbf{A}_3 = -\text{diag}\left\{\begin{bmatrix} \mathbf{0}_4 & \mathbf{h}_3 \\ \mathbf{h}_3^T & 0 \end{bmatrix}, \mathbf{0}_3, \begin{bmatrix} \mathbf{0}_2 & \mathbf{q}_{21} \\ \mathbf{q}_{21}^T & -\mathbf{q}_2(1) \end{bmatrix}\right\}$$

$$\mathbf{A}_4 = -\text{diag}\left\{\begin{bmatrix} \mathbf{0}_4 & \mathbf{h}_4 \\ \mathbf{h}_4^T & 0 \end{bmatrix}, \mathbf{0}_3, \begin{bmatrix} \mathbf{0}_2 & \mathbf{q}_{22} \\ \mathbf{q}_{22}^T & -\mathbf{q}_2(2) \end{bmatrix}\right\}$$

$$\mathbf{A}_5 = -\text{diag}\left\{\begin{bmatrix} \mathbf{0}_4 & \mathbf{0} \\ \mathbf{0} & 1 \end{bmatrix}, \mathbf{0}_3, \mathbf{0}_3\right\}$$

$$\mathbf{C} = -\text{diag}\left\{\begin{bmatrix} \mathbf{I}_4 & \mathbf{0} \\ \mathbf{0} & 0 \end{bmatrix}, \begin{bmatrix} \mathbf{I}_2 & \mathbf{0} \\ \mathbf{0} & -r_1 \end{bmatrix}, \begin{bmatrix} \mathbf{I}_2 & \mathbf{0} \\ \mathbf{0} & -r_2 \end{bmatrix}\right\}$$

where $\mathbf{h}_i$, $\mathbf{q}_{1j}$, and $\mathbf{q}_{2j}$ for $1 \leq i \leq 4$ and $1 \leq j \leq 2$ are the ith and jth columns of $\hat{\mathbf{H}}$, $\hat{\mathbf{Q}}_1$, and $\hat{\mathbf{Q}}_2$, respectively, $\mathbf{I}_k$ is the $k \times k$ identity matrix, and $\mathbf{0}_k$ is the $k \times k$ zero matrix. A strictly feasible initial set $\{\mathbf{X}_0, \mathbf{y}_0, \mathbf{S}_0\}$ can be identified as

$$\mathbf{X}_0 = \text{diag}\{\mathbf{I}_5, \mathbf{X}_{02}, \mathbf{X}_{03}\}$$
$$\mathbf{y}_0 = [1 \ 0 \ 2 \ 4 \ 20]^T$$
$$\mathbf{S}_0 = \mathbf{C} - \sum_{i=1}^{5} \mathbf{y}_0(i)\mathbf{A}_i$$

where

$$\mathbf{X}_{02} = \begin{bmatrix} 1 & 0 & -0.5 \\ 0 & 1 & 0 \\ -0.5 & 0 & 1 \end{bmatrix} \quad \text{and} \quad \mathbf{X}_{03} = \begin{bmatrix} 180 & 0 & -12 \\ 0 & 60 & -2 \\ -12 & -2 & 1 \end{bmatrix}$$

With $\gamma = 0.9$ and $\varepsilon = 10^{-3}$, it took Algorithm 14.2 six iterations and 10.73 Mflops to converge to the solution $\{\mathbf{X}^*, \mathbf{y}^*, \mathbf{S}^*\}$ where

$$\mathbf{y}^* = \begin{bmatrix} 2.044717 \\ 0.852719 \\ 2.544895 \\ 2.485678 \\ 2.916910 \end{bmatrix}$$

This corresponds to the solution points $\mathbf{r}^* \in \mathcal{R}$ and $\mathbf{s}^* \in \mathcal{S}$ (see Fig. 13.5) with

$$\mathbf{r}^* = \begin{bmatrix} 2.044717 \\ 0.852719 \end{bmatrix} \quad \text{and} \quad \mathbf{s}^* = \begin{bmatrix} 2.544895 \\ 2.485678 \end{bmatrix}$$

which yield the shortest distance between $\mathcal{R}$ and $\mathcal{S}$ as $\|\mathbf{r}^* - \mathbf{s}^*\| = 1.707845$. Note that Algorithm 14.2 usually yields a more accurate solution to the problem than Algorithm 13.6 with comparable computational complexity.

■

14.6 Projective Method of Nemirovski and Gahinet

In this section, we describe a different interior-point method for SDP problems that was proposed by Nemirovski and Gahinet in [4][14]. The name of the method, i.e., *the projective method*, comes from the fact that orthogonal projections of positive-definite matrices onto the range of a linear mapping characterized by some LMI constraint are heavily involved in the algorithm.

14.6.1 Notation and preliminaries

In the space of symmetric matrices of size $n \times n$, $\mathcal{S}^n$, we denote the set of positive-semidefinite matrices by $\mathcal{K}$ and the set of positive-definite matrices by int$\mathcal{K}$. Note that $\mathcal{K}$ is a convex cone (see Sec. 14.2.1) and the notation int$\mathcal{K}$ comes from the fact that the set of positive-definite matrices can be viewed as the interior of convex cone $\mathcal{K}$.

Given a positive-definite matrix $\mathbf{P} \in R^{n \times n}$, an inner product can be introduced in $\mathcal{S}^n$ as

$$\langle \mathbf{X}, \mathbf{Y} \rangle_P = \text{trace}(\mathbf{PXPY}) \tag{14.59}$$

which leads to the L_P norm

$$\|\mathbf{X}\|_P = [\text{trace}(\mathbf{PXPX})]^{1/2} \tag{14.60}$$

If $\mathbf{P}$ is the identity matrix, then the above norm is reduced to the Frobenius norm

$$\|\mathbf{X}\|_I = [\text{trace}(\mathbf{X}^2)]^{1/2} = \|\mathbf{X}\|_F$$

i.e., norm $\| \cdot \|_P$ in Eq. (14.60) is a generalization of the Frobenius norm $\| \cdot \|_F$.

An important concept involved in the development of the projective method is the *Dikin ellipsoid* [4] which, for a given positive-definite matrix $\mathbf{X}$, is defined as the set

$$D(\mathbf{X}) = \{\mathbf{Y} : \|\mathbf{Y} - \mathbf{X}\|_{X^{-1}}^2 < 1\} \tag{14.61}$$

Since

$$\begin{aligned} \|\mathbf{Y} - \mathbf{X}\|_{X^{-1}}^2 &= \text{trace}[\mathbf{X}^{-1}(\mathbf{Y} - \mathbf{X})\mathbf{X}^{-1}(\mathbf{Y} - \mathbf{X})] \\ &= \text{trace}[(\mathbf{X}^{-1/2}\mathbf{YX}^{-1/2} - \mathbf{I})(\mathbf{X}^{-1/2}\mathbf{YX}^{-1/2} - \mathbf{I})] \\ &= \|\mathbf{X}^{-1/2}\mathbf{YX}^{-1/2} - \mathbf{I}\|_F^2 \end{aligned}$$

the Dikin ellipsoid $D(\mathbf{X})$ can be characterized by

$$D(\mathbf{X}) = \{\mathbf{Y} : \|\mathbf{X}^{-1/2}\mathbf{YX}^{-1/2} - \mathbf{I}\|_F^2 < 1\} \tag{14.62}$$

A very useful property of the Dikin ellipsoid is that for a positive definite $\mathbf{X}$, every element in $D(\mathbf{X})$ is a positive-definite matrix (see Prob. 14.12). In other words, for an $\mathbf{X} \in \text{int}\mathcal{K}$, $D(\mathbf{X})$ is an ellipsoid centered at $\mathbf{X}$ such that the entire ellipsoid is within $\text{int}\mathcal{K}$.

The SDP problem we consider here is given by Eq. (14.9), i.e.,

$$\text{minimize} \quad \mathbf{c}^T \mathbf{x} \tag{14.63a}$$

$$\text{subject to:} \quad \mathbf{F}(\mathbf{x}) \succeq \mathbf{0} \tag{14.63b}$$

where $\mathbf{c} \in R^{p \times 1}$, $\mathbf{x} \in R^{p \times 1}$, and

$$\mathbf{F}(\mathbf{x}) = \mathbf{F}_0 + \sum_{i=1}^{p} x_i \mathbf{F}_i \tag{14.63c}$$

with $\mathbf{F}_i \in \mathcal{S}^n$ for $0 \leq i \leq p$. To start with, we need to find a strictly feasible initial point. This can be done by solving the *strict-feasibility problem* which can be stated as

$$\text{find a vector } \mathbf{x} \text{ such that } \mathbf{F}(\mathbf{x}) \succ \mathbf{0} \tag{14.64}$$

In the projective method, which is applicable to both the SDP problem in Eq. (14.63) and the strict-feasibility problem in Eq. (14.64), we consider the orthogonal projection of a positive-definite matrix $\mathbf{X}$ onto a subspace $\mathcal{E}$ of $\mathcal{S}^n$, where $\mathcal{E}$ is the range of the linear map $\mathcal{F}$ related to the LMI constraint in Eqs. (14.63b) and (14.64), i.e.,

$$\mathcal{F}\mathbf{x} = \sum_{i=1}^{p} x_i \mathbf{F}_i \tag{14.65}$$

and

$$\mathcal{E} = \{\mathbf{X} : \mathbf{X} = \mathcal{F}\mathbf{x}, \ \mathbf{x} \in R^p\} \tag{14.66}$$

The orthogonal projection of a given positive definite $\mathbf{X}$ onto subspace $\mathcal{E}$ with respect to metric $\langle, \rangle_P$ can be defined as the unique solution of the minimization problem

$$\min_{\mathbf{Y} \in \mathcal{E}} \|\mathbf{Y} - \mathbf{X}\|_P = \min_{\mathbf{x} \in R^P} \|\mathcal{F}\mathbf{x} - \mathbf{X}\|_P \tag{14.67}$$

which is a least-squares problem because

$$\begin{aligned}
\|\mathcal{F}\mathbf{x} - \mathbf{X}\|_P^2 &= \text{trace}\left[\mathbf{P}\left(\sum_{i=1}^{p} x_i \mathbf{F}_i - \mathbf{X}\right)\mathbf{P}\left(\sum_{i=1}^{p} x_i \mathbf{F}_i - \mathbf{X}\right)\right] \\
&= \text{trace}\left[\left(\sum_{i=1}^{p} x_i \hat{\mathbf{F}}_i - \hat{\mathbf{X}}\right)\left(\sum_{i=1}^{p} x_i \hat{\mathbf{F}}_i - \hat{\mathbf{X}}\right)\right] \\
&= \mathbf{x}^T \hat{\mathbf{F}} \mathbf{x} - 2\mathbf{x}^T \mathbf{v} + \kappa \tag{14.68}
\end{aligned}$$

is a quadratic function with respect to **x**, where

$$\hat{\mathbf{F}}_i = \mathbf{P}^{1/2}\mathbf{F}_i\mathbf{P}^{1/2} \tag{14.69a}$$

$$\hat{\mathbf{X}} = \mathbf{P}^{1/2}\mathbf{X}\mathbf{P}^{1/2} \tag{14.69b}$$

$$\hat{\mathbf{F}} = \{\hat{f}_{ij}, \ 1 \le i, \ j \le p\} \quad \text{with } \hat{f}_{ij} = \text{trace}(\hat{\mathbf{F}}_i\hat{\mathbf{F}}_j) \tag{14.69c}$$

$$\mathbf{v} = [v_1 \ v_2 \ \ldots \ v_p]^T \quad \text{with } v_i = \text{trace}(\hat{\mathbf{X}}\hat{\mathbf{F}}_i) \tag{14.69d}$$

$$\kappa = \text{trace}(\hat{\mathbf{X}}^2) \tag{14.69e}$$

It can be shown that if matrices $\mathbf{F}_i$ for $i = 1, 2, \ldots, p$ are linearly independent, then matrix $\hat{\mathbf{F}}$ is positive definite (see Prob. 14.13) and the unique global minimizer of the least-squares problem in Eq. (14.67) is given by

$$\mathbf{x} = \hat{\mathbf{F}}^{-1}\mathbf{v} \tag{14.70}$$

The orthogonal projection of matrix **X** onto $\mathcal{E}$ with respect to metric $\langle,\rangle_P$ is now obtained as

$$\mathbf{X}^\dagger = \sum_{i=1}^p x_i\mathbf{F}_i$$

where x_i is the ith component of vector **x** obtained from Eq. (14.70).

14.6.2 Projective method for the strict-feasibility problem

Below we assume that matrices $\mathbf{F}_i$ for $1 \le i \le p$ are linearly independent, namely, $\mathcal{F}\mathbf{x} = \mathbf{0}$ if and only if $\mathbf{x} = \mathbf{0}$, so as to assure a unique orthogonal projection of a symmetric matrix **X** onto subspace $\mathcal{E}$ with respect to metric $\langle,\rangle_P$ defined by Eq. (14.59) with **P** positive definite.

Initially we need to homogenize the LMI constraint

$$\mathcal{F}\mathbf{x} + \mathbf{F}_0 = x_1\mathbf{F}_1 + \cdots + x_p\mathbf{F}_p + \mathbf{F}_0 \succ \mathbf{0} \tag{14.71}$$

as

$$x_1\mathbf{F}_1 + \cdots + x_p\mathbf{F}_p + \tau\mathbf{F}_0 \succ \mathbf{0} \tag{14.72a}$$

$$\tau > 0 \tag{14.72b}$$

The constraints in Eq. (14.72) are equivalent to

$$\tilde{\mathcal{F}}\tilde{\mathbf{x}} = \begin{bmatrix} \mathcal{F}\mathbf{x} + \tau\mathbf{F}_0 & \mathbf{0} \\ \mathbf{0} & \tau \end{bmatrix} \succ \mathbf{0} \tag{14.73a}$$

where

$$\tilde{\mathbf{x}} = \begin{bmatrix} \mathbf{x} \\ \tau \end{bmatrix} \tag{14.73b}$$

Evidently, if vector **x** satisfies the constraint in Eq. (14.71), then $\tilde{\mathbf{x}} = [\mathbf{x}^T \ 1]^T$ satisfies the constraint in Eq. (14.73) and, conversely, if $\tilde{\mathbf{x}} = [\mathbf{x}^T \ \tau]^T$ satisfies

Eq. (14.73), then vector $\mathbf{x}/\tau$ satisfies Eq. (14.71). On the basis of the equivalence of the LMI constraints in Eqs. (14.71) and (14.73), we need to consider only the strict-feasibility problem with a homogenized LMI constraint, i.e.,

$$\text{find a vector } \mathbf{x} \text{ such that } \tilde{\mathcal{F}}\tilde{\mathbf{x}} = \sum_{i=1}^{p+1} \tilde{x}_i \tilde{\mathbf{F}}_i \succ \mathbf{0} \tag{14.74a}$$

where $\tilde{x}_{p+1} = \tau$,

$$\tilde{\mathbf{F}}_i = \begin{bmatrix} \mathbf{F}_i & \mathbf{0} \\ \mathbf{0} & 0 \end{bmatrix}_{(n+1)\times(n+1)} \qquad \text{for } 1 \leq i \leq p \tag{14.74b}$$

and

$$\tilde{\mathbf{F}}_{p+1} = \begin{bmatrix} \mathbf{F}_0 & \mathbf{0} \\ \mathbf{0} & 1 \end{bmatrix}_{(n+1)\times(n+1)} \tag{14.74c}$$

In the projective method as applied to the strict-feasibility problem in Eq. (14.74), we start with an initial point $\mathbf{X}_0 \in \text{int}\mathcal{K}$, say, $\mathbf{X}_0 = \mathbf{I}$, and generate a sequence of positive-definite matrices $\mathbf{X}_k$ in such a way that the orthogonal projection of $\mathbf{X}_k$ onto subspace $\mathcal{E}$ eventually becomes positive definite. More specifically, in the kth iteration the positive-definite matrix $\mathbf{X}_k$ is orthogonally projected onto subspace $\mathcal{E}$ with respect to metric $\langle , \rangle_{X_k^{-1}}$, and the projection obtained is denoted as $\mathbf{X}_k^\dagger$. From Eqs. (14.69) and (14.70) it follows that

$$\mathbf{X}_k^\dagger = \sum_{i=1}^{p+1} \tilde{x}_i \tilde{\mathbf{F}}_i \tag{14.75}$$

where $\tilde{x}_i$ is the ith component of vector $\tilde{\mathbf{x}}_k$ which is calculated as

$$\tilde{\mathbf{x}}_k = \hat{\mathbf{F}}^{-1}\mathbf{v} \tag{14.76a}$$

$$\hat{\mathbf{F}} = \{\hat{f}_{ij}\} \quad \text{with } \hat{f}_{ij} = \text{trace}(\mathbf{X}_k^{-1}\tilde{\mathbf{F}}_i\mathbf{X}_k^{-1}\tilde{\mathbf{F}}_j) \tag{14.76b}$$

$$\mathbf{v} = [v_1 \ v_2 \ \cdots \ v_{p+1}]^T \quad \text{with } v_i = \text{trace}(\mathbf{X}_k^{-1}\tilde{\mathbf{F}}_i) \tag{14.76c}$$

If the projection $\mathbf{X}_k^\dagger$ in Eq. (14.75) is positive definite, then the strict-feasibility problem is solved with vector $\tilde{\mathbf{x}}_k$ given by Eq. (14.76a). Otherwise, matrix $\mathbf{X}_k$ is updated according to

$$\mathbf{X}_{k+1}^{-1} = \mathbf{X}_k^{-1} - \gamma_k\mathbf{X}_k^{-1}(\mathbf{X}_k^\dagger - \mathbf{X}_k)\mathbf{X}_k^{-1} \tag{14.77a}$$

where γ_k is a positive scalar given by

$$\gamma_k = \frac{1}{1 + \rho_\infty} \tag{14.77b}$$

with

$$\rho_\infty = \max_{1 \leq i \leq n} |\lambda(\mathbf{X}_k^{-1}\mathbf{X}_k^\dagger - \mathbf{I})| \tag{14.77c}$$

In Eq. (14.77c), $\lambda(\cdot)$ denotes the eigenvalues of the matrix involved. Once $\mathbf{X}_{k+1}^{-1}$ is obtained from Eq. (14.77a), the orthogonal projection $\mathbf{X}_{k+1}^{\dagger}$ is obtained using Eqs. (14.76a)–(14.76c) with index k replaced by $k+1$, and the iteration continues until a positive-definite orthogonal projection is obtained.

To understand the updating formula in Eq. (14.77a), we first write it as

$$\mathbf{X}_{k+1}^{-1} = \mathbf{X}_k^{-1/2}(\mathbf{I} - \gamma_k \mathbf{W}_k)\mathbf{X}_k^{-1/2} \tag{14.78}$$

where

$$\mathbf{W}_k = \mathbf{X}_k^{-1/2}(\mathbf{X}_k^{\dagger} - \mathbf{X}_k)\mathbf{X}_k^{-1/2}$$

Since

$$\lambda(\mathbf{W}_k) = \lambda(\mathbf{X}_k^{-1}\mathbf{X}_k^{\dagger} - \mathbf{I})$$

we can estimate the eigenvalues of matrix $\mathbf{I} - \gamma_k \mathbf{W}_k$ as

$$\lambda(\mathbf{I} - \gamma_k \mathbf{W}_k) \geq 1 - \frac{\rho_\infty}{1 + \rho_\infty} > 0$$

which means that $\mathbf{I} - \gamma_k \mathbf{W}_k$ is a positive-definite matrix. It now follows from Eq. (14.78) that if $\mathbf{X}_k$ is positive definite, then $\mathbf{X}_{k+1}$ obtained using Eq. (14.77a) is also positive definite. Furthermore, it can be shown [4] that

$$\det(\mathbf{X}_{k+1}^{-1}) \geq \kappa \det(\mathbf{X}_k^{-1})$$

with $\kappa = e/2 \approx 1.36$, which implies that

$$\det(\mathbf{X}_k^{-1}) \geq \kappa^k \det(\mathbf{X}_0^{-1})$$

That is, if $\mathbf{X}_k^{\dagger}$ were to remain positive definite as the iterations continue, we would have

$$\det(\mathbf{X}_k^{-1}) \to \infty \qquad \text{as } k \to \infty \tag{14.79}$$

Next, we note that because $\mathbf{X}_k^{\dagger}$ is an orthogonal projection onto subspace $\mathcal{E}$, $\mathbf{X}_k^{-1}(\mathbf{X}_k^{\dagger} - \mathbf{X}_k)\mathbf{X}_k^{-1}$ is orthogonal to $\mathcal{E}$ with respect to the usual Frobenius metric. Namely, $\mathbf{X}_k^{-1}(\mathbf{X}_k^{\dagger} - \mathbf{X}_k)\mathbf{X}_k^{-1} \in \mathcal{E}^{\perp}$, the orthogonal complement of $\mathcal{E}$, with respect to the Frobenius inner product. Since the last term of the updating formula in Eq. (14.77a) is proportional to $\mathbf{X}_k^{-1}(\mathbf{X}_k^{\dagger} - \mathbf{X}_k)\mathbf{X}_k^{-1}$, we note that Eq. (14.77a) updates $\mathbf{X}_k^{-1}$ in a direction parallel to subspace $\mathcal{E}^{\perp}$. From Sec. 13.6.1, we know that $\det(\mathbf{X}_k^{-1})$ is related to the volume of the ellipsoid characterized by $\mathbf{X}_k^{-1}$ and, consequently, Eq. (14.79) implies that $\mathbf{X}_{k+1}^{-1}$ would grow in parallel to subspace $\mathcal{E}^{\perp}$ towards infinity if the iterations were not terminated. To see that this will not occur, notice that the Frobenius inner product of any two positive-semidefinite matrices is always nonnegative. In geometrical terms this means that the angle at the vertex of the convex cone $\mathcal{K}$ is exactly

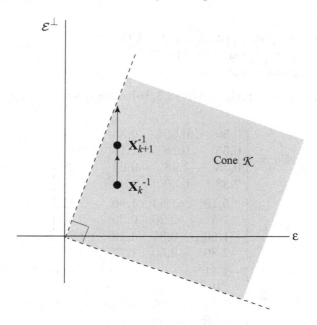

Figure 14.1. A geometrical interpretation of Eq. (14.77a).

90° [4], as illustrated in Fig. 14.1. This geometrical interpretation also suggests that if the strict-feasibility problem is solvable, i.e., cone $\mathcal{K}$ intersects with subspace $\mathcal{E}$, then $\mathcal{E}^{\perp} \cap \mathcal{K} = \{0\}$. Therefore, if the iterations do not terminate, then $\mathbf{X}_k^{-1}$ as a point in cone $\mathcal{K}$ would eventually leave the cone, i.e., becoming nonpositive definite, which obviously contradicts the fact that the matrix $\mathbf{X}_{k+1}$ updated using Eq. (14.77a) is always positive definite.

An algorithm for the solution of the homogenized strict-feasibility problem in Eq. (14.74) is as follows.

Algorithm 14.3 Projective algorithm for the homogenized strict-feasibility problem in Eq. (14.74)
Step 1
Set $k = 0$ and $\mathbf{X}_0 = \mathbf{I}$.
Step 2
Compute the orthogonal projection $\mathbf{X}_k^{\dagger}$ with respect to metric $\langle , \rangle_{X_k^{-1}}$ by using Eqs. (14.75) and (14.76).
Step 3
If $\mathbf{X}_k^{\dagger}$ is positive definite, output solution $\tilde{x}^* = \tilde{x}_k$, which is given by Eq. (14.76a), and stop; otherwise, continue with Step 4.

Step 4
Compute γ_k using Eqs. (14.77b) and (14.77c).
Update $\mathbf{X}_k^{-1}$ to $\mathbf{X}_{k+1}^{-1}$ using Eq. (14.77a).
Set $k = k + 1$ and repeat from Step 2.

Example 14.3 Applying Algorithm 14.3, solve the strict-feasibility problem in Eq. (14.64) if

$$
\mathbf{F}_0 = \begin{bmatrix} 0.50 & 0.55 & 0.33 & 2.38 \\ 0.55 & 0.18 & -1.18 & -0.40 \\ 0.33 & -1.18 & -0.94 & 1.46 \\ 2.38 & -0.40 & 1.46 & 0.17 \end{bmatrix}
$$

$$
\mathbf{F}_1 = \begin{bmatrix} 5.19 & 1.54 & 1.56 & -2.80 \\ 1.54 & 2.20 & 0.39 & -2.50 \\ 1.56 & 0.39 & 4.43 & 1.77 \\ -2.80 & -2.50 & 1.77 & 4.06 \end{bmatrix}
$$

$$
\mathbf{F}_2 = \begin{bmatrix} -1.11 & 0 & -2.12 & 0.38 \\ 0 & 1.91 & -0.25 & -0.58 \\ -2.12 & -0.25 & -1.49 & 1.45 \\ 0.38 & -0.58 & 1.45 & 0.63 \end{bmatrix}
$$

$$
\mathbf{F}_3 = \begin{bmatrix} 2.69 & -2.24 & -0.21 & -0.74 \\ -2.24 & 1.77 & 1.16 & -2.01 \\ -0.21 & 1.16 & -1.82 & -2.79 \\ -0.74 & -2.01 & -2.79 & -2.22 \end{bmatrix}
$$

$$
\mathbf{F}_4 = \begin{bmatrix} 0.58 & -2.19 & 1.69 & 1.28 \\ -2.19 & -0.05 & -0.01 & 0.91 \\ 1.69 & -0.01 & 2.56 & 2.14 \\ 1.28 & 0.91 & 2.14 & -0.75 \end{bmatrix}
$$

Solution In order to apply Algorithm 14.3, the problem at hand is first converted into the homogenized problem in Eq. (14.74) and the initial matrix $\mathbf{X}_0$ is set to $\mathbf{I}_5$. The Algorithm took four iterations and 38.8 Kflops to yield

$$
\tilde{\mathbf{x}}_4 = \begin{bmatrix} 0.214262 \\ 0.042863 \\ -0.019655 \\ -0.056181 \\ 0.078140 \end{bmatrix}
$$

which corresponds to a solution of the strict-feasibility problem in Eq. (14.64) as

$$
\begin{bmatrix} x_1 \\ x_2 \\ x_3 \\ x_4 \end{bmatrix} = \frac{\tilde{\mathbf{x}}_4(1:4)}{\tilde{\mathbf{x}}_4(5)} = \begin{bmatrix} 2.742040 \\ 0.548538 \\ -0.251537 \\ -0.718983 \end{bmatrix}
$$

where $\tilde{\mathbf{x}}_4(1 : 4)$ denotes the vector formed by using the first four components of $\tilde{\mathbf{x}}_4$. It can be verified that

$$\mathbf{F}_0 + \sum_{i=1}^{4} x_i \mathbf{F}_i$$

is a positive-definite matrix whose smallest eigenvalue is 0.1657. It is interesting to note that the sequence

$$\{\det(\mathbf{X}_k^{-1}) \text{ for } k = 0, 1, \ldots, 4\} = \{1, 2.63, 6.75, 16.58, 37.37\}$$

gives the ratio

$$\left\{ \frac{\det(\mathbf{X}_{k+1}^{-1})}{\det(\mathbf{X}_k^{-1})} \text{ for } k = 0, 1, \ldots, 3 \right\} = \{2.63, 2.57, 2.45, 2.25\}$$

This verifies that the ratio is greater than $\kappa = 1.36$.

∎

14.6.3 Projective method for SDP problems

14.6.3.1 Problem homogenization

Let us now consider the SDP problem

$$\text{minimize } \mathbf{c}^T \mathbf{x} \qquad (14.80\text{a})$$

$$\text{subject to: } \mathbf{F}(\mathbf{x}) \succeq \mathbf{0} \qquad (14.80\text{b})$$

where $\mathbf{c} \in R^{p \times 1}$, $\mathbf{x} \in R^{p \times 1}$, and

$$\mathbf{F}(\mathbf{x}) = \mathbf{F}_0 + \sum_{i=1}^{p} x_i \mathbf{F}_i \qquad (14.80\text{c})$$

with $\mathbf{F}_i \in \mathcal{S}^n$ for $0 \le i \le p$. We assume below that the problem in Eq. (14.80) is *solvable* by using an interior-point method, i.e., the interior of the feasible region described by Eq. (14.80b) is not empty, and that the objective function $\mathbf{c}^T \mathbf{x}$ has a finite lower bound in the feasible region.

As in the projective method for the strict-feasibility problem in Sec. 14.6.2, we first convert the problem at hand into the homogeneous problem

$$\underset{\mathbf{y}, \tau}{\text{minimize }} \frac{\mathbf{c}^T \mathbf{y}}{\tau} \qquad (14.81\text{a})$$

$$\text{subject to: } \mathcal{F}\mathbf{y} + \tau \mathbf{F}_0 \succeq \mathbf{0} \qquad (14.81\text{b})$$

$$\tau > 0 \qquad (14.81\text{c})$$

where

$$\mathcal{F}\mathbf{y} = \sum_{i=1}^{p} y_i \mathbf{F}_i$$

The problems in Eqs. (14.80) and (14.81) are equivalent because if vector $\mathbf{x}$ is a minimizer for the problem in Eq. (14.80), then $[\mathbf{y}^T \ \tau]^T = [\mathbf{x}^T \ 1]^T$ is a minimizer for the problem in Eq. (14.81) and, conversely, if $[\mathbf{y}^T \ \tau]^T$ is a minimizer for the problem in Eq. (14.81), then $\mathbf{x} = \mathbf{y}/\tau$ is a minimizer for the problem in Eq. (14.80). Now if we let

$$\tilde{\mathbf{c}} = \begin{bmatrix} \mathbf{c} \\ 0 \end{bmatrix}, \quad \tilde{\mathbf{d}} = \begin{bmatrix} \mathbf{0} \\ 1 \end{bmatrix}, \quad \tilde{\mathbf{x}} = \begin{bmatrix} \mathbf{y} \\ \tau \end{bmatrix} \tag{14.82a}$$

$$\tilde{\mathcal{F}}\tilde{\mathbf{x}} = \begin{bmatrix} \mathcal{F}\mathbf{y} + \tau \mathbf{F}_0 & \mathbf{0} \\ \mathbf{0} & \tau \end{bmatrix} = \sum_{i=1}^{p+1} \tilde{x}_i \tilde{\mathbf{F}}_i \tag{14.82b}$$

where

$$\tilde{\mathbf{F}}_i = \begin{bmatrix} \mathbf{F}_i & \mathbf{0} \\ \mathbf{0} & 0 \end{bmatrix}_{(n+1)\times(n+1)} \qquad \text{for } 1 \le i \le p \tag{14.82c}$$

and

$$\tilde{\mathbf{F}}_{p+1} = \begin{bmatrix} \mathbf{F}_0 & \mathbf{0} \\ \mathbf{0} & 1 \end{bmatrix}_{(n+1)\times(n+1)} \tag{14.82d}$$

then the problem in Eq. (14.81) can be expressed as

$$\text{minimize } f(\tilde{\mathbf{x}}) = \frac{\tilde{\mathbf{c}}^T \tilde{\mathbf{x}}}{\tilde{\mathbf{d}}^T \tilde{\mathbf{x}}} \tag{14.83a}$$

$$\text{subject to:} \quad \tilde{\mathcal{F}}\tilde{\mathbf{x}} \succeq \mathbf{0} \tag{14.83b}$$

$$\tilde{\mathbf{d}}^T \tilde{\mathbf{x}} \neq 0 \tag{14.83c}$$

In what follows, we describe a projective method proposed in [4][14] that applies to the SDP problem in the form of Eq. (14.83).

14.6.3.2 Solution procedure

In the projective method for the problem in Eq. (14.83), we start with a strictly feasible initial point $\tilde{\mathbf{x}}_0$ for which $\mathbf{X}_0 = \tilde{\mathcal{F}}\tilde{\mathbf{x}}_0 \succ \mathbf{0}$ and $\tilde{\mathbf{d}}^T \tilde{\mathbf{x}}_0 \neq 0$. Such an initial point $\tilde{\mathbf{x}}_0$ can be obtained by using Algorithm 14.3. In the kth iteration, the set $\{\tilde{\mathbf{x}}_k, \ \mathbf{X}_k\}$ is updated to set $\{\tilde{\mathbf{x}}_{k+1}, \ \mathbf{X}_{k+1}\}$ to achieve two goals: to reduce the objective function and to maintain strict feasibility for point $\tilde{\mathbf{x}}_{k+1}$. These goals can be achieved through the following steps:

1. Compute the orthogonal projection of $\mathbf{X}_k$ onto the subspace $\mathcal{E}$ defined by

$$\mathcal{E} = \{\mathbf{X} : \ \mathbf{X} = \tilde{\mathcal{F}}\tilde{\mathbf{x}}, \tilde{\mathbf{x}} \in R^{p+1}\} \tag{14.84}$$

and denote the orthogonal projection obtained as

$$\mathbf{X}_k^\dagger = \tilde{\mathcal{F}} \tilde{\mathbf{x}}_k$$

If $\mathbf{X}_k^\dagger \succ \mathbf{0}$, then continue with Step 2; otherwise, set

$$\mathbf{Y}_k = \mathbf{X}_k^\dagger - \mathbf{X}_k \tag{14.85}$$

and continue with Step 3.

2. Find a value of the objective function $f(\tilde{\mathbf{x}})$, f_k^*, such that

$$||\mathbf{X}_k - \mathbf{X}_k^\dagger(f_k^*)||_{X_k^{-1}} \geq 0.99 \quad \text{subject to:} \quad \mathbf{X}_k^\dagger(f_k^*) \succ \mathbf{0} \tag{14.86}$$

The matrix $\mathbf{X}_k^\dagger(f)$ in Eq. (14.86) represents the orthogonal projection of $\mathbf{X}_k$ onto $\mathcal{E}(f)$ which for a given real number f is the subspace of $\mathcal{E}$ defined by

$$\mathcal{E}(f) = \{\mathbf{X} : \mathbf{X} \in \mathcal{E} \quad \text{and} \quad (\tilde{\mathbf{c}} - f\tilde{\mathbf{d}})^T \tilde{\mathbf{x}} = 0\} \tag{14.87}$$

(see Prob. 14.14(a)). Note that $\mathcal{E}(f)$ is related to the hyperplane $\mathcal{P}(f) = \{\tilde{\mathbf{x}} : (\tilde{\mathbf{c}} - f\tilde{\mathbf{d}})^T \tilde{\mathbf{x}} = 0\}$ on which the objective function $f(\tilde{\mathbf{x}})$ assumes constant value f (see Prob. 14.14(b)).

Then compute matrix $\mathbf{Y}_k$ as

$$\mathbf{Y}_k = \mathbf{X}_k^\dagger(f_k^*) - \mathbf{X}_k \tag{14.88}$$

Details for the calculation of f_k^* and $\mathbf{X}_k^\dagger(f_k^*)$ are given in Secs. 14.6.3.4 and 14.6.3.5, respectively.

3. Update $\mathbf{X}_k$ to $\mathbf{X}_{k+1}$ as

$$\mathbf{X}_{k+1}^{-1} = \mathbf{X}_k^{-1} - \gamma_k \mathbf{X}_k^{-1} \mathbf{Y}_k \mathbf{X}_k^{-1} \tag{14.89}$$

where the step size γ_k is chosen such that $\mathbf{X}_{k+1}^{-1} \succ \mathbf{0}$ and

$$\det(\mathbf{X}_{k+1}^{-1}) \geq \kappa \det(\mathbf{X}_k^{-1})$$

for some fixed $\kappa > 1$.
Repeat from Step 1.

14.6.3.3 Choice of step size γ_k

The choice of a suitable value for γ_k in Eq. (14.89) is dependent on how the matrix $\mathbf{Y}_k$ in Eq. (14.89) is calculated. If $\mathbf{Y}_k$ is calculated using Eq. (14.85), then it means that $\mathbf{X}_k^\dagger$ obtained from Step 1 is *not* positive definite. Evidently, this is a situation similar to that in the strict-feasibility problem, and γ_k can be determined using Eqs. (14.77b) and (14.77c). On the other hand, if $\mathbf{Y}_k$ is

480

calculated using Eq. (14.88), then γ_k can be determined by using Eqs. (14.77b) and (14.77c) with $\mathbf{X}_k^\dagger$ replaced by $\mathbf{X}_k^\dagger(f_k^*)$, i.e.,

$$\gamma_k = \frac{1}{1 + \tilde{\rho}_\infty} \tag{14.90a}$$

with

$$\tilde{\rho}_\infty = \max_{1 \le i \le n} |\lambda(\mathbf{X}_k^{-1}\mathbf{X}_k^\dagger(f_k^*) - \mathbf{I})| \tag{14.90b}$$

14.6.3.4   Computation of f_k^*

For a given positive-definite matrix $\mathbf{X}_k$ and any element in $\mathcal{E}$, say, $\mathbf{X} = \tilde{\mathcal{F}}\tilde{\mathbf{x}}$, there exist two matrices $\mathbf{C}_k$ and $\mathbf{D}_k$ in $\mathcal{E}$ such that the inner products $\tilde{\mathbf{c}}^T\tilde{\mathbf{x}}$ and $\tilde{\mathbf{d}}^T\mathbf{x}$ can be represented as inner products in space $\mathcal{S}^n$, i.e.,

$$\tilde{\mathbf{c}}^T\tilde{\mathbf{x}} = \langle \mathbf{C}_k, \mathbf{X} \rangle_{X_k^{-1}} \tag{14.91a}$$

and

$$\tilde{\mathbf{d}}^T\tilde{\mathbf{x}} = \langle \mathbf{D}_k, \mathbf{X} \rangle_{X_k^{-1}} \tag{14.91b}$$

respectively (see Prob. 14.15). Consequently, we can write

$$(\tilde{\mathbf{c}} - f\tilde{\mathbf{d}})^T\tilde{\mathbf{x}} = \langle \mathbf{C}_k - f\mathbf{D}_k, \mathbf{X} \rangle_{X_k^{-1}}$$

and the linear subspace $\mathcal{E}(f)$ defined by Eq. (14.87) can be characterized by

$$\mathcal{E}(f) = \{\mathbf{X} \in \mathcal{E}, \langle \mathbf{C}_k - f\mathbf{D}_k, \mathbf{X} \rangle_{X_k^{-1}} = 0\} \tag{14.92}$$

From Fig. 14.2, it follows that the squared distance between $\mathbf{X}_k$ and $\mathbf{X}_k^\dagger(f)$ can be computed as

$$\delta^2(f) = ||\mathbf{X}_k - \mathbf{X}_k^\dagger||_{X_k^{-1}}^2 + ||\mathbf{X}_k^\dagger - \mathbf{X}_k^\dagger(f)||_{X_k^{-1}}^2 \tag{14.93}$$

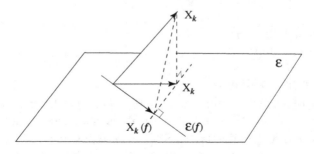

Figure 14.2.   Relation among $\mathbf{X}_k$, $\mathbf{X}_k^\dagger$, and $\mathbf{X}_k^\dagger(f)$.

Since $\mathbf{X}_k^{\dagger} - \mathbf{X}_k^{\dagger}(f)$ is orthogonal to subspace $\mathcal{E}(f)$, $\mathbf{X}_k^{\dagger} - \mathbf{X}_k^{\dagger}(f)$ can be obtained as the orthogonal projection of $\mathbf{X}_k^{\dagger}$ onto a normal of $\mathcal{E}(f)$, i.e.,

$$\mathbf{X}_k^{\dagger} - \mathbf{X}_k^{\dagger}(f) = \langle \mathbf{u}, \mathbf{X}_k^{\dagger} \rangle_{X_k^{-1}} \mathbf{u}$$

where $\mathbf{u}$ is a normal of $\mathcal{E}(f)$ with unity length with respect to metric $\langle , \rangle_{X_k^{-1}}$. This, in conjunction with the fact that $\mathbf{C}_k - f\mathbf{D}_k$ is a normal of subspace $\mathcal{E}(f)$, yields

$$||\mathbf{X}_k^{\dagger} - \mathbf{X}_k^{\dagger}(f)||^2 = \frac{\langle \mathbf{C}_k - f\mathbf{D}_k, \mathbf{X}_k^{\dagger} \rangle_{X_k^{-1}}}{||\mathbf{C}_k - f\mathbf{D}_k||^2_{X_k^{-1}}}$$

which modifies Eq. (14.93) to

$$\delta^2(f) = ||\mathbf{X}_k - \mathbf{X}_k^{\dagger}||^2_{X_k^{-1}} + \frac{\langle \mathbf{C}_k - f\mathbf{D}_k, \mathbf{X}_k^{\dagger} \rangle_{X_k^{-1}}}{||\mathbf{C}_k - f\mathbf{D}_k||^2_{X_k^{-1}}} \tag{14.94}$$

The value of f_k^* used in Step 2 in Sec. 16.6.3.2 can now be determined as follows. First, we note that the matrix $\mathbf{X}_k^{\dagger} = \tilde{\mathcal{F}} \tilde{\mathbf{x}}_k$ obtained in Step 1 is positive definite. Hence if we let $f_k = f(\tilde{\mathbf{x}}_k)$, then $\mathbf{X}_k^{\dagger}(f_k) = \mathbf{X}_k^{\dagger}$. If

$$||\mathbf{X}_k - \mathbf{X}_k^{\dagger}||_{X_k^{-1}} \geq 0.99$$

then the constraints in Eq. (14.86) are satisfied by taking $f_k^* = f_k$. Otherwise, we have

$$\delta(f_k) = ||\mathbf{X}_k - \mathbf{X}_k^{\dagger}||_{X_k^{-1}} < 0.99 \tag{14.95}$$

because $f(\tilde{\mathbf{x}}_k) = f_k$ implies that

$$\langle \mathbf{C}_k - f_k\mathbf{D}_k, \mathbf{X}_k^{\dagger} \rangle_{X_k^{-1}} = (\tilde{\mathbf{c}} - f_k\tilde{\mathbf{d}})^T \tilde{\mathbf{x}}_k = 0$$

On the other hand, the limit of $\delta(f)$ as f approaches negative infinity is equal to or larger than one. This, in conjunction with Eq. (14.95), implies the existence of an $f_k^* < f_k$ that satisfies the two constraints in Eq. (14.86). The numerical value of such an f_k^* can be determined by solving the *quadratic* equation

$$0.99 = ||\mathbf{X}_k - \mathbf{X}_k^{\dagger}||^2_{X_k^{-1}} + \frac{\langle \mathbf{C}_k - f\mathbf{D}_k, \mathbf{X}_k^{\dagger} \rangle_{X_k^{-1}}}{||\mathbf{C}_k - f\mathbf{D}_k||^2_{X_k^{-1}}} \tag{14.96}$$

for f. If f_k^* is the smaller real solution of Eq. (14.96), then we have

$$||\mathbf{X}_k - \mathbf{X}_k^{\dagger}(f_k^*)||_{X_k^{-1}} = 0.99 \tag{14.97}$$

Since $\mathbf{X}_k$ is positive definite, Eq. (14.97) indicates that $\mathbf{X}_k^{\dagger}(f_k^*)$ is located inside the Dikin ellipsoid $D(\mathbf{X}_k)$ and, therefore, $\mathbf{X}_k^{\dagger}(f_k^*)$ is positive definite.

14.6.3.5 Computation of $X_k^\dagger(f_k^*)$

By definition, $X_k^\dagger(f) = \tilde{\mathcal{F}}\tilde{x}^*$ minimizes $||X_k - \tilde{\mathcal{F}}\tilde{x}||_{X_k^{-1}}$ subject to the constraint $(\tilde{c} - f\tilde{d})^T \tilde{x} = 0$. Note that

$$||X_k - \tilde{\mathcal{F}}\tilde{x}||_{X_k^{-1}}^2 = \tilde{x}^T \hat{F}\tilde{x} - 2\tilde{x}^T v + \kappa$$

where $\hat{F}$ and v are given by Eqs. (14.76c) and (14.76d), respectively. Therefore, $\tilde{x}^*$ and $X_k^\dagger(f)$ can be obtained by solving the QP problem

$$\text{minimize } \tilde{x}^T \hat{F}\tilde{x} - 2\tilde{x}^T v + \kappa$$

$$\text{subject to: } (\tilde{c} - f\tilde{d})^T \tilde{x} = 0$$

By applying the formula in Eq. (13.11), we obtain the solution of the above QP problem as

$$q = \tilde{c} - f\tilde{d} \tag{14.98a}$$

$$\lambda^* = -\frac{q^T \hat{F}^{-1} v}{q^T \hat{F}^{-1} q} \tag{14.98b}$$

$$\tilde{x}^* = \hat{F}^{-1}(q\lambda^* + v) \tag{14.98c}$$

and the orthogonal projection of X_k onto $\mathcal{E}(f)$ is given by

$$X_k^\dagger(f) = \tilde{\mathcal{F}}\tilde{x}^* \tag{14.99}$$

where $\tilde{x}^*$ is given in Eq. (14.98c).

14.6.3.6 Algorithm

The above method can be implemented in terms of the following algorithm.

Algorithm 14.4 Projective algorithm for the homogenized SDP problem in Eq. (14.83)
Step 1
Apply Algorithm 14.3 to obtain a strictly feasible point $\tilde{x}_0$.
Evaluate $X_0 = \tilde{\mathcal{F}}\tilde{x}_0$ and compute $f_0^* = f(x_0)$.
Set $k = 0$ and initialize tolerance ε.
Select a positive integer value for L.
Step 2
Compute the orthogonal projection of X_k onto subspace $\mathcal{E}$ given by Eq. (14.84).
Denote the orthogonal projection obtained as $X_k^\dagger = \tilde{\mathcal{F}}\tilde{x}_k$.
Step 3
If $X_k^\dagger \succ 0$, continue with Step 4; otherwise, set

$$\mathbf{Y}_k = \mathbf{X}_k^\dagger - \mathbf{X}_k$$

and continue with Step 5.

Step 4

Compute $f_k = f(\tilde{\mathbf{x}}_k)$ and $\delta(f_k) = ||\mathbf{X}_k - \mathbf{X}_k^\dagger||_{\mathbf{X}_k^{-1}}$.

If

$$\delta(f_k) \geq 0.99$$

then let $f_k^* = f_k$, $\mathbf{X}_k^\dagger(f_k^*) = \mathbf{X}_k^\dagger$, and compute $\mathbf{Y}_k$ using Eq. (14.88); otherwise, determine matrices $\mathbf{C}_k$ and $\mathbf{D}_k$ in Eq. (14.91), compute f_k^* as the smallest real solution of Eq. (14.96), and obtain $(\tilde{\mathbf{x}}_k^*, \mathbf{X}_k^\dagger(f_k^*))$ using Eqs. (14.98) and (14.99) with $f = f_k^*$.

Compute $\mathbf{Y}_k$ using Eq. (14.88).

If the reduction in f_k^* during the last L iterations is consistently less than ε, output solution $\tilde{\mathbf{x}}^* = \tilde{\mathbf{x}}_k^*$ and stop; otherwise, continue with Step 5.

Step 5

Update $\mathbf{X}_k$ to $\mathbf{X}_{k+1}$ using Eq. (14.89), where parameter γ_k is determined as

$$\gamma_k = \frac{1}{1 + \rho_\infty} \quad \text{with} \quad \rho_\infty = \max_{1 \leq i \leq n} |\lambda(\mathbf{X}_k^{-1}\mathbf{Y}_k)|$$

Set $k = k + 1$ and repeat from Step 2.

An analysis on the polynomial-time convergence of the above algorithm can be found in [4][14]. The latter reference also addresses various implementation issues of the algorithm.

Example 14.4 Apply Algorithm 14.4 to solve the shortest distance problem discussed in Example 14.2.

Solution The shortest distance problem in Example 14.2 can be formulated as the SDP problem in Eq. (14.80) where $\mathbf{c} = [0\ 0\ 0\ 0\ 1]^T$ and $\mathbf{F}_i$ for $0 \leq i \leq 5$ are given by $\mathbf{F}_0 = \mathbf{C}$, $\mathbf{F}_i = -\mathbf{A}_i$ for $i = 1, 2, \ldots, 5$; on the other hand, $\mathbf{C}$ and $\mathbf{A}_i$ are defined in Example 14.2.

The problem at hand can be converted to the homogeneous SDP problem in Eq. (14.83) with

$$\tilde{\mathbf{c}} = \begin{bmatrix} 0 \\ 0 \\ 0 \\ 0 \\ 1 \\ 0 \end{bmatrix} \quad \text{and} \quad \tilde{\mathbf{d}} = \begin{bmatrix} 0 \\ 0 \\ 0 \\ 0 \\ 0 \\ 1 \end{bmatrix}$$

and $\tilde{\mathbf{F}}_i$ for $1 \leq i \leq 6$ can be determined using Eqs. (14.82c) and (14.82d). With $\varepsilon = 5 \times 10^{-5}$ and $L = 1$, it took Algorithm 14.4 36 iterations and 9.7 Mflops to converge to the solution

$$\mathbf{x}^* = \frac{\tilde{\mathbf{x}}(1:5)}{\tilde{\mathbf{x}}^*(6)} = \begin{bmatrix} 2.044301 \\ 0.852835 \\ 2.544217 \\ 2.485864 \\ 2.916757 \end{bmatrix}$$

This corresponds to the solution points $\mathbf{r}^* \in \mathcal{R}$ and $\mathbf{s}^* \in \mathcal{S}$ with

$$\mathbf{r}^* = \begin{bmatrix} 2.044301 \\ 0.852835 \end{bmatrix} \quad \text{and} \quad \mathbf{s}^* = \begin{bmatrix} 2.544217 \\ 2.485864 \end{bmatrix}$$

which yield the shortest distance between $\mathcal{R}$ and $\mathcal{S}$ as $\|\mathbf{r}^* - \mathbf{s}^*\| = 1.707835$. Note that Algorithm 14.4 generated a slightly more accurate solution than Algorithm 14.2 requiring less computation.

∎

14.7   Second-Order Cone Programming
14.7.1   Notation and definitions

The concept of convex cone has been defined in Sec. 14.2 as a convex set where any element multiplied by any nonnegative scalar still belongs to the cone (see Def. 14.1). Here we are interested in a special class of convex cones known as *second-order cones*.

Definition 14.2 A second-order cone of dimension n is defined as

$$\mathcal{K} = \left\{ \begin{bmatrix} t \\ \mathbf{u} \end{bmatrix} : t \in R, \ \mathbf{u} \in R^{n-1} \text{ for } \|\mathbf{u}\| \leq t \right\} \tag{14.100}$$

A second-order cone is also called *quadratic* or *Lorentz cone*. For $n = 1$, the second-order cone degenerates into a ray on the t axis starting from $t = 0$, as shown in Fig. 14.3a. The second-order cones for $n = 2$ and 3 are depicted in Fig. 14.3b and c, respectively.

Note that the second-order cone $\mathcal{K}$ is a convex set in R^n because for any two points in $\mathcal{K}$, $[t_1 \ \mathbf{u}_1^T]^T$ and $[t_2 \ \mathbf{u}_2^T]^T$, and $\lambda \in [0, 1]$, we have

$$\lambda \begin{bmatrix} t_1 \\ \mathbf{u}_1 \end{bmatrix} + (1-\lambda) \begin{bmatrix} t_2 \\ \mathbf{u}_2 \end{bmatrix} = \begin{bmatrix} \lambda t_1 + (1-\lambda)t_2 \\ \lambda \mathbf{u}_1 + (1-\lambda)\mathbf{u}_2 \end{bmatrix}$$

where

$$\|\lambda \mathbf{u}_1 + (1-\lambda)\mathbf{u}_2\| \leq \lambda\|\mathbf{u}_1\| + (1-\lambda)\|\mathbf{u}_2\| \leq \lambda t_1 + (1-\lambda)t_2$$

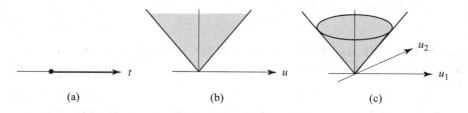

Figure 14.3. Second-order cones of dimension (a) $n = 1$, (b) $n = 2$, and (c) $n = 3$.

The primal second-order cone-programming (SOCP) problem is a constrained optimization problem that can be formulated as

$$\text{minimize} \sum_{i=1}^{q} \hat{\mathbf{c}}_i^T \mathbf{x}_i \tag{14.101a}$$

$$\text{subject to:} \quad \sum_{i=1}^{q} \hat{\mathbf{A}}_i \mathbf{x}_i = \mathbf{b} \tag{14.101b}$$

$$\mathbf{x}_i \in \mathcal{K}_i \quad \text{for } i = 1, 2, \ldots, q \tag{14.101c}$$

where $\hat{\mathbf{c}}_i \in R^{n_i \times 1}$, $\mathbf{x}_i \in R^{n_i \times 1}$, $\hat{\mathbf{A}}_i \in R^{m \times n_i}$, $\mathbf{b} \in R^{m \times 1}$, and $\mathcal{K}_i$ is the second-order cone of dimension n_i. It is interesting to note that there exists an analogy between the SOCP problem in Eq. (14.101) and the LP problem in Eq. (12.1): both problems involve a linear objective function and a linear equality constraint. While the variable vector $\mathbf{x}$ in an LP problem is constrained to the region $\{\mathbf{x} \geq \mathbf{0}, \ \mathbf{x} \in R^n\}$, which is a convex cone (see Def. 14.1), each variable vector $\mathbf{x}_i$ in an SOCP problem is constrained to the second-order cone $\mathcal{K}_i$.

The dual of the SOCP problem in Eq. (14.101) referred to hereafter as the *dual SOCP problem* can be shown to be of the form

$$\text{maximize } \mathbf{b}^T \mathbf{y} \tag{14.102a}$$

$$\text{subject to:} \quad \hat{\mathbf{A}}_i^T \mathbf{y} + \mathbf{s}_i = \hat{\mathbf{c}}_i \tag{14.102b}$$

$$\mathbf{s}_i \in \mathcal{K}_i \quad \text{for } i = 1, 2, \ldots, q \tag{14.102c}$$

where $\mathbf{y} \in R^{m \times 1}$ and $\mathbf{s}_i \in R^{n_i \times 1}$ (see Prob. 14.17). Note that a similar analogy exists between the dual SOCP problem in Eq. (14.102) and the dual LP problem in Eq. (12.2).

If we let

$$\mathbf{x} = -\mathbf{y}, \quad \hat{\mathbf{A}}_i^T = \begin{bmatrix} \mathbf{b}_i^T \\ \mathbf{A}_i^T \end{bmatrix} \quad \text{and} \quad \hat{\mathbf{c}}_i = \begin{bmatrix} d_i \\ \mathbf{c}_i \end{bmatrix} \tag{14.103}$$

where $\mathbf{b}_i \in R^{m \times 1}$ and d_i is a scalar, then the SOCP problem in Eq. (14.102) can be expressed as

$$\text{minimize } \mathbf{b}^T \mathbf{x} \tag{14.104a}$$

subject to: $\quad \|\mathbf{A}_i^T \mathbf{x} + \mathbf{c}_i\| \leq \mathbf{b}_i^T \mathbf{x} + d_i \qquad$ for $i = 1, 2, \ldots, q$

$$(14.104b)$$

(see Prob. 14.18). As we will see next, this SOCP formulation turns out to have a direct connection to many convex-programming problems in engineering and science.

14.7.2 Relations among LP, QP, SDP and SOCP Problems

The class of SOCP problems is large enough to include both LP and convex QP problems. If $\mathbf{A}_i^T = \mathbf{0}$ and $\mathbf{c}_i = \mathbf{0}$ for $i = 1, 2, \ldots, q$, then the problem in Eq. (14.104) becomes

$$\text{minimize } \mathbf{b}^T \mathbf{x}$$

subject to: $\quad \mathbf{b}_i^T \mathbf{x} + d_i \geq 0 \qquad$ for $i = 1, 2, \ldots, q$

which is obviously an LP problem.

Now consider the convex QP problem

$$\text{minimize } f(\mathbf{x}) = \mathbf{x}^T \mathbf{H} \mathbf{x} + 2\mathbf{x}^T \mathbf{p} \qquad (14.105a)$$

$$\text{subject to: } \quad \mathbf{A}\mathbf{x} \geq \mathbf{b} \qquad (14.105b)$$

where $\mathbf{H}$ is positive definite. If we write matrix $\mathbf{H}$ as $\mathbf{H} = \mathbf{H}^{T/2}\mathbf{H}^{1/2}$ and let $\tilde{\mathbf{p}} = \mathbf{H}^{-T/2}\mathbf{p}$, then the objective function in Eq. (14.105a) can be expressed as

$$f(\mathbf{x}) = \|\mathbf{H}^{1/2}\mathbf{x} + \tilde{\mathbf{p}}\|^2 - \mathbf{p}^T \mathbf{H}^{-1}\mathbf{p}$$

Since the term $\mathbf{p}^T \mathbf{H}^{-1}\mathbf{p}$ is a constant, minimizing $f(\mathbf{x})$ is equivalent to minimizing $\|\mathbf{H}^{1/2}\mathbf{x} + \tilde{\mathbf{p}}\|$ and thus the problem at hand can be converted to

$$\text{minimize } \delta \qquad (14.106a)$$

$$\text{subject to: } \quad \|\mathbf{H}^{1/2}\mathbf{x} + \tilde{\mathbf{p}}\| \leq \delta \qquad (14.106b)$$

$$\mathbf{A}\mathbf{x} \geq \mathbf{b} \qquad (14.106c)$$

where δ is an upper bound for $\|\mathbf{H}^{1/2}\mathbf{x} + \tilde{\mathbf{p}}\|$ that can be treated as an auxiliary variable of the problem. By defining

$$\tilde{\mathbf{x}} = \begin{bmatrix} \delta \\ \mathbf{x} \end{bmatrix}, \quad \tilde{\mathbf{b}} = \begin{bmatrix} 1 \\ \mathbf{0} \end{bmatrix}, \quad \tilde{\mathbf{H}} = [\mathbf{0} \ \mathbf{H}^{1/2}], \quad \tilde{\mathbf{A}} = [\mathbf{0} \ \mathbf{A}]$$

the problem becomes

$$\text{minimize } \tilde{\mathbf{b}}^T \tilde{\mathbf{x}} \qquad (14.107a)$$

$$\text{subject to: } \quad \|\tilde{\mathbf{H}}\tilde{\mathbf{x}} + \tilde{\mathbf{p}}\| \leq \tilde{\mathbf{b}}^T \tilde{\mathbf{x}} \qquad (14.107b)$$

$$\tilde{\mathbf{A}}\tilde{\mathbf{x}} \geq \mathbf{b} \qquad (14.107c)$$

which is an SOCP problem. On the other hand, it can be shown that every SOCP problem can be formulated as an SDP problem. To see this, note that the constraint $\|\mathbf{u}\| \leq t$ implies that

$$\begin{bmatrix} t\mathbf{I} & \mathbf{u} \\ \mathbf{u}^T & t \end{bmatrix} \succeq \mathbf{0}$$

(see Prob. 14.19). In other words, a second-order cone can be embedded into a cone of positive semidefinite matrices, and the SOCP problem in Eq. (14.104) can be formulated as

$$\text{minimize } \mathbf{b}^T \mathbf{x} \tag{14.108a}$$

$$\text{subject to: } \begin{bmatrix} (\mathbf{b}_i^T \mathbf{x} + d_i)\mathbf{I} & \mathbf{A}_i^T \mathbf{x} + \mathbf{c}_i \\ (\mathbf{A}_i^T \mathbf{x} + \mathbf{c}_i)^T & \mathbf{c}_i^T \mathbf{x} + d_i \end{bmatrix} \succeq 0 \tag{14.108b}$$

which is an SDP problem.

The above analysis has demonstrated that the branch of nonlinear programming known as CP can be subdivided into a series of nested branches of optimization, namely, SDP, SOCP, convex QP, and LP as illustrated in Fig. 14.4.

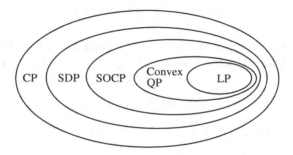

Figure 14.4. Relations among LP, convex QP, SOCP, SDP, and CP problems.

14.7.3 Examples

In this section, we present several examples to demonstrate that a variety of interesting optimization problems can be formulated as SOCP problems [15].

(i) *QP problems with quadratic constraints* A general QP problem with quadratic constrains can be expressed as

$$\text{minimize } \mathbf{x}^T \mathbf{H}_0 \mathbf{x} + 2\mathbf{p}_0^T \mathbf{x} \tag{14.109a}$$

$$\text{subject to: } \mathbf{x}^T \mathbf{H}_i \mathbf{x} + 2\mathbf{p}_i^T \mathbf{x} + r_i \leq 0 \quad \text{for } i = 1, 2, \ldots, q \tag{14.109b}$$

where $\mathbf{H}_i$ for $i = 1, 2, \ldots, q$ are assumed to be positive-definite matrices. Using the matrix decomposition $\mathbf{H}_i = \mathbf{H}_i^{T/2}\mathbf{H}_i^{1/2}$, the problem in Eq. (14.109) can be expressed as

$$\text{minimize } \|\mathbf{H}_0^{1/2}\mathbf{x} + \tilde{\mathbf{p}}_0\|^2 - \mathbf{p}_0^T \mathbf{H}_0^{-1}\mathbf{p}_0$$

$$\text{subject to: } \|\mathbf{H}_i^{1/2}\mathbf{x} + \tilde{\mathbf{p}}_i\|^2 - \mathbf{p}_i^T \mathbf{H}_i^{-1}\mathbf{p}_i + r_i \leq 0 \quad \text{for } i = 1, 2, \ldots, q$$

where $\tilde{\mathbf{p}}_i = \mathbf{H}_i^{-T/2}\mathbf{p}_i$ for $i = 0, 1. \ldots, q$. Obviously, the above problem is equivalent to the SOCP problem

$$\text{minimize } \delta \tag{14.110a}$$

$$\text{subject to:} \quad \|\mathbf{H}^{1/2}\mathbf{x} + \tilde{\mathbf{p}}_0\| \leq \delta \tag{14.110b}$$

$$\|\mathbf{H}^{1/2}\mathbf{x} + \tilde{\mathbf{p}}_i\| \leq (\mathbf{p}_i^T \mathbf{H}_i^{-1}\mathbf{p}_i - r_i)^{1/2} \text{ for } i = 1, 2, \ldots, q \tag{14.110c}$$

(ii) *Minimization of a sum of L_2 norms* Unconstrained minimization problems of the form

$$\text{minimize } \sum_{i=1}^{N} \|\mathbf{A}_i\mathbf{x} + \mathbf{c}_i\|$$

occur in a number of applications. By introducing an upper bound for each L_2-norm term in the objective function, the problem can be converted to

$$\text{minimize } \sum_{i=1}^{N} \delta_i \tag{14.111a}$$

$$\text{subject to:} \quad \|\mathbf{A}_i\mathbf{x} + \mathbf{c}_i\| \leq \delta_i \quad \text{for } i = 1, 2, \ldots, N \tag{14.111b}$$

If we define an augmented variable vector

$$\tilde{\mathbf{x}} = \begin{bmatrix} \delta_1 \\ \vdots \\ \delta_N \\ \mathbf{x} \end{bmatrix}$$

and let

$$\tilde{\mathbf{b}} = \begin{bmatrix} 1 \\ \vdots \\ 1 \\ 0 \end{bmatrix}, \quad \tilde{\mathbf{A}}_i = [\mathbf{0} \ \mathbf{A}_i], \quad \tilde{\mathbf{b}}_i = \begin{bmatrix} 0 \\ \vdots \\ 0 \\ 1 \\ 0 \\ \vdots \\ 0 \end{bmatrix} \leftarrow i\text{th component}$$

then Eq. (14.111) becomes

$$\text{minimize } \tilde{\mathbf{b}}^T \tilde{\mathbf{x}} \tag{14.112a}$$

$$\text{subject to:} \quad \|\tilde{\mathbf{A}}_i\tilde{\mathbf{x}} + \mathbf{c}_i\| \leq \tilde{\mathbf{b}}_i^T \tilde{\mathbf{x}} \quad \text{for } i = 1, 2, \ldots, N \tag{14.112b}$$

which is an SOCP problem.

Another unconstrained problem related to the problem in Eq. (14.111) is the minimax problem

$$\text{minimize } \underset{1 \leq i \leq N}{\text{maximize }} \|\mathbf{A}_i\mathbf{x} + \mathbf{c}_i\| \tag{14.113}$$

which can be re-formulated as the SOCP problem

$$\text{minimize } \delta \tag{14.114a}$$

$$\text{subject to: } \|\mathbf{A}_i\mathbf{x} + \mathbf{c}_i\| \leq \delta \quad \text{for } i = 1, 2, \ldots, N \tag{14.114b}$$

(iii) *Complex L_1-norm approximation problem* An interesting special case of the sum-of-norms problem is the complex L_1 norm approximation problem whereby a complex-valued approximate solution for the linear equation $\mathbf{A}\mathbf{x} = \mathbf{b}$ is required where $\mathbf{A}$ and $\mathbf{b}$ are complex-valued such that $\mathbf{x}$ solves the unconstrained problem

$$\text{minimize } \|\mathbf{A}\mathbf{x} - \mathbf{c}\|_1$$

where $\mathbf{A} \in C^{m \times n}$, $\mathbf{c} \in C^{m \times 1}$, $\mathbf{x} \in C^{n \times 1}$, and the L_1 norm of $\mathbf{x}$ is defined as $\|\mathbf{x}\|_1 = \sum_{k=1}^{n} |x_k|$. If we let $\mathbf{A} = [\mathbf{a}_1 \ \mathbf{a}_2 \ \cdots \ \mathbf{a}_m]^T$ and $\mathbf{c} = [c_1 \ c_2 \ \cdots \ c_m]^T$ where $\mathbf{a}_k = \mathbf{a}_{kr} + j\mathbf{a}_{ki}$, $c_k = c_{kr} + jc_{ki}$, $\mathbf{x} = \mathbf{x}_r + j\mathbf{x}_i$, and $j = \sqrt{-1}$, then we have

$$\|\mathbf{A}\mathbf{x} - \mathbf{c}\|_1 = \sum_{k=1}^{m} |\mathbf{a}_k^T\mathbf{x} - c_k|$$

$$= \sum_{k=1}^{m} [(\mathbf{a}_{kr}^T\mathbf{x}_r - \mathbf{a}_{ki}^T\mathbf{x}_i - c_{kr})^2 + (\mathbf{a}_{kr}^T\mathbf{x}_i + \mathbf{a}_{ki}^T\mathbf{x}_r - c_{ki})^2]^{1/2}$$

$$= \sum_{k=1}^{m} \left\| \underbrace{\begin{bmatrix} \mathbf{a}_{kr}^T & -\mathbf{a}_{ki}^T \\ \mathbf{a}_{ki}^T & \mathbf{a}_{kr}^T \end{bmatrix}}_{\mathbf{A}_k} \underbrace{\begin{bmatrix} \mathbf{x}_r \\ \mathbf{x}_i \end{bmatrix}}_{\hat{\mathbf{x}}} - \underbrace{\begin{bmatrix} c_{kr} \\ c_{ki} \end{bmatrix}}_{\mathbf{c}_k} \right\| = \sum_{k=1}^{m} \|\mathbf{A}_k\hat{\mathbf{x}} - \mathbf{c}_k\|$$

Hence the problem under consideration can be converted to

$$\text{minimize} \sum_{k=1}^{m} \delta_k \tag{14.115a}$$

subject to: $\|\mathbf{A}_k\hat{\mathbf{x}} - \mathbf{c}_k\| \leq \delta_k$ for $k = 1, 2, \ldots, m$

$$\tag{14.115b}$$

By letting

$$\tilde{\mathbf{x}} = \begin{bmatrix} \delta_1 \\ \vdots \\ \delta_m \\ \hat{\mathbf{x}} \end{bmatrix}, \quad \tilde{\mathbf{b}}_0 = \begin{bmatrix} 1 \\ \vdots \\ 1 \\ \mathbf{0} \end{bmatrix}, \quad \tilde{\mathbf{A}}_k = [\mathbf{0} \ \mathbf{A}_k]$$

$$\tilde{\mathbf{b}}_k = \begin{bmatrix} 0 \\ \vdots \\ 0 \\ 1 \\ 0 \\ \vdots \\ 0 \end{bmatrix} \leftarrow \text{the } k\text{th component}$$

the problem in Eq. (14.115) becomes

$$\text{minimize} \ \tilde{\mathbf{b}}^T\tilde{\mathbf{x}} \tag{14.116a}$$

subject to: $\|\tilde{\mathbf{A}}_k\tilde{\mathbf{x}} - \mathbf{c}_k\| \leq \tilde{\mathbf{b}}_k^T\tilde{\mathbf{x}}$ for $k = 1, 2, \ldots, m$

$$\tag{14.116b}$$

which is obviously an SOCP problem.

(iv) *Linear fractional problem* The linear fractional problem can be described as

$$\text{minimize} \sum_{i=1}^{p} \frac{1}{\mathbf{a}_i^T\mathbf{x} + c_i} \tag{14.117a}$$

subject to: $\mathbf{a}_i^T\mathbf{x} + c_i > 0$ for $i = 1, 2, \ldots, p$ (14.117b)

$\mathbf{b}_i^T\mathbf{x} + d_i \geq 0$ for $i = 1, 2, \ldots, q$ (14.117c)

It can be readily verified that subject to the constraints in Eq. (14.117b), each term in the objective function is convex and hence the objective function itself is also convex. It, therefore, follows that the problem in Eq. (14.117) is a CP problem. By introducing the auxiliary constraints

$$\frac{1}{\mathbf{a}_i^T\mathbf{x} + c_i} \leq \delta_i$$

i.e.,

$$\delta_i(\mathbf{z}_i^T \mathbf{x} + c_i) \geq 1$$

and

$$\delta_i \geq 0$$

the problem in Eq. (14.117) can be expressed as

$$\text{minimize} \sum_{i=1}^{p} \delta_i \tag{14.118a}$$

$$\text{subject to:} \quad \delta_i(\mathbf{a}_i^T \mathbf{x} + c_i) \geq 1 \quad \text{for } i = 1, 2, \ldots, p \tag{14.118b}$$

$$\delta_i \geq 0 \tag{14.118c}$$

$$\mathbf{b}_i^T \mathbf{x} + d_i \geq 0 \tag{14.118d}$$

Furthermore, we note that $w^2 \leq uv$, $u \geq 0$, $v \geq 0$ if and only if

$$\left\| \begin{bmatrix} 2w \\ u - v \end{bmatrix} \right\| \leq u + v$$

(see Prob. 14.20) and hence the constraints in Eqs. (14.118b) and (14.118c) can be written as

$$\left\| \begin{bmatrix} 2 \\ \mathbf{a}_i^T \mathbf{x} + c_i - \delta_i \end{bmatrix} \right\| \leq \mathbf{a}_i^T \mathbf{x} + c_i + \delta_i \qquad \text{for } i = 1, 2, \ldots, p$$

Hence the problem in Eq. (14.118) can be formulated as

$$\text{minimize} \sum_{i=1}^{p} \delta_i \tag{14.119a}$$

$$\text{subject to:} \quad \left\| \begin{bmatrix} 2 \\ \mathbf{a}_i^T \mathbf{x} + c_i - \delta_i \end{bmatrix} \right\| \leq \mathbf{a}_i^T \mathbf{x} + c_i + \delta_i \quad \text{for } i = 1, 2, \ldots, p \tag{14.119b}$$

$$\mathbf{b}_i^T \mathbf{x} + d_i \geq 0 \tag{14.119c}$$

which is an SOCP problem.

14.8   A Primal-Dual Method for SOCP Problems
14.8.1   Assumptions and KKT conditions

If we let

$$\mathbf{c} = \begin{bmatrix} \hat{\mathbf{c}}_1 \\ \hat{\mathbf{c}}_2 \\ \vdots \\ \hat{\mathbf{c}}_q \end{bmatrix}, \quad \mathbf{x} = \begin{bmatrix} \mathbf{x}_1 \\ \mathbf{x}_2 \\ \vdots \\ \mathbf{x}_q \end{bmatrix}, \quad \mathbf{s} = \begin{bmatrix} \mathbf{s}_1 \\ \mathbf{s}_2 \\ \vdots \\ \mathbf{s}_q \end{bmatrix}$$

$$\mathbf{A} = [\hat{\mathbf{A}}_1 \ \hat{\mathbf{A}}_2 \ \hat{\mathbf{A}}_2 \ \cdots \ \hat{\mathbf{A}}_q] \quad \text{and} \quad \mathcal{K} = \mathcal{K}_1 \times \mathcal{K}_2 \times \cdots \times \mathcal{K}_q$$

where $\mathcal{K}_1$, $\mathcal{K}_2$, ..., $\mathcal{K}_q$ are the second-order cones in Eqs. (14.101) and (14.102) and $\mathcal{K} = \mathcal{K}_1 \times \mathcal{K}_2 \times \cdots \times \mathcal{K}_q$ represents a second-order cone whose elements are of the form $\mathbf{x} = [x_1 \; x_2 \; \cdots \; x_q]^T$ with $x_i \in \mathcal{K}_i$ for $i = 1, 2, \ldots, q$, then the primal and dual SOCP problems in Eqs. (14.101) and (14.102) can be expressed as

$$\text{minimize } \mathbf{c}^T \mathbf{x} \tag{14.120a}$$

$$\text{subject to:} \quad \mathbf{Ax} = \mathbf{b}, \; \mathbf{x} \in \mathcal{K} \tag{14.120b}$$

and

$$\text{maximize } \mathbf{b}^T \mathbf{y} \tag{14.121a}$$

$$\text{subject to:} \quad \mathbf{A}^T \mathbf{y} + \mathbf{s} = \mathbf{c}, \; \mathbf{s} \in \mathcal{K} \tag{14.121b}$$

respectively. The feasible sets for the primal and dual SOCP problems are defined by

$$\mathcal{F}_p = \{\mathbf{x} : \; \mathbf{Ax} = \mathbf{b}, \; \mathbf{x} \in \mathcal{K}\}$$

and

$$\mathcal{F}_d = \{(\mathbf{s}, \; \mathbf{y}) : \; \mathbf{A}^T \mathbf{y} + \mathbf{s} = \mathbf{c}, \; \mathbf{s} \in \mathcal{K}\}$$

respectively. The duality gap between $\mathbf{x} \in \mathcal{F}_p$ and $(\mathbf{s}, \; \mathbf{y}) \in \mathcal{F}_d$ assumes the form

$$\delta(\mathbf{x}, \mathbf{s}, \mathbf{y}) = \mathbf{c}^T \mathbf{x} - \mathbf{b}^T \mathbf{y} = (\mathbf{A}^T \mathbf{y} + \mathbf{s})^T \mathbf{x} - \mathbf{b}^T \mathbf{y} = \mathbf{s}^T \mathbf{x} \tag{14.122}$$

A vector $\mathbf{x}_i = [t_i \; \mathbf{u}_i^T]^T$ in space $R^{n_i \times 1}$ is said to be an interior point of the second-order cone $\mathcal{K}_i$ if $\|\mathbf{u}_i\| < t_i$. If we denote the set of all interior points of $\mathcal{K}_i$ as $\mathcal{K}_i^o$ and let

$$\mathcal{K}^o = \mathcal{K}_1^o \times \mathcal{K}_2^o \times \cdots \times \mathcal{K}_q^o$$

then a strictly feasible vector for the problem in Eq. (14.120) is a vector $\mathbf{x} \in \mathcal{K}_i^o$ satisfying the constraint in Eq. (14.120b). Based on these ideas, the strictly feasible sets of the primal and dual SOCP problems are given by

$$\mathcal{F}_p^o = \{\mathbf{x} : \; \mathbf{Ax} = \mathbf{b}, \; \mathbf{x} \in \mathcal{K}^o\}$$

and

$$\mathcal{F}_d^o = \{(\mathbf{x}, \; \mathbf{y}) : \; \mathbf{A}^T \mathbf{y} + \mathbf{s} = \mathbf{c}, \; \mathbf{s} \in \mathcal{K}^o\}$$

respectively.

In the rest of the chapter, we make the following two assumptions:

1. There exists a strictly feasible point $\mathbf{x}$ for the primal problem in Eq. (14.120) and a strictly feasible pair $(\mathbf{s}, \; \mathbf{y})$ for the dual problem in Eq. (14.121), i.e., both $\mathcal{F}_p^o$ and $\mathcal{F}_d^o$ are nonempty.
2. The rows of matrix $\mathbf{A}$ are linearly independent.

Under these assumptions, solutions for the primal and dual SOCP problems exist and finding these solutions is equivalent to finding a vector set $(\mathbf{x}, \mathbf{s}, \mathbf{y}) \in \mathcal{K} \times \mathcal{K} \times R^m$ that satisfies the KKT conditions [16]

$$\mathbf{Ax} = \mathbf{b} \tag{14.123a}$$
$$\mathbf{A}^T\mathbf{y} + \mathbf{s} = \mathbf{c} \tag{14.123b}$$
$$\mathbf{x}^T\mathbf{s} = 0 \tag{14.123c}$$

where the condition in Eq. (14.123c) is referred to as the complementarity condition. From Eq. (14.123c), we note that the duality gap $\delta(\mathbf{x}, \mathbf{s}, \mathbf{y})$ at the primal and dual solution points becomes zero.

14.8.2 A primal-dual interior-point algorithm

In this section we introduce a primal-dual interior-point algorithm for SOCP, which is a slightly modified version of an algorithm proposed in [16]. In the kth iteration of the algorithm, the vector set $(\mathbf{x}_k, \mathbf{s}_k, \mathbf{y}_k)$ is updated to

$$(\mathbf{x}_{k+1}, \mathbf{s}_{k+1}, \mathbf{y}_{k+1}) = (\mathbf{x}_k, \mathbf{s}_k, \mathbf{y}_k) + \alpha_k(\Delta\mathbf{x}, \Delta\mathbf{s}, \Delta\mathbf{y}) \tag{14.124}$$

where $(\Delta\mathbf{x}, \Delta\mathbf{s}, \Delta\mathbf{y})$ is obtained by solving the linear system of equations

$$\mathbf{A}\Delta\mathbf{x} = \mathbf{b} - \mathbf{Ax} \tag{14.125a}$$
$$\mathbf{A}^T\Delta\mathbf{y} + \Delta\mathbf{s} = \mathbf{c} - \mathbf{s} - \mathbf{A}^T\mathbf{y} \tag{14.125b}$$
$$\mathbf{S}\Delta\mathbf{x} + \mathbf{X}\Delta\mathbf{s} = \sigma\mu\mathbf{e} - \mathbf{Xs} \tag{14.125c}$$

where $\mathbf{e} = [1 \; 1 \; \cdots \; 1]^T$,

$$\mathbf{X} = \text{diag}\{\mathbf{X}_1, \ldots, \mathbf{X}_q\} \quad \text{with } \mathbf{X}_i = \begin{bmatrix} t_i & \mathbf{u}_i^T \\ \mathbf{u}_i & t_i\mathbf{I}_i \end{bmatrix} \tag{14.125d}$$
$$\mathbf{S} = \text{diag}\{\mathbf{S}_1, \ldots, \mathbf{S}_q\} \tag{14.125e}$$
$$\mu = \mathbf{x}^T\mathbf{s}/q \tag{14.125f}$$

σ is a small positive scalar, and $(\mathbf{x}, \mathbf{s}, \mathbf{y})$ assumes the value of $(\mathbf{x}_k, \mathbf{s}_k, \mathbf{y}_k)$. In Eq. (14.125d), t_i and $\mathbf{u}_i$ are the first component and the remaining part of vector $\mathbf{x}_i$, respectively, and $\mathbf{I}_i$ is the identity matrix of dimension $n_i - 1$. The matrices $\mathbf{S}_i$ for $i = 1, 2, \ldots, q$ in Eq. (14.125e) are defined in a similar manner. On comparing Eq. (14.125) with Eq. (14.123), it is evident that the vector set $(\mathbf{x}_k, \mathbf{s}_k, \mathbf{y}_k)$ is updated so that the new vector set $(\mathbf{x}_{k+1}, \mathbf{s}_{k+1}, \mathbf{y}_{k+1})$ better approximates the KKT conditions in Eq. (14.123).

In Eq. (14.124), α_k is a positive scalar that is determined by the line search

$$\alpha_k = 0.75 \min(\alpha_{k1}, \alpha_{k2}, \alpha_{k3}) \tag{14.126a}$$
$$\alpha_{k1} = \max_{0 < \alpha \le 1} (\mathbf{x}_k + \alpha\Delta\mathbf{x} \in \mathcal{F}_p^o) \tag{14.126b}$$

$$\alpha_{k2} = \max_{0<\alpha\leq 1} (\mathbf{s}_k + \alpha\Delta\mathbf{s} \in \mathcal{F}_d^o) \qquad (14.126c)$$

$$\alpha_{k3} = \max_{0<\alpha\leq 1} [\mathbf{c} - \mathbf{A}^T(\mathbf{y}_k + \alpha\Delta\mathbf{y}) \in \mathcal{F}_d^o] \qquad (14.126d)$$

It follows from Eqs. (14.124) and (14.126) that the updated vector set $(\mathbf{x}_{k+1}, \mathbf{s}_{k+1}, \mathbf{y}_{k+1})$ will remain strictly feasible.

An algorithm based on the above approach is as follows.

Algorithm 14.5 Primal-dual interior-point algorithm for SOCP problems

Step 1

Input data set $(\mathbf{A}, \mathbf{b}, \mathbf{c})$, parameters q and n_i for $i = 1, 2, \ldots, q$, and tolerance ε.

Input an initial vector set $(\mathbf{x}_0, \mathbf{s}_0, \mathbf{y}_0)$ with $\mathbf{x}_0 \in \mathcal{F}_p^o$ and $(\mathbf{s}_0, \mathbf{y}_0) \in \mathcal{F}_d^o$.

Set $\mu_0 = \mathbf{x}_0^T\mathbf{s}_0/q$, $\sigma = 10^{-5}$, and $k = 0$.

Step 2

Compute the solution $(\Delta\mathbf{x}, \Delta\mathbf{s}, \Delta\mathbf{y})$ of Eqs. (14.125a)–(14.125c) where $(\mathbf{x}, \mathbf{s}, \mathbf{y}) = (\mathbf{x}_k, \mathbf{s}_k, \mathbf{y}_k)$ and $\mu = \mu_k$.

Step 3

Compute α_k using Eq. (14.126).

Step 4

Set $(\mathbf{x}_{k+1}, \mathbf{s}_{k+1}, \mathbf{y}_{k+1}) = (\mathbf{x}_k + \alpha_k\Delta\mathbf{x}, \mathbf{s}_k + \alpha_k\Delta\mathbf{s}, \mathbf{y}_k + \alpha_k\Delta\mathbf{y})$

Step 5

Compute $\mu_{k+1} = \mathbf{x}_{k+1}^T\mathbf{s}_{k+1}/q$. If $\mu_{k+1} \leq \varepsilon$, output solution $(\mathbf{x}^*, \mathbf{s}^*, \mathbf{y}^*) = (\mathbf{x}_{k+1}, \mathbf{s}_{k+1}, \mathbf{y}_{k+1})$ and stop; otherwise, set $k = k+1$ and repeat from Step 2.

Example 14.5 Apply Algorithm 14.5 to solve the shortest distance problem discussed in Example 13.5.

Solution The problem can be formulated as

$$\text{minimize } \delta$$

subject to:
$$[(x_1 - x_3)^2 + (x_2 - x_4)^2]^{1/2} \leq \delta$$

$$[x_1\ x_2] \begin{bmatrix} 1/4 & 0 \\ 0 & 1 \end{bmatrix} \begin{bmatrix} x_1 \\ x_2 \end{bmatrix} - [x_1\ x_2] \begin{bmatrix} 1/2 \\ 0 \end{bmatrix} \leq \frac{3}{4}$$

$$[x_3\ x_4] \begin{bmatrix} 5/8 & 3/8 \\ 3/8 & 5/8 \end{bmatrix} \begin{bmatrix} x_3 \\ x_4 \end{bmatrix} - [x_3\ x_4] \begin{bmatrix} 11/2 \\ 13/2 \end{bmatrix} \leq -\frac{35}{2}$$

If we let $\mathbf{x} = [\delta\ x_1\ x_2\ x_3\ x_4]^T$, the above problem can be expressed as

$$\text{minimize } \mathbf{b}^T \mathbf{x} \tag{14.127a}$$

$$\text{subject to: } \quad \|\mathbf{A}_i^T \mathbf{x} + \mathbf{c}_i\| \le \mathbf{b}_i^T \mathbf{x} + d_i \quad \text{for } i = 1, 2, 3 \tag{14.127b}$$

where

$$\mathbf{b} = [1\ 0\ 0\ 0\ 0]^T$$

$$\mathbf{A}_1^T = \begin{bmatrix} 0 & -1 & 0 & 1 & 0 \\ 0 & 0 & 1 & 0 & -1 \end{bmatrix}, \quad \mathbf{A}_2^T = \begin{bmatrix} 0 & 0.5 & 0 & 0 & 0 \\ 0 & 0 & 1 & 0 & 0 \end{bmatrix}$$

$$\mathbf{A}_3^T = \begin{bmatrix} 0 & 0 & 0 & -0.7071 & -0.7071 \\ 0 & 0 & 0 & -0.3536 & 0.3536 \end{bmatrix}$$

$$\mathbf{b}_1 = \mathbf{b}, \quad \mathbf{b}_2 = \mathbf{0}, \quad \mathbf{b}_3 = \mathbf{0}$$

$$\mathbf{c}_1 = \mathbf{0}, \quad \mathbf{c}_2 = [-0.5\ 0]^T, \quad \mathbf{c}_3 = [4.2426\ -0.7071]^T$$

$$d_1 = 0, \quad d_2 = 1, \quad \text{and} \quad d_3 = 1$$

The SOCP formulation in Eq. (14.127) is the same as that in Eq. (14.104). Hence by using Eq. (14.103), the problem at hand can be converted into the primal and dual formulations in Eq. (14.101) and Eq. (14.102).

In order to generate a strictly feasible vector set $(\mathbf{x}_0, \mathbf{s}_0, \mathbf{y}_0)$, we note that the initial point used in Example 13.5, i.e., $[x_1\ x_2\ x_3\ x_4] = [1.5\ 0.5\ 2.5\ 4]$, suggests an initial $\mathbf{y}_0 = [\beta\ -1.5\ -0.5\ -2.5\ -4]^T$ where β is a scalar to ensure that $\mathbf{s} = \mathbf{c} - \mathbf{A}^T \mathbf{y}_0 \in \mathcal{F}_d^o$. Since

$$\mathbf{s}_0 = \mathbf{c} - \mathbf{A}^T \mathbf{y}_0 = [\beta\ 1\ 3.5\ 0\ 0\ 1\ 0.25\ 0.5\ 1\ -0.3535\ -0.1767]^T$$

$n_1 = 5$, $n_2 = 3$, and $n_3 = 3$, choosing $\beta = 3.7$ guarantees that $\mathbf{s}_0 \in \mathcal{F}_d^o$. This gives

$$\mathbf{y}_0 = [3.7\ -1.5\ -0.5\ -2.5\ -4]^T$$

and

$$\mathbf{x}_0 = [3.7\ 1\ 3.5\ 0\ 0\ 1\ 0.25\ 0.5\ 1\ -0.3535\ -0.1767]^T$$

Moreover, it can be readily verified that

$$\mathbf{x}_0 = [1\ 0\ 0\ 0\ 0\ 0.1\ 0\ 0\ 0.1\ 0\ 0]^T \in \mathcal{F}_p^o$$

With $\varepsilon = 10^{-4}$, it took Algorithm 14.5 15 iterations and 2.98 Mflops to converge to vector set $(\mathbf{x}^*, \mathbf{s}^*, \mathbf{y}^*)$ where

$$\mathbf{y}^* = \begin{bmatrix} -1.707791 \\ -2.044705 \\ -0.852730 \\ -2.544838 \\ -2.485646 \end{bmatrix}$$

496

which corresponds to the solution points $\mathbf{r}^* \in \mathcal{R}$ and $\mathbf{s}^* \in \mathcal{S}$ given by

$$\mathbf{r}^* = \begin{bmatrix} 2.044705 \\ 0.852730 \end{bmatrix} \quad \text{and} \quad \mathbf{s}^* = \begin{bmatrix} 2.544838 \\ 2.485646 \end{bmatrix}$$

These points give the shortest distance between $\mathcal{R}$ and $\mathcal{S}$ as $\|\mathbf{r}^* - \mathbf{s}^*\| = 1.707790$. Compared with the results obtained in Example 13.5, we note that Algorithm 14.5 led to a more accurate solution with less computation.

■

References

1 F. Alizadeh, *Combinational Optimization with Interior Point Methods and Semidefinite Matrices*, Ph.D. thesis, University of Minnesota, Oct. 1991.

2 Y. Nesterov and A. Nemirovski, *Interior-Point Polynomial Algorithms in Convex Programming*, SIAM, Philadelphia, PA, 1994.

3 L. Vandenberghe and S. Boyd, "Semidefinite programming," *SIAM Review*, vol. 38, pp. 49–95, Mar. 1996.

4 A. Nemirovski and P. Gahinet, "The projective method for solving linear matrix inequalities," in *Proc. American Control Conference*, pp. 840–844, Baltimore, MD., June 1994.

5 C. Helmberg, F. Rendl, R. Vanderbei, and H. Wolkowicz, "An interior-point method for semidefinite programming," *SIAM J. Optim.*, vol. 6, pp. 342–361, 1996.

6 M. Kojima, S. Shindoh, and S. Hara, "Interior-point methods for the monotone linear complementarity problem in symmetric matrices," *SIAM J. Optim.*, vol. 7, pp. 86–125, 1997.

7 Y. Nesterov and M. Todd, "Primal-dual interior-point method for self-scaled cones," *SIAM J. Optim.*, vol. 8, pp. 324–364, 1998.

8 F. Alizadeh, J. A. Haeberly, and M. L. Overton, "Primal-dual interior-point methods for semidefinite programming: Convergence rates, stability and numerical results," *SIAM J. Optim.*, vol. 8, pp. 746–768, 1998.

9 R. Monteiro, "Polynomial convergence of primal-dual algorithm for semidefinite programming based on Monteiro and Zhang family of directions," *SIAM J. Optim.*, vol. 8, pp. 797–812, 1998.

10 S. Boyd, L. El Ghaoui, E. Feron, and V. Balakrishnan, *Linear Matrix Inequalities in System and Control Theory*, SIAM, Philadelphia, 1994.

11 A. S. Lewis and M. L. Overton, "Eigenvalue optimization," *Acta Numerica*, vol. 5, pp. 149–190, 1996.

12 M. Shida, S. Shindoh, and M. Kojima, "Existence and uniqueness of search directions in interior-point algorithms for the SDP and the monotone SDLCP," *SIAM J. Optim.*, vol. 8, pp. 387–398, 1998.

13 S. Mehrotra, "On the implementation of a primal-dual interior point method," *SIAM, J. Optim.*, vol. 2, pp. 575–601, 1992.

14 P. Gahinet and A. Nemirovski, "The projective method for solving linear matrix inequalities," *Math. Programming*, vol. 77, pp. 163–190, 1997.

15 M. S. Lobo, L. Vandenberghe, S. Boyd, and H. Lebret, "Applications of second-order cone programming," *Linear Algebra and Its Applications*, vol. 284, pp. 193–228, Nov. 1998.

16 R. D. C. Monteiro and T. Tsuchiya, "Polynomial convergence of primal-dual algorithms for the second-order cone program based on the MZ-family of directions," *Math. Programming*, vol. 88, pp. 61–83, 2000.

Problems

14.1 Prove that the minimization problem in Eq. (14.1) is a CP problem.

14.2 Show that the sets $\{\mathbf{X} : \mathbf{X} \in R^{n \times n}, \mathbf{X} \succeq \mathbf{0}\}$ and $\{\mathbf{x} : \mathbf{x} \in R^{n \times 1}, \mathbf{x} \geq \mathbf{0}\}$ are convex cones.

14.3 Given that $\mathbf{X} \succeq \mathbf{0}$ and $\mathbf{S} \succeq \mathbf{0}$, prove that

$$\mathbf{S} \cdot \mathbf{X} \geq 0$$

where $\mathbf{S} \cdot \mathbf{X}$ denotes the standard inner product in space $\mathcal{S}^n$ defined in Sec. 14.2.1.

14.4 (a) Prove that the inequality in Eq. (14.14) holds if and only if matrix $G(\delta, \mathbf{x})$ in Eq. (14.15) is positive semidefinite.

(b) Specify matrices $\mathbf{F}_0$ and $\mathbf{F}_i$ $(1 \leq i \leq n)$ in Eq. (14.16) so that $\mathbf{F}(\mathbf{x}) \succeq \mathbf{0}$ reformulates the constraints in Eq. (14.13c).

14.5 The problem of minimizing the maximum eigenvalue of an affine matrix can be stated as follows: given matrices $\mathbf{A}_0, \mathbf{A}_1, \ldots, \mathbf{A}_p$ in $\mathcal{S}^n$, find scalars $y_1, y_2, \ldots, y_p$ such that the maximum eigenvalue of

$$\mathbf{A}_0 + \sum_{i=1}^{p} y_i \mathbf{A}_i$$

is minimized. Formulate this optimization problem as an SDP problem.

14.6 (a) Prove that if $\mathbf{X} \succ \mathbf{0}$ or $\mathbf{S} \succ \mathbf{0}$, then

$$\mathbf{XS} = \tau \mathbf{I}$$

is equivalent to

$$\mathbf{XS} + \mathbf{SX} = 2\tau \mathbf{I}$$

(b) Give a numerical example to demonstrate that if the condition $\mathbf{X} \succ \mathbf{0}$ or $\mathbf{S} \succ \mathbf{0}$ is removed, then the equalities in part (a) are no longer equivalent.

14.7 (a) Verify the identity in Eq. (14.35).

(b) Verify the identity in Eq. (14.38).

14.8 By using Eq. (14.36) with a special symmetric matrix $\mathbf{K}$, each row of matrix $\mathbf{M} \odot \mathbf{N}$ can be determined for given matrices $\mathbf{M}$ and $\mathbf{N}$.

(a) Develop an algorithm that computes the symmetric Kronecker product $\mathbf{M} \odot \mathbf{N}$ for given $\mathbf{M}, \mathbf{N} \in R^{n \times n}$.

(b) Write a MATLAB program that implements the algorithm developed in part (a).

14.9 Prove Lemma 14.1.

14.10 Using Lemma 14.1, show that if matrices $\mathbf{M}$ and $\mathbf{N}$ commute then the solution of the Lyapunov equation

$$\mathbf{M}\mathbf{X}\mathbf{N}^T + \mathbf{N}\mathbf{X}\mathbf{M}^T = \mathbf{B}$$

where $\mathbf{B} \in \mathcal{S}^n$, $\mathbf{M}$, and $\mathbf{N}$ are given, can be expressed as

$$\mathbf{X} = \mathbf{V}\mathbf{C}\mathbf{V}^T$$

where $\mathbf{V} = [\mathbf{v}_1 \ \mathbf{v}_2 \ \cdots \ \mathbf{v}_n]$ is defined by the eigenvectors $\mathbf{v}_i$ for $1 \le i \le n$ in Lemma 14.1, and $\mathbf{C}$ is obtained by calculating $\mathbf{V}^T\mathbf{B}\mathbf{V}$ and dividing its elements by $(\alpha_i\beta_j + \beta_i\alpha_j)$ componentwise.

14.11 Verify that $\{\Delta\mathbf{x}, \ \Delta\mathbf{y}, \ \Delta\mathbf{s}\}$ obtained from Eq. (14.43) solves Eq. (14.42).

14.12 Given a positive-definite matrix $\mathbf{X}$, show that every element in the Dikin ellipsoid $D(\mathbf{X})$ defined by Eq. (14.61) is a positive-definite matrix.

14.13 Show that if matrices $\mathbf{F}_i$ for $i = 1, 2, \ldots, p$ are linearly independent, then matrix $\hat{\mathbf{F}}$ given by Eq. (14.69c) is positive definite.

14.14 (a) Show that for a given real number f, $\mathcal{E}(f)$ given by Eq. (14.87) is a linear subspace of $\mathcal{E}$.

(b) Show that there exists a one-to-one correspondence between $\mathcal{E}(f)$ and hyperplane $\{\tilde{\mathbf{x}} : \ f(\tilde{\mathbf{x}}) = f\}$.

14.15 Show that for a given $\mathbf{X}_k \succ \mathbf{0}$, there exist matrices $\mathbf{C}_k$ and $\mathbf{D}_k$ in $\mathcal{E}$ such that Eqs. (14.91a) and (14.91b) hold for any $\mathbf{X} = \tilde{\mathcal{F}}\tilde{\mathbf{x}}$ in $\mathcal{E}$.
Hint: A proof of this fact can be carried out by letting

$$\mathbf{X} = \sum_{i=1}^{p+1} x_i\tilde{\mathbf{F}}_i, \quad \mathbf{C}_k = \sum_{i=1}^{p+1} \alpha_i\tilde{\mathbf{F}}_i, \quad \mathbf{D}_k = \sum_{i=1}^{p+1} \beta_i\tilde{\mathbf{F}}_i$$

and converting Eqs. (14.91a) and (14.91b) into a linear system of equations for $\boldsymbol{\alpha} = [\alpha_1 \ \alpha_2 \ \cdots \ \alpha_{p+1}]^T$ and $\boldsymbol{\beta} = [\beta_1 \ \beta_2 \ \cdots \ \beta_{p+1}]^T$, respectively.

14.16 Applying Algorithm 14.4, solve the SDP problem

$$\text{minimize } \mathbf{c}^T\mathbf{x}$$

$$\text{subect to: } \mathbf{F}(\mathbf{x}) = \mathbf{F}_i + \sum_{i=1}^{4} x_i\mathbf{F}_i \succeq \mathbf{0}$$

where $\mathbf{c} = [1 \ 0 \ 2 \ -1]^T$ and $\mathbf{F}_i$ for $i = 0, 1, \ldots, 4$ are given in Example 14.3.

14.17 Show that the Wolfe dual of the primal SOCP problem in Eq. (14.101) assumes the form in Eq. (14.102).

14.18 Show that the SOCP problem in Eq. (14.102) is equivalent to the SOCP problem in Eq. (14.104).

14.19 Given a column vector $\mathbf{u}$ and a nonnegative scalar t such that $\|\mathbf{u}\|_2 \leq t$, the matrix

$$\begin{bmatrix} t\mathbf{I} & \mathbf{u} \\ \mathbf{u}^T & t \end{bmatrix}$$

can be constructed. Show that the matrix is positive semidefinite.

14.20 Show that $w^2 \leq uv$, $u \geq 0$, and $v \geq 0$ if and only if

$$\left\| \begin{bmatrix} 2w \\ u - v \end{bmatrix} \right\| \leq u + v$$

14.21 Solve the shortest distance problem in Prob. 13.13 by using Algorithm 14.5.

14.22 Solve the least-square minimization problem in Prob. 13.15 by using Algorithm 14.5.

Chapter 15

GENERAL NONLINEAR OPTIMIZATION PROBLEMS

15.1   Introduction

The most general class of optimization problems is the class of problems where both the objective function and the constraints are nonlinear, as formulated in Eq. (10.1). These problems can be solved by using a variety of methods such as penalty- and barrier-function methods, gradient projection methods, and sequential quadratic-programming (SQP) methods [1]. Among these methods, SQP algorithms have proved highly effective for solving general constrained problems with smooth objective and constraint functions [2]. A more recent development in nonconvex constrained optimization is the extension of the modern interior-point approaches of Chaps. 12–14 to the general class of nonlinear problems.

In this chapter, we study two SQP algorithms for nonlinear problems first with equality and then with inequality constraints. In Sec. 15.3, we modify these algorithms by including a line-search step and an updating formula such as the Broyden-Fletcher-Goldfarb-Shanno (BFGS) updating formula to estimate the Hessian of the Lagrangian. In Sec. 15.4, we study an interior-point algorithm for nonconvex constrained problems, which is a direct extension of the primal-dual interior-point methods for linear-programming (LP) and quadratic-programming (QP) problems.

15.2   Sequential Quadratic Programming Methods

SQP methods for the general nonlinear constrained optimization problem in Eq. (10.1) were first studied during the sixties by Wilson [3], and a great deal of research has been devoted to this class of methods since that time. A recent survey of SQP algorithms can be found in [4].

15.2.1 SQP problems with equality constraints

Consider the optimization problem

$$\text{minimize } f(\mathbf{x}) \tag{15.1a}$$

$$\text{subject to: } a_i(\mathbf{x}) = 0 \qquad \text{for } i = 1, 2, \ldots, p \tag{15.1b}$$

where $f(\mathbf{x})$ and $a_i(\mathbf{x})$ are continuous functions which have continuous second partial derivatives. We assume that the feasible region $\mathcal{R}$ described by Eq. (15.1b) is nonempty and that $p \leq n$. From Chap. 10, we know that the first-order necessary conditions for $\mathbf{x}^*$ to be a local minimizer of the problem in Eq. (15.1) are that there exists a $\boldsymbol{\lambda}^* \in R^p$ such that

$$\nabla \mathcal{L}(\mathbf{x}^*, \boldsymbol{\lambda}^*) = \mathbf{0} \tag{15.2}$$

where $\mathcal{L}(\mathbf{x}, \boldsymbol{\lambda})$ is the Lagrangian defined by

$$\mathcal{L}(\mathbf{x}, \boldsymbol{\lambda}) = f(\mathbf{x}) - \sum_{i=1}^{p} \lambda_i a_i(\mathbf{x})$$

and the gradient operation in Eq. (15.2) is performed with respect to $\mathbf{x}$ and $\boldsymbol{\lambda}$, i.e.,

$$\nabla = \begin{bmatrix} \nabla_x \\ \nabla_\lambda \end{bmatrix}$$

(See Sec. 10.5.2 for the details.)

If set $\{\mathbf{x}_k, \boldsymbol{\lambda}_k\}$ is the kth iterate, which is assumed to be sufficiently close to $\{\mathbf{x}^*, \boldsymbol{\lambda}^*\}$, i.e., $\mathbf{x}_k \approx \mathbf{x}^*$ and $\boldsymbol{\lambda}_k \approx \boldsymbol{\lambda}^*$, we need to find an increment $\{\boldsymbol{\delta}_x, \boldsymbol{\delta}_\lambda\}$ such that the next iterate $\{\mathbf{x}_{k+1}, \boldsymbol{\lambda}_{k+1}\} = \{\mathbf{x}_k + \boldsymbol{\delta}_x, \boldsymbol{\lambda}_k + \boldsymbol{\delta}_\lambda\}$ is closer to $\{\mathbf{x}^*, \boldsymbol{\lambda}^*\}$. If we approximate $\nabla \mathcal{L}(\mathbf{x}_{k+1}, \boldsymbol{\lambda}_{k+1})$ by using the first two terms of the Taylor series of $\nabla \mathcal{L}$ for $\{\mathbf{x}_k, \boldsymbol{\lambda}_k\}$, i.e.,

$$\nabla \mathcal{L}(\mathbf{x}_{k+1}, \boldsymbol{\lambda}_{k+1}) \approx \nabla \mathcal{L}(\mathbf{x}_k, \boldsymbol{\lambda}_k) + \nabla^2 \mathcal{L}(\mathbf{x}_k, \boldsymbol{\lambda}_k) \begin{bmatrix} \boldsymbol{\delta}_x \\ \boldsymbol{\delta}_\lambda \end{bmatrix}$$

then $\{\mathbf{x}_{k+1}, \boldsymbol{\lambda}_{k+1}\}$ is an approximation of $\{\mathbf{x}^*, \boldsymbol{\lambda}^*\}$ if the increment $\{\boldsymbol{\delta}_x, \boldsymbol{\delta}_\lambda\}$ satisfies the equality

$$\nabla^2 \mathcal{L}(\mathbf{x}_k, \boldsymbol{\lambda}_k) \begin{bmatrix} \boldsymbol{\delta}_x \\ \boldsymbol{\delta}_\lambda \end{bmatrix} = -\nabla \mathcal{L}(\mathbf{x}_k, \boldsymbol{\lambda}_k) \tag{15.3}$$

More specifically, we can write Eq. (15.3) in terms of the Hessian of the Lagrangian, $\mathbf{W}$, for $\{\mathbf{x}, \boldsymbol{\lambda}\} = \{\mathbf{x}_k, \boldsymbol{\lambda}_k\}$ and the Jacobian, $\mathbf{A}$, for $\mathbf{x} = \mathbf{x}_k$ as

$$\begin{bmatrix} \mathbf{W}_k & -\mathbf{A}_k^T \\ -\mathbf{A}_k & \mathbf{0} \end{bmatrix} \begin{bmatrix} \boldsymbol{\delta}_x \\ \boldsymbol{\delta}_\lambda \end{bmatrix} = \begin{bmatrix} \mathbf{A}_k^T \boldsymbol{\lambda}_k - \mathbf{g}_k \\ \mathbf{a}_k \end{bmatrix} \tag{15.4a}$$

where

$$\mathbf{W}_k = \nabla_x^2 f(\mathbf{x}_k) - \sum_{i=1}^{p} (\boldsymbol{\lambda}_k)_i \nabla_x^2 a_i(\mathbf{x}_k) \tag{15.4b}$$

$$\mathbf{A}_k = \begin{bmatrix} \nabla_x^T a_1(\mathbf{x}_k) \\ \nabla_x^T a_2(\mathbf{x}_k) \\ \vdots \\ \nabla_x^T a_p(\mathbf{x}_k) \end{bmatrix} \tag{15.4c}$$

$$\mathbf{g}_k = \nabla_x f(\mathbf{x}_k) \tag{15.4d}$$

$$\mathbf{a}_k = \begin{bmatrix} a_1(\mathbf{x}_k) & a_2(\mathbf{x}_k) & \cdots & a_p(\mathbf{x}_k) \end{bmatrix}^T \tag{15.4e}$$

If $\mathbf{W}_k$ is positive definite and $\mathbf{A}_k$ has full row rank, then the matrix at the left-hand side of Eq. (15.4a) is nonsingular and symmetric and the system of equations in Eq. (15.4a) can be solved efficiently for $\{\boldsymbol{\delta}_x, \boldsymbol{\delta}_\lambda\}$ as shown in Chap. 4 of [5].

Eq. (15.4a) can also be written as

$$\mathbf{W}_k \boldsymbol{\delta}_x + \mathbf{g}_k = \mathbf{A}_k^T \boldsymbol{\lambda}_{k+1} \tag{15.5a}$$

$$\mathbf{A}_k \boldsymbol{\delta}_x = -\mathbf{a}_k \tag{15.5b}$$

and these equations may be interpreted as the first-order necessary conditions for $\boldsymbol{\delta}_x$ to be a local minimizer of the QP problem

$$\text{minimize } \tfrac{1}{2}\boldsymbol{\delta}^T \mathbf{W}_k \boldsymbol{\delta} + \boldsymbol{\delta}^T \mathbf{g}_k \tag{15.6a}$$

$$\text{subject to: } \mathbf{A}_k \boldsymbol{\delta} = -\mathbf{a}_k \tag{15.6b}$$

If $\mathbf{W}_k$ is positive definite and $\mathbf{A}_k$ has full row rank, the minimizer of the problem in Eq. (15.6) can be found by using, for example, the methods discussed in Sec. 13.2. Once the minimizer, $\boldsymbol{\delta}_x$, is obtained, the next iterate is set to $\mathbf{x}_{k+1} = \mathbf{x}_k + \boldsymbol{\delta}_x$ and the Lagrange multiplier vector[1] $\boldsymbol{\lambda}_{k+1}$ is determined as

$$\boldsymbol{\lambda}_{k+1} = (\mathbf{A}_k \mathbf{A}_k^T)^{-1} \mathbf{A}_k (\mathbf{W}_k \boldsymbol{\delta}_x + \mathbf{g}_k) \tag{15.7}$$

by using Eq. (15.5a). With $\mathbf{x}_{k+1}$ and $\boldsymbol{\lambda}_{k+1}$ known, $\mathbf{W}_{k+1}$, $\mathbf{g}_{k+1}$, $\mathbf{A}_{k+1}$, and $\mathbf{a}_{k+1}$ can be evaluated. The iterations are continued until $||\boldsymbol{\delta}_x||$ is sufficiently small to terminate the algorithm. We see that the entire solution procedure consists of solving a series of QP subproblems in a sequential manner and, as a consequence, the method is often referred to as the *sequential quadratic-programming* (SQP) *method*.

From Eq. (15.3), we observe that the correct increments $\boldsymbol{\delta}_x$ and $\boldsymbol{\delta}_\lambda$ are actually the Newton direction for the Lagrangian $\mathcal{L}(\mathbf{x}, \boldsymbol{\lambda})$. For this reason, the above

---

[1]Hereafter this will be referred to as the Lagrange multiplier for the sake of simplicity.

method (along with the method for SQP problems described in Sec. 15.2.2) is sometimes referred to as the Lagrange-Newton method in the literature [1]. It should be stressed at this point that the involvement of the Lagrangian in this method is crucial. To see this more clearly, note that for an increment δ satisfying the constraints in Eq. (15.6b), we can write

$$\delta^T \mathbf{g}_k = \delta^T (\mathbf{g}_k - \mathbf{A}_k^T \boldsymbol{\lambda}_k) + \delta^T \mathbf{A}_k^T \boldsymbol{\lambda}_k$$
$$= \delta^T (\mathbf{g}_k - \mathbf{A}_k^T \boldsymbol{\lambda}_k) - \mathbf{a}_k^T \boldsymbol{\lambda}_k$$
$$= \delta^T \nabla_x \mathcal{L}(\mathbf{x}_k, \boldsymbol{\lambda}_k) + c_k$$

where c_k is independent of δ. Therefore, the problem in Eq. (15.6) can be stated as the QP problem

$$\text{minimize } \{\tfrac{1}{2}\delta^T [\nabla_x^2 \mathcal{L}(\mathbf{x}_k, \boldsymbol{\lambda}_k)]\delta + \delta^T \nabla_x \mathcal{L}(\mathbf{x}_k, \boldsymbol{\lambda}_k) + c_k\} \qquad (15.8a)$$
$$\text{subject to: } \mathbf{A}_k \delta = -\mathbf{a}_k \qquad (15.8b)$$

In effect, the QP problem in Eq. (15.6) essentially *entails minimizing the second-order approximation of the Lagrangian* $\mathcal{L}(\mathbf{x}, \boldsymbol{\lambda})$ *rather than the objective function* $f(\mathbf{x})$.

The SQP method can be implemented in terms of the following algorithm.

Algorithm 15.1 SQP algorithm for nonlinear problems with equality constraints
Step 1
Set $\{\mathbf{x}, \boldsymbol{\lambda}\} = \{\mathbf{x}_0, \boldsymbol{\lambda}_0\}$, $k = 0$, and initialize the tolerance ε.
Step 2
Evaluate $\mathbf{W}_k$, $\mathbf{A}_k$, $\mathbf{g}_k$, and $\mathbf{a}_k$ using Eqs. (15.4b) – (15.4e).
Step 3
Solve the QP problem in Eq. (15.6) for δ and compute Lagrange multiplier $\boldsymbol{\lambda}_{k+1}$ using Eq. (15.7).
Step 4
Set $\mathbf{x}_{k+1} = \mathbf{x}_k + \delta_x$. If $\|\delta_x\| \leq \varepsilon$, output $\mathbf{x}^* = \mathbf{x}_{k+1}$ and stop; otherwise, set $k = k + 1$, and repeat from Step 2.

Example 15.1 Apply Algorithm 15.1 to the minimization problem

$$\text{minimize } f(\mathbf{x}) = -x_1^4 - 2x_2^4 - x_3^4 - x_1^2 x_2^2 - x_1^2 x_3^2$$

$$\text{subject to: } a_1(\mathbf{x}) = x_1^4 + x_2^4 + x_3^4 - 25 = 0$$
$$a_2(\mathbf{x}) = 8x_1^2 + 14x_2^2 + 7x_3^2 - 56 = 0$$

Solution With $\mathbf{x}_k = [x_1\ x_2]^T$ and $\boldsymbol{\lambda}_k = [\lambda_1\ \lambda_2]^T$, Step 2 of Algorithm 15.1 gives

$$
\mathbf{W}_k =
\begin{bmatrix}
-12x_1^2 - 2x_2^2 - 2x_3^2 & -4x_1x_2 & -4x_1x_3 \\
-12\lambda_1 x_1^2 - 16\lambda_2 & & \\
& -24x_2^2 - 2x_1^2 & \\
-4x_1x_2 & -12\lambda_1 x_2^2 - 28\lambda_2 & 0 \\
& & -12x_3^2 - 2x_1^2 \\
-4x_1x_3 & 0 & -12\lambda_1 x_3^2 - 14\lambda_2
\end{bmatrix}
\tag{15.9}
$$

$$
\mathbf{g}_k =
\begin{bmatrix}
-4x_1^3 - 2x_1x_2^2 - 2x_1x_3^2 \\
-8x_2^3 - 2x_1^2 x_2 \\
-4x_3^3 - 2x_1^2 x_3
\end{bmatrix}
$$

$$
\mathbf{A}_k =
\begin{bmatrix}
4x_1^3 & 4x_2^3 & 4x_3^3 \\
16x_1 & 28x_2 & 14x_3
\end{bmatrix}
\tag{15.10}
$$

$$
\mathbf{a}_k =
\begin{bmatrix}
x_1^4 + x_2^4 + x_3^4 - 25 \\
8x_1^2 + 14x_2^2 + 7x_3^2 - 56
\end{bmatrix}
$$

With $\mathbf{x}_0 = [3\ 1.5\ 3]^T$, $\boldsymbol{\lambda}_0 = [-1\ -1]^T$, and $\varepsilon = 10^{-8}$, it took Algorithm 15.1 10 iterations to converge to

$$
\mathbf{x}^* =
\begin{bmatrix}
1.874065 \\
0.465820 \\
1.884720
\end{bmatrix}, \quad
\boldsymbol{\lambda}^* =
\begin{bmatrix}
-1.223464 \\
-0.274937
\end{bmatrix}
$$

and $f(\mathbf{x}^*) = -38.384828$. To examine whether or not $\mathbf{x}^*$ is a local minimizer, we can compute the Jacobian of the constraints in Eq. (15.10) at $\mathbf{x}^*$ and perform the QR decomposition of $\mathbf{A}^T(\mathbf{x}^*)$ as

$$
\mathbf{A}^T(\mathbf{x}^*) = \mathbf{Q}
\begin{bmatrix}
\mathbf{R} \\
\mathbf{0}
\end{bmatrix}
$$

where $\mathbf{Q} \in R^{3\times 3}$ is an orthogonal matrix. Since $\text{rank}[\mathbf{A}(\mathbf{x}^*)] = 2$, the null space of $\mathbf{A}(\mathbf{x}^*)$ is a one-dimensional subspace in E^3 which is spanned by the last column of $\mathbf{Q}$, i.e.,

$$
\mathbf{N}(\mathbf{x}^*) = \mathbf{Q}(:, 3) =
\begin{bmatrix}
-0.696840 \\
0.222861 \\
0.681724
\end{bmatrix}
$$

where $\mathbf{Q}(:, 3)$ denotes the third column of matrix $\mathbf{Q}$. This leads to

$$
\mathbf{N}^T(\mathbf{x}^*)\nabla_x^2 \mathcal{L}(\mathbf{x}^*,\ \boldsymbol{\lambda}^*)\mathbf{N}(\mathbf{x}^*) = 20.4 > 0
$$

Therefore, $\mathbf{x}^*$ is a local minimizer of the problem.

■

15.2.2 SQP problems with inequality constraints

In this section, we extend Algorithm 15.1 to the case of inequality constraints. Let us consider the general optimization problem

$$\text{minimize } f(\mathbf{x}) \tag{15.11a}$$

$$\text{subject to: } c_j(\mathbf{x}) \geq 0 \quad \text{for } j = 1, 2, \ldots, q \tag{15.11b}$$

where $f(\mathbf{x})$ and $c_j(\mathbf{x})$ are continuous and have continuous second partial derivatives, and the feasible region $\mathcal{R}$ described by Eq. (15.11b) is nonempty. Motivated by the SQP method for equality constraints studied in Sec. 15.2.1, we need to find an increment $\{\boldsymbol{\delta}_x, \boldsymbol{\delta}_\mu\}$ for the kth iterate $\{\mathbf{x}_k, \boldsymbol{\mu}_k\}$ such that the next iterate $\{\mathbf{x}_{k+1}, \boldsymbol{\mu}_{k+1}\} = \{\mathbf{x}_k + \boldsymbol{\delta}_x, \boldsymbol{\mu}_k + \boldsymbol{\delta}_\mu\}$ approximates the Karush-Kuhn-Tucker (KKT) conditions

$$\nabla_x \mathcal{L}(\mathbf{x}, \boldsymbol{\mu}) = \mathbf{0}$$

$$c_j(\mathbf{x}) \geq 0 \quad \text{for } j = 1, 2, \ldots, q$$

$$\boldsymbol{\mu} \geq \mathbf{0}$$

$$\mu_j c_j(\mathbf{x}) = 0 \quad \text{for } j = 1, 2, \ldots, q$$

in the sense that

$$\nabla_x \mathcal{L}(\mathbf{x}_{k+1}, \boldsymbol{\mu}_{k+1}) \approx \nabla_x \mathcal{L}(\mathbf{x}_k, \boldsymbol{\mu}_k) + \nabla_x^2 \mathcal{L}(\mathbf{x}_k, \boldsymbol{\mu}_k) \boldsymbol{\delta}_x$$
$$+ \nabla_{x\mu}^2 \mathcal{L}(\mathbf{x}_k, \boldsymbol{\mu}_k) \boldsymbol{\delta}_\mu = \mathbf{0} \tag{15.12a}$$

$$c_j(\mathbf{x}_k + \boldsymbol{\delta}_x) \approx c_j(\mathbf{x}_k) + \boldsymbol{\delta}_x^T \nabla_x c_j(\mathbf{x}_k) \geq 0 \text{ for } j = 1, 2, \ldots, q \tag{15.12b}$$

$$\boldsymbol{\mu}_{k+1} \geq \mathbf{0} \tag{15.12c}$$

and

$$[c_j(\mathbf{x}_k) + \boldsymbol{\delta}_x^T \nabla_x c_j(\mathbf{x}_k)](\mu_{k+1})_j = 0 \quad \text{for } j = 1, 2, \ldots, q \tag{15.12d}$$

The Lagrangian $\mathcal{L}(\mathbf{x}, \boldsymbol{\mu})$ in this case is defined as

$$\mathcal{L}(\mathbf{x}, \boldsymbol{\mu}) = f(\mathbf{x}) - \sum_{j=1}^{q} \mu_j c_j(\mathbf{x}) \tag{15.13}$$

Hence

$$\nabla_x \mathcal{L}(\mathbf{x}_k, \boldsymbol{\mu}_k) = \nabla_x f(\mathbf{x}_k) - \sum_{j=1}^{q} (\mu_k)_j \nabla_x c_j(\mathbf{x}_k) = \mathbf{g}_k - \mathbf{A}_k^T \boldsymbol{\mu}_k$$

$$\nabla_x^2 \mathcal{L}(\mathbf{x}_k, \boldsymbol{\mu}_k) = \nabla_x^2 f(\mathbf{x}_k) - \sum_{j=1}^{q} (\mu_k)_j \nabla_x^2 c_j(\mathbf{x}_k) = \mathbf{Y}_k \tag{15.14a}$$

and
$$\nabla^2_{x\mu}\mathcal{L}(\mathbf{x}_k, \boldsymbol{\mu}_k) = -\mathbf{A}^T_k$$
where $\mathbf{A}_k$ is the Jacobian of the constraints at $\mathbf{x}_k$, i.e.,

$$\mathbf{A}_k = \begin{bmatrix} \nabla^T_x c_1(\mathbf{x}_k) \\ \nabla^T_x c_2(\mathbf{x}_k) \\ \vdots \\ \nabla^T_x c_q(\mathbf{x}_k) \end{bmatrix} \tag{15.14b}$$

The approximate KKT conditions in Eq. (15.12) can now be expressed as

$$\mathbf{Y}_k\boldsymbol{\delta}_x + \mathbf{g}_k - \mathbf{A}^T_k\boldsymbol{\mu}_{k+1} = \mathbf{0} \tag{15.15a}$$

$$\mathbf{A}_k\boldsymbol{\delta}_x \geq -\mathbf{c}_k \tag{15.15b}$$

$$\boldsymbol{\mu}_{k+1} \geq \mathbf{0} \tag{15.15c}$$

$$(\boldsymbol{\mu}_{k+1})_j(\mathbf{A}_k\boldsymbol{\delta}_x + \mathbf{c}_k)_j = 0 \quad \text{for } j = 1, 2, \ldots, q \tag{15.15d}$$

where

$$\mathbf{c}_k = [c_1(\mathbf{x}_k) \; c_2(\mathbf{x}_k) \; \cdots \; c_q(\mathbf{x}_k)]^T \tag{15.16}$$

Given $(\mathbf{x}_k, \boldsymbol{\mu}_k)$, Eq. (15.15) may be interpreted as the *exact* KKT conditions of the QP problem

$$\text{minimize } \tfrac{1}{2}\boldsymbol{\delta}^T\mathbf{Y}_k\boldsymbol{\delta} + \boldsymbol{\delta}^T\mathbf{g}_k \tag{15.17a}$$

$$\text{subject to: } \mathbf{A}_k\boldsymbol{\delta} \geq -\mathbf{c}_k \tag{15.17b}$$

If $\boldsymbol{\delta}_x$ is a *regular* solution of the QP subproblem in Eq. (15.17) in the sense that the gradients of those constraints that are active at $\mathbf{x}_k$ are linearly independent, then Eq. (15.15a) can be written as

$$\mathbf{Y}_k\boldsymbol{\delta}_x + \mathbf{g}_k - \mathbf{A}^T_{ak}\hat{\boldsymbol{\mu}}_{k+1} = \mathbf{0}$$

where the rows of $\mathbf{A}_{ak}$ are those rows of $\mathbf{A}_k$ satisfying the equality $(\mathbf{A}_k\boldsymbol{\delta}_x + \mathbf{c}_k)_j = 0$ and $\hat{\boldsymbol{\mu}}_{k+1}$ denotes the associated Lagrange multiplier vector. Hence $\hat{\boldsymbol{\mu}}_{k+1}$ can be computed as

$$\hat{\boldsymbol{\mu}}_{k+1} = (\mathbf{A}_{ak}\mathbf{A}^T_{ak})^{-1}\mathbf{A}_{ak}(\mathbf{Y}_k\boldsymbol{\delta}_x + \mathbf{g}_k) \tag{15.18}$$

It follows from the complementarity condition in Eq. (15.15d) that the Lagrange multiplier $\boldsymbol{\mu}_{k+1}$ can be obtained by inserting zeros where necessary in $\hat{\boldsymbol{\mu}}_{k+1}$.

Since the key objective in the above method is to solve the QP subproblem in each iteration, the method is referred to as the *SQP method for general nonlinear minimization problems with inequality constraints*. As in the case of equality constraints, the quadratic function involved in the QP problem in Eq. (15.17) is associated with the Lagrangian $\mathcal{L}(\mathbf{x}, \boldsymbol{\mu})$ in Eq. (15.13) rather than the objective function $f(\mathbf{x})$ and, as can be seen in Eq. (15.12a), the increment $\boldsymbol{\delta}_x$ obtained by solving Eq. (15.17) is the Newton direction of the Lagrangian with respect to variable $\mathbf{x}$.

The above SQP method can be implemented in terms of the following algorithm.

Algorithm 15.2 SQP algorithm for nonlinear problems with inequality constraints

Step 1
Initialize $\{\mathbf{x}, \boldsymbol{\mu}\} = \{\mathbf{x}_0, \boldsymbol{\mu}_0\}$ where $\mathbf{x}_0$ and $\boldsymbol{\mu}_0$ are chosen such that $c_j(\mathbf{x}_0) \geq 0$ $(j = 1, 2, \ldots, q)$ and $\boldsymbol{\mu}_0 \geq 0$.
Set $k = 0$ and initialize tolerance ε.

Step 2
Evaluate $\mathbf{Y}_k$, $\mathbf{A}_k$, $\mathbf{g}_k$ and $\mathbf{c}_k$ using Eqs. (15.14a), (15.14b), (15.4d), and (15.16), respectively.

Step 3
Solve the QP problem in Eq. (15.17) for $\boldsymbol{\delta}_x$ and compute Lagrange multiplier $\hat{\boldsymbol{\mu}}_{k+1}$ using Eq. (15.18).

Step 4
Set $\mathbf{x}_{k+1} = \mathbf{x}_k + \boldsymbol{\delta}_x$. If $\|\boldsymbol{\delta}_x\| \leq \varepsilon$, output $\mathbf{x}^* = \mathbf{x}_{k+1}$ and stop; otherwise, set $k = k + 1$, and repeat from Step 2.

Example 15.2 Apply Algorithm 15.2 to solve the shortest distance problem discussed in Example 13.5.

Solution As was discussed in Example 13.5, the problem can be formulated as the constrained minimization problem

$$\text{minimize } f(\mathbf{x}) = \tfrac{1}{2}[(x_1 - x_3)^2 + (x_2 - x_4)^2]$$

$$\text{subject to:} \quad c_1(\mathbf{x}) = -[x_1 \; x_2]\begin{bmatrix} \frac{1}{4} & 0 \\ 0 & 1 \end{bmatrix}\begin{bmatrix} x_1 \\ x_2 \end{bmatrix} + [x_1 \; x_2]\begin{bmatrix} \frac{1}{2} \\ 0 \end{bmatrix} + \tfrac{3}{4} \geq 0$$

$$c_2(\mathbf{x}) = -\tfrac{1}{8}[x_3 \; x_4]\begin{bmatrix} 5 & 3 \\ 3 & 5 \end{bmatrix}\begin{bmatrix} x_3 \\ x_4 \end{bmatrix} + [x_3 \; x_4]\begin{bmatrix} \frac{11}{2} \\ \frac{13}{2} \end{bmatrix} - \tfrac{35}{2} \geq 0$$

where $\mathbf{x} = [x_1 \; x_2 \; x_3 \; x_4]^T$. Since both the objective function and the constraints are quadratic, Hessian $\mathbf{Y}_k$ is independent of $\mathbf{x}_k$ and is given by

$$\mathbf{Y}_k = \begin{bmatrix} 1 + \mu_1/2 & 0 & -1 & 0 \\ 0 & 1 + 2\mu_1 & 0 & -1 \\ -1 & 0 & 1 + 5\mu_2/4 & 3\mu_2/4 \\ 0 & -1 & 3\mu_2/4 & 1 + 5\mu_2/4 \end{bmatrix}$$

where $\boldsymbol{\mu}_k = [\mu_1 \ \mu_2]^T$, and $\mathbf{Y}_k$ is positive definite as long as $\boldsymbol{\mu}_k > 0$. With

$$\mathbf{x}_0 = \begin{bmatrix} 1.0 \\ 0.5 \\ 2.0 \\ 3.0 \end{bmatrix}, \quad \boldsymbol{\mu}_0 = \begin{bmatrix} 1 \\ 1 \end{bmatrix}, \quad \text{and} \quad \varepsilon = 10^{-5}$$

it took Algorithm 15.2 seven iterations and 49.8 Kflops to converge to the solution

$$\mathbf{x}^* = \begin{bmatrix} 2.044750 \\ 0.852716 \\ 2.544913 \\ 2.485633 \end{bmatrix} \quad \text{and} \quad \boldsymbol{\mu}^* = \begin{bmatrix} 0.957480 \\ 1.100145 \end{bmatrix}$$

Hence the solution points $\mathbf{r}^* \in \mathcal{R}$ and $\mathbf{s}^* \in \mathcal{S}$ can be obtained as

$$\mathbf{r}^* = \begin{bmatrix} 2.044750 \\ 0.852716 \end{bmatrix} \quad \text{and} \quad \mathbf{s}^* = \begin{bmatrix} 2.544913 \\ 2.485633 \end{bmatrix}$$

Therefore, the shortest distance between $\mathcal{R}$ and $\mathcal{S}$ is given by $||\mathbf{r}^* - \mathbf{s}^*||_2 = 1.707800$.

∎

15.3 Modified SQP Algorithms

The SQP algorithms described in Sec. 15.2 have good local properties. As a matter of fact, it can be shown that if (a) the initial point $\mathbf{x}_0$ is sufficiently close to a local minimizer $\mathbf{x}^*$, (b) the coefficient matrix in Eq. (15.4a) for $\{\mathbf{x}, \boldsymbol{\lambda}\} = \{\mathbf{x}_0, \boldsymbol{\lambda}_0\}$ is nonsingular, (c) the second-order sufficient conditions hold for $\{\mathbf{x}^*, \boldsymbol{\lambda}^*\}$ with rank$[\mathbf{A}(\mathbf{x}^*)] = p$, and (d) for $\{\mathbf{x}_0, \boldsymbol{\lambda}_0\}$ the QP problem has a unique solution $\boldsymbol{\delta}_x$, then Algorithm 15.1 converges quadratically (see Chap. 12 of [1]), i.e., the order of convergence is two (see Sec. 3.7). The main disadvantage of Algorithms 15.1 and 15.2 is that they may fail to converge when the initial point is not sufficiently close to $\mathbf{x}^*$. Another disadvantage is associated with the need to evaluate the Hessian of the Lagrangian which combines the Hessian of the objective function with the Hessians of the constraints. Due to the large size of the Lagrangian, these algorithms would require a large number of function evaluations even for problems of moderate size.

In this section, we present two modifications for Algorithms 15.1 and 15.2. First, we describe a line-search method proposed in [6] that enables the two SQP algorithms to converge with arbitrary initial points. Second, the Hessian of the Lagrangian is approximated using a BFGS formula proposed in [2][7]. As in the BFGS algorithm for unconstrained minimization problems (see Chap. 7), if we start with an initial approximation for the Hessian which is positive definite, then the subsequent approximations will remain positive definite under a certain mild condition. Consequently, this modification reduces the computational complexity and assures well-defined solutions of the QP subproblems involved.

15.3.1 SQP Algorithms with a Line-Search Step

Considering the constrained optimization problem in Eq. (15.11), a line-search step can by introduced in the algorithm described in Sec. 15.2.2, namely, Algorithm 15.2, by generating the $(k + 1)$th iterate as

$$\mathbf{x}_{k+1} = \mathbf{x}_k + \alpha_k \boldsymbol{\delta}_x \tag{15.19}$$

where $\boldsymbol{\delta}_x$ is the solution of the QP subproblem in Eq. (15.17), and α_k is a scalar obtained by a line search. Han [6] has proposed that scalar α_k be obtained by minimizing the one-variable *exact penalty function*

$$\psi_h(\alpha) = f(\mathbf{x}_k + \alpha \boldsymbol{\delta}_x) + r \sum_{j=1}^{q} c_j(\mathbf{x}_k + \alpha \boldsymbol{\delta}_x)_- \tag{15.20}$$

on interval $[0, \ \delta]$ where the interval length δ and parameter $r > 0$ are fixed throughout the minimization process and function $c_j(\mathbf{x}_k + \alpha \boldsymbol{\delta}_x)_-$ is given by

$$c_j(\mathbf{x}_k + \alpha \boldsymbol{\delta}_x)_- = \max[0, \ -c_j(\mathbf{x}_k + \alpha \boldsymbol{\delta}_x)] \tag{15.21}$$

Note that the second term in Eq. (15.20) is always nonnegative and contains only those constraint functions that violate the nonnegativity condition in Eq. (15.11b). The term 'penalty function' for $\psi_h(\alpha)$ in Eq. (15.20) is related to the fact that the value of $\psi_h(\alpha)$ depends partly on how many constraints in Eq. (15.11b) are violated at α and the degree of violation. By choosing an appropriate value for r, the line search as applied to $\psi_h(\alpha)$ will yield a value α_k such that the objective function at $\mathbf{x}_k + \alpha_k \boldsymbol{\delta}_x$ is reduced with fewer violated constraints and a reduced degree of violation. Because of the operation in Eq. (15.21), the penalty function is, in general, nondifferentiable and, consequently, gradient-based line searches such as those described in Chap. 7 and [1] would not work with these algorithms. However, efficient search methods for the minimization of nondifferentiable one-dimensional functions are available. See, for example, [8] for a recent survey of direct search methods.

An alternative method of determining α_k in Eq. (15.19) was proposed by Powell [7]. In this method, an inexact line-search is applied to the Lagrangian in Eq. (15.13) with $\mathbf{x} = \mathbf{x}_k + \alpha \boldsymbol{\delta}_x$ and $\boldsymbol{\mu} = \boldsymbol{\mu}_{k+1}$ to obtain the following one-variable function

$$\psi_p(\alpha) = f(\mathbf{x}_k + \alpha \boldsymbol{\delta}_x) - \sum_{j=1}^{q} (\boldsymbol{\mu}_{k+1})_j c_j(\mathbf{x}_k + \alpha \boldsymbol{\delta}_x) \tag{15.22}$$

We note that if $\boldsymbol{\mu}_{k+1} \geq 0$, the second term in Eq. (15.22) acts as a penalty term since an α with $c_j(\mathbf{x}_k + \alpha \boldsymbol{\delta}_x) < 0$ will increase the value of $\psi_p(\alpha)$ and, consequently, minimizing $\psi_p(\alpha)$ would tend to reduce the objective function along the search direction, $\boldsymbol{\delta}_x$, with fewer violated constraints and a reduced degree of violation.

As was shown by Powell, $\psi_p(\alpha)$ decreases as α varies from 0 to a small positive value, which suggests a method for determining α_k in Eq. (15.19) by minimizing $\psi_p(\alpha)$ on the interval $[0, 1]$. First, we perform a line search to find the minimizer α_1^* of $\psi_p(\alpha)$ on $[0, 1]$, i.e.,

$$\alpha_1^* = \arg \left[\min_{0 \le \alpha \le 1} \psi_p(\alpha) \right] \tag{15.23}$$

where $\arg(\cdot)$ denotes the resulting argument of the minimum of $\psi_p(\alpha)$. Unlike the L_1 exact penalty function ψ_h, function $\psi_p(\alpha)$ is differentiable and hence efficient gradient-based algorithms for inexact line searches such as those described in Sec. 4.8 can be used to find α_1^*. Next, we note that if for a particular index j we have $(\mathbf{A}_k \boldsymbol{\delta}_x + \mathbf{c}_k)_j = 0$, then the complementarity condition in Eq. (15.15d) implies that $(\boldsymbol{\mu}_{k+1})_j = 0$ and in this case the term $(\boldsymbol{\mu}_{k+1})_j c_j(\mathbf{x}_k + \alpha \boldsymbol{\delta}_x)$ is not present in Eq. (15.22). In other words, the constraints $c_j(\mathbf{x}) \ge 0$ for which $(\boldsymbol{\mu}_{k+1})_j = 0$ need to be dealt with separately. To this end, we define index set

$$\mathcal{J} = \{j : (\boldsymbol{\mu}_{k+1})_j = 0\}$$

and evaluate

$$\alpha_2^* = \min_{j \in \mathcal{J}} \max\{\alpha : c_j(\mathbf{x}_k + \alpha \boldsymbol{\delta}_x) \ge 0\} \tag{15.24}$$

The value of α_k in Eq. (15.19) is then obtained as

$$\alpha_k = 0.95 \min\{\alpha_1^*, \alpha_2^*\} \tag{15.25}$$

Once α_k is determined, the increment $\boldsymbol{\delta}_x$ in Eq. (15.18) needs to be modified to $\alpha_k \boldsymbol{\delta}_x$ in order to compute $\hat{\boldsymbol{\mu}}_{k+1}$ and $\boldsymbol{\mu}_{k+1}$. In addition, in Step 4 of Algorithm 15.2, $\mathbf{x}_{k+1}$ should be set using Eq. (15.19). With the line-search step included, the modified SQP algorithm turns out to be more robust in the sense that it converges with arbitrary initial points.

15.3.2 SQP algorithms with approximated Hessian

The BFGS updating formula discussed in Chap. 7 as applied to the Lagrangian in Eq. (15.13) is given by

$$\mathbf{Y}_{k+1} = \mathbf{Y}_k + \frac{\boldsymbol{\gamma}_k \boldsymbol{\gamma}_k^T}{\boldsymbol{\delta}_x^T \boldsymbol{\gamma}_k} - \frac{\mathbf{Y}_k \boldsymbol{\delta}_x \boldsymbol{\delta}_x^T \mathbf{Y}_k}{\boldsymbol{\delta}_x^T \mathbf{Y}_k \boldsymbol{\delta}_x} \tag{15.26}$$

where $\boldsymbol{\delta}_x = \mathbf{x}_{k+1} - \mathbf{x}_k$ and

$$\begin{aligned} \boldsymbol{\gamma}_k &= \nabla_x \mathcal{L}(\mathbf{x}_{k+1}, \boldsymbol{\mu}_{k+1}) - \nabla_x \mathcal{L}(\mathbf{x}_k, \boldsymbol{\mu}_{k+1}) \\ &= (\mathbf{g}_{k+1} - \mathbf{g}_k) - (\mathbf{A}_{k+1} - \mathbf{A}_k)^T \boldsymbol{\mu}_{k+1} \end{aligned} \tag{15.27}$$

If $\mathbf{Y}_k$ is positive definite, then $\mathbf{Y}_{k+1}$ obtained using Eq. (15.26) is also positive definite if and only if

$$\boldsymbol{\delta}_x^T \boldsymbol{\gamma}_k > 0 \tag{15.28}$$

However, this condition does not hold when the Lagrangian has a negative curvature for iterate $\{\mathbf{x}_{k+1},\ \boldsymbol{\mu}_{k+1}\}$. Powell proposed a method for overcoming this difficulty [7], which has proven quite effective. This method entails replacing the $\boldsymbol{\gamma}_k$ in Eq. (15.26) by

$$\boldsymbol{\eta}_k = \theta\boldsymbol{\gamma}_k + (1-\theta)\mathbf{Y}_k\boldsymbol{\delta}_x \tag{15.29}$$

where $\boldsymbol{\gamma}_k$ is given by Eq. (15.27) and θ is determined as

$$\theta = \begin{cases} 1 & \text{if } \boldsymbol{\delta}_x^T\boldsymbol{\gamma}_k \geq 0.2\boldsymbol{\delta}_x^T\mathbf{Y}_k\boldsymbol{\delta}_x \\ \dfrac{0.8\boldsymbol{\delta}_x^T\mathbf{Y}_k\boldsymbol{\delta}_x}{\boldsymbol{\delta}_x^T\mathbf{Y}_k\boldsymbol{\delta}_k - \boldsymbol{\delta}_x^T\boldsymbol{\gamma}_k} & \text{otherwise} \end{cases} \tag{15.30}$$

By incorporating one of the line search methods and the BFGS approximation of the Hessian into Algorithm 15.2, we obtain the modified SQP algorithm summarized below.

Algorithm 15.3 Modified SQP algorithm for nonlinear problems with inequality constraints

Step 1
Set $\{\mathbf{x},\ \boldsymbol{\mu}\} = \{\mathbf{x}_0,\ \boldsymbol{\mu}_0\}$ where $\mathbf{x}_0$ and $\boldsymbol{\mu}_0$ are chosen such that $c_j(\mathbf{x}_0) \geq 0$ for $j = 1, 2, \ldots, q$ and $\boldsymbol{\mu}_0 \geq 0$.
Set $k = 0$ and initialize the tolerance ε.
Set $\mathbf{Y}_0 = \mathbf{I}_n$.

Step 2
Evaluate $\mathbf{A}_k$, $\mathbf{g}_k$, and $\mathbf{c}_k$ using Eqs. (15.14b), (15.4d), and (15.16), respectively.

Step 3
Solve the QP problem in Eq. (15.17) for $\boldsymbol{\delta}_x$ and compute Lagrange multiplier $\hat{\boldsymbol{\mu}}_{k+1}$ using Eq. (15.18).

Step 4
Compute α_k, the value of α by either minimizing $\psi_h(\alpha)$ in Eq. (15.20) or minimizing $\psi_p(\alpha)$ in Eq. (15.22) using Eqs. (15.23)–(15.25).

Step 5
Set $\boldsymbol{\delta}_x = \alpha_k\boldsymbol{\delta}_x$ and $\mathbf{x}_{k+1} = \mathbf{x}_k + \boldsymbol{\delta}_x$.
Compute $\hat{\boldsymbol{\mu}}_{k+1}$ using Eq. (15.18).

Step 6
If $\|\boldsymbol{\delta}_x\| \leq \varepsilon$, output solution $\mathbf{x}^* = \mathbf{x}_{k+1}$ and stop; otherwise, continue with Step 7.

Step 7
Evaluate γ_k, θ, and η_k using Eqs. (15.27), (15.30), and (15.29), respectively.
Compute $\mathbf{Y}_{k+1}$ using Eq. (15.26).
Set $k = k + 1$ and repeat from Step 2.

Example 15.3 Algorithms 15.2 and 15.3 were applied to solve the shortest distance problem in Example 15.2. With $\varepsilon = 10^{-5}$, $\boldsymbol{\mu}_0 = [1 \ 1]^T$, and six different initial points in the feasible region, the results summarized in Table 15.1 were obtained where the entries x/y (e.g., 7/55) denote the number of iterations and the number of Kflops required, respectively. Symbol '×' indicates that the algorithm did not converge. As can be seen, Algorithm 15.3, which combines the idea of sequential quadratic programming with Han's or Powell's line search and Powell's version of the BFGS updating formula, is considerably more robust than Algorithm 15.2.

Table 15.1. Test Results for Example 15.3

$\mathbf{x}_0$	$\begin{bmatrix} 1.0 \\ 0.5 \\ 2.0 \\ 3.0 \end{bmatrix}$	$\begin{bmatrix} 2 \\ 0 \\ 1 \\ 5 \end{bmatrix}$	$\begin{bmatrix} 3 \\ 0 \\ 1 \\ 5 \end{bmatrix}$	$\begin{bmatrix} 0 \\ 0 \\ 1 \\ 5 \end{bmatrix}$	$\begin{bmatrix} -1 \\ 0 \\ 1 \\ 5 \end{bmatrix}$	$\begin{bmatrix} 1.0 \\ -0.5 \\ 1.0 \\ 5.0 \end{bmatrix}$
Algorithm 15.2	7/55	×	×	×	×	×
Algorithm 15.3 with Han's line search	8/69	9/83	10/88	13/116	12/109	12/108
Algorithm 15.3 with Powell's line search	8/65	10/87	11/94	12/108	12/109	11/92

■

15.3.3 SQP problems with equality and inequality constraints

Having developed SQP algorithms for nonconvex optimization with either equality or inequality constraints, it is not difficult to extend SQP to the most general case where both equality and inequality constraints are present. This optimization problem was formulated in Eq. (10.1) and is of the form

$$\text{minimize } f(\mathbf{x}) \tag{15.31a}$$

$$\text{subject to: } a_i(\mathbf{x}) = 0 \quad \text{for } i = 1, 2, \ldots, p \tag{15.31b}$$

$$c_j(\mathbf{x}) \geq 0 \quad \text{for } j = 1, 2, \ldots, q \tag{15.31c}$$

Let $\{\mathbf{x}_k, \boldsymbol{\lambda}_k, \boldsymbol{\mu}_k\}$ be the kth iterate and $\{\boldsymbol{\delta}_x, \boldsymbol{\delta}_\lambda, \boldsymbol{\delta}_\mu\}$ be a set of increment vectors such that the KKT conditions

$$\nabla_x \mathcal{L}(\mathbf{x}, \boldsymbol{\lambda}, \boldsymbol{\mu}) = \mathbf{0}$$
$$a_i(\mathbf{x}) = 0 \qquad \text{for } i = 1, 2, \ldots, p$$
$$c_j(\mathbf{x}) \geq 0 \qquad \text{for } j = 1, 2, \ldots, q$$
$$\boldsymbol{\mu} \geq \mathbf{0}$$
$$\mu_j c_j(\mathbf{x}) = 0 \qquad \text{for } j = 1, 2, \ldots, q$$

where the Lagrangian $\mathcal{L}(\mathbf{x}, \boldsymbol{\lambda}, \boldsymbol{\mu})$ is defined as

$$\mathcal{L}(\mathbf{x}, \boldsymbol{\lambda}, \boldsymbol{\mu}) = f(\mathbf{x}) - \sum_{i=1}^{p} \lambda_i a_i(\mathbf{x}) - \sum_{j=1}^{q} \mu_j c_j(\mathbf{x})$$

are satisfied approximately at the next iterate $\{\mathbf{x}_{k+1}, \boldsymbol{\lambda}_{k+1}, \boldsymbol{\mu}_{k+1}\} = \{\mathbf{x}_k + \boldsymbol{\delta}_x, \boldsymbol{\lambda}_k + \boldsymbol{\delta}_\lambda, \boldsymbol{\mu}_k + \boldsymbol{\delta}_\mu\}$. By using arguments similar to those used in Sec. 15.2.2, we obtain the approximate KKT conditions as (see Prob. 15.4)

$$\mathbf{Z}_k \boldsymbol{\delta}_x + \mathbf{g}_k - \mathbf{A}_{ek}^T \boldsymbol{\lambda}_{k+1} - \mathbf{A}_{ik}^T \boldsymbol{\mu}_{k+1} = \mathbf{0} \tag{15.32a}$$
$$\mathbf{A}_{ek} \boldsymbol{\delta}_x = -\mathbf{a}_k \tag{15.32b}$$
$$\mathbf{A}_{ik} \boldsymbol{\delta}_x \geq -\mathbf{c}_k \tag{15.32c}$$
$$\boldsymbol{\mu}_{k+1} \geq \mathbf{0} \tag{15.32d}$$
$$(\boldsymbol{\mu}_{k+1})_j (\mathbf{A}_{ik} \boldsymbol{\delta}_x + \mathbf{c}_k)_j = 0 \qquad \text{for } j = 1, 2, \ldots, q \tag{15.32e}$$

where

$$\mathbf{Z}_k = \nabla_x^2 f(\mathbf{x}_k) - \sum_{i=1}^{p} (\boldsymbol{\lambda}_k)_i \nabla_x^2 a_i(\mathbf{x}_k) - \sum_{j=1}^{q} (\boldsymbol{\mu}_k)_j \nabla_x^2 c_j(\mathbf{x}_k) \tag{15.32f}$$

$$\mathbf{g}_k = \nabla_x f(\mathbf{x}_k) \tag{15.32g}$$

$$\mathbf{A}_{ek} = \begin{bmatrix} \nabla_x^T a_1(\mathbf{x}_k) \\ \vdots \\ \nabla_x^T a_p(\mathbf{x}_k) \end{bmatrix} \tag{15.32h}$$

$$\mathbf{A}_{ik} = \begin{bmatrix} \nabla_x^T c_1(\mathbf{x}_k) \\ \vdots \\ \nabla_x^T c_q(\mathbf{x}_k) \end{bmatrix} \tag{15.32i}$$

and $\mathbf{a}_k$ and $\mathbf{c}_k$ are given by Eqs. (15.4e) and (15.16), respectively.

Given $(\mathbf{x}_k, \boldsymbol{\lambda}_k, \boldsymbol{\mu}_k)$, Eqs. (15.32a)–(15.32e) may be interpreted as the exact KKT conditions of the QP problem

$$\text{minimize } \tfrac{1}{2}\boldsymbol{\delta}^T \mathbf{Z}_k \boldsymbol{\delta} + \boldsymbol{\delta}^T \mathbf{g}_k \tag{15.33a}$$

$$\text{subject to:} \quad \mathbf{A}_{ek}\boldsymbol{\delta} = -\mathbf{a}_k \tag{15.33b}$$

$$\mathbf{A}_{ik}\boldsymbol{\delta} \geq -\mathbf{c}_k \tag{15.33c}$$

Note that if $\boldsymbol{\delta}_x$ is a regular solution of the QP subproblem in Eq. (15.33), then Eq. (15.32a) can be written as

$$\mathbf{Z}_k\boldsymbol{\delta}_x + \mathbf{g}_k - \mathbf{A}_{ek}^T \boldsymbol{\lambda}_{k+1} - \mathbf{A}_{aik}^T \hat{\boldsymbol{\mu}}_{k+1} = 0$$

where matrix $\mathbf{A}_{aik}$ is composed of those rows of $\mathbf{A}_{ik}$ that satisfy the equality $(\mathbf{A}_{ik}\boldsymbol{\delta}_x + \mathbf{c}_k)_j = 0$, and $\hat{\boldsymbol{\mu}}_{k+1}$ denotes the associated Lagrange multiplier. It follows that $\boldsymbol{\lambda}_{k+1}$ and $\hat{\boldsymbol{\mu}}_{k+1}$ can be computed as

$$\begin{bmatrix} \boldsymbol{\lambda}_{k+1} \\ \hat{\boldsymbol{\mu}}_{k+1} \end{bmatrix} = (\mathbf{A}_{ak}\mathbf{A}_{ak}^T)^{-1}\mathbf{A}_{ak}(\mathbf{Z}_k\boldsymbol{\delta}_x + \mathbf{g}_k) \tag{15.34}$$

where

$$\mathbf{A}_{ak} = \begin{bmatrix} \mathbf{A}_{ek} \\ \mathbf{A}_{aik} \end{bmatrix}$$

With $\hat{\boldsymbol{\mu}}_{k+1}$ known, the Lagrange multiplier $\boldsymbol{\mu}_{k+1}$ can be obtained by inserting zeros where necessary in $\hat{\boldsymbol{\mu}}_{k+1}$.

As for the development of Algorithm 15.3, a more robust and efficient SQP algorithm for the general nonconvex optimization problem in Eq. (15.31) can be obtained by incorporating a line-search step in the algorithm and using an approximate Hessian.

If $\boldsymbol{\delta}_x$ is the solution of the QP problem in Eq. (15.33), then the $(k+1)$th iterate assumes the form

$$\mathbf{x}_{k+1} = \mathbf{x}_k + \alpha_k \boldsymbol{\delta}_x \tag{15.35}$$

where α_k is determined as follows. First, we introduce a merit function as

$$\psi(\alpha) = f(\mathbf{x}_k + \alpha\boldsymbol{\delta}_x) + \beta \sum_{k=1}^{p} a_i^2(\mathbf{x}_k + \alpha\boldsymbol{\delta}_x) - \sum_{j=1}^{q}(\boldsymbol{\mu}_{k+1})_j c_j(\mathbf{x}_k + \alpha\boldsymbol{\delta}_x)$$

$$\tag{15.36}$$

where β is a sufficiently large positive scalar. Function $\psi(\alpha)$ is a natural generalization of function $\psi_p(\alpha)$ in Eq. (15.22) and can be obtained by including a term related to the equality constraints in Eq. (15.31b). Evidently, minimizing $\psi(\alpha)$ reduces the objective function along the search direction $\boldsymbol{\delta}_x$ and, at the same time, reduces the degree of violation for both the equality and inequality

516

constraints. Let the value of α that minimizes function $\psi(\alpha)$ in Eq. (15.36) on the interval $0 \leq \alpha \leq 1$ be

$$\alpha_1^* = \arg \left[\min_{0 \leq \alpha \leq 1} \psi(\alpha) \right] \tag{15.37}$$

Second, by following an argument similar to that in Sec. 15.3.1, we define index set

$$\mathcal{J} = \{j : (\boldsymbol{\mu}_{k+1})_j = 0\} \tag{15.38}$$

and compute

$$\alpha_2^* = \max\{\alpha : c_j(\mathbf{x}_k + \alpha\boldsymbol{\delta}_x) \geq 0, \ j \in \mathcal{J}\} \tag{15.39}$$

The value of α_k in Eq. (15.35) is then calculated as

$$\alpha_k = 0.95 \min\{\alpha_1^*, \ \alpha_2^*\} \tag{15.40}$$

Having determined $\mathbf{x}_{k+1}$ with Eq. (15.35), an approximate Hessian $\mathbf{Z}_{k+1}$ can be evaluated by using the modified BFGS updating formula as

$$\mathbf{Z}_{k+1} = \mathbf{Z}_k + \frac{\boldsymbol{\eta}_k \boldsymbol{\eta}_k^T}{\boldsymbol{\delta}_x^T \boldsymbol{\eta}_k} - \frac{\mathbf{Z}_k \boldsymbol{\delta}_x \boldsymbol{\delta}_x^T \mathbf{Z}_k}{\boldsymbol{\delta}_x^T \mathbf{Z}_k \boldsymbol{\delta}_x} \tag{15.41}$$

where

$$\boldsymbol{\eta}_k = \theta\boldsymbol{\gamma}_k + (1-\theta)\mathbf{Z}_k\boldsymbol{\delta}_x \tag{15.42}$$
$$\boldsymbol{\gamma}_k = (\mathbf{g}_{k+1} - \mathbf{g}_k) - (\mathbf{A}_{e,k+1} - \mathbf{A}_{e,k})^T \boldsymbol{\lambda}_{k+1} - (\mathbf{A}_{i,k+1} - \mathbf{A}_{i,k})^T \boldsymbol{\mu}_{k+1} \tag{15.43}$$

$$\theta = \begin{cases} 1 & \text{if } \boldsymbol{\delta}_x^T\boldsymbol{\gamma}_k \geq 0.2\,\boldsymbol{\delta}_x^T\mathbf{Z}_k\boldsymbol{\delta}_x \\ \dfrac{0.8\boldsymbol{\delta}_x^T\mathbf{Z}_k\boldsymbol{\delta}_x}{\boldsymbol{\delta}_x^T\mathbf{Z}_k\boldsymbol{\delta}_x - \boldsymbol{\delta}_x^T\boldsymbol{\gamma}_k} & \text{otherwise} \end{cases} \tag{15.44}$$

The above SQP procedure can be implemented in terms of the following algorithm.

Algorithm 15.4 SQP algorithm for nonlinear problems with equality and inequality constraints
Step 1
Set $\{\mathbf{x}, \boldsymbol{\lambda}, \boldsymbol{\mu}\} = \{\mathbf{x}_0, \boldsymbol{\lambda}_0, \boldsymbol{\mu}_0\}$ where $\mathbf{x}_0$ and $\boldsymbol{\mu}_0$ are chosen such that $c_j(\mathbf{x}_0) \geq 0$ for $j = 1, 2, \ldots, q$ and $\boldsymbol{\mu}_0 \geq \mathbf{0}$.
Set $k = 0$ and initialize the tolerance ε.
Set $\mathbf{Z}_0 = \mathbf{I}_n$.

Step 2
Evaluate $\mathbf{g}_k$, $\mathbf{A}_{ek}$, $\mathbf{A}_{ik}$, $\mathbf{a}_k$, and $\mathbf{c}_k$ using Eqs. (15.32g), (15.32h), (15.32i), (15.4e), and (15.16), respectively.
Step 3
Solve the QP problem in Eq. (15.33) for $\boldsymbol{\delta}_x$ and compute the Lagrange multipliers $\boldsymbol{\lambda}_{k+1}$ and $\hat{\boldsymbol{\mu}}_{k+1}$ using Eq. (15.34).
Step 4
Compute α_k using Eqs. (15.37), (15.39), and (15.40).
Step 5
Set $\boldsymbol{\delta}_x = \alpha_k \boldsymbol{\delta}_x$ and $\mathbf{x}_{k+1} = \mathbf{x}_k + \boldsymbol{\delta}_x$.
Step 6
If $\|\boldsymbol{\delta}_x\| \leq \varepsilon$, output solution $\mathbf{x}^* = \mathbf{x}_{k+1}$ and stop; otherwise, continue with Step 7.
Step 7
Evaluate γ_k, θ, and η_k using Eqs. (15.43), (15.44), and (15.42), respectively.
Compute $\mathbf{Z}_{k+1}$ using Eq. (15.41).
Set $k = k + 1$ and repeat from Step 2.

Example 15.4 Applying Algorithm 15.4, solve the nonlinear constrained optimization problem

$$\text{minimize } f(\mathbf{x}) = x_1^2 + x_2$$
$$\text{subject to:} \quad a_1(\mathbf{x}) = x_1^2 + x_2^2 - 9 = 0$$
$$\mathbf{Ax} \geq \mathbf{b}$$

where

$$\mathbf{A} = \begin{bmatrix} 1 & 0 \\ -1 & 0 \\ 0 & 1 \\ 0 & -1 \end{bmatrix} \quad \text{and} \quad \mathbf{b} = \begin{bmatrix} 1 \\ -5 \\ 2 \\ -4 \end{bmatrix}$$

Solution The feasible region of the problem is the part of the circle centered at the origin with radius 3 that is contained in the rectangle $1 \leq x_1 \leq 5$, $2 \leq x_2 \leq 4$. The feasible region is not a convex set and hence the problem at hand is a nonconvex problem.

Algorithm 15.4 was applied to the problem with $\varepsilon = 10^{-6}$ and $\beta = 100$ using five different initial points that satisfy the inequality constraint $\mathbf{Ax}_0 \geq \mathbf{b}$. The algorithm converged in all cases to the solution point

$$\mathbf{x}^* = \begin{bmatrix} 1 \\ 2.8284 \end{bmatrix}$$

The test results are given in Table 15.2.

518

Table 15.2 Test Results for Example 15.4

$\mathbf{x}_0$	$\begin{bmatrix} 5 \\ 4 \end{bmatrix}$	$\begin{bmatrix} 5 \\ 2 \end{bmatrix}$	$\begin{bmatrix} 1 \\ 2 \end{bmatrix}$	$\begin{bmatrix} 3 \\ 2 \end{bmatrix}$	$\begin{bmatrix} 4 \\ 3 \end{bmatrix}$
Number of iterations	6	6	2	6	6
Number of Kflops	46.9	46.0	14.8	45.5	46.1

■

15.4   Interior-Point Methods

Interior-point methods that have proven useful for LP, QP, and convex programming (CP) problems have recently been extended to nonconvex optimization problems [9]–[13]. In this section, we describe an interior-point algorithm for nonconvex optimization based on the methods described in [12][13], which is a direct extension of the primal-dual interior-point methods for LP and QP problems.

15.4.1   KKT conditions and search direction

We consider the constrained nonlinear optimization problem

$$\text{minimize } f(\mathbf{x}) \tag{15.45a}$$

$$\text{subject to: } c_j(\mathbf{x}) \geq 0 \quad \text{for } j = 1, 2, \ldots, q \tag{15.45b}$$

where $f(\mathbf{x})$ and $c_j(\mathbf{x})$ are continuous and have continuous second partial derivatives, and the feasible region $\mathcal{R}$ described by Eq. (15.45b) is nonempty.

By introducing slack variable $\mathbf{y} = [y_1 \ y_2 \ \cdots \ y_q]^T$, the above problem can be converted to

$$\text{minimize } f(\mathbf{x}) \tag{15.46a}$$

$$\text{subject to: } \mathbf{c}(\mathbf{x}) - \mathbf{y} = \mathbf{0} \tag{15.46b}$$

$$\mathbf{y} \geq \mathbf{0} \tag{15.46c}$$

where $\mathbf{c}(\mathbf{x}) = [c_1(\mathbf{x}) \ c_2(\mathbf{x}) \ \cdots \ c_q(\mathbf{x})]^T$. As in the primal Newton barrier method discussed in Sec. 12.4, the inequality constraints in Eq. (15.46c) can be incorporated into the objective function by adding a logarithmic barrier function. This yields the minimization problem

$$\text{minimize } f_\tau(\mathbf{x}) = f(\mathbf{x}) - \tau \sum_{i=1}^{q} \ln y_i \tag{15.47a}$$

$$\text{subject to: } \mathbf{c}(\mathbf{x}) - \mathbf{y} = \mathbf{0} \tag{15.47b}$$

where $\tau > 0$ is the barrier parameter. The Lagrangian for the problem in Eq. (15.47) is

$$\mathcal{L}(\mathbf{x}, \mathbf{y}, \boldsymbol{\lambda}, \tau) = f(\mathbf{x}) - \tau \sum_{i=1}^{q} \ln y_i - \boldsymbol{\lambda}^T [\mathbf{c}(\mathbf{x}) - \mathbf{y}]$$

and the KKT conditions for a minimizer of the problem in Eq. (15.47) are given by

$$\nabla_x \mathcal{L} = \nabla f(\mathbf{x}) - \mathbf{A}^T(\mathbf{x})\boldsymbol{\lambda} = \mathbf{0}$$
$$\nabla_y \mathcal{L} = -\tau \mathbf{Y}^{-1}\mathbf{e} + \boldsymbol{\lambda} = \mathbf{0}$$
$$\nabla_\lambda \mathcal{L} = \mathbf{c}(\mathbf{x}) - \mathbf{y} = \mathbf{0}$$

where

$$\mathbf{A}(\mathbf{x}) = [\nabla c_1(\mathbf{x}) \cdots \nabla c_q(\mathbf{x})]^T$$
$$\mathbf{Y} = \text{diag}\{y_1, y_2, \ldots, y_q\}$$
$$\mathbf{e} = [1 \ 1 \ \cdots \ 1]^T$$

By multiplying the equation $\nabla_y \mathcal{L} = \mathbf{0}$ by $\mathbf{Y}$, we obtain the standard primal-dual system

$$\nabla f(\mathbf{x}) - \mathbf{A}^T(\mathbf{x})\boldsymbol{\lambda} = \mathbf{0} \tag{15.48a}$$
$$-\tau \mathbf{e} + \mathbf{Y}\boldsymbol{\Lambda}\mathbf{e} = \mathbf{0} \tag{15.48b}$$
$$\mathbf{c}(\mathbf{x}) - \mathbf{y} = \mathbf{0} \tag{15.48c}$$

where $\boldsymbol{\Lambda} = \text{diag}\{\lambda_1, \lambda_2, \ldots, \lambda_q\}$.

At the kth iteration, the set of vectors $\{\mathbf{x}_k, \mathbf{y}_k, \boldsymbol{\lambda}_k\}$ is updated to $\{\mathbf{x}_{k+1}, \mathbf{y}_{k+1}, \boldsymbol{\lambda}_{k+1}\}$ as

$$\mathbf{x}_{k+1} = \mathbf{x}_k + \alpha_k \Delta \mathbf{x}_k \tag{15.49a}$$
$$\mathbf{y}_{k+1} = \mathbf{y}_k + \alpha \Delta \mathbf{y}_k \tag{15.49b}$$
$$\boldsymbol{\lambda}_{k+1} = \boldsymbol{\lambda}_k + \alpha_k \Delta \boldsymbol{\lambda}_k \tag{15.49c}$$

where α_k is a scalar to be determined using a line search, and the set of increment vectors $\{\Delta \mathbf{x}_k, \Delta \mathbf{y}_k, \Delta \boldsymbol{\lambda}_k\}$ is determined by solving the linearized equations for Eq. (15.48) as follows. First, we approximate the nonlinear terms $\nabla f(\mathbf{x})$, $\mathbf{A}(\mathbf{x})$, $\mathbf{c}(\mathbf{x})$, and $\mathbf{Y}\boldsymbol{\Lambda}\mathbf{e}$ in Eq. (15.48) at point $\mathbf{x}_{k+1}$ as

$$\mathbf{g}_{k+1} \approx \mathbf{g}_k + \nabla^2 f(\mathbf{x}_k)\Delta \mathbf{x}_k$$
$$\mathbf{A}_{k+1}^T \approx \mathbf{A}_k^T + \sum_{i=1}^{q} \nabla^2 c_i(\mathbf{x}_k)\Delta \mathbf{x}_k$$
$$\mathbf{c}_{k+1} \approx \mathbf{c}_k + \mathbf{A}_k \Delta \mathbf{x}_k$$
$$\mathbf{Y}_{k+1}\boldsymbol{\Lambda}_{k+1}\mathbf{e} \approx \mathbf{Y}_k \boldsymbol{\Lambda}_k \mathbf{e} + \boldsymbol{\Lambda}_k \Delta \mathbf{y}_k + \mathbf{Y}_k \Delta \boldsymbol{\lambda}_k$$

where $\mathbf{g}_k = \nabla f(\mathbf{x}_k)$, $\mathbf{A}_k = \mathbf{A}(\mathbf{x}_k)$, and $\mathbf{c}_k = \mathbf{c}(\mathbf{x}_k)$. The linearized system of equations for Eq. (15.48) becomes

$$
\begin{bmatrix} \mathbf{H}_k & \mathbf{0} & -\mathbf{A}_k^T \\ \mathbf{0} & \mathbf{\Lambda}_k & \mathbf{Y}_k \\ \mathbf{A}_k & -\mathbf{I} & \mathbf{0} \end{bmatrix} \begin{bmatrix} \Delta\mathbf{x}_k \\ \Delta\mathbf{y}_k \\ \Delta\boldsymbol{\lambda}_k \end{bmatrix} = \begin{bmatrix} -\mathbf{g}_k + \mathbf{A}_k^T \boldsymbol{\lambda}_k \\ \tau\mathbf{e} - \mathbf{Y}_k\mathbf{\Lambda}_k\mathbf{e} \\ -\mathbf{c}_k + \mathbf{y}_k \end{bmatrix} \tag{15.50}
$$

where $\mathbf{H}_k$ represents the Hessian of the Lagrangian, $\mathbf{H}(\mathbf{x}, \boldsymbol{\lambda})$, for $\{\mathbf{x}, \boldsymbol{\lambda}\} = \{\mathbf{x}_k, \boldsymbol{\lambda}_k\}$, i.e.,

$$
\mathbf{H}(\mathbf{x}, \boldsymbol{\lambda}) = \nabla^2 f(\mathbf{x}) - \sum_{i=1}^{q} \lambda(i)\nabla^2 c_i(\mathbf{x}) \tag{15.51}
$$

The search direction determined using Eq. (15.50) is often referred to as the Newton direction of the problem in Eq. (15.45).

Note that the matrix in Eq. (15.50) is not symmetric, but it can be made symmetric by multiplying the first equation by $-\mathbf{I}$ and the second equation by $-\mathbf{Y}_k^{-1}$. This would yield

$$
\begin{bmatrix} -\mathbf{H}_k & \mathbf{0} & \mathbf{A}_k^T \\ \mathbf{0} & -\mathbf{Y}_k^{-1}\mathbf{\Lambda}_k & -\mathbf{I} \\ \mathbf{A}_k & -\mathbf{I} & \mathbf{0} \end{bmatrix} \begin{bmatrix} \Delta\mathbf{x}_k \\ \Delta\mathbf{y}_k \\ \Delta\boldsymbol{\lambda}_k \end{bmatrix} = \begin{bmatrix} \boldsymbol{\sigma}_k \\ -\boldsymbol{\gamma}_k \\ \boldsymbol{\rho}_k \end{bmatrix} \tag{15.52a}
$$

where

$$
\boldsymbol{\sigma}_k = \mathbf{g}_k - \mathbf{A}_k^T \boldsymbol{\lambda}_k \tag{15.52b}
$$

$$
\boldsymbol{\gamma}_k = \tau\mathbf{Y}_k^{-1}\mathbf{e} - \boldsymbol{\lambda}_k \tag{15.52c}
$$

$$
\boldsymbol{\rho}_k = \mathbf{y}_k - \mathbf{c}_k \tag{15.52d}
$$

If $\{\mathbf{x}_k, \mathbf{y}_k\}$ satisfies the constraints in Eq. (15.47b), then $\boldsymbol{\rho}_k = 0$. Hence $\boldsymbol{\rho}_k$ in Eq. (15.52d) can be viewed as a measure of how far the set $\{\mathbf{x}_k, \mathbf{y}_k\}$ is from being feasible for the primal problem in Eq. (15.46). Likewise, $\boldsymbol{\sigma}_k$ in Eq. (15.52b) can be regarded as a measure of how far the set $\{\mathbf{x}_k, \boldsymbol{\lambda}_k\}$ is from being feasible for the dual problem, which is the maximization problem

$$
\text{maximize } \mathbf{y}^T\boldsymbol{\lambda}
$$
$$
\text{subject to: } \mathbf{g}(\mathbf{x}) - \mathbf{A}^T(\mathbf{x})\boldsymbol{\lambda} = 0
$$
$$
\mathbf{y} \geq 0
$$

By solving the second equation in Eq. (15.52a) for $\Delta\mathbf{y}_k$, we obtain

$$
\Delta\mathbf{y}_k = \mathbf{Y}_k\mathbf{\Lambda}_k^{-1}(-\boldsymbol{\gamma}_k - \Delta\boldsymbol{\lambda}_k) \tag{15.53}
$$

and Eq. (15.52a) is reduced to

$$
\begin{bmatrix} -\mathbf{H}_k & \mathbf{A}_k^T \\ \mathbf{A}_k & \mathbf{Y}_k\mathbf{\Lambda}_k^{-1} \end{bmatrix} \begin{bmatrix} \Delta\mathbf{x}_k \\ \Delta\boldsymbol{\lambda}_k \end{bmatrix} = \begin{bmatrix} \boldsymbol{\sigma}_k \\ \boldsymbol{\rho}_k + \mathbf{Y}_k\mathbf{\Lambda}_k^{-1}\boldsymbol{\gamma}_k \end{bmatrix} \tag{15.54}
$$

Explicit formulas for the solution of Eq. (15.52a) can be obtained as

$$\Delta \mathbf{x}_k = -\mathbf{N}_k^{-1}\mathbf{g}_k + \tau \mathbf{N}_k^{-1}\mathbf{A}_k^T\mathbf{Y}_k^{-1}\mathbf{e} + \mathbf{N}_k^{-1}\mathbf{A}_k^T\mathbf{Y}_k^{-1}\mathbf{\Lambda}_k\boldsymbol{\rho}_k \quad (15.55a)$$

$$\Delta \mathbf{y}_k = -\mathbf{A}_k\mathbf{N}_k^{-1}\mathbf{g}_k + \tau \mathbf{A}_k\mathbf{N}_k^{-1}\mathbf{A}_k^T\mathbf{Y}_k^{-1}\mathbf{e}$$
$$- (\mathbf{I} - \mathbf{A}_k\mathbf{N}_k^{-1}\mathbf{A}_k^T\mathbf{Y}_k^{-1}\mathbf{\Lambda}_k)\boldsymbol{\rho}_k \quad (15.55b)$$

$$\Delta \boldsymbol{\lambda}_k = \mathbf{Y}_k^{-1}\mathbf{\Lambda}_k(\boldsymbol{\rho}_k - \mathbf{A}_k\Delta\mathbf{x}_k) + \boldsymbol{\gamma}_k \quad (15.55c)$$

where $\mathbf{N}_k$ is the so-called *dual normal matrix*

$$\mathbf{N}(\mathbf{x},\ \mathbf{y},\ \boldsymbol{\lambda}) = \mathbf{H}(\mathbf{x},\ \boldsymbol{\lambda}) + \mathbf{A}^T(\mathbf{x})\mathbf{Y}^{-1}\mathbf{\Lambda}\mathbf{A}(\mathbf{x}) \quad (15.56)$$

evaluated for $\{\mathbf{x},\ \mathbf{y},\ \boldsymbol{\lambda}\} = \{\mathbf{x}_k,\ \mathbf{y}_k,\ \boldsymbol{\lambda}_k\}$.

As will be seen below, for convex problems the search direction given by Eq. (15.55) works well and a step length, α_k, can be determined by minimizing a suitable merit function. For nonconvex problems, however, the above search direction needs to be modified so as to assure a descent direction for the merit function.

15.4.2   A merit function for convex problems

A suitable merit function is one whose minimization along a search direction leads to progress towards finding a local minimizer. In this regard, the penalty functions described in Eqs. (15.20) and (15.22) can be regarded as merit functions evaluated along search direction $\mathbf{d}_k$ at point $\mathbf{x}_k$. A suitable merit function for convex problems is

$$\psi_{\beta,\tau}(\mathbf{x},\ \mathbf{y}) = f(\mathbf{x}) - \tau \sum_{i=1}^{q} \ln y_i + \frac{\beta}{2}\|\mathbf{y} - \mathbf{c}(\mathbf{x})\|^2$$

which can also be expressed as

$$\psi_{\beta,\tau}(\mathbf{x},\ \mathbf{y}) = f_\tau(\mathbf{x}) + \frac{\beta}{2}\|\boldsymbol{\rho}(\mathbf{x},\ \mathbf{y})\|^2 \quad (15.57)$$

where $f_\tau(\mathbf{x})$ is given by Eq. (15.47), $\boldsymbol{\rho}(\mathbf{x},\ \mathbf{y}) = \mathbf{y} - \mathbf{c}(\mathbf{x})$, and $\beta \geq 0$ is a parameter to be determined later.

Evidently, this merit function is differentiable with respect to the elements of $\mathbf{x}$ and $\mathbf{y}$. With a sufficiently large β, minimizing $\psi_{\beta,\tau}$ at $\{\mathbf{x}_k + \alpha\Delta\mathbf{x}_k,\ \mathbf{y}_k + \alpha\Delta\mathbf{y}_k\}$ reduces the objective function $f_\tau(\mathbf{x})$ and, at the same time, the new point is closer to the feasible region because of the presence of the term $\beta\|\boldsymbol{\rho}(\mathbf{x},\ \mathbf{y})\|^2/2$.

Let $\{\Delta\mathbf{x}_k,\ \Delta\mathbf{y}_k,\ \Delta\boldsymbol{\lambda}_k\}$ given by Eq. (15.55) be the search direction at the kth iteration. Using Eqs. (15.55) and (15.57), we can verify that

$$s_k = \begin{bmatrix} \nabla_x\psi_{\beta,\tau}(\mathbf{x}_k,\ \mathbf{y}_k) \\ \nabla_y\psi_{\beta,\tau}(\mathbf{x}_k,\ \mathbf{y}_k) \end{bmatrix}^T \begin{bmatrix} \Delta\mathbf{x}_k \\ \Delta\mathbf{y}_k \end{bmatrix}$$

$$= -\boldsymbol{\xi}_k^T \mathbf{N}_k^{-1} \boldsymbol{\xi}_k + \tau \mathbf{e}^T \mathbf{Y}_k^{-1} \boldsymbol{\rho}_k + \boldsymbol{\xi}_k^T \mathbf{N}_k^{-1} \mathbf{A}_k^T \mathbf{Y}_k^{-1} \boldsymbol{\Lambda}_k \boldsymbol{\rho}_k$$
$$- \beta \|\boldsymbol{\rho}_k\|^2 \tag{15.58}$$

where $\boldsymbol{\xi}_k = \mathbf{g}_k - \tau \mathbf{A}_k^T \mathbf{Y}_k^{-1} \mathbf{e}$ (see Prob. 15.8). If the dual normal matrix $\mathbf{N}(\mathbf{x}, \mathbf{y}, \boldsymbol{\lambda})$ is positive definite for $\{\mathbf{x}, \mathbf{y}, \boldsymbol{\lambda}\} = \{\mathbf{x}_k, \mathbf{y}_k, \boldsymbol{\lambda}_k\}$ and $\{\mathbf{x}_k, \mathbf{y}_k\}$ is not feasible, i.e., $\boldsymbol{\rho}_k \neq \mathbf{0}$, then from Eq. (15.58) it follows that $\{\Delta\mathbf{x}_k, \Delta\mathbf{y}_k\}$ is not a descent direction for merit function $\psi_{\beta,\tau}(\mathbf{x}, \mathbf{y})$ for $\{\mathbf{x}, \mathbf{y}\} = \{\mathbf{x}_k, \mathbf{y}_k\}$ only if

$$\tau \mathbf{e}^T \mathbf{Y}_k^{-1} \boldsymbol{\rho}_k + \boldsymbol{\xi}_k^T \mathbf{N}_k^{-1} \mathbf{A}_k^T \mathbf{Y}_k^{-1} \boldsymbol{\Lambda}_k \boldsymbol{\rho}_k > 0 \tag{15.59}$$

In such a case, we can choose a β which is greater than or equal to $\beta_{\min}$ where

$$\beta_{\min} = (-\boldsymbol{\xi}_k^T \mathbf{N}_k^{-1} \boldsymbol{\xi}_k + \tau \mathbf{e}^T \mathbf{Y}_k^{-1} \boldsymbol{\rho}_k + \boldsymbol{\xi}_k^T \mathbf{N}_k^{-1} \mathbf{A}_k^T \mathbf{Y}_k^{-1} \boldsymbol{\Lambda}_k \boldsymbol{\rho}_k) / \|\boldsymbol{\rho}_k\|^2 \tag{15.60}$$

to ensure that the inner product s_k in Eq. (15.58) is negative and, therefore, $\{\Delta\mathbf{x}_k, \Delta\mathbf{y}_k\}$ is a descent direction for $\psi_{\beta,\tau}(\mathbf{x}, \mathbf{y})$ for $\{\mathbf{x}, \mathbf{y}\} = \{\mathbf{x}_k, \mathbf{y}_k\}$. In practice, β is initially set to zero and remains unchanged as long as $\{\Delta\mathbf{x}_k, \Delta\mathbf{y}_k\}$ is a descent direction for $\psi_{\beta,\tau}$. If, with $\beta = 0$, s_k in Eq. (15.58) is nonnegative, then $\beta_{\min}$ is calculated using Eq. (15.60) and β is set to $10\beta_{\min}$ so as to ensure that s_k is negative. Note that the above analysis was carried out under the assumption that the dual normal matrix $\mathbf{N}(\mathbf{x}, \mathbf{y}, \boldsymbol{\lambda})$ defined by Eq. (15.56) is positive definite. If the objective function $f(\mathbf{x})$ is convex and the constraint functions $c_j(\mathbf{x})$ for $j = 1, 2, \ldots, q$ are all concave, i.e., if the optimization problem in Eq. (15.45) is a CP problem, then Eq. (15.48b) in conjunction with the fact that $\tau > 0$ and $\mathbf{Y}$ is positive definite implies that $\boldsymbol{\lambda} > \mathbf{0}$ and, therefore, matrix $\mathbf{N}(\mathbf{x}, \mathbf{y}, \boldsymbol{\lambda})$ is positive definite. In other words, we have shown that for a CP problem there exists a β that causes the search direction in Eq. (15.55) to be a descent direction for merit function $\psi_{\beta,\tau}(\mathbf{x}, \mathbf{y})$.

Once the search direction $\{\Delta\mathbf{x}_k, \Delta\mathbf{y}_k, \boldsymbol{\lambda}_k\}$ is computed and an appropriate β is chosen, the scalar α_k in Eq. (15.49) is calculated in two steps: (a) Find $\alpha_{\max}$ such that $\mathbf{y}_k + \alpha_{\max}\Delta\mathbf{y}_k > \mathbf{0}$ and $\boldsymbol{\lambda}_k + \alpha_{\max}\Delta\boldsymbol{\lambda}_k > \mathbf{0}$; for $\mathbf{y}_k > \mathbf{0}$ and $\boldsymbol{\lambda}_k > \mathbf{0}$, $\alpha_{\max}$ can be computed as

$$\alpha_{\max} = 0.95 \left[\max_{1 \le i \le q} \left(-\frac{\Delta\mathbf{y}_k(i)}{\mathbf{y}_k(i)}, -\frac{\Delta\boldsymbol{\lambda}_k(i)}{\boldsymbol{\lambda}_k(i)} \right) \right]^{-1} \tag{15.61}$$

(b) Perform a line search on interval $[0, \alpha_{\max}]$ to find α_k, the value of α that minimizes the one-variable function $\psi_{\beta,\tau}(\mathbf{x}_k + \alpha\Delta\mathbf{x}_k, \mathbf{y}_k + \alpha\Delta\mathbf{y}_k)$.

As in the Newton barrier method for LP and QP problems, the value of the barrier parameter τ is fixed in the subproblem in Eq. (15.47). Once a minimizer of this subproblem is obtained, it can serve as the initial point for the same subproblem with a reduced barrier parameter τ. This procedure is continued until the difference in norm between two consecutive minimizers is less than a given tolerance and, at that time, the minimizer of the corresponding subproblem is deemed to be a solution for the problem in Eq. (15.45). From Eq. (15.48b), it is quite obvious that an appropriate value of τ should be proportional to $\mathbf{y}_k^T \boldsymbol{\lambda}_k / q$, as in the case of LP and QP problems (see, for example, Eq. (12.50)). In [11], the use of the formula

$$\tau_{k+1} = \delta \left\{ \min \left[(1-r) \frac{1-\zeta}{\zeta}, 2 \right] \right\}^3 \frac{\mathbf{y}_k^T \boldsymbol{\lambda}_k}{q} \qquad (15.62)$$

was proposed for the update of τ, where q is the number of constraints involved,

$$\zeta = \frac{q \min_{1 \le i \le q} [\mathbf{y}_k(i) \boldsymbol{\lambda}_k(i)]}{\mathbf{y}_k^T \boldsymbol{\lambda}_k}$$

and r and δ are parameters which are set to 0.95 and 0.1, respectively. The interior-point algorithm for convex programming problems can now be summarized as follows.

Algorithm 15.5 Interior-point algorithm for CP problems with inequality constraints

Step 1
Input an initial set $\{\mathbf{x}_0, \ \mathbf{y}_0, \ \boldsymbol{\lambda}_0\}$ with $\mathbf{y}_0 > 0$ and $\boldsymbol{\lambda}_0 > 0$ and an initial barrier parameter τ_0. Set $l = 0$, $\{\mathbf{x}_0^*, \ \mathbf{y}_0^*, \ \boldsymbol{\lambda}_0^*\} = \{\mathbf{x}_0, \ \mathbf{y}_0, \ \boldsymbol{\lambda}_0\}$, and initialize the outer-loop tolerance ε_{outer}.

Step 2
Set $k = 0$, $\tau = \tau_l$, and initialize the inner-loop tolerance ε_{inner}.

Step 3

Step 3.1
Set $\beta = 0$ and evaluate $\{\Delta\mathbf{x}_k, \ \Delta\mathbf{y}_k, \ \Delta\boldsymbol{\lambda}_k\}$ using Eq. (15.55) and s_k using Eq. (15.58).

Step 3.2
If $s_k \ge 0$, compute $\beta_{\min}$ using Eq. (15.60) and set $\beta = 10\beta_{\min}$; otherwise, continue with Step 3.3.

Step 3.3
Compute $\alpha_{\max}$ using Eq. (15.61) and perform a line search to find the value of α_k that minimizes $\psi_{\beta,\tau}(\mathbf{x}_k + \alpha\Delta\mathbf{x}_k, \ \mathbf{y}_k + \alpha\Delta\mathbf{y}_k)$ on $[0, \ \alpha_{\max}]$.

Step 3.4
Set $\{\mathbf{x}_{k+1}, \ \mathbf{y}_{k+1}, \ \boldsymbol{\lambda}_{k+1}\}$ using Eq. (15.49).

Step 3.5

If $||\alpha_k \Delta \mathbf{x}_k|| + ||\alpha_k \Delta \mathbf{y}_k|| + ||\alpha_k \Delta \boldsymbol{\lambda}_k|| < \varepsilon_{inner}$, set
$\{\mathbf{x}_{l+1}^*, \mathbf{y}_{l+1}^*, \boldsymbol{\lambda}_{l+1}^*\} = \{\mathbf{x}_{k+1}, \mathbf{y}_{k+1}, \boldsymbol{\lambda}_{k+1}\}$ and continue with
Step 4; otherwise, set $k = k + 1$ and repeat from Step 3.1.

Step 4

If $||\mathbf{x}_l^* - \mathbf{x}_{l+1}^*|| + ||\mathbf{y}_l^* - \mathbf{y}_{l+1}^*|| + ||\boldsymbol{\lambda}_l^* - \boldsymbol{\lambda}_{l+1}^*|| < \varepsilon_{outer}$, output
$\{\mathbf{x}^*, \mathbf{y}^*, \boldsymbol{\lambda}^*\} = \{\mathbf{x}_l^*, \mathbf{y}_l^*, \boldsymbol{\lambda}_l^*\}$ and stop; otherwise, calculate τ_{l+1}
using Eq. (15.62), set $\{\mathbf{x}_0, \mathbf{y}_0, \boldsymbol{\lambda}_0\} = \{\mathbf{x}_l^*, \mathbf{y}_l^*, \boldsymbol{\lambda}_l^*\}$, $l = l + 1$, and
repeat from Step 2.

In Step 3.3 an inexact line search based on the Goldstein conditions in
Eqs. (4.55) and (4.56) can be applied with $\rho = 0.1$. Initially, α_0 in these
inequalities is set to α_{max}. If the inequalities are not satisfied then the value of
α_0 is successively halved until they are satisfied.

Example 15.5 Apply Algorithm 15.5 to the shortest-distance problem in Example 13.5 with $\varepsilon_{inner} = 10^{-3}$, $\varepsilon_{outer} = 10^{-5}$, $\tau_0 = 0.001$, and an initial set
$\{\mathbf{x}_0, \mathbf{y}_0, \boldsymbol{\lambda}_0\}$ with $\mathbf{x}_0 = [0\ 1\ 2\ 2]^T$, $\mathbf{y}_0 = [2\ 2]^T$, and $\boldsymbol{\lambda}_0 = [1\ 1]^T$. Note that
$\mathbf{x}_0$ violates both constraints $c_1(\mathbf{x}) \geq 0$ and $c_2(\mathbf{x}) \geq 0$.

Solution The minimization problem at hand is a CP problem and Algorithm
15.4 is, therefore, applicable. To apply the algorithm, we compute

$$\mathbf{g}(\mathbf{x}) = \begin{bmatrix} x_1 - x_3 \\ x_2 - x_4 \\ x_3 - x_1 \\ x_4 - x_2 \end{bmatrix}$$

$$\mathbf{A}(\mathbf{x}) = \begin{bmatrix} \frac{1-x_1}{2} & -2x_2 & 0 & 0 \\ 0 & 0 & \frac{-5x_3-3x_4+22}{4} & \frac{-3x_3-5x_4+26}{4} \end{bmatrix}$$

$$\mathbf{H}(\mathbf{x}, \boldsymbol{\lambda}) = \begin{bmatrix} 1+\frac{\lambda_1}{2} & 0 & -1 & 0 \\ 0 & 1+2\lambda_1 & 0 & -1 \\ -1 & 0 & 1+\frac{5\lambda_2}{4} & \frac{3\lambda_2}{4} \\ 0 & -1 & \frac{3\lambda_2}{4} & 1+\frac{5\lambda_2}{4} \end{bmatrix}$$

where λ_1 and λ_2 are the first and second components of $\boldsymbol{\lambda}$, respectively. It took
Algorithm 15.4 three outer-loop iterations and 22.4 Kflops to converge to the
solution

$$\mathbf{x}^* = \begin{bmatrix} 2.044750 \\ 0.852716 \\ 2.544913 \\ 2.485633 \end{bmatrix}, \ \mathbf{y}^* = \begin{bmatrix} 0.131404 \\ 0.114310 \end{bmatrix} \times 10^{-6}, \ \text{and } \boldsymbol{\lambda}^* = \begin{bmatrix} 0.957480 \\ 1.100145 \end{bmatrix}$$

Therefore, we obtain

$$\mathbf{r}^* = \begin{bmatrix} 2.044750 \\ 0.852716 \end{bmatrix} \quad \text{and} \quad \mathbf{s}^* = \begin{bmatrix} 2.544913 \\ 2.485633 \end{bmatrix}$$

Hence the shortest distance between $\mathcal{R}$ and $\mathcal{S}$ is $||\mathbf{r}^* - \mathbf{s}^*|| = 1.707800$.

∎

15.4.3 Algorithm modifications for nonconvex problems

If the problem in Eq. (15.45) is not a CP problem, the dual normal matrix $\mathbf{N}(\mathbf{x}, \mathbf{y}, \boldsymbol{\lambda})$ in Eq. (15.56) may be indefinite and in such a case, the algorithm developed in Sec. 15.4.2 needs to be modified to deal with the indefiniteness of matrix $\mathbf{N}(\mathbf{x}, \mathbf{y}, \boldsymbol{\lambda})$.

A simple and effective way to fix the problem is to modify the Hessian of the Lagrangian as

$$\hat{\mathbf{H}}(\mathbf{x}, \boldsymbol{\lambda}) = \mathbf{H}(\mathbf{x}, \boldsymbol{\lambda}) + \eta \mathbf{I} \tag{15.63}$$

where $\eta \geq 0$ is chosen to yield a modified dual normal matrix

$$\mathbf{N}(\mathbf{x}, \mathbf{y}, \boldsymbol{\lambda}) = \hat{\mathbf{H}}(\mathbf{x}, \boldsymbol{\lambda}) + \mathbf{A}^T(\mathbf{x})\mathbf{Y}^{-1}\boldsymbol{\Lambda}\mathbf{A}(\mathbf{x}) \tag{15.64}$$

which is positive definite. With this modification, the search direction $\{\Delta\mathbf{x}_k, \Delta\mathbf{y}_k, \Delta\boldsymbol{\lambda}_k\}$ given by Eq. (15.55) remains a descent direction for merit function $\psi_{\beta,\tau}$, and Algorithm 15.5 applies.

For problems of moderate size, a suitable value of η can be determined as follows. First, we examine the eigenvalues of $\mathbf{H}(\mathbf{x}, \boldsymbol{\lambda})$. If they are all positive, then we set $\eta = 0$. Otherwise, we use $\bar{\eta} = 1.2\eta_0$ as an upper bound of η where η_0 is the magnitude of the most negative eigenvalue of $\mathbf{H}(\mathbf{x}, \boldsymbol{\lambda})$. Evidently, with $\eta = \bar{\eta}$ in Eq. (15.63), the modified $\mathbf{N}(\mathbf{x}, \mathbf{y}, \boldsymbol{\lambda})$ in Eq. (15.64) is positive definite. Next, we successively halve η and test the positive definiteness of $\mathbf{N}(\mathbf{x}, \mathbf{y}, \boldsymbol{\lambda})$ with $\eta = 2^{-m}\bar{\eta}$ for $m = 1, 2, \ldots$ until an $\eta = 2^{-m^*}\bar{\eta}$ is reached for which $\mathbf{N}(\mathbf{x}, \mathbf{y}, \boldsymbol{\lambda})$ fails to be positive definite. The value of η in Eq. (15.63) is then taken as

$$\eta = 2^{-(m^*-1)}\bar{\eta} \tag{15.65}$$

A computationally more economical method for finding a suitable value of η, which is based on matrix factorization of the reduced KKT matrix in Eq. (15.54), can be found in [12].

The modified version of Algorithm 15.5 is as follows.

526

Algorithm 15.6 Interior-point algorithm for nonconvex problems with inequality constraints

Step 1
Input an initial set $\{\mathbf{x}_0,\ \mathbf{y}_0,\ \boldsymbol{\lambda}_0\}$ with $\mathbf{y}_0 > 0$, $\boldsymbol{\lambda}_0 > 0$, and an initial barrier parameter τ_0.
Set $l = 0$, $\{\mathbf{x}_0^*,\ \mathbf{y}_0^*,\ \boldsymbol{\lambda}_0^*\} = \{\mathbf{x}_0,\ \mathbf{y}_0,\ \boldsymbol{\lambda}_0\}$, and initialize the outer-loop tolerance ε_{outer}.

Step 2
Set $k = 0$, $\tau = \tau_l$, and initialize the inner-loop tolerance ε_{inner}.

Step 3

Step 3.1
Evaluate the eigenvalues of $\mathbf{H}(\mathbf{x}_k,\ \boldsymbol{\lambda}_k)$. If they are all positive, continue with Step 3.2; otherwise, set $\bar{\eta} = 1.2\eta_0$ where η_0 is the magnitude of the most negative eigenvalue of $\mathbf{H}(\mathbf{x}_k,\ \boldsymbol{\lambda}_k)$; test the positive definiteness of $\mathbf{N}(\mathbf{x}_k,\ \mathbf{y}_k,\ \boldsymbol{\lambda}_k)$ in Eq. (15.64) with $\eta = 2^{-m}\bar{\eta}$ for $m = 1,\ 2,\ \dots$ until a value $\eta = 2^{-m^*}\bar{\eta}$ is obtained for which $\mathbf{N}(\mathbf{x}_k,\ \mathbf{y}_k,\ \boldsymbol{\lambda}_k)$ fails to be positive definite; evaluate

$$\hat{\mathbf{H}}(\mathbf{x}_k,\ \boldsymbol{\lambda}_k) = \mathbf{H}(\mathbf{x}_k,\ \boldsymbol{\lambda}_k) + \eta\mathbf{I}$$
$$\mathbf{N}_k = \hat{\mathbf{H}}(\mathbf{x}_k,\ \boldsymbol{\lambda}_k) + \mathbf{A}^T(\mathbf{x}_k)\mathbf{Y}_k^{-1}\boldsymbol{\Lambda}_k\mathbf{A}(\mathbf{x}_k)$$

with $\eta = 2^{-(m^*-1)}\bar{\eta}$.

Step 3.2
Set $\beta = 0$ and evaluate $\{\Delta\mathbf{x}_k,\ \Delta\mathbf{y}_k,\ \Delta\boldsymbol{\lambda}_k\}$ using Eq. (15.55) and s_k using Eq. (15.58).

Step 3.3
If $s_k \geq 0$, compute $\beta_{\min}$ using Eq. (15.60) and set $\beta = 10\beta_{\min}$; otherwise, continue with Step 3.4.

Step 3.4
Compute $\alpha_{\max}$ using Eq. (15.61) and perform a line search to find α_k, the value of α that minimizes $\psi_{\beta,\tau}(\mathbf{x}_k + \alpha\Delta\mathbf{x}_k,\ \mathbf{y}_k + \alpha\Delta\mathbf{y}_k)$ on $[0,\ \alpha_{\max}]$.

Step 3.5
Set $\{\mathbf{x}_{k+1},\ \mathbf{y}_{k+1},\ \boldsymbol{\lambda}_{k+1}\}$ using Eq. (15.49).

Step 3.6
If $||\alpha_k\Delta\mathbf{x}_k|| + ||\alpha_k\Delta\mathbf{y}_k|| + ||\alpha_k\Delta\boldsymbol{\lambda}_k|| < \varepsilon_{inner}$, set $\{\mathbf{x}_{l+1}^*,\ \mathbf{y}_{l+1}^*,\ \boldsymbol{\lambda}_{l+1}^*\} = \{\mathbf{x}_{k+1},\ \mathbf{y}_{k+1},\ \boldsymbol{\lambda}_{k+1}\}$ and continue with Step 4; otherwise, set $k = k + 1$ and repeat from Step 3.1.

Step 4
If $||\mathbf{x}_l^* - \mathbf{x}_{l+1}^*|| + ||\mathbf{y}_l^* - \mathbf{y}_{l+1}^*|| + ||\boldsymbol{\lambda}_l^* - \boldsymbol{\lambda}_{l+1}^*|| < \varepsilon_{outer}$, output $\{\mathbf{x}^*,\ \mathbf{y}^*,\ \boldsymbol{\lambda}^*\} = \{\mathbf{x}_l^*,\ \mathbf{y}_l^*,\ \boldsymbol{\lambda}_l^*\}$ and stop; otherwise, calculate τ_{l+1} using Eq. (15.62), set $\{\mathbf{x}_0,\ \mathbf{y}_0,\ \boldsymbol{\lambda}_0\} = \{\mathbf{x}_l^*,\ \mathbf{y}_l^*,\ \boldsymbol{\lambda}_l^*\}$, $l = l + 1$, and repeat from Step 2.

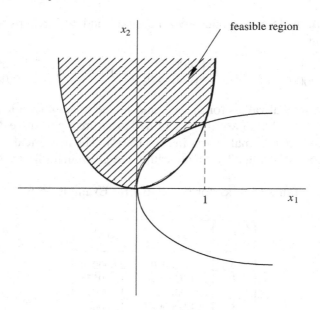

Figure 15.1. Feasible region of the problem in Example 15.5.

Example 15.6 Applying Algorithm 15.6, solve the nonconvex optimization problem

$$\text{minimize } f(\mathbf{x}) = (x_1 - 2)^2 + (x_2 - 1)^2$$
$$\text{subject to:} \quad c_1(\mathbf{x}) = -x_1^2 + x_2 \geq 0$$
$$c_2(\mathbf{x}) = -x_1 + x_2^2 \geq 0$$

Solution The feasible region of this problem, shown as the shaded area in Fig. 15.1, is obviously nonconvex. To apply the algorithm, we compute

$$\mathbf{g}(\mathbf{x}) = \begin{bmatrix} 2(x_1 - 2) \\ 2(x_2 - 1) \end{bmatrix}, \quad \mathbf{A}(\mathbf{x}) = \begin{bmatrix} -2x_1 & 1 \\ -1 & 2x_2 \end{bmatrix}$$
$$\mathbf{H}(\mathbf{x}, \boldsymbol{\lambda}) = \begin{bmatrix} 2 + 2\lambda_1 & 0 \\ 0 & 2 - 2\lambda_2 \end{bmatrix}$$

where λ_1 and λ_2 are the first and second components of $\boldsymbol{\lambda}$, respectively. Since $\boldsymbol{\lambda} > \mathbf{0}$, $\mathbf{H}(\mathbf{x}, \boldsymbol{\lambda})$ becomes indefinite if the second component of $\boldsymbol{\lambda}$ is greater than 1.

With

$$\mathbf{x}_0 = \begin{bmatrix} -1 \\ 2 \end{bmatrix}, \quad \mathbf{y}_0 = \begin{bmatrix} 20 \\ 20 \end{bmatrix}, \quad \boldsymbol{\lambda}_0 = \begin{bmatrix} 1 \\ 2 \end{bmatrix}$$
$$\tau_0 = 0.001, \quad \varepsilon_{inner} = 5 \times 10^{-4}, \quad \text{and} \quad \varepsilon_{outer} = 10^{-5}$$

528

it took Algorithm 15.5 four outer-loop iterations and 34.7 Kflops to converge to the solution

$$\mathbf{x}^* = \begin{bmatrix} 1.000004 \\ 1.000007 \end{bmatrix}, \quad \mathbf{y}^* = \begin{bmatrix} 0.000009 \\ 0.891272 \end{bmatrix} \times 10^{-5}, \quad \boldsymbol{\lambda}^* = \begin{bmatrix} 0.999983 \\ 0.000007 \end{bmatrix}$$

The above numerical values for $\mathbf{x}_0$, $\mathbf{y}_0$, and $\boldsymbol{\lambda}_0$ led to an indefinite $\mathbf{N}(\mathbf{x}_0, \mathbf{y}_0, \boldsymbol{\lambda}_0)$, but an $\eta = 0.600025$ was then identified to assure the positive definiteness of the modified dual normal matrix in Eq. (15.64). The numerical values of the four iterates, $\mathbf{x}_k$, generated by the algorithm are given in Table 15.3.

Table 15.3 $\mathbf{x}_k$ for $k = 0$ to 4 for Example 15.6

k	$\mathbf{x}_k(1)$	$\mathbf{x}_k(2)$
0	−1.000000	2.000000
1	1.034101	1.080555
2	1.000013	1.000013
3	1.000004	1.000007
4	1.000004	1.000007

∎

References

1 R. Fletcher, *Practical Methods of Optimization*, 2nd ed., Wiley, New York, 1987.
2 P. E. Gill, W. Murray, and M. A. Saunders, "SNOPT: An SQP algorithm for large-scale constrained optimization," Research Report NA 97–2, Dept. of Mathematics, Univ. of California, San Diego, 1997.
3 R. B. Wilson, *A Simplicial Method for Concave Programming*, Ph.D. dissertation, Graduate School of Business Administration, Harvard University, Cambridge, MA., 1963.
4 P. T. Boggs and J. W. Tolle, "Sequential quadratic programming," *Acta numerica*, vol. 4, pp. 1–52, 1995.
5 G. H. Golub and C. F. Van Loan, *Matrix Computation*, 2nd ed., Baltimore, The Johns Hopkins University Press, MD, 1989.
6 S. P. Han, "A globally convergent method for nonlinear programming," *J. Optimization Theory and Applications*, vol. 22, pp. 297–309, July 1977.
7 M. J. D. Powell, "Algorithms for nonlinear constraints that use Lagrangian functions," *Math. Programming*, vol. 14, pp. 224–248, 1978.
8 M. H. Wright, "Direct search methods: Once scorned, now respectable," in *Numerical Analysis 1995*, D. F. Griffiths and G. A. Watson eds., pp. 191–208, Addison Wesley Longman, UK.
9 A. El-Bakry, R. Tapia, T. Tsuchiya, and Y. Zhang, "On the formulation and theory of the Newton interior-point method for nonlinear programming," *J. Optimization Theory and Applications*, vol. 89, pp. 507–541, 1996.
10 A. Forsgren and P. Gill, "Primal-dual interior methods for nonconvex nonlinear programming," Technical Report NA-96-3, Dept. of Mathematics, Univ. of California, San Diego, 1996.

11 D. M. Gay, M. L. Overton, and M. H. Wright, "A primal-dual interior method for nonconvex nonlinear programming," *Proc. 1996 Int. Conf. on Nonlinear Programming*, Kluwer Academic Publishers, 1998.

12 R. J. Vanderbei and D. F. Shanno, "An interior-point algorithm for nonconvex nonlinear programming," Research Report SOR-97-21 (revised), Statistics and Operations Res., Princeton University, June 1998.

13 D. F. Shanno and R. J. Vanderbei, "Interior-point methods for nonconvex nonlinear programming: Orderings and high-order methods," Research Report SOR-99-5, Statistics and Operations Res., Princeton University, May 1999.

Problems

15.1 The Lagrange multiplier λ_{k+1} can be computed using Eq. (15.7) if $\mathbf{A}_k$ has full row rank. Modify Eq. (15.7) so as to make it applicable to the case where $\mathbf{A}_k$ does not have full row rank.

15.2 Apply Algorithm 15.1 to the problem

$$\text{minimize } f(\mathbf{x}) = \ln(1 + x_1^2) + x_2$$

$$\text{subject to:} \quad (1 + x_1^2)^2 + x_2^2 - 4 = 0$$

15.3 Apply Algorithm 15.2 or Algorithm 15.3 to the problem

$$\text{minimize } f(\mathbf{x}) = 0.01x_1^2 + x_2^2$$

$$\text{subject to:} \quad \begin{aligned} c_1(\mathbf{x}) &= x_1 x_2 - 25 \geq 0 \\ c_2(\mathbf{x}) &= x_1^2 + x_2^2 - 25 \geq 0 \\ c_3(\mathbf{x}) &= x_1 \geq 2 \end{aligned}$$

15.4 Derive the approximate KKT conditions in Eqs. (15.32a)–(15.32e).

15.5 Apply Algorithm 15.4 to the nonconvex problem

$$\text{minimize } f(\mathbf{x}) = 2x_1^2 + 3x_2$$

$$\text{subject to:} \quad \begin{aligned} a_1(\mathbf{x}) &= x_1^2 + x_2^2 - 16 = 0 \\ \mathbf{Ax} &\geq \mathbf{b} \end{aligned}$$

where

$$\mathbf{A} = \begin{bmatrix} 1 & 0 \\ -1 & 0 \\ 0 & 1 \\ 0 & -1 \end{bmatrix} \quad \text{and} \quad \mathbf{b} = \begin{bmatrix} 2 \\ -5 \\ 1 \\ -5 \end{bmatrix}$$

15.6 Using Eqs. (15.52)–(15.54), derive the formulas in Eq. (15.55).

15.7 Show that if the Hessian, $\mathbf{H}(\mathbf{x}, \lambda)$, is positive definite, then the dual normal matrix $\mathbf{N}(\mathbf{x}, \mathbf{y}, \lambda)$ in Eq. (15.56) is also positive definite.

15.8 Using Eqs. (15.55) and (15.57), derive the expression of s_k in Eq. (15.58).

15.9 Show that the inner product s_k is negative if $\beta \geq \beta_{\min}$ where $\beta_{\min}$ is given by Eq. (15.60).

15.10 Apply Algorithm 15.5 to the CP problem

$$\text{minimize } f(\mathbf{x}) = (x_1 - 2)^2 + (x_2 - 1)^2$$
$$\text{subject to:} \quad c_1(\mathbf{x}) = -x_1^2 + x_2 \geq 0$$
$$c_2(\mathbf{x}) = x_1 - x_2^2 \geq 0$$

15.11 Apply Algorithm 15.6 to the nonconvex problem

$$\text{minimize } f(\mathbf{x}) = -x_1 x_2$$
$$\text{subject to:} \quad c_1(\mathbf{x}) = 1 - x_1^2 - x_2^2 \geq 0$$

15.12 Apply Algorithm 15.6 to the nonconvex problem in Prob. 15.3, and compare the solution with that obtained in Prob. 15.3.

15.13 Consider the nonlinear constrained problem

$$\text{minimize } f(\mathbf{x})$$
$$\text{subject to:} \quad 0 \leq c_i(\mathbf{x}) \leq r_i \qquad \text{for } i = 1, 2, \ldots, q$$

(a) Show that the problem just described can be converted into the problem

$$\text{minimize } f(\mathbf{x})$$
$$\text{subject to:} \quad c_i(\mathbf{x}) - y_i = 0$$
$$y_i + p_i = r_i \qquad \text{for } i = 1, 2, \ldots, q$$
$$y_i \geq 0, \ p_i \geq 0$$

(b) Use the method outlined in part (a) to deal with the nonlinear constrained problem

$$\text{minimize } f(\mathbf{x})$$
$$\text{subject to:} \quad a_i(\mathbf{x}) = 0 \qquad \text{for } i = 1, 2, \ldots, p$$

15.14 Consider the constrained problem

$$\text{minimize } f(\mathbf{x})$$
$$\text{subject to:} \quad l_i \leq x_i \leq u_i \qquad \text{for } i = 1, 2, \ldots, n$$

where l_i and u_i are constants.

(a) Convert the inequality constraints to equality constraints by introducing slack variables.

(b) Follow the development in Secs. 15.4.1 and 15.4.2, derive a system of linear equations similar to Eq. (15.52a) for the search direction.

(c) Using the system of linear equations in part (b), derive a reduced KKT system similar to that in Eq. (15.54).

15.15 Convert the constrained problem

$$\text{minimize } f(\mathbf{x}) = 100(x_1^2 - x_2)^2 + (x_1 - 1)^2 + 90(x_3^2 - x_4)^2$$
$$+(x_3 - 1)^2 + 10.1[(x_2 - 1)^2 + (x_4 - 1)^2]$$
$$+19.8(x_2 - 1)(x_4 - 1)$$

$$\text{subject to:} \quad -10 \le x_i \le 10 \quad \text{for } i = 1, 2, 3, 4$$

into the form in Eq. (15.46) and solve it using Algorithm 15.6.

Chapter 16

APPLICATIONS OF CONSTRAINED OPTIMIZATION

16.1   Introduction

Constrained optimization provides a general framework in which a variety of design criteria and specifications can be readily imposed on the required solution. Usually, a multivariable objective function that quantifies a performance measure of a design can be identified. This objective function may be linear, quadratic, or highly nonlinear, and usually it is differentiable so that its gradient and sometimes Hessian can be evaluated. In a real-life design problem, the design is carried out under certain physical limitations with limited resources. If these limitations can be quantified as equality or inequality constraints on the design variables, then a constrained optimization problem can be formulated whose solution leads to an optimal design that satisfies the limitations imposed. Depending on the degree of nonlinearity of the objective function and constraints, the problem at hand can be a linear programming (LP), quadratic programming (QP), convex programming (CP), semidefinite programming (SDP), second-order cone programming (SOCP), or general nonlinear constrained optimization problem.

This chapter is devoted to several applications of some of the constrained optimization algorithms studied in Chaps. 11–15 in the areas of digital signal processing, control, robotics, and telecommunications. In Sec. 16.2, we show how constrained algorithms of the various types, e.g., LP, QP, CP, SDP algorithms, can be utilized for the design of digital filters. The authors draw from their extensive research experience on the subject [1][2]. Section 16.3 introduces several models for uncertain dynamic systems and develops an effective control strategy known as model predictive control for this class of systems, which involves the use of SDP. In Sec. 16.4, LP and SDP are applied to solve a problem that entails optimizing the grasping force distribution for dextrous robotic hands. In Sec. 16.5, an SDP-based method for multiuser detection and a CP approach to minimize bit-error rate for wireless communication systems is described.

16.2 Design of Digital Filters

16.2.1 Design of linear-phase FIR filters using QP

In many applications of digital filters in communication systems, it is often desirable to design linear-phase finite-duration impulse response (FIR) digital filters with a specified maximum passband error, δ_p, and/or a specified maximum stopband gain, δ_a [3] (see Sec. B.9.1). FIR filters of this class can be designed relatively easily by using a QP approach as described below.

For the sake of simplicity, we consider the problem of designing a linear-phase lowpass FIR filter of even order N (odd length $N+1$) with normalized passband and stopband edges ω_p and ω_a, respectively (see Sec. B.9.2). The frequency response of such a filter can be expressed as

$$H(e^{j\omega}) = e^{-j\omega N/2} A(\omega)$$

as in Eq. (9.33) and the desired amplitude response, $A_d(\omega)$, can be assumed to be of the form given by Eq. (9.39). If we use the piecewise-constant weighting function defined by Eq. (9.40), then the objective function $e_l(\mathbf{x})$ in Eq. (9.35a) becomes

$$e_l(\mathbf{x}) = \int_0^{\omega_p} [A(\omega) - 1]^2 \, d\omega + \gamma \int_{\omega_a}^{\pi} A^2(\omega) \, d\omega \qquad (16.1\text{a})$$

$$= \mathbf{x}^T \mathbf{Q}_l \mathbf{x} - 2\mathbf{x}^T \mathbf{b}_l + \kappa \qquad (16.1\text{b})$$

where $\mathbf{x}$ is given by Eq. (9.35b), and $\mathbf{Q}_l = \mathbf{Q}_{l1} + \mathbf{Q}_{l2}$ and $\mathbf{b}_l$ are given by Eq. (9.37). If the weight γ in Eq. (16.1a) is much greater than 1, then minimizing $e_l(\mathbf{x})$ would yield an FIR filter with a minimized least-squares error in the stopband but the passband error would be left largely unaffected. This problem can be fixed by imposing the constraint

$$|A(\omega) - 1| \leq \delta_p \qquad \text{for } \omega \in [0, \, \omega_p] \qquad (16.2)$$

where δ_p is the upper bound on the amplitude of the passband error. With $A(\omega) = \mathbf{c}_l^T(\omega)\mathbf{x}$ where $\mathbf{c}_l(\omega)$ is defined by Eq. (9.36a), Eq. (16.2) can be written as

$$\mathbf{c}_l^T(\omega)\mathbf{x} \leq 1 + \delta_p \qquad \text{for } \omega \in [0, \, \omega_p] \qquad (16.3\text{a})$$

and

$$-\mathbf{c}_l^T(\omega)\mathbf{x} \leq -1 + \delta_p \qquad \text{for } \omega \in [0, \, \omega_p] \qquad (16.3\text{b})$$

Note that the frequency variable ω in Eq. (16.3) can assume an infinite set of values in the range 0 to ω_p. A realistic way to implement these constraints is to impose the constraints on a finite set of sample frequencies $\mathcal{S}_p = \{\omega_i^{(p)} : i = 1, 2, \ldots, M_p\}$ in the passband. Under these circumstances, the above constraints can be expressed in matrix form as

$$\mathbf{A}_p \mathbf{x} \leq \mathbf{b}_p \tag{16.4a}$$

where

$$\mathbf{A}_p = \begin{bmatrix} \mathbf{c}_l^T(\omega_1^{(p)}) \\ \vdots \\ \mathbf{c}_l^T(\omega_{M_p}^{(p)}) \\ -\mathbf{c}_l^T(\omega_1^{(p)}) \\ \vdots \\ -\mathbf{c}_l^T(\omega_{M_p}^{(p)}) \end{bmatrix} \quad \text{and} \quad \mathbf{b}_p = \begin{bmatrix} 1+\delta_p \\ \vdots \\ 1+\delta_p \\ -1+\delta_p \\ \vdots \\ -1+\delta_p \end{bmatrix} \tag{16.4b}$$

Additional constraints can be imposed to ensure that the maximum stopband gain, δ_a, is also well controlled. To this end, we impose the constraint

$$|A(\omega)| \leq \delta_a \quad \text{for } \omega \in [\omega_a, \pi] \tag{16.5}$$

A discretized version of Eq. (16.5) is given by

$$\mathbf{c}_l^T(\omega)\mathbf{x} \leq \delta_a \quad \text{for } \omega \in \mathcal{S}_a \tag{16.6a}$$
$$-\mathbf{c}_l^T(\omega)\mathbf{x} \leq \delta_a \quad \text{for } \omega \in \mathcal{S}_a \tag{16.6b}$$

where $\mathcal{S}_a = \{\omega_i^{(a)} : i = 1, 2, \ldots, M_a\}$ is a set of sample frequencies in the stopband. The inequality constraints in Eq. (16.6) can be expressed in matrix form as

$$\mathbf{A}_a \mathbf{x} \leq \mathbf{b}_a \tag{16.7a}$$

where

$$\mathbf{A}_a = \begin{bmatrix} \mathbf{c}_l^T(\omega_1^{(a)}) \\ \vdots \\ \mathbf{c}_l^T(\omega_{M_a}^{(a)}) \\ -\mathbf{c}_l^T(\omega_1^{(a)}) \\ \vdots \\ -\mathbf{c}_l^T(\omega_{M_a}^{(a)}) \end{bmatrix} \quad \text{and} \quad \mathbf{b}_a = \delta_a \begin{bmatrix} 1 \\ \vdots \\ 1 \end{bmatrix} \tag{16.7b}$$

The design problem can now be formulated as the optimization problem

$$\text{minimize } e(\mathbf{x}) = \mathbf{x}^T \mathbf{Q}_l \mathbf{x} - 2\mathbf{b}_l \mathbf{x} + \kappa \tag{16.8a}$$

$$\text{subject to: } \begin{bmatrix} \mathbf{A}_p \\ \mathbf{A}_a \end{bmatrix} \mathbf{x} \leq \begin{bmatrix} \mathbf{b}_p \\ \mathbf{b}_a \end{bmatrix} \tag{16.8b}$$

There are $(N + 2)/2$ design variables in vector $\mathbf{x}$ and $2(M_p + M_a)$ linear inequality constraints in Eq. (16.8b). Since matrix $\mathbf{Q}_l$ is positive definite, the problem under consideration is a convex QP problem that can be solved using the algorithms studied in Chap. 13.

Example 16.1 Applying the above method, design a linear-phase lowpass FIR digital filter that would satisfy the following specifications: passband edge = 0.45π, stopband edge = 0.5π, maximum passband error $\delta_p = 0.025$, minimum stopband attenuation = 40 dB. Assume idealized passband and stopband gains of 1 and 0, respectively, and a normalized sampling frequency of 2π.

Solution The design was carried out by solving the QP problem in Eq. (16.8) using Algorithm 13.1. We have used a weighting constant $\gamma = 3 \times 10^3$ in Eq. (16.1a) and $\delta_p = 0.025$ in Eq. (16.2). The maximum stopband gain, δ_a, in Eq. (16.5) can be deduced from the minimum stopband attenuation, A_a, as

$$\delta_a = 10^{-0.05A_a} = 10^{-2}$$

(see Sec. B.9.1). We assumed 80 uniformly distributed sample frequencies with respect to the passband $[0, 0.45\pi]$ and 10 sample frequencies in the lower one-tenth of the stopband $[0.5\pi, \pi]$, which is usually the most critical part of the stopband, i.e., $M_p = 80$ and $M_a = 10$ in sets $\mathcal{S}_p$ and $\mathcal{S}_a$, respectively.

Unfortunately, there are no analytical methods for predicting the filter order N that would yield a filter which would meet the required specifications but a trial-and-error approach can often be used. Such an approach has resulted in a filter order of 84.

The amplitude of the passband ripple and the minimum stopband attenuation achieved were 0.025 and 41.65 dB, respectively. The amplitude response of the filter is plotted in Fig. 16.1. It is interesting to note that an equiripple error has been achieved with respect to the passband, which is often a desirable feature.

∎

16.2.2 Minimax design of FIR digital filters using SDP

Linear-phase FIR filters are often designed very efficiently using the so-called *weighted-Chebyshev* method which is essentially a minimax method based on the *Remez exchange algorithm* [1, Chap. 15]. These filters can also be designed using a minimax method based on SDP, as will be illustrated in this section. In fact, the SDP approach can be used to design FIR filters with arbitrary amplitude and phase responses including certain types of filters that cannot be designed with the weighted-Chebyshev method [4].

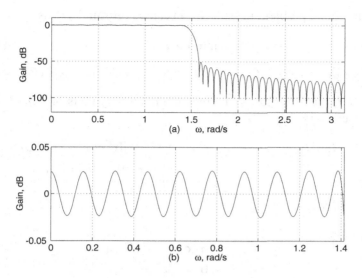

Figure 16.1. Amplitude response of the filter for Example 16.1: (a) For baseband $0 \leq \omega \leq \pi$, (b) for passband $0 \leq \omega \leq \omega_p$.

Consider an FIR filter of order N characterized by the general transfer function

$$H(z) = \sum_{n=0}^{N} h_n z^{-n} \tag{16.9}$$

The frequency response of such a filter can be expressed as

$$H(e^{j\omega}) = \sum_{n=0}^{N} h_n e^{-jn\omega} = \mathbf{h}^T[\mathbf{c}(\omega) - j\mathbf{s}(\omega)] \tag{16.10}$$

where $\mathbf{c}(\omega)$ and $\mathbf{s}(\omega)$ are given by Eqs. (9.26a) and (9.26b), respectively, and $\mathbf{h} = [h_0 \ h_1 \ \cdots \ h_N]^T$. Let $H_d(\omega)$ be the desired frequency response and assume a normalized sampling frequency of 2π. In a minimax design, we need to find a coefficient vector $\mathbf{h}$ that solves the optimization problem

$$\underset{\mathbf{h}}{\text{minimize}} \ \max_{\omega \in \Omega}[W(\omega)|H(e^{j\omega}) - H_d(\omega)|] \tag{16.11}$$

where Ω is a frequency region of interest over the positive half of the baseband $[0, \pi]$ and $W(\omega)$ is a given weighting function.

If δ denotes the upper bound of the squared weighted error in Eq. (16.11), i.e.,

$$W^2(\omega)|H(e^{j\omega}) - H_d(\omega)|^2 \leq \delta \qquad \text{for } \omega \in \Omega \tag{16.12}$$

538

then the minimax problem in Eq. (16.11) can be reformulated as

$$\text{minimize } \delta \tag{16.13a}$$

$$\text{subject to:} \quad W^2(\omega)|H(e^{j\omega}) - H_d(\omega)|^2 \leq \delta \quad \text{for } \omega \in \Omega \tag{16.13b}$$

Now let $H_r(\omega)$ and $H_i(\omega)$ be the real and imaginary parts of $H_d(\omega)$, respectively. We can write

$$\begin{aligned}
W^2(\omega)|H(e^{j\omega}) - H_d(\omega)|^2 &= W^2(\omega)\{[\mathbf{h}^T\mathbf{c}(\omega) - H_r(\omega)]^2 \\
&\quad + [\mathbf{h}^T\mathbf{s}(\omega) + H_i(\omega)]^2\} \\
&= \alpha_1^2(\omega) + \alpha_2^2(\omega)
\end{aligned} \tag{16.14}$$

where

$$\begin{aligned}
\alpha_1(\omega) &= \mathbf{h}^T\mathbf{c}_w(\omega) - H_{rw}(\omega) \\
\alpha_2(\omega) &= \mathbf{h}^T\mathbf{s}_w(\omega) + H_{iw}(\omega) \\
\mathbf{c}_w(\omega) &= W(\omega)\mathbf{c}(\omega) \\
\mathbf{s}_w(\omega) &= W(\omega)\mathbf{s}(\omega) \\
H_{rw}(\omega) &= W(\omega)H_r(\omega) \\
H_{iw}(\omega) &= W(\omega)H_i(\omega)
\end{aligned}$$

Using Eq. (16.14), the constraint in Eq. (16.13b) becomes

$$\delta - \alpha_1^2(\omega) - \alpha_2^2(\omega) \geq 0 \quad \text{for } \omega \in \Omega \tag{16.15}$$

It can be shown that the inequality in Eq. (16.15) holds if and only if

$$\mathbf{D}(\omega) = \begin{bmatrix} \delta & \alpha_1(\omega) & \alpha_2(\omega) \\ \alpha_1(\omega) & 1 & 0 \\ \alpha_2(\omega) & 0 & 1 \end{bmatrix} \succeq 0 \quad \text{for } \omega \in \Omega \tag{16.16}$$

(see Prob. 16.3) i.e., $\mathbf{D}(\omega)$ is positive definite for the frequencies of interest. If we write

$$\mathbf{x} = \begin{bmatrix} \delta \\ \mathbf{h} \end{bmatrix} = \begin{bmatrix} x_1 \\ x_2 \\ \vdots \\ x_{N+2} \end{bmatrix} \tag{16.17}$$

where $x_1 = \delta$ and $[x_2\ x_3\ \cdots\ x_{N+2}]^T$, then matrix $\mathbf{D}(\omega)$ is *affine* with respect to $\mathbf{x}$. If $\mathcal{S} = \{\omega_i : i = 1, 2, \ldots, M\} \subset \Omega$ is a set of frequencies which is sufficiently dense on Ω, then a discretized version of Eq. (16.16) is given by

$$\mathbf{F}(\mathbf{x}) \succeq 0 \tag{16.18a}$$

where

$$\mathbf{F}(\mathbf{x}) = \text{diag}\{\mathbf{D}(\omega_1), \ \mathbf{D}(\omega_2), \ \ldots, \ \mathbf{D}(\omega_M)\} \tag{16.18b}$$

and the minimization problem in Eq. (16.13) can be converted into the optimization problem

$$\text{minimize } \mathbf{c}^T \mathbf{x} \tag{16.19a}$$

$$\text{subject to: } \mathbf{F}(\mathbf{x}) \succeq \mathbf{0} \tag{16.19b}$$

where $\mathbf{c} = [1 \ 0 \ \cdots \ 0]^T$. Upon comparing Eq. (16.19) with Eq. (14.9), we conclude that this problem belongs to the class of SDP problems studied in Chap. 14.

Example 16.2 Assuming idealized passband and stopband gains of 1 and 0, respectively, and a normalized sampling frequency of 2π, apply the SDP-based minimax approach described in Sec. 16.2.2 to design a lowpass FIR filter of order 84 with a passband edge $\omega_p = 0.45\pi$ and a stopband edge $\omega_a = 0.5\pi$.

Solution The design was carried out by solving the SDP problem in Eq. (16.19) using Algorithm 14.1. The desired specifications can be achieved by assuming an idealized frequency response of the form

$$H_d(\omega) = \begin{cases} e^{-j42\omega} & \text{for } \omega \in [0, \ \omega_p] \\ 0 & \text{for } \omega \in [\omega_a, \ \omega_s/2] \end{cases}$$

$$= \begin{cases} e^{-j42\omega} & \text{for } \omega \in [0, \ 0.45\pi] \\ 0 & \text{for } \omega \in [0.5\pi, \ \pi] \end{cases}$$

For a filter order $N = 84$, the variable vector $\mathbf{x}$ has 86 elements as can be seen in Eq. (16.17). We assumed 300 sample frequencies that were uniformly distributed in $\Omega = [0, \ 0.45] \cup [0.5\pi, \ \pi]$, i.e., $M = 300$ in set S. Consequently, matrix $\mathbf{F}(\mathbf{x})$ in Eq. (16.19b) is of dimension 900×900. Using a piecewise constant representation for the weighting function $W(\omega)$ defined in Eq. (9.40) with $\gamma = 1.5$, a filter was obtained that has an equiripple amplitude response as can be seen in the plots of Fig. 16.2a and b. The maximum passband error and minimum stopband attenuation were 0.0098 and 43.72 dB, respectively.

The existence of a unique equiripple linear-phase FIR-filter design for a given set of amplitude-response specifications is guaranteed by the so-called *alternation theorem* (see p. 677 of [1]). This design has a constant group delay of $N/2$ s. Interestingly, the FIR filter designed here has a constant group delay of $N/2 = 42$, as can be seen in the delay characteristic of Fig. 16.2c, and this feature along with the equiripple amplitude response achieved suggests that the SDP minimax approach actually obtained the unique best equiripple linear-phase design. The SDP approach is much more demanding than the Remez exchange algorithm in terms of computation effort. However, it can be used to design FIR filter types that cannot be designed with the Remez exchange algorithm, for example, low-delay FIR filters with approximately constant passband group delay.

■

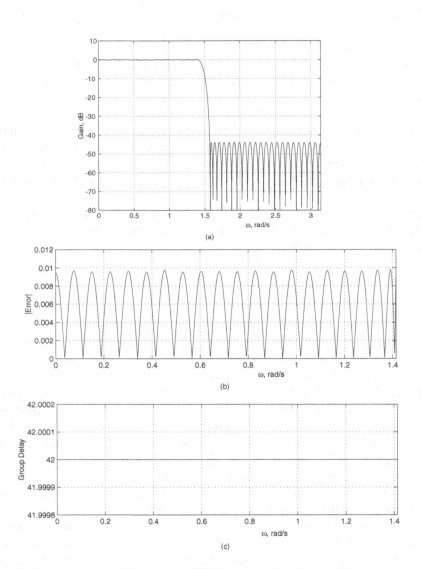

Figure 16.2. Performance of the lowpass FIR filter designed in Example 16.2: (a) Amplitude response, (b) passband error, (c) group-delay characteristic.

16.2.3   Minimax design of IIR digital filters using SDP

16.2.3.1   Introduction

Infinite-duration impulse response (IIR) digital filters offer improved selectivity and computational efficiency and reduced system delay compared to what can be achieved using FIR filters of comparable approximation accuracy [1]. The major drawbacks of an IIR design are that linear phase response can be achieved only approximately and the design must deal with the stability problem which does not exist in the FIR case.

A linear phase response is often required in digital filters for the purpose of avoiding phase distortion in the signals to be processed. Since signal components transmitted through stopbands are usually heavily attenuated, the linearity of the phase response is typically unimportant in stopbands. Consequently, IIR filters which have an approximately linear phase response in passbands and possibly a nonlinear phase response in stopbands are often entirely satisfactory particularly if they are also more economical in terms of computational complexity. Several methods are available for the design of IIR filters with approximately linear phase response in passbands [5]–[8].

The stability problem can be handled in several ways, see, for example, [1][5]–[8]. A popular approach is to impose *stability constraints* that establish a class of stable IIR filters from which the best solution for the design problem can be obtained. Obviously, this leads to a constrained optimization formulation for the design. However, technical difficulties can often occur if we attempt to implement a stability constraint that is explicitly related to the design variables. This is because the locations of the poles of the transfer function, which determine the stability of the filter, are related to the filter coefficients in a highly nonlinear and implicit way even for filters of moderate orders. Linear stability constraints that depend on the design variables affinely were proposed in [6][8]. These constraints depend on the frequency variable ω which can vary from 0 to π. Their linearity makes it possible to formulate the design of stable IIR filters as LP or convex QP problems. It should be mentioned, however, that constraints of this class are *sufficient* conditions for stability and are often too restrictive to permit a satisfactory design, especially the linear constraint proposed in [6].

Below, we formulate the design of an IIR filter as an SDP problem. The stability of the filter is assured by using a *single* linear matrix inequality (LMI) constraint, which fits nicely into an SDP formulation and does not depend on continuous parameters other than the design variables.

The transfer function of the IIR filter to be designed is assumed to be of the form

$$H(z) = \frac{A(z)}{B(z)} \qquad (16.21a)$$

where

$$A(z) = \sum_{i=0}^{N} a_i z^{-i} \qquad (16.21b)$$

$$B(z) = 1 + \sum_{i=1}^{K} b_i z^{-i} \qquad (16.21c)$$

and K is an integer between 1 and N. The particular form of the denominator polynomial $B(z)$ in Eq. (16.20c) has $N - K$ poles at the origin which, as was recently observed in [8], can be beneficial in the design of certain types of digital filters.

16.2.3.2   LMI constraint for stability

The stability of a filter represented by transfer function $H(z)$ such as that in Eq. (16.20a) is guaranteed if the zeros of polynomial $B(z)$ in Eq. (16.20c) are strictly inside the unit circle as was stated earlier. It can be shown that the zeros of $B(z)$ are the eigenvalues of the matrix

$$\mathbf{D} = \begin{bmatrix} -b_1 & -b_2 & \cdots & -b_{N-1} & -b_N \\ 1 & 0 & \cdots & 0 & 0 \\ 0 & 1 & \cdots & 0 & 0 \\ \vdots & \vdots & \ddots & \vdots & \vdots \\ 0 & 0 & \cdots & 1 & 0 \end{bmatrix} \qquad (16.22)$$

(see Prob. 16.4(a)). Consequently, the filter is stable if the moduli of the eigenvalues are all strictly less than one. The well-known Lyapunov theory [9] states that $\mathbf{D}$ represents a stable filter if and only if there exists a positive definite matrix $\mathbf{P}$ such that $\mathbf{P} - \mathbf{D}^T \mathbf{P} \mathbf{D}$ is positive definite, i.e.,

$$\mathcal{F} = \{\mathbf{P} : \mathbf{P} \succ \mathbf{0} \text{ and } \mathbf{P} - \mathbf{D}^T \mathbf{P} \mathbf{D} \succ \mathbf{0}\} \qquad (16.23)$$

is nonempty. Using simple linear algebraic manipulations, it can be verified that the set $\mathcal{F}$ in Eq. (16.22) can be characterized by

$$\mathcal{F} = \left\{\mathbf{P} : \begin{bmatrix} \mathbf{P}^{-1} & \mathbf{D} \\ \mathbf{D}^T & \mathbf{P} \end{bmatrix} \succ \mathbf{0}\right\} \qquad (16.24)$$

(see Prob. 16.4). Note that unlike the constraints in Eq. (16.22), matrix $\mathbf{D}$ in Eq. (16.23) (hence the coefficients of $B(z)$) appears *affinely*.

16.2.3.3   SDP formulation of the design problem

Given a desired frequency response $H_d(\omega)$, a minimax design of a stable IIR filter can be obtained by finding a transfer function $H(z)$ such as that in

Eq. (16.20) which solves the constrained optimization problem

$$\text{minimize } \max_{0 \le \omega \le \pi} \left[W(\omega) | H(e^{j\omega}) - H_d(\omega) | \right] \qquad (16.25a)$$

$$\text{subject to: } B(z) \neq 0 \quad \text{for } |z| \ge 1 \qquad (16.25b)$$

(see Sec. B.7). The frequency response of the filter can be expressed as

$$H(e^{j\omega}) = \frac{A(\omega)}{B(\omega)} \qquad (16.26)$$

where

$$A(\omega) = \sum_{n=0}^{N} a_n e^{-jn\omega} = \mathbf{a}^T \mathbf{c}(\omega) - j\mathbf{a}^T \mathbf{s}(\omega)$$

$$B(\omega) = 1 + \sum_{n=1}^{K} b_n e^{-jn\omega} = 1 + \mathbf{b}^T \hat{\mathbf{c}}(\omega) - j\mathbf{b}^T \hat{\mathbf{s}}(\omega)$$

$$\mathbf{a} = [a_0 \ a_1 \ \cdots \ a_N]^T$$
$$\mathbf{b} = [b_1 \ b_2 \ \cdots \ b_K]^T$$
$$\mathbf{c}(\omega) = [1 \ \cos \omega \ \cdots \ \cos N\omega]^T$$
$$\mathbf{s}(\omega) = [0 \ \sin \omega \ \cdots \ \sin N\omega]^T$$
$$\hat{\mathbf{c}}(\omega) = [\cos \omega \ \cos 2\omega \ \cdots \ \cos K\omega]^T$$
$$\hat{\mathbf{s}}(\omega) = [\sin \omega \ \sin 2\omega \ \cdots \ \sin K\omega]^T$$

and $H_d(\omega)$ can be written as

$$H_d(\omega) = H_r(\omega) + jH_i(\omega) \qquad (16.27)$$

where $H_r(\omega)$ and $H_i(\omega)$ denote the real and imaginary parts of $H_d(\omega)$, respectively.

Following the reformulation step in the FIR case (see Sec. 16.2.2), the problem in Eq. (16.24) can be expressed as

$$\text{minimize } \delta \qquad (16.28a)$$

$$\text{subject to: } \quad W^2(\omega) | H(e^{j\omega}) - H_d(\omega) |^2 \le \delta \quad \text{for } \omega \in \Omega$$
$$(16.28b)$$
$$B(z) \neq 0 \quad \text{for } |z| \ge 1 \qquad (16.28c)$$

where $\Omega = [0, \pi]$, we can write

$$W^2(\omega) | H(e^{j\omega}) - H_d(\omega) | = \frac{W^2(\omega)}{|B(\omega)|^2} |A(\omega) - B(\omega)H_d(\omega)|^2 \qquad (16.29)$$

which suggests the following iterative scheme: In the kth iteration, we seek to find polynomials $A_k(z)$ and $B_k(z)$ that solve the constrained optimization problem

$$\text{minimize } \delta \qquad (16.30\text{a})$$

subject to: $\quad \dfrac{W^2(\omega)}{|B_{k-1}(\omega)|^2}|A(\omega) - B(\omega)H_d(\omega)|^2 \leq \delta \quad \text{for } \omega \in \Omega$

$$(16.30\text{b})$$

$$B(z) \neq 0 \qquad \text{for } |z| \geq 1 \qquad (16.30\text{c})$$

where $B_{k-1}(\omega)$ is obtained in the $(k-1)$th iteration. An important difference between the problems in Eqs. (16.27) and (16.29) is that the constraint in Eq. (16.27b) is highly nonlinear because of the presence of $B(\omega)$ as the denominator of $H(e^{j\omega})$ while the constraint in Eq. (16.29b) is a *quadratic* function with respect to the components of $\mathbf{a}$ and $\mathbf{b}$ and $W^2(\omega)/|B_{k-1}(\omega)|^2$ is a weighting function.

Using arguments similar to those in Sec. 16.2.2, it can be shown that the constraint in Eq. (16.29) is equivalent to

$$\mathbf{\Gamma}(\omega) \succeq \mathbf{0} \qquad \text{for } \omega \in \Omega \qquad (16.31)$$

where

$$\mathbf{\Gamma}(\omega) = \begin{bmatrix} \delta & \alpha_1(\omega) & \alpha_2(\omega) \\ \alpha_1(\omega) & 1 & 0 \\ \alpha_2(\omega) & 0 & 1 \end{bmatrix}$$

with

$$\alpha_1(\omega) = \hat{\mathbf{x}}^T \mathbf{c}_k - H_{rw}(\omega)$$
$$\alpha_2(\omega) = \hat{\mathbf{x}}^T \mathbf{s}_k + H_{iw}(\omega)$$
$$\hat{\mathbf{x}} = \begin{bmatrix} \mathbf{a} \\ \mathbf{b} \end{bmatrix}, \quad \mathbf{c}_k = \begin{bmatrix} \mathbf{c}_w \\ \mathbf{u}_w \end{bmatrix}, \quad \mathbf{s}_k = \begin{bmatrix} \mathbf{s}_w \\ \mathbf{v}_w \end{bmatrix}$$
$$w_k = \frac{W(\omega)}{|B_{k-1}(\omega)|}$$
$$\mathbf{c}_w = w_k \mathbf{c}(\omega)$$
$$\mathbf{s}_w = w_k \mathbf{s}(\omega)$$
$$H_{rw}(\omega) = w_k H_r(\omega)$$
$$H_{iw}(\omega) = w_k H_i(\omega)$$
$$\mathbf{u}_w = w_k[-H_i(\omega)\hat{\mathbf{s}}(\omega) - H_r(\omega)\hat{\mathbf{c}}(\omega)]$$
$$\mathbf{v}_w = w_k[-H_i(\omega)\hat{\mathbf{c}}(\omega) + H_r(\omega)\hat{\mathbf{s}}(\omega)]$$

As for the stability constraint in Eq. (16.29c), we note from Sec. 16.2.3.2 that for a stable filter there exists a $\mathbf{P}_{k-1} \succ \mathbf{0}$ that solves the Lyapunov equation [9]

$$\mathbf{P}_{k-1} - \mathbf{D}_{k-1}^T \mathbf{P}_{k-1} \mathbf{D}_{k-1} = \mathbf{I} \qquad (16.32)$$

where $\mathbf{I}$ is the $K \times K$ identity matrix and $\mathbf{D}_{k-1}$ is a $K \times K$ matrix of the form in Eq. (16.21) with $-\mathbf{b}_{k-1}^T$ as its first row. Eq. (16.23) suggests a stability constraint for the digital filter as

$$\begin{bmatrix} \mathbf{P}_{k-1}^{-1} & \mathbf{D} \\ \mathbf{D}^T & \mathbf{P}_{k-1} \end{bmatrix} \succ \mathbf{0} \qquad (16.33)$$

or

$$\mathbf{Q}_k = \begin{bmatrix} \mathbf{P}_{k-1}^{-1} - \tau\mathbf{I} & \mathbf{D} \\ \mathbf{D}^T & \mathbf{P}_{k-1} - \tau\mathbf{I} \end{bmatrix} \succeq \mathbf{0} \qquad (16.34)$$

where $\mathbf{D}$ is given by Eq. (16.21) and $\tau > 0$ is a scalar that can be used to control the stability margin of the IIR filter. We note that (a) $\mathbf{Q}_k$ in Eq. (16.33) depends on $\mathbf{D}$ (and hence on $\hat{\mathbf{x}}$) affinely; and (b) because of Eq. (16.31), the positive definite matrix $\mathbf{P}_{k-1}$ in Eq. (16.33) is *constrained*. Consequently, Eq. (16.33) is a *sufficient* (but not necessary) constraint for stability. However, if the iterative algorithm described above converges, then the matrix sequence $\{\mathbf{D}_k\}$ also converges. Since the existence of a $\mathbf{P}_{k-1} \succ \mathbf{0}$ in Eq. (16.31) is a necessary and sufficient condition for the stability of the filter, the LMI constraint in Eq. (16.33) becomes less and less restrictive as the iterations continue.

Combining a discretized version of Eq. (16.30) with the stability constraint in Eq. (16.33), the constrained optimization problem in Eq. (16.29) can now be formulated as

$$\text{minimize } \mathbf{c}^T\mathbf{x} \qquad (16.35a)$$

$$\text{subject to: } \begin{bmatrix} \mathbf{\Gamma}_k & \mathbf{0} \\ \mathbf{0} & \mathbf{Q}_k \end{bmatrix} \succeq \mathbf{0} \qquad (16.35b)$$

where

$$\mathbf{x} = \begin{bmatrix} \delta \\ \hat{\mathbf{x}} \end{bmatrix} = \begin{bmatrix} \delta \\ \mathbf{a} \\ \mathbf{b} \end{bmatrix}, \quad \mathbf{c} = \begin{bmatrix} 1 \\ 0 \\ \vdots \\ 0 \end{bmatrix}$$

and

$$\mathbf{\Gamma}_k = \text{diag}\{\mathbf{\Gamma}(\omega_1), \ \mathbf{\Gamma}(\omega_2), \ \dots, \ \mathbf{\Gamma}(\omega_M)\}$$

In the above equation, $\{\omega_i : 1 \leq i \leq M\}$ is a set of frequencies in the range of interest. Since both $\mathbf{\Gamma}_k$ and $\mathbf{Q}_k$ depend on variable vector $\mathbf{x}$ affinely, the problem in Eq. (16.34) is an SDP problem.

16.2.3.4 Iterative SDP algorithm

Given a desired frequency response $H_d(\omega)$, a weighting function $W(\omega)$, and the orders of $A(z)$ and $B(z)$, namely, N and K, respectively, we can start the design with an initial point $\hat{\mathbf{x}}_0 = [\mathbf{a}_0^T \ \mathbf{b}_0^T]^T$ with $\mathbf{b}_0 = \mathbf{0}$. Coefficient vector $\mathbf{a}_0$

is obtained by designing an FIR filter assuming a desired frequency response $H_d(\omega)$ using a routine design algorithm [1]. The SDP problem formulated in Eq. (16.34) is solved for $k = 1$. If $||\hat{\mathbf{x}}_k - \hat{\mathbf{x}}_{k-1}||$ is less than a prescribed tolerance ε, then $\hat{\mathbf{x}}_k$ is deemed to be a solution for the design problem. Otherwise, the SDP in Eq. (16.34) is solved for $k = 2$, etc. This algorithm is illustrated by the following example.

Example 16.3 Assuming idealized passband and stopband gains of 1 and 0, respectively, and a normalized sampling frequency of 2π, apply the above iterative minimax approach to design an IIR lowpass digital filter that would meet the following specifications: passband edge $\omega_p = 0.5\pi$, stopband edge $\omega_a = 0.6\pi$, maximum passband error $\delta_p \leq 0.02$, minimum stopband attenuation $A_a \geq 34$ dB, group delay in passband = 9 s with a maximum deviation of less than 1 s.

Solution The required IIR filter was designed by solving the SDP problem in Eq. (16.34) using Algorithm 14.1. The desired specifications were achieved by using an idealized frequency of the form

$$H_d(\omega) = \begin{cases} e^{-j9\omega} & \text{for } \omega \in [0,\ \omega_p] \\ 0 & \text{for } \omega \in [\omega_a,\ \omega_s/2] \end{cases}$$
$$= \begin{cases} e^{-j9\omega} & \text{for } \omega \in [0,\ 0.5\pi] \\ 0 & \text{for } \omega \in [0.6\pi,\ \pi] \end{cases}$$

along with $N = 12$, $K = 6$, $W(\omega) = 1$ on $[0,\ 0.5\pi] \cup [0.6\pi, \pi]$ and zero elsewhere, $\tau = 10^{-4}$, and $\varepsilon = 5 \times 10^{-3}$. The constraint $\Gamma(\omega) \succeq 0$ was discretized over a set of 240 equally-spaced sample frequencies on $[0,\ \omega_p] \cup [\omega_a,\ \pi]$. It took the algorithm 50 iterations to converge to a solution. The poles and zeros of the filter obtained are given in Table 16.1, and $a_0 = 0.00789947$. The largest pole magnitude is 0.944.

The performance of the filter obtained can be compared with that of an alternative design reported by Deczky as Example 1 in [5], which has the same passband and stopband edges and filter order. As can be seen in Table 16.2 and Fig. 16.3, the present design offers improved performance as well as a reduced group delay. In addition, the present filter has only six nonzero poles, which would lead to reduced computational complexity in the implementation of the filter.

■

Table 16.1 Zeros and poles of the transfer function for Example 16.3

Zeros	Poles
−2.12347973 −1.22600378 1.49482238 ± j0.55741991 0.75350472 ± j1.37716837 −0.89300316 ± j0.65496710 −0.32277491 ± j0.93626367 −0.49091195 ± j0.86511412	−0.15960464 ± j0.93037014 −0.03719150 ± j0.55679595 0.24717453 ± j0.18656749 Plus another 6 poles at the origin

Table 16.2 Performance comparisons for Example 16.3

Filter	Iterative SDP Design	Deczky's Design [5]
Maximum passband error in magnitude	0.0171	0.0549
Minimum stopband attenuation, dB	34.7763	31.5034
Maximum ripple of group delay in passband, s	0.8320	1.3219

16.3 Model Predictive Control of Dynamic Systems

One of the challenges encountered in modeling and control of real-life dynamic systems is the development of controllers whose performance remains robust against various uncertainties that exist due to modeling errors, sensor noise, power-supply interference, and finite word length effects of the controller itself. Model predictive control (MPC) is a popular open-loop control methodology that has proven effective for the control of slow-varying dynamic systems such as process control in chemical, oil refinement, and pulp and paper industries [10][11]. At each control instant, a model predictive controller performs online optimization to generate an optimal control input based on a model that describes the dynamics of the system to be controlled and the available input and output measurements. In [11], it was shown that robust MPC that takes into account model uncertainty and various constraints on the

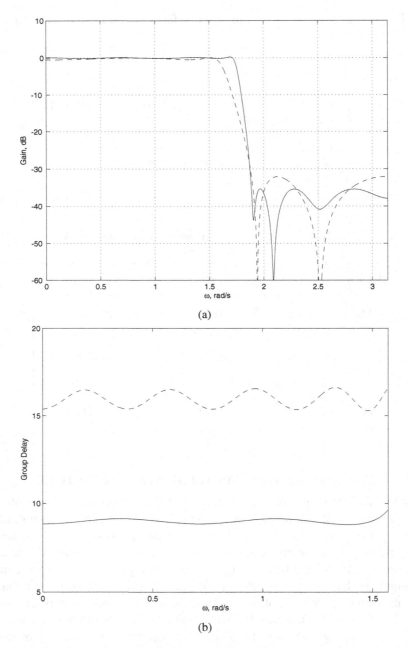

Figure 16.3. Performance of the IIR filter designed (solid lines) and the filter in [5] (dashed lines) for Example 16.3: (a) Amplitude response, (b) passband group delay characteristic.

control input and plant output can be designed using SDP techniques. In this section, we follow the methodology used in [11] to illustrate several SDP-based techniques for the design of robust model predictive controllers.

16.3.1 Polytopic model for uncertain dynamic systems

A linear discrete-time time-varying dynamic system can be modeled in terms of a state-space formulation as [9]

$$\mathbf{x}(k+1) = \mathbf{A}(k)\mathbf{x}(k) + \mathbf{B}(k)\mathbf{u}(k) \tag{16.36a}$$
$$\mathbf{y}(k) = \mathbf{C}\mathbf{x}(k) \tag{16.36b}$$

where $\mathbf{y}(k) \in R^{m \times 1}$ is the output vector, $\mathbf{u}(k) \in R^{p \times 1}$ is the input vector, and $\mathbf{x}(k) \in R^{n \times 1}$ is the state vector at time instant k. The matrices $\mathbf{A}(k) \in R^{n \times n}$ and $\mathbf{B}(k) \in R^{n \times m}$ are *time dependent*. The time dependence of the system matrix $\mathbf{A}(k)$ and the input-to-state matrix $\mathbf{B}(k)$ can be utilized to describe systems whose dynamic characteristics vary with time. In order to incorporate modeling uncertainties into the model in Eq. (16.35), the pair $[\mathbf{A}(k)\ \mathbf{B}(k)]$ is allowed to be a member of the polytope $\mathcal{M}$ defined by

$$\mathcal{M} = \text{Co}\{[\mathbf{A}_1\ \mathbf{B}_1],\ [\mathbf{A}_2\ \mathbf{B}_2],\ \ldots,\ [\mathbf{A}_L\ \mathbf{B}_L]\}$$

where Co denotes the convex hull spanned by $[\mathbf{A}_i\ \mathbf{B}_i]$ for $1 \le i \le L$, which is defined as

$$\mathcal{M} = \{[\mathbf{A}\ \mathbf{B}] : [\mathbf{A}\ \mathbf{B}] = \sum_{i=1}^{L} \lambda_i [\mathbf{A}_i\ \mathbf{B}_i],\ \lambda_i \ge 0,\ \sum_{i=1}^{L} \lambda_i = 1\} \tag{16.37}$$

(see Sec. A.16).

The linear model in Eq. (16.35) subject to the constraint $[\mathbf{A}(k)\ \mathbf{B}(k)] \in \mathcal{M}$ can be used to describe a wide variety of real-life dynamic systems. As an example, consider the angular positioning system illustrated in Fig. 16.4 [12]. The control problem is to use the input voltage to the motor to rotate the antenna such that the antenna angle, θ, relative to some reference tracks the angle of the moving target, θ_r. The discrete-time equation of the motion of the antenna can be derived from its continuous-time counterpart by discretization using a sampling period of 0.1 s and a first-order approximation of the derivative as

$$\mathbf{x}(k+1) = \begin{bmatrix} \theta(k) \\ \dot{\theta}(k+1) \end{bmatrix} = \begin{bmatrix} 1 & 0.1 \\ 0 & 1 - 0.1\alpha(k) \end{bmatrix} \mathbf{x}(k) + \begin{bmatrix} 0 \\ 0.1\eta \end{bmatrix} u(k)$$
$$= \mathbf{A}(k)\mathbf{x}(k) + \mathbf{B}u(k) \tag{16.38a}$$
$$y(k) = [1\ 0]\mathbf{x}(k) = \mathbf{C}\mathbf{x}(k) \tag{16.38b}$$

where $\eta = 0.787$. The parameter $\alpha(k)$ in matrix $\mathbf{A}(k)$ is proportional to the coefficient of viscous friction in the rotating parts of the antenna, and is assumed

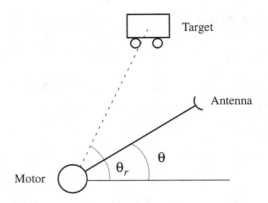

Figure 16.4. Angular positioning system.

to be arbitrarily time-varying in the range $0.1 \leq \alpha(k) \leq 10$. It follows that Eq. (16.37) is a polytopic model with $\mathbf{A}(k) \in \mathrm{Co}\,\{\mathbf{A}_1,\ \mathbf{A}_2\}$ where

$$\mathbf{A}_1 = \begin{bmatrix} 1 & 0.10 \\ 0 & 0.99 \end{bmatrix}, \quad \mathbf{A}_2 = \begin{bmatrix} 1 & 0.1 \\ 0 & 0 \end{bmatrix} \tag{16.37c}$$

Below, we deal with several aspects of MPCs.

16.3.2   Introduction to robust MPC

At sampling instant k, a robust MPC uses plant measurements and a model, such as the polytopic model in Eq. (16.35), to predict future outputs of the system. These measurements are utilized to compute m control inputs, $\mathbf{u}(k + i|k)$ for $i = 0,\ 1,\ \ldots,\ m-1$, by solving the minimax optimization problem

$$\underset{\mathbf{u}(k+i|k),\ 0 \leq i \leq m-1}{\text{minimize}} \quad \underset{[\mathbf{A}(k+i)\ \mathbf{B}(k+i)] \in \mathcal{M},\, i \geq 0}{\text{max}} \quad J_p(k) \tag{16.38}$$

The notation $\mathbf{u}(k + i|k)$ denotes the control decision made at instant $k + i$ based on the measurements available at instant k; $J(k)$ is an objective function that measures system performance and is given by

$$J(k) = \sum_{i=0}^{\infty} [\mathbf{x}^T(k + i|k)\mathbf{Q}\mathbf{x}(k + i|k) + \mathbf{u}^T(k + i|k)\mathbf{R}\mathbf{u}(k + i|k)] \tag{16.39}$$

where $\mathbf{Q} \succeq \mathbf{0}$ and $\mathbf{R} \succ \mathbf{0}$ are constant weighting matrices; $\mathbf{x}(k + i|k)$ denotes the system state at instant $k + i$, which is predicted using the measurements at instant k; and $\mathbf{u}(k + i|k)$ denotes the control input at instant $k + i$ obtained by solving the problem in Eq. (16.38). It follows that the control inputs obtained by solving the problem in Eq. (16.38) take into account the system's uncertainty by minimizing the worst-case value of $J(k)$ among all possible plant models included in set $\mathcal{M}$. The control inputs so computed are, therefore, *robust* against model uncertainties. At any given sampling instant k, the solution of the optimization problem in Eq. (16.38) provides a total of m control actions $\mathbf{u}(k|k)$, $\mathbf{u}(k+1|k)$, $\ldots$, $\mathbf{u}(k+m-1|k)$, but in a model predictive controller only the first control action, $\mathbf{u}(k|k)$, is implemented. At the next sampling instant, new measurements are obtained based on which the problem in Eq. (16.38) is solved again to provide a new set of m control actions; the first one is then implemented.

Frequently, the solution of the above minimax problem is computationally too demanding to implement and in the MPC literature the problem in Eq. (16.38) has been addressed by deriving an upper bound of the objective function $J(k)$ and then minimizing this upper bound with a constant state-feedback control law

$$\mathbf{u}(k+i|k) = \mathbf{F}\mathbf{x}(k+i|k) \qquad \text{for } i \geq 0 \qquad (16.40)$$

Let us assume that there exists a quadratic function $V(\mathbf{x}) = \mathbf{x}^T\mathbf{P}\mathbf{x}$ with $\mathbf{P} \succ \mathbf{0}$ such that for all $\mathbf{x}(k+i|k)$ and $\mathbf{u}(k+i|k)$ satisfying Eq. (16.35) and for $[\mathbf{A}(k+i)\ \mathbf{B}(k+i)] \in \mathcal{M}$, $i \geq 0$, $V(\mathbf{x})$ satisfies the inequality

$$V[\mathbf{x}(k+i+1|k)] - V[\mathbf{x}(k+i|k)] \leq -[\mathbf{x}^T(k+i|k)\mathbf{Q}\mathbf{x}(k+i|k)$$

$$+\mathbf{u}^T(k+i|k)\mathbf{R}\mathbf{u}(k+i|k)] \qquad (16.41)$$

If the objective function is finite, then the series in Eq. (16.39) must converge and, consequently, we have $\mathbf{x}(\infty|k) = \mathbf{0}$ which implies that $V[\mathbf{x}(\infty|k)] = 0$. By summing the inequality in Eq. (16.41) from $i = 0$ to ∞, we obtain $J(k) \leq V[\mathbf{x}(k|k)]$. This means that $V[\mathbf{x}(k|k)]$ is an upper bound of the objective function, which is considerably easier to deal with than $J(k)$. In the next section, we study the condition under which a positive definite matrix $\mathbf{P}$ exists such that $V(\mathbf{x})$ satisfies the condition in Eq. (16.41); we then formulate a modified optimization problem that can be solved by SDP algorithms.

16.3.3   Robust unconstrained MPC by using SDP

If γ is an upper bound of $V[\mathbf{x}(k|k)]$, namely,

$$V[\mathbf{x}(k|k)] = \mathbf{x}^T(k|k)\mathbf{P}\mathbf{x}(k|k) \leq \gamma \qquad (16.42)$$

then minimizing $V[\mathbf{x}(k|k)]$ is equivalent to minimizing γ. If we let

$$\mathbf{S} = \gamma\mathbf{P}^{-1} \qquad (16.43)$$

then $\mathbf{P} \succ \mathbf{0}$ implies that $\mathbf{S} \succ \mathbf{0}$ and Eq. (16.42) becomes

$$1 - \mathbf{x}^T(k|k)\mathbf{S}^{-1}\mathbf{x}(k|k) \geq 0$$

which is equivalent to

$$\begin{bmatrix} 1 & \mathbf{x}^T(k|k) \\ \mathbf{x}(k|k) & \mathbf{S} \end{bmatrix} \succeq \mathbf{0} \tag{16.44}$$

At sampling instant k, the state vector $\mathbf{x}(k|k)$ is assumed to be a known measurement which is used in a state feedback control $\mathbf{u}(k|k) = \mathbf{Fx}(k|k)$ (see Eq. (16.40)). Recall that for $V[\mathbf{x}(k|k)]$ to be an upper bound of $J(k)$, $V(k|k)$ is required to satisfy the condition in Eq. (16.41). By substituting Eqs. (16.40) and (16.35) into Eq. (16.41), we obtain

$$\mathbf{x}^T(k+i|k)\mathbf{W}\mathbf{x}(k+i|k) \leq 0 \qquad \text{for } i \geq 0 \tag{16.45}$$

where

$$\mathbf{W} = [\mathbf{A}(k+i) + \mathbf{B}(k+i)\mathbf{F}]^T \mathbf{P}[\mathbf{A}(k+i) + \mathbf{B}(k+i)\mathbf{F}] \\ -\mathbf{P} + \mathbf{F}^T\mathbf{R}\mathbf{F} + \mathbf{Q}$$

Evidently, Eq. (16.45) holds if $\mathbf{W} \preceq \mathbf{0}$. Now if we let $\mathbf{Y} = \mathbf{FS}$ where $\mathbf{S}$ is related to matrix $\mathbf{P}$ by Eq. (16.43), then based on the fact that the matrix inequality

$$\begin{bmatrix} \mathbf{D} & \mathbf{F} \\ \mathbf{H}^T & \mathbf{G} \end{bmatrix} \succ \mathbf{0} \tag{16.46}$$

is equivalent to

$$\mathbf{G} \succ \mathbf{0} \quad \text{and} \quad \mathbf{D} - \mathbf{HG}^{-1}\mathbf{H}^T \succ \mathbf{0} \tag{16.47}$$

or

$$\mathbf{D} \succ \mathbf{0} \quad \text{and} \quad \mathbf{G} - \mathbf{H}^T\mathbf{D}^{-1}\mathbf{H} \succ \mathbf{0} \tag{16.48}$$

it can be shown that $\mathbf{W} \preceq \mathbf{0}$ is equivalent to

$$\begin{bmatrix} \mathbf{S} & \mathbf{SA}_{k+i}^T + \mathbf{Y}^T\mathbf{B}_{k+i}^T & \mathbf{SQ}^{1/2} & \mathbf{Y}^T\mathbf{R}^{1/2} \\ \mathbf{A}_{k+i}\mathbf{S} + \mathbf{B}_{k+i}\mathbf{Y} & \mathbf{S} & 0 & 0 \\ \mathbf{Q}^{1/2}\mathbf{S} & 0 & \gamma\mathbf{I}_n & 0 \\ \mathbf{R}^{1/2}\mathbf{Y} & 0 & 0 & \gamma\mathbf{I}_p \end{bmatrix} \succeq \mathbf{0} \tag{16.49}$$

where $\mathbf{A}_{k+i}$ and $\mathbf{B}_{k+i}$ stand for $\mathbf{A}(k+i)$ and $\mathbf{B}(k+i)$, respectively (see Probs. 16.6 and 16.7). Since the matrix inequality in Eq. (16.49) is affine with respect to $[\mathbf{A}(k+i)\ \mathbf{B}(k+i)]$, Eq. (16.49) is satisfied for all $[\mathbf{A}(k+i)\ \mathbf{B}(k+i)] \in \mathcal{M}$ defined by Eq. (16.36) if there exist $\mathbf{S} \succ \mathbf{0}$, $\mathbf{Y}$, and scalar γ such that

$$\begin{bmatrix} \mathbf{S} & \mathbf{SA}_j^T + \mathbf{Y}^T\mathbf{B}_j^T & \mathbf{SQ}^{1/2} & \mathbf{Y}^T\mathbf{R}^{1/2} \\ \mathbf{A}_j\mathbf{S} + \mathbf{B}_j\mathbf{Y} & \mathbf{S} & 0 & 0 \\ \mathbf{Q}^{1/2}\mathbf{S} & 0 & \gamma\mathbf{I}_n & 0 \\ \mathbf{R}^{1/2}\mathbf{Y} & 0 & 0 & \gamma\mathbf{I}_p \end{bmatrix} \succeq \mathbf{0} \tag{16.50}$$

for $j = 1, 2, \ldots, L$. Therefore, the unconstrained robust MPC can be formulated in terms of the constrained optimization problem

$$\underset{\gamma, \mathbf{S}, \mathbf{Y}}{\text{minimize}} \ \gamma \tag{16.51a}$$

$$\text{subject to: constraints in Eqs. (16.44) and (16.50)} \tag{16.51b}$$

There are a total of $L+1$ matrix inequality constraints in this problem in which the variables γ, $\mathbf{S}$, and $\mathbf{Y}$ are present affinely. Therefore, this is an SDP problem and the algorithms studied in Chap. 14 can be used to solve it. Once the optimal matrices $\mathbf{S}^*$ and $\mathbf{Y}^*$ are obtained, the optimal feedback matrix can be computed as

$$\mathbf{F}^* = \mathbf{Y}^* \mathbf{S}^{*-1} \tag{16.52}$$

Example 16.4 Design a robust MPC for the angular positioning system discussed in Sec. 16.3.1. Assume that the initial angular position and angular velocity of the antenna are $\theta(0) = 0.12$ rad and $\dot{\theta}(0) = -0.1$ rad/s, respectively. The goal of the MPC is to steer the antenna to the desired position $\theta_r = 0$. The weighting matrix $\mathbf{R}$ in $J(k)$ in this case is a scalar and is set to $R = 2 \times 10^{-5}$.

Solution Note that $\theta(k)$ is related to $\mathbf{x}(k)$ through the equation $\theta(k) = [1 \ 0]\mathbf{x}(k)$; hence Eq. (16.37b) implies that $y(k) = \theta(k)$, and

$$y^2(k + i|k) = \mathbf{x}^T(k + i|k) \begin{bmatrix} 1 & 0 \\ 0 & 0 \end{bmatrix} \mathbf{x}(k + i|k)$$

The objective function can be written as

$$J(k) = \sum_{i=0}^{\infty} [y^2(k + i|k) + Ru^2(k + i|k)]$$

$$= \sum_{i=0}^{\infty} [\mathbf{x}^T(k + i|k)\mathbf{Q}\mathbf{x}(k + i|k) + Ru^2(k + i|k)]$$

where

$$\mathbf{Q} = \begin{bmatrix} 1 & 0 \\ 0 & 0 \end{bmatrix} \quad \text{and} \quad R = 2 \times 10^{-5}$$

Since the control system under consideration has only one scalar input, namely, the voltage applied to the motor, $u(k + i|k)$ is a scalar. Consequently, the feedback gain $\mathbf{F}$ is a row vector of dimension 2. Other known quantities in the constraints in Eqs. (16.44) and (16.50) are

$$\mathbf{x}(0|0) = \begin{bmatrix} 0.12 \\ -0.10 \end{bmatrix}, \quad \mathbf{Q}^{1/2} = \begin{bmatrix} 1 & 0 \\ 0 & 0 \end{bmatrix}, \quad R^{1/2} = 0.0045$$

$$\mathbf{B}_j = \begin{bmatrix} 0 \\ 0.0787 \end{bmatrix} \quad \text{for } j = 1, 2$$

and matrices $\mathbf{A}_1$ and $\mathbf{A}_2$ are given by Eq. (16.37c).

With the above data and a sampling period of 0.1 s, the solution of the SDP problem in Eq. (16.51), $\{\mathbf{Y}^*, \mathbf{S}^*\}$, can be obtained using Algorithm 14.1, and by using Eq. (16.52) $\mathbf{F}^*$ can be deduced. The optimal MPC can then be computed using the state feedback control law

$$u(k+1|k) = \mathbf{F}^*\mathbf{x}(k+1|k) \qquad \text{for } k = 0, 1, \ldots \qquad (16.53)$$

The state $\mathbf{x}(k+1|k)$ in Eq. (16.53) is calculated using the model in Eq. (16.37), where $\mathbf{A}(k)$ is selected randomly from the set

$$\mathcal{M} = \text{Co}\{\mathbf{A}_1, \mathbf{A}_2\}$$

Figure 16.5a and b depicts the angular position $\theta(k)$ and velocity $\dot{\theta}(k)$ obtained over the first 2 seconds, respectively. It is observed that both $\theta(k)$ and $\dot{\theta}(k)$ are steered to the desired value of zero within a second. The corresponding MPC profile $u(k)$ for $k = 1, 2, \ldots, 20$ is shown in Fig. 16.5c.

■

16.3.4 Robust constrained MPC by using SDP

Frequently, it is desirable to design an MPC subject to certain constraints on the system's input and/or output. For example, constraints on the control input may become necessary in order to represent limitations on control equipment (such as value saturation in a process control scenario). The need for constraints can be illustrated in terms of Example 16.4. In Fig. 16.5c, we observe that at instants 0.1, 0.2, and 0.3 s, the magnitude of the control voltage exceeds 2 V. In such a case, the controller designed would become nonfeasible if the maximum control magnitude were to be limited to 2 V. In the rest of the section we develop robust model-predictive controllers with L_2 norm and componentwise input constraints using SDP.

16.3.4.1 L_2-norm input constraint

As in the unconstrained MPC studied in Sec. 16.3.3, the objective function considered here is also the upper bound γ in Eq. (16.42), and the state feedback control $\mathbf{u}(k+i|k) = \mathbf{F}\mathbf{x}(k+i|k)$ is assumed throughout this section. From Sec. 16.3.3, we know that the matrix inequality constraint in Eq. (16.50) implies the inequality in Eq. (16.49) which, in conjunction with Eq. (16.35), leads to

$$\mathbf{x}^T(k+i+1|k)\mathbf{P}\mathbf{x}(k+i+1|k) - \mathbf{x}^T(k+i|k)\mathbf{P}\mathbf{x}(k+i|k)$$
$$\leq -\mathbf{x}^T(k+i|k)(\mathbf{F}^T\mathbf{R}\mathbf{F} + \mathbf{S})\mathbf{x}(k+i|k) < 0$$

Hence

$$\mathbf{x}^T(k+i+1|k)\mathbf{P}\mathbf{x}(k+i+1|k) < \mathbf{x}^T(k+i|k)\mathbf{P}\mathbf{x}(k+i|k)$$

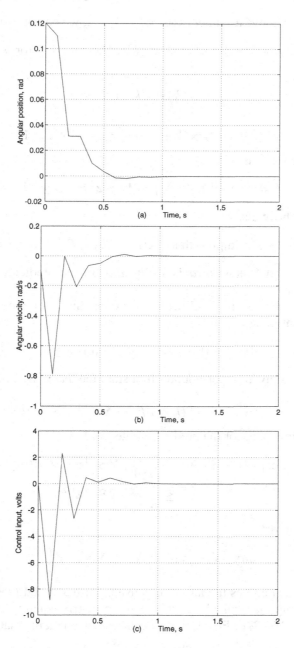

Figure 16.5. Performance of MPC in Example 16.4 with $R = 2 \times 10^{-5}$: (a) Angular position $\theta(k)$, (b) angular velocity $\dot{\theta}(k)$, (c) profile of the MPC.

By repeating the above argument for $i = 0, 1, \ldots,$ we conclude that

$$\mathbf{x}^T(k + i|k)\mathbf{Px}(k + i|k) < \mathbf{x}^T(k|k)\mathbf{Px}(k|k) \qquad \text{for } i \geq 1$$

Therefore,

$$\mathbf{x}^T(k|k)\mathbf{Px}(k|k) \leq \gamma$$

implies that

$$\mathbf{x}^T(k + i|k)\mathbf{Px}(k + i|k) \leq \gamma \qquad \text{for } i \geq 1$$

So if we define set $\mathcal{E}$ as

$$\mathcal{E} = \{\mathbf{z} : \mathbf{z}^T\mathbf{Pz} \leq \gamma\} = \{\mathbf{z} : \mathbf{z}^T\mathbf{S}^{-1}\mathbf{z} \leq 1\} \tag{16.54}$$

then from the above analysis

$$\mathbf{x}(k|k) \in \mathcal{E} \quad \text{implies that} \quad \mathbf{x}(k + i|k) \in \mathcal{E} \qquad \text{for } i \geq 1$$

In other words, set $\mathcal{E}$ is an *invariant ellipsoid* for the predicted states of the uncertain system.

Now let us consider the Euclidean norm constraint on the control input at sampling instant k, i.e.,

$$||\mathbf{u}(k + i|k)|| \leq u_{max} \qquad \text{for } i \geq 0 \tag{16.55}$$

where u_{max} is a given upper bound. In a state feedback MPC, the control is given by

$$\mathbf{u}(k + i|k) = \mathbf{Fx}(k + i|k) = \mathbf{YS}^{-1}\mathbf{x}(k + i|k)$$

Since set $\mathcal{E}$ is invariant for the predicted state, we have

$$\max_{i \geq 0}||\mathbf{u}(k + i|k)||^2 = \max_{i \geq 0}||\mathbf{YS}^{-1}\mathbf{x}(k + i|k)||^2$$

$$\leq \max_{\mathbf{z} \in \mathcal{E}}||\mathbf{YS}^{-1}\mathbf{z}||^2 \tag{16.56}$$

It can be shown that

$$\max_{\mathbf{z} \in \mathcal{E}}||\mathbf{YS}^{-1}\mathbf{z}||^2 = \lambda_{max}(\mathbf{YS}^{-1}\mathbf{Y}^T) \tag{16.57}$$

where $\lambda_{max}(\mathbf{M})$ denotes the largest eigenvalue of matrix $\mathbf{M}$ (see Prob. 16.8). Further, by using the equivalence between the matrix equality in Eq. (16.46) and that in Eq. (16.47) or Eq. (16.48), it can be shown that the matrix inequality

$$\begin{bmatrix} u_{max}^2\mathbf{I} & \mathbf{Y} \\ \mathbf{Y}^T & \mathbf{S} \end{bmatrix} \succeq \mathbf{0} \tag{16.58}$$

implies that

$$\lambda_{max}(\mathbf{YS}^{-1}\mathbf{Y}^T) \leq u_{max}^2 \tag{16.59}$$

(see Prob. 16.9). Therefore, the L_2-norm input constraint in Eq. (16.55) holds if the matrix inequality in Eq. (16.58) is satisfied. Thus a robust MPC that allows an L_2-norm input constraint can be formulated by adding the constraint in Eq. (16.58) to the SDP problem in Eq. (16.51) as

$$\underset{\gamma, \mathbf{S}, \mathbf{Y}}{\text{minimize}} \ \gamma \qquad (16.60\text{a})$$

subject to: constraints in Eqs. (16.44), (16.50), and (16.58) $\qquad$ (16.60b)

Since variables $\mathbf{S}$ and $\mathbf{Y}$ are present in Eq. (16.58) affinely, this is again an SDP problem.

16.3.4.2 Componentwise input constraints

Another type of commonly used input constraint is an upper bound for the magnitude of each component of the control input, i.e.,

$$|u_j(k+i|k)| \le u_{j,max} \qquad \text{for } i \ge 0, \ j = 1, 2, \ldots, p \qquad (16.61)$$

It follows from Eq. (16.56) that

$$\max_{i \ge 0} |u_j(k+i|k)|^2 = \max_{i \ge 0} |[\mathbf{Y}\mathbf{S}^{-1}\mathbf{x}(k+i|k)]_j|^2$$

$$\le \max_{\mathbf{z} \in \mathcal{E}} |(\mathbf{Y}\mathbf{S}^{-1}\mathbf{z})_j|^2 \quad \text{for } j = 1, 2, \ldots, p$$

$$(16.62)$$

Note that the set $\mathcal{E}$ defined in Eq. (16.54) can be expressed as

$$\mathcal{E} = \{\mathbf{z} : \mathbf{z}^T\mathbf{S}^{-1}\mathbf{z} \le 1\} = \{\mathbf{w} : ||\mathbf{w}|| \le 1, \ \mathbf{w} = \mathbf{S}^{-1/2}\mathbf{z}\}$$

which, in conjunction with the use of the Cauchy-Schwarz inequality, modifies Eq. (16.61) into

$$\max_{i \ge 0} |u_j(k+i|k)|^2 \le \max_{||\mathbf{w}|| \le 1} |(\mathbf{Y}\mathbf{S}^{-1/2}\mathbf{w})_j|^2 \le ||(\mathbf{Y}\mathbf{S}^{-1/2})_j||^2$$

$$= (\mathbf{Y}\mathbf{S}^{-1}\mathbf{Y})_{i,j} \qquad \text{for } j = 1, 2, \ldots, p \quad (16.63)$$

It can be readily verified that if there exists a symmetric matrix $\mathbf{X} \in R^{p \times p}$ such that

$$\begin{bmatrix} \mathbf{X} & \mathbf{Y} \\ \mathbf{Y}^T & \mathbf{S} \end{bmatrix} \succeq \mathbf{0} \qquad (16.64\text{a})$$

where the diagonal elements of $\mathbf{X}$ satisfy the inequalities

$$X_{jj} \le u_{j,max}^2 \qquad \text{for } j = 1, 2, \ldots, p \qquad (16.64\text{b})$$

then

$$(\mathbf{Y}^T\mathbf{S}^{-1}\mathbf{Y})_{jj} \le u_{j,max}^2 \qquad \text{for } j = 1, 2, \ldots, p$$

which, by virtue of Eq. (16.63), implies the inequalities in Eq. (16.61) (see Prob. 16.10). Therefore, the componentwise input constraints in Eq. (16.61) hold if there exists a symmetric matrix $\mathbf{X}$ that satisfies the inequalities in Eqs. (16.64a) and (16.64b). Hence a robust MPC with the input constraints in Eq. (16.61) can be formulated by modifying the SDP problem in Eq. (16.51) to

$$\underset{\gamma, \mathbf{S}, \mathbf{X}, \mathbf{Y}}{\text{minimize}} \ \gamma \qquad (16.65a)$$

subject to the constraints in Eqs. (16.44), (16.50), and (16.64) $\qquad$ (16.65b)

Example 16.5 Design an MPC for the angular positioning system discussed in Sec. 16.3.1 with input constraint

$$|u(k|k + i)| \le 2 \qquad \text{for } i \ge 0$$

The initial state $\mathbf{x}(0)$ and other parameters are the same as in Example 16.4.

Solution Since $u(k|k + i)$ is a scalar, $||u(k + i|k)|| = |u(k + i|k)|$. Hence the L_2-norm input and the componentwise input constraints become identical. With $u_{max} = 2$ V, the MPC can be obtained by solving the SDP problem in Eq. (16.60) using Algorithm 14.1 for each sampling instant k. The angular position $\theta(k)$ and velocity $\dot{\theta}(k)$ over the first 2 seconds are plotted in Fig. 16.6a and b, respectively. The control profile is depicted in Fig. 16.6c where we note that the magnitude of the control voltage has been kept within the 2-V bound.

It is interesting to note that the magnitude of the MPC commands can also be reduced by using a larger value of weighting factor R in $J(k)$. For the angular positioning system in question with $R = 0.0035$, the unconstrained MPC developed in Example 16.4 generates the control profile shown in Fig. 16.7a, which obviously satisfies the constraint $|u(k|k + i)| \le 2$. However, the corresponding $\theta(k)$ and $\dot{\theta}(k)$ plotted in Fig. 16.7b and c, respectively, indicate that such an MPC takes a longer time to steer the system from the same initial state to the desired zero state compared to what is achieved by the constrained MPC. ∎

16.4 Optimal Force Distribution for Robotic Systems with Closed Kinematic Loops

Because of their use in a wide variety of applications ranging from robotic surgery to space exploration, robotic systems with closed kinematic loops such as multiple manipulators handling a single workload, dextrous hands with fingers closed[1] through the object grasped (see Fig. 16.8), and multilegged vehicles

---

[1] A kinematic finger/chain is said to be closed if both ends of the finger/chain are mechanically constrained.

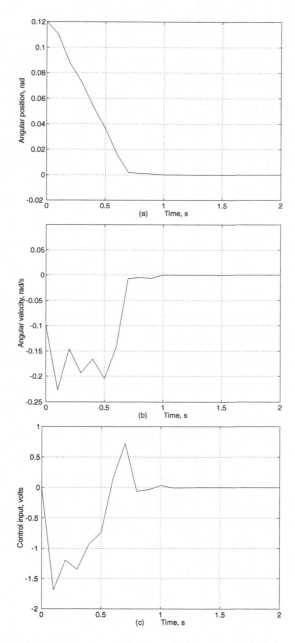

Figure 16.6. Performance of MPC in Example 16.5. (a) Angular position $\theta(k)$, (b) angular velocity $\dot{\theta}(k)$, (c) profile of the constrained MPC.

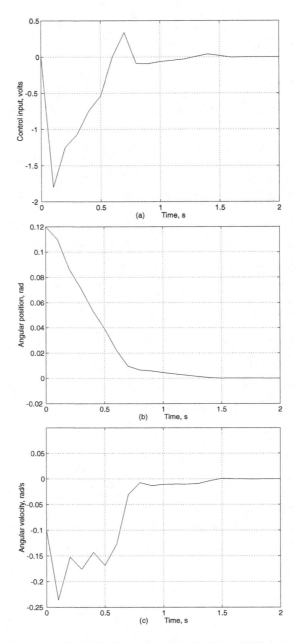

Figure 16.7. Performance of MPC in Example 16.4 with $R = 0.0035$ (a) Profile of the MPC, (b) angular position $\theta(k)$, (c) angular velocity $\dot{\theta}(k)$.

with kinematic chains closed through the body (see Fig. 16.9) have become an increasingly important subject of study in the past several years [13]–[18]. An issue of central importance for this class of robotic systems is the *force distribution* that determines the joint torques and forces to generate the desired motion of the workload [14].

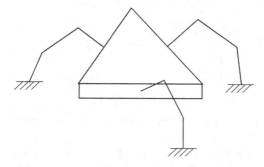

Figure 16.8. Three coordinated manipulators (also known as a three-finger dextrous hand) grasping an object.

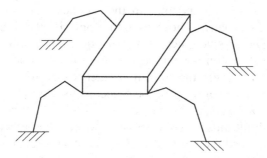

Figure 16.9. Multilegged vehicle.

In Sec. 16.4.1, the force distribution problem for multifinger dextrous hands is described and two models for the contact forces are studied. The optimal force distribution problem is then formulated and solved using LP and SDP in Secs. 16.4.2 and 16.4.3, respectively.

16.4.1 Force distribution problem in multifinger dextrous hands

Consider a dextrous hand with m fingers grasping an object such as that depicted in Fig. 16.10 for $m = 3$. The contact force c_i of the ith finger is supplied by the finger's n_j joint torques τ_{ij} for $j = 1, 2, \ldots, n_j$, and f_{ext} is an external force exerted on the object. The force distribution problem is to find

562

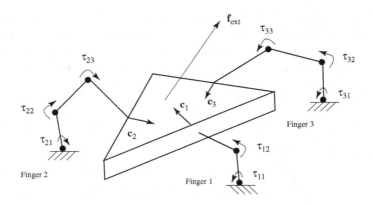

Figure 16.10. A three-finger hand grasping an object.

the contact forces c_i for $i = 1, 2, \ldots, m$ that would balance the external force $f_{ext} \in R^6$ so as to assure a stable grasp. The dynamics of the system can be represented by the equation

$$\mathbf{W}\mathbf{c} = -\mathbf{f}_{ext} \qquad (16.66)$$

where $\mathbf{c}$ is a vector whose components are the m contact forces c_i for $1 \leq i \leq m$ and $\mathbf{W} \in R^{6 \times 3m}$ is a matrix whose columns comprise the directions of the m contact forces. The product vector $\mathbf{W}\mathbf{c}$ in Eq. (16.66) is a six-dimensional vector whose first three components represent the overall contact force and last three components represent the overall contact torque relative to a frame of reference with the center of mass of the object as its origin [14][17].

To maintain a stable grasp, the contact forces whose magnitudes are within the friction force limit must remain positive towards the object surface. There are two commonly used models to describe a contact force, namely, the *point-contact* and *soft-finger contact model*. In the point-contact model, the contact force c_i has three components, a component c_{i1} that is orthogonal and two components c_{i2} and c_{i3} that are tangential to the object surface as shown in Fig. 16.11a. In the soft-finger contact model, c_i has an additional component c_{i4}, as shown in Fig. 16.11b, that describes the torsional moment around the normal on the object surface [17].

Friction force plays an important role in stable grasping. In a point-contact model, the friction constraint can be expressed as

$$\sqrt{c_{i2}^2 + c_{i3}^2} \leq \mu_i c_{i1} \qquad (16.67)$$

where c_{i1} is the normal force component, c_{i2} and c_{i3} are the tangential components of the contact force c_i, and $\mu_i > 0$ denotes the friction coefficient at the contact point. It follows that for a given friction coefficient $\mu_i > 0$, the constraint in Eq. (16.67) describes a *friction cone* as illustrated in Fig. 16.12.

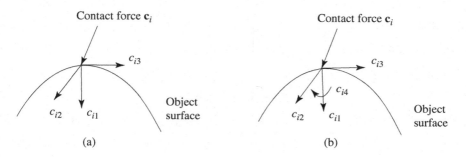

Figure 16.11. (a) Point-contact model, (b) soft-finger contact model.

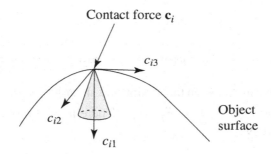

Figure 16.12. Friction cone as a constraint on contact force $\mathbf{c}_i$.

Obviously, the friction force modeled by Eq. (16.67) is nonlinear: for a fixed μ_i and c_{i1}, the magnitude of the tangential force is constrained to within a circle of radius $\mu_i c_{i1}$. A linear constraint for the friction force can be obtained by approximating the circle with a square as shown in Fig. 16.13. The approximation involved can be described in terms of the linear constraints [14]

$$c_{i1} \geq 0 \tag{16.68a}$$

$$-\frac{\mu_i}{\sqrt{2}} c_{i1} \leq c_{i2} \leq \frac{\mu_i}{\sqrt{2}} c_{i1} \tag{16.68b}$$

$$-\frac{\mu_i}{\sqrt{2}} c_{i1} \leq c_{i3} \leq \frac{\mu_i}{\sqrt{2}} c_{i1} \tag{16.68c}$$

The friction limits in a soft-finger contact model depend on both the torsion and shear forces, and can be described by a linear or an elliptical approximation [17]. The linear model is given by

$$\frac{1}{\mu_i} f_t + \frac{1}{\hat{\mu}_{ti}} |c_{i4}| \leq c_{i1} \tag{16.69}$$

564

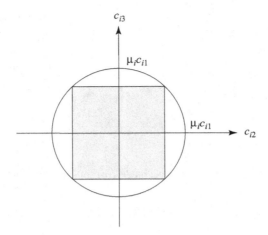

Figure 16.13. Linear approximation for friction cone constraint.

where $\hat{\mu}_{ti}$ is a constant between the torsion and shear limits, μ_i is the tangential friction coefficient, and $f_t = \sqrt{c_{i2}^2 + c_{i3}^2}$. The elliptical model, on the other hand, is described by

$$c_{i1} \geq 0 \qquad (16.70\text{a})$$

$$\frac{1}{\mu_i}(c_{i2}^2 + c_{i3}^2) + \frac{1}{\mu_{ti}}c_{i4}^2 \leq c_{i1}^2 \qquad (16.70\text{b})$$

where μ_{ti} is a constant.

16.4.2 Solution of optimal force distribution problem by using LP

The problem of finding the optimal force distribution of an m-finger dextrous hand is to find the contact forces c_i for $1 \leq i \leq m$ that optimize a performance index subject to the force balance constraint in Eq. (16.66) and friction-force constraints in one of Eqs. (16.67)–(16.70).

A typical performance measure in this case is the weighted sum of the m normal force components c_{i1} ($1 \leq i \leq m$), i.e.,

$$p = \sum_{i=1}^{m} w_i c_{i1} \qquad (16.71)$$

If we employ the point-contact model and let

$$\mathbf{c} = \begin{bmatrix} \mathbf{c}_1 \\ \vdots \\ \mathbf{c}_m \end{bmatrix}, \quad \mathbf{c}_i = \begin{bmatrix} c_{i1} \\ c_{i2} \\ c_{i3} \end{bmatrix}, \quad \mathbf{w} = \begin{bmatrix} \mathbf{w}_1 \\ \vdots \\ \mathbf{w}_m \end{bmatrix}, \quad \mathbf{w}_i = \begin{bmatrix} w_i \\ 0 \\ 0 \end{bmatrix}$$

then the objective function in Eq. (16.71) can be expressed as

$$p(\mathbf{c}) = \mathbf{w}^T \mathbf{c} \tag{16.72}$$

and the friction-force constraints in Eq. (16.68) can be written as

$$\mathbf{Ac} \geq \mathbf{0} \tag{16.73}$$

where

$$\mathbf{A} = \begin{bmatrix} \mathbf{A}_1 & & \mathbf{0} \\ & \ddots & \\ \mathbf{0} & & \mathbf{A}_m \end{bmatrix} \quad \text{and} \quad \mathbf{A}_i = \begin{bmatrix} 1 & 0 & 0 \\ \mu_i/\sqrt{2} & -1 & 0 \\ \mu_i/\sqrt{2} & 1 & 0 \\ \mu_i/\sqrt{2} & 0 & -1 \\ \mu_i/\sqrt{2} & 0 & 1 \end{bmatrix}$$

Obviously, the problem of minimizing function $p(\mathbf{c})$ in Eq. (16.72) subject to the linear inequality constraints in Eq. (16.73) and linear equality constraints

$$\mathbf{Wc} = -\mathbf{f}_{ext} \tag{16.74}$$

is an LP problem and many algorithms studied in Chaps. 11 and 12 are applicable. In what follows, the above LP approach is illustrated using a four-finger robot hand grasping a rectangular object. The same robot hand was used in [17] to demonstrate a gradient-flow-based optimization method.

Example 16.6 Find the optimal contact forces $\mathbf{c}_i$ for $i = 1, 2, \ldots, 4$, that minimize the objective function in Eq. (16.73) subject to the constraints in Eqs. (16.72) and (16.74) for a four-finger robot hand grasping the rectangular object illustrated in Fig. 16.14.

Solution The input data of the problem is given by

$$\mathbf{W}^T = \begin{bmatrix} 0 & 1 & 0 & 0 & 0 & -a_1 \\ 1 & 0 & 0 & 0 & 0 & b \\ 0 & 0 & 1 & -b & a_1 & 0 \\ 0 & 1 & 0 & 0 & 0 & a_2 \\ 1 & 0 & 0 & 0 & 0 & b \\ 0 & 0 & 1 & -b & -a_2 & 0 \\ 0 & -1 & 0 & 0 & 0 & -a_3 \\ 1 & 0 & 0 & 0 & 0 & -b \\ 0 & 0 & 1 & b & -a_3 & 0 \\ 0 & -1 & 0 & 0 & 0 & a_4 \\ 1 & 0 & 0 & 0 & 0 & -b \\ 0 & 0 & 1 & b & a_4 & 0 \end{bmatrix} \tag{16.75}$$

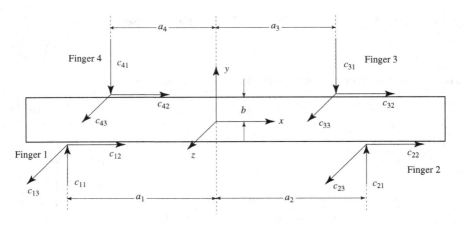

Figure 16.14. Grasping a rectangular object with four fingers.

where $a_1 = 0.1$, $a_2 = 0.15$, $a_3 = 0.05$, $a_4 = 0.065$, and $b = 0.02$. The weights, μ_i, and $\mathbf{f}_{ext}$ are given by $w_i = 1$, $\mu_i = 0.4$ for $1 \leq i \leq 4$, and

$$\mathbf{f}_{ext} = [0\ 0\ -1\ 0\ 0\ 0]^T \tag{16.76}$$

The rank of matrix $\mathbf{W}$ is 6; hence the solutions of Eq. (16.74) can be characterized by the equation

$$\mathbf{c} = -\mathbf{W}^+\mathbf{f}_{ext} + \mathbf{V}_\eta\phi \tag{16.77}$$

where $\mathbf{W}^+$ denotes the Moore-Penrose pseudoinverse of $\mathbf{W}$, $\mathbf{V}_\eta$ is the matrix formed using the last 6 columns of $\mathbf{V}$ obtained from the singular-value decomposition $\mathbf{W} = \mathbf{U}\boldsymbol{\Sigma}\mathbf{V}^T$, and $\phi \in R^{6\times 1}$ is the free parameter vector (see Sec. 10.4). Using Eq. (16.77), the above LP problem is reduced to

$$\text{minimize } \hat{\mathbf{w}}^T\phi \tag{16.78a}$$

$$\text{subject to } \hat{\mathbf{A}}\phi \geq \hat{\mathbf{b}} \tag{16.78b}$$

where

$$\hat{\mathbf{w}} = \mathbf{V}_\eta^T\mathbf{w}, \quad \hat{\mathbf{A}} = \mathbf{A}\mathbf{V}_\eta, \quad \hat{\mathbf{b}} = \mathbf{A}\mathbf{W}^+\mathbf{f}_{ext}$$

The reduced LP problem was solved by using Algorithm 11.1. In [14], this solution method is referred to as the *compact LP method*. If ϕ^* is the minimizer of the LP problem in Eq. (16.78), then the minimizer of the original LP problem is given by

$$\mathbf{c}^* = -\mathbf{W}^+\mathbf{f}_{ext} + \mathbf{V}_\eta\phi^*$$

which leads to

$$\mathbf{c}^* = \begin{bmatrix} \mathbf{c}_1^* \\ \mathbf{c}_2^* \\ \mathbf{c}_3^* \\ \mathbf{c}_4^* \end{bmatrix}$$

with

$$\mathbf{c}_1^* = \begin{bmatrix} 1.062736 \\ 0.010609 \\ 0.300587 \end{bmatrix}, \quad \mathbf{c}_2^* = \begin{bmatrix} 0.705031 \\ 0.015338 \\ 0.199413 \end{bmatrix}$$

$$\mathbf{c}_3^* = \begin{bmatrix} 1.003685 \\ -0.038417 \\ 0.283885 \end{bmatrix}, \quad \mathbf{c}_4^* = \begin{bmatrix} 0.764082 \\ 0.012470 \\ 0.216115 \end{bmatrix}$$

The minimum value of $p(\mathbf{c})$ at $\mathbf{c}^*$ was found to be 3.535534. ∎

16.4.3   Solution of optimal force distribution problem by using SDP

The LP-based solution discussed in Sec. 16.4.2 is an approximate solution because it was obtained for the case where the quadratic friction-force constraint in Eq. (16.67) is approximated using a linear model. An improved solution can be obtained by formulating the problem at hand as an SDP problem. To this end, we need to convert the friction-force constraints into linear matrix inequalities [17].

For the point-contact case, the friction-force constraint in Eq. (16.67) yields

$$\mu_i c_{i1} \geq 0$$

and

$$\mu_i^2 c_{i1}^2 - (c_{i2}^2 + c_{i3}^2) \geq 0$$

Hence Eq. (16.67) is equivalent to

$$\mathbf{P}_i = \begin{bmatrix} \mu_i c_{i1} & 0 & c_{i2} \\ 0 & \mu_i c_{i1} & c_{i3} \\ c_{i2} & c_{i3} & u_1 c_{i1} \end{bmatrix} \succeq \mathbf{0} \qquad (16.79)$$

(see Prob. 16.11a). For an m-finger robot hand, the constraint on point-contact friction forces is given by

$$\mathbf{P}(\mathbf{c}) = \begin{bmatrix} \mathbf{P}_1 & & \mathbf{0} \\ & \ddots & \\ \mathbf{0} & & \mathbf{P}_m \end{bmatrix} \succeq \mathbf{0} \qquad (16.80)$$

where $\mathbf{P}_i$ is defined by Eq. (16.79). Similarly, the constraint on the soft-finger friction forces of an m-finger robot hand can be described by Eq. (16.80) where matrix $\mathbf{P}_i$ is given by

$$\mathbf{P}_i = \left[\begin{array}{cccc|ccc}
c_{i1} & 0 & 0 & 0 & & & \\
0 & \alpha_i & 0 & c_{i2} & & \mathbf{0} & \\
0 & 0 & \alpha_i & c_{i3} & & & \\
0 & c_{i2} & c_{i3} & \alpha_i & & & \\
\hline
& & & & \beta_i & 0 & c_{i2} \\
& \mathbf{0} & & & 0 & \beta_i & c_{i3} \\
& & & & c_{i2} & c_{i3} & \beta_i
\end{array}\right] \tag{16.81}$$

with $\alpha_i = \mu_i(c_{i1} + c_{i4}/\hat{\mu}_{ti})$ and $\beta_i = \mu_i(c_{i1} - c_{i4}/\hat{\mu}_{ti})$ for the linear model in Eq. (16.69) or

$$\mathbf{P}_i = \left[\begin{array}{cccc}
c_{i1} & 0 & 0 & \alpha_i c_{i2} \\
0 & c_{i1} & 0 & \alpha_i c_{i3} \\
0 & 0 & c_{i1} & \beta_i c_{i4} \\
\alpha_i c_{i2} & \alpha_i c_{i3} & \beta_i c_{i4} & c_{i1}
\end{array}\right] \tag{16.82}$$

with $\alpha_i = 1/\sqrt{\mu_i}$ and $\beta_i = 1/\sqrt{\mu_{ti}}$ for the elliptical model in Eq. (16.70) (see Prob. 16.11(b) and (c)).

Note that matrix $\mathbf{P}(\mathbf{c})$ for both point-contact and soft-finger models is *linear* with respect to parameters c_{i1}, c_{i2}, c_{i3}, and c_{i4}.

The optimal force distribution problem can now be formulated as

$$\text{minimize } p = \mathbf{w}^T \mathbf{c} \tag{16.83a}$$

$$\text{subject to:} \quad \mathbf{W}\mathbf{c} = -\mathbf{f}_{ext} \tag{16.83b}$$

$$\mathbf{P}(\mathbf{c}) \succeq \mathbf{0} \tag{16.83c}$$

where $\mathbf{c} = [\mathbf{c}_1^T \ \mathbf{c}_2^T \ \cdots \ \mathbf{c}_m^T]^T$ with $\mathbf{c}_i = [c_{i1} \ c_{i2} \ c_{i3}]^T$ for the point-contact case or $\mathbf{c}_i = [c_{i1} \ c_{i2} \ c_{i3} \ c_{i4}]^T$ for the soft-finger case, and $\mathbf{P}(\mathbf{c})$ is given by Eq. (16.80) with $\mathbf{P}_i$ defined by Eq. (16.79) for the point-contact case or Eq. (16.82) for the soft-finger case. By using the variable elimination method discussed in Sec. 10.4, the solutions of Eq. (16.83b) can be expressed as

$$\mathbf{c} = \mathbf{V}_\eta \phi + \mathbf{c}_0 \tag{16.84}$$

with $\mathbf{c}_0 = -\mathbf{W}^+ \mathbf{f}_{ext}$ where $\mathbf{W}^+$ is the Moore-Penrose pseudo-inverse of $\mathbf{W}$. Thus the problem in Eq. (16.83) reduces to

$$\text{minimize } \hat{p} = \hat{\mathbf{w}}^T \phi \tag{16.85a}$$

$$\text{subject to:} \quad \mathbf{P}(\mathbf{V}_\eta \phi + \mathbf{c}_0) \succeq \mathbf{0} \tag{16.85b}$$

Since $\mathbf{P}(\mathbf{V}_\eta \phi + \mathbf{c}_0)$ is affine with respect to vector ϕ, the optimization problem in Eq. (16.85) is a standard SDP problem of the type studied in Chap. 14.

Example 16.7 Find the optimal contact forces $\mathbf{c}_i$ for $1 \leq i \leq 4$ that would solve the minimization problem in Eq. (16.83) for the 4-finger robot hand grasping the rectangular object illustrated in Fig. 16.14, using the soft-finger model in Eq. (16.70) with $\mu_i = 0.4$ and $\mu_{ti} = \sqrt{0.2}$ for $1 \leq i \leq 4$.

Solution The input data are given by

$$\mathbf{w} = \begin{bmatrix} 1 & 0 & 0 & 0 & 1 & 0 & 0 & 0 & 1 & 0 & 0 & 0 & 1 & 0 & 0 & 0 \end{bmatrix}^T$$

$$\mathbf{f}_{ext} = \begin{bmatrix} 1 & 1 & -1 & 0 & 0.5 & 0.5 \end{bmatrix}^T$$

$$\mathbf{W}^T = \begin{bmatrix}
0 & 1 & 0 & 0 & 0 & -a_1 \\
1 & 0 & 0 & 0 & 0 & b \\
0 & 0 & 1 & -b & a_1 & 0 \\
0 & 0 & 0 & 0 & -1 & 0 \\
0 & 1 & 0 & 0 & 0 & a_2 \\
1 & 0 & 0 & 0 & 0 & b \\
0 & 0 & 1 & -b & -a_2 & 0 \\
0 & 0 & 0 & 0 & -1 & 0 \\
0 & -1 & 0 & 0 & 0 & -a_3 \\
1 & 0 & 0 & 0 & 0 & -b \\
0 & 0 & 1 & b & -a_3 & 0 \\
0 & 0 & 0 & 0 & 1 & 0 \\
0 & -1 & 0 & 0 & 0 & a_4 \\
1 & 0 & 0 & 0 & 0 & -b \\
0 & 0 & 1 & b & a_4 & 0 \\
0 & 0 & 0 & 0 & 1 & 0
\end{bmatrix}$$

where the numerical values of a_1, a_2, a_3, and b are the same as in Example 16.6.

By applying Algorithm 14.1 to the SDP problem in Eq. (16.85), the minimizer ϕ^* was found to be

$$\phi^* = \begin{bmatrix}
-2.419912 \\
-0.217252 \\
3.275539 \\
0.705386 \\
-0.364026 \\
-0.324137 \\
-0.028661 \\
0.065540 \\
-0.839180 \\
0.217987
\end{bmatrix}$$

Eq. (16.84) then yields

$$\mathbf{c}^* = \mathbf{V}_\eta \phi^* + \mathbf{c}_0 = \begin{bmatrix}
\mathbf{c}_1^* \\
\mathbf{c}_2^* \\
\mathbf{c}_3^* \\
\mathbf{c}_4^*
\end{bmatrix}$$

where

$$\mathbf{c}_1^* = \begin{bmatrix} 2.706396 \\ -1.636606 \\ 0.499748 \\ -0.015208 \end{bmatrix}, \quad \mathbf{c}_2^* = \begin{bmatrix} 0.003041 \\ -0.000633 \\ 0.000252 \\ -0.000172 \end{bmatrix}$$

$$\mathbf{c}_3^* = \begin{bmatrix} 3.699481 \\ 0.638543 \\ 0.500059 \\ -0.541217 \end{bmatrix}, \quad \mathbf{c}_4^* = \begin{bmatrix} 0.009955 \\ -0.001303 \\ -0.000059 \\ 0.000907 \end{bmatrix}$$

The minimum value of $p(\mathbf{c})$ at $\mathbf{c}^*$ is $p(\mathbf{c})^* = 6.418873$. ∎

16.5 Multiuser Detection in Wireless Communication Channels

Multiuser communication systems are telecommunication systems where several users can transmit information through a common channel [19] as illustrated in Fig. 16.15. A typical system of this type is a cellular communication system where a number of mobile users in a cell send their information to the receiver at the base station of the cell.

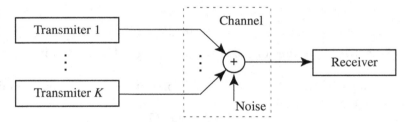

Figure 16.15. A multiuser communication system.

There are three basic multiple access methods for multiuser communication, namely, *frequency-division multiple access* (FDMA), *time-division multiple access* (TDMA), and *code-division multiple access* (CDMA). In FDMA, the available channel bandwidth is divided into a number of nonoverlapping subchannels and each subchannel is assigned to a user. In TDMA, the unit time duration known as frame duration is divided into several nonoverlapping time intervals, and each time interval is assigned to a user. In CDMA, each user is assigned a distinct *code sequence* which spreads the user's information signal across the assigned frequency band. These code sequences have small cross-correlation with each other so that signals from different users can be separated at the receiver using a bank of match filters, each performing cross-correlation of the received signal with a particular code sequence.

In order to accommodate asynchronous users in CDMA channels, practical code sequences are not orthogonal [20]. This nonorthogonality leads to nonzero cross-correlation between each pair of code sequences. Therefore, users interfere with each other and any interferer with sufficient power at the receiver can cause significant performance degradation. Multiuser detection is a demodulation technique that can perform quite effectively in the presence of multiple access interference. The purpose of this section is to demonstrate that several multiuser detection problems can be addressed using modern optimization methods. In Sec. 16.5.1, the CDMA channel model and the maximum-likelihood (ML) multiuser detector [21] is reviewed. A near-optimal multiuser detector for direct sequence (DS)-CDMA channels using SDP relaxation [22] is described in Sec. 16.5.2. In digital communication systems, performance is usually measured in terms of the probability that a signal bit is in error at the receiver output, and this probability is referred to as the *bit-error rate* (BER). In Sec. 16.5.3, we describe a linear multiuser detection algorithm based on minimizing the BER subject to a set of reasonable constraints [23].

16.5.1 Channel model and ML multiuser detector

16.5.1.1 CDMA channel model

We consider a DS-CDMA system where K users transmit information bits through a common channel. The bit interval of each user is T_b seconds and each information bit belongs to the set $\{1, -1\}$. Each signal is assigned a signature waveform $s(t)$, often called *spreading sequence*, given by

$$s(t) = \sum_{i=1}^{N} (-1)^{c_i} p_{T_c}[t - (i-1)T_c] \qquad \text{for } t \in [0, T_b] \qquad (16.86)$$

where $p_{T_c}(t)$ is a rectangular pulse which takes the value of one for $0 \le t \le T_c$ and zero elsewhere, $\{c_1, c_2, \ldots, c_N\}$ is a binary sequence, $N = T_b/T_c$ is the length of the signature waveform, which is often referred to as the *spreading gain*. Typically, the waveform of $p_{T_c}(t)$ is common to all the users, and it is the binary sequence $\{c_1, c_2, \ldots, c_N\}$ assigned to each user that distinguishes the different signature waveforms. One of the commonly used binary sequences is the Gold sequence which has low crosscorrelations for all possible cyclic shifts [19]. The signature waveforms are normalized to have unit energy, i.e., $\|s_k(t)\|^2 = 1$ for $1 \le k \le K$. The received baseband signal is given by

$$y(t) = \sum_{i=0}^{\infty} \sum_{k=1}^{K} A_k^i b_k^i s_k(t - iT_b - \tau_k) + n(t) \qquad (16.87)$$

where b_k^i is an information bit, τ_k is the transmission delay, A_k^i is the signal amplitude of the kth user, and $n(t)$ is additive white Gaussian noise (AWGN) with variance σ^2. A DS-CDMA system is said to be *synchronous* if τ_k in Eq. (16.87) is zero for $1 \le k \le K$, and thus

$$y(t) = \sum_{k=1}^{K} A_k b_k s_k(t) + n(t) \tag{16.88}$$

where t can assume values in the bit interval $[0, T_b]$.

Demodulation is achieved by filtering the received signal $y(t)$ with a bank of matched filters. The filter bank consists of K filters, each matched to each signature waveform, and the filtered signals are sampled at the end of each bit interval. The outputs of the matched filters are given by

$$y_k = \int_0^{T_b} y(t) s_k(t)\, dt \qquad \text{for } 1 \le k \le K \tag{16.89}$$

Using Eq. (16.88), Eq. (16.89) can be expressed as

$$y_k = A_k b_k + \sum_{j \ne k} A_j b_j \rho_{jk} + n_k \qquad \text{for } 1 \le k \le K \tag{16.90}$$

where

$$\rho_{jk} = \int_0^{T_b} s_j(t) s_k(t)\, dt$$

and

$$n_k = \int_0^{T_b} n(t) s_k(t)\, dt$$

The discrete-time synchronous model in Eq. (16.90) can be described in matrix form as

$$\mathbf{y} = \mathbf{RAb} + \mathbf{n} \tag{16.91}$$

where $\mathbf{y} = [y_1\ y_2\ \cdots\ y_K]^T$, $\mathbf{A} = \text{diag}\{A_1,\ A_2,\ \ldots,\ A_K\}$, $\mathbf{b} = [b_1\ b_2\ \cdots\ b_K]^T$, $\mathbf{R}_{ij} = \rho_{ij}$, and $\mathbf{n} = [n_1\ n_2\ \cdots\ n_K]^T$. Since $n(t)$ in Eq. (16.88) is an AWGN with variance σ^2, the term $\mathbf{n}$ in Eq. (16.91) is a zero-mean Gaussian noise vector with covariance matrix $\sigma^2 \mathbf{R}$.

If we consider an ideal channel which is free of background noise and the signature waveforms are orthogonal to each other, then Eq. (16.89) assumes the form $y_k = A_k b_k$. In such a case the information bit b_k can be perfectly detected based on the output of the kth matched filter, y_k. In a realistic CDMA channel, however, the signature waveforms are nonorthogonal [20] and hence the second term at the right-hand side of Eq. (16.90), which quantifies the multiple access interference (MAI), is always nonzero. The MAI in conjunction with the noise represented by term n_k in Eq. (16.90) can in many cases be so large that it is difficult to estimate the transmitted information based on the outputs of the matched filters without further processing. A *multiuser detector* is essentially a digital signal processing algorithm or processor that takes **y** as its input to estimate the transmitted information vector **b** such that a low probability of error is achieved.

16.5.1.2 ML multiuser detector

The goal of the optimal multiuser detector is to generate an estimate of the information vector **b** in Eq. (16.91) that maximizes the log-likelihood function defined by

$$f(\mathbf{b}) = \exp\left(-\frac{1}{2\sigma^2}\int_0^{T_b}[y(t) - \sum_{k=1}^{K} A_k b_k s_k(t)]^2\, dt\right) \qquad (16.92)$$

which is equivalent to maximizing the quadratic function

$$\Omega(\mathbf{b}) = \left[\sum_{k=1}^{K} A_k b_k s_k(t)\right]^2 = 2\mathbf{b}^T\mathbf{A}\mathbf{y} - \mathbf{b}^T\mathbf{A}\mathbf{R}\mathbf{A}\mathbf{b} \qquad (16.93)$$

By defining the unnormalized crosscorrelation matrix as $\mathbf{H} = \mathbf{A}\mathbf{R}\mathbf{A}$ and letting $\mathbf{p} = -2\mathbf{A}\mathbf{y}$, the ML detector is characterized by the solution of the combinatorial optimization problem [21]

$$\text{minimize } \mathbf{x}^T\mathbf{H}\mathbf{x} + \mathbf{x}^T\mathbf{p} \qquad (16.94a)$$

$$\text{subject to:} \quad \mathbf{x}_i \in \{1,\, -1\} \quad \text{for } i = 1,\, 2,\, \ldots,\, K \qquad (16.94b)$$

Because of the binary constraints in Eq. (16.94b), the optimization problem in Eq. (16.94) is an integer programming (IP) problem. Its solution can be obtained by exhaustive evaluation of the objective function over 2^K possible values of **x**. However, the amount of computation involved becomes prohibitive even for a moderate number of users.

16.5.2 Near-optimal multiuser detector using SDP relaxation

The near-optimal multiuser detector described in [22] is based on a relaxation of the so-called *MAX-CUT problem* as detailed below.

16.5.2.1 SDP relaxation of MAX-CUT problem

We begin by examining the MAX-CUT problem which is a well-known IP problem in graph theory. It can be formulated as

$$\text{maximize } \tfrac{1}{2}\sum_{i<j}\sum w_{ij}(1 - x_i x_j) \qquad (16.95a)$$

$$\text{subject to:} \quad x_i \in \{1, -1\} \qquad \text{for } 1 \le i \le n \qquad (16.95\text{b})$$

where w_{ij} denotes the weight from node i to node j in the graph. The constraints in Eq. (16.95b) can be expressed as $x_i^2 = 1$ for $1 \le i \le n$. If we define a symmetric matrix $\mathbf{W} = \{w_{ij}\}$ with $w_{ii} = 0$ for $1 \le i \le n$, then the objective function in Eq. (16.95a) can be expressed as

$$\tfrac{1}{2}\sum_{i<j}\sum w_{ij}(1 - x_i x_j) = \tfrac{1}{4}\sum\sum w_{ij} - \tfrac{1}{4}\mathbf{x}^T \mathbf{W}\mathbf{x}$$

$$= \tfrac{1}{4}\sum\sum w_{ij} - \tfrac{1}{4}\text{trace}(\mathbf{W}\mathbf{X})$$

where $\text{trace}(\cdot)$ denotes the trace of the matrix and $\mathbf{X} = \mathbf{x}\mathbf{x}^T$ with $\mathbf{x} = [x_1 \ x_2 \ \cdots \ x_n]^T$ (see Sec. A.7). Note that the set of matrices $\{\mathbf{X} : \mathbf{X} = \mathbf{x}\mathbf{x}^T \text{ with } x_i^2 = 1$ for $1 \le i \le n\}$ can be characterized by $\{\mathbf{X} : x_{ii} = 1 \text{ for } 1 \le i \le n, \mathbf{X} \succeq \mathbf{0},$ and $\text{rank}(\mathbf{X}) = 1\}$ where x_{ii} denotes the ith diagonal element of $\mathbf{X}$. Hence the problem in Eq. (16.95) can be expressed as

$$\text{minimize trace}(\mathbf{W}\mathbf{X}) \qquad (16.96\text{a})$$

$$\text{subject to:} \quad \mathbf{X} \succeq \mathbf{0} \qquad (16.96\text{b})$$

$$x_{ii} = 1 \qquad \text{for } 1 \le i \le n \qquad (16.96\text{c})$$

$$\text{rank}(\mathbf{X}) = 1 \qquad (16.96\text{d})$$

In [24], Geomans and Williamson proposed a relaxation of the above problem by removing the rank constraint in Eq. (16.96d), which leads to

$$\text{minimize trace}(\mathbf{W}\mathbf{X}) \qquad (16.97\text{a})$$

$$\text{subject to :} \quad \mathbf{X} \succeq \mathbf{0} \qquad (16.97\text{b})$$

$$x_{ii} = 1 \qquad \text{for } 1 \le i \le n \qquad (16.97\text{c})$$

Note that the objective function in Eq. (16.97a) is a linear function of $\mathbf{X}$ and the constraints in Eqs. (16.97b) and (16.97c) can be combined into an LMI as

$$\sum_{i>j}\sum x_{ij}\mathbf{F}_{ij} + \mathbf{I} \succeq \mathbf{0}$$

where, for each (i, j) with $i > j$, $\mathbf{F}_{ij}$ is a symmetric matrix whose (i, j)th and (j, i)th components are one and zero elsewhere. The problem in Eq. (16.97) fits into the formulation in Eq. (14.9) and, therefore, is an SDP problem. For this reason, the problem in Eq. (16.97) is known as an *SDP relaxation* of the IP problem in Eq. (16.95) and, equivalently, of the problem in Eq. (16.96).

If we denote the minimum values of the objective functions in the problems of Eqs. (16.96) and (16.97) as μ^* and ν^*, respectively, then since the feasible region of the problem in Eq. (16.96) is a subset of the feasible region of the problem in Eq. (16.97), we have $\nu^* \leq \mu^*$. Further, it has been shown that if the weights w_{ij} are all nonnegative, then $\nu^* \geq 0.87856\mu^*$ [25]. Therefore, we have

$$0.87856\mu^* \leq \nu^* \leq \mu^* \qquad (16.98)$$

This indicates that the solution of the SDP problem in Eq. (16.97) is in general a good approximation of the solution of the problem in Eq. (16.96). It is the good quality of the approximation in conjunction with the SDP's polynomial-time computational complexity that makes the Geomans-Williamson SDP relaxation an attractive optimization tool for combinatorial minimization problems. As a consequence, this approach has found applications in graph optimization, network management, and scheduling [26][27]. In what follows, we present an SDP-relaxation-based algorithm for multiuser detection.

16.5.2.2 An SDP-relaxation-based multiuser detector

Let

$$\hat{\mathbf{X}} = \begin{bmatrix} \mathbf{xx}^T & \mathbf{x} \\ \mathbf{x}^T & 1 \end{bmatrix} \quad \text{and} \quad \mathbf{C} = \begin{bmatrix} \mathbf{H} & \mathbf{p}/2 \\ \mathbf{p}^T/2 & 1 \end{bmatrix} \qquad (16.99)$$

By using the property that trace($\mathbf{AB}$) = trace($\mathbf{BA}$), the objective function in Eq. (16.94) can be expressed as

$$\mathbf{x}^T\mathbf{H}\mathbf{x} + \mathbf{x}^T\mathbf{p} = \text{trace}(\mathbf{C}\hat{\mathbf{X}}) \qquad (16.100)$$

(see Prob. 16.13(*a*)). Using an argument similar to that in Sec. 16.5.2.1, the constraint in Eq. (16.94b) can be converted to

$$\hat{\mathbf{X}} \succeq 0, \quad \hat{x}_{ii} = 1 \quad \text{for } 1 \leq i \leq K \qquad (16.101a)$$
$$\text{rank } (\hat{\mathbf{X}}) = 1 \qquad (16.101b)$$

where $\hat{x}_{ii}$ denotes the ith diagonal element of $\hat{\mathbf{X}}$ (see Prob. 16.13(*b*)). By removing the rank constraint in Eq. (16.101b), we obtain an SDP relaxation of the optimization problem in Eq. (16.94) as

$$\text{minimize trace}(\mathbf{C}\hat{\mathbf{X}}) \qquad (16.102a)$$

$$\text{subject to:} \quad \hat{\mathbf{X}} \succeq 0 \qquad (16.102b)$$
$$\hat{x}_{ii} = 1 \quad \text{for } i = 1, 2, \ldots, K+1 \qquad (16.102c)$$

The variables in the original problem in Eq. (16.94) assume only the values of 1 or -1 while the variable $\hat{\mathbf{X}}$ in the SDP minimization problem (16.102) has real-valued components. In what follows, we describe two approaches that can be used to generate a binary solution for the problem in Eq. (16.94) based on the solution $\hat{\mathbf{X}}$ of the SDP problem in Eq. (16.102).

Let the solution of the problem in Eq. (16.102) be denoted as $\hat{\mathbf{X}}^*$. It follows from Eq. (16.99) that $\hat{\mathbf{X}}^*$ is a $(K+1) \times (K+1)$ symmetric matrix of the form

$$\hat{\mathbf{X}}^* = \begin{bmatrix} \mathbf{X}^* & \mathbf{x}^* \\ \mathbf{x}^{*T} & 1 \end{bmatrix} \qquad (16.103)$$

with

$$\hat{x}_{ii}^* = 1 \qquad \text{for } i = 1, 2, \ldots, K.$$

In view of Eq. (16.103), our first approach is simply to apply operator sgn($\cdot$) to $\mathbf{x}^*$ in Eq. (16.103), namely,

$$\hat{\mathbf{b}} = \text{sgn}(\mathbf{x}^*) \qquad (16.104)$$

where $\mathbf{x}^*$ denotes the vector formed by the first K components in the last column of $\hat{\mathbf{X}}^*$.

At the cost of more computation, a better binary solution can be obtained by using the eigendecomposition of matrix $\hat{\mathbf{X}}^*$, i.e., $\hat{\mathbf{X}}^* = \mathbf{U}\mathbf{S}\mathbf{U}^T$, where $\mathbf{U}$ is an orthogonal and $\mathbf{S}$ is a diagonal matrix with the eigenvalues of $\hat{\mathbf{X}}^*$ as its diagonal components in decreasing order (see Sec. A.9). It is well known that an optimal rank-one approximation of $\hat{\mathbf{X}}^*$ in the L_2 norm sense is given by $\lambda_1 \mathbf{u}_1 \mathbf{u}_1^T$, where λ_1 is the largest eigenvalue of $\hat{\mathbf{X}}^*$ and $\mathbf{u}_1$ is the eigenvector associated with λ_1 [28]. If we denote the vector formed by the first K components of $\mathbf{u}_1$ as $\tilde{\mathbf{u}}$, and the last component of $\mathbf{u}_1$ by u_{K+1}, i.e.,

$$\mathbf{u}_1 = \begin{bmatrix} \tilde{\mathbf{u}} \\ u_{K+1} \end{bmatrix}$$

then the optimal rank-one approximation of $\hat{\mathbf{X}}^*$ can be written as

$$\hat{\mathbf{X}}^* \approx \lambda_1 \mathbf{u}_1 \mathbf{u}_1^T = \lambda_1 \begin{bmatrix} \tilde{\mathbf{u}}\tilde{\mathbf{u}}^T & u_{K+1}\tilde{\mathbf{u}} \\ u_{K+1}\tilde{\mathbf{u}}^T & u_{K+1}^2 \end{bmatrix}$$

$$= \frac{\lambda_1}{u_{K+1}^2} \begin{bmatrix} \tilde{\mathbf{x}}_1 \tilde{\mathbf{x}}_1^T & \tilde{\mathbf{x}}_1 \\ \tilde{\mathbf{x}}_1^T & 1 \end{bmatrix} \qquad (16.105)$$

where $\tilde{\mathbf{x}}_1 = u_{K+1}\tilde{\mathbf{u}}$. Since $\lambda_1 > 0$, on comparing Eqs. (16.103) and (16.105) we note that the signs of the components of vector $\tilde{\mathbf{x}}_1$ are likely to be the same as the signs of the corresponding components in vector $\mathbf{x}^*$. Therefore, a binary solution of the problem in Eq. (16.94) can be generated as

$$\hat{\mathbf{b}} = \begin{cases} \text{sgn}(\tilde{\mathbf{u}}) & \text{if } u_{K+1} > 0 \\ -\text{sgn}(\tilde{\mathbf{u}}) & \text{if } u_{K+1} < 0 \end{cases} \qquad (16.106)$$

16.5.2.3 Solution suboptimality

Because of the relaxation involved, the detector described is *suboptimal* but, as mentioned in Sec. 16.5.2.1, the SDP relaxation of the MAX-CUT problem yields a good suboptimal solution. However, there are two important differences between the SDP problems in Eqs. (16.97) and (16.102): The diagonal components of $\mathbf{W}$ in Eq. (16.97) are all zero whereas those of $\mathbf{C}$ in Eq. (16.102) are all strictly positive; and although the off-diagonal components in $\mathbf{W}$ are assumed to be nonnegative, matrix $\mathbf{C}$ may contain negative off-diagonal components. Consequently, the bounds in Eq. (16.98) do not always hold for the SDP problem in Eq. (16.102). However, as will be demonstrated in terms of some experimental results presented below, the near-optimal detector offers comparable performance to that of the optimal ML detector.

In the next section, we describe an alternative but more efficient SDP-relaxation-based detector.

16.5.2.4 Efficient-relaxation-based detector via duality

Although efficient interior-point algorithms such as those in [27][29] (see Secs. 14.4–14.5) can be applied to solve the SDP problem in Eq. (16.102), numerical difficulties can arise because the number of variables can be quite large even for the case of a moderate number of users. For example, if $K = 20$, the dimension of vector $\mathbf{x}$ in Eq. (16.99) is 20 and the number of variables in $\hat{\mathbf{X}}$ becomes $K(K+1)/2 = 210$. In this section, we present a more efficient approach for the solution of the SDP problem under consideration. Essentially, we adopt an indirect approach by first solving the dual SDP problem, which involves a much smaller number of variables, and then convert the solution of the dual problem to that of the primal SDP problem.

We begin by rewriting the SDP problem in Eq. (16.102) as

$$\text{minimize trace}(\mathbf{C}\hat{\mathbf{X}}) \qquad (16.107a)$$

$$\text{subject to:} \quad \hat{\mathbf{X}} \succeq 0 \qquad (16.107b)$$

$$\text{trace}(\mathbf{A}_i\mathbf{X}) = 1 \qquad \text{for } i = 1, 2, \ldots, K+1 \qquad (16.107c)$$

where $\mathbf{A}_i$ is a diagonal matrix whose diagonal components are all zero except for the ith component which is 1. It follows from Chap. 14 that the dual of the problem in Eq. (16.107) is given by

$$\text{minimize } -\mathbf{b}^T\mathbf{y} \qquad (16.108a)$$

$$\text{subject to:} \quad \mathbf{S} = \mathbf{C} - \sum_{i=1}^{K+1} y_i \mathbf{A}_i \qquad (16.108b)$$

$$\mathbf{S} \succeq 0 \qquad (16.108c)$$

where $\mathbf{y} = [y_1\ y_2\ \cdots\ y_{K+1}]^T$ and $\mathbf{b} = [1\ 1\ \cdots\ 1]^T \in C^{(K+1)\times 1}$. Evidently, the dual problem in Eq. (16.108) involves only $K+1$ variables and it is, therefore, much easier to solve then the primal problem. Any efficient interior-point algorithm can be used for the solution such as the projective algorithm proposed by Nemirovski and Gahinet [30] (see Sec. 14.6).[2]

In order to obtain the solution of the primal SDP problem in Eq. (16.107), we need to carry out some analysis on the Karush-Kuhn-Tucker (KKT) conditions for the solutions of the problems in Eqs. (16.107) and (16.108). The KKT conditions state that the set $\{\hat{\mathbf{X}}^*, \mathbf{y}^*\}$ solves the problems in Eqs. (16.107) and (16.108) if and only if they satisfy the conditions

$$\sum_{i=1}^{K+1} y_i^* \mathbf{A}_i + \mathbf{S}^* = \mathbf{C} \tag{16.109a}$$

$$\text{trace}(\mathbf{A}_i \hat{\mathbf{X}}^*) = 1 \quad \text{for } i = 1, 2, \ldots, K+1 \tag{16.109b}$$

$$\mathbf{S}^* \hat{\mathbf{X}}^* = \mathbf{0} \tag{16.109c}$$

$$\hat{\mathbf{X}}^* \succeq \mathbf{0} \text{ and } \mathbf{S}^* \succeq \mathbf{0} \tag{16.109d}$$

From Eq. (16.109a), we have

$$\mathbf{S}^* = \mathbf{C} - \sum_{i=1}^{K+1} y_i^* \mathbf{A}_i \tag{16.110}$$

Since the solution $\mathbf{y}^*$ is typically obtained by using an *iterative* algorithm, e.g., the projective algorithm of Nemirovski and Gahinet, $\mathbf{y}^*$ can be a good approximate solution only of the problem in Eq. (16.109), which means that $\mathbf{y}^*$ is in the *interior* of the feasible region. Consequently, matrix $\mathbf{S}^*$ remains *positive definite*. Therefore, the set $\{\mathbf{y}^*, \mathbf{S}^*, \hat{\mathbf{X}}^*\}$ can be regarded as a point in the feasible region that is sufficiently close to the limiting point of the central path for the problems in Eqs. (16.107) and (16.108). Recall that the central path is defined as a parameterized set $\{\mathbf{y}(\tau), \mathbf{S}(\tau), \hat{\mathbf{X}}(\tau)$ for $\tau > 0\}$ that satisfies the modified KKT conditions

$$\sum_{i=1}^{K+1} y_i(\tau) \mathbf{A}_i + \mathbf{S}(\tau) = \mathbf{C} \tag{16.111a}$$

$$\text{tr}(\mathbf{A}_i \hat{\mathbf{X}}(\tau)) = 1 \quad \text{for } i = 1, 2, \ldots, K+1 \tag{16.111b}$$

$$\mathbf{S}(\tau)\hat{\mathbf{X}}(\tau) = \tau \mathbf{I} \tag{16.111c}$$

$$\hat{\mathbf{X}}(\tau) \succeq \mathbf{0} \text{ and } \mathbf{S}(\tau) \succeq \mathbf{0} \tag{16.111d}$$

[2]The projective method has been implemented in the MATLAB LMI Control Toolbox for solving a variety of SDP problems [31].

The relation between Eqs. (16.109) and (16.111) becomes transparent since the entire central path defined by Eq. (16.111) lies in the interior of the feasible region and as $\tau \rightarrow 0$, the path converges to the solution set $\{\mathbf{y}^*, \mathbf{S}^*, \hat{\mathbf{X}}^*\}$ that satisfies Eq. (16.109).

From Eq. (16.111c), it follows that

$$\hat{\mathbf{X}}(\tau) = \tau \mathbf{S}^{-1}(\tau) \qquad (16.112)$$

which suggests an approximate solution of (16.107) as

$$\hat{\mathbf{X}} = \tau (\mathbf{S}^*)^{-1} \qquad (16.113)$$

for some sufficiently small $\tau > 0$, where $\mathbf{S}^*$ is given by Eq. (16.110). In order for matrix $\hat{\mathbf{X}}$ in Eq. (16.113) to satisfy the equality constraints in Eq. (16.107c), $\hat{\mathbf{X}}$ needs to be slightly modified using a scaling matrix $\mathbf{\Pi}$ as

$$\hat{\mathbf{X}}^* = \mathbf{\Pi}(\mathbf{S}^*)^{-1}\mathbf{\Pi} \qquad (16.114a)$$

where

$$\mathbf{\Pi} = \text{diag}\{\xi_1^{1/2} \; \xi_2^{1/2} \; \cdots \; \xi_{K+1}^{1/2}\} \qquad (16.114b)$$

and ξ_i is the ith diagonal component of $(\mathbf{S}^*)^{-1}$. In Eq. (16.114a) we have pre- and post-multiplied $(\mathbf{S}^*)^{-1}$ by $\mathbf{\Pi}$ so that matrix $\hat{\mathbf{X}}^*$ remains *symmetric* and *positive definite*. It is worth noting that by imposing the equality constraints in Eq. (16.107c) on $\hat{\mathbf{X}}$, the parameter τ in Eq. (16.113) is absorbed in the scaling matrix $\mathbf{\Pi}$.

In summary, an approximate solution $\hat{\mathbf{X}}$ of the SDP problem in Eq. (16.107) can be efficiently obtained by using the following algorithm.

Algorithm 16.1 SDP-relaxation algorithm based on dual problem
Step 1
Form matrix $\mathbf{C}$ using Eq. (16.99).
Step 2
Solve the dual SDP problem in Eq. (16.108) and let its solution be $\mathbf{y}^*$.
Step 3
Compute $\mathbf{S}^*$ using Eq. (16.110).
Step 4
Compute $\hat{\mathbf{X}}^*$ using Eq. (16.114).
Step 5
Compute $\hat{\mathbf{b}}$ using Eq. (16.104) or (16.106).

We conclude this section with two remarks on the computational complexity of the above algorithm and the accuracy of the solution obtained. To a large extent, the mathematical complexity of the algorithm is determined by Steps 2 and

4 where a $(K + 1)$-variable SDP problem is solved and a $(K + 1) \times (K + 1)$ positive definite matrix is inverted, respectively. Consequently, the dual approach reduces the amount of computation required considerably compared to that required to solve the $K(K + 1)/2$-variable SDP problem in Eq. (16.107) directly. Concerning the accuracy of the solution, we note that it is the binary solution that determines the performance of the multiuser detector. Since the binary solution is the output of the sign operation (see Eqs. (16.104) and (16.106)), the approximation introduced in Eq. (16.114) is expected to have an insignificant negative effect on the solution.

Example 16.8 Apply the primal and dual SDP-relaxation-based multiuser detectors to a six-user synchronous system and compare their performance with that of the ML detector described in Sec. 16.5.1.2 in terms of bit-error rate (BER) and computational complexity.

Solution For the sake of convenience, we refer to the detectors based on the primal and dual problems of Sec. 16.5.2.2 and Sec. 16.5.2.4 as the SDPR-P and SDPR-D detectors, respectively. The SDP problems in Eqs. (16.102) and (16.108) for the SDPR-P and SDPR-D detectors were solved by using Algorithms 14.1 and 14.4, respectively. The user signatures used in the simulations were 15-chip Gold sequences. The received signal power of the six users were set to 5, 3, 1.8, 0.6, 0.3, and 0.2, respectively. The last (weakest) user with power 0.2 was designated as the desired user. The average BERs for the SDPR-P, SDPR-D, and ML detectors are plotted versus the signal-to-noise ratio (SNR) in Fig. 16.16, and as can be seen the demodulation performance of the SDPR-P and SDPR-D detectors is consistently very close to that of the ML detector.

The computational complexity of the detectors was evaluated in terms of CPU time and the results for the SDPR-P, SDPR-D, and ML detectors are plotted in Fig. 16.17 versus the number of active users. As expected, the amount of computation required by the ML detector increases exponentially with the number of users as shown in Fig. 16.17a. The SDPR detectors reduce the amount of computation to less than 1 percent and between the SDPR-P and SDPR-D detectors, the latter one, namely, the one based on the dual problem, is significantly more efficient as can be seen in Fig. 16.17b.

16.5.3 A constrained minimum-BER multiuser detector
16.5.3.1 Problem formulation

Although the SDP-based detectors described in Sec. 16.5.2 achieve near optimal performance with reduced computational complexity compared to that of the ML detector, the amount of computation they require is still too large for real-time applications. A more practical solution is to develop *linear* multiuser

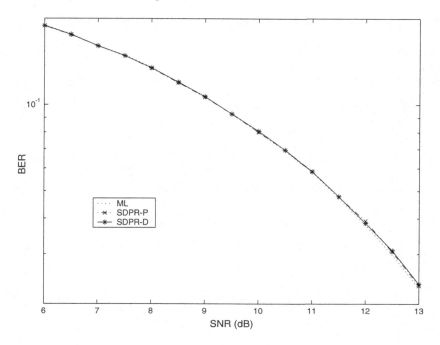

Figure 16.16. BER of six-user synchronous DS-CDMA system in AWGN channel.

detectors that estimate the users' information bits by processing the observation data with an FIR digital filter. Several linear detectors with satisfactory performance have been recently developed [20]. However, in general, these detectors do not provide the lowest BER and, therefore, it is of interest to develop a constrained minimum-BER detector that minimizes BER directly.

We consider a DS-CDMA channel with K synchronous users whose continuous-time model is given by Eq. (16.88). Within the observation window, the critically sampled version of the received signal $\mathbf{r} = [y(0)\ y(\Delta)\ \cdots\ y[(N-1)\Delta]]^T$, where Δ denotes the sampling period, can be expressed as

$$\mathbf{r} = \mathbf{Sb} + \mathbf{n} \qquad (16.115)$$

where

$$\mathbf{S} = [A_1\mathbf{s}_1\ A_2\mathbf{s}_2\ \cdots\ A_K\mathbf{s}_K]$$
$$\mathbf{s}_k = [s_k(0)\ s_k(\Delta)\ \cdots\ s_k[(N-1)\Delta]]^T$$
$$\mathbf{b} = [b_1\ b_2\ \cdots\ b_K]^T$$
$$\mathbf{n} = [n(0)\ n(\Delta)\ \cdots\ n[(N-1)\Delta]]^T$$

In Eq. (16.115), $\mathbf{n}$ is an AWGN signal with zero mean and variance $\sigma^2\mathbf{I}$, and $\mathbf{s}_k \in R^{N \times 1}$ is the signature signal of the kth user.

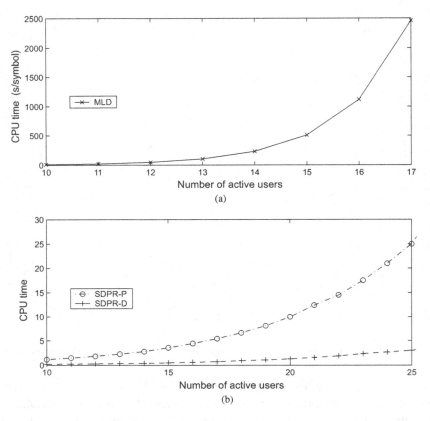

Figure 16.17. Computational complexity of (a) ML detector and (b) SDPR-P and SDPR-D detectors.

■

The linear multiuser detector to be investigated in this section can be regarded as an FIR filter of length N that is characterized by its coefficient vector $\mathbf{c} \in R^{N \times 1}$. From the channel model in Eq. (16.115), it follows that the output of the detector is given by

$$\mathbf{c}^T \mathbf{r} = \mathbf{c}^T \mathbf{S} \mathbf{b} + \mathbf{c}^T \mathbf{n}$$

Let the kth user be the desired user. We want to detect its information bit with minimum error regardless of the information bits sent by the other $K - 1$ users. If $\hat{\mathbf{b}}_i$ for $1 \leq i \leq 2^{K-1}$ are the possible information vectors with their kth entry $b_k = 1$, then for each $\hat{\mathbf{b}}_i$ the output of the detector is given by $\mathbf{c}^T \mathbf{r} = \mathbf{c}^T \hat{\mathbf{v}}_i + \mathbf{c}^T \mathbf{n}$ where $\hat{\mathbf{v}}_i = \mathbf{S} \hat{\mathbf{b}}_i$. The BER of the kth user can be shown to be [19]

$$P(\mathbf{c}) = \frac{1}{2^{K-1}} \sum_{i=1}^{2^{K-1}} Q\left(\frac{\mathbf{c}^T \hat{\mathbf{v}}_i}{||\mathbf{c}||\sigma}\right) \qquad (16.116)$$

with

$$Q(x) = \frac{1}{\sqrt{2\pi}} \int_x^\infty d^{-v^2/2} \, dv \qquad (16.117)$$

A detector whose coefficient vector $\mathbf{c}^*$ minimizes $P(\mathbf{c})$ in Eq. (16.116) can be referred to as a *constrained minimum-BER* (CMBER) *detector*. An optimal linear detector is an unconstrained optimization algorithm that minimizes the BER objective function $P(\mathbf{c})$ in Eq. (16.116). A difficulty associated with the above unconstrained problem is that function $P(\mathbf{c})$ is highly nonlinear and there may exist more than one local minimum. Consequently, convergence to $\mathbf{c}^*$ cannot be guaranteed for most optimization algorithms. In what follows, we present a *constrained optimization* formulation of the problem that can be used to implement a CMBER detector [23].

It can be shown that any local minimizer of the BER objective function in Eq. (16.116) subject to constraints

$$\mathbf{c}^T \hat{\mathbf{v}}_i \geq 0 \qquad \text{for } 1 \leq i \leq 2^{K-1} \qquad (16.118)$$

is a global minimizer. Furthermore, with the constraint $||\mathbf{c}|| = 1$, the global minimizer is unique (see Prob. 16.14).

Before proceeding to the problem formulation, it should be mentioned that the constraints in Eq. (16.118) are reasonable in the sense that they will not exclude good local minimizers. This can be seen from Eqs. (16.116) and (16.117) which indicate that nonnegative inner products $\mathbf{c}^T \mathbf{v}_i$ for $1 \leq i \leq 2^{K-1}$ tend to reduce $P(\mathbf{c})$ compared with negative inner products. Now if we define the set

$$I = \{\mathbf{c} : \mathbf{c} \text{ satisfies Eq. (16.118) and } ||\mathbf{c}|| = 1\} \qquad (16.119)$$

then it can be readily shown that as long as vectors $\{\mathbf{s}_i : 1 \leq i \leq K\}$ are linearly independent, set I contains an infinite number of elements (see Prob. 16.15). Under these circumstances, we can formulate the multiuser detection problem at hand as the constrained minimization problem

$$\text{minimize } P(\mathbf{c}) \qquad (16.120a)$$

$$\text{subject to: } \quad \mathbf{c} \in I \qquad (16.120b)$$

16.5.3.2 Conversion of the problem in Eq. (16.120) into a CP problem

We start with a simple conversion of the problem in Eq. (16.120) into the following problem

584

$$\text{minimize } P(\mathbf{c}) = \frac{1}{2^{K-1}} \sum_{i=1}^{2^{K-1}} Q(\mathbf{c}^T \mathbf{v}_i) \qquad (16.121\text{a})$$

$$\text{subject to: } \mathbf{c}^T \mathbf{v}_i \geq 0 \quad \text{for } 1 \leq i \leq 2^{K-1} \qquad (16.121\text{b})$$

$$\|\mathbf{c}\| = 1 \qquad (16.121\text{c})$$

where

$$\mathbf{v}_i = \frac{\hat{\mathbf{v}}_i}{\sigma} \quad \text{for } 1 \leq i \leq w^{K-1}$$

Note that the problem in Eq. (16.121) is *not* a CP problem because the feasible region characterized by Eqs. (16.121b) and (16.121c) is not convex. However, it can be readily verified that the solution of Eq. (16.121) coincides with the solution of the constrained optimization problem

$$\text{minimize } P(\mathbf{c}) \qquad (16.122\text{a})$$

$$\text{subject to: } \mathbf{c}^T \mathbf{v}_i \geq 0 \quad \text{for } 1 \leq i \leq 2^{K-1} \qquad (16.122\text{b})$$

$$\|\mathbf{c}\| \leq 1 \qquad (16.122\text{c})$$

This is because for any $\mathbf{c}$ with $\|\mathbf{c}\| < 1$, we always have $P(\hat{\mathbf{c}}) \leq P(\mathbf{c})$ where $\hat{\mathbf{c}} = \mathbf{c}/\|\mathbf{c}\|$. In other words, the minimizer $\mathbf{c}^*$ of the problem in Eq. (16.122) always satisfies the constraint $\|\mathbf{c}^*\| = 1$. A key distinction between the problems in Eqs. (16.121) and (16.122) is that the latter one is a CP problem for which a number of efficient algorithms are available (see Chap. 13).

16.5.3.3 Newton-barrier method

The optimization algorithm described below fits into the class of barrier function methods studied in Chap. 12 but it has several additional features that are uniquely associated with the present problem. These include a closed-form formula for evaluating the Newton direction and an efficient line search.

By adopting a barrier function approach, we can drop the nonlinear constraint in Eq. (16.122c) and convert the problem in Eq. (16.122) into the form

$$\text{minimize } F_\mu(\mathbf{c}) = P(\mathbf{c}) - \mu \ln(1 - \mathbf{c}^T \mathbf{c}) \qquad (16.123\text{a})$$

$$\text{subject to: } \mathbf{c}^T \mathbf{v}_i \geq 0 \quad \text{for } 1 \leq i \leq 2^{K-1} \qquad (16.123\text{b})$$

where $\mu > 0$ is the barrier parameter. With a strictly feasible initial point $\mathbf{c}_0$, which strictly satisfies the constraints in Eqs. (16.122b) and (16.122c), the logarithmic term in Eq. (16.123a) is well defined. The gradient and Hessian of $F_\mu(\mathbf{c})$ are given by

$$\nabla F_\mu(\mathbf{c}) = -\sum_{i=1}^{M} \frac{1}{M} e^{-\beta_i^2/2} \mathbf{v}_i + \frac{2\mu \mathbf{c}}{1 - \|\mathbf{c}\|^2} \qquad (16.124)$$

$$\nabla^2 F_\mu(\mathbf{c}) = \sum_{i=1}^{M} \frac{1}{M} e^{-\beta_i^2/2} \beta_i \mathbf{v}_i \mathbf{v}_i^T + \frac{2\mu}{1 - \|\mathbf{c}\|^2} \mathbf{I} \qquad (16.125)$$

$$+ \frac{4\mu}{(1 - \|\mathbf{c}\|)^2} \mathbf{c} \mathbf{c}^T \qquad (16.126)$$

where $M = 2^{K-1}$ and $\beta_i = \mathbf{c}^T \mathbf{v}_i$ for $1 \leq i \leq M$. Note that the Hessian in the interior of the feasible region, i.e., $\mathbf{c}$ with $\beta_i = \mathbf{c}^T \mathbf{v}_i > 0$ and $\|\mathbf{c}\| < 1$, is positive definite. This suggests that at the $(k + 1)$th iteration, $\mathbf{c}_{k+1}$ can be obtained as

$$\mathbf{c}_{k+1} = \mathbf{c}_k + \alpha_k \mathbf{d}_k \qquad (16.127)$$

where the search direction $\mathbf{d}_k$ is given by

$$\mathbf{d}_k = -[\nabla^2 F_\mu(\mathbf{c}_k)]^{-1} \nabla F_\mu(\mathbf{c}_k) \qquad (16.128)$$

The positive scalar α_k in Eq. (16.127) can be determined by using a line search as follows. First, we note that the one-variable function $F_\mu(\mathbf{c}_k + \alpha \mathbf{d}_k)$ is strictly convex on the interval $[0, \bar{\alpha}]$ where $\bar{\alpha}$ is the largest positive scalar such that $\mathbf{c}_k + \alpha \mathbf{d}_k$ remains feasible for $0 \leq \alpha \leq \bar{\alpha}$. Once $\bar{\alpha}$ is determined, $F_\mu(\mathbf{c}_k + \alpha \mathbf{d}_k)$ is a unimodal function on $[0, \bar{\alpha}]$ and the search for the minimizer of the function can be carried out using one of the well known methods such as quadratic or cubic interpolation or the Golden-section method (see Chap. 4). To find $\bar{\alpha}$, we note that a point $\mathbf{c}_k + \alpha \mathbf{d}_k$ satisfies the constraints in Eq. (16.122b) if

$$(\mathbf{c}_k + \alpha \mathbf{d}_k)^T \mathbf{v}_i \geq 0 \qquad \text{for } 1 \leq i \leq M \qquad (16.129)$$

Since $\mathbf{c}_k$ is feasible, we have $\mathbf{c}_k^T \mathbf{v}_i \geq 0$ for $1 \leq i \leq M$. Hence for those indices i such that $\mathbf{d}_k^T \mathbf{v}_i \geq 0$, any nonnegative α will satisfy Eq. (16.129). In other words, only those constraints in Eq. (16.122b) whose indices are in the set

$$\mathcal{I}_k = \{i : \mathbf{d}_k^T \mathbf{v}_i < 0\} \qquad (16.130)$$

will affect the largest value of α that satisfies Eq. (16.129), and that value of α can be computed as

$$\bar{\alpha}_1 = \min_{i \in \mathcal{I}_k} \left(\frac{\mathbf{c}_k^T \mathbf{v}_i}{-\mathbf{d}_k^T \mathbf{v}_i} \right) \qquad (16.131)$$

In order to satisfy the constraint in Eq. (16.122c), we solve the equation

$$\|\mathbf{c}_k + \alpha \mathbf{d}_k\|^2 = 1$$

to obtain the solution

$$\alpha = \bar{\alpha}_2 = \frac{[(\mathbf{c}_k^T \mathbf{d}_k)^2 - \|\mathbf{d}_k\|^2(\|\mathbf{c}_k\|^2 - 1)]^{1/2} - \mathbf{c}_k^T \mathbf{d}_k}{\|\mathbf{d}_k\|^2} \qquad (16.132)$$

586

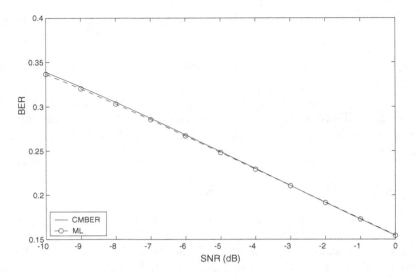

Figure 16.18. Performance comparison of CMBER and ML detectors for a system with 10 equal-power users.

The value of $\bar{\alpha}$ can then be taken as $\min(\bar{\alpha}_1, \bar{\alpha}_2)$. In practice, we must keep the next iterate strictly inside the feasible region to ensure that the barrier function in Eq. (16.123a) is well defined. To this end we can use

$$\bar{\alpha} = 0.99 \min(\bar{\alpha}_1, \bar{\alpha}_2) \tag{16.133}$$

The above iterative optimization procedure is continued until the difference between two successive solutions is less than a prescribed tolerance. For a strictly feasible initial point, the Newton-barrier method described above always converges to the global minimizer for an arbitrary positive μ. However, the value of μ does affect the behavior of the algorithm. A small μ may lead to an ill-conditioned Hessian while a large μ may lead to slow convergence. A μ in the interval $[0.001, 0.1]$ would guarantee a well-conditioned Hessian and allow a fast convergence.

The BER performance of the CMBER detector is compared with that of the ML detector in Fig. 16.18 for a system with 10 equal-power users. As can be seen, the performance of the CMBER is practically the same as that of the ML detector.

References

1 A. Antoniou, *Digital Singal Processing: Signals, Systems, and Filters*, McGraw-Hill, New York, 2005.
2 W.-S. Lu and A. Antoniou, *Two-Dimensional Digital Filters*, Marcel Dekker, New York, 1992.

3 J. W. Adams, "FIR digital filters with least-squares stopbands subject to peak-gain constraints," *IEEE Trans. Circuits Syst.*, vol. 38, pp. 376–388, April 1991.

4 W.-S. Lu, "Design of nonlinear-phase FIR digital filters: A semidefinite programming approach," *IEEE Int. Symp. on Circuits and Systems,* vol. III, pp. 263–266, Orlando, FL., May 1999.

5 A. G. Deczky, "Synthesis of recursive digital filters using the minimum p-error criterion," *IEEE Trans. Audio and Electroacoustics*, vol. 20, pp. 257–263, 1972.

6 A. T. Chottra and G. A. Jullien, "A linear programming approach to recursive digital filter design with linear phase," *IEEE Trans. Circuits Syst.*, vol. 29, pp. 139–149, Mar. 1982.

7 W.-S. Lu, S.-C. Pei, and C.-C. Tseng, "A weighted least-squares method for the design of stable 1-D and 2-D IIR filters," *IEEE Trans. Signal Processing*, vol. 46, pp. 1–10, Jan. 1998.

8 M. Lang, "Weighted least squares IIR filter design with arbitrary magnitude and phase responses and specified stability margin," *IEEE Symp. on Advances in Digital Filtering and Signal Processing*, pp. 82–86, Victoria, BC, June 1998.

9 T. Kailath, *Linear Systems*, Englewood Cliffs, Prentice-Hall, NJ., 1981.

10 C. E. Garcia, D. M. Prett, and M. Morari, "Model predictive control: Theory and practice — a survey," *Automatica*, vol. 25, pp. 335–348, 1989.

11 M. V. Kothare, V. Balakrishnan, and M. Morari, "Robust constrained model predictive control using linear matrix inequalities," *Automatica*, vol. 32, pp. 1361–1379, 1996.

12 H. Kwakernaak and R. Sivan, *Linear Optimal Control Systems*, Wiley, New York, 1972.

13 J. Kerr and B. Roth, "Analysis of multifingered hands," *Int. J. Robotics Research*, vol. 4, no. 4, pp. 3–17, Winter 1986.

14 D. E. Orin and F.-T. Cheng, "General dynamic formulation of the force distribution equations," *Proc. 4th Int. Conf. on Advanced Robotics*, pp. 525–546, Columbus, Ohio, June 13-15, 1989.

15 F.-T. Cheng and D. E. Orin, "Efficient algorithm for optimal force distribution — The compact-dual LP method," *IEEE Trans. Robotics and Automation*, vol. 6, pp. 178–187, April 1990.

16 E. S. Venkaraman and T. Iberall, *Dextrous Robot Hands*, Springer Verlag, New York, 1990.

17 M. Buss, H. Hashimoto, and J. B. Moore, "Dextrous hand grasping force optimization," *IEEE Trans. Robotics and Automation*, vol. 12, pp. 406-418, June 1996.

18 K. Shimoga, "Robot grasp synthesis algorithms: A survey," *Int. J. Robotics Research*, vol. 15, pp. 230–266, June 1996.

19 J. G. Proakis, *Digial Communications*, 3rd ed., McGraw-Hill, New York, 1995.

20 S. Verdú, *Multiuser Detection*, Cambridge University Press, New York, 1998.

21 S. Verdú, "Minimum probability of error for asynchronous Gaussian multiple-access channels," *IEEE Trans. Inform. Theory*, vol. 32, pp. 85–96, Jan. 1986.

22 X. M. Wang, W.-S. Lu, and A. Antoniou, "A near-optimal multiuser detector for CDMA channels using semidefinite programming relaxation," *Proc. Int. Symp. Circuits Syst.*, Sydney, Australia, June 2001.

23 X. F. Wang, W.-S. Lu, and A. Antoniou, "Constrained minimum-BER multiuser detection," *IEEE Trans. Signal Processing*, vol. 48, pp. 2903–2909, Oct. 2000.

24 M. X. Geomans and D. P. Williamson, "Improved approximation algorithms for maximum cut and satisfiability problem using semidefinite programming," *J. ACM*, vol. 42, pp. 1115–1145, 1995.

25 M. X. Geomans and D. P. Williamson, ".878-approximation algorithm for MAX-CUT and MAX-2SAT," *Proc. 26th ACM Symp. Theory of Computing*, pp. 422–431, 1994.

26 L. Vandenberghe and S. Boyd, "Semidefinite programming," *SIAM Review*, vol. 38, pp. 49–95, 1996.

27 H. Wolkowicz, R. Saigal, and L. Vandenberghe, *Handbook on Semidefinite Programming*, Kluwer Academic, MA, 2000.

28 G. W. Stewart, *Introduction to Matrix Computations*, New York, Academic Press, 1973.

29 K. C. Toh, R. H. Tütüncü, and M. J. Todd, "On the implementation of SDPT3 version 3.1 — a MATLAB software package for semidefinite-quadratic-linear Programming," *Proc. IEEE Conf. on Computer-Aided Control System Design*, Sept. 2004.

30 A. Nemirovski and P. Gahinet, "The projective method for solving linear matrix inequalities," *Math. Programming*, Series B, vol. 77, pp. 163–190, 1997.

31 P. Gahinet, A. Nemirovski, A. J. Laub, and M. Chilali, *Manual of LMI Control Toolbox*, Natick: MathWorks Inc., May 1990.

Problems

16.1 Write a MATLAB program to implement the constrained optimization algorithm described in Sec. 16.2, and use it to obtain the design in Example 16.1.

16.2 Derive the expression in Eq. (16.14).

16.3 Show that the inequality in Eq. (16.15) holds if and only if matrix $\mathbf{D}(\omega)$ in Eq. (16.16) is positive semidefinite.

16.4 (a) Show that the zeros of polynomial $z^K B(z)$ in Eq. (16.20c) are the eigenvalues of matrix $\mathbf{D}$ in Eq. (16.21).

     (b) Show that if the matrix inequality in Eq. (16.23) holds for some matrix $\mathbf{P} \succ \mathbf{0}$, then the largest modulus of the eigenvalues of matrix $\mathbf{D}$ in Eq. (16.21) is strictly less than one.

16.5 Show that the constraint in Eq. (16.29b) is equivalent to the matrix equality in Eq. (16.30).

16.6 Show that the matrix inequality in Eq. (16.46) is equivalent to the matrix inequalities in Eq. (16.47) or those in Eq. (16.48).

16.7 Using the results of Prob. 16.6, show that matrix $\mathbf{W}$ in Eq. (16.45) is negative semidefinite if and only if the matrix inequality in Eq. (16.49) holds.

16.8 Assuming that matrix $\mathbf{S} \in R^{n \times n}$ is positive definite, $\mathbf{Y} \in R^{p \times n}$, and $\mathcal{E}$ is defined by Eq. (16.54), show that the formula in Eq. (16.57) is valid.

16.9 Using the result of Prob. 16.6, show that the matrix inequality in Eq. (16.58) implies the inequality in Eq. (16.59).

16.10 Show that if there exists a symmetric matrix $\mathbf{X} \in R^{p \times p}$ such that the conditions in Eqs. (16.64a) and (16.64b) are satisfied, then $(\mathbf{Y}\mathbf{S}^{-1}\mathbf{Y})_{jj} \leq u_{j,\max}^2$.

16.11 (a) Show that Eq. (16.67) is equivalent to Eq. (16.79).

     (b) Show that the constraint in Eq. (16.69) assures the positive semidefiniteness of matrix $\mathbf{P}_i$ in Eq. (16.81).

     (c) Show that the constraints in Eq. (16.70) assure to the positive semidefiniteness of matrix $\mathbf{P}_i$ in Eq. (16.82).

16.12 (*a*) It has been shown that the smallest eigenvalue $\lambda_{\min}$ of matrix $\mathbf{P}(\mathbf{c})$ in Eq. (16.83c) can be viewed as a measure of the strictest friction and by how much the contact forces are away from slippage [17]. Modify the constraint in Eq. (16.83c) such that $\lambda_{\min}$ of $\mathbf{P}(\mathbf{c})$ is no less than a given threshold, say, ε, and the modified problem in Eq. (16.83) remains an SDP problem.

(*b*) Solve the optimal force distribution problem in Example 16.7 with the additional requirement that $\lambda_{\min}$ of $\mathbf{P}(\mathbf{c})$ be no less than $\varepsilon = 0.05$.

16.13 (*a*) Show that the objective function in Eq. (16.94a) can be expressed as

$$\text{trace}(\mathbf{C}\hat{\mathbf{X}})$$

where

$$\hat{\mathbf{X}} = \begin{bmatrix} \mathbf{x}\mathbf{x}^T & \mathbf{x} \\ \mathbf{x}^T & 1 \end{bmatrix} \quad \text{and} \quad \mathbf{C} = \begin{bmatrix} \mathbf{H} & \mathbf{p}/2 \\ \mathbf{p}^T/2 & 1 \end{bmatrix}$$

(*b*) Using the results obtained in part (*a*), show that the optimization problem in Eq. (16.94) can be reformulated as a problem which is identical with that in Eq. (16.101).

16.14 (*a*) Show that any local minimizer of the BER cost function in Eq. (16.116) subject to the constraints in Eq. (16.118) is a global minimizer.

(*b*) Show that with an additional constraint $\|\mathbf{c}\| = 1$, the global minimizer for the problem in part (*a*) is unique.

16.15 Show that if the signature vectors $\{s_k : 1 \le i \le K\}$ in Eq. (16.115) are linearly independent, then set I defined by Eq. (16.119) contains an infinite number of elements.

16.16 Show that the constrained problem in Eq. (16.122) is a CP problem.

Appendix A
Basics of Linear Algebra

A.1   Introduction

In this appendix we summarize some basic principles of linear algebra [1]–[4] that are needed to understand the derivation and analysis of the optimization algorithms and techniques presented in the book. We state these principles without derivations. However, a reader with an undergraduate-level linear-algebra background should be in a position to deduce most of them without much difficulty. Indeed, we encourage the reader to do so as the exercise will contribute to the understanding of the optimization methods described in this book.

In what follows, R^n denotes a vector space that consists of all column vectors with n real-valued components, and C^n denotes a vector space that consists of all column vectors with n complex-valued components. Likewise, $R^{m \times n}$ and $C^{m \times n}$ denote spaces consisting of all $m \times n$ matrices with real-valued and complex-valued components, respectively. Evidently, $R^{m \times 1} \equiv R^m$ and $C^{m \times 1} \equiv C^m$. Boldfaced uppercase letters, e.g., $\mathbf{A}$, $\mathbf{M}$, represent matrices, and boldfaced lowercase letters, e.g., $\mathbf{a}$, $\mathbf{x}$, represent column vectors. $\mathbf{A}^T$ and $\mathbf{A}^H = (\mathbf{A}^*)^T$ denote the transpose and complex-conjugate transpose of matrix $\mathbf{A}$, respectively. $\mathbf{A}^{-1}$ (if it exists) and $\det(\mathbf{A})$ denote the inverse and determinant of square matrix $\mathbf{A}$, respectively. The identity matrix of dimension n is denoted as $\mathbf{I}_n$. Column vectors will be referred to simply as vectors henceforth for the sake of brevity.

A.2 Linear Independence and Basis of a Span

A number of vectors $\mathbf{v}_1$, $\mathbf{v}_2$, $\ldots$, $\mathbf{v}_k$ in R^n are said to be *linearly independent* if

$$\sum_{i=1}^{k} \alpha_i \mathbf{v}_i = 0 \tag{A.1}$$

only if $\alpha_i = 0$ for $i = 1, 2, \ldots, k$. Vectors $\mathbf{v}_1$, $\mathbf{v}_2$, $\ldots$, $\mathbf{v}_k$ are said to be *linearly dependent* if there exit real scalars α_i for $i = 1, 2, \ldots, k$, with at least one nonzero α_i, such that Eq. (A.1) holds.

A subspace $\mathcal{S}$ is a subset of R^n such that $\mathbf{x} \in \mathcal{S}$ and $\mathbf{y} \in \mathcal{S}$ imply that $\alpha \mathbf{x} + \beta \mathbf{y} \in \mathcal{S}$ for any real scalars α and β. The set of all linear combinations of vectors $\mathbf{v}_1$, $\mathbf{v}_2$, $\ldots$, $\mathbf{v}_k$ is a subspace called the *span* of $\{\mathbf{v}_1, \mathbf{v}_2, \ldots, \mathbf{v}_k\}$ and is denoted as span$\{\mathbf{v}_1, \mathbf{v}_2, \ldots, \mathbf{v}_k\}$.

Given a set of vectors $\{\mathbf{v}_1, \mathbf{v}_2, \ldots, \mathbf{v}_k\}$, a subset of r vectors $\{\mathbf{v}_{i_1}, \mathbf{v}_{i_2}, \ldots, \mathbf{v}_{i_r}\}$ is said to be a maximal linearly independent subset if (a) vectors $\mathbf{v}_{i_1}$, $\mathbf{v}_{i_2}$, $\ldots$, $\mathbf{v}_{i_r}$ are linearly independent, and (b) any vector in $\{\mathbf{v}_1, \mathbf{v}_2, \ldots, \mathbf{v}_k\}$ can be expressed as a linear combination of $\mathbf{v}_{i_1}$, $\mathbf{v}_{i_2}$, $\ldots$, $\mathbf{v}_{i_r}$. In such a case, the vector set $\{\mathbf{v}_{i_1}, \mathbf{v}_{i_2}, \ldots, \mathbf{v}_{i_r}\}$ is called a *basis* for span$\{\mathbf{v}_1, \mathbf{v}_2, \ldots, \mathbf{v}_k\}$ and integer r is called the *dimension* of the subspace The dimension of a subspace $\mathcal{S}$ is denoted as $\dim(\mathcal{S})$.

Example A.1 Examine the linear dependence of vectors

$$\mathbf{v}_1 = \begin{bmatrix} 1 \\ -1 \\ 3 \\ 0 \end{bmatrix}, \quad \mathbf{v}_2 = \begin{bmatrix} 0 \\ 2 \\ 1 \\ -1 \end{bmatrix}, \quad \mathbf{v}_3 = \begin{bmatrix} 3 \\ -7 \\ 7 \\ 2 \end{bmatrix}, \quad \text{and} \quad \mathbf{v}_4 = \begin{bmatrix} -1 \\ 5 \\ -1 \\ -2 \end{bmatrix}$$

and obtain a basis for span$\{\mathbf{v}_1, \mathbf{v}_2, \mathbf{v}_3, \mathbf{v}_4\}$.

Solution We note that

$$3\mathbf{v}_1 + 2\mathbf{v}_2 - 2\mathbf{v}_3 - 3\mathbf{v}_4 = 0 \tag{A.2}$$

Hence vectors $\mathbf{v}_1$, $\mathbf{v}_2$, $\mathbf{v}_3$, and $\mathbf{v}_4$ are linearly dependent. If

$$\alpha_1 \mathbf{v}_1 + \alpha_2 \mathbf{v}_2 = 0$$

then

$$\begin{bmatrix} \alpha_1 \\ -\alpha_1 + 2\alpha_2 \\ 3\alpha_1 \\ -\alpha_2 \end{bmatrix} = 0$$

which implies that $\alpha_1 = 0$ and $\alpha_2 = 0$. Hence $\mathbf{v}_1$ and $\mathbf{v}_2$ are linearly independent. We note that

$$\mathbf{v}_3 = 3\mathbf{v}_1 - 2\mathbf{v}_2 \tag{A.3}$$

and by substituting Eq. (A.3) into Eq. (A.2), we obtain

$$-3\mathbf{v}_1 + 6\mathbf{v}_2 - 3\mathbf{v}_4 = \mathbf{0}$$

i.e.,

$$\mathbf{v}_4 = -\mathbf{v}_1 + 2\mathbf{v}_2 \tag{A.4}$$

Thus vectors $\mathbf{v}_3$ and $\mathbf{v}_4$ can be expressed as linear combinations of $\mathbf{v}_1$ and $\mathbf{v}_2$. Therefore, $\{\mathbf{v}_1, \mathbf{v}_2\}$ is a basis of span$\{\mathbf{v}_1, \mathbf{v}_2, \mathbf{v}_3, \mathbf{v}_4\}$.

■

A.3 Range, Null Space, and Rank

Consider a system of linear equations

$$\mathbf{A}\mathbf{x} = \mathbf{b} \tag{A.5}$$

where $\mathbf{A} \in R^{m \times n}$ and $\mathbf{b} \in R^{m \times 1}$. If we denote the ith column of matrix $\mathbf{A}$ as $\mathbf{a}_i \in R^{m \times 1}$, i.e.,

$$\mathbf{A} = [\mathbf{a}_1 \ \mathbf{a}_2 \ \cdots \ \mathbf{a}_n]$$

and let

$$\mathbf{x} = [x_1 \ x_2 \ \ldots \ x_n]^T$$

then Eq. (A.5) can be written as

$$\sum_{i=1}^{n} x_i \mathbf{a}_i = \mathbf{b}$$

It follows from the above expression that Eq. (A.5) is solvable if and only if

$$\mathbf{b} \in \text{span}\{\mathbf{a}_1, \mathbf{a}_2, \ \ldots, \ \mathbf{a}_n\}$$

The subspace span$\{\mathbf{a}_1, \mathbf{a}_2, \ \ldots, \ \mathbf{a}_n\}$ is called the *range* of $\mathbf{A}$ and is denoted as $\mathcal{R}(\mathbf{A})$. Thus, Eq. (A.5) has a solution if and only if vector $\mathbf{b}$ is in the range of $\mathbf{A}$.

The dimension of $\mathcal{R}(\mathbf{A})$ is called the *rank* of $\mathbf{A}$, i.e., $r = \text{rank}(\mathbf{A}) = \dim[\mathcal{R}(\mathbf{A})]$. Since $\mathbf{b} \in \text{span}\{\mathbf{a}_1, \mathbf{a}_2, \ \ldots, \ \mathbf{a}_n\}$ is equivalent to

$$\text{span}\{\mathbf{b}, \mathbf{a}_1, \ \ldots, \ \mathbf{a}_n\} = \text{span}\{\mathbf{a}_1, \mathbf{a}_2, \ \ldots, \ \mathbf{a}_n\}$$

we conclude that Eq. (A.5) is solvable if and only if

$$\text{rank}(\mathbf{A}) = \text{rank}([\mathbf{A} \ \mathbf{b}]) \tag{A.6}$$

It can be shown that $\text{rank}(\mathbf{A}) = \text{rank}(\mathbf{A}^T)$. In other words, *the rank of a matrix is equal to the maximum number of linearly independent columns or rows.*

Another important concept associated with a matrix $\mathbf{A} \in R^{m \times n}$ is the *null space* of $\mathbf{A}$, which is defined as

$$\mathcal{N}(\mathbf{A}) = \{\mathbf{x} : \mathbf{x} \in R^n, \ \mathbf{A}\mathbf{x} = \mathbf{0}\}$$

It can be readily verified that $\mathcal{N}(\mathbf{A})$ is a subspace of R^n. If $\mathbf{x}$ is a solution of Eq. (A.5) then $\mathbf{x} + \mathbf{z}$ with $\mathbf{z} \in \mathcal{N}(\mathbf{A})$ also satisfies Eq. (A.5). Hence Eq. (A.5) has a unique solution only if $\mathcal{N}(\mathbf{A})$ contains just one component, namely, the zero vector in R^n. Furthermore, it can be shown that for $\mathbf{A} \in R^{m \times n}$

$$\text{rank}(\mathbf{A}) + \dim[\mathcal{N}(\mathbf{A})] = n \qquad (A.7)$$

(see [2]). For the important special case where matrix $\mathbf{A}$ is square, i.e., $n = m$, the following statements are equivalent: (*a*) there exists a unique solution for Eq. (A.5); (*b*) $\mathcal{N}(\mathbf{A}) = \{\mathbf{0}\}$; (*c*) $\text{rank}(\mathbf{A}) = n$.

A matrix $\mathbf{A} \in R^{m \times n}$ is said to have full column rank if $\text{rank}(\mathbf{A}) = n$, i.e., the n column vectors of $\mathbf{A}$ are linearly independent, and $\mathbf{A}$ is said to have full row rank if $\text{rank}(\mathbf{A}) = m$, i.e., the m row vectors of $\mathbf{A}$ are linearly independent.

Example A.2 Find the rank and null space of matrix

$$\mathbf{V} = \begin{bmatrix} 1 & 0 & 3 & -1 \\ -1 & 2 & -7 & 5 \\ 3 & 1 & 7 & -1 \\ 0 & -1 & 2 & -2 \end{bmatrix}$$

Solution Note that the columns of $\mathbf{V}$ are the vectors $\mathbf{v}_i$ for $i = 1, 2, \ldots, 4$ in Example A.1. Since the maximum number of linearly independent columns is 2, we have $\text{rank}(\mathbf{V}) = 2$. To find $\mathcal{N}(\mathbf{V})$, we write $\mathbf{V} = [\mathbf{v}_1 \ \mathbf{v}_2 \ \mathbf{v}_3 \ \mathbf{v}_4]$; hence the equation $\mathbf{V}\mathbf{x} = \mathbf{0}$ becomes

$$x_1\mathbf{v}_1 + x_2\mathbf{v}_2 + x_3\mathbf{v}_3 + x_4\mathbf{v}_4 = \mathbf{0} \qquad (A.8)$$

Using Eqs. (A.3) and (A.4), Eq. (A.8) can be expressed as

$$(x_1 + 3x_3 - x_4)\mathbf{v}_1 + (x_2 - 2x_3 + 2x_4)\mathbf{v}_2 = \mathbf{0}$$

which implies that

$$x_1 + 3x_3 - x_4 = 0$$
$$x_2 - 2x_3 + 2x_4 = 0$$

i.e.,

$$x_1 = -3x_3 + x_4$$
$$x_2 = 2x_3 - 2x_4$$

Hence any vector $\mathbf{x}$ that can be expressed as

$$\mathbf{x} = \begin{bmatrix} x_1 \\ x_2 \\ x_3 \\ x_4 \end{bmatrix} = \begin{bmatrix} -3x_3 + x_4 \\ 2x_3 - 2x_4 \\ x_3 \\ x_4 \end{bmatrix} = \begin{bmatrix} -3 \\ 2 \\ 1 \\ 0 \end{bmatrix} x_3 + \begin{bmatrix} 1 \\ -2 \\ 0 \\ 1 \end{bmatrix} x_4$$

with arbitrary x_3 and x_4 satisfies $\mathbf{Ax} = \mathbf{0}$. Since the two vectors in the above expression, namely,

$$\mathbf{n}_1 = \begin{bmatrix} -3 \\ 2 \\ 1 \\ 0 \end{bmatrix} \quad \text{and} \quad \mathbf{n}_2 = \begin{bmatrix} 1 \\ -2 \\ 0 \\ 1 \end{bmatrix}$$

are linearly independent, we have $\mathcal{N}(\mathbf{V}) = \text{span}\{\mathbf{n}_1, \mathbf{n}_2\}$. ∎

A.4 Sherman-Morrison Formula

The Sherman-Morrison formula [4] states that given matrices $\mathbf{A} \in C^{n \times n}$, $\mathbf{U} \in C^{n \times p}$, $\mathbf{W} \in C^{p \times p}$, and $\mathbf{V} \in C^{n \times p}$, such that $\mathbf{A}^{-1}$, $\mathbf{W}^{-1}$ and $(\mathbf{W}^{-1} + \mathbf{V}^H \mathbf{A}^{-1} \mathbf{U})^{-1}$ exist, then the inverse of $\mathbf{A} + \mathbf{U} \mathbf{W} \mathbf{V}^H$ exists and is given by

$$(\mathbf{A} + \mathbf{U} \mathbf{W} \mathbf{V}^H)^{-1} = \mathbf{A}^{-1} - \mathbf{A}^{-1} \mathbf{U} \mathbf{Y}^{-1} \mathbf{V}^H \mathbf{A}^{-1} \qquad (\text{A.9})$$

where

$$\mathbf{Y} = \mathbf{W}^{-1} + \mathbf{V}^H \mathbf{A}^{-1} \mathbf{U} \qquad (\text{A.10})$$

In particular, if $p = 1$ and $\mathbf{W} = 1$, then Eq. (A.9) assumes the form

$$(\mathbf{A} + \mathbf{u}\mathbf{v}^H)^{-1} = \mathbf{A}^{-1} - \frac{\mathbf{A}^{-1} \mathbf{u}\mathbf{v}^H \mathbf{A}^{-1}}{1 + \mathbf{v}^H \mathbf{A}^{-1} \mathbf{u}} \qquad (\text{A.11})$$

where $\mathbf{u}$ and $\mathbf{v}$ are vectors in $C^{n \times 1}$. Eq. (A.11) is useful for computing the inverse of a rank-one modification of $\mathbf{A}$, namely, $\mathbf{A} + \mathbf{u}\mathbf{v}^H$, if $\mathbf{A}^{-1}$ is available.

Example A.3 Find $\mathbf{A}^{-1}$ for

$$\mathbf{A} = \begin{bmatrix} 1.04 & 0.04 & \cdots & 0.04 \\ 0.04 & 1.04 & \cdots & 0.04 \\ \vdots & \vdots & & \vdots \\ 0.04 & 0.04 & \cdots & 1.04 \end{bmatrix} \in \mathcal{R}^{10 \times 10}$$

Solution Matrix $\mathbf{A}$ can be treated as a rank-one perturbation of the identity matrix:

$$\mathbf{A} = \mathbf{I} + \mathbf{p}\mathbf{p}^T$$

where $\mathbf{I}$ is the identity matrix and $\mathbf{p} = [0.2\ 0.2\ \cdots\ 0.2]^T$. Using Eq. (A.11), we can compute

$$\mathbf{A}^{-1} = (\mathbf{I} + \mathbf{p}\mathbf{p}^T)^{-1} = \mathbf{I} - \frac{\mathbf{p}\mathbf{p}^T}{1 + \mathbf{p}^T\mathbf{p}} = \mathbf{I} - \frac{1}{1.4}\mathbf{p}\mathbf{p}^T$$

$$= \begin{bmatrix} 0.9714 & -0.0286 & \cdots & -0.0286 \\ -0.0286 & 0.9714 & \cdots & -0.0286 \\ \vdots & \vdots & & \vdots \\ -0.0286 & -0.0286 & \cdots & 0.9714 \end{bmatrix}$$

A.5 Eigenvalues and Eigenvectors

The *eigenvalues* of a matrix $\mathbf{A} \in C^{n \times n}$ are defined as the n roots of its so-called *characteristic equation*

$$\det(\lambda\mathbf{I} - \mathbf{A}) = 0 \tag{A.12}$$

If we denote the set of n eigenvalues $\{\lambda_1, \lambda_2, \ldots, \lambda_n\}$ by $\lambda(\mathbf{A})$, then for a $\lambda_i \in \lambda(\mathbf{A})$, there exists a nonzero vector $\mathbf{x}_i \in C^{n \times 1}$ such that

$$\mathbf{A}\mathbf{x}_i = \lambda_i\mathbf{x}_i \tag{A.13}$$

Such a vector is called an *eigenvector* of $\mathbf{A}$ associated with eigenvalue λ_i.

Eigenvectors are not unique. For example, if $\mathbf{x}_i$ is an eigenvector of matrix $\mathbf{A}$ associated with eigenvalue λ_i and c is an arbitrary nonzero constant, then $c\mathbf{x}_i$ is also an eigenvector of $\mathbf{A}$ associated with eigenvalue λ_i.

If $\mathbf{A}$ has n distinct eigenvalues $\lambda_1, \lambda_2, \ldots, \lambda_n$ with associated eigenvectors $\mathbf{x}_1, \mathbf{x}_2, \ldots, \mathbf{x}_n$, then these eigenvectors are linearly independent; hence we can write

$$\mathbf{A}[\mathbf{x}_1\ \mathbf{x}_2\ \cdots\ \mathbf{x}_n] = [\mathbf{A}\mathbf{x}_1\ \mathbf{A}\mathbf{x}_2\ \cdots\ \mathbf{A}\mathbf{x}_n] = [\lambda_1\mathbf{x}_1\ \lambda_2\mathbf{x}_2\ \cdots\ \lambda_n\mathbf{x}_n]$$

$$= [\mathbf{x}_1\ \mathbf{x}_2\ \cdots\ \mathbf{x}_n] \begin{bmatrix} \lambda_1 & & \mathbf{0} \\ & \ddots & \\ \mathbf{0} & & \lambda_n \end{bmatrix}$$

In effect,

$$\mathbf{A}\mathbf{X} = \mathbf{X}\mathbf{\Lambda}$$

or

$$\mathbf{A} = \mathbf{X}\mathbf{\Lambda}\mathbf{X}^{-1} \tag{A.14}$$

with

$$\mathbf{X} = [\mathbf{x}_1\ \mathbf{x}_2\ \cdots\ \mathbf{x}_n] \quad \text{and} \quad \mathbf{\Lambda} = \text{diag}\{\lambda_1, \lambda_1, \ldots, \lambda_n\}$$

where diag$\{\lambda_1, \lambda_2, \ldots, \lambda_n\}$ represents the diagonal matrix with components $\lambda_1, \lambda_2, \ldots, \lambda_n$ along its diagonal. The relation in (A.14) is often referred to as an *eigendecomposition* of $\mathbf{A}$.

A concept that is closely related to the eigendecomposition in Eq. (A.14) is that of similarity transformation. Two square matrices $\mathbf{A}$ and $\mathbf{B}$ are said to be *similar* if there exists a nonsingular $\mathbf{X}$, called a *similarity transformation*, such that

$$\mathbf{A} = \mathbf{XBX}^{-1} \tag{A.15}$$

From Eq. (A.14), it follows that if the eigenvalues of $\mathbf{A}$ are distinct, then $\mathbf{A}$ is similar to $\mathbf{\Lambda} = \text{diag}\{\lambda_1, \lambda_2, \ldots, \lambda_n\}$ and the similarity transformation involved, $\mathbf{X}$, is composed of the n eigenvectors of $\mathbf{A}$. For arbitrary matrices with repeated eigenvalues, the eigendecomposition becomes more complicated. The reader is referred to [1]–[3] for the theory and solution of the eigenvalue problem for the general case.

Example A.4 Find the diagonal matrix $\mathbf{\Lambda}$, if it exists, that is similar to matrix

$$\mathbf{A} = \begin{bmatrix} 4 & -3 & 1 & 1 \\ 2 & -1 & 1 & 1 \\ 0 & 0 & 1 & 2 \\ 0 & 0 & 2 & 1 \end{bmatrix}$$

Solution From Eq. (A.12), we have

$$\begin{aligned} \det(\lambda \mathbf{I} - \mathbf{A}) &= \det \begin{bmatrix} \lambda - 4 & 3 \\ -2 & \lambda + 1 \end{bmatrix} \cdot \det \begin{bmatrix} \lambda - 1 & -2 \\ -2 & \lambda - 1 \end{bmatrix} \\ &= (\lambda^2 - 3\lambda + 2)(\lambda^2 - 2\lambda - 3) \\ &= (\lambda - 1)(\lambda - 2)(\lambda + 1)(\lambda - 3) \end{aligned}$$

Hence the eigenvalues of $\mathbf{A}$ are $\lambda_1 = 1$, $\lambda_2 = 2$, $\lambda_3 = -1$, and $\lambda_4 = 3$. An eigenvector $\mathbf{x}_i$ associated with eigenvalue λ_i satisfies the relation

$$(\lambda_i \mathbf{I} - \mathbf{A})\mathbf{x}_i = 0$$

For $\lambda_1 = 1$, we have

$$\lambda_1 \mathbf{I} - \mathbf{A} = \begin{bmatrix} -3 & 3 & -1 & -1 \\ -2 & 2 & -1 & -1 \\ 0 & 0 & 0 & -2 \\ 0 & 0 & -2 & 0 \end{bmatrix}$$

It is easy to verify that $\mathbf{x}_1 = [1\ 1\ 0\ 0]^T$ satisfies the relation

$$(\lambda_1 \mathbf{I} - \mathbf{A})\mathbf{x}_1 = 0$$

Similarly, $\mathbf{x}_2 = [3\ 2\ 0\ 0]^T$, $\mathbf{x}_3 = [0\ 0\ 1\ -1]^T$, and $\mathbf{x}_4 = [1\ 1\ 1\ 1]^T$ satisfy the relation

$$(\lambda_i \mathbf{I} - \mathbf{A})\mathbf{x}_i = 0 \qquad \text{for } i = 2,\ 3,\ 4$$

If we let

$$\mathbf{X} = [\mathbf{x}_1\ \mathbf{x}_2\ \mathbf{x}_3\ \mathbf{x}_4] = \begin{bmatrix} 1 & 3 & 0 & 1 \\ 1 & 2 & 0 & 1 \\ 0 & 0 & 1 & 1 \\ 0 & 0 & -1 & 1 \end{bmatrix}$$

then we have

$$\mathbf{A}\mathbf{X} = \mathbf{\Lambda}\mathbf{X}$$

where

$$\mathbf{\Lambda} = \text{diag}\{1,\ 2,\ -1,\ 3\}$$

∎

A.6   Symmetric Matrices

The matrices encountered most frequently in numerical optimization are symmetric. For these matrices, an elegant eigendecomposition theory and corresponding computation methods are available. If $\mathbf{A} = \{a_{ij}\} \in R^{n \times n}$ is a symmetric matrix, i.e., $a_{ij} = a_{ji}$, then there exists an orthogonal matrix $\mathbf{X} \in R^{n \times n}$, i.e., $\mathbf{X}\mathbf{X}^T = \mathbf{X}^T\mathbf{X} = \mathbf{I}_n$, such that

$$\mathbf{A} = \mathbf{X}\mathbf{\Lambda}\mathbf{X}^T \tag{A.16}$$

where $\mathbf{\Lambda} = \text{diag}\{\lambda_1,\ \lambda_2,\ \ldots,\ \lambda_n\}$. If $\mathbf{A} \in C^{n \times n}$ is such that $\mathbf{A} = \mathbf{A}^H$, then $\mathbf{A}$ is referred to as a *Hermitian matrix*. In such a case, there exists a so-called *unitary matrix* $\mathbf{U} \in C^{n \times n}$ for which $\mathbf{U}\mathbf{U}^H = \mathbf{U}^H\mathbf{U} = \mathbf{I}_n$ such that

$$\mathbf{A} = \mathbf{U}\mathbf{\Lambda}\mathbf{U}^H \tag{A.17}$$

In Eqs. (A.16) and (A.17), the diagonal components of $\mathbf{\Lambda}$ are eigenvalues of $\mathbf{A}$, and the columns of $\mathbf{X}$ and $\mathbf{U}$ are corresponding eigenvectors of $\mathbf{A}$.

The following properties can be readily verified:

(a) A square matrix is nonsingular if and only if all its eigenvalues are nonzero.

(b) The magnitudes of the eigenvalues of an orthogonal or unitary matrix are always equal to unity.

(c) The eigenvalues of a symmetric or Hermitian matrix are always real.

(d) The determinant of a square matrix is equal to the product of its eigenvalues.

A symmetric matrix $\mathbf{A} \in R^{n \times n}$ is said to be *positive definite, positive semidefinite, negative semidefinite, negative definite* if $\mathbf{x}^T\mathbf{A}\mathbf{x} > 0$, $\mathbf{x}^T\mathbf{A}\mathbf{x} \geq 0$, $\mathbf{x}^T\mathbf{A}\mathbf{x} \leq 0$, $\mathbf{x}^T\mathbf{A}\mathbf{x} < 0$, respectively, for all nonzero $\mathbf{x} \in R^{n \times 1}$.

Using the decomposition in Eq. (A.16), it can be shown that matrix $\mathbf{A}$ is positive definite, positive semidefinite, negative semidefinite, negative definite, if and only if its eigenvalues are positive, nonnegative, nonpositive, negative, respectively. Otherwise, $\mathbf{A}$ is said to be indefinite. We use the shorthand notation $\mathbf{A} \succ, \succeq, \preceq, \prec \mathbf{0}$ to indicate that $\mathbf{A}$ is positive definite, positive semidefinite, negative semidefinite, negative definite throughout the book.

Another approach for the characterization of a square matrix $\mathbf{A}$ is based on the evaluation of the *leading principal minor determinants*. A *minor determinant*, which is usually referred to as a *minor*, is the determinant of a submatrix obtained by deleting a number of rows and an equal number of columns from the matrix. Specifically, a minor of order r of an $n \times n$ matrix $\mathbf{A}$ is obtained by deleting $n - r$ rows and $n - r$ columns. For example, if

$$\mathbf{A} = \begin{bmatrix} a_{11} & a_{12} & a_{13} & a_{14} \\ a_{21} & a_{22} & a_{23} & a_{24} \\ a_{31} & a_{32} & a_{33} & a_{34} \\ a_{41} & a_{42} & a_{43} & a_{44} \end{bmatrix}$$

then

$$\Delta_3^{(123,123)} = \begin{vmatrix} a_{11} & a_{12} & a_{13} \\ a_{21} & a_{22} & a_{23} \\ a_{31} & a_{32} & a_{33} \end{vmatrix}, \quad \Delta_3^{(134,124)} = \begin{vmatrix} a_{11} & a_{12} & a_{14} \\ a_{31} & a_{32} & a_{34} \\ a_{41} & a_{42} & a_{44} \end{vmatrix}$$

and

$$\Delta_2^{(12,12)} = \begin{vmatrix} a_{11} & a_{12} \\ a_{21} & a_{22} \end{vmatrix}, \quad \Delta_2^{(13,14)} = \begin{vmatrix} a_{11} & a_{14} \\ a_{31} & a_{34} \end{vmatrix}$$

$$\Delta_2^{(24,13)} = \begin{vmatrix} a_{21} & a_{23} \\ a_{41} & a_{43} \end{vmatrix}, \quad \Delta_2^{(34,34)} = \begin{vmatrix} a_{33} & a_{34} \\ a_{43} & a_{44} \end{vmatrix}$$

are third-order and second-order minors, respectively. An nth-order minor is the determinant of the matrix itself and a first-order minor, i.e., if $n-1$ rows and $n - 1$ columns are deleted, is simply the value of a single matrix component.[1]

If the indices of the deleted rows are the same as those of the deleted columns, then the minor is said to be a *principal minor*, e.g., $\Delta_3^{(123,123)}$, $\Delta_2^{(12,12)}$, and $\Delta_2^{(34,34)}$ in the above examples.

Principal minors $\Delta_3^{(123,123)}$ and $\Delta_2^{(12,12)}$ in the above examples can be represented by

$$\Delta_3^{(1,2,3)} = \det \mathbf{H}_3^{(1,2,3)}$$

and

$$\Delta_2^{(1,2)} = \det \mathbf{H}_2^{(1,2)}$$

---

[1] The zeroth-order minor is often defined to be unity.

respectively. An arbitrary principal minor of order i can be represented by

$$\Delta_i^{(l)} = \det \mathbf{H}_i^{(l)}$$

where

$$\mathbf{H}_i^{(l)} = \begin{bmatrix} a_{l_1 l_1} & a_{l_1 l_2} & \cdots & a_{l_1 l_i} \\ a_{l_2 l_1} & a_{l_2 l_2} & \cdots & a_{l_2 l_i} \\ \vdots & \vdots & & \vdots \\ a_{l_i l_1} & a_{l_i l_2} & \cdots & a_{l_i l_i} \end{bmatrix}$$

and $l \in \{l_1, l_2, \ldots, l_i\}$ with $1 \le l_1 < l_2 < \cdots < l_i \le n$ is the set of rows (and columns) retained in submatrix $\mathbf{H}_i^{(l)}$.

The specific principal minors

$$\Delta_r = \begin{vmatrix} a_{11} & a_{12} & \cdots & a_{1r} \\ a_{21} & a_{22} & \cdots & a_{2r} \\ \vdots & \vdots & & \vdots \\ a_{r1} & a_{r2} & \cdots & a_{rr} \end{vmatrix} = \det \mathbf{H}_r$$

for $1 \le r \le n$ are said to be the *leading principal minors* of an $n \times n$ matrix. For a 4×4 matrix, the complete set of leading principal minors is as follows:

$$\Delta_1 = a_{11}, \quad \Delta_2 = \begin{vmatrix} a_{11} & a_{12} \\ a_{21} & a_{22} \end{vmatrix}$$

$$\Delta_3 = \begin{vmatrix} a_{11} & a_{12} & a_{13} \\ a_{21} & a_{22} & a_{23} \\ a_{31} & a_{32} & a_{33} \end{vmatrix}, \quad \Delta_4 = \begin{vmatrix} a_{11} & a_{12} & a_{13} & a_{14} \\ a_{21} & a_{22} & a_{23} & a_{24} \\ a_{31} & a_{32} & a_{33} & a_{34} \\ a_{41} & a_{42} & a_{43} & a_{44} \end{vmatrix}$$

The leading principal minors of a matrix $\mathbf{A}$ or its negative $-\mathbf{A}$ can be used to establish whether the matrix is positive or negative definite whereas the principal minors of $\mathbf{A}$ or $-\mathbf{A}$ can be used to establish whether the matrix is positive or negative semidefinite. These principles are stated in terms of Theorem 2.9 in Chap. 2 and are often used to establish the nature of the Hessian matrix in optimization algorithms.

The fact that a nonnegative real number has positive and negative square roots can be extended to the class of positive semidefinite matrices. Assuming that matrix $\mathbf{A} \in R^{n \times n}$ is positive semidefinite, we can write its eigendecomposition in Eq. (A.16) as

$$\mathbf{A} = \mathbf{X}\mathbf{\Lambda}\mathbf{X}^T = \mathbf{X}\mathbf{\Lambda}^{1/2}\mathbf{W}\mathbf{W}^T\mathbf{\Lambda}^{1/2}\mathbf{X}^T$$

where $\mathbf{\Lambda}^{1/2} = \text{diag}\{\lambda_1^{1/2}, \lambda_2^{1/2}, \ldots, \lambda_n^{1/2}\}$ and $\mathbf{W}$ is an arbitrary orthogonal matrix, which leads to

$$\mathbf{A} = \mathbf{A}^{1/2}(\mathbf{A}^{1/2})^T \tag{A.18}$$

where $\mathbf{A}^{1/2} = \mathbf{X}\mathbf{\Lambda}^{1/2}\mathbf{W}$ and is called an *asymmetric square root* of $\mathbf{A}$. Since matrix $\mathbf{W}$ can be an arbitrary orthogonal matrix, an infinite number of asymmetric square roots of $\mathbf{A}$ exist. Alternatively, since $\mathbf{X}$ is an orthogonal matrix, we can write

$$\mathbf{A} = (\alpha\mathbf{X}\mathbf{\Lambda}^{1/2}\mathbf{X}^T)(\alpha\mathbf{X}\mathbf{\Lambda}^{1/2}\mathbf{X}^T)$$

where α is either 1 or -1, which gives

$$\mathbf{A} = \mathbf{A}^{1/2}\mathbf{A}^{1/2} \tag{A.19}$$

where $\mathbf{A}^{1/2} = \alpha\mathbf{X}\mathbf{\Lambda}^{1/2}\mathbf{X}^T$ and is called a *symmetric square root* of $\mathbf{A}$. Again, because α can be either 1 or -1, more than one symmetric square roots exist. Obviously, the symmetric square roots $\mathbf{X}\mathbf{\Lambda}^{1/2}\mathbf{X}^T$ and $-\mathbf{X}\mathbf{\Lambda}^{1/2}\mathbf{X}^T$ are positive semidefinite and negative semidefinite, respectively.

If $\mathbf{A}$ is a complex-valued positive semidefinite matrix, then *non-Hermitian* and *Hermitian square roots* of $\mathbf{A}$ can be obtained using the eigendecomposition in Eq. (A.17). For example, we can write

$$\mathbf{A} = \mathbf{A}^{1/2}(\mathbf{A}^{1/2})^H$$

where $\mathbf{A}^{1/2} = \mathbf{U}\mathbf{\Lambda}^{1/2}\mathbf{W}$ is a non-Hermitian square root of $\mathbf{A}$ if $\mathbf{W}$ is unitary. On the other hand,

$$\mathbf{A} = \mathbf{A}^{1/2}\mathbf{A}^{1/2}$$

where $\mathbf{A}^{1/2} = \alpha\mathbf{U}\mathbf{\Lambda}^{1/2}\mathbf{U}^H$ is a Hermitian square root if $\alpha = 1$ or $\alpha = -1$.

Example A.5 Verify that

$$\mathbf{A} = \begin{bmatrix} 2.5 & 0 & 1.5 \\ 0 & \sqrt{2} & 0 \\ 1.5 & 0 & 2.5 \end{bmatrix}$$

is positive definite and compute a symmetric square root of $\mathbf{A}$.

Solution An eigendecomposition of matrix $\mathbf{A}$ is

$$\mathbf{A} = \mathbf{X}\mathbf{\Lambda}\mathbf{X}^T$$

with

$$\mathbf{\Lambda} = \begin{bmatrix} 4 & 0 & 0 \\ 0 & 2 & 0 \\ 0 & 0 & 1 \end{bmatrix} \quad \text{and} \quad \mathbf{X} = \begin{bmatrix} \sqrt{2}/2 & 0 & -\sqrt{2}/2 \\ 0 & -1 & 0 \\ \sqrt{2}/2 & 0 & \sqrt{2}/2 \end{bmatrix}$$

Since the eigenvalues of $\mathbf{A}$ are all positive, $\mathbf{A}$ is positive definite. A symmetric square root of $\mathbf{A}$ is given by

$$\mathbf{A}^{1/2} = \mathbf{X}\mathbf{\Lambda}^{1/2}\mathbf{X}^T = \begin{bmatrix} 1.5 & 0 & 0.5 \\ 0 & \sqrt{2} & 0 \\ 0.5 & 0 & 1.5 \end{bmatrix}$$

∎

A.7 Trace

The trace of an $n \times n$ square matrix, $\mathbf{A} = \{a_{ij}\}$, is *the sum of its diagonal components*, i.e.,

$$\text{trace}(\mathbf{A}) = \sum_{i=1}^{n} a_{ii}$$

It can be verified that the trace of a square matrix $\mathbf{A}$ with eigenvalues λ_1, λ_2, ..., λ_n is equal to the sum of its eigenvalues, i.e.,

$$\text{trace}(\mathbf{A}) = \sum_{i=1}^{n} \lambda_i$$

A useful property pertaining to the product of two matrices is that the trace of a square matrix $\mathbf{AB}$ is equal to the trace of matrix $\mathbf{BA}$, i.e.,

$$\text{trace}(\mathbf{AB}) = \text{trace}(\mathbf{BA}) \tag{A.20}$$

By applying Eq. (A.20) to the quadratic form $\mathbf{x}^T \mathbf{H} \mathbf{x}$, we obtain

$$\mathbf{x}^T \mathbf{H} \mathbf{x} = \text{trace}(\mathbf{x}^T \mathbf{H} \mathbf{x}) = \text{trace}(\mathbf{H} \mathbf{x} \mathbf{x}^T) = \text{trace}(\mathbf{H} \mathbf{X})$$

where $\mathbf{X} = \mathbf{x} \mathbf{x}^T$. Moreover, we can write a general quadratic function as

$$\mathbf{x}^T \mathbf{H} \mathbf{x} + 2 \mathbf{p}^T \mathbf{x} + \kappa = \text{trace}(\hat{\mathbf{H}} \hat{\mathbf{X}}) \tag{A.21}$$

where

$$\hat{\mathbf{H}} = \begin{bmatrix} \mathbf{H} & \mathbf{p} \\ \mathbf{p}^T & \kappa \end{bmatrix} \quad \text{and} \quad \hat{\mathbf{X}} = \begin{bmatrix} \mathbf{x} \mathbf{x}^T & \mathbf{x} \\ \mathbf{x}^T & 1 \end{bmatrix}$$

A.8 Vector Norms and Matrix Norms

A.8.1 Vector norms

The L_p norm of a vector $\mathbf{x} \in C^n$ for $p \geq 1$ is given by

$$\|\mathbf{x}\|_p = \left(\sum_{i=1}^{n} |x_i|^p \right)^{1/p} \tag{A.22}$$

where p is a positive integer and x_i is the ith component of $\mathbf{x}$. The most popular L_p norms are $\| \cdot \|_1$, $\| \cdot \|_2$, and $\| \cdot \|_\infty$, where the infinity norm $\| \cdot \|_\infty$ can easily be shown to satisfy the relation

$$\|\mathbf{x}\|_\infty = \lim_{p \to \infty} \left(\sum_{i=1}^{n} |x_i|^p \right)^{1/p} = \max_i |x_i| \tag{A.23}$$

For example, if $\mathbf{x} = [1 \; 2 \; \cdots \; 100]^T$, then $\|\mathbf{x}\| = 581.68$, $\|\mathbf{x}\|_{10} = 125.38$, $\|\mathbf{x}\|_{50} = 101.85$, $\|\mathbf{x}\|_{100} = 100.45$, $\|\mathbf{x}\|_{200} = 100.07$ and, of course, $\|\mathbf{x}\|_\infty = 100$.

The important point to note here is that for an even p, the L_p norm of a vector is a *differentiable* function of its components but the L_∞ norm is *not*. So when the L_∞ norm is used in a design problem, we can replace it by an L_p norm (with p even) so that powerful calculus-based tools can be used to solve the problem. Obviously, the results obtained can only be *approximate* with respect to the original design problem. However, as indicated by Eq. (9.23), the difference between the approximate and exact solutions becomes insignificant if p is sufficiently large.

The inner product of two vectors $\mathbf{x},\ \mathbf{y} \in C^n$ is a scalar given by

$$\mathbf{x}^H\mathbf{y} = \sum_{i=1}^{n} x_i^* y_i$$

where x_i^* denotes the complex-conjugate of x_i. Frequently, we need to estimate the absolute value of $\mathbf{x}^H\mathbf{y}$. There are two well-known inequalities that provide tight upper bounds for $|\mathbf{x}^H\mathbf{y}|$, namely, the *Hölder inequality*

$$|\mathbf{x}^H\mathbf{y}| \leq \|\mathbf{x}\|_p\|\mathbf{y}\|_q \tag{A.24}$$

which holds for any $p \geq 1$ and $q \geq 1$ satisfying the equality

$$\frac{1}{p} + \frac{1}{q} = 1$$

and the *Cauchy-Schwartz inequality* which is the special case of the Hölder inequality with $p = q = 2$, i.e.,

$$|\mathbf{x}^H\mathbf{y}| \leq \|\mathbf{x}\|_2\|\mathbf{y}\|_2 \tag{A.25}$$

If vectors $\mathbf{x}$ and $\mathbf{y}$ have unity lengths, i.e., $\|\mathbf{x}\|_2 = \|\mathbf{y}\|_2 = 1$, then Eq. (A.25) becomes

$$|\mathbf{x}^H\mathbf{y}| \leq 1 \tag{A.26}$$

A geometric interpretation of Eq. (A.26) is that for unit vectors $\mathbf{x}$ and $\mathbf{y}$, the inner product $\mathbf{x}^H\mathbf{y}$ is equal to $\cos\theta$, where θ denotes the angle between the two vectors, whose absolute value is always less than one.

Another property of the L_2 norm is its invariance under orthogonal or unitary transformation. That is, if $\mathbf{A}$ is an orthogonal or unitary matrix, then

$$\|\mathbf{A}\mathbf{x}\|_2 = \|\mathbf{x}\|_2 \tag{A.27}$$

The L_p norm of a vector $\mathbf{x}$, $\|\mathbf{x}\|_p$, is monotonically decreasing with respect to p for $p \geq 1$. For example, we can relate $\|\mathbf{x}\|_1$ and $\|\mathbf{x}\|_2$ as

$$\|\mathbf{x}\|_1^2 = \left(\sum_{i=1}^{n}|x_i|\right)^2$$
$$= |x_1|^2 + |x_2|^2 + \cdots + |x_n|^2 + 2|x_1x_2| + \cdots + 2|x_{n-1}x_n|$$
$$\geq |x_1|^2 + |x_2|^2 + \cdots + |x_n|^2 = \|\mathbf{x}\|_2^2$$

604

which implies that

$$\|\mathbf{x}\|_1 \geq \|\mathbf{x}\|_2$$

Furthermore, if $\|\mathbf{x}\|_\infty$ is numerically equal to $|x_k|$ for some index k, i.e.,

$$\|\mathbf{x}\|_\infty = \max_i |x_i| = |x_k|$$

then we can write

$$\|\mathbf{x}\|_2 = (|x_1|^2 + \cdots + |x_n|^2)^{1/2} \geq (|x_k|^2)^{1/2} = |x_k| = \|\mathbf{x}\|_\infty$$

i.e.,

$$\|\mathbf{x}\|_2 \geq \|\mathbf{x}\|_\infty$$

Therefore, we have

$$\|\mathbf{x}\|_1 \geq \|\mathbf{x}\|_2 \geq \|\mathbf{x}\|_\infty$$

In general, it can be shown that

$$\|\mathbf{x}\|_1 \geq \|\mathbf{x}\|_2 \geq \|\mathbf{x}\|_3 \geq \cdots \geq \|\mathbf{x}\|_\infty$$

A.8.2 Matrix norms

The L_p norm of matrix $\mathbf{A} = \{a_{ij}\} \in C^{m \times n}$ is defined as

$$\|\mathbf{A}\|_p = \max_{\mathbf{x} \neq 0} \frac{\|\mathbf{Ax}\|_p}{\|\mathbf{x}\|_p} \qquad \text{for } p \geq 1 \tag{A.28}$$

The most useful matrix L_p norm is the L_2 norm

$$\|\mathbf{A}\|_2 = \max_{\mathbf{x} \neq 0} \frac{\|\mathbf{Ax}\|_2}{\|\mathbf{x}\|_2} = \left[\max_i \left|\lambda_i(\mathbf{A}^H\mathbf{A})\right|\right]^{1/2} = \left[\max_i \left|\lambda_i(\mathbf{AA}^H)\right|\right]^{1/2} \tag{A.29}$$

which can be easily computed as the square root of the largest eigenvalue magnitude in $\mathbf{A}^H\mathbf{A}$ or $\mathbf{AA}^H$. Some other frequently used matrix L_p norms are

$$\|\mathbf{A}\|_1 = \max_{\mathbf{x} \neq 0} \frac{\|\mathbf{Ax}\|_1}{\|\mathbf{x}\|_1} = \max_{1 \leq j \leq n} \sum_{i=1}^{m} |a_{ij}|$$

and

$$\|\mathbf{A}\|_\infty = \max_{\mathbf{x} \neq 0} \frac{\|\mathbf{Ax}\|_\infty}{\|\mathbf{x}\|_\infty} = \max_{1 \leq i \leq m} \sum_{j=1}^{n} |a_{ij}|$$

Another popular matrix norm is the Frobenius norm which is defined as

$$\|\mathbf{A}\|_F = \left(\sum_{i=1}^{m}\sum_{j=1}^{n} |a_{ij}|^2\right)^{1/2} \tag{A.30}$$

which can also be calculated as

$$\|\mathbf{A}\|_F = [\text{trace}(\mathbf{A}^H \mathbf{A})]^{1/2} = [\text{trace}(\mathbf{A}\mathbf{A}^H)]^{1/2} \quad\quad (A.31)$$

Note that the matrix L_2 norm and the Frobenius norm are *invariant* under orthogonal or unitary transformation, i.e., if $\mathbf{U} \in C^{n \times n}$ and $\mathbf{V} \in C^{m \times m}$ are unitary or orthogonal matrices, then

$$\|\mathbf{U}\mathbf{A}\mathbf{V}\|_2 = \|\mathbf{A}\|_2 \quad\quad (A.32)$$

and

$$\|\mathbf{U}\mathbf{A}\mathbf{V}\|_F = \|\mathbf{A}\|_F \qu\quad (A.33)$$

Example A.6 Evaluate matrix norms $\|\mathbf{A}\|_1$, $\|\mathbf{A}\|_2$, $\|\mathbf{A}\|_\infty$, and $\|\mathbf{A}\|_F$ for

$$\mathbf{A} = \begin{bmatrix} 1 & 5 & 6 & 3 \\ 0 & 4 & -7 & 0 \\ 3 & 1 & 4 & 1 \\ -1 & 1 & 0 & 1 \end{bmatrix}$$

Solution

$$\|\mathbf{A}\|_1 = \max_{1 \le j \le 4} \left(\sum_{i=1}^{4} |a_{ij}| \right) = \max\{5,\ 11,\ 17,\ 5\} = 17$$

$$\|\mathbf{A}\|_\infty = \max_{1 \le i \le 4} \left(\sum_{j=1}^{4} |a_{ij}| \right) = \max\{15,\ 11,\ 9,\ 3\} = 15$$

$$\|\mathbf{A}\|_F = \left(\sum_{i=1}^{4} \sum_{j=1}^{4} |a_{ij}|^2 \right)^{1/2} = \sqrt{166} = 12.8841$$

To obtain $\|\mathbf{A}\|_2$, we compute the eigenvalues of $\mathbf{A}^T \mathbf{A}$ as

$$\lambda(\mathbf{A}^T \mathbf{A}) = \{0.2099,\ 6.9877,\ 47.4010,\ 111.4014\}$$

Hence

$$\|\mathbf{A}\|_2 = [\max_i |\lambda_i(\mathbf{A}^T \mathbf{A})|]^{1/2} = \sqrt{111.4014} = 10.5547$$

∎

A.9 Singular-Value Decomposition

Given a matrix $\mathbf{A} \in C^{m \times n}$ of rank r, there exist unitary matrices $\mathbf{U} \in C^{m \times m}$ and $\mathbf{V} \in C^{n \times n}$ such that

$$\mathbf{A} = \mathbf{U} \boldsymbol{\Sigma} \mathbf{V}^H \tag{A.34}$$

where

$$\boldsymbol{\Sigma} = \begin{bmatrix} \mathbf{S} & \mathbf{0} \\ \mathbf{0} & \mathbf{0} \end{bmatrix}_{m \times n} \tag{A.34}$$

and

$$\mathbf{S} = \text{diag}\{\sigma_1, \ \sigma_2, \ \ldots, \ \sigma_r\} \tag{A.34}$$

with $\sigma_1 \geq \sigma_2 \geq \cdots \geq \sigma_r > 0$.

The matrix decomposition in Eq. (A.34a) is known as the *singular-value decomposition* (SVD) of $\mathbf{A}$. It has many applications in optimization and elsewhere. If $\mathbf{A}$ is a real-valued matrix, then $\mathbf{U}$ and $\mathbf{V}$ in Eq. (A.34a) become orthogonal matrices and $\mathbf{V}^H$ becomes $\mathbf{V}^T$. The positive scalars σ_i for $i = 1, 2, \ldots, r$ in Eq. (A.34c) are called the *singular values* of $\mathbf{A}$. If $\mathbf{U} = [\mathbf{u}_1 \ \mathbf{u}_2 \ \cdots \ \mathbf{u}_m]$ and $\mathbf{V} = [\mathbf{v}_1 \ \mathbf{v}_2 \ \cdots \ \mathbf{v}_n]$, vectors $\mathbf{u}_i$ and $\mathbf{v}_i$ are called the *left* and *right singular vectors* of $\mathbf{A}$, respectively. From Eq. (A.34), it follows that

$$\mathbf{A}\mathbf{A}^H = \mathbf{U} \begin{bmatrix} \mathbf{S}^2 & \mathbf{0} \\ \mathbf{0} & \mathbf{0} \end{bmatrix}_{m \times m} \mathbf{U}^H \tag{A.35}$$

and

$$\mathbf{A}^H \mathbf{A} = \mathbf{V} \begin{bmatrix} \mathbf{S}^2 & \mathbf{0} \\ \mathbf{0} & \mathbf{0} \end{bmatrix}_{n \times n} \mathbf{V}^H \tag{A.35}$$

Therefore, the singular values of $\mathbf{A}$ are the positive square roots of the nonzero eigenvalues of $\mathbf{A}\mathbf{A}^H$ (or $\mathbf{A}^H\mathbf{A}$), the ith left singular vector $\mathbf{u}_i$ is the ith eigenvector of $\mathbf{A}\mathbf{A}^H$, and the ith right singular vector $\mathbf{v}_i$ is the ith eigenvector of $\mathbf{A}^H\mathbf{A}$.

Several important applications of the SVD are as follows:

(a) The L_2 norm and Frobenius norm of a matrix $\mathbf{A} \in C^{m \times n}$ of rank r are given, respectively, by

$$\|\mathbf{A}\|_2 = \sigma_1 \tag{A.36}$$

and

$$\|\mathbf{A}\|_F = \left(\sum_{i=1}^{r} \sigma_i^2 \right)^{1/2} \tag{A.37}$$

(b) The *condition number* of a nonsingular matrix $\mathbf{A} \in C^{n \times n}$ is defined as

$$\text{cond}(\mathbf{A}) = \|\mathbf{A}\|_2 \|\mathbf{A}^{-1}\|_2 = \frac{\sigma_1}{\sigma_n} \tag{A.38}$$

(c) The range and null space of a matrix $\mathbf{A} \in C^{m \times n}$ of rank r assume the forms

$$\mathcal{R}(\mathbf{A}) = \text{span}\{\mathbf{u}_1, \mathbf{u}_2, \ldots, \mathbf{u}_r\} \tag{A.39}$$

$$\mathcal{N}(\mathbf{A}) = \text{span}\{\mathbf{v}_{r+1}, \mathbf{v}_{r+2}, \ldots, \mathbf{v}_n\} \tag{A.40}$$

(d) Properties and computation of Moore-Penrose pseudo-inverse:

The Moore-Penrose pseudo-inverse of a matrix $\mathbf{A} \in C^{m \times n}$ is defined as the matrix $\mathbf{A}^+ \in C^{n \times m}$ that satisfies the following four conditions:

(i) $\mathbf{A}\mathbf{A}^+\mathbf{A} = \mathbf{A}$

(ii) $\mathbf{A}^+\mathbf{A}\mathbf{A}^+ = \mathbf{A}^+$

(iii) $(\mathbf{A}\mathbf{A}^+)^H = \mathbf{A}\mathbf{A}^+$

(iv) $(\mathbf{A}^+\mathbf{A})^H = \mathbf{A}^+\mathbf{A}$

Using the SVD of $\mathbf{A}$ in Eq. (A.34), the Moore-Penrose pseudo-inverse of $\mathbf{A}$ can be obtained as

$$\mathbf{A}^+ = \mathbf{V}\boldsymbol{\Sigma}^+\mathbf{U}^H \tag{A.41}$$

where

$$\boldsymbol{\Sigma}^+ = \begin{bmatrix} \mathbf{S}^{-1} & \mathbf{0} \\ \mathbf{0} & \mathbf{0} \end{bmatrix}_{n \times m} \tag{A.41}$$

and

$$\mathbf{S}^{-1} = \text{diag}\{\sigma_1^{-1}, \sigma_2^{-1}, \ldots, \sigma_r^{-1}\} \tag{A.41}$$

Consequently, we have

$$\mathbf{A}^+ = \sum_{i=1}^{r} \frac{\mathbf{v}_i \mathbf{u}_i^H}{\sigma_i} \tag{A.42}$$

(e) For an underdetermined system of linear equations

$$\mathbf{A}\mathbf{x} = \mathbf{b} \tag{A.43}$$

where $\mathbf{A} \in C^{m \times n}$, $\mathbf{b} \in C^{m \times 1}$ with $m < n$, and $\mathbf{b} \in \mathcal{R}(\mathbf{A})$, all the solutions of Eq. (A.43) are characterized by

$$\mathbf{x} = \mathbf{A}^+\mathbf{b} + \mathbf{V}_r\boldsymbol{\phi} \tag{A.44}$$

where $\mathbf{A}^+$ is the Moore-Penrose pseudo-inverse of $\mathbf{A}$,

$$\mathbf{V}_r = [\mathbf{v}_{r+1} \ \mathbf{v}_{r+2} \ \cdots \ \mathbf{v}_n] \tag{A.44}$$

is a matrix of dimension $n \times (n-r)$ composed of the last $n-r$ columns of matrix $\mathbf{V}$ which is obtained by constructing the SVD of $\mathbf{A}$ in Eq. (A.34), and $\phi \in C^{(n-r) \times 1}$ is an *arbitrary* $(n-r)$-dimensional vector. Note that the first term in Eq. (A.44a), i.e., $\mathbf{A}^+\mathbf{b}$, is a solution of Eq. (A.43) while the second term, $\mathbf{V}_r\phi$, belongs to the null space of $\mathbf{A}$ (see Eq. (A.40)). Through vector ϕ, the expression in Eq. (A.44) parameterizes all the solutions of an underdetermined system of linear equations.

Example A.7 Perform the SVD of matrix

$$\mathbf{A} = \begin{bmatrix} 2.8284 & -1 & 1 \\ 2.8284 & 1 & -1 \end{bmatrix}$$

and compute $\|\mathbf{A}\|_2$, $\|\mathbf{A}\|_F$, and $\mathbf{A}^+$.

Solution To compute matrix $\mathbf{V}$ in Eq. (A.34a), from Eq. (A.35b) we obtain

$$\mathbf{A}^T\mathbf{A} = \begin{bmatrix} 16 & 0 & 0 \\ 0 & 2 & -2 \\ 0 & -2 & 2 \end{bmatrix} = \mathbf{V} \begin{bmatrix} 16 & 0 & 0 \\ 0 & 4 & 0 \\ 0 & 0 & 0 \end{bmatrix} \mathbf{V}^T$$

where

$$\mathbf{V} = \begin{bmatrix} 1 & 0 & 0 \\ 0 & 0.7071 & -0.7071 \\ 0 & -0.7071 & -0.7071 \end{bmatrix} = [\mathbf{v}_1 \ \mathbf{v}_2 \ \mathbf{v}_3]$$

Hence the nonzero singular values of $\mathbf{A}$ are $\sigma_1 = \sqrt{16} = 4$ and $\sigma_2 = \sqrt{4} = 2$. Now we can write (A.34a) as $\mathbf{U\Sigma} = \mathbf{AV}$, where

$$\mathbf{U\Sigma} = [\sigma_1\mathbf{u}_1 \ \sigma_2\mathbf{u}_2 \ \mathbf{0}] = [4\mathbf{u}_1 \ 2\mathbf{u}_2 \ \mathbf{0}]$$

and

$$\mathbf{AV} = \begin{bmatrix} 2.8284 & -1.4142 & 0 \\ 2.8284 & 1.4142 & 0 \end{bmatrix}$$

Hence

$$\mathbf{u}_1 = \frac{1}{4}\begin{bmatrix} 2.8284 \\ 2.8284 \end{bmatrix} = \begin{bmatrix} 0.7071 \\ 0.7071 \end{bmatrix}, \quad \mathbf{u}_2 = \frac{1}{2}\begin{bmatrix} -1.4142 \\ 1.4142 \end{bmatrix} = \begin{bmatrix} -0.7071 \\ 0.7071 \end{bmatrix}$$

and

$$\mathbf{U} = [\mathbf{u}_1 \ \mathbf{u}_2] = \begin{bmatrix} 0.7071 & -0.7071 \\ 0.7071 & 0.7071 \end{bmatrix}$$

On using Eqs. (A.36) and (A.37), we have

$$\|\mathbf{A}\|_2 = \sigma_1 = 4 \quad \text{and} \quad \|\mathbf{A}\|_F = (\sigma_1^2 + \sigma_2^2)^{1/2} = \sqrt{20} = 4.4721$$

Now from Eq. (A.42), we obtain

$$\mathbf{A}^+ = \frac{\mathbf{v}_1\mathbf{u}_1^T}{\sigma_1} + \frac{\mathbf{v}_2\mathbf{u}_2^T}{\sigma_2} = \begin{bmatrix} 0.1768 & 0.1768 \\ -0.2500 & 0.2500 \\ 0.2500 & -0.2500 \end{bmatrix}$$

∎

A.10   Orthogonal Projections

Let S be a subspace in C^n. Matrix $\mathbf{P} \in C^{n \times n}$ is said to be an orthogonal projection matrix onto S if $\mathcal{R}(\mathbf{P}) = S$, $\mathbf{P}^2 = \mathbf{P}$, and $\mathbf{P}^H = \mathbf{P}$, where $\mathcal{R}(\mathbf{P})$ denotes the range of transformation $\mathbf{P}$ (see Sec. A.3), i.e., $\mathcal{R}(\mathbf{P}) = \{\mathbf{y} : \mathbf{y} = \mathbf{P}\mathbf{x}, \ \mathbf{x} \in C^n\}$. The term 'orthogonal projection' originates from the fact that if $\mathbf{x} \in C^n$ is a vector outside S, then $\mathbf{P}\mathbf{x}$ is a vector in S such that $\mathbf{x} - \mathbf{P}\mathbf{x}$ is orthogonal to every vector in S and $\|\mathbf{x} - \mathbf{P}\mathbf{x}\|$ is the minimum distance between $\mathbf{x}$ and s, i.e., $\min \|\mathbf{x} - \mathbf{s}\|$, for $\mathbf{s} \in S$, as illustrated in Fig. A.1.

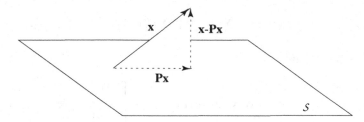

Figure A.1.   Orthogonal projection of $\mathbf{x}$ onto subspace S.

Let $\{\mathbf{s}_1, \mathbf{s}_2, \ldots, \mathbf{s}_k\}$ be a basis of a subspace S of dimension k (see Sec. A.2) such that $\|\mathbf{s}_i\| = 1$ and $\mathbf{s}_i^T \mathbf{s}_j = 0$ for $i, j = 1, 2, \ldots, k$ and $i \neq j$. Such a basis is called *orthonormal*. It can be readily verified that an orthogonal projection matrix onto S can be explicitly constructed in terms of an orthonormal basis as

$$\mathbf{P} = \mathbf{S}\mathbf{S}^H \qquad (A.45)$$

where

$$\mathbf{S} = [\mathbf{s}_1 \ \mathbf{s}_2 \ \cdots \ \mathbf{s}_k] \qquad (A.45)$$

It follows from Eqs. (A.39), (A.40), and (A.45) that $[\mathbf{u}_1 \ \mathbf{u}_2 \ \cdots \ \mathbf{u}_r] \cdot [\mathbf{u}_1 \ \mathbf{u}_2 \ \cdots \ \mathbf{u}_r]^H$ is the orthogonal projection onto $\mathcal{R}(\mathbf{A})$ and $[\mathbf{v}_{r+1} \ \mathbf{v}_{r+2} \ \cdots \ \mathbf{v}_n] \cdot [\mathbf{v}_{r+1} \ \mathbf{v}_{r+2} \ \cdots \ \mathbf{v}_n]^H$ is the orthogonal projection onto $\mathcal{N}(\mathbf{A})$.

Example A.8 Let $S = \mathrm{span}\{\mathbf{v}_1, \mathbf{v}_2\}$ where

$$\mathbf{v}_1 = \begin{bmatrix} 1 \\ 1 \\ 1 \end{bmatrix} \quad \text{and} \quad \mathbf{v}_2 = \begin{bmatrix} -1 \\ 1 \\ 1 \end{bmatrix}$$

Find the orthogonal projection onto S.

Solution First, we need to find an orthonormal basis $\{\mathbf{s}_1, \mathbf{s}_2\}$ of subspace S. To this end, we take

$$\mathbf{s}_1 = \frac{\mathbf{v}_1}{\|\mathbf{v}_1\|} = \begin{bmatrix} 1/\sqrt{3} \\ 1/\sqrt{3} \\ 1/\sqrt{3} \end{bmatrix}$$

610

Then we try to find vector $\hat{s}_2$ such that $\hat{s}_2 \in \mathcal{S}$ and $\hat{s}_2$ is orthogonal to s_1. Such an $\hat{s}_2$ must satisfy the relation

$$\hat{s}_2 = \alpha_1 v_1 + \alpha_2 v_2$$

for some α_1, α_2 and

$$\hat{s}_2^T s_1 = 0$$

Hence we have

$$(\alpha_1 v_1^T + \alpha_2 v_2^T)s_1 = \alpha_1 v_1^T s_1 + \alpha_2 v_2^T s_1 = \sqrt{3}\alpha_1 + \frac{1}{\sqrt{3}}\alpha_2 = 0$$

i.e., $\alpha_2 = -3\alpha_1$. Thus

$$\hat{s}_2 = \alpha_1 v_1 - 3\alpha_1 v_2 = \alpha_1 \begin{bmatrix} 4 \\ -2 \\ -2 \end{bmatrix}$$

where α_1 is a parameter that can assume an arbitrary nonzero value.

By normalizing vector $\hat{s}_2$, we obtain

$$s_2 = \frac{\hat{s}_2}{\|\hat{s}_2\|} = \frac{1}{\sqrt{4^2 + (-2)^2 + (-2)^2}} \begin{bmatrix} 4 \\ -2 \\ -2 \end{bmatrix} = \frac{1}{\sqrt{6}} \begin{bmatrix} 2 \\ -1 \\ -1 \end{bmatrix}$$

It now follows from Eq. (A.45) that the orthogonal projection onto $\mathcal{S}$ can be characterized by

$$\mathbf{P} = [s_1 \ s_2][s_1 \ s_2]^T = \begin{bmatrix} 1 & 0 & 0 \\ 0 & 0.5 & 0.5 \\ 0 & 0.5 & 0.5 \end{bmatrix}$$

∎

A.11   Householder Transformations and Givens Rotations
A.11.1   Householder transformations

The *Householder transformation* associated with a nonzero vector $\mathbf{u} \in R^{n \times 1}$ is characterized by the symmetric orthogonal matrix

$$\mathbf{H} = \mathbf{I} - 2\frac{\mathbf{u}\mathbf{u}^T}{\|\mathbf{u}\|^2} \tag{A.46}$$

If

$$\mathbf{u} = \mathbf{x} - \|\mathbf{x}\|\mathbf{e}_1 \tag{A.47}$$

where $e_1 = [1 \ 0 \ \cdots \ 0]^T$, then the Householder transformation will convert vector x to coordinate vector e_1 to within a scale factor $\|x\|$, i.e.,

$$Hx = \|x\| \begin{bmatrix} 1 \\ 0 \\ \vdots \\ 0 \end{bmatrix} \qquad (A.48)$$

Alternatively, if vector u in Eq. (A.46) is chosen as

$$u = x + \|x\|e_1 \qquad (A.49)$$

then

$$Hx = -\|x\| \begin{bmatrix} 1 \\ 0 \\ \vdots \\ 0 \end{bmatrix} \qquad (A.50)$$

From Eqs. (A.47) and (A.49), we see that the transformed vector Hx contains $n - 1$ zeros. Furthermore, since H is an orthogonal matrix, we have

$$\|Hx\|^2 = (Hx)^T Hx = x^T H^T Hx = x^T x = \|x\|^2$$

Therefore, Hx preserves the length of x. For the sake of numerical robustness, a good choice of vector u between Eqs. (A.47) and (A.49) is

$$u = x + \text{sign}(x_1)\|x\|e_1 \qquad (A.51)$$

because the alternative choice, $u = x - \text{sign}(x_1)\|x\|e_1$, may yield a vector u whose magnitude becomes too small when x is close to a multiple of e_1.

Given a matrix $A \in R^{n \times n}$, the matrix product HA is called a *Householder update* of A and it can be evaluated as

$$HA = \left(I - \frac{2uu^T}{\|u\|^2} \right) A = A - uv^T \qquad (A.52)$$

where

$$v = \alpha A^T u, \quad \alpha = -\frac{2}{\|u\|^2} \qquad (A.52)$$

We see that a Household update of A is actually a rank-one correction of A, which can be obtained by using a matrix-vector multiplication and then an outer product update. In this way, a Householder update can be carried out efficiently without requiring matrix multiplication explicitly.

By successively applying the Householder update with appropriate values of u, a given matrix A can be transformed to an upper triangular matrix. To see

612

this, consider a matrix $\mathbf{A} \in R^{n \times n}$ and let $\mathbf{H}_i$ be the ith Householder update such that after $k - 1$ successive applications of $\mathbf{H}_i$ for $i = 1, 2, \ldots, k - 1$ the transformed matrix becomes

$$
\mathbf{A}^{(k-1)} = \mathbf{H}_{k-1} \cdots \mathbf{H}_1 \mathbf{A} =
\begin{bmatrix}
* & * & \cdots & * & & & * \\
0 & * & & * & & & \\
& & \ddots & & & & \\
0 & 0 & & * & & & \\
\hdashline
0 & 0 & \cdots & 0 & & & \\
\vdots & \vdots & \cdots & \vdots & \mathbf{a}_k^{(k-1)} & \cdots & \mathbf{a}_n^{(k-1)} \\
0 & 0 & \cdots & 0 & & &
\end{bmatrix}
\tag{A.53}
$$

The next Householder update is characterized by

$$
\mathbf{H}_k = \mathbf{I} - 2 \frac{\mathbf{u}_k \mathbf{u}_k^T}{\|\mathbf{u}_k\|^2} =
\begin{bmatrix} \mathbf{I}_{k-1} & \mathbf{0} \\ \mathbf{0} & \tilde{\mathbf{H}}_k \end{bmatrix}
\tag{A.54}
$$

where

$$
\mathbf{u}_k =
\begin{bmatrix} 0 \\ \vdots \\ 0 \\ \mathbf{u}_k^{(k-1)} \end{bmatrix}
\Big\} k - 1
\quad, \quad
\mathbf{u}_k^{(k-1)} = \mathbf{a}_k^{(k-1)} + \mathrm{sign}[\mathbf{a}_k^{(k-1)}(1)] \|\mathbf{a}_k^{(k-1)}\| \mathbf{e}_1
$$

$$
\tilde{\mathbf{H}}_k = \mathbf{I}_{n-k+1} - 2 \frac{\mathbf{u}_k^{(k-1)} (\mathbf{u}_k^{(k-1)})^T}{\|\mathbf{u}_k^{(k-1)}\|^2}
$$

and $\mathbf{a}_k^{(k-1)}(1)$ represents the first component of vector $\mathbf{a}_k^{(k-1)}$.

Evidently, premultiplying $\mathbf{A}^{(k-1)}$ by $\mathbf{H}_k$ alters only the lower right block of $\mathbf{A}^{(k-1)}$ in Eq. (A.53) thereby converting its first column $\mathbf{a}_k^{(k-1)}$ to $[* \, 0 \, \cdots \, 0]^T$. Proceeding in this way, all the entries in the lower triangle will become zero.

Example A.9 Applying a series of Householder transformations, reduce matrix

$$
\mathbf{A} = \begin{bmatrix} 1 & 0 & 3 & -1 \\ -1 & 2 & -7 & 5 \\ 3 & 1 & 7 & -1 \\ 0 & -1 & 2 & -2 \end{bmatrix}
$$

to an upper triangular matrix.

Solution Using Eq. (A.51), we compute vector $\mathbf{u}_1$ as

$$\mathbf{u}_1 = \begin{bmatrix} 1 \\ -1 \\ 3 \\ 0 \end{bmatrix} + \sqrt{11} \begin{bmatrix} 1 \\ 0 \\ 0 \\ 0 \end{bmatrix} = \begin{bmatrix} 1 + \sqrt{11} \\ -1 \\ 3 \\ 0 \end{bmatrix}$$

The associated Householder transformation is given by

$$\mathbf{H} = \mathbf{I} - \frac{2\mathbf{u}_1\mathbf{u}_1^T}{\|\mathbf{u}_1\|^2} = \begin{bmatrix} -0.3015 & 0.3015 & -0.9045 & 0 \\ 0.3015 & 0.9302 & 0.2095 & 0 \\ -0.9045 & 0.2095 & 0.3714 & 0 \\ 0 & 0 & 0 & 1 \end{bmatrix}$$

The first Householder update is found to be

$$\mathbf{A}^{(1)} = \mathbf{H}_1\mathbf{A} = \begin{bmatrix} -3.3166 & -0.3015 & -9.3469 & 2.7136 \\ 0 & 2.0698 & -4.1397 & 4.1397 \\ 0 & 0.7905 & -1.5809 & 1.5809 \\ 0 & -1 & 2 & -2 \end{bmatrix}$$

From Eq. (A.53), we obtain

$$\mathbf{a}_2^{(1)} = \begin{bmatrix} 2.0698 \\ 0.7905 \\ -1 \end{bmatrix}$$

Using Eq. (A.54), we can compute

$$\mathbf{u}_2^{(1)} = \begin{bmatrix} 4.5007 \\ 0.7905 \\ -1 \end{bmatrix}$$

$$\tilde{\mathbf{H}}_2 = \begin{bmatrix} -0.8515 & -0.3252 & 0.4114 \\ -0.3252 & 0.9429 & 0.0722 \\ 0.4114 & 0.0722 & 0.9086 \end{bmatrix}$$

and

$$\mathbf{H}_2 = \begin{bmatrix} 1 & \mathbf{0} \\ \mathbf{0} & \tilde{\mathbf{H}}_2 \end{bmatrix}$$

By premultiplying matrix $\mathbf{H}_1\mathbf{A}$ by $\mathbf{H}_2$, we obtain the required upper triangular matrix in terms of the second Householder update as

$$\mathbf{H}_2\mathbf{H}_1\mathbf{A} = \begin{bmatrix} -3.3166 & -0.3015 & -9.3469 & 2.7136 \\ 0 & -2.4309 & 4.8617 & -4.8617 \\ 0 & 0 & 0 & 0 \\ 0 & 0 & 0 & 0 \end{bmatrix}$$

614

A.11.2   Givens rotations

Givens rotations are rank-two corrections of the identity matrix and are characterized by

$$
\mathbf{G}_{ik}(\theta) =
\begin{bmatrix}
1 & \cdots & 0 & \cdots & 0 & \cdots & 0 \\
\vdots & \ddots & \vdots & & \vdots & & \vdots \\
0 & \cdots & c & \cdots & s & \cdots & 0 \\
\vdots & & \vdots & \ddots & \vdots & & \vdots \\
0 & \cdots & -s & \cdots & c & \cdots & 0 \\
\vdots & & \vdots & & \vdots & \ddots & \vdots \\
0 & \cdots & 0 & \cdots & 0 & \cdots & 1
\end{bmatrix}
\begin{matrix} \\ \\ i \\ \\ k \\ \\ \\ \end{matrix}
$$

$$
\begin{matrix} & i & & k \end{matrix}
$$

for $1 \le i,\ k \le n$, where $c = \cos\theta$ and $s = \sin\theta$ for some θ. It can be verified that $\mathbf{G}_{ik}(\theta)$ is an orthogonal matrix and $\mathbf{G}_{ik}^T(\theta)\mathbf{x}$ only affects the ith and kth components of vector $\mathbf{x}$, i.e.,

$$
\mathbf{y} = \mathbf{G}_{ik}^T(\theta)\mathbf{x} \quad \text{with} \quad y_l =
\begin{cases}
cx_i - sx_k & \text{for } l = i \\
sx_i + cx_k & \text{for } l = k \\
x_l & \text{otherwise}
\end{cases}
$$

By choosing an appropriate θ such that

$$
sx_i + cx_k = 0 \tag{A.55}
$$

the kth component of vector $\mathbf{y}$ is forced to zero. A numerically stable method for determining suitable values for s and c in Eq. (A.55) is described below, where we denote x_i and x_k as a and b, respectively.

   (a) If $b = 0$, set $c = 1,\ s = 0$.
   (b) If $b \ne 0$, then
      (i) if $|b| > |a|$, set

$$
\tau = -\frac{a}{b}, \quad s = \frac{1}{\sqrt{1+\tau^2}}, \quad c = \tau s
$$

   (ii) otherwise, if $|b| \le |a|$, set

$$
\tau = -\frac{b}{a}, \quad c = \frac{1}{\sqrt{1+\tau^2}}, \quad s = \tau c
$$

Note that when premultiplying matrix $\mathbf{A}$ by $\mathbf{G}_{ik}^T(\theta)$, matrix $\mathbf{G}_{ik}^T(\theta)\mathbf{A}$ alters only the ith and kth rows of $\mathbf{A}$. The application of Givens rotations is illustrated by the following example.

Appendix A: Basics of Linear Algebra

Example A.10 Convert matrix $\mathbf{A}$ given by

$$\mathbf{A} = \begin{bmatrix} 3 & -1 \\ -3 & 5 \\ 2 & 1 \end{bmatrix}$$

into an upper triangular matrix by premultiplying it by an orthogonal transformation matrix that can be obtained using Givens rotations.

Solution To handle the first column, we first use $\mathbf{G}_{2,3}^T(\theta)$ to force its last component to zero. In this case, $a = -3$ and $b = 2$, hence

$$\tau = \frac{2}{3}, \quad c = \frac{1}{\sqrt{1+\tau^2}} = 0.8321, \quad \text{and} \quad s = \tau c = 0.5547$$

Therefore, matrix $\mathbf{G}_{2,3}(\theta_1)$ is given by

$$\mathbf{G}_{2,3}(\theta_1) = \begin{bmatrix} 1 & 0 & 0 \\ 0 & 0.8321 & 0.5547 \\ 0 & -0.5547 & 0.8321 \end{bmatrix}$$

which leads to

$$\mathbf{G}_{2,3}^T(\theta_1)\mathbf{A} = \begin{bmatrix} 3 & -1 \\ -3.6056 & 3.6056 \\ 0 & 3.6056 \end{bmatrix}$$

In order to apply $\mathbf{G}_{1,2}^T(\theta_2)$ to the resulting matrix to force the second component of its first column to zero, we note that $a = 3$ and $b = -3.6056$; hence

$$\tau = \frac{3}{3.6056}, \quad s = \frac{1}{\sqrt{1+\tau^2}} = 0.7687, \quad \text{and} \quad c = \tau s = 0.6396$$

Therefore, matrix $\mathbf{G}_{1,2}(\theta_2)$ is given by

$$\mathbf{G}_{1,2}(\theta_2) = \begin{bmatrix} 0.6396 & 0.7687 & 0 \\ -0.7687 & 0.6396 & 0 \\ 0 & 0 & 1 \end{bmatrix}$$

and

$$\mathbf{G}_{1,2}^T(\theta_2)\mathbf{G}_{2,3}^T(\theta_1)\mathbf{A} = \begin{bmatrix} 4.6904 & -3.4112 \\ 0 & 1.5374 \\ 0 & 3.6056 \end{bmatrix}$$

Now we can force the last component of the second column of the resulting matrix to zero by applying $\mathbf{G}_{2,3}^T(\theta_3)$. With $a = 1.5374$ and $b = 3.6056$, we compute

$$\tau = \frac{1.5374}{3.6056}, \quad s = \frac{1}{\sqrt{1+\tau^2}} = 0.9199, \quad \text{and} \quad c = \tau s = 0.3922$$

Therefore, matrix $\mathbf{G}_{2,3}(\theta_3)$ is given by

$$\mathbf{G}_{2,3}(\theta_3) = \begin{bmatrix} 1 & 0 & 0 \\ 0 & -0.3922 & 0.9199 \\ 0 & -0.9199 & -0.3922 \end{bmatrix}$$

which yields

$$\mathbf{G}_{2,3}^T(\theta_3)\mathbf{G}_{1,2}^T(\theta_2)\mathbf{G}_{2,3}^T(\theta_1)\mathbf{A} = \begin{bmatrix} 4.6904 & -3.4112 \\ 0 & -3.9196 \\ 0 & 0 \end{bmatrix}$$

∎

A.12   QR Decomposition
A.12.1   Full-rank case

A QR decomposition of a matrix $\mathbf{A} \in R^{m \times n}$ is given by

$$\mathbf{A} = \mathbf{QR} \qquad \text{(A.56)}$$

where $\mathbf{Q} \in R^{m \times m}$ is an orthogonal matrix and $\mathbf{R} \in R^{m \times n}$ is an upper triangular matrix.

In general, more than one QR decompositions exist. For example, if $\mathbf{A} = \mathbf{QR}$ is a QR decomposition of $\mathbf{A}$, then $\mathbf{A} = \tilde{\mathbf{Q}}\tilde{\mathbf{R}}$ is also a QR decomposition of $\mathbf{A}$ if $\tilde{\mathbf{Q}} = \mathbf{Q}\tilde{\mathbf{I}}$ and $\tilde{\mathbf{R}} = \tilde{\mathbf{I}}\mathbf{R}$ and $\tilde{\mathbf{I}}$ is a diagonal matrix whose diagonal comprises a mixture of 1's and -1's. Obviously, $\tilde{\mathbf{Q}}$ remains orthogonal and $\tilde{\mathbf{R}}$ is a triangular matrix but the signs of the rows in $\tilde{\mathbf{R}}$ corresponding to the -1's in $\tilde{\mathbf{I}}$ are changed compared with those in $\mathbf{R}$.

For the sake of convenience, we assume in the rest of this section that $m \geq n$. This assumption implies that $\mathbf{R}$ has the form

$$\mathbf{R} = \begin{bmatrix} \hat{\mathbf{R}} \\ \mathbf{0} \end{bmatrix} \begin{matrix} \}n \text{ rows} \\ \}m - n \text{ rows} \end{matrix}$$

$$\underbrace{\qquad\qquad}_{n \text{ columns}}$$

where $\hat{\mathbf{R}}$ is an upper triangular square matrix of dimension n, and that Eq. (A.56) can be expressed as

$$\mathbf{A} = \hat{\mathbf{Q}}\hat{\mathbf{R}} \qquad \text{(A.57)}$$

where $\hat{\mathbf{Q}}$ is the matrix formed by the first n columns of $\mathbf{Q}$. Now if we let $\hat{\mathbf{Q}} = [\mathbf{q}_1 \ \mathbf{q}_2 \ \cdots \ \mathbf{q}_n]$, Eq. (A.57) yields

$$\mathbf{Ax} = \hat{\mathbf{Q}}\hat{\mathbf{R}}\mathbf{x} = \hat{\mathbf{Q}}\hat{\mathbf{x}} = \sum_{i=1}^{n} \hat{x}_i \mathbf{q}_i$$

In other words, if $\mathbf{A}$ has full column rank n, then the first n columns in $\mathbf{Q}$ form an orthogonal basis for the range of $\mathbf{A}$, i.e., $\mathcal{R}(\mathbf{A})$.

As discussed in Sec. A.11.1, a total of n successive applications of the Householder transformation can convert matrix $\mathbf{A}$ into an upper triangular matrix, $\mathbf{R}$, i.e.,

$$\mathbf{H}_n \cdots \mathbf{H}_2 \mathbf{H}_1 \mathbf{A} = \mathbf{R} \tag{A.58}$$

Since each $\mathbf{H}_i$ in Eq. (A.58) is orthogonal, we obtain

$$\mathbf{A} = (\mathbf{H}_n \cdots \mathbf{H}_2 \mathbf{H}_1)^T \mathbf{R} = \mathbf{Q}\mathbf{R} \tag{A.59}$$

where $\mathbf{Q} = (\mathbf{H}_n \cdots \mathbf{H}_2 \mathbf{H}_1)^T$ is an orthogonal matrix and, therefore, Eqs. (A.58) and (A.59) yield a QR decomposition of $\mathbf{A}$. This method requires $n^2(m - n/3)$ multiplications [3].

An alternative approach for obtaining a QR decomposition is to apply Givens rotations as illustrated in Sec. A.11.2. For a general matrix $\mathbf{A} \in R^{m \times n}$ with $m \geq n$, a total of $mn - n(n + 1)/2$ Givens rotations are required to convert $\mathbf{A}$ into an upper triangular matrix and this Givens-rotation-based algorithm requires $1.5n^2(m - n/3)$ multiplications [3].

A.12.2 QR decomposition for rank-deficient matrices

If the rank of a matrix $\mathbf{A} \in R^{m \times n}$ where $m \geq n$ is less than n, then there is at least one zero component in the diagonal of $\mathbf{R}$ in Eq. (A.56). In such a case, the conventional QR decomposition discussed in Sec. A.12.1 does not always produce an orthogonal basis for $\mathcal{R}(\mathbf{A})$. For such rank-deficient matrices, however, the Householder-transformation-based QR decomposition described in Sec. A.12.1 can be modified as

$$\mathbf{A}\mathbf{P} = \mathbf{Q}\mathbf{R} \tag{A.60}$$

where $\text{rank}(\mathbf{A}) = r < n$, $\mathbf{Q} \in R^{m \times m}$ is an orthogonal matrix,

$$\mathbf{R} = \begin{bmatrix} \mathbf{R}_{11} & \mathbf{R}_{12} \\ 0 & 0 \end{bmatrix} \tag{A.60}$$

where $\mathbf{R}_{11} \in R^{r \times r}$ is a triangular and nonsingular matrix, and $\mathbf{P} \in R^{n \times n}$ assumes the form

$$\mathbf{P} = [\mathbf{e}_{s_1} \ \mathbf{e}_{s_2} \ \cdots \ \mathbf{e}_{s_n}]$$

where $\mathbf{e}_{s_i}$ denotes the s_ith column of the $n \times n$ identity matrix and index set $\{s_1, s_2, \ldots, s_n\}$ is a permutation of $\{1, 2, \ldots, n\}$. Such a matrix is said to be *a permutation matrix* [1].

To illustrate how Eq. (A.60) is obtained, assume that $k - 1$ (with $k - 1 < r$) Householder transformations and permutations have been applied to $\mathbf{A}$ to obtain

$$\mathbf{R}^{(k-1)} = (\mathbf{H}_{k-1} \cdots \mathbf{H}_2 \mathbf{H}_1)\mathbf{A}(\mathbf{P}_1 \mathbf{P}_2 \cdots \mathbf{P}_{k-1})$$

$$= \begin{bmatrix} \mathbf{R}_{11}^{(k-1)} & \mathbf{R}_{12}^{(k-1)} \\ \mathbf{0} & \mathbf{R}_{22}^{(k-1)} \end{bmatrix} \begin{matrix} \}k-1 \\ \}m-k+1 \end{matrix} \qquad (\text{A.61})$$

$$\underbrace{}_{k-1} \quad \underbrace{}_{n-k+1}$$

where $\mathbf{R}_{11}^{(k-1)} \in R^{(k-1)\times(k-1)}$ is upper triangular and $\text{rank}(\mathbf{R}_{11}^{(k-1)}) = k-1$. Since $\text{rank}(\mathbf{A}) = r$, block $\mathbf{R}_{22}^{(k-1)}$ is nonzero. Now we postmultiply Eq. (A.61) by a permutation matrix $\mathbf{P}_k$ which rearranges the last $n - k + 1$ columns of $\mathbf{R}^{(k-1)}$ such that the column in $\mathbf{R}_{22}^{(k-1)}$ with the largest L_2 norm becomes its first column. A Householder matrix $\mathbf{H}_k$ is then applied to obtain

$$\mathbf{H}_k \mathbf{R}^{(k-1)} \mathbf{P}_k = \begin{bmatrix} \mathbf{R}_{11}^{(k)} & \mathbf{R}_{12}^{(k)} \\ \mathbf{0} & \mathbf{R}_{22}^{(k)} \end{bmatrix} \begin{matrix} \}k \\ \}m-k \end{matrix}$$

$$\underbrace{}_{k} \quad \underbrace{}_{n-k}$$

where $\mathbf{R}_{11}^{(k)} \in R^{k\times k}$ is an upper triangular nonsingular matrix. If $r = k$, then $\mathbf{R}_{22}^{(k)}$ must be a zero matrix since $\text{rank}(\mathbf{A}) = r$; otherwise, $\mathbf{R}_{22}^{(k)}$ is a nonzero block, and we proceed with postmultiplying $\mathbf{R}^{(k)}$ by a new permutation matrix $\mathbf{P}_{k+1}$ and then premultiplying by a Householder matrix $\mathbf{H}_{k+1}$. This procedure is continued until the modified QR decomposition in Eq. (A.60) is obtained where

$$\mathbf{Q} = (\mathbf{H}_r \mathbf{H}_{r-1} \cdots \mathbf{H}_1)^T \qquad \text{and} \qquad \mathbf{P} = \mathbf{P}_1 \mathbf{P}_2 \cdots \mathbf{P}_r$$

The decomposition in Eq. (A.60) is called the *QR decomposition of matrix* **A** with *column pivoting*. It follows from Eq. (A.60) that the first r columns of matrix **Q** form an orthogonal basis for the range of **A**.

Example A.11 Find a QR decomposition of the matrix

$$\mathbf{A} = \begin{bmatrix} 1 & 0 & 3 & -1 \\ -1 & 2 & -7 & 5 \\ 3 & 1 & 7 & -1 \\ 0 & -1 & 2 & 2 \end{bmatrix}$$

Solution In Example A.9, two Householder transformation matrices

$$\mathbf{H}_1 = \begin{bmatrix} -0.3015 & 0.3015 & -0.9045 & 0 \\ 0.3015 & 0.9302 & 0.2095 & 0 \\ -0.9045 & 0.2095 & 0.3714 & 0 \\ 0 & 0 & 0 & 1 \end{bmatrix}$$

$$\mathbf{H}_2 = \begin{bmatrix} 1 & 0 & 0 & 0 \\ 0 & -0.8515 & -0.3252 & 0.4114 \\ 0 & -0.3252 & 0.9429 & 0.0722 \\ 0 & 0.4114 & 0.0722 & 0.9086 \end{bmatrix}$$

were obtained that reduce matrix $\mathbf{A}$ to the upper triangular matrix

$$\mathbf{R} = \mathbf{H}_2\mathbf{H}_1\mathbf{A} = \begin{bmatrix} -3.3166 & -0.3015 & -9.3469 & 2.7136 \\ 0 & -2.4309 & 4.8617 & -4.8617 \\ 0 & 0 & 0 & 0 \\ 0 & 0 & 0 & 0 \end{bmatrix}$$

Therefore, a QR decomposition of $\mathbf{A}$ can be obtained as $\mathbf{A} = \mathbf{QR}$ where $\mathbf{R}$ is the above upper triangular matrix.

$$\mathbf{Q} = (\mathbf{H}_2\mathbf{H}_1)^{-1} = \mathbf{H}_1^T\mathbf{H}_2^T = \begin{bmatrix} -0.3015 & 0.0374 & -0.9509 & 0.0587 \\ 0.3015 & -0.8602 & -0.1049 & 0.3978 \\ -0.9045 & -0.2992 & 0.2820 & 0.1130 \\ 0 & 0.4114 & 0.0722 & 0.9086 \end{bmatrix}$$

∎

A.13 Cholesky Decomposition

For a symmetric positive-definite matrix $\mathbf{A} \in \mathcal{R}^{n \times n}$, there exists a unique lower triangular matrix $\mathbf{G} \in R^{n \times n}$ with positive diagonal components such that

$$\mathbf{A} = \mathbf{GG}^T \tag{A.62}$$

The decomposition in Eq. (A.62) is known as the *Cholesky decomposition* and matrix $\mathbf{G}$ as the *Cholesky triangle*.

One of the methods that can be used to obtain the Cholesky decomposition of a given positive-definite matrix is based on the use of the outer-product updates [1] as illustrated below.

A positive-definite matrix $\mathbf{A} \in R^{n \times n}$ can be expressed as

$$\mathbf{A} = \begin{bmatrix} a_{11} & \mathbf{u}^T \\ \mathbf{u} & \mathbf{B} \end{bmatrix} \tag{A.63}$$

where a_{11} is a positive number. It can be readily verified that with

$$\mathbf{T} = \begin{bmatrix} \frac{1}{\sqrt{a_{11}}} & \mathbf{0} \\ -\mathbf{u}/a_{11} & \mathbf{I}_{n-1} \end{bmatrix} \tag{A.64}$$

we have

$$\mathbf{TAT}^T = \begin{bmatrix} 1 & \mathbf{0} \\ \mathbf{0} & \mathbf{B} - \mathbf{uu}^T/a_{11} \end{bmatrix} \equiv \begin{bmatrix} 1 & \mathbf{0} \\ \mathbf{0} & \mathbf{A}_1 \end{bmatrix} \tag{A.65}$$

which implies that

$$\mathbf{A} = \begin{bmatrix} \sqrt{a_{11}} & \mathbf{0} \\ \mathbf{u}/\sqrt{a_{11}} & \mathbf{I}_{n-1} \end{bmatrix} \begin{bmatrix} 1 & \mathbf{0} \\ \mathbf{0} & \mathbf{B} - \mathbf{uu}^T/a_{11} \end{bmatrix} \begin{bmatrix} \sqrt{a_{11}} & \mathbf{u}/\sqrt{a_{11}} \\ \mathbf{0} & \mathbf{I}_{n-1} \end{bmatrix}$$
$$\equiv \mathbf{G}_1 \begin{bmatrix} 1 & \mathbf{0} \\ \mathbf{0} & \mathbf{A}_1 \end{bmatrix} \mathbf{G}_1^T \tag{A.66}$$

where $\mathbf{G}_1$ is a lower triangular matrix and $\mathbf{A}_1 = \mathbf{B} - \mathbf{uu}^T/a_{11}$ is an $(n-1) \times (n-1)$ symmetric matrix. Since $\mathbf{A}$ is positive definite and $\mathbf{T}$ is nonsingular, it follows from Eq. (A.65) that matrix $\mathbf{A}_1$ is positive definite; hence the above procedure can be applied to matrix $\mathbf{A}_1$. In other words, we can find an $(n-1) \times (n-1)$ lower triangular matrix $\mathbf{G}_2$ such that

$$\mathbf{A}_1 = \mathbf{G}_2 \begin{bmatrix} 1 & \mathbf{0} \\ \mathbf{0} & \mathbf{A}_2 \end{bmatrix} \mathbf{G}_2^T \tag{A.67}$$

where $\mathbf{A}_2$ is an $(n-2) \times (n-2)$ positive-definite matrix. By combining Eqs. (A.66) and (A.67), we obtain

$$\mathbf{A} = \begin{bmatrix} \sqrt{a_{11}} & \mathbf{0} \\ \mathbf{u}/\sqrt{a_{11}} & \mathbf{G}_2 \end{bmatrix} \begin{bmatrix} \mathbf{I}_2 & \mathbf{0} \\ \mathbf{0} & \mathbf{A}_2 \end{bmatrix} \begin{bmatrix} \sqrt{a_{11}} & \mathbf{u}^T/\sqrt{a_{11}} \\ \mathbf{0} & \mathbf{G}_2^T \end{bmatrix}$$
$$\equiv \mathbf{G}_{12} \begin{bmatrix} \mathbf{I}_2 & \mathbf{0} \\ \mathbf{0} & \mathbf{A}_2 \end{bmatrix} \mathbf{G}_{12}^T \tag{A.68}$$

where $\mathbf{I}_2$ is the 2×2 identity matrix and $\mathbf{G}_{12}$ is lower triangular. The above procedure is repeated until the second matrix at the right-hand side of Eq. (A.68) is reduced to the identity matrix $\mathbf{I}_n$. The Cholesky decomposition of $\mathbf{A}$ is then obtained.

Example A.12 Compute the Cholesky triangle of the positive-definite matrix

$$\mathbf{A} = \begin{bmatrix} 4 & -2 & 1 \\ -2 & 7 & -1 \\ 1 & -1 & 1 \end{bmatrix}$$

Solution From Eq. (A.66), we obtain

$$\mathbf{G}_1 = \begin{bmatrix} 2 & 0 & 0 \\ -1 & 1 & 0 \\ 0.5 & 0 & 1 \end{bmatrix}$$

and

$$\mathbf{A}_1 = \begin{bmatrix} 7 & -1 \\ -1 & 1 \end{bmatrix} - \frac{1}{4} \begin{bmatrix} -2 \\ 1 \end{bmatrix} [2\ 1] = \begin{bmatrix} 6 & -0.50 \\ -0.50 & 0.75 \end{bmatrix}$$

Now working on matrix $\mathbf{A}_1$, we get

$$\mathbf{G}_2 = \begin{bmatrix} \sqrt{6} & 0 \\ -0.5/\sqrt{6} & 1 \end{bmatrix}$$

and

$$\mathbf{A}_2 = 0.75 - (-0.5)^2/6 = 0.7083$$

In this case, Eq. (A.66) becomes

$$\mathbf{A} = \begin{bmatrix} 2 & 0 & 0 \\ -1 & \sqrt{6} & 0 \\ 0.5 & -0.5/\sqrt{6} & 1 \end{bmatrix} \begin{bmatrix} 1 & 0 & 0 \\ 0 & 1 & 0 \\ 0 & 0 & 0.7083 \end{bmatrix} \begin{bmatrix} 2 & 0 & 0 \\ -1 & \sqrt{6} & 0 \\ 0.5 & -0.5/\sqrt{6} & 1 \end{bmatrix}^T$$

Finally, we use

$$\mathbf{G}_3 = \sqrt{0.7083} \approx 0.8416$$

to reduce $\mathbf{A}_2$ to $\mathbf{A}_3 = 1$, which leads to the Cholesky triangle

$$\mathbf{G} = \begin{bmatrix} 2 & 0 & 0 \\ -1 & \sqrt{6} & 0 \\ 0.5 & -0.5/\sqrt{6} & \sqrt{0.7083} \end{bmatrix} \approx \begin{bmatrix} 2 & 0 & 0 \\ -1 & 2.4495 & 0 \\ 0.5 & -0.2041 & 0.8416 \end{bmatrix}$$

∎

A.14 Kronecker Product

Let $\mathbf{A} \in R^{p \times m}$ and $\mathbf{B} \in R^{q \times n}$. The Kronecker product of $\mathbf{A}$ and $\mathbf{B}$, denoted as $\mathbf{A} \otimes \mathbf{B}$, is a $pq \times mn$ matrix defined by

$$\mathbf{A} \otimes \mathbf{B} = \begin{bmatrix} a_{11}\mathbf{B} & \cdots & a_{1m}\mathbf{B} \\ \vdots & & \vdots \\ a_{p1}\mathbf{B} & \cdots & a_{pm}\mathbf{B} \end{bmatrix} \tag{A.69}$$

where a_{ij} denotes the (i, j)th component of $\mathbf{A}$ [5]. It can be verified that

(i) $(\mathbf{A} \otimes \mathbf{B})^T = \mathbf{A}^T \otimes \mathbf{B}^T$

(ii) $(\mathbf{A} \otimes \mathbf{B}) \cdot (\mathbf{C} \otimes \mathbf{D}) = \mathbf{A}\mathbf{C} \otimes \mathbf{B}\mathbf{D}$ where $\mathbf{C} \in R^{m \times r}$ and $\mathbf{D} \in R^{n \times s}$

(iii) If $p = m$, $q = n$, and $\mathbf{A}$, $\mathbf{B}$ are nonsingular, then

$$(\mathbf{A} \otimes \mathbf{B})^{-1} = \mathbf{A}^{-1} \otimes \mathbf{B}^{-1}$$

(iv) If $\mathbf{A} \in R^{m \times m}$ and $\mathbf{B} \in R^{n \times n}$, then the eigenvalues of $\mathbf{A} \otimes \mathbf{B}$ and $\mathbf{A} \otimes \mathbf{I}_n + \mathbf{I}_m \otimes \mathbf{B}$ are $\lambda_i \mu_j$ and $\lambda_i + \mu_j$, respectively, for $i = 1, 2, \ldots, m$ and $j = 1, 2, \ldots, n$, where λ_i and μ_j are the ith and jth eigenvalues of $\mathbf{A}$ and $\mathbf{B}$, respectively.

The Kronecker product is useful when we are dealing with matrix variables. If we use nvec($\mathbf{X}$) to denote the column vector obtained by stacking the column vectors of matrix $\mathbf{X}$, then it is easy to verify that for $\mathbf{M} \in R^{p \times m}$, $\mathbf{N} \in R^{q \times n}$ and $\mathbf{X} \in R^{n \times m}$, we have

$$\text{nvec}(\mathbf{NXM}^T) = (\mathbf{M} \otimes \mathbf{N})\text{nvec}(\mathbf{X}) \tag{A.70}$$

In particular, if $p = m = q = n$, $\mathbf{N} = \mathbf{A}^T$, and $\mathbf{M} = \mathbf{I}_n$, then Eq. (A.70) becomes

$$\text{nvec}(\mathbf{A}^T\mathbf{X}) = (\mathbf{I}_n \otimes \mathbf{A}^T)\text{nvec}(\mathbf{X}) \tag{A.71}$$

Similarly, we have

$$\text{nvec}(\mathbf{XA}) = (\mathbf{A}^T \otimes \mathbf{I}_n)\text{nvec}(\mathbf{X}) \tag{A.71}$$

For example, we can apply Eq. (A.71) to the Lyapunov equation [5]

$$\mathbf{A}^T\mathbf{P} + \mathbf{PA} = -\mathbf{Q} \tag{A.72}$$

where matrices $\mathbf{A}$ and $\mathbf{Q}$ are given and $\mathbf{Q}$ is positive definite. First, we write Eq. (A.72) in vector form as

$$\text{nvec}(\mathbf{A}^T\mathbf{P}) + \text{nvec}(\mathbf{PA}) = -\text{nvec}(\mathbf{Q}) \tag{A.73}$$

Using Eq. (A.71), Eq. (A.73) becomes

$$(\mathbf{I}_n \otimes \mathbf{A}^T)\text{nvec}(\mathbf{P}) + (\mathbf{A}^T \otimes \mathbf{I}_n)\text{nvec}(\mathbf{P}) = -\text{nvec}(\mathbf{Q})$$

which can be solved to obtain nvec($\mathbf{P}$) as

$$\text{nvec}(\mathbf{P}) = -(\mathbf{I}_n \otimes \mathbf{A}^T + \mathbf{A}^T \otimes \mathbf{I}_n)^{-1}\text{nvec}(\mathbf{Q}) \tag{A.74}$$

Example A.13 Solve the Lyapunov equation

$$\mathbf{A}^T\mathbf{P} + \mathbf{PA} = -\mathbf{Q}$$

for matrix $\mathbf{P}$ where

$$\mathbf{A} = \begin{bmatrix} -2 & -2 \\ 1 & 0 \end{bmatrix} \quad \text{and} \quad \mathbf{Q} = \begin{bmatrix} 1 & -1 \\ -1 & 2 \end{bmatrix}$$

Solution From Eq. (A.69), we compute

$$\mathbf{I}_2 \otimes \mathbf{A}^T + \mathbf{A}^T \otimes \mathbf{I}_2 = \begin{bmatrix} -4 & 1 & 1 & 0 \\ -2 & -2 & 0 & 1 \\ -2 & 0 & -2 & 1 \\ 0 & -2 & -2 & 0 \end{bmatrix}$$

Since

$$\text{nvec}(\mathbf{Q}) = \begin{bmatrix} 1 \\ -1 \\ -1 \\ 2 \end{bmatrix}$$

Eq. (A.74) gives

$$\text{nvec}(\mathbf{P}) = -(\mathbf{I}_2 \otimes \mathbf{A}^T + \mathbf{A}^T \otimes \mathbf{I}_2)^{-1}\text{nvec}(\mathbf{Q})$$

$$= - \begin{bmatrix} -4 & 1 & 1 & 0 \\ -2 & -2 & 0 & 1 \\ -2 & 0 & -2 & 1 \\ 0 & -2 & -2 & 0 \end{bmatrix}^{-1} \begin{bmatrix} 1 \\ -1 \\ -1 \\ 2 \end{bmatrix} = \begin{bmatrix} 0.5 \\ 0.5 \\ 0.5 \\ 3 \end{bmatrix}$$

from which we obtain

$$\mathbf{P} = \begin{bmatrix} 0.5 & 0.5 \\ 0.5 & 3 \end{bmatrix}$$

∎

A.15 Vector Spaces of Symmetric Matrices

Let $\mathcal{S}^n$ be the vector space of real symmetric $n \times n$ matrices. As in the n-dimensional Euclidean space where the inner product is defined for two vectors, the *inner product* for matrices $\mathbf{A}$ and $\mathbf{B}$ in $\mathcal{S}^n$ is defined as

$$\mathbf{A} \cdot \mathbf{B} = \text{trace}(\mathbf{AB})$$

If $\mathbf{A} = (a_{ij})$ and $\mathbf{B} = (b_{ij})$, then we have

$$\mathbf{A} \cdot \mathbf{B} = \text{trace}(\mathbf{AB}) = \sum_{i=1}^{n}\sum_{j=1}^{n} a_{ij}b_{ij} \qquad (A.75)$$

The norm $\|\mathbf{A}\|_{\mathcal{S}^n}$ associated to this inner product is

$$\|\mathbf{A}\|_{\mathcal{S}^n} = \sqrt{\mathbf{A} \cdot \mathbf{A}} = \left[\sum_{i=1}^{n}\sum_{j=1}^{n} a_{ij}^2\right]^{1/2} = \|\mathbf{A}\|_F \qquad (A.76)$$

where $\|\mathbf{A}\|_F$ denotes the Frobenius norm of $\mathbf{A}$ (see Sec. A.8.2).

An important set in space $\mathcal{S}^n$ is the set of all positive-semidefinite matrices given by

$$\mathcal{P} = \{\mathbf{X} : \mathbf{X} \in \mathcal{S}^n \text{ and } \mathbf{X} \succeq \mathbf{0}\} \qquad (A.77)$$

A set $\mathcal{K}$ in a vector space is said to be a *convex cone* if $\mathcal{K}$ is a convex set such that $\mathbf{v} \in \mathcal{K}$ implies $\alpha\mathbf{v} \in \mathcal{K}$ for any nonnegative scalar α. It is easy to verify that set $\mathcal{P}$ forms a convex cone in space $\mathcal{S}^n$.

Let matrices $\mathbf{X}$ and $\mathbf{S}$ be two components of $\mathcal{P}$, i.e., $\mathbf{X} \succeq \mathbf{0}$ and $\mathbf{S} \succeq \mathbf{0}$. The eigendecomposition of $\mathbf{X}$ gives

$$\mathbf{X} = \mathbf{U}\mathbf{\Lambda}\mathbf{U}^T \tag{A.78}$$

where $\mathbf{U} \in R^{n \times n}$ is orthogonal and $\mathbf{\Lambda} = \mathrm{diag}\{\lambda_1, \lambda_2, \ldots, \lambda_n\}$. The decomposition in Eq. (A.78) can be expressed as

$$\mathbf{X} = \sum_{i=1}^{n} \lambda_i \mathbf{u}_i \mathbf{u}_i^T$$

where $\mathbf{u}_i$ denotes the ith column of $\mathbf{U}$. By using the property that

$$\mathrm{trace}(\mathbf{A}\mathbf{B}) = \mathrm{trace}(\mathbf{B}\mathbf{A})$$

(see Eq. (A.20)), we can compute the inner product $\mathbf{X} \cdot \mathbf{S}$ as

$$\mathbf{X} \cdot \mathbf{S} = \mathrm{trace}(\mathbf{X}\mathbf{S}) = \mathrm{trace}\left(\sum_{i=1}^{n} \lambda_i \mathbf{u}_i \mathbf{u}_i^T \mathbf{S}\right) = \sum_{i=1}^{n} \lambda_i \, \mathrm{trace}(\mathbf{u}_i \mathbf{u}_i^T \mathbf{S})$$

$$= \sum_{i=1}^{n} \lambda_i \, \mathrm{trace}(\mathbf{u}_i^T \mathbf{S}\mathbf{u}_i) = \sum_{i=1}^{n} \lambda_i \mu_i \tag{A.79}$$

where $\mu_i = \mathbf{u}_i^T \mathbf{S}\mathbf{u}_i$. Since both $\mathbf{X}$ and $\mathbf{S}$ are positive semidefinite, we have $\lambda_i \geq 0$ and $\mu_i \geq 0$ for $i = 1, 2, \ldots, n$. Therefore, Eq. (A.79) implies that

$$\mathbf{X} \cdot \mathbf{S} \geq 0 \tag{A.80}$$

In other words, the inner product of two positive-semidefinite matrices is always nonnegative.

A further property of the inner product on set $\mathcal{P}$ is that if $\mathbf{X}$ and $\mathbf{S}$ are positive semidefinite and $\mathbf{X} \cdot \mathbf{S} = 0$, then the product matrix $\mathbf{X}\mathbf{S}$ must be the zero matrix, i.e.,

$$\mathbf{X}\mathbf{S} = \mathbf{0} \tag{A.81}$$

To show this, we can write

$$\mathbf{u}_i^T \mathbf{X}\mathbf{S}\mathbf{u}_j = \mathbf{u}_i^T \left(\sum_{k=1}^{n} \lambda_k \mathbf{u}_k \mathbf{u}_k^T\right) \mathbf{S}\mathbf{u}_j = \lambda_i \mathbf{u}_i^T \mathbf{S}\mathbf{u}_j \tag{A.82}$$

Using the Cauchy-Schwartz inequality (see Eq. (A.25)), we have

$$|\mathbf{u}_i^T \mathbf{S}\mathbf{u}_j|^2 = |(\mathbf{S}^{1/2}\mathbf{u}_i)^T (\mathbf{S}^{1/2}\mathbf{u}_j)|^2 \leq \|\mathbf{S}^{1/2}\mathbf{u}_i\|^2 \|\mathbf{S}^{1/2}\mathbf{u}_j\|^2 = \mu_i \mu_j \tag{A.83}$$

Now if $\mathbf{X} \cdot \mathbf{S} = 0$, then Eq. (A.79) implies that

$$\sum_{i=0}^{n} \lambda_i \mu_i = 0 \tag{A.84}$$

Since λ_i and μ_i are all nonnegative, Eq. (A.84) implies that $\lambda_i \mu_i = 0$ for $i = 1, 2, \ldots, n$; hence for each index i, either $\lambda_i = 0$ or $\mu_i = 0$. If $\lambda_i = 0$, Eq. (A.82) gives

$$\mathbf{u}_i^T \mathbf{X} \mathbf{S} \mathbf{u}_j = 0 \tag{A.85}$$

If $\lambda_i \neq 0$, then μ_i must be zero and Eq. (A.83) implies that $\mathbf{u}_i^T \mathbf{S} \mathbf{u}_j = 0$ which, in conjunction with Eq. (A.82) also leads to Eq. (A.85). Since Eq. (A.85) holds for any i and j, we conclude that

$$\mathbf{U}^T \mathbf{X} \mathbf{S} \mathbf{U} = \mathbf{0} \tag{A.86}$$

Since $\mathbf{U}$ is nonsingular, Eq. (A.86) implies Eq. (A.81).

Given $p + 1$ symmetric matrices $\mathbf{F}_0$, $\mathbf{F}_1$, $\ldots$, $\mathbf{F}_p$ in space $\mathcal{S}^n$, and a p-dimensional vector $\mathbf{x} = [x_1 \ x_2 \ \cdots \ x_p]^T$, we can generate a symmetric matrix

$$\mathbf{F}(\mathbf{x}) = \mathbf{F}_0 + x_1 \mathbf{F}_1 + \cdots + x_p \mathbf{F}_p = \mathbf{F}_0 + \sum_{i=1}^{p} x_i \mathbf{F}_i \tag{A.87}$$

which is said to be *affine* with respect to $\mathbf{x}$. Note that if the constant term $\mathbf{F}_0$ were a zero matrix, then $\mathbf{F}(\mathbf{x})$ would be a *linear* function of vector $\mathbf{x}$, i.e., $\mathbf{F}(\mathbf{x})$ would satisfy the condition $\mathbf{F}(\alpha\mathbf{x} + \beta\mathbf{y}) = \alpha\mathbf{F}(\mathbf{x}) + \beta\mathbf{F}(\mathbf{y})$ for any vectors $\mathbf{x}, \mathbf{y} \in R^p$ and any scalars α and β. However, because of the presence of $\mathbf{F}_0$, $\mathbf{F}(\mathbf{x})$ in Eq. (A.87) is *not* linear with respect to $\mathbf{x}$ in a strict sense and the term 'affine' is often used in the literature to describe such a class of matrices. In effect, the affine property is a somewhat relaxed version of the linearity property.

In the context of linear programming, the concept of an *affine manifold* is sometimes encountered. A manifold is a subset of the Euclidean space that satisfies a certain structural property of interest, for example, a set of vectors satisfying the relation $\mathbf{x}^T \mathbf{c} = \beta$. Such a set of vectors may possess the affine property, as illustrated in the following example.

Example A.14 Describe the set of n-dimensional vectors $\{\mathbf{x} : \mathbf{x}^T \mathbf{c} = \beta\}$ for a given vector $\mathbf{c} \in R^{n \times 1}$ and a scalar β as an affine manifold in the n-dimensional Euclidean space E^n.

Solution Obviously, the set of vectors $\{\mathbf{x} : \mathbf{x}^T \mathbf{c} = \beta\}$ is a subset in E^n. If we denote $\mathbf{x} = [x_1 \ x_2 \ \cdots \ x_n]^T$ and $\mathbf{c} = [c_1 \ c_2 \ \cdots \ c_n]^T$, then equation $\mathbf{x}^T \mathbf{c} = \beta$ can be expressed as $F(\mathbf{x}) = 0$ where

$$F(\mathbf{x}) = -\beta + x_1 c_1 + x_2 c_2 + \cdots + x_n c_n \tag{A.88}$$

By viewing $-\beta$, c_1, c_2, $\ldots$, c_n as one-dimensional symmetric matrices, $F(\mathbf{x})$ in Eq. (A.88) assumes the form in Eq. (A.87), which is affine with respect to $\mathbf{x}$. Therefore the set $\{\mathbf{x} : \mathbf{x}^T \mathbf{c} = \beta\}$ is an affine manifold in E^n.

Example A.15 Convert the following constraints

$$\mathbf{X} = (x_{ij}) \succeq \mathbf{0} \qquad \text{for } i, j = 1, 2, 3 \tag{A.89}$$

and

$$x_{ii} = 1 \qquad \text{for } i = 1, 2, 3 \tag{A.89}$$

into a constraint of the type

$$\mathbf{F}(\mathbf{x}) \succeq \mathbf{0} \tag{A.90}$$

for some vector variable $\mathbf{x}$ where $\mathbf{F}(\mathbf{x})$ assumes the form in Eq. (A.87).

Solution The constraints in Eqs. (A.89a) and (A.89b) can be combined into

$$\mathbf{X} = \begin{bmatrix} 1 & x_{12} & x_{13} \\ x_{12} & 1 & x_{23} \\ x_{13} & x_{23} & 1 \end{bmatrix} \succeq \mathbf{0} \tag{A.91}$$

Next we write matrix $\mathbf{X}$ in (A.91) as

$$\mathbf{X} = \mathbf{F}_0 + x_{12}\mathbf{F}_1 + x_{13}\mathbf{F}_2 + x_{23}\mathbf{F}_3$$

where $\mathbf{F}_0 = \mathbf{I}_3$ and

$$\mathbf{F}_1 = \begin{bmatrix} 0 & 1 & 0 \\ 1 & 0 & 0 \\ 0 & 0 & 0 \end{bmatrix}, \quad \mathbf{F}_2 = \begin{bmatrix} 0 & 0 & 1 \\ 0 & 0 & 0 \\ 1 & 0 & 0 \end{bmatrix}, \quad \mathbf{F}_3 = \begin{bmatrix} 0 & 0 & 0 \\ 0 & 0 & 1 \\ 0 & 1 & 0 \end{bmatrix}$$

Hence the constraint in Eq. (A.89) can be expressed in terms of Eq. (A.90) with $\mathbf{F}_i$ given by the above equations and $\mathbf{x} = [x_{12} \ x_{13} \ x_{23}]^T$. ∎

A.16 Polygon, Polyhedron, Polytope, and Convex Hull

A *polygon* is a closed plane figure with an arbitrary number of sides. A polygon is said to be convex if the region inside the polygon is a convex set (see Def. 2.7). A convex polygon with m sides can be described in terms of m linear inequalities which can be expressed in matrix form as

$$\mathcal{P}_y = \{\mathbf{x} : \mathbf{A}\mathbf{x} \geq \mathbf{b}\} \tag{A.92}$$

where $\mathbf{A} \in R^{m \times 2}$, $\mathbf{x} \in R^{2 \times 1}$, and $\mathbf{b} \in R^{m \times 1}$.

A *convex polyhedron* is an n-dimensional extension of a convex polygon. A convex polyhedron can be described by the equation

$$\mathcal{P}_h = \{\mathbf{x} : \mathbf{A}\mathbf{x} \geq \mathbf{b}\} \tag{A.93}$$

where $\mathbf{A} \in R^{m \times n}$, $\mathbf{x} \in R^{n \times 1}$, and $\mathbf{b} \in R^{m \times 1}$. For example, a 3-dimensional convex polyhedron is a 3-dimensional solid which consists of several polygons, usually joined at their edges such as that shown Fig. 11.4.

A polyhedron may or may not be bounded depending on the numerical values of $\mathbf{A}$ and $\mathbf{b}$ in Eq. (A.93). A bounded polyhedron is called a *polytope*.

Given a set of points $S = \{p_1, p_2, \ldots, p_L\}$ in an n-dimensional space, the *convex hull* spanned by S is defined as the smallest convex set that contains S. It can be verified that the convex hull is characterized by

$$\text{Co}\{p_1, p_2, \ldots, p_L\} = \{p : p = \sum_{i=1}^{L} \lambda_i p_i, \ \lambda_i \geq 0, \sum_{i=1}^{L} \lambda_i = 1\} \quad \text{(A.94)}$$

In the above definition, each point p_i represents an abstract n-dimensional point. For example, if point p_i is represented by an n-dimensional vector, say, $\mathbf{v}_i$, then the convex hull spanned by the L vectors $\{\mathbf{v}_1, \mathbf{v}_2, \ldots, \mathbf{v}_L\}$ is given by

$$\text{Co}\{\mathbf{v}_1, \mathbf{v}_2, \ldots, \mathbf{v}_L\} = \{\mathbf{v} : \mathbf{v} = \sum_{i=1}^{L} \lambda_i \mathbf{v}_i, \ \lambda_i \geq 0, \ \sum_{i=1}^{L} \lambda_i = 1\} \quad \text{(A.95)}$$

Alternatively, if point p_i is represented by a pair of matrices $[\mathbf{A}_i \ \mathbf{B}_i]$ with $\mathbf{A}_i \in R^{n \times n}$ and $\mathbf{B}_i \in R^{n \times m}$, then the convex hull spanned by $\{[\mathbf{A}_i \ \mathbf{B}_i] \text{ for } i = 1, 2, \ldots, L\}$ is given by

$$\text{Co}\{[\mathbf{A}_1 \ \mathbf{B}_1], [\mathbf{A}_2 \ \mathbf{B}_2], \ldots, [\mathbf{A}_L \ \mathbf{B}_L]\} =$$
$$\{[\mathbf{A} \ \mathbf{B}] : [\mathbf{A} \ \mathbf{B}] = \sum_{i=1}^{L} \lambda_i [\mathbf{A}_i \ \mathbf{B}_i], \ \lambda_i \geq 0, \ \sum_{i=1}^{L} \lambda_i = 1\}$$

References

1 G. H. Golub and C. F. Van Loan, *Matrix Computations*, 2nd ed., The Johns Hopkins University Press, Baltimore, 1989.
2 R. A. Horn and C. R. Johnson, *Matrix Analysis*, Cambridge University Press, New York, 1991.
3 G. W. Stewart, *Introduction to Matrix Computations*, Academic Press, New York, 1973.
4 P. E. Gill, W. Murray, and M. H. Wright, *Numerical Linear Algebra and Optimization*, vol. 1, Addison-Wesley, New York, 1991.
5 S. Barnett, *Polynomials and Linear Control Systems*, Marcel Dekker, New York, 1983.

Appendix B
Basics of Digital Filters

B.1   Introduction

Several of the unconstrained and constrained optimization algorithms described in this book have been illustrated in terms of examples taken from the authors' research on the application of optimization algorithms for the design of digital filters. To enhance the understanding of the application of the algorithms presented to the design of digital filters, we provide in this appendix a concise introduction to the basic concepts and principles of digital filters as well as typical design problems associated with these systems. A detailed treatment of the subject can be found in [1].

B.2   Characterization

Digital filters are digital systems that can be used to process discrete-time signals. A single-input single-output digital filter can be represented by a block diagram as shown in Fig. B.1a where $x(nT)$ and $y(nT)$ are the excitation (input) and response (output), respectively. The excitation and response are sequences of numbers such as those illustrated in Fig B.1b and c. In the most general case, the response of a digital filter at instant nT is a function of a number of values of the excitation $x[(n+K)T]$, $x[(n+K-1)T]$, ..., $x(nT)$, $x[(n-1)T]$, ..., $x[(n-M)T]$ and a number of values of the response $y[(n-1)T]$, $y[(n-2)T]$, ..., $y[(n-N)T]$ where M, K and N are positive integers, i.e.,

$$y(nT) = \mathcal{R}x(nT) = f\{x[(n+K)T], \ldots, x[(n-M)T],$$
$$y[(n-1)T], \ldots, y[(n-N)T]\} \tag{B.1}$$

where $\mathcal{R}$ is an operator that can be interpreted as "*is the response produced by the excitation*".

Digital filters can be linear or nonlinear, time invariant or time dependent, causal or noncausal. In a linear digital filter, the response of the filter to a linear

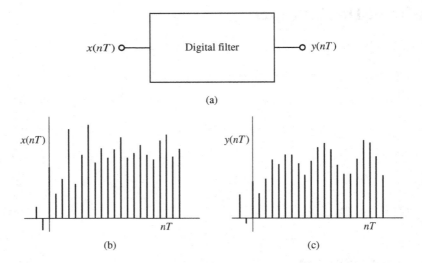

Figure B.1. (a) Block diagram representation of a digital filter, (b) excitation, (c) response.

combination of two signals $x_1(nT)$ and $x_2(nT)$ satisfies the relation

$$\mathcal{R}[\alpha x_1(nT) + \beta x_2(nT)] = \alpha \mathcal{R}x_1(nT) + \beta \mathcal{R}x_2(nT)$$

In an initially relaxed time-invariant digital filter, we have

$$\mathcal{R}x(nT - kT) = y(nT - kT)$$

This relation states, in effect, that a delayed excitation will produce a delayed but otherwise unchanged response. In this context, an *initially relaxed* digital filter is one whose response $y(nT)$ is zero for $nT < 0$ if $x(nT) = 0$ for all $nT < 0$. A causal digital filter, on the other hand, is a filter whose response at instant nT depends only on values of the input at instants nT, $(n-1)T$, …, $(n-M)T$, i.e., it is independent of future values of the excitation.

In a general linear, time-invariant, causal digital filter, Eq. (B.1) assumes the form of a linear recursive difference equation, i.e.,

$$y(nT) = \sum_{i=0}^{N} a_i x(nT - iT) - \sum_{i=1}^{N} b_i y(nT - iT) \qquad \text{(B.2)}$$

where a_i for $0 \le i \le N$ and b_i for $1 \le i \le N$ are constants. Some of these constants can be zero. A digital filter characterized by Eq. (B.2) is said to be *recursive*, since the response depends on a number of values of the excitation as well as a number of values of the response. Integer N, namely, the order of the difference equation, is said to be the *order* of the digital filter.

If the response of a digital filter at instant nT is independent of all the previous values of the response, i.e., $y(nT - T)$, $y(nT - 2T)$, ..., then Eq. (B.2) reduces to the nonrecursive equation

$$y(nT) = \sum_{i=0}^{N} a_i x(nT - iT) \tag{B.3}$$

which characterizes an Nth-order *nonrecursive filter* (see Secs. 4.2 and 4.3 of [1]). The number of coefficients in the difference equation, namely, $N + 1$ is said to be the *length* of the nonrecursive filter.

B.3 Time-Domain Response

The time-domain response of a digital filter to some excitation is often required and to facilitate the evaluation of time-domain responses, a number of standard signals are frequently used. Typical signals of this type are the unit impulse, unit step, and unit sinusoid which are defined in Table B.1 (see Sec. 4.3 of [1]).

Table B.1 Discrete-time standard signals

Function	Definition
Unit impulse	$\delta(nT) = \begin{cases} 1 & \text{for } n = 0 \\ 0 & \text{for } n \neq 0 \end{cases}$
Unit step	$u(nT) = \begin{cases} 1 & \text{for } n \geq 0 \\ 0 & \text{for } n < 0 \end{cases}$
Unit sinusoid	$u(nT) \sin \omega nT$

From Eq. (B.3), the impulse response of an arbitrary Nth-order nonrecursive filter, denoted as $h(nT)$, is given by

$$h(nT) \equiv y(nT) = \mathcal{R}\delta(nT)$$

$$= \sum_{i=0}^{N} a_i \delta(nT - iT)$$

Now from the definition of the unit impulse in Table B.1, we can readily show that ..., $h(-2T) = 0$, $h(-T) = 0$, $h(0) = a_0$, $h(T) = a_1$, ..., $h(NT) = a_N$, $h[(N + 1)T] = 0$, $h[(N + 2)T] = 0$, ..., i.e.,

$$h(nT) = \begin{cases} a_i & \text{for } n = i \\ 0 & \text{otherwise} \end{cases} \tag{B.4}$$

632

In effect, the impulse response in nonrecursive digital filters is of finite duration and for this reason these filters are also known as *finite-duration impulse response (FIR) filters*. On the other hand, the use of Eq. (B.2) gives the impulse response of a recursive filter as

$$y(nT) = h(nT) = \sum_{i=0}^{N} a_i \delta(nT - iT) - \sum_{i=1}^{N} b_i y(nT - iT)$$

and if the filter is initially relaxed, we obtain

$$
\begin{aligned}
y(0) = h(0) &= a_0 \delta(0) + a_1 \delta(-T) + a_2 \delta(-2T) + \cdots \\
&\quad -b_1 y(-T) - b_2 y(-2T) - \cdots = a_0 \\
y(T) = h(T) &= a_0 \delta(T) + a_1 \delta(0) + a_2 \delta(-T) + \cdots \\
&\quad -b_1 y(0) - b_2 y(-T) - \cdots = a_1 - b_1 a_0 \\
y(2T) = h(2T) &= a_0 \delta(2T) + a_1 \delta(T) + a_2 \delta(0) + \cdots \\
&\quad -b_1 y(T) - b_2 y(0) - \cdots \\
&= a_2 - b_1(a_1 - b_1 a_0) - b_2 a_0
\end{aligned}
$$

$$\vdots$$

Evidently, in this case the impulse response is of infinite duration since the response at instant nT depends on previous values of the response which are always finite. Hence recursive filters are also referred to as *infinite-duration impulse response (IIR) filters*.

Other types of time-domain response are the unit-step and the sinusoidal responses. The latter is of particular importance because it leads to a frequency-domain characterization for digital filters.

Time-domain responses of digital filters of considerable complexity can be deduced by using the z transform. The z transform of a signal $x(nT)$ is defined as

$$\mathcal{Z}x(nT) = X(z) = \sum_{n=-\infty}^{\infty} x(nT)z^{-n} \tag{B.5}$$

where z is a complex variable. The conditions for the convergence of $X(z)$ can be found in Sec. 3.3 of [1].

B.4 Stability Property

A digital filter is said to be *stable* if any bounded excitation will produce a bounded response. In terms of mathematics, a digital filter is stable if and only if any input $x(nT)$ such that

$$|x(nT)| \leq P < \infty \qquad \text{for all } n$$

will produce an output $y(nT)$ that satisfies the condition

$$|y(nT)| \leq Q < \infty \qquad \text{for all } n$$

where P and Q are positive constants.

A necessary and sufficient condition for the stability of a causal digital filter is that its impulse response be absolutely summable over the range $0 \leq nT \leq \infty$, i.e.,

$$\sum_{n=0}^{\infty} |h(nT)| \leq R < \infty \tag{B.6}$$

(see Sec. 4.7 of [1] for proof).

Since the impulse response of FIR filters is always of finite duration, as can be seen in Eq. (B.4), it follows that it is absolutely summable and, therefore, these filters are always stable.

B.5 Transfer Function

The analysis and design of digital filters is greatly simplified by representing the filter in terms of a transfer function. This can be derived from the difference equation or the impulse response and it can be used to find the time-domain response of a filter to an arbitrary excitation or its frequency-domain response to an arbitrary linear combination of sinusoidal signals.

B.5.1 Definition

The transfer function of a digital filter can be defined as the ratio of the z transform of the response to the z transform of the excitation, i.e.,

$$H(z) = \frac{Y(z)}{X(z)} \tag{B.7}$$

From the definition of the z transform in Eq. (B.5), it can be readily shown that

$$\mathcal{Z}x(nT - kT) = z^{-k}X(z)$$

and

$$\mathcal{Z}[\alpha x_1(nT) + \beta x_2(nT)] = \alpha X_1(z) + \beta X_2(z)$$

Hence if we apply the z transform to both sides of Eq. (B.2), we obtain

$$Y(z) = \mathcal{Z}y(nT) = \mathcal{Z}\sum_{i=0}^{N} a_i x(nT - iT) - \mathcal{Z}\sum_{i=1}^{N} b_i y(nT - iT)$$

$$= \sum_{i=0}^{N} a_i z^{-i} \mathcal{Z}x(nT) - \sum_{i=1}^{N} b_i z^{-i} \mathcal{Z}y(nT)$$

$$= \sum_{i=0}^{N} a_i z^{-i} X(z) - \sum_{i=1}^{N} b_i z^{-i} Y(z) \tag{B.8}$$

Therefore, from Eqs. (B.7) and (B.8), we have

$$H(z) = \frac{Y(z)}{X(z)} = \frac{\sum_{i=0}^{N} a_i z^{-i}}{1 + \sum_{i=1}^{N} b_i z^{-i}} = \frac{\sum_{i=0}^{N} a_i z^{N-i}}{z^N + \sum_{i=1}^{N} b_i z^{N-i}} \qquad (B.9)$$

The transfer function happens to be the z transform of the impulse response, i.e., $H(z) = \mathcal{Z} h(nT)$ (see Sec. 5.2 of [1]).

In FIR filters, $b_i = 0$ for $1 \leq i \leq N$ and hence the transfer function in Eq. (B.9) assumes the form

$$H(z) = \sum_{i=0}^{N} a_i z^{-i}$$

Since coefficients a_i are numerically equal to the impulse response values $h(iT)$ for $0 \leq i \leq N$, as can be seen in Eq. (B.4), the transfer function for FIR filters is often expressed as

$$H(z) = \sum_{i=0}^{N} h(iT) z^{-i} \quad \text{or} \quad \sum_{n=0}^{N} h_n z^{-n} \qquad (B.10)$$

where h_n is a simplified representation of $h(nT)$.

B.5.2   Zero-pole form

By factorizing the numerator and denominator polynomials in Eq. (B.9), the transfer function can be expressed as

$$H(z) = \frac{A(z)}{B(z)} = \frac{H_0 \prod_{i=1}^{Z} (z - z_i)^{m_i}}{\prod_{i=1}^{P} (z - p_i)^{n_i}} \qquad (B.11)$$

where $z_1, z_2, \ldots, z_Z$ and $p_1, p_2, \ldots, p_P$ are the zeros and poles of $H(z)$, m_i and n_i are the orders of zero z_i and pole p_i, respectively, $\sum_i^{Z} m_i = \sum_i^{P} n_i = N$, and H_0 is a multiplier constant. Evidently, the zeros, poles, and multiplier constant describe the transfer function and, in turn, the digital filter, completely. Typically, the zeros and poles of digital filters are simple, i.e., $m_i = n_i = 1$ for $1 \leq i \leq N$, and in such a case $Z = P = N$.

From Eq. (B.10), the transfer function of an FIR filter can also be expressed as

$$H(z) = \frac{1}{z^N} \sum_{n=0}^{N} h_n z^{N-n}$$

and, in effect, all the poles in an FIR filter are located at the origin of the z plane.

B.6   Time-Domain Response Using the Z Transform

The time-domain response of a digital filter to an arbitrary excitation $x(nT)$ can be readily obtained from Eq. (B.7) as

$$y(nT) = \mathcal{Z}^{-1}[H(z)X(z)] \qquad (B.12)$$

i.e., we simply obtain the inverse-z transform of $H(z)X(z)$ (see Sec. 5.4 of [1]).

B.7 Z-Domain Condition for Stability

The stability condition in Eq. (B.6), namely, the requirement that the impulse response be absolutely summable over the range $0 \leq nT \leq \infty$ is difficult to apply in practice because it requires complete knowledge of the impulse response over the specified range. Fortunately, this condition can be converted into a corresponding z-domain condition that is much easier to apply as follows: *A digital filter is stable if and only if all the poles of the transfer function are located strictly inside the unit circle of the z plane.* In mathematical terms, a digital filter with poles

$$p_i = r_i e^{j\psi_i}$$

where $r_i = |p_i|$ and $\psi_i = \arg p_i$ for $1 \leq i \leq N$ is stable if and only if

$$r_i < 1$$

The z-plane areas of stability and instability are illustrated in Fig. B.2.

From the above discussion we note that for a stable digital filter, the denominator of the transfer function, $B(z)$, must *not* have zeros on or outside the unit circle $|z| = 1$ and, therefore, an alternative way of stating the z-domain stability condition is

$$B(z) \neq 0 \qquad \text{for } |z| \geq 1 \qquad (B.13)$$

The poles of the transfer function are the zeros of polynomial $B(z)$ in Eq. (B.11) and it can be easily shown that these are numerically equal to the eigenvalues of matrix

$$\mathbf{D} = \begin{bmatrix} -b_1 & -b_2 & \cdots & -b_{N-1} & -b_N \\ 1 & 0 & \cdots & 0 & 0 \\ 0 & 1 & \cdots & 0 & 0 \\ \vdots & \vdots & \ddots & \vdots & \vdots \\ 0 & 0 & \cdots & 1 & 0 \end{bmatrix}$$

(see Prob. 16.4(a)). Consequently, *an IIR filter is stable if and only if the moduli of the eigenvalues of matrix* $\mathbf{D}$ *are all strictly less than one.*

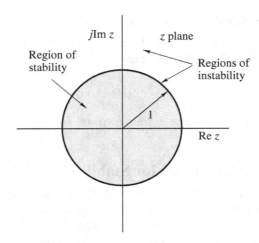

Figure B.2. Z-plane areas of stability and instability in IIR filters.

B.8   Frequency, Amplitude, and Phase Responses

A most important type of time-domain response for a digital filter is its steady-state sinusoidal response which leads to a frequency-domain characterization for the filter.

As shown in Sec. 5.5.1 of [1], the steady-state sinusoidal response of a stable digital filter can be expressed as

$$\tilde{y}(nT) = \lim_{n \to \infty} \mathcal{R}[u(nT) \sin \omega nT]$$
$$= M(\omega) \sin[\omega nT + \theta(\omega)]$$

where

$$M(\omega) = |H(e^{j\omega T})| \quad \text{and} \quad \theta(\omega) = \arg H(e^{j\omega T}) \tag{B.14}$$

Thus, the steady-state effect of a digital filter on a sinusoidal excitation is to introduce a *gain* $M(\omega)$ and a phase shift $\theta(\omega)$, which can be obtained by evaluating the transfer function $H(z)$ on the unit circle $z = e^{j\omega T}$ of the z plane. $H(e^{j\omega T})$, $M(\omega)$, and $\theta(\omega)$ in Eq. (B.14) as functions of ω are known as the *frequency response*, *amplitude response*, and *phase response*, respectively.

The frequency response of a digital filter characterized by a transfer function of the form given in Eq. (B.11) can be obtained as

$$H(z)|_{z \to e^{j\omega T}} = H(e^{j\omega T}) = M(\omega)e^{j\theta(\omega)} \tag{B.15}$$

$$= \frac{H_0 \prod_{i=1}^{Z}(e^{j\omega T} - z_i)^{m_i}}{\prod_{i=1}^{P}(e^{j\omega T} - p_i)^{n_i}} \tag{B.16}$$

and by letting

$$e^{j\omega T} - z_i = M_{z_i}e^{j\psi_{z_i}} \tag{B.17}$$

$$e^{j\omega T} - p_i = M_{p_i} e^{j\psi_{p_i}} \tag{B.18}$$

Eqs. (B.14)–(B.18) give

$$M(\omega) = \frac{|H_0| \prod_{i=1}^{Z} M_{z_i}^{m_i}}{\prod_{i=1}^{P} M_{p_i}^{n_i}} \tag{B.19}$$

$$\theta(\omega) = \arg H_0 + \sum_{i=1}^{Z} m_i \psi_{z_i} - \sum_{i=1}^{P} n_i \psi_{p_i} \tag{B.20}$$

where $\arg H_0 = \pi$ if H_0 is negative. Thus the gain and phase shift of a digital filter at some frequency ω can be obtained by calculating the magnitudes and angles of the complex numbers in Eqs. (B.17) and (B.18) and then substituting these values in Eqs. (B.19) and (B.20). These calculations are illustrated in Fig. B.3 for the case of a transfer function with two zeros and two poles. The vectors from the zeros and poles to point B represent the complex numbers in Eqs. (B.17) and (B.18), respectively.

The amplitude and phase responses of a digital filter can be plotted by evaluating the gain and phase shift for a series of frequencies $\omega_1, \omega_2, \ldots, \omega_K$ over the frequency range of interest.

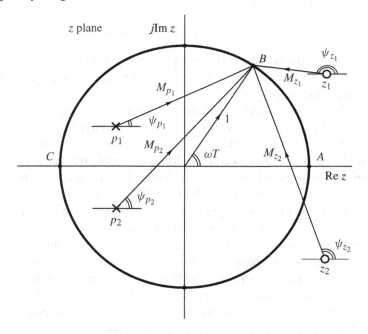

Figure B.3. Calculation of gain and phase shift of a digital filter.

Point A in Fig. B.3 corresponds to $\omega = 0$, i.e., zero frequency, and one complete revolution of vector $e^{j\omega T}$ about the origin corresponds to an increase in frequency of $\omega_s = 2\pi/T$ rad/s; this is known as the *sampling frequency*. Point C, on the other hand, corresponds to an increase in frequency of π/T, i.e., half the sampling frequency, which is often referred to as the *Nyquist frequency*. In the design of digital filters, a normalized sampling frequency of 2π rad/s is usually used, for the sake of simplicity, which corresponds to a Nyquist frequency of π and a normalized sampling period, $T = 2\pi/\omega_s$, of 1 s.

If vector $e^{j\omega T}$ in Fig. B.3 is rotated k complete revolutions starting at some arbitrary point, say, point B, it will return to its original position and the values of $M(\omega)$ and $\theta(\omega)$ will be the same as before according to Eqs. (B.19) and (B.20). Therefore,

$$H(e^{j(\omega+k\omega_s)T}) = H(e^{j\omega T})$$

In effect, the *frequency response* of a digital filter is a *periodic function* of frequency with a period ω_s. It, therefore, follows that knowledge of the frequency response of a digital filter over the base period $-\omega_s/2 \leq \omega \leq \omega_s/2$ provides a complete frequency-domain characterization for the filter. This frequency range is often referred to as the *baseband*.

Assuming real transfer-function coefficients, the amplitude response of a digital filter can be easily shown to be an even function and the phase response an odd function of frequency, i.e.,

$$M(-\omega) = M(\omega) \quad \text{and} \quad \theta(-\omega) = -\theta(\omega)$$

Consequently, a frequency-domain description of a digital filter over the positive half of the baseband, i.e., $0 \leq \omega \leq \omega_s/2$, constitutes a complete frequency-domain description of the filter.

Digital filters can be used in a variety of applications for example to pass low and reject high frequencies (lowpass filters), to pass high and reject low frequencies (highpass filters), or to pass or reject a range of frequencies (bandpass or bandstop filters). In this context, low and high frequencies are specified in relation to the positive half of the baseband, e.g., frequencies in the upper part of the baseband are deemed to be high frequencies. A frequency range over which the digital filter is required to pass or reject frequency components is said to be a *passband* or *stopband* as appropriate.

In general, the amplitude response is required to be close to unity in passbands and approach zero in stopbands. A constant passband gain close to unity is required to ensure that the different sinusoidal components of the signal are subjected to the same gain. Otherwise, so-called *amplitude distortion* will occur. The gain is required to be as small as possible in stopbands to ensure that undesirable signals are as far as possible rejected.

Note that the gain of a filter can vary over several orders of magnitude and for this reason it is often represented in terms of decibels as

$$M(\omega)_{\text{dB}} = 20 \log_{10} M(\omega)$$

In passbands, $M(\omega) \approx 1$ and hence we have $M(\omega)_{\mathrm{dB}} \approx 0$. On the other hand, in stopbands the gain is a small fraction and hence $M(\omega)_{\mathrm{dB}}$ is a negative quantity. To avoid this problem, stopbands are often specified in terms of *attenuation* which is defined as the reciprocal of the gain in decibels, i.e.,

$$A(\omega) = 20 \log_{10} \frac{1}{M(\omega)} \qquad (B.21)$$

Phase shift in a signal is associated with a delay and the delay introduced by a digital filter is usually measured in terms of the *group delay* which is defined as

$$\tau(\omega) = -\frac{d\theta(\omega)}{d\omega} \qquad (B.22)$$

If different sinusoidal components of the signal with different frequencies are delayed by different amounts, a certain type of distortion known as *phase distortion* (or *delay distortion*) is introduced, which is sometimes objectionable. This type of distortion, can be minimized by ensuring that the group delay is as far as possible constant in passbands, and a constant group delay corresponds to a linear phase response as can be readily verified by using Eq. (B.22) (see Sec. 5.7 of [1]).

B.9 Design

The design of digital filters involves four basic steps as follows:

- Approximation

- Realization

- Implementation

- Study of the effects of roundoff errors

The *approximation* step is the process of deducing the transfer function coefficients such that some desired amplitude or phase response is achieved. *Realization* is the process of obtaining a digital network that has the specified transfer function. *Implementation* is the process of constructing a system in hardware or software form based on the transfer function or difference equation characterizing the digital filter. Digital systems constructed either in terms of special- or general-purpose hardware are implemented using finite arithmetic and there is, therefore, a need to proceed to the fourth step of the design process, namely, the study of the effects of roundoff errors on the performance of the digital filter (see Sec. 8.1 of [1]). In the context of optimization, the design of a digital filter is usually deemed to be the solution of just the approximation problem.

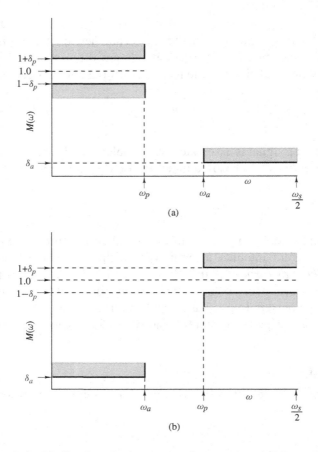

Figure B.4. Idealized amplitude responses for lowpass and highpass filters.

B.9.1 Specified amplitude response

The design of an FIR or IIR filter that would satisfy certain amplitude-response requirements starts with an idealized amplitude response such as those depicted in Fig. B.4a to d for lowpass, highpass, bandpass, and bandstop filters, respectively. In lowpass and highpass filters, parameters ω_p and ω_a are the *passband* and *stopband edges*. In bandpass and bandstop filters, on the other hand, ω_{p1} and ω_{p2} are the lower and upper passband edges, and ω_{a1} and ω_{a2} are the lower and upper stopband edges. The frequency bands between passbands and stopbands, e.g., the range $\omega_p < \omega < \omega_a$ in a lowpass or highpass filter, are called *transition bands*, for obvious reasons. The idealized passband gain is usually assumed to be unity but some other value could be used if necessary.

The objective of design in digital filters is to find a set of transfer function coefficients which would yield an amplitude response that falls within the passband and stopband templates shown in Fig. B.4a to d. For example, in the case

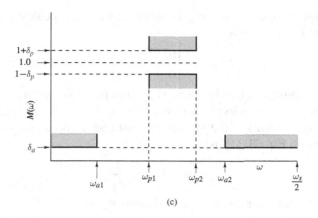

(c)

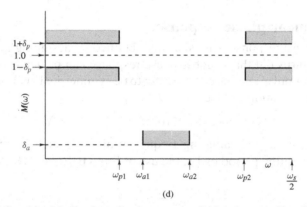

(d)

Figure B.4 Cont'd. Idealized amplitude responses for bandpass and bandstop filters.

of a lowpass filter, we require that

$$1 - \delta_p \leq M(\omega) \leq 1 + \delta_p$$

with respect to the passband and

$$M(\omega) \leq \delta_a$$

with respect to the stopband.

The peak-to-peak passband error in Fig. B.4a to d is often expressed in decibels as

$$A_p = 20\log_{10}(1 + \delta_p) - 20\log_{10}(1 - \delta_p) = 20\log_{10}\frac{1 + \delta_p}{1 - \delta_p} \qquad (B.23)$$

642

which is often referred to as the *passband ripple*. The peak error δ_p can be deduced from Eq. (B.23) as

$$\delta_p = \frac{10^{0.05A_p} - 1}{10^{0.05A_p} + 1}$$

Since the maximum stopband gain δ_a is a small fraction which corresponds to a negative quantity when expressed in decibels, the stopband specification is often expressed in terms of the minimum stopband attenuation, A_a, which can be obtained from Eq. (B.21) as

$$A_a = 20 \log_{10} \frac{1}{\delta_a}$$

The peak stopband error can be deduced from A_a as

$$\delta_a = 10^{-0.05A_a}$$

B.9.2   Linear phase response

A linear phase response is most easily obtained by designing the filter as an FIR filter. It turns out that a linear phase response can be achieved by simply requiring the impulse response of the filter to be symmetrical or antisymmetrical with respect to its midpoint (see Sec. 9.2 of [1]), i.e.,

$$h_n = h_{N-n} \qquad \text{for } n = 0, 1, \ldots, N \qquad (B.24)$$

Assuming a normalized sampling frequency of 2π, which corresponds to a normalized sampling period of 1 s, the use of Eqs. (B.10) and (B.24) will show that

$$H(e^{j\omega}) = e^{-j\omega N/2} A(\omega)$$

where

$$A(\omega) = \sum_{n=0}^{N/2} a_n \cos n\omega$$

$$a_n = \begin{cases} h_N/2 & \text{for } n = 0 \\ 2h_{N/2-n} & \text{for } n \neq 0 \end{cases}$$

for even N and

$$A(\omega) = \sum_{n=0}^{(N-1)/2} a_n \cos[(n + 1/2)\omega]$$

$$a_n = 2h_{(N-1)/2-n}$$

for odd N. The quantity $A(\omega)$ is called the *gain function* and, in fact, its magnitude is the gain of the filter, i.e.,

$$M(\omega) = |A(\omega)| \qquad (B.25)$$

B.9.3 Formulation of objective function

Let us assume that we need to design an Nth-order IIR filter with a transfer function such as that in Eq. (B.9) whose amplitude response is required to approach one of the idealized amplitude responses shown Fig. B.4a to d. Assuming a normalized sampling frequency of 2π rad/s, the amplitude response of such a filter can be expressed as

$$|H(e^{j\omega})| = M(\mathbf{x}, \omega) = \left| \frac{\sum_{i=0}^{N} a_i e^{-j\omega i}}{1 + \sum_{i=1}^{N} b_i e^{-j\omega i}} \right|$$

where

$$\mathbf{x} = [a_0 \ a_1 \ \cdots \ a_N \ b_1 \ b_2 \ \cdots b_N]^T$$

is the parameter vector. An error function can be constructed as

$$e(\mathbf{x}, \omega) = M(\mathbf{x}, \omega) - M_0(\omega) \tag{B.26}$$

where $M_0(\omega)$ is the required idealized amplitude response, for example,

$$M_0(\omega) = \begin{cases} 1 & \text{for } 0 \leq \omega \leq \omega_p \\ 0 & \text{otherwise} \end{cases}$$

in the case of a lowpass filter. An objective function can now be constructed in terms of one of the standard norms of the error function, e.g., the L_2 norm

$$F = \int_{\omega \in \Omega} |e(\mathbf{x}, \omega)|^2 d\omega \tag{B.27}$$

where Ω denotes the positive half of the normalized baseband $[0, \pi]$. Minimizing F in Eq. (B.27) would yield a *least-squares* solution. Alternatively, we can define the objective function as

$$F = \max_{\omega \in \Omega} |e(\mathbf{x}, \omega)| = \lim_{p \to \infty} \int_{\omega \in \Omega} |e(\mathbf{x}, \omega)|^p d\omega \tag{B.28}$$

where p is a positive integer, which would yield a minimax solution.

A more general design can be accomplished by forcing the frequency response of the filter, $H(e^{j\omega})$, to approach some desired idealized frequency response $H_d(\omega)$ by minimizing the least-pth objective function

$$F = \max_{\omega \in \Omega} |e(\mathbf{x}, \omega)| = \lim_{p \to \infty} \int_{\omega \in \Omega} |H(e^{j\omega T}) - H_d(\omega)|^p d\omega \tag{B.29}$$

As before, we can assign $p = 2$ to obtain a least-squares solution or let $p \to \infty$ to obtain a minimax solution.

Discretized versions of the objective functions in Eqs. (B.27)–(B.29) can be deduced by sampling the error in Eq. (B.26), $e(\mathbf{x}, \omega)$, at frequencies ω_1, ω_2, ..., ω_K, and thus the vector

$$\mathbf{E}(\mathbf{x}) = [e_1(\mathbf{x})\, e_2(\mathbf{x})\, \cdots\, e_K(\mathbf{x})]^T$$

can be formed where

$$e_i(\mathbf{x}) = e(\mathbf{x}, \omega_i)$$

for $i = 1, 2, \ldots, K$. At this point, an objective function can be constructed in terms of the L_p norm of $\mathbf{E}\,(\mathbf{x})$, as

$$F = ||\mathbf{E}(\mathbf{x})||_p = \left[\sum_{i=1}^{K} |e_i(\mathbf{x})|^p \right]^{1/p}$$

where we can assign $p = 2$ for a least-squares solution or $p \to \infty$ for a minimax solution.

The above objective functions can be readily applied for the design of FIR filters by setting the denominator coefficients of the transfer function, b_i for $1 \leq i \leq N$, to zero. If a linear phase response is also required, it can be readily achieved by simply forcing the coefficients $a_i \equiv h_i$ for $1 \leq i \leq N$ to satisfy the symmetry property in Eq. (B.24) and this can be accomplished by using the amplitude response given by Eq. (B.25).

Reference

1 A. Antoniou, *Digital Signal Processing: Signals, Systems, and Filters*, McGraw-Hill, New York, 2005.

Index

648

658

666

Printed in the United States
By Bookmasters